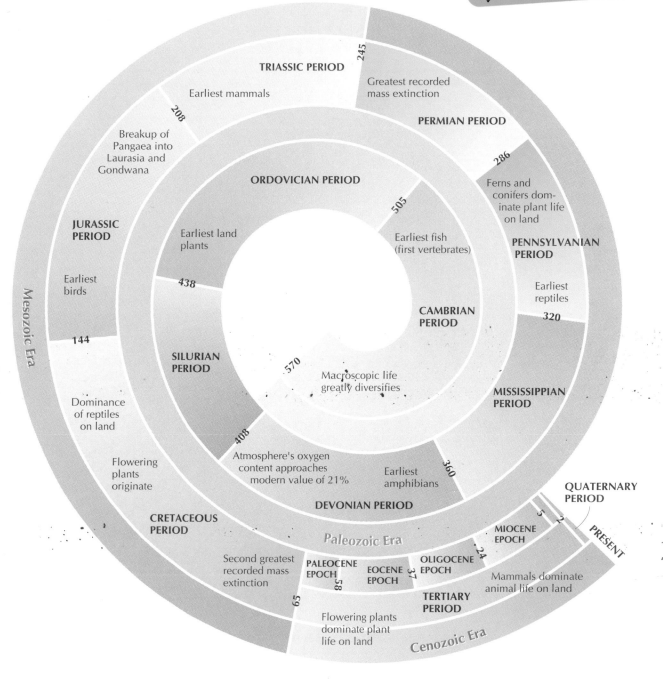

# Physical Science
## A Unified Approach

# Physical Science
## A Unified Approach

**Jerry Schad**
San Diego Mesa College

Brooks/Cole Publishing Company
I(T)P™ An International Thomson Publishing Company

Pacific Grove • Albany • Bonn • Boston • Cincinnati • Detroit • London • Madrid
Melbourne • Mexico City • New York • Paris • San Francisco
Singapore • Tokyo • Toronto • Washington

Sponsoring Editor: *Harvey Pantzis*
Project Development Editor: *Casey FitzSimons*
Marketing Team: *Connie Jirovsky, Jean Thompson*
Editorial Associate: *Beth Wilbur*
Production Editor: *Nancy L. Shammas*
Production Assistant: *Tessa McGlasson*
Manuscript Editor: *Carol Beal*
Permissions Editor: *May Clark*

Interior and Cover Design: *Roy R. Neuhaus*
Interior Illustration: *Precision Graphics*
Cover Photo: *Joe Viesti/Viesti Associates*
Art Editor: *Lisa Torri*
Photo Editor: *Larry Molmud*
Photo Researcher: *Sue C. Howard*
Indexer: *Do Mi Stauber*
Typesetting: *TECHarts*
Cover Printing: *Color Dot Graphics, Inc.*
Printing and Binding: *Quebecor/Hawkins*

*For more information, contact:*

BROOKS/COLE PUBLISHING COMPANY
511 Forest Lodge Road
Pacific Grove, CA 93950
USA

International Thomson Publishing Europe
Berkshire House 168-173
High Holborn
London WC1V 7AA
England

Thomas Nelson Australia
102 Dodds Street
South Melbourne, 3205
Victoria, Australia

Nelson Canada
1120 Birchmount Road
Scarborough, Ontario
Canada M1K 5G4

International Thomson Editores
Campos Eliseos 385, Piso 7
Col. Polanco
11560 México D. F. México

International Thomson Publishing GmbH
Königswinterer Strasse 418
53227 Bonn
Germany

International Thomson Publishing Asia
221 Henderson Road
#05-10 Henderson Building
Singapore 0315

International Thomson Publishing Japan
Hirakawacho Kyowa Building, 3F
2-2-1 Hirakawacho
Chiyoda-ku, Tokyo 102
Japan

Printed in the United States of America

3   4   5   6   7   06   05   04   03

**Library of Congress Cataloging-in-Publication Data**
Schad, Jerry.
    Physical science : a unified approach / Jerry Schad.
        p.      cm.
    Includes bibliographical references and index.
    ISBN 0-534-19248-3
    1. Physical sciences.    I. Title.
Q158.5.S3   1996                                    95-21545
500.2—dc20                                          CIP

Slinky is a registered trademark of Corestates Bank, N.A.

*This book is dedicated to my father, Jack Schad, amateur radio operator W6JSF since 1930, former electronics instructor, and former electronic parts merchant. His guidance sparked my youthful interest in electricity and magnetism.*

# Preface

*Physical Science: A Unified Approach* surveys the realm of the physical sciences—physics, chemistry, earth science, and astronomy—in an entirely new and engaging way. The goals of the book are bold and comprehensive: to examine all the major revelations of contemporary physical science, to learn how scientists acquire information and synthesize knowledge, and to discover connections among the various scientific disciplines.

Until now, virtually all college courses in introductory physical science (and the textbooks written for them) have treated each physical science discipline as an insular subject. Again and again, freshman students have been abruptly cast into the murky waters of pure physics during the opening weeks, myopically focusing on details, all the while losing sight of the big picture. This boot camp method has been replaced in this book by a kinder, more informative approach.

In Part I (the first four chapters) of *Physical Science: A Unified Approach*, the student is introduced to the fundamental components of our physical world: space, time, matter, and energy. Topics such as measurement, units,

**COSMIC ZOOM** The cosmic zoom in Chapter 2 explores the microscopic, macroscopic, and supermacroscopic realms of our universe.

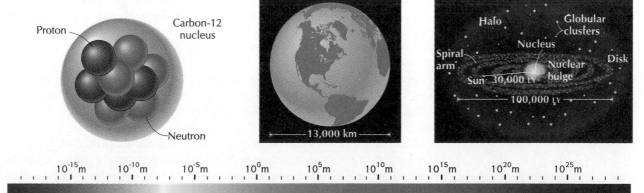

**TIMELINE** Chapter 3's timeline surveys 15 billion years of astronomical, geological, biological, and technological history—right up to today's information revolution.

and the scientific method are covered as well. Two major elements within Part I are attention-getters: Chapter 2's visually rich cosmic zoom takes students on a mind-expanding spatial journey, beginning in the microscopic realm of subatomic particles and moving toward the supermacroscopic world of billions of galaxies. More than merely an exposition of the very small and very large, this zoom feature introduces students to forces and phenomena that apply to each level of size. Chapter 3's graphic timeline accomplishes a similar goal, except that the focus is time rather than space. Starting with the big bang, the timeline's descriptive and visual elements guide students through the major astronomical, geological, and biological episodes that have led to humanity's existence on Earth. The cosmic zoom, the timeline, and other elements in Part I serve as a friendly introduction to many great ideas in science. These introductions are followed up by more detailed coverage of the same ideas in subsequent chapters.

In the middle of the book, Parts II and III, topics in physics take the lead. The four chapters of Part II cover the relatively tangible physical phenomena of motion, gravity, energy, heat, and waves. Applications of these phenomena in sciences other than physics (primarily chemistry, geology, meteorology, and astronomy) are introduced at opportune times to strengthen the student's grasp of physical principles. The four chapters of Part III focus on the somewhat less tangible topics of electricity, magnetism, light, and nuclear energy. Again, these chapters contain many examples and applications drawn from sciences other than physics.

The five chapters of Part IV conclude this book and highlight five overarching concepts—reigning intellectual paradigms or important organizational tools—associated with the physical sciences. Relativity and quantum mechanics are two paradigms that uphold contemporary physics. The periodic table serves as a central organizing tool in chemistry and much of physics. Plate tectonics has become the accepted paradigm by which geologists and other earth scientists gain a global understanding of our dynamic planet. The final chapter, on cosmic evolution, explains how the universe and its contents have evolved over time.

# Features

Apart from the distinctive organization of this book, and apart from the cosmic zoom and timeline sections in Part I, several other features make this book user-friendly for both the instructor and the student. The first two items in the following list are unique pedagogical features.

- The opening two-page spread in each chapter contains a diagram indicating where we are going with the subject matter in that chapter. The title of each major section or unit of boxed material is listed in the diagram. By reading the diagrams from left to right and top to bottom, instructors and students can get a sense of the order in which the major topics of each chapter are presented. More important, any close relationship between topics is shown by overlapping rectangles. A similar diagram, summarizing the subject matter of this entire book, appears on page xiii.
- Small icons appear alongside major sections of text in each chapter and also with boxed sections of text. These icons let the student know which branch of science is being discussed. There are five icons.

 The orbiting bodies symbolize **physics.**

 The graduated flask symbolizes **chemistry.**

 Planet Earth symbolizes **earth science.**

 The spiral galaxy symbolizes **astronomy.**

 The DNA segment symbolizes **biology** and **biochemistry.**

- The material in this book is presented at a primarily descriptive and conceptual level, with mathematics held to a minimum. Mathematical formulas are introduced when necessary to succinctly and precisely express certain physical laws and relationships. Sample problems are sprinkled throughout the book, and there are several problems involving mathematical solutions at the ends of most chapters.
- Considerable flexibility is built into this textbook. Part I was designed to provide students having little or no background in science with a rudimentary literacy in all of the major disciplines of physical science. For other students, this material will be review. The instructor may wish to skip certain portions of Part I or assign them as supplementary reading.
- Parts II and III follow a backbone-of-physics approach to understanding physical science and may contain topics deemed more important or less important for particular audiences. Some of the specialized material included in the boxed sections can be skipped, assigned as supplementary reading, or passed over in favor of instructor-provided materials. Students often appreciate learning about the physical environment of their own region; thus, supple-

mental material on local geology and meteorology should be worked into the course where appropriate.

The five chapters of Part IV offer a great deal of flexibility. These shorter chapters—spotlighting relativity, quantum mechanics, the periodic table, plate tectonics, and cosmic evolution—help round out the student's view of the physical world. Any or all of these chapters can be skipped, though most instructors will probably want to cover at least some aspects of the periodic table and plate tectonics.

- Nearly 600 drawings and photographs are included in this text. Each was designed or chosen to illustrate a salient point; none were inserted to fill space or belabor a point. The pedagogical use of color in the drawings is precise and transparent.

- The style of writing in this textbook is friendly, accessible, and engaging, yet not patronizing or excessively informal. I have tried to convey my enthusiasm for the subject matter and the sense of wonder that comes from synthesizing myriad bits of information about the universe into an elegant conceptual whole.

- The end-of-chapter material includes a summary of important concepts as well as exercises for the student. The exercises consist of *multiple-choice questions* (the answer key is provided at the end of each chapter), *short-answer questions, problems,* and *questions for thought.* Answers to the odd-numbered short-answer questions and the odd-numbered problems are given in Appendix E. Many of the questions for thought are open-ended and serve as excellent starting points for class discussion.

- Scientific terminology is set off in the text in italic or boldface type. The more important terms are indicated by boldface wherever they are defined in the text and also appear in the glossary at the end of the book.

## Ancillaries

The following ancillary items are available with this text to assist the instructor in preparing for course lectures and exams.

- Instructor's Manual
  The Instructor's Manual provides help in transforming a traditionally compartmentalized course to an integrated one. Alternative ways to use this text are presented. Classroom demonstrations, media aids, topics for further investigation, and solutions to each chapter's even questions and problems are provided.

- Test Items
  Approximately 500 multiple-choice, true-false, and short-answer test items are provided and are available on DOS, Windows, and Macintosh platforms.

- Transparencies
  One hundred four-color transparency acetates for overhead projection are available.

- Art and Animations CD-ROM
  Key figures from the text come alive with animation in this CD-ROM available to the instructor as a teaching tool. It is available for both

Macintosh and Windows and includes easy-to-use search features, key terms, and an image index. The CD-ROM also contains still versions of approximately 75 diagrams in the text.

# Acknowledgments

The successful conclusion of the more than three years of effort in writing this textbook would not have been possible without the help of many people who have contributed to the development and refinement of my ideas. Key parts of the manuscript were critiqued by my colleagues Chuck Corum, Fred Jappe, and Sam Scharber at Mesa College. Fred, who has taught humanities courses as well as chemistry, crafted most of the material in Chapter 17's feature on theology and science. Karen Featherby read the entire manuscript and provided much constructive criticism from a student perspective.

During the tenure of this project, it was a real pleasure to work with the team at Brooks/Cole Publishing Company. My sincere thanks to president Bill Roberts and executive editor Harvey Pantzis for lending support to my vision and providing the resources necessary to create this textbook. Editors Maureen Allaire and Lisa Moller, editorial associate Beth Wilbur, and consultant Casey FitzSimons worked diligently to keep the project on track during the manuscript phase. They facilitated the review process, curbed my worst excesses, and consistently provided the encouragement I needed in order to keep going on this large and difficult undertaking. For nearly a year, production editor Nancy Shammas flawlessly directed the transformation of manuscript text, art, photographs, and captions into the finished product you now hold in your hands. Designer Roy Neuhaus masterminded the book's extraordinarily friendly and colorful design. Art editor Lisa Torri's guidance of the exceptional art program and photo researcher Sue Howard's efforts at tracking down photo illustrations were essential in producing a book with a remarkable visual impact.

While I have had the pleasure of closely working with all of the above-mentioned people, many others associated with Brooks/Cole lent their efforts and ideas to the project: Carol Beal, Ellen Brownstein, May Clark, Connie Jirovsky, Barbara Kimmel, Larry Molmud, Elizabeth Rammel, Howard Perry, and Do Mi Stauber. In addition, I would like to acknowledge and express my appreciation to the following reviewers from institutions other than my own:

Robert Backes, Pittsburg State University; Franklin Brown, Tallahassee Community College; Carl Davis, Danville Area Community College; Ted Erickson, Ricks College; William Faissler, Northeastern University; Ray Glienna, Glendale College; Stan Hirschi, Central Michigan University; Keith Honey, West Virginia Institute of Technology; Charles Perrino, California State University at Hayward; Russell Roy, Santa Fe Community College; Jubran Wakim, Middle Tennessee State University; Linda Wilson, Middle Tennessee State University; and Larry Weaver, Kansas State University.

*Jerry Schad*

# WHERE WE ARE GOING IN THIS BOOK

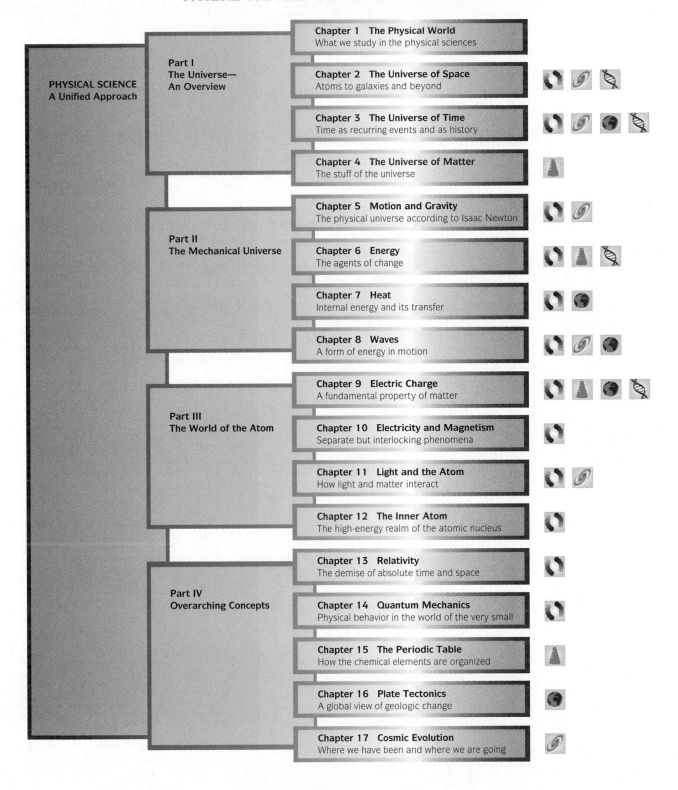

# Brief Contents

# Contents

**PART II**

# The Mechanical Universe 180

**Chapter 8**

# Waves   352

# PART III
# The World of the Atom
# 392

**Chapter 12**   **The Inner Atom   508**

# Preface

*Physical Science: A Unified Approach,* surveys the realm of the physical sciences—physics, chemistry, earth science, and astronomy—in an entirely new and engaging way. The goals of the book are bold and comprehensive: to examine all the major revelations of contemporary physical science, to learn how scientists acquire information and synthesize knowledge, and to discover connections among the various scientific disciplines.

Until now, virtually all college courses in introductory physical science (and the textbooks written for them) have treated each physical science discipline as an insular subject. Again and again, freshman students have been abruptly cast into the murky waters of pure physics during the opening weeks, myopically focusing on details, all the while losing sight of the big picture. This boot camp method has been replaced in this book by a kinder, more informative approach.

In Part I (the first four chapters) of *Physical Science: A Unified Approach,* the student is introduced to the fundamental components of our physical world: space, time, matter, and energy. Topics such as measurement, units,

**COSMIC ZOOM** The cosmic zoom in Chapter 2 explores the microscopic, macroscopic, and supermacroscopic realms of our universe.

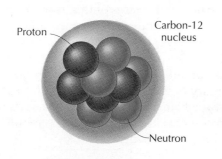

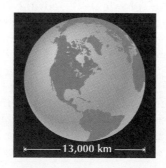

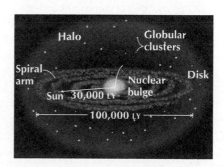

The physical sciences belong to a field of knowledge and inquiry called natural science. **Natural science** deals with phenomena that can be observed and measured in some concrete way: the motion of the planets, water rushing down a stream, lightning, and the beating of a heart. The behavioral and social sciences (like psychology and history) and pure mathematics are considered to lie outside the purview of natural science. They deal primarily with abstractions or phenomena that cannot be measured quantitatively (using numbers).

Natural science has as its subject the entire universe. It is customary, academically at least, to divide natural science into the study of living things (the biological sciences) and nonliving things (the physical sciences). The **physical sciences** include physics, chemistry, astronomy, and the earth sciences (primarily geology and meteorology), which are the areas covered in this book.

**Physics** is concerned with the most basic aspects of nature, such as matter, energy, motion, and force. **Chemistry** builds on physics by investigating the composition, properties, and transformations of matter. The biological sciences, in turn, flow from chemistry. (Except for brief coverage of some important biological structures and processes, biology topics are not included in this book.)

**Geology** is the study of Earth's surface and interior, and **meteorology** deals with the atmosphere and weather. **Astronomy** is the study of the universe and its contents, including our home planet. All three of these sciences borrow heavily from physics and chemistry.

Part I of this textbook has four chapters. We begin in Chapter 1 by asking what is "physical" about the world we will study in this course. We then discuss the importance of measurement in science and describe the *scientific method*, the method of inquiry that facilitates scientific advancement. Chapters 2, 3, and 4 offer broad insights into the relative sizes of things in our universe, the enormous range of cosmic time, and the fundamental arrangements of atoms and molecules.

Part I is not only a survey of the universe but also a summary of the entire scope of natural science. It lays a firm foundation for the increasingly narrower physical science topics found in the succeeding chapters of this book.

# PART I

# The Universe— An Overview

# CHAPTER 1
# The Physical World

*Scientists search for order and regularity in natural phenomena. Some of the things we study in the physical sciences are dynamic and beautifully symmetric, such as the wave emanating from the site of a fallen water droplet.*

**WHERE WE ARE GOING IN CHAPTER 1**

**The physical world**
What we study in the physical sciences

**What is physical?**
Matter, energy, space, and time
make up the physical world

**Properties and measurements**
What we measure and how we express it

**Science and its methodology**
All scientific explanations are subject
to further verification

In principle, everything that exists in the physical world—past, present, and future—is appropriate for inquiry within the physical sciences. But what, exactly, belongs to the physical world?

## What Is Physical?

Pay attention to what your senses are telling you right now. Notice the light reaching you from various sources. Judge the brightness, color, and direction of the light. Notice each sound coming to your ears. Consider its loudness and distinctive characteristics. Feel the texture and weight of this textbook in your hands. Smell the faint odor of its paper and inks. Touch the tip of your tongue to a clean page to detect its acidity, or sour taste.

When you probe an object with your senses, you discover the *physical characteristics* or *properties* of that object. When you look at a cloud in the sky, you consciously or subconsciously judge its shape, size, color, bright-

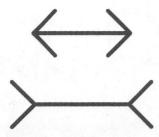

**FIGURE 1.1** A famous optical illusion. The lower horizontal line looks longer. Use a ruler to confirm that the two horizontal lines are equal in length.

ness, and other characteristics and properties, and you thereby recognize it as a cloud. When you hold a cool glass of water in your hand, you sense the temperature, weight, and rigidity of the glass. The taste and smell of the water itself may tell you something about its chemical composition. For instance, bottled spring water contains trace amounts of certain minerals known to have a subtle, pleasant taste. Municipal tap water may contain disinfecting chemicals that give it an unpleasant taste and smell.

All information about the physical world comes to us through our senses. The information arrives through direct contact with the skin or taste buds, through airborne molecules intercepted by the organs of smell, through light waves entering the eyes, and through sound waves entering the ears. The brain learns of this information from electrical impulses sent through the nervous system. Light and sound are two forms of **energy,** something that can cause change. Tangible things—a glass of water, for instance—we call **matter.** Both matter and energy are physical; they are part of the physical world.

Human sensations are a starting point for identifying what belongs to the physical world. Our sensations, however, do not always lead us to what is physically "real." Most people have experienced mild hallucinations due to illness or extreme fatigue. Under these circumstances, false sensations can be evoked in our brains. We can see, hear, or otherwise "sense" things that have nothing to do with the messages sent by our sense organs.

Even when hallucinations are not involved, we are often mistaken about the nature of physical things. A case in point is the optical illusion shown in Figure 1.1. Another illusion you can easily demonstrate for yourself shows how subjective the sense of touch is: Immerse one hand in a container of hot water and the other hand in ice-cold water, and leave them there for 10 seconds or so. Then remove both hands and place them in a third container filled with cool or lukewarm water. How would you describe the temperature of the water in the third container?

In a world full of sensory information and illusions, how can we find objectivity and truth? We can resolve the optical illusion of Figure 1.1 with the help of a ruler. With this measuring tool or instrument, we can make a measurement—an evaluation of length—that is less subject to perceptual

**PHOTO 1.1** The full moon rising over San Diego looks large in this photograph because the photograph was shot with a telephoto lens. The scene appears magnified, as if viewed through binoculars. In magnifying a scene, however, we miss what we would ordinarily see with our peripheral vision. The mental processing that goes on in normal vision can give rise to a similar effect: We may subconsciously ignore all but the arresting details of a real-world scene and thus miss everything else around it. The arresting detail here is the moon looming over the tall buildings. That makes the moon *seem* bigger. Actually, the buildings in the photograph were 4 miles away from the camera; they do not appear large to the naked eye at that distance. If the buildings are really small when seen at that distance, then so, too, is the moon.

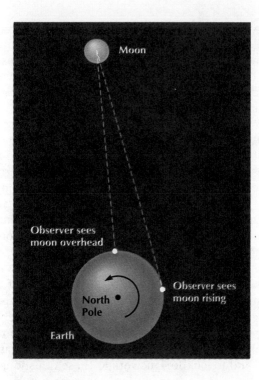

**FIGURE 1.2** The moon is slightly closer, and therefore appears larger in the sky, when it is overhead. Diagram not to scale.

distortions. For the hot- and cold-water experiment, we can make an objective measurement of temperature by using a thermometer.

Today, we routinely dispel illusions by using instruments that measure or record physical properties. Have you ever seen a huge, bubblelike full moon hovering above the horizon? Most people claim that the moon is bigger when it is near the horizon than when it is high in the sky (see Photo 1.1). We can dispel this "moon illusion" by repeatedly photographing the moon as it moves upward from the horizon. Amazingly, the moon's image on film actually *grows* slightly as it rises. Figure 1.2 shows why.

Instruments can assist us in making measurements that are more precise and accurate than those made by the senses alone. **Precision** refers to how consistent repeated measurements of the same quantity are. **Accuracy** refers to how close the measurement of a quantity comes to the accepted or true value of that quantity. Any instrument must be adjusted properly, or *calibrated,* in order to give accurate measurements. (You will deal with precision and accuracy in measurement if you take a laboratory course in the physical sciences.)

Instruments do more than help remove distortions of perception; many operate as supersensitive extensions of our senses. For example, telescopes, especially those equipped with electronic cameras, have broadened our range of sight dramatically. The latest microscopes allow us to detect, measure, and even manipulate objects thousands of times smaller than a grain of sand. Microphones and electronic amplifying equipment have greatly extended our natural hearing abilities. Electrochemical sensors can detect certain odors or airborne chemicals far better than humans—though in many cases not as well as bloodhounds or other animals with an extremely acute sense of smell.

Although the physical world consists of many things with measurable properties, our knowledge of these things is sometimes far from complete. We may not have the proper instruments necessary to investigate a particular object or phenomenon. We may not have the theoretical or mathematical tools to make sense of what is going on. Or we may not have the mental agility to answer certain questions.

We sometimes encounter difficulties when we attempt to trace past events or to make predictions about the future. Some of the most profound questions we can ask about the physical world are related to events of the far past or the projected future. How (and more audaciously, why) did the universe begin? How did life arise and develop on Earth, and has it come into being elsewhere? What is the ultimate fate of the universe? Fortunately, there is no shortage of clues available to help us answer these questions. The very structure of matter in our world today is a consequence of events in the earliest history of the universe. Biological fossils by the billions wait for their discovery within the rock layers of Earth's crust. "Fossil light" comes to us from distant parts of the universe after journeys lasting billions of years. Nevertheless, the important questions about our existence may never be resolved conclusively or to the satisfaction of everyone. If the past is a guide to the future, however, we will gain a better understanding of the physical world as long as we continue to ask questions and to look for answers to these questions.

## Order from Chaos

Chaos implies disorder: a state in which chance rules. Order implies consistency and predictability. The world around us has plenty of both. When we consider how powerless we are in the face of unpredictable or poorly predicted disasters such as hurricanes, earthquakes, and even social and political upheavals, the world seems chaotic. At the same time, some things never change: We can always depend on the certainty of night following day and day following night.

One of the primary tasks of scientists who study the physical world is to extract order from disorder—to recognize some kind of regularity within apparent chaos. They often proceed in this task by dividing complex structures into simpler parts and by describing complex phenomena in terms of more fundamental processes, or *laws of nature*.

In the ancient Greek culture, natural philosophers (counterparts of today's scientists) searched for order amid chaos, just as modern scientists do. The Greeks, however, divided their universe into two parts: the heavens above, where the perfectly formed sun, moon, planets, and stars moved with stately precision; and Earth beneath, with all its imperfect messiness. The heavens and Earth were considered to be separate realms, subject to different laws.

About four centuries ago, during the scientific revolution that accompanied the Renaissance in Europe, Nicolaus Copernicus, Johannes Kepler, Galileo Galilei, Isaac Newton, and others introduced the idea that Earth, like other planets, is merely a part of the universe. They showed that the same physical laws that operate on Earth apply everywhere else as well (as we shall emphasize in the chapters ahead). The germ of this idea was not entirely new; many cultures on other continents accepted the same general notion. The European scientists, however, were the first to link mathematics, a nonphysical system of thought, to the new theory of physical world order.

Today, mathematics is the universal language of science. Most of the fundamental laws of physics can be expressed in terms of relatively simple mathematical formulas, a few of which you will study in this book. Scientists have shown repeatedly that order can be drawn from chaos in many familiar kinds of phenomena—yet no one still knows *why*.

Chemists of the scientific revolution era, such as Antoine Lavoisier and John Dalton (Chapter 4), revived earlier ideas about the atomic makeup of matter and showed how tiny particles called *atoms* form the various substances we see with our eyes, hold in our hands, and measure with the help of analytical instruments. Today, the atom is recognized as an aggregate of smaller particles, some of which are further divided into still-smaller and more basic particles. Furthermore, physicists now are more confident than ever that all natural processes operating at present in the universe are controlled by no more than four **fundamental forces,** or *fundamental interactions.* These four—the strong, weak, electromagnetic, and gravitational forces—will be introduced in Chapter 2 and referred to many times in this book.

## Space and Time

We noted earlier that matter and energy belong to the physical world. But what about space and time? Do they have measurable characteristics that would make them just as "physical" as matter and energy? Yes! In the modern view, space is not merely a vacuum (the absence of matter). Both matter and energy can move through space. The behavior of anything moving in space depends on a characteristic called the *geometry of space* (the shape of space), and the particular geometry in turn depends on the presence of matter, near or far. Similarly, our measurement of any interval of time depends on such factors as the motion of the device keeping time and the presence of matter in the space surrounding the timekeeping device. So since space and time have characteristics that can change and be measured, they, too, are part of the physical world. To summarize: *Space, time, matter, and energy constitute what we call the physical world.*

Ways of defining space, time, matter, and energy—or anything else physical, for that matter—usually involve some kind of circular explanation. For example, any useful definition of time must involve the movement of something (matter or energy) through space. What about energy? Energy implies change, and change is associated with time. Rather than concern ourselves with absolute definitions of time, space, matter, and energy, it is more fruitful simply to use them as tools for measurement. For example, we can make a definitive statement that a certain basketball player is 2 meters tall, without worrying about what a meter (or any other measure of space) really is. We simply take a meterstick and use it to measure the player's height.

Albert Einstein, one of the most gifted and perceptive physicists of the 20th century, showed how space and time can have different characteristics for different observers. He also demonstrated that matter and energy are equivalent to each other, that space and time are inseparably linked, and that matter and energy together are connected to space and time. His theories of relativity, which are based on a few simple assumptions about nature, and the physical consequences drawn from these conclusions unified the disparate concepts of space, time, matter, and energy. For most of this book, we will not be concerned with these deep interrelationships, although we will touch on them occasionally.

## Properties and Measurements

Since physical things have properties, we must have a clear idea of what these properties are, and we must be able to measure them. Properties associated with material bodies include weight, hardness, temperature, and speed. Properties associated with light include color (a rough measure of what physicists specify as wavelength) and intensity. It is remarkable that every physical property you can imagine can in some way be expressed in terms of one or more of only four **fundamental properties:** length, time, mass, and charge. These properties are fundamental in that they cannot be expressed in simpler terms.

## The Four Fundamental Properties

**Length** (also called distance) is measured in one dimension of space at a time: right to left, north to south, or in any direction you choose. **Time** is also one-dimensional (the past is behind us, the future is ahead of us, and there are no side paths as we march through time). **Mass** can be defined as quantity of matter. On Earth, mass is closely related to weight. In addition to mass, matter may also have **charge** (electric charge). Combinations of these fundamental properties give rise to all the other properties, which are called **derived properties.**

## Derived Properties

Properties like area and volume are not fundamental, because they can be broken down into two or more separate measurements of length. Imagine a rectangular tile. A single measurement of length does not allow you to determine the area of the tile's surface. You must know both how long and how wide the tile is. The property called area thus depends on *two* measurements of length. A tile's area is specified in terms of two applications of length, such as square inches or square feet.

Now, think of a rectangular block. *Three* measurements of length are required to determine its volume. A block's volume is specified in terms of three applications of length, such as cubic inches or cubic feet.

Speed, a familiar characterization of bodies that move, relies on measurements of length and time. Density—by definition, mass divided by volume—relies on measurements of volume (which is entirely derived from

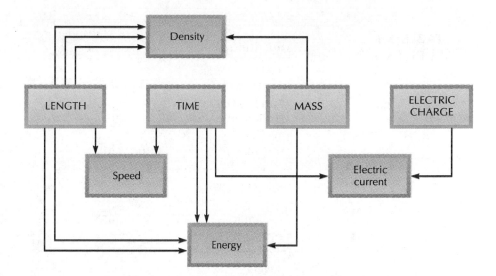

**FIGURE 1.3** All physical properties can be specified in terms of one or more of the four fundamental properties: length, time, mass, and electric charge. Included in this diagram are several derived properties, such as speed, which is defined in terms of length (or distance) and time. Electric current is defined in terms of electric charge and time. Density depends on mass plus the three dimensions of space (or length) that are used to define volume. Energy can be defined in terms of one measurement of mass, two measurements of length, and two measurements of time.

length) and of mass. Energy can be expressed in terms of mass, length, and time. Area, volume, speed, density, and energy are all derived properties because each one incorporates two or more applications of the fundamental properties (see Figure 1.3). In the coming chapters of this book, we will take a closer look at each of the fundamental properties and many of the derived properties.

# Units and the SI

Measurements of any physical property, fundamental or derived, must be expressed in terms of a number and a unit. A **unit** is a well-defined and agreed-upon value of a measurable property. Inches, feet, miles, meters, and kilometers are examples of units in common use for measuring length or distance. When specifying a physical property, be sure to include the units that go with the measurement. It is meaningless to say that your height is 2 but quite reasonable to say that your height is 2 meters—especially if you're a basketball player.

Although the old British system of units (feet, pounds, quarts, and so on) remains in common use in the United States, scientists worldwide have adopted the **International System of Units** (Système Internationale d'Unités), denoted as SI in all languages. The SI is the modern form of the metric system first proposed more than two centuries ago in France. The SI is organized so that multiples or fractions of any unit can be easily expressed in terms of powers of ten. Prefixes (or prefix abbreviations) are commonly used in place of the powers-of-ten notation, as noted in Table 1.1.

| TABLE 1.1 | Unit Multiplier | Unit Multiplier (Power of 10)* | Prefix | Prefix Abbreviation |
|---|---|---|---|---|
| SI Prefixes | 1,000,000,000,000 (one trillion) | $10^{12}$ | tera | T |
| | 1,000,000,000 (one billion) | $10^{9}$ | giga | G |
| | 1,000,000 (one million) | $10^{6}$ | mega | M |
| | 1000 (one thousand) | $10^{3}$ | kilo | k |
| | 1/10 (one-tenth) | $10^{-1}$ | deci | d |
| | 1/100 (one-hundreth) | $10^{-2}$ | centi | c |
| | 1/1000 (one-thousandth) | $10^{-3}$ | milli | m |
| | 1/1,000,000 (one-millionth) | $10^{-6}$ | micro | $\mu$ |
| | 1/1,000,000,000 (one-billionth) | $10^{-9}$ | nano | n |
| | 1/1,000,000,000,000 (one-trillionth) | $10^{-12}$ | pico | p |

*Powers of ten are explained in Appendix A.

The SI defines seven base units, three of which are applicable to the material presented in Parts I and II of this book. The base unit for length is the **meter** (m), which is equivalent to about four times the length of this textbook. Originally, the meter was expressed as one-ten-millionth of the distance between Earth's equator and the North Pole. Today, it is defined to be "the distance traveled by light in a vacuum during 1/299,792,459 second." This way of defining the meter is useful because the speed of light in a vacuum is constant. Some common multiples of the meter you will see in this book are the kilometer (km), centimeter (cm), millimeter (mm), and nanometer (nm).

For mass, the base SI unit is the **kilogram** (kg). This textbook has a mass of about 1.5 kg. Officially, the kilogram is defined to be the exact mass of a prototype object stored at the International Bureau of Weights and Measures in France. Second-generation 1-kg prototypes used by other countries have had their mass carefully compared with that of the original in France. Earlier, the gram (1/1000 kg) was considered the base unit of mass in the metric system.

The **second** (s) is the base SI unit for time. Originally, the second was defined as 1/86,400 day (there are $60 \times 60 \times 24 = 86,400$ s in a day). After Earth's rotation was shown to vary slightly over time, it became necessary to adopt a stricter standard. Today, precise measurements of time are accomplished by atomic clocks. Officially, the second is now defined as the duration of a specified number of vibrations made by a specific type of

**PHOTO 1.2** This second-generation standard kilogram is housed in a double bell jar at the U.S. National Institute of Standards and Technology.

atom. Atomic clocks and the new definition of the second are discussed in Chapter 3.

We will frequently use metric units in this book. To give you a good sense for the magnitude of some metric units when they are introduced, we will occasionally compare them with the equivalent British system units. If you wish to make conversions from one system to another, see the formulas and tables in Appendix C.

# Science and Its Methodology

Science is a systematized body of knowledge pertaining to the natural world. Scientists are interested in the whats, wheres, whens, hows, and whys of nature. The term *science* also applies to the *process* of seeking answers to questions posed about nature. The methodology used to acquire these answers is commonly called the **scientific method** of inquiry. Scientific methodology includes the processes of gathering and analyzing information, developing some kind of interpretation, and testing the interpretation. These steps are diagramed in Figure 1.4.

## Scientific Methodology

The gathering of information is an essential feature of scientific methodology. This is done by *observation* or by performing an *experiment* and then observing the results. The results can be qualitative (such as the discovery of a new phenomenon) or quantitative. When quantitative measurements are involved, the results are known as *data*.

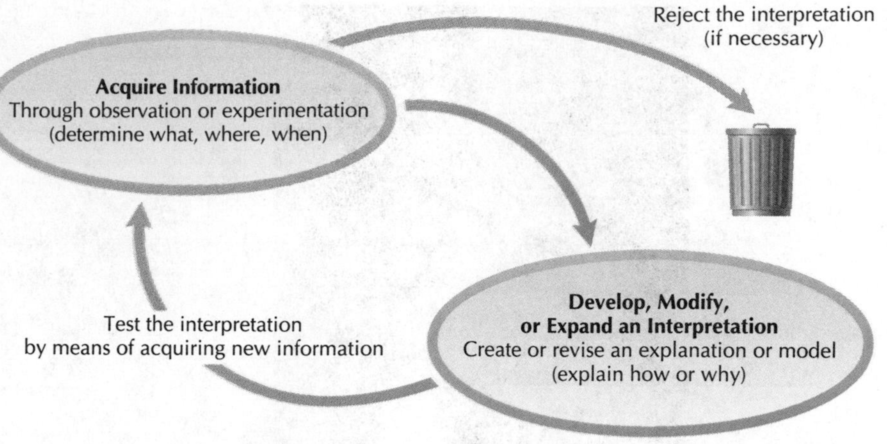

**Acquire Information**
Through observation or experimentation
(determine what, where, when)

Reject the interpretation
(if necessary)

Test the interpretation
by means of acquiring new information

**Develop, Modify,
or Expand an Interpretation**
Create or revise an explanation or model
(explain how or why)

**FIGURE 1.4**   Scientific methodology (often referred to as the scientific method) involves relentless testing of any interpretation developed for the purpose of explaining natural phenomena. An interpretation that fails any stage of testing must be either revised or rejected. An interpretation supported by extensive testing may gain consensus among the scientific community, but it is never considered an absolute truth. An absolute truth, such as the mathematical statement $2 + 2 = 4$, needs no further testing. No scientific interpretation leaves the "loop" above, unless it is rejected.

Scientists typically gather data in laboratories, where experiments can take place under controlled (or at least known) conditions. This is especially true of the work chemists and physicists do. Sometimes, direct experimentation is not possible and observation alone must suffice. No one, for instance, has ever removed material from a star and transported it to Earth for analysis. Astronomers, however, use telescopes to collect starlight. An amazing amount of information can be extracted from a star's light by using a technique called spectral analysis, or spectroscopy. No one has ever brought a piece of Earth's core to the surface for analysis in the laboratory either, but geologists have gained some knowledge of what is in the core by means of observing and analyzing seismic (earthquake) waves traveling through Earth.

The process of acquiring information, by experimentation or by observation, often leads to the discovery of something new. Discoveries can result from a researcher's deliberate search for something he or she suspects exists, or a discovery may be unanticipated or purely accidental. Many discoveries are made during the course of surveys (the wholesale gathering of data). Thus, sky surveys have revealed entirely new types of celestial objects and phenomena never witnessed before. Comprehensive studies of the deep ocean floor have revealed structures that have helped revolutionize the earth sciences.

Scientists are always anxious to acquire the use of new and improved equipment—bigger telescopes, more powerful lasers, or bigger particle accelerators. Better tools such as these are better able to detect and measure faint sources of light or subtle phenomena. Much of the ordinary work in science, however, is done behind the cutting edge of discovery. Initial findings often involve weak or uncertain measurements, and further measurements are usually needed to confirm earlier results. Sometimes, follow-up work can be done with less than state-of-the-art equipment.

In the early stages of a scientific quest, observation and experimentation are most helpful in answering the questions *what, where,* and *when*:

- The question *what* is often addressed by means of classification. Observable things can be grouped by similarities, and labels can be applied to each group. For instance, matter can be classified as being in a solid, liquid, or gaseous state. Many decades ago, astronomers separated galaxies into four basic types—spiral, barred spiral, elliptical, and irregular—long before they knew much more about galaxies than their shape. The 19th-century chemists tried to make sense out of the tremendous diversity of the known elements by recognizing that certain elements could be grouped together by similar chemical behavior. Their efforts eventually led to today's periodic table of the elements (which we shall discuss in detail in Chapter 15). The question *what* can be addressed quantitatively, as well. When a geologist wishes to identify a certain mineral substance, the process may involve precise measurements of the mineral's density and other quantifiable properties.

- The issue of *where* can often be settled by defining a coordinate system and then specifying where a body is located relative to some reference point. (Various coordinate systems are introduced in Chapter 2.) Our knowledge of *where* depends on the yardsticks (or metersticks)—literally or figuratively—we use for measuring distance. Generally, the farther away a body is, the more difficult it is to measure its distance. This is especially true in astronomy. We can precisely measure the distance to neighboring bodies like the moon, but distances to remote galaxies are really *estimates* based on certain commonly accepted assumptions. These assumptions are far from arbitrary; they rest on the belief that nature exhibits consistent patterns of behavior both near and far.

- The question *when* is more difficult, because it often involves extrapolation into the past or future. For example, our confidence in predicting that Earth will be on the opposite side of the sun six months from now rests on the fact that Earth has always made half a turn around the sun in every half year during the past. But how can we be absolutely sure that some massive interloping object might not swoop into the Solar System unexpectedly and bump Earth from its present orbit? (The probability of this event happening is real but vanishingly small.) Just as measurements of where something is become more uncertain with increasing distance, scenarios of the past or future tend to become less certain over greater time intervals. Our knowledge of past events in geologic and cosmic history rests on evidence that cannot always be interpreted clearly.

Scientific methodology enters its most interesting phase after the basic issues of *what, where,* and *when* are taken care of, at least at some level of detail. The deeper questions to be addressed are *how* and *why*. Answers to *how* and *why* questions involve some kind of interpretation that is subject to further testing by experimentation or observation. This can take several forms: The simplest is a **hypothesis,** a provisional explanation—sometimes only an educated guess—that is often used as a guide to further investigation. A hypothesis must be rejected (or modified) if new evidence contradicts it.

A hypothesis can become a scientific **principle** (or *law*) if it survives the rigors of extensive testing through further experimentation or observa-

**PHOTO 1.3**  Sir Isaac Newton (1642–1727) advanced our understanding of motion, gravity, heat, light, and optics. He also shares credit for the invention of calculus, a mathematical technique that underlies most of the detailed computations of physical science today. Many of Newton's findings and conclusions were published in the *Principia*, which many scholars have singled out as the most important book in the literature of science.

tion. Often, a scientist testing an interpretation will devise an experiment and make predictions about its results. If the predicted results are found to be true, the interpretation is supported. On the other hand, the failure of a prediction either proves the interpretation false or forces the scientist to modify it.

Principles and laws usually describe *how* something works. Newton's law of gravitation (Chapter 5) describes *how* masses attract each other by a specific kind of force called gravity. Archimedes' principle (Chapter 7) tells something about *how* objects behave when immersed in a fluid. (A principle or a law is often named after the scientist who first discovers or develops it. *Newton's laws* refer to several important laws governing motion and gravity developed in the 17th century by Sir Isaac Newton, arguably the greatest scientist in history. *Archimedes' principle* is named after the ancient Greek philosopher-scientist Archimedes.) Principles and laws can be expressed in different ways. For example, Archimedes' principle is commonly written as a single, succinct sentence. Newton's law of gravitation is most clearly stated as a mathematical equation. Often, scientific laws can be represented by using the framework of a simple graph.

While a principle or a law may be helpful in explaining how something works, a **theory** often goes one step further. A scientific theory is a detailed explanation covering some general aspect of nature. Most scientific theories are so broad and have so many consequences that several people contribute to their development (and sometimes to their demise).

Theories strive to explain *why* (in a mechanical sense, not a philosophical sense), not just *how*. Newton's law of gravitation is an excellent description of *how* gravity works, but today we have an even better description of gravity, known as Einstein's general theory of relativity (Chapter 13). The general theory of relativity explains *why* gravity (in the sense that Newton pictured it) works in terms of certain basic and very general relationships among mass, space, and time.

Theories often take the form of scientific **models,** or metaphors that we can readily understand. A model can be a simplified version of reality, devised to bring out the basic character of a complex or multifaceted phenomenon. One example is the *heliocentric* (sun-centered) *model* of the Solar System advanced by the 16th-century astronomer Nicolaus Copernicus. As we shall see in Box 3.4, the heliocentric model eventually replaced a much older and widely accepted model in which Earth occupied the center of the universe. In time, the basic character of the Copernicus model was shown to be correct, though some of its details had to be modified in order to agree with observations that became more and more refined.

In practice, the distinctions between a hypothesis and a law, or a law and a theory, are often blurry. Sometimes, hypotheses quickly become known as laws after substantial (but not exhaustive) testing. When people theorize from the results of certain experiments or observations, they may really be doing nothing more than inventing a new hypothesis or comparing their results with already accepted laws, principles, or theories. Sometimes, a powerful theory will spring from the mind of a highly creative individual without, at first, much evidence to support it. Einstein's relativity theories were intuitive creations based on "what-if" scenarios, or "thought experiments." Decades' worth of subsequent experimentation and observation have shown that these revolutionary theories were not just clever but also correct—as far as we know today.

To summarize: Observational and experimental testing can eliminate an incorrect scientific interpretation or render it as an approximate model of some better interpretation. But testing can never prove anything as absolute truth. Consistent, positive results in tests only imply that the interpretation is provisionally correct. A viable scientific interpretation can never exit the loop shown in Figure 1.4.

# The Limits of Science

Science deals with measurable and quantifiable aspects of nature, but it cannot encompass all of what we perceive as reality. Human beings experience love, joy, sadness, grief, beauty, faith, and courage. These emotions and abstractions, which cannot be measured or quantified, are difficult or impossible to deal with in science. They lie within the realm of the arts and humanities. Similarly, science fails to answer ultimate *why* questions: Why do we humans exist? Why does the universe exist? Questions like these belong to the realms of philosophy and religion.

Science may not have all the answers, but it has changed and will continue to change the world in profound ways. Basic science (science for its own sake) will continue to drive tomorrow's applied science and technology, and that affects us all. Technological innovations bring us both benefits and harm. The same marvel of technology, the laser, that speeds our passage through supermarket checkout counters and eases the trauma of surgery has the capability of being used as a weapon in war. It is up to all of us—not just scientists—as citizens of the world to make the right decisions that will steer scientific knowledge in beneficial directions. The arts, humanities, philosophy, and religion can help us in that effort.

## CHAPTER 1
# Summary

The physical world consists of matter, energy, space, and time—all of which have measurable properties or characteristics.

The properties of physical things can be quantified in terms of one or more of the following basic concepts: mass, length, time, and electric charge.

Any measurement of a physical property can be expressed as a number coupled with a standard unit. The metric (SI) system of units is the adopted standard of measurement in the physical sciences.

The goal of scientific methodology is to arrive at better explanations of how nature works. These explanations (which can be called hypotheses, laws, principles, and theories) are never taken to represent absolute truth; rather, they are ideas subject to further testing and possible revision or even rejection. Sometimes, to simplify and bring out the basic character of complex phenomena, scientists devise models, or simplified explanations.

## CHAPTER 1
# Questions

## Multiple Choice

1. A description of any aspect of the physical world must always involve
   a) things humans can detect with the five senses
   b) mathematical theorems
   c) the measurement of one or more physical properties
   d) the characteristics of matter

2. Any physical measurement must be expressed in terms of
   a) a decimal number
   b) one or more units
   c) a number and a unit
   d) a number and one of the basic units of length, mass, and time

3. The four fundamental physical properties are
   a) matter, energy, space, and time
   b) length, time, mass, and electric charge
   c) earth, air, fire, and water
   d) length, width, depth, and time

4. The SI units for length, mass, and time are the
   a) meter, gram, and second
   b) kilometer, kilogram, and second
   c) centimeter, gram, and second
   d) meter, kilogram, and second

5. A scientific model is a (an)
   a) exact description of some natural process
   b) simplified description of some natural process
   c) scale model of any structure of any size in the universe
   d) serious-looking person wearing clothes for sale

6. The scientific method often involves
   a) providing many alternative explanations for the same phenomenon
   b) proving mathematical theorems
   c) devising testable predictions about physical behavior
   d) the attainment of absolute truth

## Questions

1. All scientific hypotheses or theories explain something about nature or the physical world. What other essential quality of a hypothesis or theory makes it scientific?

2. Which of the various branches of physical science relies most on pure observation (as opposed to experimentation)?

## Problems

1. How many nanometers (nm) are there in 1 kilometer (km)?

2. If a computer hard disk has a capacity of 230 megabytes of information, how many kilobytes of information does it have? How many bytes of information does it have? (A byte of information is equivalent to 8 bits. Each bit is the equivalent of the answer to a single yes-no question.)

## Questions for Thought

1. Pick up an object of your own choosing, and make a list of as many physical properties as you can think of that pertain to it. How would you go about measuring each of its properties?

2. In commerce and manufacturing, as well as in science, almost all of the world's countries have adopted the metric system. In the United States, the effort of making the conversion from the old-style, awkward British units to metric units has stalled. Why do you think it has?

3. Why is the statement $2 + 2 = 4$ not considered a scientific hypothesis?

## Answers to Multiple-Choice Questions

1. c     2. c     3. b     4. d     5. b     6. c

# The Universe of Space

*Our universe is built on a vast hierarchy of scale. Atoms are a thousand million million times smaller in diameter than Earth. The universe is more than a thousand million million million times larger in diameter than Earth. Here, Earth is rising over the pocked, utterly barren surface of our nearest celestial neighbor, the moon—a "mere" 240,000 miles away.*

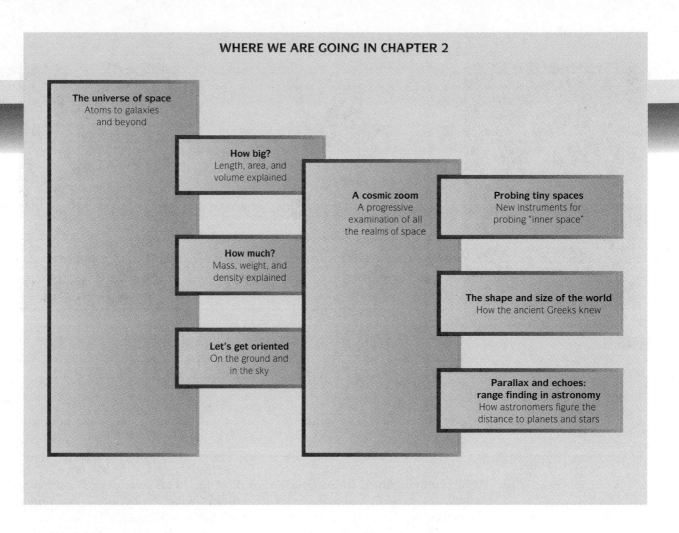

**WHERE WE ARE GOING IN CHAPTER 2**

**The universe of space**
Atoms to galaxies
and beyond

**How big?**
Length, area, and
volume explained

**How much?**
Mass, weight, and
density explained

**Let's get oriented**
On the ground and
in the sky

**A cosmic zoom**
A progressive
examination of all
the realms of space

**Probing tiny spaces**
New instruments for
probing "inner space"

**The shape and size of the world**
How the ancient Greeks knew

**Parallax and echoes:
range finding in astronomy**
How astronomers figure the
distance to planets and stars

Space, as Webster's dictionary defines it, is "a boundless, continuous expanse extending in all directions or in three dimensions, within which all material things are contained." In this chapter, we will study many realms of space, small and large, and become familiar with the small and large units of matter that make up our universe.

Space is a kind of stage on which changes are played out. The planets of our Solar System whirl in an orderly way through a space that is practically devoid of matter. In the air we breathe, tiny particles jiggle and jostle about, darting for small fractions of a second through what is basically empty space before colliding with neighboring particles. Inside a solid piece of rock, where it may seem that there is no room at all for anything to move, tiny units of matter called *atoms* seethe and vibrate. Within the atom itself, subatomic particles of matter whirl and dance in a confined region of space.

The space we experience in everyday life—which may span a city, a state or even a continent—strikes a middle ground in terms of size. We know how matter behaves on *macroscopic scales* like these. We are

a

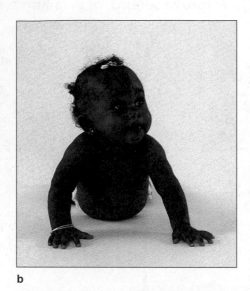

b

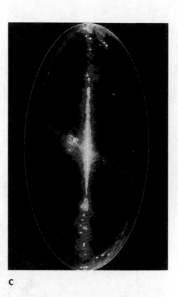

c

**PHOTO 2.1**    Anything smaller than an aphid (a) can be considered microscopic in size. Using our five senses alone, we humans can explore the macroscopic world of ordinary size (b). The supermacroscopic, or astronomical, scale of the Milky Way galaxy (c) lies mostly beyond our direct experience.

acquainted with natural behaviors we can see or feel: Heavy objects, when dropped, always fall. Sunlight warms the day. Liquids and gases flow, but solids do not. Other natural behaviors, not directly accessible to our senses, exist on *microscopic scales* and on *supermacroscopic* (very large, or astronomical) *scales.* Some of these behaviors satisfy our intuitive sensibilities about how things are supposed to work. Other kinds of behavior, especially that of very tiny units of matter, confound what we take to be common sense.

Our overview of space, which constitutes the central theme of this chapter, starts with a realm smaller than the atom and ends with the universe itself. We will present this overview by means of a "cosmic zoom," an imaginary zoom through microscopic, macroscopic, and supermacroscopic space. The range of length or distance we are covering in the zoom is so enormous that it is awkward to use an ordinary, linear scale of length or distance to specify and illustrate the relative sizes of things. Instead, we will utilize an exponential scale based on powers of ten. A look at Figure 2.1 will reveal the "power" of a powers-of-ten scale—its ability to encompass and easily distinguish among a wide range of numbers or quantities.

Scientists frequently employ powers of ten for rough estimates, especially when they need to express very large numbers or compare two quantities very different in magnitude. A scientist, for example, would say that a baseball is approximately $10^9$ times larger in diameter than a carbon atom. The $10^9$ factor is equivalent to 1 followed by nine zeros: 1,000,000,000, or 1 billion.

For precise specifications, scientists often make use of *scientific notation,* a shorthand technique of writing very large or very small numbers. Scientific notation, along with powers of ten, is reviewed in Appendix A. Since both are used fairly frequently in this book, you should be familiar with them.

This chapter's preliminary material includes the sections "How Big?," "How Much?," and "Let's Get Oriented." By going over this material, you

**FIGURE 2.1** On the linear scale (top), each interval to the right represents a fixed increase of the same number of units—in this case 5 m per interval. On the exponential scale based on powers of ten (bottom), each interval to the right represents an increase of 10 times as much. The linear scale allows you to picture differences between heights of a few meters to tens of meters. The exponential scale covers a range of thousandths of a meter to thousands of meters in height. Its beauty is that you can easily make comparisons between small things (mice and ants) and also between large things (airplanes and buildings) on the same scale.

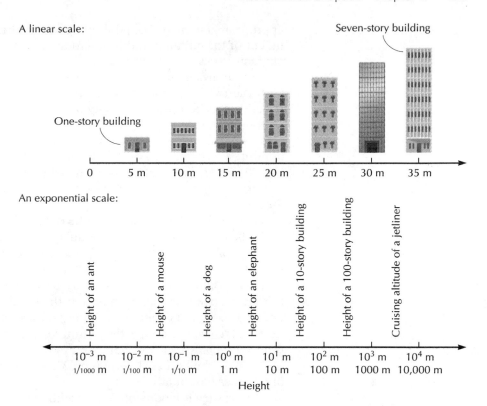

A linear scale:

Seven-story building

One-story building

0   5 m   10 m   15 m   20 m   25 m   30 m   35 m

An exponential scale:

Height of an ant

Height of a mouse

Height of a dog

Height of an elephant

Height of a 10-story building

Height of a 100-story building

Cruising altitude of a jetliner

| $10^{-3}$ m | $10^{-2}$ m | $10^{-1}$ m | $10^{0}$ m | $10^{1}$ m | $10^{2}$ m | $10^{3}$ m | $10^{4}$ m |
| 1/1000 m | 1/100 m | 1/10 m | 1 m | 10 m | 100 m | 1000 m | 10,000 m |

Height

will gain a solid understanding of some of the basics of measurement relating to size and location.

## How Big?

Measurements of size always involve the fundamental property of length. In simple terms, we could describe the size of the glass box skyscraper of Figure 2.2 as being only its height, $h$. (Height in this case is essentially a single measurement of length.) This one-dimensional viewpoint is certainly

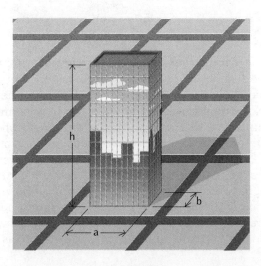

**FIGURE 2.2** The size of this building can be expressed in terms of height, area, or volume.

appropriate for an airline pilot flying at low altitude over the city, since the height of the building in this situation takes precedence over its other size characteristics.

A mapmaker, on the other hand, may be concerned not with height but with the two horizontal dimensions (lengths $a$ and $b$) of the building that constitute its "footprint" on a city block. The building would appear simply as a flat rectangle on a city street map. When $a$ and $b$ are perpendicular (at right angles) to each other, as assumed in this case, the *area* ($A$) of the building's footprint can be easily calculated: $A = ab$. For example, a building 30 meters (m) long and 20 m wide would have a footprint area of (30 m)(20 m) = 600 $m^2$. The unit $m^2$ (read "square meters") is the SI unit for area.

The engineers responsible for heating or cooling the building would naturally be concerned with the amount of air contained within it. They would need to consider *volume* ($V$), which takes into account all three of the building's dimensions $a$, $b$, and $h$. For our box-shaped building, where the $a$, $b$, and $h$ measurements lie in mutually perpendicular directions, the calculation of volume is simply $V = abh$. If the building is 30 m long, 20 m wide, and 100 m high, its volume is (30 m)(20 m)(100 m) = 60,000 $m^3$. The unit $m^3$ (read "cubic meters") is the SI unit for volume. Other metric units of volume, commonly used by chemists, are the liter ($10^{-3}$ $m^3$) and the milliliter ($10^{-6}$ $m^3$). One liter is slightly more than one U.S. quart, and a milliliter is one-thousandth of a liter.

Interesting relationships occur when we compare the length, area, and volume of one object with the same properties of another object having the same shape but different size. Using cubes for simplicity, we will investigate the relationships among length, area, and volume. For each of three cubes, 1 m, 2 m, and 3 m on a side (pictured in Figure 2.3), let us tabulate the cross-sectional area (the area of one side), the surface area (the total area of all six sides), and the volume (see the table).

| Length of One Side | Cross-sectional Area | Surface Area | Volume |
|---|---|---|---|
| 1 m | 1 $m^2$ | 6 $m^2$ | 1 $m^3$ |
| 2 m | 4 $m^2$ | 24 $m^2$ | 8 $m^3$ |
| 3 m | 9 $m^2$ | 54 $m^2$ | 27 $m^3$ |

A careful look at the table reveals certain patterns. Both area (either type listed) and volume increase faster than length for bigger and bigger cubes. If we double the length of a cube, the new area becomes 4 times greater than it was before, and the new volume becomes 8 times greater than it was before. If we triple the length of a cube, the area becomes 9 times greater, and the volume becomes 27 times greater.

We can generalize these trends, as follows: *Any area measurement of a cube increases in proportion to the second power (square) of the cube's length.* In symbols, this relationship can be written as $A \propto l^2$. The symbol $l$ means length, and $\propto$ means "is proportional to." Also: *The volume of a cube increases in proportion to the third power (cube) of the cube's length.* In symbols, this relationship can be written as $V \propto l^3$. These relationships

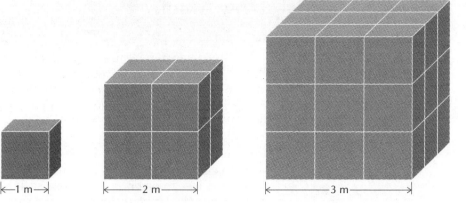

**FIGURE 2.3** Count the cubes. Eight 1-m cubes fit into the 2-m cube, and twenty-seven 1-m cubes fit into the 3-m cube.

work with other three-dimensional shapes as well. For spheres we can write $A \propto r^2$ and $V \propto r^3$, where $r$ is the radius (the distance between the center and the surface) of the sphere.

---

**Problem 2.1**

Cube A is 8 cm long and cube B is 4 cm long. How many times larger is the surface area of cube A than the surface area of cube B?

**Solution**

We will use the relationship $A \propto l^2$. Cube A's length is two times greater than cube B's length. Since $l$ is two times greater, $l^2$ is four times greater ($2^2 = 4$). The surface area $A$ is proportional to $l^2$; so cube A's surface area is four times greater than cube B's area.

---

**Problem 2.2**

Sphere X has a radius of 10 m and sphere Y has a radius of 1 m. How many times more surface area and volume does sphere X have than sphere Y?

**Solution**

For the surface area comparison, we will use the relationship $A \propto r^2$. Sphere X's radius $r$ is 10 times larger than sphere Y's radius. Since $r$ is 10 times greater, $r^2$ is 100 times greater ($10^2 = 100$). Area is proportional to $r^2$, so sphere X has 100 times greater surface area than sphere Y.

For the volume comparison, we use the relationship $V \propto r^3$. Since $r$ is 10 times greater, $r^3$ is 1000 times greater ($10^3 = 1000$). Volume is proportional to $r^3$, so sphere X has 1000 times greater volume than sphere Y.

---

These examples illustrate that when an object is scaled up in size, its surface area increases faster than its length across one dimension, and its volume increases faster than its area.

# How Much?

We can use volume to express how much material a body contains, but a more fundamental measure of the amount of something is its mass. As mentioned earlier, the mass of a body is the quantity of matter that body has. If you could tear down any body to the level of its subatomic particles and count all of them, you would have a good measure of mass. (In Chapter 5, we will define mass in two other ways.)

Mass, incidentally, is not the same as weight. Weight is the downward force exerted by a mass in a given gravitational environment. When you weigh yourself on a bathroom scale, the scale is really measuring force. If you are a 60-kg person, a bathroom scale shows your weight to be about 132 pounds (lb). If you, along with the same scale, were transported to the moon, where the surface gravity is 1/6 as much as on Earth, your weight according to the scale would be a paltry 22 lb. If you and the scale were floating around in the "weightless" environment of an orbiting space shuttle, you would not press against the scale at all, and its dial would register zero. You— and your mass of 60 kg—still exist, of course. The substance of your body does not disappear just because you have traveled to a different environment.

For homogeneous materials (materials that are the same throughout), mass and volume are proportional to each other. For example, 2 m³ of dry sand on a given beach has twice the mass of 1 m³ of another sample of the same kind of sand.

In common language we say that a brick is "heavy" and air is "light." However, if all the air contained in a typical classroom were compressed into the volume of an average-sized red brick and placed on a scale, the "brick" of compressed air would far outweigh a brick of clay. Even a gold brick of the sort kept at Fort Knox would fail to equal the weight of a classroom's worth of air. When we say that things are heavy and light, we often

**PHOTO 2.2** These two objects have the same mass. The metal weight has greater density because its volume is smaller.

| TABLE 2.1 | Substance | Density (kg/m³) | Density (g/cm³) |
|---|---|---|---|
| Densities of | Air (at sea level) | 1.29 | 0.00129 |
| Common | Ice | 917 | 0.917 |
| Substances | Water | 1,000 | 1.000 |
| | Aluminum | 2,700 | 2.70 |
| | Iron | 7,850 | 7.85 |
| | Gold | 19,300 | 19.3 |

are speaking of *density*. When scientists refer to *light gases* like hydrogen and helium, and *heavy metals* like lead and gold, they are referring to the relative densities of these substances.

**Density** is defined as mass divided by volume. Symbolically, this can be written as

$$\rho = m/V \hspace{4cm} \text{Equation 2.1}$$

where $\rho$ (the Greek letter rho) stands for density, $m$ stands for mass, and $V$ stands for volume. Since the SI units for mass and volume are kg and m³, respectively, the SI unit for density is kg/m³. Another metric unit for density in common use is grams per centimeter cubed (g/cm³). Table 2.1 gives the densities (rounded off ) of several common substances. Note from the figures given in the table that 1 g/cm³ = 1000 kg/m³. Notice, also, that water has a density of 1 g/cm³. This is no coincidence. The gram was originally defined to be the mass of 1 cm³ of water (at 4 °C temperature). Water in ice form is less dense than in liquid form—which is why ice floats in water. Gold, with a density of approximately 19 g/cm³, is the densest familiar substance.

**Problem 2.3**
If 4.4 metric tons of sand fills a bulldozer bucket having a volume capacity of 2 m³, what is the sand's density? (*Note*: 1 metric ton = 1000 kg.)

**Solution**
$\rho = m/V$ = 4400 kg/2 m³ = 2200 kg/m³. (This density can also be expressed as 2.2 g/cm³.) Sand sinks in water because it is denser than water.

## Let's Get Oriented

Whether we wish to navigate through the space of city streets or map the space that lies outside Earth, two kinds of information are needed: distance and direction. Distances must be measured or specified relative to some reference point, a starting point that is often assumed to be the observer. Directions are measured or specified relative to some agreed-upon reference direction, such as north, east, up, or down. The distance and direction of a body determined in this way give us a clear picture of the body's position in space. Let us examine some of the traditional ways of expressing position.

## Flat Grids

Old-fashioned towns often have a rectangular layout of streets and avenues. Figure 2.4 shows a flat, rectangular coordinate grid—a *Cartesian grid*—superimposed on a town map. The grid in our illustration was chosen so that the origin ($x = 0$, $y = 0$) coincides with the center of the town square, and it is arranged so that the positive $y$-axis points north and the positive $x$-axis points east.

Assuming that the town is flat, any point on any street (or the ground floor of any building) can be uniquely specified by just two numbers. For the manhole cover shown, the coordinates are $x = 3.5$, $y = 3$. Since the streets themselves are numbered, we can also say it is "in the middle of 3rd Street, halfway between 3rd Avenue and 4th Avenue."

How might we specify the exact location of, say, a certain pipe junction in a buried sewer line, or maybe a person in an office on the fourth floor of

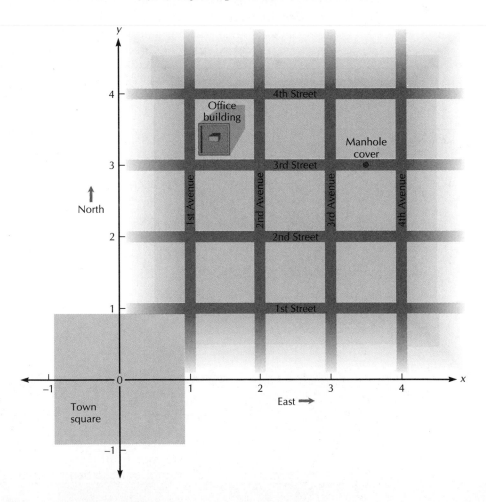

**FIGURE 2.4** Street grids of old-fashioned towns lend themselves well to Cartesian coordinates.

a building at 3rd Street and 1st Avenue? For that, we need a third coordinate (a $z$-coordinate) perpendicular to the $x$- and $y$-coordinates, on which we can specify depth or height.

## Polar Grids

On small scales, Earth's surface is considered to be flat, so flat maps and Cartesian grids work fine. When our planet is viewed from afar (Figure 2.5), however, it is clear that a flat Cartesian grid cannot be fit onto its curved surface. By distorting the grid, though, we can successfully wrap it completely around the planet. The center of curvature of the grid is taken to be Earth's center. (Picture the hub of a wheel being at the center of the wheel's rim. In a similar manner, Earth's center is surrounded by the grid that wraps around Earth's surface.) Any point on the grid can now be specified in terms of a set of two angles. The two poles of the grid are set on the north and south geographic poles of Earth, which are defined by Earth's rotational axis.

Earth's center serves as the apex of angles that mark **latitude.** Several parallels of latitude (east-west—trending circles) are shown in Figure 2.5. Earth spins parallel to these parallels of latitude. Earth's equator, which divides the globe into Northern and Southern hemispheres, is the largest parallel of latitude. Latitude values increase from 0° at the equator, to 90°N at the North Pole, and to 90°S at the South Pole.

Meridians of **longitude** run in the north-south direction and always cross parallels of latitude at right angles. Thus all meridians of longitude converge at both poles. On world maps and globes, longitude meridians are typically drawn in 15° increments because Earth turns on its axis 15° for every hour (abbreviated h) of time (15° multiplied by the 24 h in a day gives one complete 360° rotation). Note that these meridians are 15° apart, as measured from Earth's center, on the equator only. The meridians are more closely spaced near the poles, a consequence of the distortion that unavoidably results when bending a flat grid so that it fits the surface of

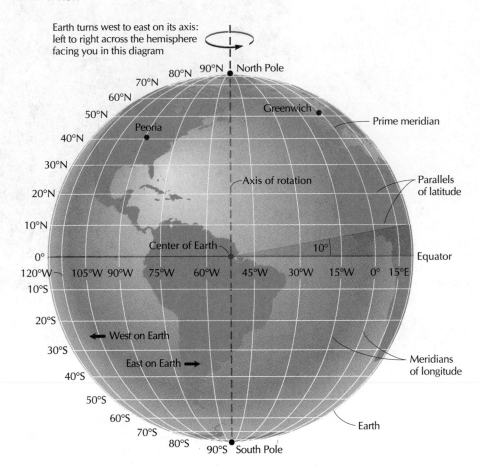

**FIGURE 2.5**  Earth's polar grid of latitude parallels and longitude meridians is based on Earth's equator and poles.

a sphere. Equal increments of longitude do not necessarily translate to equal increments of distance: Meridians 1° of longitude apart are separated by approximately 111 km at the equator, but the distance between the same meridians approaches zero near the poles.

Longitude is measured east and west from the so-called prime meridian (0° meridian) that passes through the Greenwich Observatory in England. England ruled the seas at the time that the latitude/longitude system was invented for purposes of navigation, so it is not surprising that its national observatory would be chosen as a point of reference. Longitude increases to a maximum of 180° both east and west of the prime meridian. Longitudes of 180° E and 180° W are identical. The international date line runs along or near this longitude meridian.

With a single ordered pair of numbers, one number for latitude and the other for longitude, any location on Earth's surface can be specified precisely. Peoria, Illinois, for example, has rough coordinates of Lat = 40° N, Long = 90° W. A more precise specification of location for a particular house in Peoria would be something like Lat = 40°41'27"N, Long = 89°42'09"W. The symbols ' and " refer to "arcminutes" and "arcseconds," respectively (1° = 60' and 1' = 60").

As noted earlier, a third coordinate would be needed to completely describe the location of something above or below Earth's surface. Mea-

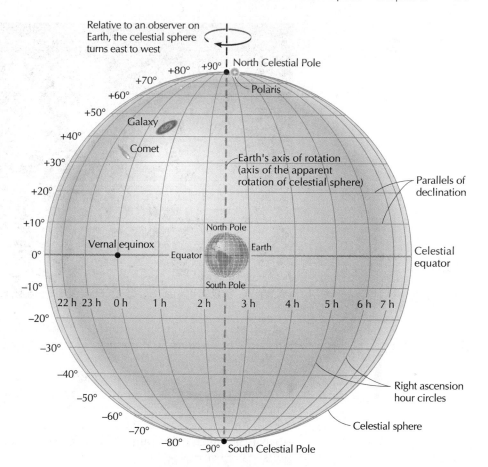

Relative to an observer on
Earth, the celestial sphere
turns east to west

North Celestial Pole

Polaris

Earth's axis of rotation
(axis of the apparent
rotation of celestial sphere)

Parallels of
declination

North Pole

Vernal equinox

Equator          Earth

South Pole

Celestial
equator

+90°  +80°  +70°  +60°  +50°  +40°  +30°  +20°  +10°  0°  −10°  −20°  −30°  −40°  −50°  −60°  −70°  −80°  −90°

Galaxy

Comet

22 h  23 h  0 h  1 h  2 h  3 h  4 h  5 h  6 h  7 h

Right ascension
hour circles

Celestial sphere

South Celestial Pole

**FIGURE 2.6** The projection
of Earth's polar coordinate grid
onto the celestial sphere provides the framework for the
celestial coordinates of right
ascension and declination.

surements on this coordinate could be in reference to sea level, to the center of Earth, or to some other convenient zero point.

For the purpose of mapping celestial objects, astronomers frequently
utilize a coordinate system derived from the terrestrial (earthly) latitude
and longitude system. The celestial coordinate **right ascension** (RA) is
analogous to longitude, and **declination** (Dec) is analogous to latitude.
Both coordinates are mapped onto an imaginary spherical shell, called the
**celestial sphere,** as shown in Figure 2.6. Viewed from Earth, the celestial
sphere represents our perspective of everything in the sky around us. (Of
course, only half of the whole sky can be seen at any one time, since Earth
itself blocks the other half.) Star maps, which are typically printed on flat
surfaces, depict large or small portions of the celestial sphere. Such maps
show only the directions of stars as seen from Earth, not their actual locations in three-dimensional space. Flat star maps are subject to some
amount of distortion, just as flat maps of Earth's surface are.

Since ancient times, stars have been grouped together in certain parts
of the sky and given names for their resemblance (or supposed resemblance) to the shapes of mythological characters, animals, or objects. These
"picture patterns" of stars are called **constellations.** The stars of a given
constellation usually do not lie close to each other in space; some stars may
be relatively near and some relatively far from Earth. Astronomers today

**PHOTO 2.4** The seven stars of the Big Dipper lie over moon-lit Mount Rainier, Washington.

use constellations to refer to different regions of the sky. We make use of a similar kind of mental mapping when we refer to various parts of a nation by its states or provinces.

As shown on the celestial sphere of Figure 2.6, the **celestial equator** is the projection of Earth's equatorial plane outward. Earth's rotational axis points in opposite directions toward the north and south **celestial poles.** A moderately bright star, Polaris (the North Star), lies very nearly in the direction of the north celestial pole, but the south celestial pole has no conspicuous star near it.

By convention, declination uses + and − symbols to represent north and south. Right ascension employs angle units called *hours* (1 h = 15°, so there are 24 h of right ascension around the entire sky). Right ascension values increase going east in reference to a particular point in the sky called the vernal equinox (the significance of the vernal equinox will be discussed in Chapter 3).

The RA/Dec system does not rotate with Earth; instead, it stays fixed with respect to the distant stars. Consequently, when viewed by observers on the surface of the spinning Earth, the celestial sphere, with its RA/Dec grid affixed, seems to be in motion, like a hollow sphere overhead turning east to west.

When an astronomer wishes to communicate the approximate direction of some object (such as a comet) in the sky, he or she might say that it is "in" (in the direction of) Orion, Leo, or some other constellation. (In much the same manner, someone might refer to Peoria as being in Illinois.) Right ascension and declination figures would be given if more precision were desired. The comet's location thus might be reported as something like RA = 23 h 24 min, Dec = +33° 19′. This particular comet location on the celestial sphere (direction in the sky) is marked in Figure 2.6.

Keep in mind that RA/Dec celestial coordinates are based on the celestial equator and the celestial poles. These features and the coordinate grid are fixed relative to the stars; they do not depend on the observer's local horizon. No two observers at different places on Earth's spherical surface see exactly the same sky overhead at the same time. Also, celestial objects move across the observer's sky as Earth rotates. A star's RA/Dec coordinates alone do not tell exactly where in the sky a given star should be. For that, the observer's time and latitude must be specified.

Also keep in mind that RA/Dec celestial coordinates really specify directions, not actual locations in space. The location of a celestial body in three dimensions cannot be known unless its distance, measured from the observer, is known.

Technically, the celestial coordinates are *not* exactly "fixed" to the stars, as we stated earlier. One reason is that the random motions of stars in space (which are exceedingly slow as observed from Earth) cause them to change their positions on the celestial sphere, even to the extent of altering the patterns of the familiar constellations. Without a telescope, it would be hard to notice such changes unless they could be observed over thousands of years. Another reason is that Earth's rotational axis *precesses*, or wobbles, much as a toy top wobbles as it turns, only much slower. Earth's axis traces out a cone in space over a leisurely period of about 26,000 years. Since the celestial coordinate system is based on the orientation of that axis, the imaginary grid lines shift slowly with time. To keep up with these changes, astronomers commonly update star atlases and catalogs every 50 years. Coordinates for the year 2000 are given in nearly all the current astronomical publications.

## Diurnal Motion

Because Earth spins west to east, the stars (and everything else in the sky) seem to move in the opposite direction—east to west. For this reason, celestial objects viewed from Earth are in constant apparent motion: They move upward in the east half of the sky and sink in the west half. The apparent paths followed by celestial objects as a consequence of Earth's rotation over a day are called **diurnal circles.** A portion of a diurnal circle is called a *diurnal arc*.

Figure 2.7 will help you visualize how the stars move along diurnal circles, as seen from three different latitudes on Earth. For any observer, the whole celestial sphere seems to turn on an axis (a dashed line in the drawings) connecting the north celestial pole (NCP) and the south celestial pole (SCP). This is the axis of the sky's apparent rotation, and it is parallel to Earth's axis of rotation. The orientation of the axis in the sky depends on the observer's latitude: In Figure 2.7(a), the observer at Earth's equator sees the NCP on the north horizon, the SCP on the south horizon, and the stars rising and setting at right angles to the horizon. In Figure 2.7(b), the observer at latitude 45°N (corresponding to the northern tier of states across the United States) sees the stars rising and setting at oblique angles to the horizon. In Figure 2.7(c), the observer at Earth's North Pole sees the NCP at the zenith (Z), or straight-up position in the sky. The SCP is straight down. The stars, as seen from the North Pole (and also the South Pole), travel on paths parallel to the horizon.

A close look at Figure 2.7 reveals several interesting relationships. In Figure 2.7(b), any star closer to the NCP than the diurnal circle with star A

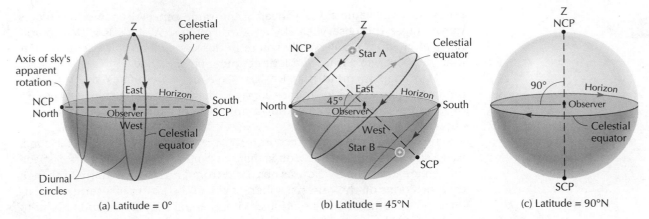

(a) Latitude = 0°

(b) Latitude = 45°N

(c) Latitude = 90°N

**FIGURE 2.7** These drawings of the sky as seen from different latitudes on Earth suggest the apparent motions of the stars. The celestial sphere is shown for observers located at (a) Earth's equator, (b) a midnorthern latitude, and (c) Earth's North Pole. In each drawing, the observer stands in the middle of a patch of Earth's surface that appears to be flat. The rim of this circular patch is the observer's horizon.

on it always stays above the horizon. These objects are said to be *circumpolar*. In the same diagram, any object close to the SCP never rises. Star B touches the horizon once each day, but never rises. The circumpolar and hidden regions are equal in size, and the size of both depends on the observer's latitude. Polaris and many of the constellations that lie near it are circumpolar for North America.

Notice, too, how the **altitude,** or angle above the horizon, of the NCP is the same as the observer's north latitude. We see in Figure 2.7 that when observers at latitudes 0°, 45°N, and 90°N look toward the NCP, they look at altitudes of 0°, 45°, and 90°, respectively. The farther north the observer, the higher the NCP (or Polaris) lies in the sky. Mariners have used this relationship as a navigational aid for thousands of years.

**PHOTO 2.5** The camera was aimed toward the north celestial pole during this 7-h time exposure of the night sky above Delicate Arch in Arches National Park, Utah. The brightest track inside the arch was produced by Polaris, the North Star.

**PHOTO 2.6** This all-night exposure, taken at Tyee Lakes in California's Sierra Nevada range, shows stars rising from the northeast (left) and east (right) horizons. Some star reflections can be seen on the lake's surface, which remained almost glassy smooth because of a lack of winds during the 9-h exposure.

The diurnal arcs of celestial bodies ("star trails" or "star tracks") can be photographed by placing a camera on a fixed tripod and taking a time exposure lasting many minutes or several hours. See Photos 2.5 and 2.6 for two spectacular examples.

## A Cosmic Zoom

Now that we have covered some of the essentials of spatial measurent, we are ready to embark on an imaginary journey through the universe. Our "zoom" through the universe starts with an illustration of some of the smallest known entities in nature, called quarks, and ends with the huge bubble-and-void structures that seem to pervade our universe on the largest scales astronomers can presently discern. This is an expansion through 40 powers of ten. In the pages that follow, we will open 19 "windows" that reveal the larger and larger things being discussed along the powers-of-ten continuum. Each window is accompanied by its own section of text explaining what we see. Even at this swift rate, it will take 38 pages to pass through the microscopic and macroscopic worlds to the supermacroscopic world of billions of galaxies. When looking at any window, you can keep in touch with your approximate location on the spatial scale by referring to the graphic bar depicting the powers of ten at the bottom of the page.

Along the way of the zoom, you will discover many important structures in nature, from atoms to superclusters of galaxies. You will learn something about the forces that hold matter together at many different scales. You will also learn how forces acting at a microscopic scale give rise to the forms, shapes, and other characteristics of materials we deal with every day in the macroscopic world.

# Quarks

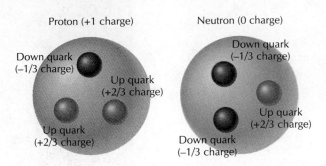

Proton (+1 charge)     Neutron (0 charge)

Down quark (−1/3 charge)
Up quark (+2/3 charge)
Up quark (+2/3 charge)

Down quark (−1/3 charge)
Up quark (+2/3 charge)
Down quark (−1/3 charge)

**WINDOW A** The fundamental building blocks of matter consist largely of tiny particles called quarks.

We begin our cosmic zoom at the vanishingly small scale of about $10^{-15}$ m. One meter is 15 powers of ten, or 1 quadrillion times greater than $10^{-15}$ m. If an object $10^{-15}$ m wide were enlarged to the size of a small grain of sand, then a meter scaled up in the same way would stretch from Earth to the sun. We will use nanometers, or billionths of a meter (1 nm = $10^{-9}$ m), to specify the dimensions of various objects in this window and in the next few windows.

**Quarks,** represented here as grayish spheres, are some of the smallest and most fundamental entities believed to exist in nature. Quarks are thought to comprise 99.9% of the ordinary matter in the universe. Curiously, no one has yet been able to isolate and examine a lone quark. This has not been for lack of trying; quarks have a strong tendency to bind together in twos and threes to form many of the tiny units of matter, called *subatomic particles,* that physicists have been studying for decades.

As illustrated in this window, trios of quarks make up the particles of ordinary matter we call protons and neutrons. The remaining, roughly 0.1% of ordinary matter is in the form of electrons. No electrons are portrayed in either this window or the next, since electrons ordinarily travel in paths well away from any quarks. All **electrons** possess a tiny but fixed electric charge, often written −1. The 1 denotes what was earlier

thought to be the smallest possible unit of charge, and the minus indicates the type of charge (there are only two kinds of charge: positive and negative). On the basis of the electron's charge, quarks bear charges of exactly +2/3 or −1/3.

There are several kinds of quarks, including the common ones denoted as *up* and *down*. Each up quark has a charge of +2/3, and each down quark has a charge of −1/3. Two up quarks and one down quark combine to make a **proton,** which has an overall charge of +1. Two down quarks and one up quark combine to make a **neutron,** which has a zero (or neutral) charge. Up and down quarks have nearly identical mass, so the masses of the proton and the neutron are nearly the same.

Quarks bind together by means of the strongest of the four fundamental forces (interactions) of nature: the **strong force.** The strong force is also responsible for holding together protons and neutrons to form atomic nuclei (Window B).

Experiments seem to indicate that quarks may be as small as 1/1000 the size of a proton, which itself is about a millionth of a nanometer ($10^{-15}$ m). Since no one has actually seen a quark (and the concept of seeing anything that small is probably meaningless anyway), the quarks in Window A are shown as nondescript grayish spheres.

YOU ARE HERE

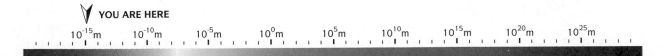

$10^{-15}$m   $10^{-10}$m   $10^{-5}$m   $10^{0}$m   $10^{5}$m   $10^{10}$m   $10^{15}$m   $10^{20}$m   $10^{25}$m

# The Atomic Nucleus

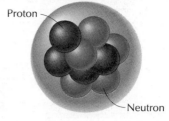

Carbon-12 nucleus

Proton

Neutron

**WINDOW B** An atomic nucleus is made up of a cluster of protons and neutrons.

In Window B, six protons and six neutrons cluster together to form a compact structure called an **atomic nucleus** (the nucleus, or central part, of an atom). This particular nucleus belongs to a carbon atom. The strong force, which binds quarks together inside protons and neutrons, also binds protons and neutrons together to form atomic nuclei.

Within the nucleus, the strong force succeeds in overcoming a weaker force, the **electromagnetic force,** which tries to push protons apart from each other. (You may remember the rule of thumb about electric charges: like charges repel; unlike charges attract. Each proton has a positive charge, so two or more protons always repel each other by virtue of their like charges.) Despite its great strength inside the nucleus, the strong force falls to zero immediately outside the nucleus. Electrons, under the influence of the weaker but far-reaching attractive electromagnetic force, whirl around the nucleus at a comparatively large distance.

A nucleus of protons and neutrons plus enough electrons circulating around the nucleus to match the number of protons (one for one) make up the unit of matter we call an **atom.** (The one exception to this description is the typical hydrogen atom, which has one proton, one electron, and no neutrons.) Atoms are electrically neutral, because they are composed of equal numbers of protons (+1 charge) and electrons (−1 charge).

More than a century ago, atoms were thought to be indivisible. Now we know that atoms are aggregates of smaller particles: protons, neutrons, and electrons. Protons and neutrons are made of quarks, but electrons (as far as we know at present) are indivisible.

A substance consisting solely of atoms with the same number of protons is called an **element.** About 100 elements exist on Earth and in the rest of the universe. Carbon, oxygen, and hydrogen are three examples of elements (many more will be described in the chapters to come). Pure carbon consists of atoms whose nuclei all have exactly six protons. Pure oxygen is nothing but oxygen atoms, all of which have eight protons. Hydrogen is the simplest element; every hydrogen nucleus contains a single proton.

All atoms of a given element closely resemble each other, but this does not mean they must be identical in all respects. Atoms of the same element can have different weights, based on the number of neutrons contained in their nuclei. Hydrogen, for example, comes in three forms: ordinary hydrogen, with a nucleus of one proton; deuterium (or heavy hydrogen), with one proton and one neutron; and tritium, with one proton and two neutrons. Atoms of different weight but of the same element are called **isotopes.** Typically, for a given element, one isotope is common and the rest are relatively uncommon. Deuterium and tritium atoms, for example, are rarer than ordinary hydrogen atoms.

The carbon nucleus shown in Window B is the common, stable isotope of carbon, carbon-12. (The number 12 refers to the total number of protons and neutrons). At least seven other isotopes of carbon exist, but most are unstable. A carbon-14 nucleus, for example, contains the required 6 protons (which *makes* it carbon) and 8 neutrons. But too many neutrons are present in a carbon-14 nucleus to ensure its stability indefinitely. There is a 50–50 chance that within a period of about 5700 years, any given carbon-14 nucleus will undergo a change called *beta decay* and transform itself into the nucleus of another element (nitro-

## TABLE 2.2  The Four Fundamental Forces in Nature

| Name | Strength | Range | Behavior |
|------|----------|-------|----------|
| Strong force (strong nuclear force) | About $10^{38}$ times stronger than the gravitational force | Less than $10^{-14}$ m (no more than the diameter of an atomic nucleus) | Binds quarks together into protons and neutrons; binds protons and neutrons together into atomic nuclei |
| Electromagnetic force (electromagnetism) | About $10^{36}$ times stronger than the gravitational force | Infinite | Causes attraction or repulsion between charged bodies; strength of the force decreases with increasing distance; binds electrons to nuclei; holds atoms together in molecules and crystalline solids; governs ordinary electronic and magnetic phenomena |
| Weak force (weak nuclear force) | About $10^{29}$ times stronger than the gravitational force | Less than $10^{-15}$ m (roughly the distance between nucleons) | Causes beta decay in some unstable nuclei |
| Gravitational force (gravitation) | The weakest by far of the four forces | Infinite | Causes attraction between masses; strength of the force decreases with increasing distance; not important for small masses, yet dominates interactions between bodies with large masses |

*Note*: Particle accelerator experiments have shown that under conditions of extremely high energy, the electromagnetic and weak forces behave indistinguishably—the two act as one "electroweak" force. It is possible that at still higher levels of energy, the fundamental forces could unify to become two forces or perhaps even one "superforce." This superforce may have pervaded the universe in its earliest moments. In today's natural world, the four forces listed here behave independently.

gen). During beta decay, the strong force is overcome by a disruptive force within the nucleus called the **weak force.** An electron is produced and then ejected from the nucleus. The remaining nucleus is that of a nitrogen atom. Carbon-14 is said to be a *radioactive isotope* because the emitted electron carries away (radiates) energy.

In general, heavy nuclei are less stable than lighter nuclei, because the range of the strong force is extremely short: The strong force may lose its grip on the particles of the larger nuclei. One of the heaviest stable nuclei is that of lead-206, with 82 protons and 124 neutrons (82 + 124 = 206). Uranium, a heavy element with 92 protons, has several radioactive isotopes.

An atom can undergo another kind of change having nothing to do with its nucleus. By any of several processes, an atom may either lose one or more of its electrons or capture one or more extra electrons. If it does, the atom acquires a net charge—a positive charge if electrons are lost by the atom, and a negative charge if electrons are gained. (Remember that electrons are negatively charged; if an atom gains an electron it acquires an extra amount of negative charge.) Whenever an atom possesses a net charge, it is known as an **ion.**

So far, we have introduced three of the four fundamental forces of nature. The strong force and the weak force exist only in or very near atomic nuclei (for this reason, they are also known as the *strong nuclear force* and the *weak nuclear force*). The electromagnetic force, roughly 100 times weaker than the strong force, is unlimited in range. The fourth, the **gravitational force,** or gravitation, affects all particles that have mass. Despite its infinite reach, gravitation's influence on small bits of matter like atomic nuclei or atoms is negligible; it is significant only when large amounts of mass interact. Table 2.2 summarizes several important characteristics of the four fundamental forces. We will refer to these forces frequently in succeeding chapters.

# A Carbon Atom

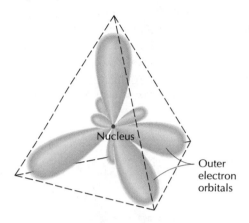

**WINDOW C**   This carbon atom, like all atoms, consists of a dense, positively charged nucleus surrounded by negatively charged electrons that move about in particular patterns.

The nucleus of a carbon atom has six electrons moving around it, just enough to equal the number of protons. The electrons of any atom travel through a volume of space vastly larger than the nucleus of that atom. Each is held in place by electromagnetic forces—specifically, the attraction between the positively charged nucleus and each negatively charged electron. If the electromagnetic force did not exist, the rapidly moving electrons could not bind to atomic nuclei. Then our universe would have no atoms and nothing with which to build larger, cohesive structures.

Window C, with its roughly 10,000-fold expansion in scale compared to Window B, reveals the structure of an entire carbon atom. The size of the carbon nucleus in Window C is greatly exaggerated; if drawn to the scale of the electron orbits shown, it would be invisible to the eye. Despite its relatively small size, the nucleus possesses more than 99.9% of the mass of a typical atom. The electrons, which whirl around the nucleus at incredible speeds, account for less than 0.1% of the mass.

Electrons in atoms move in patterns called *orbitals*. In a lone hydrogen atom, a single electron fills an orbital in the shape of a spherical shell centered on the nucleus. In a lone carbon atom, a total of six electrons arrange themselves in two spherical orbitals and two dumbbell-like orbitals. Carbon atoms, however, are almost never isolated from others—at least on Earth. Carbon atoms easily bind to themselves and to other atoms by means of electromagnetic forces. Chemists call the attachments between atoms *bonds*. When atoms of any kind bond together they can form a molecule. A **molecule** can be defined as an aggregate of atoms having a net charge of zero.

One of the simplest molecules is methane (the principal constituent of natural gas). The methane molecule consists of a single carbon atom surrounded by four hydrogen atoms. The single carbon atom in Window C has its outer four electron orbitals pictured *as if* the atom were linked to the four hydrogen atoms in a methane molecule. These hydrogen atoms (not shown in the illustration) would lie at the ends of the four, long, pronglike electron orbitals extending out toward the corners of an imaginary tetrahedron, or pyramid.

Each of the four long prongs pictured in Window C can be thought of as a **chemical bond**—a potential linkage to other atoms. In the methane molecule, the four long orbital prongs are straight. In other molecular arrangements involving carbon, the prongs are often bent. Carbon

**V** YOU ARE HERE

$10^{-15}$m   $10^{-10}$m   $10^{-5}$m   $10^{0}$m   $10^{5}$m   $10^{10}$m   $10^{15}$m   $10^{20}$m   $10^{25}$m

atoms bond easily to other carbon atoms and to many other kinds of atoms. For this reason, chemists consider carbon to be the most versatile element. Carbon atoms form the basic structure of molecules as simple as methane and as complex as the DNA molecules that reside in every living cell. A single DNA molecule can contain more than $10^{11}$ carbon atoms!

Atoms are almost entirely empty space, because the parts that make up an atom are tiny compared to the volume of the whole atom. If a typical atomic nucleus were enlarged to the size of a cherry pit and placed in the center of a football stadium, then the outermost electrons belonging to that nucleus would spend most of their time outside the stadium. The electrons are quite insubstantial; it takes nearly 2000 of them to equal the mass of either the proton or the neutron.

No ordinary microscope has ever rendered a clear picture of an atom; in fact, atoms are far too

## BOX 2.1
## Probing Tiny Spaces

### Microscopes—Traditional and Modern

 Viewed at micrometer scale (1 $\mu$m = $10^{-6}$ m, or one millionth of a meter), the head of a pin looms large as an ocean liner, and dust motes seem as big as the passengers on board. At nanometer scale (1 nm = $10^{-9}$ m, or one-billionth of a meter), individual atoms become tangible entities, and you could spend days exploring the atomic peaks and hollows on the surface of single dust mote.

Beginning with the invention of the ordinary light microscope in the mid-1600s, scientists have developed an increasingly diverse toolbox of sophisticated instruments capable of probing inner spaces such as these. During the past decade, the resolution (ability to render sharp, clear images) of the newer kinds of microscopes has increased dramatically. Researchers now have the ability to map the surfaces of solid materials on the atomic level.

Some of the most advanced instruments that are used to peer into the world of the very small are also capable of moving individual atoms or molecules from place to place. The emerging field of *nanotechnology*, which deals with precise manipulations of matter on the nanometer and somewhat larger scales, promises futuristic dividends, some of which are described next.

**Optical microscope** The resolution of ordinary optical microscopes, which gather and focus light from small objects, is limited by the wave nature of light itself. No object smaller than the characteristic length, or wavelength, of the light used to illuminate the object can be seen clearly. Since visible light has a wavelength of somewhat less than 1 $\mu$m (1/1000 mm), nothing smaller than about 1 $\mu$m in diameter can be seen distinctly, even in the best optical microscopes. Magnification helps, but only up to a certain point. At magnifications greater than about 1000 power (1000 X), microscope images viewed in visible light become indistinct and very dim. Several techniques, including the use of shorter-wavelength ultraviolet light, can be used to marginally improve the resolution of light microscopes.

**Transmitting electron microscope (TEM)** In the TEM, a beam of high-speed electrons (which behave like tiny wavelike packets having a wavelength of about 0.01 nm) is directed into a thin slice of an object being examined. The electrons penetrate the less dense portions of the slice, and they are deflected or absorbed by the more dense parts of the slice. The transmitted electrons go on to strike a phosphorescent target screen or a photographic

small to the seen with the light our eyes respond to. Our knowledge of atoms and molecules has traditionally come from indirect physical and chemical experiments. Now, however, several new instruments (see Box 2.1) are delivering increasingly vivid and detailed images of the world of the small, including the fuzzy outlines of atoms themselves.

Since atoms are mostly empty space and ordinary matter is made of atoms, then everything we see on Earth is literally full of holes—on the atomic scale. Elsewhere in the universe, highly compressed forms of matter can be found. In the collapsed bodies known as neutron stars, for example, protons, neutrons, and electrons merge to form a material trillions of times denser than ordinary matter. If a teaspoon's worth of material from a neutron star could be brought to Earth's surface, it would weigh about a billion tons.

In the next two windows of the cosmic zoom,

plate, thus building up a picture consisting of light and dark areas. Magnifications of several hundred thousand can be obtained, but the image is purely two-dimensional—it lacks depth. The vague outlines of individual atoms can be observed with the most powerful TEMs.

**Scanning electron microscope (SEM)**  The SEM sacrifices much in resolution in return for its ability to map the surface of an object in three dimensions. In this instrument, an exceedingly thin beam of electrons scans back and forth and up or down across a small object. Electrons scattered off the object at various angles during the scans are collected by a bank of detectors, and information from the detectors is correlated by computer. The resulting lifelike image is displayed on a television screen or computer monitor.

**X-ray microscope**  Several techniques can be employed to focus or scatter X-rays from small objects. Since the impinging X-rays have wavelengths much smaller than that of visible light, the resolution obtained is at least ten times better than that of light microscopes.

**Scanning tunneling microscope (STM)**  The STM works by means of a sharp-pointed tungsten needle, only several atoms wide at its tip, that scans back and forth (like someone mowing a lawn) across the surface of a solid. If the space between an atom on the surface and the tip is small enough (less than 1 nm), electrons from the surface can "tunnel" or migrate across and produce a tiny electric current. By means of a feedback mechanism, the scanning tip is moved up and down to maintain constant current. The undulations of the tip are recorded along each scan line, and many parallel scan lines make up an image that is displayed on a computer screen. The resolution obtained in these surface mappings is less than 1 nm, which is roughly equivalent to a magnification of many millions.

If just the right amounts of charge are applied to the needle, the STM can be used to pull individual atoms (or molecules) from a surface, move them, and reposition them elsewhere. The STM's manipulative abilities may well be used for a dazzling array of practical applications in the future. In several well-publicized stunts, researchers have created company logos and pictures out of individual atoms or clumps of atoms. Extensions of this technology could result in data storage on almost unimaginably small scales. If individual atoms placed on a surface could represent the zeros and ones of computer language, then theoretically the information content of the entire U.S. Library of Congress could be placed on a single 30-cm-wide disk. Current technologies require over 100,000 disks of the same size.

Chemists have used STMs to fashion miniature "beakers" of atoms in which they can observe chemical reactions on an atomic scale. Someday, chemists using STM technology may routinely assemble molecules atom by atom, just as we can     *(continued)*

we will deal with whole atoms and molecules rather than parts of atoms. Interactions within atomic nuclei are primarily the concern of nuclear physics. Interactions among atoms and molecules are the primary focus of chemistry. Material on basic chemistry, starting with simple molecules, is covered in Chapter 4 and in chapters beyond that. Our zoom in this chapter, though, is going to take us through the complex DNA molecule and then through larger structures associated with a living organism.

*(Box 2.1 continued)* construct ball-and-stick models of molecules by fitting the pieces together with our hands.

**Atomic force microscope (AFM)** The scanning tip of an AFM (Photo 2.7), which closely resembles the STM, reads the surface of a small object by touching it (in much the same way that a blind person reads Braille). The weak electromagnetic force applied to the tip can be so small that the tip may glide over surfaces as delicate as the outer membrane of a biological cell. A slight increase in the force transforms the AFM into a precision scalpel capable of dissecting a living cell, thereby exposing its biochemical operations to scrutiny and manipulation.

## Micromachines and Virtual Reality

Recent advances in microtechnology have included the invention of *micromachines*—motors and gears with rotors smaller than the width of a hair and actuators that move like inchworms. They are being developed for manipulation of matter on micrometer scales. Micromachines can barely move themselves at present, but further advances could lead to a wide variety of futuristic devices. Someday, micromachines might direct pulses of light in fiber-optic systems, repair faulty circuits in computer chips, and assist in the process of reconnecting severed nerves or regenerating damaged nerves.

*Virtual reality*—a computer-driven, artificial environment of sights, sounds, and other sensations—is already realized in airplane flight simulators and in advanced video games. Virtual reality systems coupled to STMs or AFMs may someday allow researchers to "feel" as well as "see" their way around the surfaces of atoms and molecules. Users of such a machine would see by means of a video

**PHOTO 2.7** The atomic force microscope.

screen and feel by means of special gloves capable of amplifying and transmitting tiny forces to the user's fingers and hands. Conversely, an individual molecule could be built or altered by a researcher manipulating pressure-sensitive gloves so as to direct the movement of an STM or AFM.

# A Nucleotide

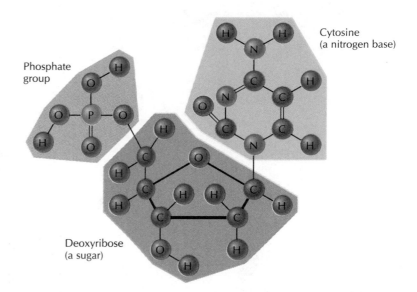

Phosphate group

Cytosine (a nitrogen base)

Deoxyribose (a sugar)

**WINDOW D** Nucleotides, which consist of many atoms bonded together, are fundamental units of the immensely complex DNA molecule.

Most of the molecules associated with biological activity are exceedingly complex. Perhaps none is so fascinating in its intricacy as *deoxyribonucleic acid*, or *DNA*. Strands of DNA reside in the nucleus, or central part of nearly every living cell. (This cell nucleus is not the same as an atomic nucleus.)

The DNA molecule consists of enormously long chains of relatively small molecular units called *nucleotides*. One of the four kinds of nucleotides that make up DNA is shown in Window D. The spheres labeled C, H, O, N, and P refer to atoms of carbon, hydrogen, oxygen, nitrogen, and phosphorus. Lines connecting the spheres represent chemical bonds. The atoms join to form multiatom units, and three of these

units make a nucleotide. In molecular drawings like this, there is some attempt to show three-dimensional structure: The pentagonal arrangement of carbon and oxygen atoms in the sugar unit is viewed as if it were tilted, with the bond on the near side of the pentagon drawn as a thick line.

Each nucleotide of DNA consists of a sugar (deoxyribose), a phosphate group, and any one of four nitrogen bases, which are called adenine, thymine, guanine, and cytosine (symbolized by A, T, G, and C). The nitrogen bases distinguish one nucleotide from another. In the next window, we see how nucleotides join to form the famous *double-helix*, or twisted ladder, structure of DNA.

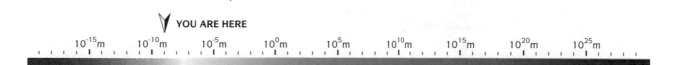

YOU ARE HERE

$10^{-15}$m  $10^{-10}$m  $10^{-5}$m  $10^{0}$m  $10^{5}$m  $10^{10}$m  $10^{15}$m  $10^{20}$m  $10^{25}$m

# A DNA Strand

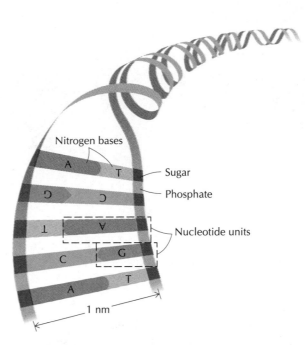

Nitrogen bases

Sugar

Phosphate

Nucleotide units

1 nm

**WINDOW E**   Millions of nucleotide pairs make up the "rungs" of the ladderlike DNA molecule.

In this somewhat broader view of part of a DNA strand, each nucleotide, with its A, T, G, or C base pointing inward, forms a link on a ladderlike chain. The twisted "rails" of the chain are made of alternating sugar and phosphate units, and the "rungs" of the ladder are pairs of nitrogen bases. Stable bonds are formed between the A and T bases and between the G and C bases. No other combinations are possible using these four bases, so the rungs are of essentially two types.

A typical human strand of DNA, if stretched out in a straight line, would span about a meter and contain several hundred million base pairs. If we moved along *one* rail of the ladder, a small portion of the base sequence might read something like "...ACTGGTTCAGATC...." The four different letters can be thought of as symbols in the genetic code of the cell to which the DNA molecule belongs. The exact sequence of these

letters (actually, base pairs) directs the life and function of the cell and determines what role the cell will play if it is part of a larger organism.

When it is time for a cell to divide, the DNA within it "unzips" itself down the middle, becoming two separate strands having unpaired bases. Individual nucleotides of the proper type floating around in the cell medium bind with the unpaired bases on each single strand: A binds with T, T binds with A, G binds with C, and C binds with G. The final result, ideally at least, is two complete and identical DNA molecules, each one consisting of one "old" rail and one just-assembled "new" rail. Both copies of the DNA molecule separate, and each one eventually resides in the nucleus of a daughter cell. Each daughter cell can, in turn, divide and form two more daughter cells housing identical DNA molecules.

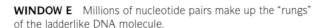

**YOU ARE HERE**

$10^{-15}$m   $10^{-10}$m   $10^{-5}$m   $10^{0}$m   $10^{5}$m   $10^{10}$m   $10^{15}$m   $10^{20}$m   $10^{25}$m

# A Chromosome

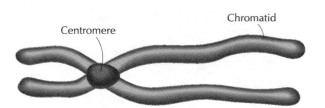

Centromere

Chromatid

**WINDOW F** Many chromosomes, each of which contains a coiled-up DNA molecule, inhabit the nucleus of each biological cell.

Each long strand of DNA, which coils itself up in an almost unbelievably intricate way, constitutes a *chromosome*. Within each chromosome are thousands of genes, consisting of about 10,000 base pairs each. Each gene is a set of instructions for some particular trait of an organism. Forty-six chromosomes reside in the nucleus of every normal human cell. This complete set of DNA in those chromosomes contains all the genetic information needed to build a human being, though only a fraction of the total information is used by any specific kind of cell for its own function.

Collectively, the complete set of genetic information for an organism is called a *genome*. Biologists are now engaged in a long-term research effort, called the Human Genome Project, to map the entire human genome of some 3 billion base pairs. If the human genome were compared to a book, with the base pairs represented by the four letters A, T, G, and C, the "book of human life" would consist of about 3 billion letters—enough to

fill about 2000 volumes the size of this textbook. Of the 100,000 or more genes belonging to the human genome, only about 100 genes distinguish one individual from another. Genetically speaking, almost all human beings are 99.9% alike.

If or when the human genome is fully deciphered, the knowledge gained would allow more widespread application of the recombinant DNA (genetic-engineering) techniques that are already in wide use today for a variety of pharmaceutical and agricultural purposes. Potentially, manipulation of the human genome could help prevent or alleviate nearly every hereditary disease that plagues humanity. Some believe that within decades (or perhaps sooner), genetic engineering techniques could be used to "create" custom-designed living beings, humans or otherwise. The ethical issues raised by such radical meddling in what has long been considered the affairs of "nature" or "God" will likely be the focus of intense debate over the next several years.

**YOU ARE HERE**

$10^{-15}$m   $10^{-10}$m   $10^{-5}$m   $10^{0}$m   $10^{5}$m   $10^{10}$m   $10^{15}$m   $10^{20}$m   $10^{25}$m

# A Neuron

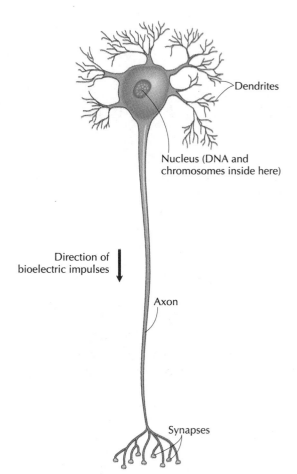

Dendrites

Nucleus (DNA and chromosomes inside here)

Direction of bioelectric impulses

Axon

Synapses

**WINDOW G** Neurons, or nerve cells, are only a little smaller than what normal human vision can detect.

The adult human body contains about $10^{13}$ cells of a wide variety of types and sizes. Roughly 1%, or $10^{11}$, of them are *neurons*, or nerve cells, residing in the brain and throughout the nervous system. Window G centers on a neuron in the brain. Radiating from the cell body are intricately branching dendrites and a long, thin axon. Neurons interconnect at sites called synapses, where there is a small gap between the axon of one neuron and the dendrite of another.

Neurons are relatively large cells that bridge the gap between the microscopic and macroscopic world. The tiny, fiberlike axons of brain cells are about a millimeter long. Neurons in the spinal cord can have axons longer than a meter.

Impulses travel into a neuron cell body by way of the dendrites and out of it by way of the axon. The impulses are electrical, which means that they involve moving charges. Unlike electricity in a wire (where charge is carried by electrons), these *bioelectric impulses* involve movements of sodium and potassium ions. Bioelectric

impulses travel at about 100 m/s (about 200 miles per hour, mi/h), much slower than electrical impulses in a wire. Like all living cells, neurons require a constant input of energy to continue functioning. That energy is provided by nutrients taken into the body.

Outside the brain, neurons (nerves) connect to muscle and other tissues. Inside the brain, neurons link together in an immensely complex pattern. The dendrites and axons of brain neurons are thought to form as many as $10^{14}$ connections among themselves. The structural complexity of the most advanced computer chips produced

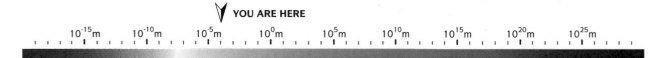

**YOU ARE HERE**

$10^{-15}$m $\quad$ $10^{-10}$m $\quad$ $10^{-5}$m $\quad$ $10^{0}$m $\quad$ $10^{5}$m $\quad$ $10^{10}$m $\quad$ $10^{15}$m $\quad$ $10^{20}$m $\quad$ $10^{25}$m

**PHOTO 2.8**  These interconnected nerve cells (stained with silver to increase their visibility) represent a tiny portion of the web of neurons in the human brain.

today is primitive in comparison to that of the human brain.

The brain's complexity lies partly in the great diversity of its neurons but mostly in the intricate ways the neurons are connected. In an effort to study the brain's function, neuroscientists and computer scientists are busy designing computer circuits (hardware) and programs (software) that mimic the neural networks of our brain. The computer models now in use are crude compared to the brains of even lower animals, but they do seem capable of "learning" from prior experience. These efforts, in addition to benefiting the health field, may lead to important advances in computer science, because current computers, which rely on sequential instructions, cannot handle certain kinds of tasks easily managed by the brain.

Like some kinds of cells in the body, neurons proliferate very early in life but after birth have almost no ability to divide or regenerate. Existing neurons, however, can make new connections with one another. New experiences and learning of all kinds fosters this process. By taking this course, you are (hopefully) increasing the complexity of your brain.

# A Human Being

**WINDOW H**   The typical adult human is somewhat less than two meters tall.

The brain is but one of many organs and tissues that make up a human body. Arriving at Window H in our cosmic zoom, we are within the macroscopic realm of size. It is in this realm, more than any other, that we can use our five senses alone in an effort to understand our immediate environment.

To greatly sharpen our perception of length or distance on the human scale—and make quantitative measurements—we can employ some kind of measuring tool, like the measuring stick, 1 m long, shown in Window H. Metersticks have convenient subdivisions of 100 centimeters (cm) and 1000 millimeters (mm). Persons with good eyesight can often interpolate between the tiny millimeter marks and make estimates of length having a precision of one- or two-tenths of a millimeter.

Today's scientific enterprise depends on accurate measurements using a variety of instruments far more sophisticated than metersticks. Even so, without the use of any measuring tool whatsoever, humans have historically made sense of worlds much larger than themselves. Humans have kept track of the stately progression of days, waxing and waning moons, changing seasons, and other celestial cycles for thousands of years. They have counted their paces or relied on times of travel to estimate distances across empires. Employing only observation, the ancient Greeks worked out the mechanism of eclipses of the sun and moon and discovered that our planet is spherical. With only the crudest of instruments, plus the power of geometry and reasoning, they were also able to estimate the relative distances of the sun and moon and to measure Earth's circumference (see Box 2.2).

Macroscopic behavior on a human scale often contains clues to underlying microscopic behavior. The Roman poet Lucretius (96–55 B.C.) noted that we never see the flecks of gold disintegrating from a ring that grows thin over a lifetime, because, he thought, the ring must be made of particles (atoms) too tiny to see. Similarly, the way that a rock or a mineral crystal breaks when subjected to a sharp blow tells us something about its inner structure.

**Y YOU ARE HERE**

$10^{-15}$ m    $10^{-10}$ m    $10^{-5}$ m    $10^{0}$ m    $10^{5}$ m    $10^{10}$ m    $10^{15}$ m    $10^{20}$ m    $10^{25}$ m

On the human scale, only one of the four fundamental forces (gravitation) seems to act in obvious ways. Nearly every object we see is affected by the downward pull of our planet's gravity. We are often unaware, however, of the many hidden roles played by electromagnetic interactions.

Consider light, for example. Light is a form of electromagnetic energy that moves at the incredibly high speed of 300,000 kilometers per second (km/s). Consider, too, chemical reactions, all of which are fundamentally electromagnetic. Combustion (fire) is one type of chemical reaction. When paper, wood, gasoline, and other flammable substances burn, they combine rapidly with the oxygen in the air. Heat and light are released when the reaction takes place. Also consider life, which is itself a complex web of chemical reactions. Biochemical reactions, like ordinary chemi-

BOX 2.2

# The Shape and Size of the World

The idea that Earth is not flat can be traced as far back as ancient Greece. Several of the ancient Greek philosophers deduced that Earth is a spherical body, much like the moon. They arrived at this conclusion by making several observations.

First, when a ship sails out to sea on a clear day, a shoreline observer watches it progressively disappear from view, bottom first, over the horizon. This observation is consistent with an ocean surface having a convex curvature. Second, the constellations shift northward in the sky as one travels south and shift southward in the sky as one travels north. This observation supports the idea that the traveler moves along a curved surface, not a flat surface. Third, during lunar eclipses, the boundary between light and dark on the moon is invariably a circular arc. The Greeks knew that lunar eclipses are caused by the moon passing through Earth's shadow, and they recognized that only a sphere could always cast a circular shadow.

Once the spherical nature of Earth was established, scientists could measure its size. Eratosthenes, a Greek astronomer living in Alexandria, Egypt, made a remarkable measurement of Earth's size in the third century B.C. Eratosthenes utilized simple geometry and two basic assumptions: (1) Earth is spherical, and (2) the sun is so far away from us that its rays are virtually parallel. Refer to Figure 2.8 as you follow Eratosthenes' procedure and reasoning.

Eratosthenes knew that at noon on a certain date, the sun shone vertically so as to illuminate the bottom of a deep well at Syene, a town some distance south of Alexandria. Eratosthenes also knew that in Alexandria, at the same time and date, the sun did not shine vertically. Eratosthenes set up a gnomon (vertical pole) in Alexandria at noon on the proper date and measured, by means of the gnomon's shadow, the small angle between the straight-up direction of the gnomon (the zenith as seen from Alexandria) and the direction of the sun's rays. The angle turned out to be a bit more than 7°, or almost exactly 1/50 of a circle. Since the two lines representing the sun's light rays in Figure 2.8 are parallel, the line *CZ* cutting across them has interior angles that are both 1/50 of a circle. Therefore, the distance between Alexandria and Syene, along Earth's curved surface, is equal to 1/50 of Earth's circumference.

A pacer had been employed to count the number of strides (of a known length) between Syene and Alexandria, so Eratosthenes knew the Syene-Alexandria distance: Translated into modern units of length it is approximately 800 km. Earth's circumference, then, is 50 times 800 km, which is 40,000 km. Dividing the circumference by $\pi$ gives the diameter of the spherical Earth: about 13,000 km. The distance units Eratosthenes used are not known with certainty. According *(continued)*

cal reactions, either give off or absorb energy. All exchanges of energy due to chemical reactions can be traced to the behavior of electrons in the atomic and molecular realm.

As its name implies, the electromagnetic force embodies aspects of two different phenomena that are often studied separately: electricity and magnetism. Both have been put to use in modern life to an astonishing degree. Today, we spend much of our time using devices that are electromagnetic. Most people would find it hard to imagine life without radios, televisions, computers, dishwashers, hair dryers, and all the other appliances that plug into wall sockets or run on batteries.

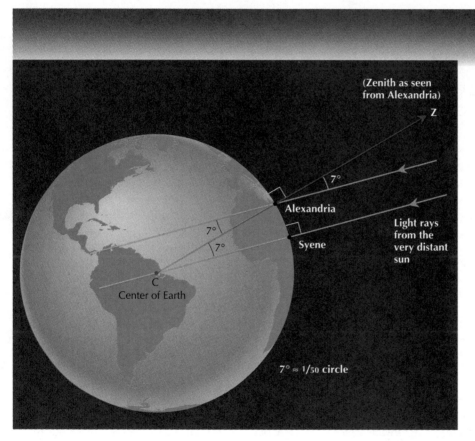

**FIGURE 2.8** In this diagram (not to scale), the sun's rays shine vertically downward as seen at Syene, but not at Alexandria. The 7° angle between the zenith and the sun's direction at Alexandria means that there is a 7° arc along Earth's surface between Syene and Alexandria.

*(Box 2.2 continued)*    to some scholars, his dimensions for Earth are within 1% of the values we accept today.

Despite the work of Eratosthenes, it would be almost two millennia before the concept of a flat Earth would be abandoned, at least in the public consciousness. Columbus and other navigators of the 15th century accepted the notion of a spherical Earth, though at that time the arrangement of the continents and oceans was poorly understood. The voyage of Magellan, which circumnavigated the entire globe in the years 1519–1522, finally provided irrefutable evidence of Earth's true shape.

# A Building

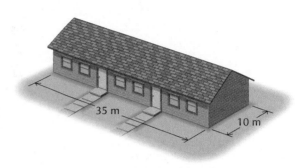

35 m

10 m

**WINDOW I** Structures that are tens of meters wide, such as this school building, must be built to withstand the pull of Earth's gravity.

As we approach the size of, say, a classroom building, gravity becomes increasingly important. When designing a building or other large structure, an engineer's main task is to counter the downward pull of Earth's gravity, which affects every one of the building's components. Buildings must also be designed to accommodate sideways forces such as wind loads and (in some geographic areas) the shaking of earthquakes.

Electromagnetic forces are responsible for supporting the building, but only on a microscopic scale. A piece of steel or concrete rigid enough to help support a building owes its solidity to attractive and repulsive electromagnetic forces. These forces keep atoms and molecules from shifting about in solid materials.

On larger and larger macroscopic scales, electromagnetic interactions become less significant. Why? Are there not a huge number of charged particles—protons and electrons—in an object as massive as a classroom building? Of course there are, but (1) matter in bulk tends to contain *equal* numbers of protons and electrons, so the net charge of all the particles is zero; and (2) under normal circumstances, there is no tendency for

electrons or protons to migrate one way or another. Within, say, the walls of a building, there is little or no *separation* of charges over macroscopic scales of distance; and therefore, attractive or repulsive forces are negligible over such distances.

We are ignoring in this discussion unusual circumstances such as lightning striking the building. A lightning strike could cause a sudden surge of charge down through the walls. We are also ignoring the electrical currents, driven by distant generators, that flow through the wiring in a building's shell.

When we consider measuring the size of large objects like buildings, *rounding off* to whole numbers of meters (or yards, or feet) usually gives us a clear enough mental picture of the building's dimensions. Knowing the building's dimensions to the nearest millimeter does not increase our basic understanding of the building's size. Rounded-off values of many quantities will be utilized in this book to simplify conceptual understanding. For example, performing simple calculations with the speed of light is simpler when we use the rounded value 300,000 km/s rather than the precise value 299,792.458 km/s.

YOU ARE HERE

$10^{-15}$m    $10^{-10}$m    $10^{-5}$m    $10^{0}$m    $10^{5}$m    $10^{10}$m    $10^{15}$m    $10^{20}$m    $10^{25}$m

# A College Campus

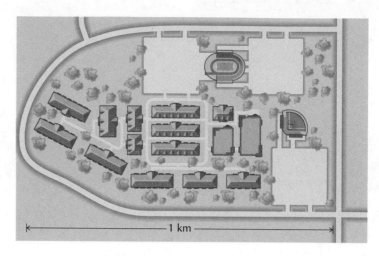

**WINDOW J** A great deal of energy must be expended to construct a college campus, and even more energy is needed to maintain it over time.

The various components of a campus—buildings, students, faculty, staff, departments, and administration—are analogous to (though much simpler than) a higher biological organism such as a human being. Despite their many differences, biological systems and artificial systems (like college campuses) depend on a constant influx of energy. This is not to say that the material (atoms) belonging to either system would disappear without added energy; but each system would become more and more *disorganized* if deprived of the energy it needs to function properly.

For example, consider what happens when a human being is deprived of oxygen (oxygen plus food yields the metabolic energy we need to function). Without oxygen, our marvelously intricate biological cells die and eventually decompose. Organization degrades into disorganization.

Most of the energy infused into a typical campus comes by way of electrical transmission lines. Some of the energy (for heating, especially) may be generated on site by burning natural gas, coal, or oil. Additional energy is provided by humans, in the form of both physical labor and thought (thinking requires energy, because it involves electrical activity among neurons). Now imagine a campus without these influxes of energy—no one to push brooms, to fix leaky roofs, or to shuffle papers; no electricity to power lighting fixtures or run computers; no one thinking. Disorder reigns.

In this situation, physicists and chemists refer to a quantity called entropy. **Entropy** is a measure of the *disorder* in a physical system. We say, for example, that a given mass of water in a liquid state has more entropy than the same mass of water in a solid state (ice). Molecules in the water slip and slide over one another in a nearly random fashion, whereas molecules of the ice are locked in place in an organized pattern. We could also say that the entropy of a well-maintained campus is a lot less than that of a campus that has been abandoned and left to decay.

Every kind of experiment ever devised to evaluate order and disorder confirms that interacting things (atoms, molecules, and anything bigger) have a natural tendency to become more disorderly in their behavior with time. Translated into the jargon of science, this tendency can be expressed as an important law: *The entropy of any physical system, isolated from the rest of the universe, cannot decrease.* The law says that, at best, an isolated system of interacting particles or units *might* maintain a given measure of order forever. This, however, is extremely unlikely to happen. Ordinarily, isolated systems with many interacting parts tend to become more disorderly with time. The tendency toward disorder can be reversed, but only by influxes of energy into a system from outside the system.

▼ YOU ARE HERE

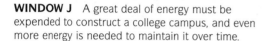

$10^{-15}$m    $10^{-10}$m    $10^{-5}$m    $10^{0}$m    $10^{5}$m    $10^{10}$m    $10^{15}$m    $10^{20}$m    $10^{25}$m

# A City

Window K shows a larger, but still organized, structure spanning several kilometers—a small city. A typical city receives an enormous amount of energy in the form of sunlight, though little or none of it is used in any direct sense to keep the city organized. Most of the energy consumed for manufacturing, transportation, housing, and other urban uses in North America comes from burning fossil fuels such as coal, oil, and natural gas. Large metropolitan areas such as Los Angeles (Photo 2.9) require prodigious amounts of energy to keep functioning. A large fraction of that energy is electricity delivered to users by means of long-distance electric transmission lines.

Fossil fuels are the most versatile and widely used sources of practical energy in the world. They are employed to run motorized vehicles, heat buildings, and provide most of the energy necessary to run electric generators at electric generating stations. Additional practical energy is derived from a variety of other sources: Falling water (hydroelectric power) is used to run machinery or generate electricity. Nuclear power is used almost exclusively to run electric generators. Wind turbines that harness the force of mov-

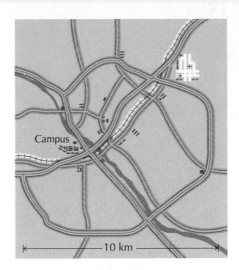

**WINDOW K** Coordinated entities as large as a city would not function without the availability of easily exploited sources of practical energy.

ing air are used to generate electricity. Direct solar energy is employed for heating and for generating electricity. Many other, experimental methods of harnessing practical energy have been conceived, and some have been implemented for small-scale applications.

Several phenomena of concern to earth scientists occur on the size scale of a small city or larger. Changes in the atmosphere, such as winds or storms, arise because of temperature differences from place to place. Large temperature differences are apt to occur over distances of hundreds of kilometers or more, but along coastlines and in mountainous terrain, temperature—and therefore, the local weather or climate—may vary significantly over as little as a kilometer. On or just below Earth's surface, geologists recognize rock *formations* (units of rock of a common nature) that may have dimensions as small as those of a city block or as large as those of a mountain range. By analyzing the chemical and mineral composition of the rocks belonging to a formation, and by noting the relative positions of formations, geologists can shed light on the geologic history of entire regions. Catastrophic phenomena such as earthquakes, volcanoes, and violent storms also tend to take place within geographic areas at least as broad as a small city.

**PHOTO 2.9** One of the world's largest cities, both in area and in population, the Los Angeles metropolitan area requires an average of about $10^{10}$ watts, or 10,000 megawatts of electric power.

**YOU ARE HERE**

$10^{-15}$m $\quad$ $10^{-10}$m $\quad$ $10^{-5}$m $\quad$ $10^{0}$m $\quad$ $10^{5}$m $\quad$ $10^{10}$m $\quad$ $10^{15}$m $\quad$ $10^{20}$m $\quad$ $10^{25}$m

# A State

**WINDOW L** On viewing a large enough patch of Earth's surface, we realize it is not flat but has a curvature.

If you were an astronaut riding a low-orbiting spacecraft passing over California (pictured in Window L), the ground below would not seem flat but noticeably curved. A big state like California spans about 9° of latitude; its curvature is roughly that of a 1-inch-wide patch on a beach ball. North America (say from Alaska to Florida) curves through a greater arc on Earth's surface: It spans some 60°, or one-sixth of the entire circumference of the globe.

Low-orbiting spacecraft race over the surface of Earth at about 8 km/s and at heights of about 200 km. At that height, the vast majority of Earth's atmosphere lies below, and outer space lies above. The term *outer space* can be taken to mean the space that contains everything else in the universe besides Earth and its atmosphere.

As we continue to expand toward outer space in the succeeding windows and view larger and larger bodies and structures, we will find almost nothing resembling the "flat" Earth we perceive over short distances. Spherical and disklike bodies predominate at larger scales, as a consequence of two things: gravity and rotation. The self-gravitation of a large body tends to pull it into a spherical shape, but rotation can alter that shape. Virtually everything in the universe rotates at some rate. Rotation causes a body to bulge at its equator; the faster the rotation, the more prominent the bulge. Also, when diffuse material orbits a central body, the material tends to settle into a disk. Saturn's ring system is an example.

Similarly, when we look at the motion of bodies, like planets, moving in space, we find that straight-line motion is rare. The paths, or orbits, taken by celestial bodies are closely approximated by curves called *conic sections*. They include circles and ellipses, which are closed curves, and parabolas and hyperbolas, which are open-ended curves. These curves make perfect sense to physicists and astronomers: They are the paths of bodies moving under gravitational interactions.

Looking down on the serene landscape of California, we may find it hard to imagine the magnitude of solar energy falling on it. In full sunlight, more than 100 trillion ($10^{14}$) watts (W) of solar power—equivalent to the continuous energy release of a trillion 100-W lightbulbs—falls on California. Plants capture some of the energy in this sunlight and convert it to energy stored in sugar or starch. Some of this stored energy is appropriated by plant-eating animals. Other animals eat the plant eaters. Humans exist at the top of the food chain; they skim off energy from the sun that has been gathered by a wide variety of organisms. Essentially, humans live on solar energy!

**YOU ARE HERE**

$10^{-15}$m   $10^{-10}$m   $10^{-5}$m   $10^{0}$m   $10^{5}$m   $10^{10}$m   $10^{15}$m   $10^{20}$m   $10^{25}$m

# Planet Earth

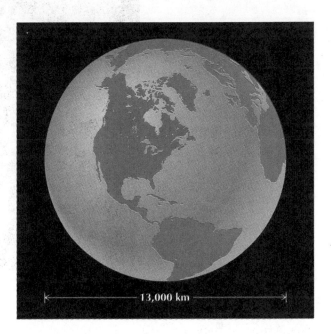

**WINDOW M** Earth is a self-contained sphere. The sun supplies the energy needed by living organisms that thrive on or near its surface.

13,000 km

The spherical nature of our planet is fully revealed from the perspective of an observer about 34,000 km above Earth's surface. This is the orbital radius of a *geostationary satellite*—a satellite orbiting above the equator and having a revolution period of 24 h. Geostationary satellites revolve at the same rate that Earth turns, so they remain fixed in the sky with respect to observers on the ground beneath them.

There is virtually no air in the space traversed by satellites, but it is a mistake to think that gravity disappears there. The presence or absence of an atmosphere has nothing to do with gravity being present or not. Earth exerts a gravitational pull on any material body 34,000 km away, but its comparative strength there is about 40 times weaker than on the ground.

Gravity is responsible for Earth's shape. Every particle within Earth attracts every other particle, so Earth has pulled itself together into the smallest possible shape: a sphere. Careful measurements show that Earth is not quite a perfect sphere but, rather, an oblate spheroid: It bulges slightly at the equator due to its rotation.

As viewed from the vantage point of any high-altitude satellite, Earth's atmosphere appears as a thin, hazy blanket. About two-thirds of the atmosphere's mass lies between sea level and the 9-km altitude of Earth's highest mountain, Mount Everest. The atmosphere is made up of mostly nitrogen and oxygen, both transparent gases; but here and there water vapor in the atmosphere condenses into tiny droplets or ice crystals, forming clouds.

Without clouds, Earth's continents down below are easily distinguished from the oceans, mainly by color. If we could somehow witness geological changes in enormously sped-up time from a geostationary satellite, we would be amazed to see the continents apparently drifting relative to each other, somewhat like lily pads on a wind-ruffled pond. Unlike lily pads on a pond,

**YOU ARE HERE**

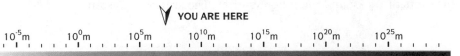

$10^{-15}$m    $10^{-10}$m    $10^{-5}$m    $10^{0}$m    $10^{5}$m    $10^{10}$m    $10^{15}$m    $10^{20}$m    $10^{25}$m

**PHOTO 2.10** Earth is a restless planet. Its fluid oceans and atmosphere are in a constant state of flux, so that each molecule of water or air typically drifts at a speed of less than a meter to several meters of distance every second. Driven by slow-moving currents of molten rock in its interior, Earth's continents slip and lurch along at an average rate of several centimeters of distance per year.

however, Earth's continents do not float on the oceans, nor do they physically plow through the rock that underlies the oceans. Instead, Earth's **lithosphere**—its solid outer shell—is cracked into large and small pieces called *plates*. These lithospheric plates can contain pieces of ocean bottom as well as continents or parts of continents. The plates are propelled horizontally (hundreds of kilometers of distance over millions of years) by slow-moving currents in the hot, soft rock of Earth's interior, on which they float.

The plates interact with one other in a variety of ways: Wherever plates move apart from each other, molten material from Earth's interior comes up, solidifies, and fills in the gap. Where two plates collide, their edges crumple, or the edge of one plate gets forced under the other. In some places, plate edges move laterally and grind against each other. The comprehensive theory encompassing these interactions, called *plate tectonics,* has revolutionized geology and the earth sciences, and it has its own chapter (Chapter 16) in this book.

Internally, Earth shares a kinship with the moon, Mars, Venus, and Mercury. All have dense interiors of rocky and metallic material. The sim-

ilarities, however, abruptly end at the surfaces of these bodies. Earth's atmosphere, hydrosphere (oceans and other bodies of water), and biosphere are unique among its neighbors. The biosphere—the intricate web of life on or near Earth's surface—includes nearly 6 billion people and countless millions of other biological species. No other planets are, at present, known to harbor life.

Some 30 Earth diameters away from us orbits our own natural satellite, the moon. During the six manned missions to the moon mounted by the United States from 1969 to 1972, astronauts explored the moon's surface on foot and by means of a lunar rover car. Roughly 382 kilograms (kg) of lunar rock and soil were brought back by these missions, much of it still being analyzed. While visiting the moon, the astronauts were separated from Earth by no more than about a quarter million miles of space, or about twice the distance traveled by an average car before it winds up in a junkyard. No human yet has traveled beyond the moon, but dozens of unmanned spacecraft, bristling with cameras and other instruments, have been launched on missions to other planets in the spacious realm of the Solar System.

# The Inner Solar System

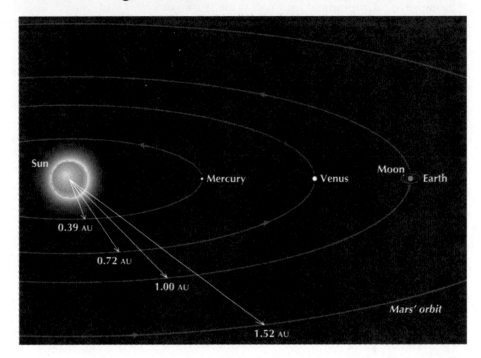

**WINDOW N**  The Solar System is incredibly spacious. Earth and its planetary neighbors are of trivial mass when compared to the sun.

The **Solar System** encompasses the sun, nine planets, and assorted smaller bodies such as moons, asteroids, and comets. The planets revolve around the sun in the same direction as the sun rotates, and almost all of them travel in near-circular orbits that lie in or near a common plane. The farther a planet is from the sun, the slower it moves, the farther it has to travel to complete one orbit, and the longer it takes to complete its own "year," which is one revolution.

Moons or satellites revolve around most of the planets. Asteroids and comets, which are small to very small in size compared with planets, follow their own independent orbits around the sun (see the next window, Window O).

The empty spaces of the Solar System, known as *interplanetary space*, are not really empty. Interplanetary space is filled with a smattering of dust particles and a tenuous (very thin) gas called *plasma*. The plasma consists of fast-moving, charged particles emanated by the sun.

In Window N, we look obliquely upon the plane of the Solar System as if we, the observers, were far above Earth's midnorthern latitudes. We see the orbits of the four innermost planets. They, like all the planets, travel counterclockwise around the sun when viewed from the north. If shown at true scale, Earth, the moon, Venus, and Mercury would be invisible, so each is drawn as if it were 100 times larger in diameter. The diameter of the sun, and also the moon's orbit, are exaggerated by a factor of 10. The relative distances between the sun and each of the planets are shown correctly.

For clarity, Window N shows the sun and the first three planets lying in a straight line. Only very rarely would all four bodies actually line up like this.

Earth travels around the sun in an orbit that is not quite circular and that has an average radius of approximately 150,000,000 km. Astronomers refer to this distance (more precisely,

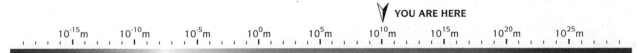

149,597,870 km) as the **astronomical unit** (AU). Although the astronomical unit is not an SI unit, it is a convenient yardstick for expressing distances in and around the solar system. In Windows N and O, the average radii of planet orbits are given in astronomical units. Today, these dimensions are known with great precision, owing largely to the technique of bouncing radio waves (see Box 2.3) off the surfaces of planets.

The sun, a rather typical star, is special to us only because it is relatively close, only 1 AU distant. The nearest "other" star is some 300,000 AU away. The sun is a sphere of hot gas more than 100 times larger in diameter than Earth and more than 1 million times larger in volume ($100^3 = 1,000,000$). This is equivalent to saying that more than a million Earths would fit inside the sun. The sun's mass, however, is only about 300,000 times

## BOX 2.3
## Parallax and Echoes: Range Finding in Astronomy

By living in the modern, mobile world, nearly all of us (sooner or later) end up having traveled distances comparable to the Earth-moon separation—approximately 384,000 km (240,000 mi). Airline pilots and frequent flyers cover similar distances in a matter of months. Astronauts have traveled to the moon in a matter of days. The moon has become not just an abstraction in the night sky but a real place separated from us by a distance most people can reasonably comprehend.

The distances to the sun and planets—hundreds and thousands of times the Earth-moon distance—are harder to fathom. Comprehending the distances to the stars requires a stretch of the imagination. Distances to the stars are so great that a gargantuan unit of distance, called the *light-year* (LY), equivalent to about 10 trillion km (6 trillion mi) is often used. How can we measure such mind-boggling distances?

One way is through a geometrical method called **trigonometric parallax.** Parallax is the change in the apparent direction of an object owing to a change in the vantage point of an observer. The trigonometric parallax method relies on observations of the same object made from two places separated from each other by a known distance, or *baseline.*

Human vision exploits the parallax effect, because humans use two eyes (separated by about 7 cm of distance) to view the world. To understand how parallax works, hold this book up so that surrounding objects, near and far, can be seen behind it. Cover one eye and notice how all objects in view tend to collapse to a single plane. Uncover the hidden eye and notice how much your depth perception (your sense of near and far) is improved. With both eyes open, your brain combines two images coming from two eyes separated by a baseline of about 7 cm.

The closer to your eyes an object is, the more significant is the effect of parallax. Hold a pencil upright at arm's length and line it up, using one eye only, with a distant object such as a utility pole or a tree. Focus on the distant object and then switch eyes, back and forth. Watch what happens when you move the pencil closer to your face: It seems to jump across a greater distance. The closer the nearby object, the greater the parallactic shift. The farther the object, the less the parallactic shift.

If a pencil were viewed from a distance greater than a few tens of meters, the shift would become undetectable if it were viewed across the same baseline of 7 cm. Of course, you could cheat a bit and move your head around laterally, thereby widening the baseline and increasing your ability to notice parallax and judge distance.

With a large enough baseline, even the moon can be seen in two different positions with respect to a more distant background. In the second century B.C., the Greek astronomer Hipparchus estimated the distance to the moon by comparing the positions of the moon (relative to the sun behind it) noted by two widely separated observers during a solar eclipse. He knew the length of the short side (baseline) of a long, thin triangle, and he also knew one

that of Earth, so its average density is considerably less than Earth's. This result is not unexpected for a ball of mostly low-density hydrogen and helium gas. Earth (and similarly Mercury, Venus, and Mars) consists almost entirely of rocky and metallic materials that are rich in elements heavier than hydrogen and helium.

The sun, like all stars, emits plenty of light, heat, and other electromagnetic radiation. The light we can see from the planets in the night sky is not emitted by the planets themselves; it is merely the reflected light of the sun. The starlike glimmers of Venus, Jupiter, and sometimes Mars are bright enough to outshine the brightest stars of the night sky. Seen close up through the cameras of space probes, the planets seem to glow against the dark background of interplanetary space (see Photo 2.11).

interior angle (the small angle between the two separate lines of sight toward the moon). A simple calculation gives the length of either long side of the triangle, which is the moon's distance. Hipparchus' estimate was within about 20% of the correct value—amazing, considering that no telescope or modern instrument was used.

After the telescope was invented in the 17th century, astronomers could compute the distances to the closer planets by comparing simultaneous telescopic observations of a given planet made by observers stationed on nearly opposite sides of Earth (a baseline of 10,000 km or more). Today, trigonometric parallax is no longer used to figure distances in the Solar System (radar ranging works much better, as we will describe momentarily). Trigonometric parallax, however, remains the fundamental tool for determining the distances to the nearer stars.

## Trigonometric Parallax as Applied to the Stars

As we look out toward the stars, our vantage point continually changes because we (and Earth itself) are in constant motion around the sun. When six months have gone by, we are halfway around the sun. We find ourselves displaced by 2 AU from our original position, because Earth's orbit has a radius of 1 AU (1 AU ≈ 150,000,000 km). A baseline of 2 AU is not large compared with star distances, but it is broad enough to produce measurable parallactic shifts in some stars.

**Stellar parallax** is the name given to the trigonometric parallax method by which astronomers measure distances to nearby stars. As illustrated in Figure 2.9, a relatively close star, star A, has a relatively large parallactic shift, or *(continued)*

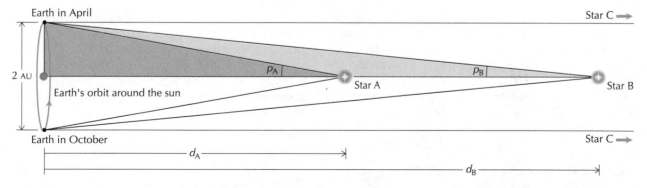

**FIGURE 2.9** The distance to star B ($d_B$) is greater than the distance to star A ($d_A$), because parallax angle $p_B$ is smaller than parallax angle $p_A$. The diagram is not to scale; even the closest stars are hundreds of thousands of astronomical units distant from Earth and the sun.

Inside the sun's enormously hot and compressed core, nuclear reactions liberate energy at a colossal rate. This energy slowly seeps out to the cooler surface, where it is released mostly in the form of light, heat, and ultraviolet radiation. Some of the gaseous material on the sun's surface gets thrown upward into a superhot, low-density region, where it becomes ionized (that is, the atoms of the gas lose one or more electrons and become ions). The resulting halo of ionized gas,

**PHOTO 2.11**   This reconstruction of Mars, based on photographic data obtained from several unmanned space missions, simulates the planet's appearance from space. A string of dormant volcanoes can be seen at left. Across the bottom sprawls the canyonlike Valles Marineris (Valley of the Mariner), which is about as long as the North American continent is wide. The thin and usually transparent atmosphere of Mars allows us to see its surface clearly from space.

*(Box 2.3 continued)*     parallax angle ($p$). Star B, farther away, has a smaller parallax angle. The very distant star C, which represents the vast majority of stars seen in the typical field of view of a telescope, has no detectable shift at all. Lines of sight toward it are essentially parallel (and $p \approx 0$).

The official parallax angle of any star as it is listed in star catalogs ($p_A$ and $p_B$ for the stars in Figure 2.9) is defined to be its parallax angle using a baseline of exactly 1 AU, rather than the 2 AU baseline employed for actual measurements. This baseline is used because the radius, not the diameter, of Earth's orbit is defined as the astronomical unit, and the astronomical unit has long been one of astronomy's most useful yardsticks.

The trigonometric calculation used to determine a star's distance is simple. For long, thin triangles only, such as those encountered in stellar parallax, the long side of the triangle (the distance to the star, $d$), is inversely proportional to the small angle $p$ between the two lines of sight. We may write this as $d \propto 1/p$.

For a baseline of 1 unit and a parallax angle of 1 arcsecond ($1'' = 1/3600°$), the distance along either long side of the long, thin triangle is 206,265 units (see Figure 2.10). A star with a parallax of $1''$ is 206,265 AU away. The distance 206,265 AU has been given a special name: 1 **parsec.** Parsec, meaning "parallax-second," is abbreviated as pc. The parsec unit is favored over the light-year unit by most professional astronomers. You will see the expressions kpc (1 kiloparsec = 1000 pc) and Mpc (1 megaparsec = 1,000,000 pc) later in this book. One parsec is approximately equal to 3.26 LY. For $d$ in parsecs and $p$ in arcseconds, the relationship $d \propto 1/p$ becomes

$$d = 1/p$$

No parallax of a star (other than, of course, the sun) exceeds $1''$. That is, none is closer than 1 pc (about 3 LY). The most distant stars measurable by parallax technique using current telescopes have $p \approx 0.01''$ and $d \approx 100$ pc $\approx 300$ LY. Nearly 10,000 stars out to this limit have had their parallaxes measured so far.

**FIGURE 2.10**   If the short side (baseline) of a long, thin triangle is 1 unit and the small angle $p$ is $1''$, then the triangle's length $d$ is 206,265 units.

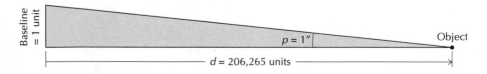

Baseline = 1 unit

$p = 1''$

Object

$d = 206,265$ units

or plasma, is called the *solar corona*. The corona's ghostly glow becomes visible from Earth whenever the bright disk of the sun is blocked by the moon during a total solar eclipse (see Photo 2.12).

Fast-moving ions and electrons spew forth from the corona to form the *solar wind*. The solar wind and its associated magnetic fields pervade the entire Solar System, waxing and waning according to the sun's variable activity.

**PHOTO 2.12** The sun emits a prodigious amount of light, only a tiny fraction of which emanates from the thin, superhot plasma making up the solar corona. The corona can be observed by the naked eye only during a total solar eclipse, like this one photographed from Baja California, Mexico, on July 11, 1991.

# Radar Ranging in the Solar System

Trigonometric parallax, when applied in astronomy, often relies on relatively inaccurate measurements of tiny angles. Any uncertainty in angular measurement results in the same uncertainty in distance. Within the realm of the Solar System, astronomers today use a more accurate technique: **radar ranging** (or simply *radar*). The technique relies on the time it takes radio waves to travel to and from any hard-surfaced object, such as one of the inner (terrestrial) planets or the moon. Radio waves, which are fundamentally the same type of energy as light, travel at a precisely known speed: the speed of light.

Radar ranging is similar to the technique of *echolocation* used by bats. Bats can fly perfectly well in total darkness because they navigate by listening to the echoes of the high-pitched squeaks they produce. Humans can practice echolocation, too. Imagine, for example, that a flat wall is facing you at some unknown distance. By clapping your hands and waiting for the echo to return, you can roughly gauge the distance to the wall. If you know the speed of sound (340 m/s on a cool day; slightly faster on a warm day) and the exact time between the transmission of the clapping sound and its return as an echo, you can calculate the wall's distance with great precision.

Figure 2.11 shows successive positions of the pulse of sound propagating from a hand clap. The pulse moves outward in an ever-widening arc and strikes the wall, where it is reflected back, again in an ever-widening arc. If the echo reaches your ears after 1 s, then the round-trip distance traveled by the sound (assuming 340 m/s as the speed of sound) is 340 m. The *one-way* distance to the wall is half that, or 170 m. For a bat zipping along in the dark, emitting a squeak, and listening for *(continued)*

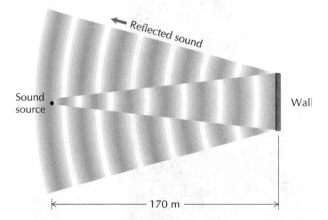

**FIGURE 2.11** If sound travels at 340 m/s on a particular day, it will take exactly 1 s to go the distance to the wall and back. The time of travel is a direct measure of the distance to the wall. (*Note:* The sound gets rapidly weaker as it travels outward along the ever-widening series of arcs drawn here.)

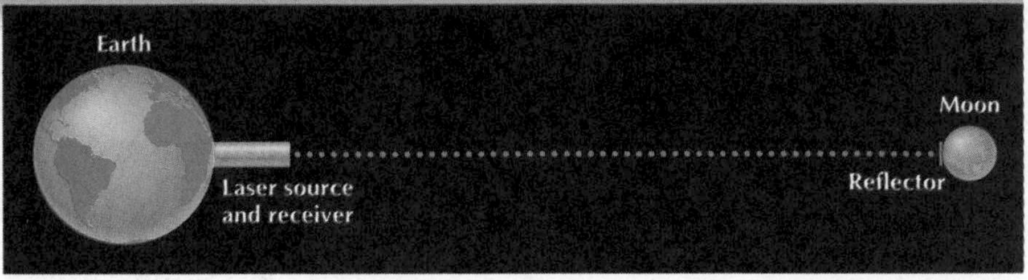

FIGURE 2.12 The distance between the source of light on Earth and the reflector on the moon is one-half the product of the speed of light and the round-trip travel time of the light.

*(Box 2.3 continued)* echoes, a 1-s delay would indicate that the source of the echo—perhaps an obstacle to be avoided—is a comfortable 170 m away. If the echo delay is 0.1 s, then the sound waves have gone only one-tenth as far, and the obstacle lies only 17 m away.

Precision echolocation and radar ranging rely on the following relationship:

$$\text{distance} = \text{speed} \times \text{time}$$

For radar ranging, astronomers use short-duration pulses of radio energy, focused and beamed by a dish-shaped antenna toward a target body. The same antenna later receives the returning radio echo, and an electronic clock measures the exact intervening time. For nearer targets such as the

moon, light emitted by a laser and sent through a telescope can be used instead of radio waves. Laser beam ranging works well for the moon, since the moon has on its surface several reflecting prism arrays left by astronauts during the Apollo moon landings. Both light and radio waves propagate through empty space unimpeded and at a speed of about 300,000 km/s.

Let us figure the distance to the moon (see Figure 2.12), assuming that a pulse of either light or radio energy sent from Earth returns in exactly 2.6 s:

$$\begin{aligned}
\text{distance} &= \text{speed} \times \text{time} \\
&= (300,000 \text{ km/s})(2.6 \text{ s}) \\
&= 780,000 \text{ km}
\end{aligned}$$

This is for the round-trip. The one-way trip (the distance to the moon) covers *half* the round-trip distance, or 390,000 km. What is really measured in this instance is the distance between a specific point on Earth and a specific point on the moon. Corrections must be applied in order to obtain the more meaningful distance between the *centers* of the two bodies. The moon's distance also varies greatly during the course of each orbit; but with enough observations over time, we can figure the average distance.

When large, dish-shaped antennas and powerful radio transmitters and receivers are used, radio echoes are discernible from all the nearby planets, despite the great distances involved. The lag time is significant: For Venus, the time between the transmission of a radio pulse and the detection of its faint echo is somewhere between about 5 min (300 s) and about 28 min (1700 s). These delays correspond to Venus being at its near and far positions relative to Earth.

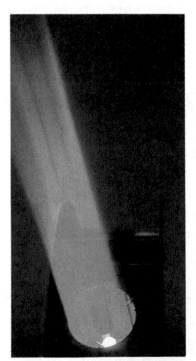

PHOTO 2.13 This laser beam (actually a series of very short-duration pulses), emerging from a telescope in France, is being used to measure the moon's distance to an accuracy of 3 cm.

# The Outer Solar System

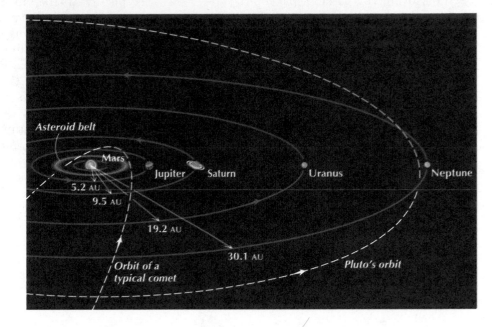

Mercury, Venus, Earth, and Mars constitute the **terrestrial planets** (Earth-like planets). The terrestrial planets have average densities of about 5 g/cm$^3$, a figure consistent with their rock and metal composition. Beyond Mars is a gap where no large planets exist. Most of the several thousand known asteroids (or minor planets) occupy this region. **Asteroids** are bodies rich in rocky or metallic material, not unlike the material that makes up the terrestrial planets. Asteroids range in size up to about a twentieth of the diameter of Earth.

Beyond the asteroids lie the widely spaced orbits of the **Jovian planets** (giant planets): Jupiter, Saturn, Uranus, and Neptune. The term *Jovian* refers to their prototype, Jupiter, which is the largest of the group. Jupiter's diameter is more than 11 times that of Earth's. Jupiter has roughly 1400 times Earth's volume but only 318 times Earth's mass. All the Jovian planets have relatively low densities, roughly 1 g/cm$^3$. In this feature they are quite similar to the sun—which is

not surprising, because they too are rich in the lightest elements, hydrogen and helium. Jupiter itself has been called a "failed" star. If Jupiter could be scaled up 1000-fold in mass, it would become a star very similar to the sun.

Distances are vast in the outer part of the Solar System, so the diameters of each planet shown in Window O are exaggerated by a factor of 500. The sun is enlarged by a factor of 50.

All four of the Jovian planets are known to possess ring systems, but only Saturn has a spectacular one. Saturn's ring system, which is really a swarm of billions of rock- and boulder-sized ice fragments, is one of the flattest structures in the Solar System. Its diameter measures some 140,000 km, but its average thickness is less than 0.1 km—razor-thin, to say the least!

Only three satellites, or moons, reside in the inner Solar System: Earth has one large one, and Mars has two very small ones. Each of the Jovian planets, however, possesses an extensive family of satellites: 59 at last count for all four Jovian

**YOU ARE HERE**

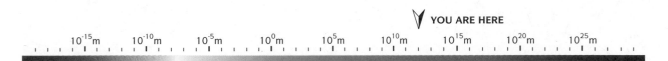

planets. More than half of these satellites have been discovered by passing spacecraft.

Pluto is an unusual planet, smaller in size than several of the Jovian satellites and with a composition rich in ice. Pluto's elongated path carries it more than half again as far from the sun as Neptune and then back inward to a point slightly closer to the sun than Neptune. Currently (from 1979 to 1999), Pluto is closer to the sun than Neptune.

Pluto is probably better classified as an asteroid than a planet. According to one theory, it is but one of a swarm of icy bodies, up to thousands of kilometers across, lying well beyond Neptune's orbit. Several such bodies, a few hundred kilometers across, have already been located; and further systematic searches are expected to reveal more.

There is good evidence for the existence of large numbers of comets in the cold depths of space surrounding our sun. **Comets** are small (typically 1–10 km across), ice-rich bodies believed to be left over from the formation of the Solar System. Billions or even trillions of them are thought to surround the sun in a huge, spherical cloud (called the Oort comet cloud), out to perhaps 50,000 AU. Occasionally, a comet from the Oort cloud strays into the inner Solar System and swings around the sun at relatively close range. If it passes within about 1 AU of the sun and similarly close to Earth, it may glow bright enough to be seen by the naked eyes of observers on Earth. After days or weeks as a conspicuous nighttime object, the comet withdraws to the outermost margins of the Solar System and quickly fades from view. A few so-called periodic comets, such as Comet Halley, return within a human lifetime, but most known comets have orbital periods of thousands or millions of years.

# Neighboring Stars

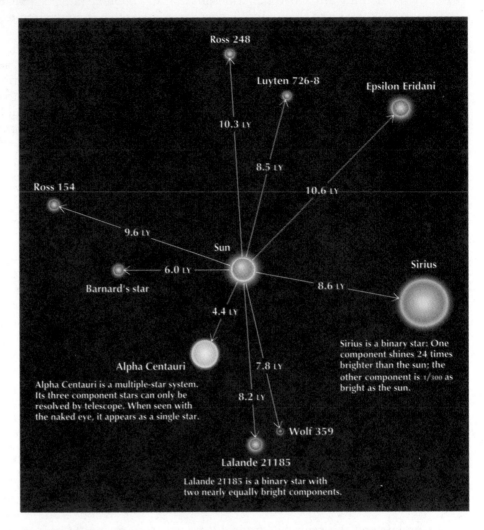

Ross 248

Luyten 726-8

Epsilon Eridani

10.3 LY

8.5 LY

10.6 LY

Ross 154

9.6 LY

Sun

6.0 LY

Sirius

8.6 LY

Barnard's star

4.4 LY

**Sirius is a binary star:** One component shines 24 times brighter than the sun; the other component is 1/300 as bright as the sun.

Alpha Centauri

7.8 LY

**Alpha Centauri is a multiple-star system.** Its three component stars can only be resolved by telescope. When seen with the naked eye, it appears as a single star.

8.2 LY

Wolf 359

Lalande 21185

**Lalande 21185 is a binary star with two nearly equally bright components.**

**WINDOW P** The sun is separated from its nearest stellar neighbors by distances of at least several light-years, which is many trillions of miles.

Window P places us in *interstellar space,* the realm of space associated with the stars. It is difficult to conceive of the leap in scale—more than a factor of 10,000 compared to the previous illustration—we are now making. At this scale, infinitesimally tiny stars float in a sea of practically nothing. If they were drawn in Window P in proportion to their true sizes, the dots of ink representing them would be not much larger than atoms. (Remember, of course, that stars are still *large* compared to the size of Earth.)

In Window P, a schematic illustration of the handful of the nearest stars in the so-called solar neighborhood, distance relationships are shown correctly, but star positions are not. The symbols for the stars are disks, not the pointed symbols many people associate with stars. The pointed appearance of bright stars in the night sky is caused by distortions in human vision. Stars also seem to fluctuate in brightness, or twinkle. The twinkling effect is caused by the way that Earth's continually moving atmosphere distorts the light

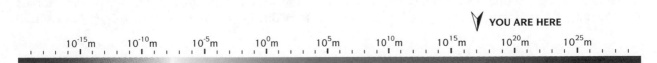

YOU ARE HERE

$10^{-15}$m    $10^{-10}$m    $10^{-5}$m    $10^{0}$m    $10^{5}$m    $10^{10}$m    $10^{15}$m    $10^{20}$m    $10^{25}$m

coming through it. Astronauts never see stars twinkling while they are above the atmosphere.

The sizes of the disks shown in Window P are related to the true brightness of the stars plotted, not to their physical size. The brightness scale is greatly compressed, though: The brightest star, Sirius, is actually about a million times brighter than the dimmest star plotted, Wolf 359. The color coding of the disks indicates each star's color and also its surface temperature: Red stars are relatively cool, white stars are hot, and blue stars are hotter still. (Star colors are actually quite pale. What astronomers refer to as red or blue stars are really reddish white and bluish white.) Alpha Centauri and Sirius, both conspicuous in Earth's nighttime sky, are rather hot and bright. All the other stars shown in this window are much dimmer and cooler; in fact, most are invisible from Earth without the use of a large telescope. When we catalog the properties of the stars, we find that dim and cool stars are by far the most common.

Quite possibly, several of the stars of the solar neighborhood could possess planets, as our sun does. The direct detection of such planets is difficult and so far unrealized. With its sharpened optics, the Hubble space telescope may soon be put to this task. Meanwhile, studies are underway to detect gravitational perturbations (irregularities) in the motions of various stars. Already, slight oscillations in the motions of some stars have provided convincing evidence of a planet or planets exerting gravitational tugs on their parent stars.

To understand the scale of Window P, imagine that you could travel outward from the sun's center at the speed of light, 300,000 km/s. After only 2.3 s you would penetrate the sun's surface. Earth's orbit would slip by 8.3 minutes (min) later. It would take roughly 4 h to reach Neptune or Pluto at its current position. To reach Alpha Centauri at light speed would take 4.4 years!

Astronomers, wishing to express astronomical distances in less-than-astronomical numbers, often use the unit called a **light-year** (LY). The light-year is *not* a unit of time; it is simply the distance light travels in one year. We say that Alpha Centauri is 4.4 LY away because its light takes 4.4 years to reach us. For the record, 1 LY is approximately equal to 64,000 AU and $10^{13}$ km, which is equivalent to about 6 trillion miles.

Light that we now see coming from the direction of Alpha Centauri left that star 4.4 years ago, so in a sense we are seeing quite deeply into the past. Actually, everything we see is a bit "old." You see the person next to you as he or she was a few nanoseconds ago, and the sun's light shows us how it appeared 8.3 minutes ago. We will refer to this delay as the *look-back time*.

Another distance unit commonly used by astronomers is the *parsec* (pc), which is approximately equal to 3.26 LY. The parsec was defined by means of stellar parallax (Box 2.3), the most common method of measuring the distances of nearby stars.

If we could view a time-lapse animation of the neighboring stars that surround us, we would notice their positions slowly changing in the sky over thousands of years. Some will drift out of our neighborhood in a million years' time; others will drift in. But this starry world is not totally aimless. The sun and its neighbors share a common movement in the galaxy we belong to (seen in Window Q). Also, half or more of all the stars are locked in stable gravitational embraces. Most of them are **binary stars**, two stars that endlessly revolve about each other. Sirius is a well-known binary, though its two components are greatly mismatched in size and brightness. Multiple-star systems can also form, with three or more stars traveling in more or less stable orbits.

When examined closely, Alpha Centauri turns out to be a triple-star system: a binary pair in a small orbit plus an exceedingly faint third companion, called Proxima Centauri. Technically speaking, Proxima (about 4.3 LY away), not Alpha, is the closest known star to our sun.

Worth noting at this point is that the falling stars or shooting stars that look like streaks of light in the night sky are not stars and have nothing to do with stars. They are meteors, which are small bits of interplanetary debris that plunge into Earth's upper atmosphere and burn up owing to friction. If the sky is dark enough, a particle need only be the size of a grain of sand for its fiery disintegration to be visible from the ground.

# The Milky Way Galaxy

**WINDOW Q**  Hundreds of billions of stars are contained within our galaxy, the Milky Way galaxy, whose principal parts extend across 100,000 light-years of space.

Another big leap in scale gives us Window Q, with a view of our home galaxy: the **Milky Way galaxy** (also referred to as "our galaxy"). This enormous, disk-shaped structure contains some 200 billion stars. Our knowledge of how many stars are really in it depends on a class of stars we know little about—the cool, exceedingly faint stars we find difficult or impossible to detect at large distances. Some astronomers believe our galaxy could have as many as 1 trillion ($10^{12}$) stars. (Let us say with certainty that even with millions of hamburgers sold daily, the giant McDonald's chain of restaurants has so far succeeded in serving only a fraction as many hamburgers as there are stars in our galaxy!)

Astronomers are less certain of the total number of planets harbored in the Milky Way galaxy. A large body of indirect evidence supports the idea that many (perhaps nearly half) of all stars are accompanied by planets. If this is true, then the number of planets in our galaxy could exceed the number of stars. Only a minuscule fraction of these planets would bear any close resemblance to Earth.

So numerous are the stars pictured in Window Q that none can be seen as an individual entity. Here and there, a fuzzy bright spot indicates where stars are clumped together in star clusters. Other bright spots, irregular in shape, indicate nebulas, or glowing clouds of interstellar gas.

Our sun and its neighboring stars, shrunk to microscopic size on the scale of this illustration, lie more than halfway out (about 30,000 LY) from the center to the ill-defined rim of the prominent disk of our galaxy. The sun, along with the rest of the Solar System, travels within this disk on a

**YOU ARE HERE**

$10^{-15}$m    $10^{-10}$m    $10^{-5}$m    $10^{0}$m    $10^{5}$m    $10^{10}$m    $10^{15}$m    $10^{20}$m    $10^{25}$m

roughly circular orbit around the galactic center at the rate of approximately 250 km/s. (Someone moving at this speed on Earth could whiz from Baltimore to New York City in 1 s.) Even at that rate, it takes our sun and Solar System a staggering 250 million years to complete a single orbit.

Like commuters on one side of a busy, multi-lane freeway, stars in the disk travel more or less parallel to each other as they circle the galactic center. Some are moving a bit faster than average, some slower; and there are many "lane changes" going on. Occasionally, a star or a cluster of stars hurtles sideways or obliquely through the disk, much like cars speeding by on overpasses or underpasses. These stars belong to a separate part of our galaxy, called the *halo*. Despite the cross-traffic, collisions between stars are exceedingly rare, because the sizes of the stars are tiny compared to the spaces between them. That any two stars would collide in our region of the galaxy is about as likely as two gnats, flying randomly in the Grand Canyon, colliding with each other. (Here we assume the gnats are not of the opposite sex!)

Currently, our galaxy is recognized as having four distinct parts:

1. The *nuclear bulge* at the center consists of a flattened sphere of closely spaced, older, redder stars. The stars are spaced only about 1 LY apart on average near the center of the bulge. Inside this bulge is the galactic nucleus, a seething region of stars and interstellar gas and dust, just 16 LY across, radiating some 100 million times as much energy as the sun.

2. The round, flat *disk,* stretching out some 50,000 LY from the center, is less than 10,000 LY thick, with most of its stars concentrated into a thinner disk in the central plane only 2000 LY thick. Within that central plane, the stars are separated by an average of about 4 LY.

The oblique view of our galaxy illustrated in Window Q shows the spiral, pinwheel structure of the disk. A wide variety of stars inhabit the disk and contribute to its brightness; however, the rare superluminous ones, most of which are young

and blue, contribute the majority of the light. These young stars tend to lie along the curved features we call spiral arms. Older, redder, dimmer stars are present everywhere in the disk, but their contribution to the total light of the galaxy is scant. Our sun lies near the disk's central plane, on the edge of a spiral arm. Recent observations suggest that the spiral arms of the Milky Way galaxy may trail from both ends of a pronounced barlike structure across the center. If so, our galaxy probably could be classified as a rarer type called barred spiral.

The disk of our galaxy also contains diffuse clouds of gas and dust, amounting to about 4% of the disk's total mass. The gas is quite transparent, but the dust, which tends to accumulate in the disk's central plane, greatly impedes our ability to see distant objects in that plane. For this reason, no one has been able to "see" the galactic nucleus, except with the help of instruments such as radio telescopes that pick up radiation other than visible light.

3. The sparsely populated *halo* extends outward in all directions from the nuclear bulge, out to perhaps 100,000 LY from the galaxy's center. It consists almost exclusively of old stars, many of which belong to **globular clusters.** These clusters, which contain tens to hundreds of thousands of stars densely packed into a ball, are the oldest and most stable structures within the galaxy.

4. The *galactic corona* consists of "dark" mass or matter of an unknown nature spread out in all directions to a radius of 300,000 LY or more from our galaxy's center. As much as 90% of the galaxy's total mass could reside in this mysterious feature. From the measured motions of stars in the galaxy's outer disk, which are controlled by gravitational forces, astronomers conclude that a massive corona must exist—though no one has identified the nature of its contents. Huge numbers of very faint stars and large amounts of diffuse material are considered current possibilities.

The interior view of our galaxy is not difficult to visualize. We do it by gazing upward on a dark, clear summer evening (far away from the glare of city lights). Amid a sky peppered with bright and dim stars, a milky white, diffuse band of light stretches overhead from horizon to horizon. It is known as the **Milky Way**. If we could view the entire band, including the part below the horizon, we would recognize it as a great circle with us at the center. The combined light of millions of distant stars in our galaxy's disk makes up the Milky Way. (The term *Milky Way galaxy*, borrowed from the earlier notion of the Milky Way in the sky, can lead to some confusion. Virtually everything we can see in the sky without the use of a telescope, toward *and* away from the Milky Way, belongs to the Milky Way galaxy.)

The brightness of the Milky Way increases toward the constellation Sagittarius, which hovers low in the southern sky on midsummer evenings as seen from the Northern Hemisphere. (Sagittarius is pictured hovering over the treetops in the photograph on p. 3.) The galactic nucleus lies in this direction, far behind many dusty, light-absorbing interstellar clouds that block the view.

When we look in directions *away* from the Milky Way, we can see deep into our galaxy's sparsely populated halo, where globular clusters follow a variety of elongated orbits around the galactic center. We can also see well beyond the halo into intergalactic space, where the light from billions of galaxies travels freely.

Considered as a unit, our galaxy is neither expanding nor contracting. Gravity, which tries to pull everything together, constrains stars that would otherwise drift away because of their incessant motion. Electromagnetism is thought to play only a minor role in phenomena existing on a galactic scale. For example, a weak magnetic field pervading the galaxy seems to control the paths of the rapidly moving, charged particles known as *cosmic rays.*

The Milky Way galaxy, a **spiral galaxy,** is but one of a vast number of similar galaxies spotted and cataloged by astronomers so far. Other kinds of galaxies, classified according to shape and structure, exist as well. Strip away the disk component of a spiral galaxy and you have what is essentially an **elliptical galaxy,** a featureless blob of mostly old stars and globular clusters with an overall round or elliptical shape. Ellipticals are somewhat more common than spirals. Take the disk component of a spiral galaxy, scramble its components so that the galaxy becomes rather disorganized and amorphous, and you have an **irregular galaxy,** the rarest of the three principal types. Each of these galaxy types encompasses an astounding range of size and other properties. "Giant" ellipticals, for example, can contain up to 10 trillion ($10^{13}$) stars; "dwarf" ellipticals may have only a few million ($10^6$–$10^7$) stars.

# Clusters of Galaxies

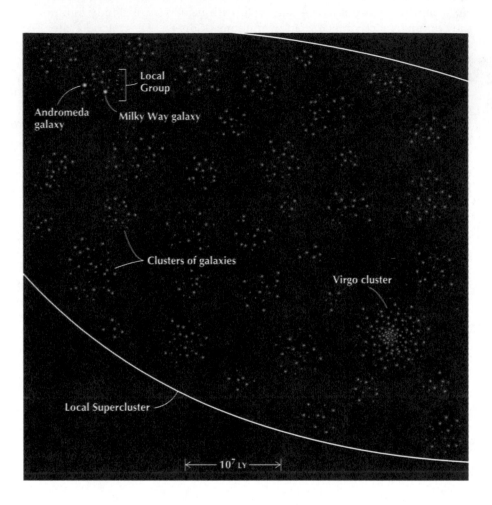

**WINDOW R**  Galaxies clump into clusters, and clusters of galaxies are often included within still larger entities called superclusters of galaxies.

From our vantage point within the Milky Way galaxy, billions of galaxies seem to stretch outward for billions of light-years in all directions. When we survey these galaxies by mapping their distances and directions, we find that they have a tendency to clump together. Most galaxies belong to loosely defined units called *clusters* of galaxies.

Our Milky Way galaxy belongs to what is commonly known as the **Local Group.** The Local

Group, with fewer than 30 known member galaxies, is considered a rather poor cluster. The relatively large and massive Milky Way galaxy anchors one end, and the even more ponderous Andromeda galaxy (also named M31), another spiral galaxy, anchors the opposite end. A small spiral galaxy called M33, plus two dozen or more dwarf elliptical and irregular galaxies, round out the Local Group.

**YOU ARE HERE**

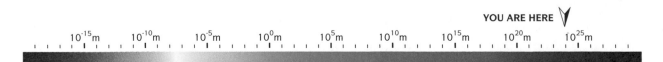

**PHOTO 2.14** The Andromeda galaxy is thought to bear a close resemblance to our own Milky Way galaxy, though it is somewhat larger and more massive.

Two of the dwarf irregulars, the Large Magellanic Cloud and the Small Magellanic Cloud, are located only about 170,000 LY away from our sun. Both are considered satellites of our own galaxy. Their disorganized structures have probably been shaped by the gravitational pull of the Milky Way galaxy. Studies of the Large Magellanic Cloud's present motion indicate that it will eventually slide into the Milky Way galaxy and lose its identity as a separate galaxy. Its stars and diffuse gas and dust will gradually disperse and merge with our galaxy's disk, thereby thickening it. Another satellite galaxy, dimly seen through the thick clouds of interstellar dust in the direction of Sagittarius, has recently been discovered only 50,000 LY from the Milky Way galaxy's center. It, too, seems to be merging with our galaxy.

In contrast to individual stars, which almost never collide with each other, galaxies (in the Local Group and elsewhere) occasionally graze or pass through one another. Such galactic collisions (or "galactic cannibalism," as it is known when a small galaxy is consumed by a large one) may profoundly affect the shape and structure of the surviving galaxy or galaxies. Thus a galaxy of a given type, such as spiral or elliptical, may not necessarily stay that way forever.

The Andromeda galaxy, the sole example of a galaxy (outside our own) visible to the naked eye from the Northern Hemisphere, lies some 2 million LY away. North Americans can easily spot it (under dark skies) on late fall evenings, when it lies high overhead amid the stars of the constellation Andromeda. Look for a hazy, oval patch of light, unlike all the sharply defined stars around it. Ponder that its dimly seen light, the combined contribution of perhaps half a trillion stars, is only now arriving on Earth after a journey lasting 2 million years!

If we define the macroscopic world as the range of distances directly accessible to our senses, then the Andromeda galaxy is surely the farthest outer limit. Beyond lies what undoubtedly must be considered a supermacroscopic realm. Today, the scope of this realm expands with every new advance in light-gathering (telescope) technology.

Looking beyond the Local Group, we find, at about 50 million LY out, the Virgo cluster of galaxies. The roughly 1000 members of this rich cluster swarm about the single largest galaxy in the cluster's center—the giant elliptical galaxy M87. Many astronomers assume that M87's great size and mass is the result of numerous past encounters with other galaxies.

Surveys seem to indicate that clusters of galaxies, rich and poor, fill the universe. The largest ones identified so far contain nearly 10,000 galaxies. These same surveys have also revealed a higher order of clustering: superclustering. The Virgo cluster and about 100 other poor clusters of galaxies, including our own Local Group, belong to the **Local Supercluster.** Its pancake-shaped form stretches some 100 million LY, more than filling Window R.

# The Universe at Large

Cosmic voids

Local Supercluster

$10^9$ LY

**WINDOW S**  Billions of galaxies inhabit the universe of our present perception. On the largest scales, galaxies seem to be concentrated along the walls of huge cosmic voids, each of which is hundreds of millions of light-years wide.

The *universe* can be defined as everything that exists. One of the goals of astronomy (and of physics) is to accumulate a body of knowledge about the universe: its size, its contents, and the laws that govern it. Today, with the help of giant telescopes, sensitive detectors, inventive methods of analyzing data, and sophisticated mathematical tools, scientists visualize an almost incomprehensibly vast universe, full of objects ranging from mundane to exotic. Wherever scientists have looked, on any spatial scale, they have found patterns. How, then, is matter organized on the very largest scales?

Window S illustrates the texture of the universe at large—or at least our current conception of it. We see what look like bubbles enclosing empty spaces, not unlike the structure of soap suds or the head on a freshly poured mug of beer. The majority of the billions of galaxies in our region of the universe seem to lie along fuzzy, bubblelike walls that enclose immense *cosmic voids*. The voids are not empty; they are thought to contain about 10-20% as many galaxies as the walls do. The largest wall, or boundary between voids, identified so far has been named the Great Wall. It extends for a distance of at least 250 million LY. The galaxies in the walls have an irregular, filamentary, and somewhat lumpy distribution. Our Local Supercluster is but one lump.

The interpretation of the universe at large in Window S is based on detailed mappings of thousands of galaxies closer than a billion light-years.

**YOU ARE HERE**

$10^{-15}$m  $10^{-10}$m  $10^{-5}$m  $10^0$m  $10^5$m  $10^{10}$m  $10^{15}$m  $10^{20}$m  $10^{25}$m

At farther distances, structural details are more difficult to discern and to map; nonetheless, the foamlike texture is thought to dominate at very great distances as well.

Every advance in telescope technology has widened our cosmic horizons in every direction we look. We can speak of an "observed" universe containing things we have seen, and we can imagine various models of the real universe that are either marginally bigger or much bigger than the part of it we are now capable of seeing. So far, however, we have no conclusive evidence that the universe is either finite or infinite in extent.

How large is the part of the universe within our observational grasp? New discoveries are made and distance records fall nearly every year. Recently, individual galaxies have been spotted at estimated distances of up to about 13 or 14 billion LY, which is thought to be about 90% of the distance to the theoretical edge of the observable universe—all that is potentially viewable. The observable universe is finite for reasons that will be explained in Chapter 17.

As human minds have comprehended a larger and larger universe, Earth has become more and more provincial in the cosmic scheme of things. Earth's fall from cosmic glory began several centuries ago, when the astronomer Copernicus proposed that the sun and not Earth lies at the center of our Solar System. By the 19th century there was no doubt: Parallax (see Box 2.3) was detected in nearby stars, indicating that Earth moves about the sun.

During the latter 19th century, the sun's nature as a sphere of hot gas was conclusively revealed by means of analyzing the sun's spectrum of light. Analysis of the light from the stars showed that they, too, are suns—and equivalently, that the sun is a star. By the end of the 19th century, our Solar System was pictured as being near the center of a huge, disk-shaped cluster of stars: the Milky Way galaxy.

Great strides in understanding were made in the 1920s and 1930s as astronomers put to full use the first generation of truly giant telescopes. They mapped the Milky Way galaxy in detail and demonstrated that our Solar System is not at the galaxy's center. They also showed that our galaxy is only one inhabiting a universe containing at least millions of other galaxies.

In the late 20th century, we recognize our Earth as being one of nine small, cool bodies orbiting a stable, unremarkable star. Our star orbits the center of a rather large, but otherwise ordinary, spiral galaxy. Our galaxy inhabits a small region somewhere within the currently observed universe, and the observed universe contains a *minimum* of 100 billion ($10^{11}$) other galaxies!

At present, we cannot conclude that our local region of the universe is at or near any central point. Indeed, the universe may have neither a center nor an edge. The universe may be "unbounded" in the same sense that Earth's *surface* is unbounded: You could travel forever upon Earth's surface and never reach a center or an edge.

By multiplying the minimum number of galaxies projected to exist by the average number of stars per galaxy (roughly $10^{11}$ also), we can conclude that there are at least $10^{11} \times 10^{11} = 10^{22}$ stars in the universe. If planetary systems commonly accompany stars, as most astronomers think, then there could be even more planets than stars in the universe. (If this perspective is too broad for your taste, making Earth and therefore humanity seem trivial, console yourself with the knowledge that astronomers and biologists alike agree that life of any sort on any given planet is probably rare. Intelligent life may be vanishingly rare.)

Our cosmic zoom sequence has allowed us to visualize a universe built on a hierarchy of scales. The bubbles and voids illustrated here are more than 43 powers of ten ($10^{43}$) times larger than quarks. As this book goes to press, we have no indication of matter being organized on any scale beyond the range we have reviewed in this chapter. Historically, however, investigations at the cutting edge of physics and astronomy have consistently led us deeper and deeper into the inner space of tiny particles and into the outer space of the universe at large.

## CHAPTER 2
# Summary

When an object of a given shape is scaled up in all three dimensions, its cross-sectional area and its surface area increase faster than (proportional to the square of) the object's width. The object's volume increases in proportion to the cube of the object's width.

Mass is a measurement of quantity of matter. Weight is the downward force an object applies when resting on Earth (or on another planet). Density is defined as mass divided by volume.

Measurements of space can be in one, two, and three dimensions. Various two- and three-dimensional coordinate systems have been devised to specify locations on Earth's surface, directions in the sky, and positions of objects relative to Earth's surface.

No matter what the spatial scale, the distribution of matter in the universe is seldom smooth. On microscopic scales, subatomic particles (protons, neutrons, and electrons) clump together to form atoms; and atoms often join together to form molecules. There are about 100 kinds of atoms (elements) and many possible molecular combinations of atoms. Immense numbers of atoms and molecules make up the macroscopic objects we can see.

Nine planets and assorted smaller bodies orbit the sun, which is a rather typical star. Galaxies, such as our own Milky Way galaxy, contain billions of stars clustered together. Galaxies themselves cluster in small and large groups. Clusters of galaxies, numbering in the millions, are distributed across the universe in a way that is not uniformly smooth.

All natural phenomena we observe in the universe today are governed by four basic interactions: the strong, weak, electromagnetic, and gravitational forces.

The strong force binds subatomic particles in the nuclei of atoms.

The weak force mediates certain forms of instability (radioactive decay) within atomic nuclei.

The electromagnetic force binds electrons to atomic nuclei, producing stable atoms; bonds atoms together into molecules; and holds assemblages of atoms or molecules together in microscopic and macroscopic structures.

The gravitational force dominates phenomena involving bodies of large mass. Gravitation is responsible for the spherical shape of stars and planets; it creates the gravity we experience on Earth's surface; and it controls the motions of planets, stars, and galaxies throughout the universe.

## CHAPTER 2
# Questions

## Multiple Choice

1. If you assume that the radius of Jupiter is 11 times greater than the radius of Earth, what would be the ratio between Jupiter's volume and Earth's volume?
   a) 11 to 1
   b) 33 to 1
   c) 121 to 1
   d) 1331 to 1

2. One kilogram of steel and 1 kg of feathers do not differ from each other in
   a) mass
   b) volume
   c) density
   d) the number of atoms they contain

3. If the mass of your body is 78 kg as measured on Earth, how much would your mass be on the moon's surface?
   a) 13 kg
   b) 13 lb
   c) 78 kg
   d) 0 kg

4. Two cubic meters of a certain type of dry soil has a mass of 6 metric tons (1 metric ton = 1000 kg). What is the density of this soil?
   a) $300 \text{ kg/m}^3$
   b) $1000 \text{ kg/m}^3$
   c) $3000 \text{ kg/m}^3$
   d) $6000 \text{ kg/m}^3$

5. The nucleus of an atom
   a) is about 1 nm in diameter
   b) always contains protons and neutrons
   c) is electrically neutral
   d) contains most of the atom's mass

6. The force that holds the particles of an atomic nucleus together is the
   a) strong force
   b) weak force
   c) electromagnetic force
   d) gravitational force

7. Which of the four fundamental forces is responsible for the formation of molecules?
   a) electromagnetic
   b) gravitational
   c) strong
   d) weak

8. The most fundamental units of matter are now thought to include the tiny particles called
   a) quarks
   b) protons
   c) atoms
   d) photons

9. An atom consists of a
   a) small object of positively charged material imbedded with one or more negatively charged electrons
   b) small, positively charged nucleus surrounded at a distance by negatively charged electrons
   c) small, negatively charged nucleus surrounded at a distance by positively charged protons
   d) nucleus of neutrons surrounded by protons

10. An atom of the carbon-13 isotope contains which and how many particles in its nucleus?
   a) 13 protons
   b) 13 protons, 13 neutrons, and 13 electrons
   c) 6 protons and 7 neutrons
   d) 6 protons, 7 neutrons, and 6 electrons

11. Nearly all the volume occupied by ordinary matter consists of
   a) neutrons
   b) protons
   c) electrons
   d) nothing (empty space)

12. The force that keeps the planets from drifting away from the sun is the
   a) strong force
   b) weak force
   c) electromagnetic force
   d) gravitational force

13. A double-helix structure characterizes the
   a) orbits of electrons around atoms
   b) DNA molecule
   c) axons and dendrites of neurons
   d) disk of the Milky Way galaxy

14. In the Solar System,
   a) the planets travel in the same direction around the sun

b) the spaces between planetary orbits are roughly the same

c) all the planets are roughly the same size

d) the outer planets travel slightly faster in their orbits than the inner planets do

15. Earth
    a) completes one orbit around the center of the Milky Way galaxy in one year
    b) along with the sun takes approximately 250,000,000 years to circle the center of the Milky Way galaxy
    c) is the outermost of the four terrestrial planets
    d) is the innermost of the four Jovian planets

# Questions

1. Assuming that Earth's diameter is four times that of the moon (it is actually a little less than that), how much more surface area does Earth have than the moon? How much more volume does Earth have than the moon?

2. Look up the longitude and latitude of your city, town, or school on a map or atlas.

3. How many coordinates (numbers) are required to specify the position of a point along a line—a highway, for example? How many are required for a plane—say, a soccer field? How many are required to specify one's exact position within Lake Michigan?

4. Compare the size of a typical atom to the size of its nucleus.

5. What are the approximate RA/Dec celestial coordinates of the galaxy plotted on the sky map of Figure 2.6?

6. Where does most of the mass of an atom reside? Which particles in atoms have by far the least mass?

7. What is an isotope? How are isotopes of a given element similar to each other? How do they differ?

8. What basic force (or forces) organizes matter on the subatomic level? What force is responsible for the structure of the atom and for the structure of materials we can hold in our hands? What force organizes matter on astronomical scales of distance?

9. If the electromagnetic force is approximately $10^{36}$ times stronger than the gravitational force (as Table 2.2 indicates), why does gravity seem so much more important in the macroscopic world of our everyday experience?

10. What is the human genome?

11. What does the term *entropy* mean?

12. How did Eratosthenes measure the size of Earth?

13. Describe Earth's position with respect to the Solar System, the Milky Way galaxy, the Local Group, the Local Supercluster, and the universe at large.

# Problems

1. Approximately 1000 Jupiter-sized bodies would fit into the volume of the sun. What, then, is the ratio between the sun's diameter and Jupiter's diameter?

2. If 5400 kg of aluminum scrap metal is melted down and cast into a solid mass, how much volume would that mass have? (*Hint:* You will need to refer to Table 2.1.)

3. Assuming that Earth's circumference is 40,000 km, how much distance along Earth's surface does 1° of latitude correspond to? How much distance does 1° of longitude correspond to?

4. For an observer at latitude 45°N on Earth, part of the celestial sphere is always blocked by the south horizon, because that part of the sky lies more or less over the South Pole. What declination would the southernmost visible stars have if they were viewed from latitude 45°N?

5. Using the radar technique to determine the distance to Mars, a radio astronomer measures a time lag of exactly 1000 s between the emission of a radio pulse through her antenna and the reception of the same pulse as reflected by Mars. What was the distance to Mars, in kilometers and in astronomical units, at the time of the experiment?

6. How many seconds does it take light to travel between the sun and Earth? (Assume that 1 AU = 150,000,000 km.) What is this figure in minutes?

7. Assuming that the sun is moving in a circular orbit around the center of the Milky Way galaxy at a constant speed of 250 km/s, and that it takes

250 million years to complete one orbit, what is the circumference of the sun's orbit in kilometers? (*Hint:* Remember that distance equals speed multiplied by time.) What is that circumference in light-years? (Recall that 1 LY $= 10^{13}$ km.) If you divide the circumference by $2\pi$, or 6.28, and obtain the radius of the sun's orbit in light-years, do you come up with a figure consistent with the dimensions of the Milky Way galaxy shown in Window Q?

## Questions for Thought

1. What is space? Is space the same as nothing? Does space have any properties or characteristics that we can describe? Can we say what *anything* really is?

2. If you could stand on Earth's South Pole, which way would be east? Which way would be west? Which way would be north? What are the latitude/longitude coordinates of the South Pole?

3. What does it mean to say that the gravitational and electromagnetic forces are "infinite" in range?

4. What would the universe be like (on large and small scales) if there was no electromagnetic force? What would it be like if there was no gravitational force?

5. If there is always a 50–50 chance that within a period of 5700 years a carbon-14 nucleus will decay and transform itself into a nucleus of another element, how much of an original 20-milligram (mg) sample of pure carbon-14 will be left intact after 5700 years? How much of the original sample will remain after 11,400 years?

6. There are approximately $10^{22}$ gas molecules in a liter of air, and there are roughly $10^{22}$ liters of air in Earth's atmosphere. Roughly how many gas molecules are in the atmosphere? How does this number compare to the estimated number of stars in the known universe? How does it compare (roughly at least) to the number of neurons in the brains of all humans now on Earth, recall-

ing that the present human population exceeds 5 billion?

7. If Earth's atmosphere could be condensed into a liquid equal in density to water, then this liquid would cover the entire globe to a depth of about 10 m. Using the results of Question 6, try to come up with a rough estimate of the number of molecules contained in the world's oceans. For simplicity, assume that the depth of the ocean water, if it could be spread uniformly around the globe, would be 1 km. Also assume that molecules in the atmosphere have, on average, the same mass as molecules in the oceans.

8. How do, or how have, increases in entropy affected your life?

9. How far can we probe into the microscopic world with any of our five senses alone? What is the farthest object we can detect with any of our five senses (without the use of instruments)?

10. If we could make parallax measurements of the nearby stars from various vantage points along Jupiter's orbit, how much more accurate would these measurements be than those obtained by our present observatories on Earth? A proposed spacecraft dubbed *Hipparchus* may one day make parallax measurements from the outer regions of the Solar System, roughly 40 AU from Earth. Currently, we can measure parallaxes of stars as far as about 100 pc. With a 40 AU baseline, what would be the distance of the farthest stars we could apply the parallax method to?

11. What would the Milky Way (the visible band of light in the night sky) look like if our Solar System inhabited the outer rim of the Milky Way galaxy's disk?

## Answers to Multiple-Choice Questions

| | | | | | |
|---|---|---|---|---|---|
| 1. d | 2. a | 3. c | 4. c | 5. d | 6. a |
| 7. a | 8. a | 9. b | 10. c | 11. d | 12. d |
| 13. b | 14. a | 15. b | | | |

# CHAPTER 3
# The Universe of Time

*The swinging pendulum of a clock measures time. All clocks measure change of some kind. Without change, time is impossible to measure. Most astronomers claim that time's reach into the past is finite; whether time's reach into the future is finite or infinite is unknown.*

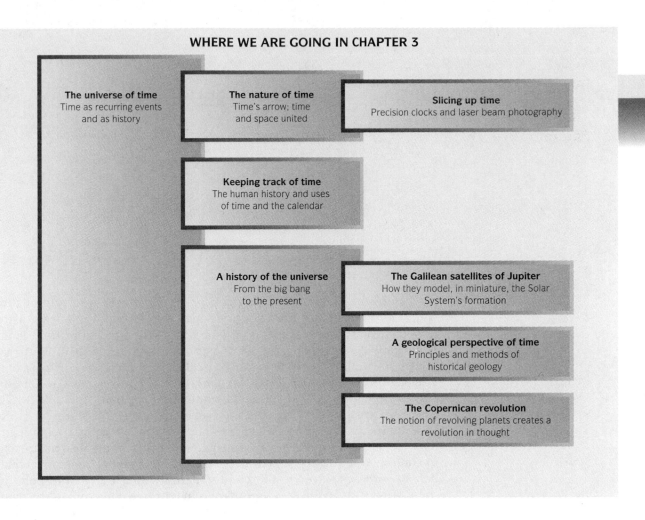

The word *time,* with all its mysterious and ambiguous meanings, merits as many as 30 separate definitions in some English language dictionaries. In the physical sciences, time loses some of its ambiguity, but not its mystery. Time helps us understand change in the physical world, and change lies at the heart of every physical process.

Although we cannot confidently say what time (or anything physical, for that matter) really *is,* we can offer an "operational" definition of time—that is, a practical definition based on some kind of action related to it: *Time is what a clock measures.*

What is a clock? A clock can be any mechanism based on regularly recurring actions. The swinging pendulum of a grandfather clock has a regularly recurring action. So does the balance wheel that ticks to and fro in an old-fashioned windup clock or wristwatch. Clocks are fundamentally counting machines. A clock's hands (or its digital display) show the translation between the counted recurring events and whatever time units are appropriate: seconds, minutes, or hours.

| TABLE 3.1 Some Time Intervals in Nature | Recurrent Event | Approximate Time Interval (s) | Time Unit |
|---|---|---|---|
| | Sun's revolution around the center of our galaxy | $7 \times 10^{15}$ | Galactic Year* |
| | Earth's revolution around the sun | $3 \times 10^7$ | Year |
| | Earth's rotation on its axis | $9 \times 10^4$ | Day |
| | Normal human heartbeat | 0.8 | |
| | Vibration of an atom in a solid | $10^{-13}$ | |
| | Revolution of an electron around a hydrogen nucleus | $10^{-16}$ | Atomic Year* |

*These terms are informal expressions.

Before the invention of mechanical clocks, longer intervals of time could be reckoned (as they still are today) by naturally recurring astronomical events—natural clocks, if you will. A day's worth of time is the equivalent of the sun circling the celestial sphere once, as we on Earth see it. The sun's diurnal motion is caused by Earth's rotation, so Earth is a kind of clock that "ticks" once a day. Similarly, Earth revolves once around the sun in the period we call a year. The time intervals listed in Table 3.1 are based on a wide variety of clock mechanisms existing in nature.

The Italian astronomer Galileo Galilei was one of the first scientists to exploit the connection between time and recurring events that we are able to control. While sitting in a drafty cathedral, Galileo noticed how a chandelier overhead swayed in short and sometimes longer arcs in response to puffs of wind. It seemed to him that no matter how long an arc the chandelier swept out, it would always complete one oscillation (back-and-forth movement) in the same amount of time. Later, under controlled conditions, Galileo set up similar arrangements (pendulums) using weights swinging on the ends of long strings. He discovered that a pendulum's period depends only on its length. Shorter pendulums oscillate more frequently, while longer pendulums oscillate less frequently. A pendulum of fixed length, he also found, oscillates with a fixed period, independent of how far it swings to either side.

Galileo's discovery ushered in the development of giant pendulum clocks, which soon became the central fixture of town squares throughout Europe. These clocks were phenomenally precise compared with earlier timekeeping devices; and they could be calibrated (adjusted) to run at slower or more rapid rates, simply by shortening or lengthening the pendulum. The era of standardized timekeeping was underway. This chapter's "Keeping Track of Time" section traces the origins of modern timekeeping and draws parallels between cyclical astronomical activity (such as seasons and moon phases) and the calendar. As for clocks themselves, they evolved from the bulky, mechanical devices of centuries past to today's elaborate electronic devices that keep count of the very steady vibrations of atoms (as we shall see in Box 3.1).

Besides being what a clock measures, time has another, broader meaning useful for the purpose of this chapter. Time is the totality of all moments—past, present, and future. Time can be thought of as a continu-

**FIGURE 3.1** A timeline is a one-dimensional graph of time. Unlike any of the three dimensions of space in the universe in which we live, we cannot reverse our direction along the one dimension of time. "Now" on this graph constantly slides to the right.

Past ⟵————————————⟶ Future

Now

um (a continuous, unbroken sequence) of moments on which events can be mapped. Space can be thought of as a continuum, too: While traveling through space, as from one city to another, we can map as many intermediate points along the way as we desire.

Time's continuum is one-dimensional and it flows in one direction only: away from the past and toward the future. We can visualize time in this sense by using a one-dimensional space—a line—as a metaphor for time. We may draw a timeline (see Figure 3.1), with the present time at the center, the past to the left, and the future to the right. "Now" is an infinites-

BOX 3.1

# Slicing Up Time

Galileo's method of counting the swings of a pendulum is reasonably precise for time intervals of seconds, minutes, or longer. But for accurate measurements of shorter periods, one needs a recurring mechanism that is both very regular and very rapid. Cesium electronic clocks, which precisely monitor the passage of time today, exploit a mechanism that has proven to be one of the most stable and rapid in nature: atomic vibrations. These clocks, which are accurate to within about three-millionths of a second per year, count the oscillations of light waves given off by undisturbed cesium-133 atoms. By the modern standard of time, it takes exactly 9,192,631,770 such vibrations to make one second (1 s).

Much of today's science and technology depends upon processes that take place not merely in seconds but in tiny fractions of a second. For example, the key microprocessor chip in many modern desktop computers runs at a clock speed of about 100 megahertz (100 MHz), which means that it can execute 100 million switching operations per second.

Just as microscopes have illuminated the world of inner space, electronic clocks (used as high-speed stopwatches), high-speed cameras, and pulsed laser beams have allowed us to probe the details of short-duration events. In the late 1800s, the photographer Eadweard Muybridge developed a camera having a shutter that could open and close in just 1/500 s (0.002 s, or 2 ms). One of its first uses was to show that at some point in its stride, a galloping horse has all four hooves off the ground—a conjecture that no one could prove beforehand because of the limitations of human vision (see Photo 3.1).

The time resolution of human vision is only about 1/20 s, which means that, at best, no more than 20 sequential images can be discerned by the eye and brain as being separate from     *(continued)*

**PHOTO 3.1** In 1878, Eadweard Muybridge's "automatic electro-photographic" camera produced this series of stop-action images. About 1/25 s passes between adjacent images.

imally small slice of time caught between the past and the future, and it never stops moving to the right. Timelines, of course, are commonly employed by historians for visualizing a sequence of historical events and by planners as an aid to scheduling future projects.

Unfolding across many pages in the second half of this chapter is a timeline encompassing the broadest possible history of the universe, beginning with the event known as the big bang. According to current scientific thought, the big bang represents the birth of the universe and the beginning of time, some 15 billion years ago. The timeline ends at our time, the close of the twentieth century. Important periods or events of cosmic history, geologic history, and human history (insofar as it pertains to science and technology) are summarized in each of several text boxes keyed to the appropriate places on the timeline.

*(Box 3.1 continued)*    one another over 1 s of time. Old-fashioned jerky movies seem disjointed because those films were shot at less than 20 frames per second. Standard television transmissions deliver one complete picture on your screen every 1/30 s, so motion depicted on TV looks reasonably smooth and continuous.

The electronic flash units built into today's inexpensive "point-and-shoot" cameras fire with a duration of about 1/1000 s. Common flash photography, then, freezes single events with a time resolution of 1 ms or better. Some specialized flash units can emit pulses a hundred times shorter than that—as short as $10^{-5}$ s.

Today, laser technology has progressed to the point where laser light can be produced in pulses that last only a few femtoseconds (1 fs = $10^{-15}$ s). By comparison, cesium atomic clocks are practically slowpokes; they count events that recur about every $10^{-10}$ s, or 100,000 fs.

Femtosecond pulses of laser light are also remarkable for their size. A laser turned on for 1 s produces a beam of light 300,000 km long if nothing stands in the way of the outgoing beam. When laser light is reduced to a pulse that lasts just 3 fs, however, it becomes a beam only about 1 $\mu$m ($10^{-6}$ m) long, which is the diameter of a typical bacterium.

By using these tiny pulses of laser light in a technique called *femtosecond photography,* chemists today can reconstruct events on the atomic scale that take place trillions of times faster than the human eye can follow. During a typical chemical reaction involving individual molecules or atoms, there is a transitional phase during which bonds between atoms are broken and new bonds are formed. After the reaction is over, the same number of atoms are present, but the arrangement of the atoms is different. Until recently, chemists could not observe the transitional phase because it is so brief, typically less than 1 picosecond (1 ps = $10^{-12}$ s). In order to "see," or resolve, transitional events, chemists must use flashes of illumination that are much briefer than the events themselves—hence the need for femtosecond lasers. Femtosecond photography may give chemists insights on how to precisely control chemical reactions, which could lead to a variety of applications, both expected and unforeseen.

**PHOTO 3.2**    Some specialized electronic flash units are capable of pulses lasting only $10^{-5}$ s (10 $\mu$s). This is short enough to "stop" an arrow or a speeding bullet in flight.

## The Nature of Time

Unlike movements through space, "travel" through time is strictly a one-way affair. You can travel either way in any dimension of space (north or south, east or west, or up or down on Earth), but time travel in the past direction is forbidden. The flow of time from past to future through the present makes sense to us because, generally speaking, events of the past and present directly affect what will happen in the future, never the other way around.

We must clarify the last sentence by saying that it is impossible to predict exactly what will happen in the future even if our knowledge of past and present events is exact. There are at least two reasons for this: quantum indeterminacy (Chapter 14), which is all-important in the realm of the atom; and chaotic behavior (Chapter 5), which is thought to apply in varying degrees to systems of interacting bodies on both microscopic and macroscopic scales. Note that, for human beings at least, the *anticipation* of future events can indirectly affect the present. For example, your present action of taking a physical science course is probably in response to some goal, such as earning an academic degree.

### Time's Arrow

The past-to-future flow of time may seem perfectly sensible, but inquiring minds ask, "*Why* is it?" Why does the "arrow of time" point from the past to the future? This may be a simple question, but it is not trivial. We may never know exactly why, but time's inexorable march into the future seems to be closely related to the fact that *entropy* (a measure of randomness or disorder in physical systems) has a strong tendency to increase as time goes on. The notion of increasing entropy was introduced in the previous chapter, and we will become acquainted with it again in Chapter 7 in connection with heat engines (devices that convert heat to other forms of energy).

Let us imagine that we *could* reverse the arrow of time. Absurd sequences of events, like those viewed on a movie played backward, could then really happen: Spilled milk and shards of glass could gather themselves together into a glass of milk that rises to a tabletop and comes to rest as a child's hand withdraws from it.

The direction of time's arrow, it seems, points in the "direction" of spontaneously increasing entropy. This is equivalent to saying that a relatively disordered system (spilled milk and glass shards) must *come after* the same system being in a more orderly state (milk in a glass on the tabletop). The reverse sequence of events not only seems nonsensical, but also never occurs spontaneously. Of course, you could meticulously gather all the glass shards and glue them back together, and you could suction up the milk and place it in the reassembled glass. But that would involve a large effort, or influx of energy, which is hardly spontaneous.

Scientists believe that the tendency for entropy to increase with time is a characteristic of the universe as a whole. As an isolated system containing all matter and all energy, the universe is pictured as steadily evolving from being in a relatively organized state toward one in which there is more and more random activity.

# Space and Time United

Prior to the early 1900s, most scientists were comfortable with the concept of *absolute time,* or the idea that time ticks along at the same rate everywhere and for everyone. However, time possesses ambiguities even more profound than those suggested by its many definitions in the dictionary. Albert Einstein's 1905 paper on special relativity and his 1916 paper on general relativity asserted that time is relative—that two different observers could disagree about the rate at which the same clock is ticking. Other things once believed to be absolute and unchanging, like length and mass, also turned out to be relative.

In Einstein's relativistic way of looking at things, the rate at which time passes at a given place depends on the observer's movement relative to that place. It also depends on whether the observer is being accelerated and on the presence of any massive body or bodies nearby. Note that relativistic effects are exceedingly small and undetectable with regard to the physical phenomena we experience in everyday life. However, relativity theory is essential for understanding phenomena involving very high speeds, very strong gravity, and other unusual circumstances.

An important consequence of the relativity of time is that no two events occurring in different places can be said to happen simultaneously for all observers. Even though two events in separate places might be observed as being simultaneous from one person's point of view, other people observing from other points of view might disagree. (Relativity theory will be dealt with further in Chapter 13; we simply have mentioned it here in connection with the unity of space and time.)

In Chapter 2, we saw how one, two, or three dimensions of space can be mapped by using grids or coordinates. It takes three coordinates to specify the location of a point in three-dimensional space, the space we live in. As mentioned earlier, time can be thought of as a single coordinate, and we can visualize it by means of an analogy with space, as in a timeline. In relativity theory, space and time are intimately connected, and everything in the universe exists on a *space-time continuum*—three dimensions of space joined with one dimension of time. Each instant of time at a given place in space is a unique "point" in space-time.

Complete space-time models are impossible to construct, or even visualize, if we insist on trying to plot all four dimensions of space-time as dimensions of space alone, because space *has* only three dimensions. Much simpler models of space-time can be made, however, if we choose to ignore one or two of the dimensions of space. One such model, with only one dimension of space plotted against the single dimension of time, is the simple Cartesian graph shown in Figure 3.2.

To understand Figure 3.2, imagine a simplified space in which you travel only along a north-south line. You leave home; drive to school (15 mi north of your home, half an hour away); attend classes until noon; walk at a brisk 4-mi/h clip to an off-campus restaurant 2 mi north of campus; eat; walk back; attend more classes; and so on. All through the day, you arrive at new points in space-time, *whether or not* you were actually in motion. Every point in the simplified space-time you experience during the day is unique and unique to you alone, unless you share the same point or points of space-time with another student by, for example, embracing him or her.

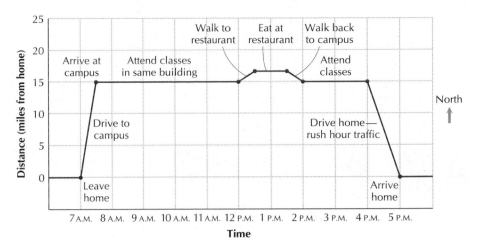

**FIGURE 3.2**  On this journey through space-time on a typical school day, you never "stand still." Time's inexorable march into the future carries you along whether or not you move through space.

Relativity theory says that each of us experiences the world around us in a different way whenever we are apart from each other, moving with respect to each other, or subject to different gravitational environments.

Most scientific work can be done without regard to the notion of the space-time continuum. Chemists performing laboratory experiments, engineers applying the laws of physics to design a structure, and geologists dating rock formations can consider space and time as being separate from each other.

Relativity and the unity of space and time often *is* important in astronomy, though. In part, it is important because astronomers deal with both vast distances and very long time intervals. In 1987, an exploding star, dubbed Supernova 1987A, suddenly appeared in the Large Magellanic Cloud in the southern part of the sky. Those who looked at it witnessed light born during the cataclysmic death of a star some 160,000 LY distant. Since the light took 160,000 years to reach Earth, these astronomers hardly observed the event itself, no more than any of us have observed the Civil War because we have examined photographs of it. Rather, a small slice of space-time (the events in and around the exploding star), very much removed from the space-time the astronomers inhabited, finally became accessible to them. Light carried information from the exploding star to the eyes of the astronomers. Light is a carrier of information that can bind together any two points in the universe. Two points close together in space can share information quickly, while two widely spaced points are separated by a big gulf of time.

In another sense, astronomers depend on Einstein's relativity theory for the concept of a *black hole* in space, which is a theoretical construct of general relativity. The notion of a black hole can explain many observed astronomical phenomena that formerly could not be explained in any satisfactory manner. Both space and time are severely distorted near a black hole.

Black holes and supernovas illustrate that today's conception of the universe is far more grand in time and space than what we might have learned from using our five senses alone. In this chapter and in the previous chapter, we are trying to understand a universe that is unimaginably large and intricate. Today, scientists have an impressive array of tools that can pry into the secret recesses of the natural world. They look inward toward the realm of the atom with the help of speeding subatomic particles. They look outward toward the galaxies and back into time with huge collectors of light and radio waves that focus their collected data upon arrays of electronic detectors. They sift through Earth's crust in search of evidence of the terrestrial past. They reach through time and space more deeply than ever before.

# Keeping Track of Time

An average human life lasts about 3 billion heartbeats, or some 2.4 billion seconds. Imagine how insufferably boring our lives would be if we marked time only by units as brief as that. Fortunately, we live in a world with longer rhythms—days, months, and years—imposed on us by nature. Each of these rhythms is rooted in cyclical events taking place in the sky above.

## The Day and the Time of Day

As we saw in Chapter 2, the daily, or *diurnal,* cycle is a consequence of Earth's rotation. (Rotation and revolution, by the way, are not the same thing. A body rotates by spinning on an axis; it revolves by circling another body.) One rotation of Earth equals one day, but a problem arises when we ask just to what this rotation refers. When measured with respect to the distant stars, the rotation period (measured by the time units on our watches) is approximately 23 h, 56 min—some 4 min shorter than a "normal" 24-h day. This shorter period, called the *sidereal day,* is the basis for **sidereal time,** or star time. Sidereal time is reckoned by the positions of stars in the sky, which is chiefly of interest to astronomers. If, on the other hand, we measure how long it takes Earth to rotate once relative to the sun, we get the normal 24 h of a *solar day.* The difference between the solar day and the sidereal day is illustrated in Figure 3.3.

Sidereal time is used by astronomers whenever it is important to locate, relative to the horizon, any celestial body whose celestial coordinates are known. Sidereal time is divided into hours, minutes, and seconds in the same way that solar time is, but the sidereal time units are slightly shorter than the hour, minute, and second units of solar time. If you look at Figure 2.6 in Chapter 2 again, you will understand how the "hour" units of right ascension are related to the hours of sidereal time. An observer on Earth sees the celestial sphere turning at a rate of 1 hour of angle (15°) for every 1 h of sidereal time.

The 4-min difference between sidereal time and solar time means that stars rise, set, or otherwise arrive at any given position in the sky about 4 min earlier per day as reckoned by the normal solar time we live by. For example, if the star Sirius rises at 9:00 P.M. tonight, then tomorrow night it will rise at 8:56 P.M. The 4-min difference accumulates to nearly half an hour after 1 week and to about 2 hours after 1 month. If you wanted to see Sirius in the sky earlier than, say, 7:00 P.M., you would have to wait for at least a month. After 6 months, the 4-min per day difference accumulates to

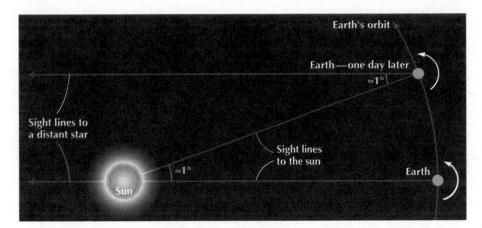

**FIGURE 3.3** If Earth did not revolve around the sun, Earth would stay in the lower position (at right), and one turn on its axis relative to the stars would be equal to one turn relative to the sun. But Earth does move around the sun once (360°) in about 365 days, a rate of about 1° per day. Since Earth travels to the upper position after one day, it must turn on its axis an extra 1° (which takes approximately 4 min) to complete one rotation relative to the sun. Thus, the solar day is about 4 min longer than the sidereal (star) day. In solar time units, 1 sidereal day is 23 h, 56 min.

12 h, which is equivalent to half a turn of Earth. Therefore, opposite halves of the celestial sphere are seen at the same clock time for dates that are 6 months apart (see Figure 3.4).

Solar time is based on the angular position of the sun in the sky relative to the **celestial meridian,** the imaginary great circle dividing the east and west halves of the sky. During the A.M. (*ante meridiem,* "before midday") hours, the sun lies east of the celestial meridian. During the P.M. (*post meridiem,* "after midday") hours, it lies west of the celestial meridian. If the sun happens to be 15° west of the celestial meridian, the time is 1 P.M.; if the sun is 30° west, it is 2 P.M. Ideally, the designation 12 noon corresponds to the sun's midday position when it transits (crosses) the observer's celestial meridian (see Figure 3.5).

Strictly speaking, our measure of solar time as described is **apparent solar time**—time based on the sun's observed angular position in the sky— which is not quite the time we keep on our watches. For one thing, the time between two successive meridian transits of the sun is seldom exactly 24 h. Because of factors related to the shape of Earth's orbit and the tilt of its axis, the sun may transit the meridian either early or late as reckoned by a uniformly ticking clock. These irregularities may amount to as much as

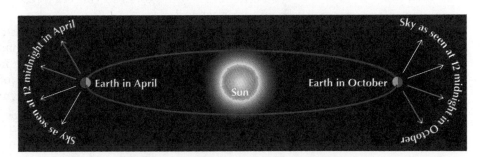

**FIGURE 3.4** Two opposite parts of the sky can be seen at the same solar time for dates that are 6 months apart.

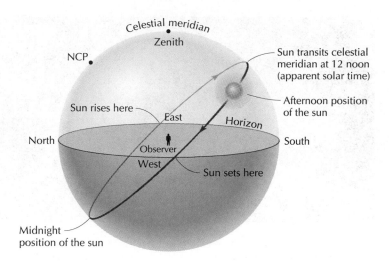

**FIGURE 3.5** Solar time is measured in reference to the celestial meridian, which passes from the north point to the south point over the zenith. When the sun transits (crosses) the celestial meridian, as seen from any given location, the apparent solar time there is 12 noon. Various corrections must be applied to obtain standard time from apparent time. (NCP is the north celestial pole.)

16 min. Nonetheless, there is a certain average, or *mean,* period associated with intervals of time between successive transits, and this mean solar day is what is taken as the basis for **mean solar time.** Since Earth repeats its orbital path around the sun year after year, the irregularities are repeated; and therefore, timekeepers can easily make corrections for "early" or "late" arrivals of the sun on any day of the year.

During most of the 19th century, each community kept its own mean solar time. As it became common for people to be able to travel more quickly and farther by train or automobile, time zones were introduced. In this scheme, which has been adopted by nearly all countries of the world today, Earth is theoretically divided into 24 longitudinal strips, each 15° wide. The strips, or *standard time zones,* are centered on *standard meridians* of longitude 15° apart. Across the contiguous United States, the standard time zones called Eastern, Central, Mountain, and Pacific are based on the standard meridians at 75° W, 90° W, 105° W, and 120° W, respectively. As shown in Figure 3.6, the actual time zone boundaries are quite irregular.

Within any standard time zone, all clocks are properly set when they agree with the mean solar time on the standard meridian associated with that zone. This setting is what is called **standard time.** If you move east from a standard meridian, astronomical events (including sunrise and sunset) occur earlier by 4 min on your watch for every degree of longitude, because Earth turns 15° of longitude for every hour (60 min). If you travel west, astronomical events occur later. If you move far enough east or west, you enter a new time zone, so you must reset your watch 1 h ahead or 1 h behind, respectively.

Daylight saving time (or daylight time) further complicates our already complex system of keeping time. Despite its name, no time is saved. When we set our clocks 1 h ahead of standard time in the spring, we are essentially rescheduling events 1 h earlier. We are forced to get up an hour earlier in the morning and retire an hour earlier in the evening. By sacrificing an hour of early-morning sunshine (or twilight, as the case may be), we apparently gain an hour of extra light at the day's end. There is, in fact, a small saving in that less fuel is burned at generating stations to meet the demand for electric lighting in the evening.

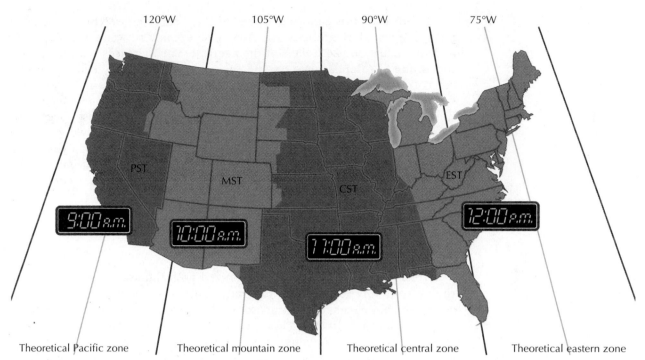

120°W    105°W    90°W    75°W

PST

MST

CST

EST

9:00 a.m.    10:00 a.m.    11:00 a.m.    12:00 p.m.

Theoretical Pacific zone    Theoretical mountain zone    Theoretical central zone    Theoretical eastern zone

**FIGURE 3.6**  Pacific standard time (PST) is defined to be the mean solar time at the standard longitude meridian of 120°W. Mountain standard time (MST) is the mean solar time at the 105°W meridian. Central standard time (CST) is the mean solar time at the 90°W meridian. Eastern standard time (EST) is the mean solar time at the 75°W meridian. The actual time zones are close to but not coincident with the theoretical time zones. From April through October, almost all states within the United States observe daylight time (PDT, MDT, CDT, and EDT). If EST were 12 noon, as shown, then EDT would be 1:00 P.M.

# The Year

One revolution of Earth around the sun is equivalent to 1 year. But again, as in our discussion of the day, we must specify just what this revolution refers to. A single revolution of Earth with respect to the stars, it turns out, is not exactly equal to the seasonal period. This small difference is explained by the very slow *precession* (wobble) of Earth's axis over a period of about 26,000 years—a phenomenon discussed in Chapter 2 in connection with celestial coordinates. Since it is more meaningful to have a year and a yearly calendar that keeps pace with the seasons and not star positions, the *average* length of our year is indeed defined with respect to the cycle of the seasons.

The average year as defined by the seasons is equivalent to 365.2422 mean solar days. We run into another complication: Clearly, we cannot have years that last exactly 365.2422 days! Note that the fraction 0.2422 day is nearly equal to 0.2500 day, or 1/4 day. A calendar consisting only of 365-day years cannot synchronize with the seasons. After 4 years, it would be about 1 day off. After 8 years, it would be 2 days off; and so on. To correct for this discrepancy, we can have a string of three normal years followed by a **leap year** of 366 days. The average length of this 4-year cycle is $365\frac{1}{4}$ days. Indeed, if we consider the contemporary years 1996, 2000, 2004, 2008, and so on, we find they are all leap years.

If, however, this pattern of three 365-day years followed by one 366-day year is repeated indefinitely, the small discrepancy between 365.2500 days and the actual 365.2422 days of the average year increases with time. Our reckoning of calendar dates would then very slowly get ahead of the seasons. To correct for this difference, we must occasionally omit a leap year. Our modern calendar, the **Gregorian calendar,** accomplishes this adjustment on a regular basis by omitting certain leap years at prescribed intervals. The omitted leap years are those century years not evenly divisible by 400. Thus, the century years 1700, 1800, 1900, 2100 are not leap years; the century years 1600, 2000, 2400 (all divisible by 400) *are* leap years. The Gregorian calendar has an error of 1 day in 3300 years. It is accepted the world over for keeping track of civil events.

## The Seasonal Cycle

Today's calendar is well synchronized with the seasons. But what are the seasons, anyway? Let us note that we are discussing astronomical seasons here, not necessarily the seasons related to a particular climate or geographical location. In Canada's far-northern forests, for example, autumn leaves start falling in August or September. In many regions near Earth's equator, seasonal effects on the environment are minor and the local weather is likely to be more or less the same the year round. Regardless of how they affect or do not affect the local environment, astronomical seasons always involve substantial changes in the diurnal path of the sun across the sky.

The changes are due to the fact that Earth's axis of rotation is not perpendicular to the plane defined by Earth's revolution around the sun—this plane is known as the **ecliptic plane.** Rather, Earth's axis is tilted through an angle of $23\frac{1}{2}°$. Equivalently, we can say that the ecliptic plane is inclined $23\frac{1}{2}°$ with respect to Earth's rotational, or equatorial, plane. These relationships are illustrated in Figure 3.7.

In Figure 3.7 we see how the Northern Hemisphere receives the most sunlight, and the Southern Hemisphere the least sunlight, on June 21 or 22 (the exact date varies because of the leap year cycle). This occasion, called the **summer solstice,** signals the beginning of summer in the Northern Hemisphere and the beginning of winter in the Southern Hemisphere. During the **winter solstice,** around December 21, the situation is reversed. Both hemispheres get equal amounts of sunlight during both the **vernal equinox** (spring equinox) and the **autumnal equinox** (fall equinox) on or near, respectively, March 21 and September 22.

The seasonal cycle can also be demonstrated by the changing position of the sun on the celestial sphere, as illustrated in Figure 3.8. The sun follows an apparent path west to east on the celestial sphere that arises because of Earth's motion around the sun. This path, the **ecliptic,** is a great circle tilted $23\frac{1}{2}°$ with respect to the celestial equator. The tilt is not surprising when you realize that Earth's equatorial plane, which produces the celestial equator in the sky, is distinct from the ecliptic plane defined by Earth's revolution around the sun (as we saw in Figure 3.7).

The sun reaches four important "milestones" in its apparent annual journey along the ecliptic, as seen in Figure 3.8. At the vernal equinox the

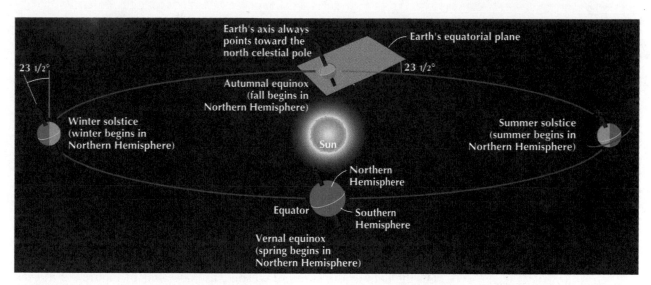

**FIGURE 3.7**  At the vernal equinox, Earth's Northern and Southern Hemispheres are illuminated equally by the sun (in the perspective shown here, we see only the night half of Earth). At the autumnal equinox, both hemispheres are also illuminated equally (we see only the daylight side of Earth in the drawing). At the summer solstice, the Northern Hemisphere is tilted toward the sun and the Southern Hemisphere is tilted away from the sun. The Northern Hemisphere is beginning its summer season, while the Southern Hemisphere is beginning its winter season. At the winter solstice, just the opposite relationships are occurring.

sun is crossing the celestial equator (declination 0°) moving obliquely north. At the summer solstice the sun is farthest north in the sky (declination $+23\frac{1}{2}°$). At the autumnal equinox the sun is again crossing the celestial equator, this time moving south. At the winter solstice the sun is farthest south (declination $-23\frac{1}{2}°$).

Figure 3.9 shows the consequences of the sun's seasonal migration in the sky as viewed by observers at different latitudes on Earth. In Figure 3.9(a), the observer is situated at a mid–Northern hemisphere latitude of 45°N. On either the vernal or autumnal equinox, the sun rises due east, attains a maximum altitude of 45° as it transits the meridian at midday, and

**FIGURE 3.8**  The sun's apparent journey (eastward, as shown by arrows) along the ecliptic includes four milestones at intervals of one-fourth of a year. The equinoxes and solstices refer to both the exact instant of the beginning of the seasons and the exact position of the sun at those instants. The declination of the sun changes rapidly near the equinoxes and slowly near the solstices. Earth, which belongs at the center of the celestial sphere drawn here, is omitted for clarity.

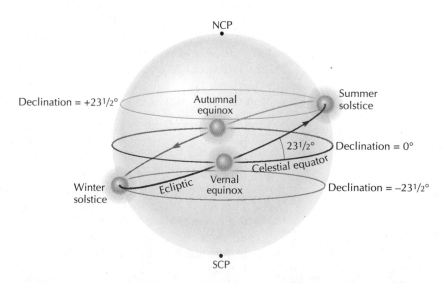

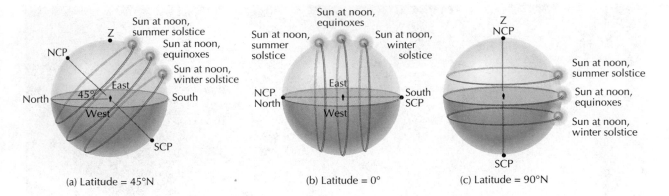

(a) Latitude = 45°N          (b) Latitude = 0°          (c) Latitude = 90°N

**FIGURE 3.9**  The sun's diurnal paths across the sky are shown for solstice and equinox days. The paths are illustrated for observers at (a) 45°N latitude, (b) 0° latitude, and (c) 90°N latitude.

finally sets due west. Exactly half the sun's diurnal circle lies above the horizon, so daytime lasts 12 h or half a day. (The term *equinox* means "equal night," referring to equal day and night lengths.)

On the date of the summer solstice, the observer in Figure 3.9(a) sees the sun rise at a point considerably north of east, transit at altitude $68\frac{1}{2}°$ ($68\frac{1}{2}° = 45° + 23\frac{1}{2}°$), and set at a point well north of west. On that day, the sun is above the horizon for about $15\frac{1}{2}$ h, and below the horizon for about $8\frac{1}{2}$ h. At the winter solstice, the sun comes up roughly in the southeast, transits at an altitude of only $21\frac{1}{2}°$ and sets roughly southwest. At that time, the sun lies above the horizon for only about $8\frac{1}{2}$ hours. (The word *solstice* means "stand still." For several days around either solstice, the sun rises or sets at horizon directions that are more or less fixed. This is the same as saying that the sun's declination changes very little around the solstices.)

For observers at Earth's equator, Figure 3.9(b), the sun's diurnal arc across the sky always amounts to a half circle. Day and night hours are equal for every day of the year.

For observers at Earth's North Pole, Figure 3.9(c), the seasonal changes are extreme. On the day of the summer solstice, the sun moves in a circle parallel to the horizon and $23\frac{1}{2}°$ above it. In the 3 months that follow, the sun gradually spirals downward until it just skims the horizon on the day of the autumnal equinox. During the next 6 months, the sun remains below the horizon, until it comes up again on the next vernal equinox. The 24-h "day," as seen from either of Earth's poles, is of little significance. Rather, the annual changes in the declination of the sun are all-important in controlling the light-and-dark cycle.

By astronomical considerations alone, the weather should be warm in the Northern Hemisphere around the time of summer solstice and cool around the time of winter solstice. Let us see why:

1. The farther north the sun is, the longer is its diurnal arc above the horizon, and the longer the sun shines each day.
2. The farther north the sun is, the greater is its altitude at any given time during the day. Midday sunlight is more intense at or near summer solstice than at or near winter solstice because of effects illustrated and noted in Figure 3.10. The more intense the sunlight, the warmer the ground gets. The lower atmosphere is largely heated from the ground up; so the warmer the ground, the higher the temperature of the air above it.

**FIGURE 3.10** In the Northern Hemisphere, a sunbeam's "footprint" spreads over a smaller area of Earth's surface in June than in December. Consequently, more solar energy per unit area of Earth's surface is absorbed in June than in December. Also, more energy in a given-size sunbeam can reach the surface in June because it traverses a shorter path through Earth's absorbing atmosphere.

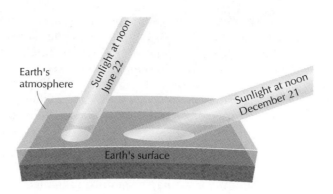

## The Zodiac

The major planets of the Solar System (except for Pluto), along with Earth's moon, have orbits that are nearly in the same plane. Consequently, the moon and planets as we see them in the sky are always found within a belt about 18° wide and centered on the ecliptic. This belt, called the **zodiac,** passes through 12 constellations specially designated as constellations of the zodiac. In its yearly apparent path around the sky, the sun enters a new zodiacal constellation every month or so (see Figure 3.11).

As seen by the naked eye, seven celestial bodies—all of them conspicuous to some extent—appear to drift through the constellations of the zodiac. These seven are the sun, the moon, Mercury, Venus, Mars, Jupiter, and Saturn. In ancient times, each of these seven was known as a *planet,* meaning "wanderer" in Greek. Today, the meaning of the word *planet* has changed somewhat, and we no longer call the sun or moon planets.

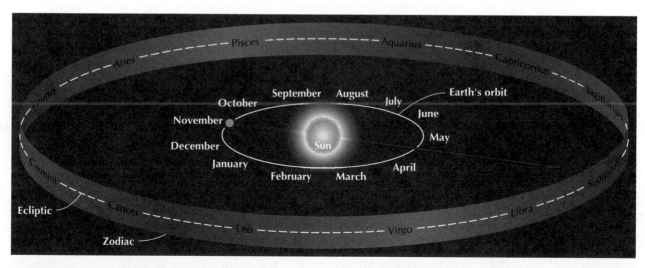

**FIGURE 3.11** The zodiac is a narrow band of the sky centered on the ecliptic. The 12 constellations of the zodiac occupy 12 roughly equally spaced zones along this band. They represent *directions* in the sky, not places. During most of November, the sun lies "in" (in the direction of) Scorpius as shown by the arrow. During most of December, the sun lies in Sagittarius. At that time, it is impossible to see the stars that make up Sagittarius (unless a total solar eclipse is occurring), but we could deduce the direction of the sun by noting that Gemini can be seen in the nighttime sky directly opposite (180° from) the sun.

In most of the Romance languages (Spanish, French, Italian, Portuguese, and others), the names of the days of the week derive from the seven original celestial wanderers. In English, only 3 of the 7 days bear a close resemblance in name: Sunday (sun), Monday (moon), and Saturday (Saturn). The choice of 7 days per weekly cycle was an arbitrary one. It is probable that our modern week would consist of 8 days if in ancient times there were eight wanderers visible to the naked eye instead of seven.

## The Moon's Phases and the Month

The moon's shape seems to change from night to night, *waxing* (becoming more full) for a little more than 2 weeks and then *waning* (becoming less full) for another 2-plus weeks. If you observe the moon's shape, or *phase,* on several successive nights, you will notice that changes in the moon's phase are related to changes in the moon's position relative to the sun.

The changing phases of the moon are the result of viewing the sunlit side of the moon from different angles. As seen in Figure 3.12, only one half the moon (the right side, as illustrated) can be illuminated by the sun. When the moon lies exactly or almost exactly between Earth and the sun, its phase is *new.* At this phase, the moon is in the sky at the same time as the sun, because it is in the same, or almost the same, direction. Note in Figure 3.12 that you would have to be somewhere on Earth's daytime hemisphere, 6 A.M. through 6 P.M., in order for the new moon to be in the sky above. You do not normally see the new moon, however, because (1) very little, if any, part of the illuminated half of the moon is facing Earth, and (2) the moon is hidden in the sun's glare. In one exceptional case you can see the new moon: During a solar eclipse, the new moon appears as a dark silhouette against the sun.

Careful observers can spot the thin, waxing crescent moon, low in the west after sunset, when its "age" (time after new phase) is between 1 and 2 days. On succeeding evenings, the moon's crescent shape gets thicker until the *first-quarter* phase is attained at an age of 7 or 8 days.

Appearing half-lit in the sky, the first-quarter moon rises at approximately 12 noon, transits the celestial meridian at about 6 P.M. (around sunset), and sets around 12 midnight. You can think of the first-quarter moon as being up only during the P.M. hours. (The exact rise, transit, and set times for any phase of the moon are greatly influenced by seasonal factors and by the peculiarities of keeping standard or daylight time at the observer's location on Earth.)

After first quarter, waxing gibbous (not quite full) phases continue. When the moon becomes *full,* about 15 days after new, the moon rises at sunset, transits at midnight, and sets at sunrise. It lies opposite the sun in the sky and opposite where the new moon was about 2 weeks earlier. In the event of a lunar eclipse, the full moon passes either partially or wholly through Earth's shadow.

Waning gibbous phases lead toward the *third-quarter* moon, which lies opposite where the first quarter moon was seen about two weeks earlier. Notice in Figure 3.12 that you would have to be in the 12 midnight through 12 noon position on Earth to see the third quarter moon in the sky. The moon at third-quarter phase, then, is "up" only during the A.M. hours.

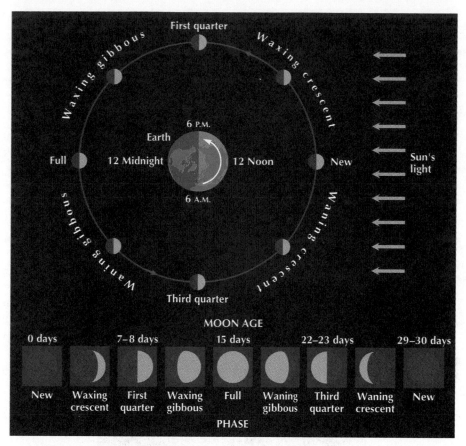

**FIGURE 3.12** The lunar phases we see from Earth are a consequence of the way we look at the moon at different times during the lunar cycle. Starting from the new phase, the cycle is divided into quarters: When the moon has traveled one-quarter of the way around Earth (with respect to the sun), its phase is first quarter. We see exactly half the moon's sunlit side at that time. At the full phase (representing the second quarter), the moon has traveled halfway around Earth and appears fully illuminated. The third-quarter (also called last-quarter) phase is next. Completion of the fourth and final quarter occurs at the next new phase. Various waxing and waning crescent and gibbous phases occur between the quarter phases. The entire cycle lasts an average of about $29\frac{1}{2}$ days. (*Note*: The actual orbit of the moon is not circular, as shown, but elliptical, with Earth off center with respect to the orbit. Consequently, the Earth-moon distance varies slightly, by about 10%. During each lunar cycle, the apparent size of the moon grows and shrinks by about 10% as a result of the changing distance.)

Finally, a sequence of waning crescent phases, visible in the early-morning sky, leads to the next new moon, which begins the cycle anew.

If you flip through a monthly calendar with the moon's phases printed, you will discover that any particular lunar phase occurs, on the average, just a bit earlier each successive month. This difference occurs because the cycle of phases lasts an average of about $29\frac{1}{2}$ days, while the average length of the various months, February excepted, is about $30\frac{1}{2}$ days.

As it turns out, there are approximately 12.4 lunar phase cycles per year. That is, the moon travels about 12.4 times around Earth, with respect to the sun, for every one revolution of Earth around the sun. Early calendar makers, whose conventions we have adopted in our modern calendar, were more concerned with fitting an integral number of months (12 months) into the year than with attempting to precisely synchronize the months to the moon's phases. This is a shame because a 12.4-month calendar might well have concluded with a short month consisting of a string of 11 days that could have been declared holidays!

## Solar and Lunar Eclipses

Eclipses involving the sun or the moon range from barely noticeable to dramatic and almost terrifying. **Solar eclipses** are those in which the sun is partially or totally hidden by the moon—an event that can occur only when the moon has a new phase. **Lunar eclipses** involve the moon's passage through the shadow cast by Earth—an event that can occur only when the moon has a full phase.

Two types of shadows are cast by either Earth or the moon: an *umbra,* or dark, inner shadow in which all the sun's light is absent; and a surrounding *penumbra,* in which only some of the sun's light is missing. In Figure 3.13, which illustrates the geometry of a solar eclipse, the umbra is shown as barely reaching Earth. This situation, which gives rise to a *total solar eclipse* (pictured in Photo 3.3), is typical of many but not all solar eclipses. Sometimes, during the new moon, the umbra misses Earth by passing either too far north or too far south (above or below the plane of the page in Figure 3.13), yet the penumbra alone may graze some portion of Earth's surface. This situation produces a *partial solar eclipse.*

**PHOTO 3.3** On February 26, 1979, the moon's umbra swept across the northwestern states of the United States and some of Canada's provinces, producing a widely observed total solar eclipse. This photograph was taken at 8 A.M. from eastern Oregon, where totality lasted for only 2 min.

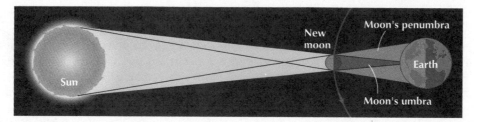

**FIGURE 3.13**  This diagram (not to scale) illustrates the geometry of a solar eclipse. The eclipse occurs when the moon's umbra and penumbra sweep across Earth (from top to bottom in this view) as the moon orbits Earth. Because the width of the umbra is relatively narrow, only a small fraction of Earth experiences totality during any eclipse. Totality is brief for any observer in the umbra's path because the umbra moves roughly as fast as the moon itself moves in its own orbit, approximately 1 km/s (2000 mi/h). The much more widespread areas on Earth swept by the moon's penumbra see a partial solar eclipse. Within the penumbra, the farther the observer is from the umbra's edge, the smaller the "bite" taken out of the sun by the moon's silhouette.

Sometimes, the new moon's umbra fails to reach Earth's surface, not because there is poor alignment but because the moon may at that time lie at or near the far point in its noncircular orbit around Earth. In that particular situation, the umbra tapers to a point somewhere above Earth's surface. The result, for observers who are located directly below the umbra, is an eclipse in which the outer rim of the sun appears as a thin, intensely bright *annulus,* or ring around the silhouette of the moon. Such an *annular solar eclipse* is pictured in Photo 3.4.

**PHOTO 3.4**  Southern California witnessed an annular solar eclipse at sunset on January 4, 1992.

Total solar eclipses are among the rarest and most spectacular phenomena a person can experience. Unlike earthquakes, volcanoes, tornadoes, or hurricanes, eclipses are both harmless and astoundingly predictable. Total solar eclipses occur, on the average, about every year and a half *somewhere* on Earth. However, because the moon's shadow is so narrow (its cross section at Earth's distance never exceeds 270 km), an observer at any *given* location on Earth's surface has to wait an average of about 360 years to see one, assuming perpetually cloudless skies. Totality for any fixed observer never lasts more than $7\frac{1}{2}$ min.

Total solar eclipses start in a covert manner, with a series of partial phases leading to totality. Sunlight fades, slowly at first, in response to the moon's ever-bigger bite in the solar disk. As the moon moves over the last remaining sliver of the exposed sun, the remaining daylight fades in a matter of seconds to an eerie pseudo-twilight. Some of the brighter stars and planets may become visible. The sun's corona, normally overwhelmed a millionfold by the bright glare of the sun's surface, radiates outward from what appears to be the perfectly round and perfectly black disk of the moon. All too soon, totality ends and daylight resumes as the moon begins uncovering the sun.

Lunar eclipses (Figure 3.14) are of three varieties. *Total lunar eclipses* occur whenever the moon passes entirely inside Earth's umbra. Often, the totally eclipsed moon is not entirely dark, because some of the sun's light refracts (bends) through Earth's upper atmosphere and into the umbra, thereby illuminating the moon with a faint, reddish glow (Photos 3.5 and 3.6). A sequence of partial phases precedes and follows totality, but the term *partial lunar eclipse* is reserved for events in which the moon grazes Earth's umbra and fails to enter it completely. *Penumbral lunar eclipses,* which are often unnoticeable, occur when the moon enters Earth's penumbral shadow but not the umbral shadow.

The moon does not eclipse the sun during every new moon, nor does it pass through any part of Earth's umbra during every full moon. The occurrence of eclipses is regulated by the tilt of the moon's orbit, approximately 5° with respect to the ecliptic plane (a relationship illustrated in Figure 3.15). As a result of the tilt, eclipses can occur only during either of two periods each year, each lasting about 3 weeks, called *eclipse seasons*.

For another way of visualizing when eclipses can occur, refer to Figure 3.16. In this diagram, the sun does not lie at a *node,* or intersection between

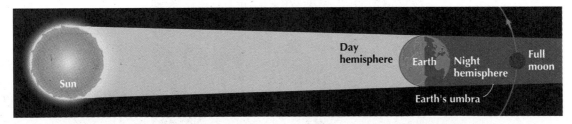

**FIGURE 3.14** Lunar eclipses occur when the full moon sweeps through Earth's generously wide shadow (only the umbra is shown here). A total lunar eclipse results whenever the moon enters Earth's umbra completely. If the moon grazes the umbra (by being above or below the plane of this page), the eclipse is partial. Lunar eclipses are more widely observable than solar eclipses, because they are visible (weather permitting) to anyone on Earth's night hemisphere. Lunar eclipse totality can last an hour or more as the moon makes its way through the shadow.

**PHOTO 3.5** The moon is seen passing through Earth's umbra (at 5-min intervals) in this multiple-exposure photograph of the widely observed total lunar eclipse of November 28, 1993. The diurnal motion of both the moon and the background stars is apparent, because the camera was placed on a fixed tripod and was carried along by Earth's rotation.

the moon's path in the sky and the sun's path (the ecliptic). When the moon passes in front of the sun at a spot on the zodiac that is not near a node, there is a significant misalignment among sun, Earth, and moon, and eclipses do not occur. When, however, Earth's motion around the sun carries the sun to a position near or at a node, *and* the moon is either new or full at the same time, eclipses can occur. At that time, all three bodies—sun, Earth, and moon—lie along a straight line.

Eclipse seasons occur somewhat earlier in the year with each successive year, as suggested by the dates of future eclipses listed in Table 3.2. The changing dates occur because the moon's orbit precesses (wobbles)

**PHOTO 3.6** A close-up of the November 28, 1993, lunar eclipse. During totality, the moon appeared approximately 10,000 times dimmer than a normal full moon.

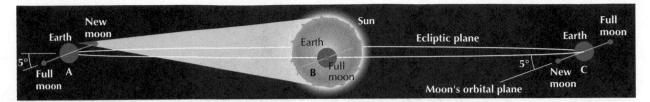

**FIGURE 3.15** Eclipse seasons are explained by the tilt of the moon's orbit, approximately 5° with respect to the ecliptic. When Earth is at or near position A, a solar eclipse cannot occur because the new moon's shadow misses Earth. Similarly at position A, Earth's shadow (not shown) cannot engulf the full moon. Three months after position A, when Earth lies at position B, a straight line passes through the sun, Earth, and either the full moon (shown) or the new moon (hidden behind Earth in this view). Eclipses of either kind can occur for about 10 days before and 10 days after Earth is at position B. After another 3 months, Earth is at position C and no eclipses can occur. Eclipses can occur again when Earth lies opposite position B (behind the sun in this view). Sizes, distances, and angles in this diagram are not to scale.

with a period of 18.6 years. The nodes of the moon's orbit slide westward along the ecliptic, completing one circuit in 18.6 years.

Today, the average person on the street could tell you virtually nothing about the solar and lunar rhythms in plain sight overhead. Many people of the world's so-called primitive past cultures, however, kept track of the equinoxes, the solstices, and the seasonal patterns of stars at night, because their survival often depended on the ability to anticipate and plan for seasonal changes. Native Americans took special note of the equinoxes and

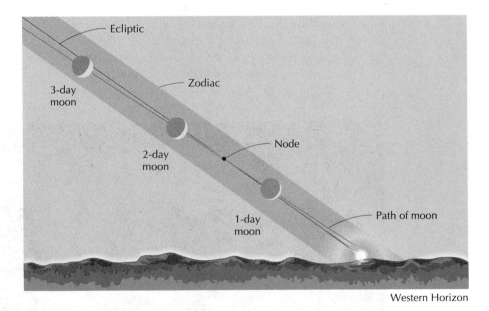

Western Horizon

**FIGURE 3.16** In the days following new moon, the moon drifts east of (away from) the sun along the zodiac while its crescent phase thickens. The diagram shows the location of the moon in the sky, at the time of sunset, 1, 2, and 3 days after new moon. The sun's apparent path along the ecliptic takes it about 1° farther east each day along the ecliptic, while the moon moves east about 13° per day. The plane of the moon's orbit is tilted 5° with respect to the ecliptic. As a consequence, the moon is never farther than 5° from the ecliptic (always well within the zodiac band) and crosses it twice during each lunar cycle. The two crossover points, called nodes, lie opposite each other on the celestial sphere. The apparent sizes of the sun and the moon have been exaggerated in this diagram.

**TABLE 3.2**
Solar and
Lunar Eclipses,
1996 through 2006

| Date | Event | Totality/Annularity Visible from North America |
|---|---|---|
| 1996, April 4 | Total lunar eclipse | Partially |
| 1996, September 27 | Total lunar eclipse | Yes |
| 1997, March 9 | Total solar eclipse | No |
| 1997, September 16 | Total lunar eclipse | No |
| 1998, February 26 | Total solar eclipse | No |
| 1998, August 22 | Annular solar eclipse | No |
| 1999, February 16 | Annular solar eclipse | No |
| 1999, August 11 | Total solar eclipse | No |
| 2000, January 21 | Total lunar eclipse | Yes |
| 2000, July 16 | Total lunar eclipse | No |
| 2001, January 9 | Total lunar eclipse | No |
| 2001, June 21 | Total solar eclipse | No |
| 2001, December 14 | Annular solar eclipse | No |
| 2002, June 10 | Annular solar eclipse | No |
| 2002, December 4 | Total solar eclipse | No |
| 2003, May 16 | Total lunar eclipse | Yes |
| 2003, May 31 | Annular solar eclipse | No |
| 2003, November 9 | Total lunar eclipse | Yes |
| 2003, November 23 | Total solar eclipse | No |
| 2004, May 4 | Total lunar eclipse | No |
| 2004, October 28 | Total lunar eclipse | Yes |
| 2005, April 8 | Annular-total solar eclipse | No |
| 2005, October 3 | Annular solar eclipse | No |
| 2006, March 29 | Total solar eclipse | No |
| 2006, September 22 | Annular solar eclipse | No |

*Note*: This list does not include the relatively uninteresting partial solar eclipses and partial and penumbral lunar eclipses.

solstices, and some tribes marked these events by means of inscriptions chiseled into or painted on rock. Inscriptions on a sandstone butte in Chaco Culture National Historic Park, New Mexico, show that the Anasazi people of the American Southwest (circa 1000 A.D.) had a keen awareness of various astronomical cycles. In addition to following the sun's equinox-solstice cycle, the Anasazi were apparently aware of the subtle, 18.6-year lunar cycle we have mentioned in connection with eclipses.

# A History of the Universe

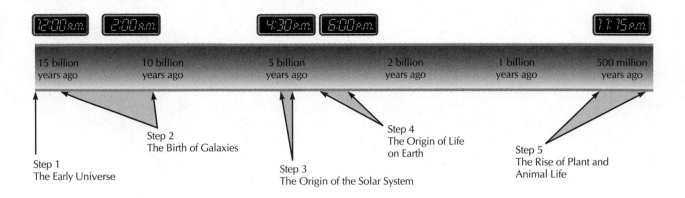

The remainder of this chapter traces the history of the universe from its earliest describable moment to the present. As a visual aid to the unfolding of time described in this section, a graphic timeline is included, in condensed form above and in expanded form on the following pages, noting several important events or epochs (called steps) in cosmic history. We begin our timeline, as we should, with the definitive event known as the *big bang*. As the name big bang suggests, the universe (or at least what we can trace of it) is thought to have originated some 15 billion years ago in an event that featured a sudden, rapid expansion of everything that exists now from a very small and very hot state.

The big bang is Step 1 in our history of the unfolding universe. Subsequent steps will take us through the formative stages of galaxies—and of our galaxy in particular. With the passing of billions of years, we will examine the history of our own neighborhood within the Milky Way galaxy: our sun and Solar System, the planet Earth, the emergence of life on Earth, and finally our own species, *Homo sapiens sapiens*.

Because astronomical and geological events are often specified according to how long ago those events took place, we have calibrated our timeline in units of "years ago." As an aid to understanding the passage of time in a relative sense, we have also included digital clock symbols placed above the timeline. These clock symbols track time in the forward direction as if the entire 15-billion-year history of the universe were compressed into a single 24-hour day. Our digital clock reads 12 midnight at the time of the big bang, 12 noon at half the present age of the universe, and 12 midnight again on January 1, 2000—a date arbitrarily chosen to represent "now" at the far end of the timeline. Using this metaphor, the first glimmerings of modern science, which traces its roots back to ancient Greece, would have occurred at around 11:59:59.99 P.M.—just 1/100 second before midnight.

## The Early Universe

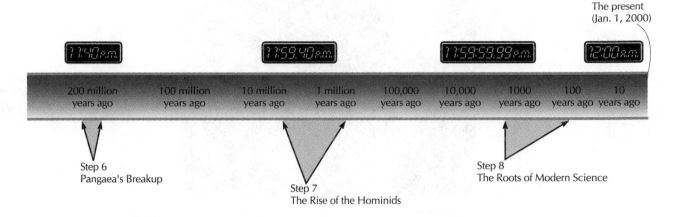

200 million years ago · 100 million years ago · 10 million years ago · 1 million years ago · 100,000 years ago · 10,000 years ago · 1000 years ago · 100 years ago · 10 years ago

The present (Jan. 1, 2000)

Step 6
Pangaea's Breakup

Step 7
The Rise of the Hominids

Step 8
The Roots of Modern Science

## Step 1: The Early Universe

The **big bang theory** supposes that the universe originated in a state of extremely high density and extremely high temperature. Since the time of its origin, the universe has been expanding and cooling. The cooling of matter in the universe eventually allowed particles of matter to accumulate, mainly by gravity, into large structures such as stars, planets, and galaxies. Propelled by the initial impetus of the big bang, the universe persists in an overall expansion today.

Much of what you read in these pages about the big bang may seem unlikely or even fanciful. Some of the details as currently envisioned are undoubtedly incorrect, and it is possible that the theory is wrong—as scientists may one day discover when they are wiser about the workings of the universe. Nevertheless, nearly all scientists who study the universe subscribe to the general idea of a big bang origin.

The big bang idea is a convincing one because it can explain what we presently see in the universe, and it can explain what we see of the past universe as well. (Recall that we look deeply into the past when we look deeply into space, because it takes time for light to travel from faraway places to Earth.) The primary piece of evidence in support of a big bang origin is the observation that virtually all other galaxies are moving away from our galaxy. These other galaxies appear to be receding from us with velocities proportional to their distances. Thus, the universe is undergoing a uniform expansion. By running time rapidly *backward* in our minds, we can easily imagine the universe collapsing to a point.

Since its introduction into the scientific mainstream in the late 1940s and 1950s, the big bang theory has evolved from a relatively simple idea to one that grew in complexity and sophistication as better observations were made and knowledge was gained. Today, the major contributors of new knowledge about the universe are astronomers who study the origin and

**PHOTO 3.7** Although the term *big bang* immediately conjures up the image of a huge explosion, a better image is that of a creation moment followed by an unfolding of all space, time, and matter. This computer-generated pattern, of a type called a *fractal,* illustrates one aspect of this unfolding. Contained in the rapidly expanding early universe were regions of space, large and small, where matter would eventually clump together to form galaxies, stars, and other aggregates of matter down to molecules in size.

evolution of the universe at large (the field called *cosmology*), and physicists who study the behavior of subatomic particles (the field called *particle physics*). There is much collaboration between scientists in these two fields, because the behavior of matter and energy in the universe's infancy is closely mimicked during the collisions between high-speed particles in machines that physicists use called *particle accelerators.*

When particles collide at high speed (and therefore at conditions of high energy) in particle accelerators, they give rise to showers of short-lived, fragmentary particles having unusual properties. These same particles were thought to have existed during the big bang's earliest phases, because similar conditions of energy prevailed then. By studying these short-lived particles, scientists gain insights into how the universe worked during its earliest and hottest moments.

In Chapter 17, which is devoted to cosmology and the big bang theory, you will learn in detail how scientists came to the conclusion that a big bang inaugurated the universe and why the universe is thought to have an age of approximately 15 billion years. Taking these conclusions for granted in this chapter, we will describe, in chronological order, some important events that are thought to have taken place during the first few moments of the universe's existence.

In the beginning, everything in the universe was confined to a single, infinitesimally small point in space-time. The "fabric" of space was compressed to essentially zero volume. Time was virtually the same everywhere in the universe because all places were essentially the same. We will call this time "time zero" ($t = 0$), the earliest moment of the universe's existence. (It is natural to ask what happened *before* $t = 0$. We do not know. The question may or may not be answerable. Time may not have existed before $t = 0$. If so, then the question is meaningless.)

Time zero represents the initiation of the expansion of the universe. The expansion, which was originally propelled by a short-lived force (or

**FIGURE 3.17**   Like the surface of a polka-dot balloon being blown up, the space of the universe gets larger with time. An ant crawling on the surface would find that as time goes on, there is more space between the dots.

forces), continues today. It is a mistake to think that some kind of force is needed to *keep* the universe expanding. The impetus for the expansion came during the first fleeting moments of the universe's existence; and since then, the matter in the universe has been merely coasting under its own momentum. Astronomical observations show that the universe continues to expand, and no reversal in this trend is expected any time soon.

It is also a mistake to think of the big bang as having taken place within a preexisting space and then filling that space, because everything that now exists in the universe—matter, energy, space, and time—was locked up in the then very small universe. Indeed, the mechanics of the big bang do not resemble at all a conventional explosion, like that of a bomb or firecracker. Picture, instead, the analogy of a spherical balloon, painted with many dots, being blown up. As the surface area of the balloon expands, the distance between any two dots painted on the balloon's surface increases. Any creature confined to this surface (like an ant) could measure the increasing distances between the dots and conclude that the "space" that it lives in is expanding (see Figure 3.17).

The surface area of a balloon is a two-dimensional space. Points on it can be described with coordinates having only two numbers. The "real" space of the universe is three-dimensional, requiring coordinates with three numbers. Real space also expands so that the distance between any two places increases with time. The travel time of light between these places increases as well.

As our universe expands, the matter in it is becoming less dense overall. Also, the energy in it is becoming "thinned out"; that is, the average temperature of the universe is decreasing. The very early, rapidly expanding universe has, some 15 billion years later, evolved into a still-expanding universe with an average temperature of only a few degrees above absolute zero. The temperature of *absolute zero* is denoted as 0 Kelvin (0 K) on the absolute, or Kelvin, temperature scale; it is the lowest possible tempera-

**PHOTO 3.8** Inside particle detectors such as this one, physicists produce "little bangs" that replicate, on a tiny scale, the enormously energetic conditions that existed during the big bang. Each "little bang" is created by a subatomic particle, moving at very close to the speed of light, smashing into a fixed target or colliding with another particle moving in the opposite direction. Each collision releases a spray of short-lived, exotic particles, and these particles undergo further interactions and transformations. Processes such as pair production and pair annihilation are routinely observed.

ture. Another important change is that the matter of the universe, once almost perfectly uniform in distribution, has become quite lumpy with galaxies and the stars that compose galaxies.

We cannot describe with great confidence what went on in the universe from $t = 0$ up to about $t = 10^{-4}$ s. Theorists have certainly not shrunk from this task: There has been much speculation about what went on as early as about $t = 10^{-43}$ s. Then, they surmise, matter, energy, space, and time were indistinguishable, and all four of today's fundamental forces were united in one "superforce." The world's most powerful particle accelerators can now simulate, on a very small scale, some of the same conditions that existed at about $t = 10^{-12}$ s in the early universe, so the theorists have at least some experimental evidence to support or disprove their ideas.

The theoretical picture becomes clearer from about $t = 10^{-6}$ s onward. At $t = 10^{-6}$ s, the known laws of physics (extrapolated backward in time) indicate that the universe had a temperature of about $10^{13}$ K (Kelvin units of temperature are essentially equivalent to °C, or Celsius units, in the higher ranges of temperature) and a density of about $10^{15}$ g/cm$^3$. Just before this time, the universe was a hot, dense soup of subatomic particles—protons, neutrons, electrons, certain other exotic particles (many of which are unstable in the present universe) and *gamma rays.* Gamma rays are massless **photons,** or tiny units of pure energy. Photons, not to be with confused with *protons,* are the particlelike entities that make up the various forms of radiant energy we call gamma rays, X-rays, ultraviolet radiation, visible light, infrared radiation, and radio waves. We will discuss various aspects of radiant energy in several chapters to come.

When gamma rays of sufficient energy collide, they can vanish, leaving in their wake two particles having the same mass but opposite electric charge (if they have any charge at all) and opposite magnetic characteristics. This process is called *pair production.* One particle produced is of normal matter and the other particle (the so-called antiparticle) is of a type

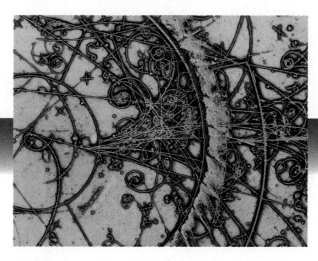

**PHOTO 3.9** Particle-detector tracings, such as this one from Europe's CERN research facility, illustrate the complexity of high-energy subatomic particle collisions and suggest how violent a place the universe was during its earliest moments.

called *antimatter*. Possible matter-antimatter pairs include the proton-antiproton pair, the neutron-antineutron pair, and the electron-positron pair (a positron can also be called an antielectron).

Antimatter particles are routinely observed in high-energy, particle accelerator experiments. Once a particle of antimatter collides with a particle of matter, however, the two immediately disappear in a flash of energy: a gamma ray. This process is called *pair annihilation*.

Annihilation quickly takes place at any temperature, but pair production occurs only at extremely high temperatures. The temperature required for pair production depends on the masses of the particles produced: Much higher temperatures are needed to produce proton-antiproton or neutron-antineutron pairs than electron-positron pairs. These last-named pairs require lower temperatures because electrons and positrons are about 1800 times less massive than the other particles and antiparticles.

After $t = 10^{-6}$ s, with the universe cooling to less than $10^{13}$ K, the pair-production process could no longer produce proton-antiproton and neutron-antineutron pairs. However, it did keep producing, for a while longer, the lighter electrons and positrons. As the universe continued to cool, wholesale annihilation took place among the protons and neutrons and their antiparticles, but the annihilation could not quite proceed to completion. Presumably (and for reasons that seem to be confirmed by detailed particle accelerator experiments today), a slight excess of matter—one particle of matter for every 100 million or so matter-antimatter particle pairs—survived the annihilation. Gamma ray photons at that stage would have outnumbered the protons and neutrons by a factor of about 100 million to 1. Surveys of the universe at large seem to support this theory: Photons still outnumber protons and neutrons by about this ratio.

Superimposed on the seething mix of photons, particles, and antiparticles already mentioned were particles called *neutrinos*. Neutrinos have no charge and little or no mass. They remain abundant in today's universe, but

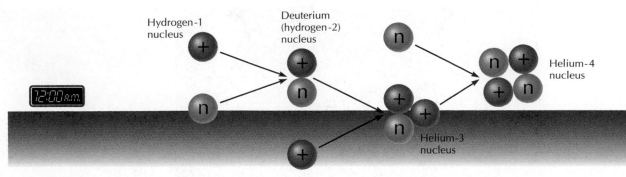

**15 billion years ago**

**FIGURE 3.18** This series of simplified steps traces a major route toward the production of helium in the very early universe. In this process two protons, or hydrogen nuclei (labeled +), and two neutrons (labeled n) join to become a helium nucleus. The temperature at this stage was far too hot for any nuclei to capture electrons and thereby become whole atoms.

they seldom interact with ordinary matter, which makes them difficult, but not impossible, to detect. There may also have been other hypothetical, ghostlike particles, whimsically dubbed WIMPs (*w*eakly *i*nteracting *m*assive *p*articles). According to some theories, accumulations of these exotic particles would later play a role in the formation of galaxies.

By the time the universe was about 4 s old, corresponding to a temperature of about $10^{10}$ K and a density of about $10^4$ g/cm$^3$, virtually all of the particles that make up today's atoms had been, by one process or another, already fashioned.

Until about $t = 3$ min, none of the constituents of today's ordinary atoms could stick together, because gamma rays would simply knock them apart. From about $t = 3$ to $t = 30$ min, gamma rays were steadily getting weaker (or stretched out in wavelength) as the space of the universe continued to expand. By the end of that period, gamma rays had so weakened that some protons and neutrons could link together, by means of the strong force, into simple combination. After this period was over, about a fourth of all the protons and neutrons had been caught up into small bundles of two protons and two neutrons each, forming the nuclei of what would eventually become helium. A very small fraction became bundles of one proton and one neutron, which are the nuclei of deuterium, a rare hydrogen isotope.

One such nucleus-building process involves the following simplified steps, shown in Figure 3.18. First, a proton and a neutron collide and stick, by means of the stong force, to form a deuterium, or hydrogen-2 nucleus. Next, the deuterium nucleus plus another proton collide and form a helium-3 nucleus. Finally, another neutron is added to form a stable helium-4 nucleus.

After about $t = 30$ min, further reactions of this sort could not take place because of the rapidly declining temperature. Because helium-3 nuclei are unstable, they eventually broke apart, leaving only hydrogen-1 and helium-4, plus small traces of hydrogen-2, plus miniscule traces of one or two elements heavier than helium.

**PHOTO 3.10** About 30 minutes after the beginning of time, the temperature of the universe stood at approximately 100 million K, about as hot as a hydrogen bomb explosion moments after detonation.

If the "fires" of the big bang had remained hot enough for a much longer time, then many heavier nuclei (of elements like carbon, oxygen, and iron) could have been formed. But by $t = 30$ min, the temperature had sunk to $10^8$ K—too cool to allow protons and neutrons, or combinations thereof, to stick. *Nucleosynthesis,* the buildup of atomic nuclei from simpler components, stopped. Approximately three-fourths of the universe's ordinary matter ended up in the form of unattached protons, which are ordinary hydrogen nuclei. The remainder was overwhelmingly helium nuclei, with small traces of deuterium and perhaps even tinier traces of nuclei belonging to two elements just heavier than helium: lithium and beryllium. For this reason, when we look at the make up of the universe today, we find that it is composed primarily of hydrogen and helium in a ratio of about 75 to 25% by mass.

Today, obviously, the universe contains a substantial amount of the elements heavier than hydrogen and helium. Witness Earth and the other terrestrial planets, whose crusts and interiors are made of elements such as iron, aluminum, silicon, and oxygen. These heavier elements, we have good reason to believe, were produced deep within stars, a process that started no earlier than a few hundred million years after the big bang. This process, called stellar nucleosynthesis, will be dealt with in Chapter 17.

At $t = 30$ min, electrons—which are of very low mass and consequently are greatly affected by gamma rays—had no chance of joining with any of the available hydrogen or helium nuclei to form stable atoms. For the next million years or so, this situation prevailed. The universe was a steadily cooling soup of bare hydrogen and helium nuclei and *free,* or unattached, electrons buffeted by photons. Conversely, the photons could not escape the soup of nuclei and electrons (even though they incessantly travel at the speed of light) because of their constant deflection by the free electrons. Matter and energy were bound together.

At about $t = 1$ million years, the situation changed. By then, the universe's temperature had dropped to only 3000 K (about as hot as a typical

12:01 a.m.

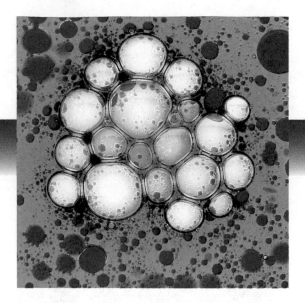

**PHOTO 3.11**  Just as bubbles tend to coagulate on the surface of a liquid, regions of slightly greater density in the early universe attracted each other through gravity and formed lumps of larger mass.

incandescent lightbulb) and the density had decreased to about $10^{-16}$ g/cm$^3$. Because of the tremendous expansion and cooling since very early times, photons that used to be gamma rays were now stretched to longer wavelengths and weakened to the point that they became photons of ordinary light. These photons of lower energy finally lost their ability to perturb the free electrons. The electrons settled down into orbits around the hydrogen and helium nuclei, where they remained free (for the most part) from disturbance by photons. At last, complete atoms—the building blocks of ordinary matter—could exist.

With the capture of electrons by atomic nuclei, photons could now travel freely in all directions. The universe, which was formerly opaque, became transparent, and light flooded everything. This event is often referred to as the *dawn of light*.

Many of the same photons released during the dawn of light still travel the universe today. By now they are stretched and weakened to the point that they are approximately equivalent in wavelength and energy to the microwave photons typically emitted inside microwave ovens. As we shall emphasize in Chapter 17, the detection of these photons, which constitute what is called the **cosmic background radiation,** lends crucial support to the big bang theory.

Before the dawn of light, the strong and electromagnetic forces controlled the universe's overall structure. After the dawn of light, matter was free to move under the influence of gravity. More and more, gravitation assumed the major role in determining the universe's structure.

For millions of years after the dawn of light, the universe consisted of an expanding, luminous, ever-thinning cloud of hydrogen and helium gas. Eventually, however, some regions of the universe that, by chance or otherwise, were slightly denser than other regions began to pull themselves together under gravitational attraction. This was the beginning of the process that ultimately led to stars and galaxies.

**13 or 14 billion years ago**

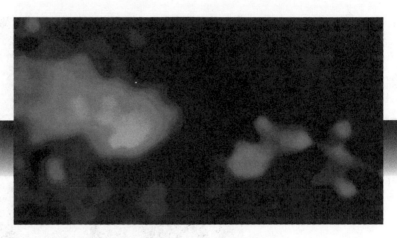

**PHOTO 3.12** This photograph taken by the Hubble space telescope shows the core region of the most distant known galaxy, dubbed 4C41.17. Its light, now arriving on Earth, is some 14 billion years old. Part of this galaxy's light is being obscured by intervening dust. The bright clumps are thought to be forming star clusters, each containing enough mass to make several billion stars. This picture may depict a stage in our own galaxy's development, perhaps a billion years after the big bang.

## Step 2: The Birth of Galaxies

The distribution of matter in the universe today is lumpy, with galaxies arranged in clusters and the galaxies themselves containing anywhere from millions to trillions of small, dense, hot lumps called stars. Surveys of nearby galaxies and galaxy clusters have given us this general picture. When we look billions of light-years away (and thus billions of years back in time), our knowledge is less complete. Only the brightest, most massive galaxies are visible to us, and the lesser galaxies are too faint to see. At the farthest limits of what we can perceive by light, perhaps 13 billion or 14 billion light-years distant, we see some galaxies that consist of luminous gases and fully formed stars. Thus, some 14 billion years ago—only about 1 billion years after the big bang—at least some galaxies were already in existence.

Most theories of the origins of galaxies presume that *gravitational instabilities,* or density fluctuations, on one scale or another, were present in the early universe. If the conditions were right, the denser regions could start pulling themselves together, thereby working against the overall expansion of matter in the universe. One of the central problems in cosmology over the past decade has been to explain (theoretically at least) how a universe of ordinary matter (hydrogen and helium) that was apparently very smooth at the time of the dawn of light could have developed distinct and dense lumps over what is generally considered too short a time, a "mere" 1 billion years.

In the early 1990s, NASA's COBE (*Co*smic *B*ackground *E*xplorer) satellite measured faint ripples in the intensity of the cosmic background radiation released at the dawn of light (see Photo 3.13). The results showed large-scale variations of only 1 part in about 50,000, which correspond to very tiny density fluctuations in regions that would later become as large as the cosmic voids and bubbles seen today. It is anticipated that future,

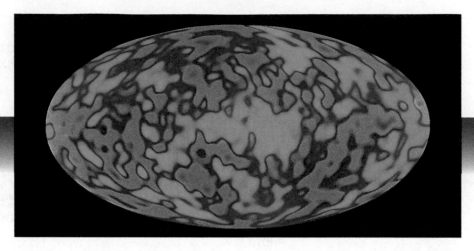

**2:00 a.m.**

**13 or 14 billion
years ago**

**PHOTO 3.13** This false-color (color-coded) map of microwave energy coming from all parts of the sky shows irregularities in the cosmic background radiation. The mean temperature of the background radiation, 2.73 K, is shown as deep blue. Pink and red areas are slightly warmer, and light blue areas are slightly cooler. This image represents a summary of data gathered by NASA's COBE satellite.

more meticulous measurements of the cosmic background radiation will reveal smaller density fluctuations that could explain clumpiness on smaller scales, like those of galaxies.

So far, the COBE measurements support cosmological scenarios that include a phase of sudden *inflation* (apart from the more uniform expansion associated with earlier big bang models) in the earliest moments of the universe and also support theories that call for a large amount of dark matter. The inflation model will be discussed in Chapter 17. **Dark matter** is material, so far undetected by any direct means, that appears to be exerting large gravitational forces on luminous objects we can see, such as stars in the outer parts of galaxies and whole galaxies within clusters of galaxies. There are strong indications that as much as 99% of all matter in the universe may be in this nonluminous form. Visible matter, such as the luminous stars and the diffuse interstellar matter inside galaxies, may be immersed in a much more massive, diffuse sea of dark matter. If dark matter had accumulated into well-defined denser units before the dawn of light, then those units could have served as "seeds" that would have accelerated the formation of galaxies and larger structures.

Another issue concerns the order in which the large units of the universe formed. Did the galaxies form from the "top down"—that is, did supercluster or larger-sized clouds of hydrogen and helium fragment into a myriad of smaller clouds that would become galaxies? Or did they form from the "bottom up," whereby the individual galaxies would form first and then draw together by gravitational attraction into clusters or superclusters of galaxies? We do not yet know.

Scientists think they have a clearer picture of how individual galaxies like ours formed. Here is a possible scenario of the birth and early history of the Milky Way.

Interactions among the protogalactic clouds in the early universe set

**PHOTO 3.14** The great globular cluster Omega Centauri, lying some 20,000 LY from Earth, is the largest and richest globular cluster in the Milky Way galaxy. The several hundred thousand stars contained in it were formed during the same early epoch of cosmic time, that is, one or two billion years after the big bang. Such a huge assemblage of stars tends to remain stable over billions of years, even as its member stars move about, like bees in a swarm, at many different speeds and in many different directions.

virtually all of them spinning slowly. One of them would become the Milky Way galaxy. As the proto–Milky Way contracted toward its present dimensions, it spun faster and gradually started to assume the shape of a disk. Before the entire body could flatten out, hundreds of large condensations (with denser matter) within it began to rapidly contract. Each contracting condensation fragmented into a swarm of hundreds of thousands of smaller clouds, and these smaller clouds collapsed to become individual stars. Thus the **globular clusters,** which now contain the oldest stars of the galaxy, were born. The estimated ages of these stars (13 or 14 billion years) can be taken as the rough age of the Milky Way galaxy.

The densely packed, spherical-shaped globular clusters, you may remember from Chapter 2, today inhabit the spherical halo of our galaxy. Their eccentric orbits around the galaxy's center, which are not confined to the galaxy's disk, emulate, to some extent, the motion of material in the protogalactic cloud before it had a chance to flatten out.

What about the vast majority of the material that did not get caught up in globular clusters? Some of it fell toward the center of the cloud, triggering an episode of star formation there; and the rest gradually settled into the shape of our galaxy's present disk. Star formation proceeded at a somewhat slower pace in the disk, and indeed, it still continues today. There we find older stars, middle-aged stars, young stars, embryonic stars, and quite a lot of diffuse material that will come together to form future stars. Our sun, born 4.6 billion years ago out of a cloud of interstellar gas and dust, can be considered middle-aged. During its lifetime, the sun has circled the galactic center about 20 times.

Our cosmic history will soon center on the sun, the Solar System, and Earth and its living inhabitants. From a less provincial viewpoint, however, we can make two statements about what has occurred up to the present in the universe at large.

# The Origin of The Solar System

**5 billion years ago**

**PHOTO 3.15** The Orion nebula, which appears to the naked eye as a faint, fuzzy star in the sword of the constellation Orion, the Hunter, lies at an estimated distance of 1500 ʟʏ. This and other glowing nebulas, or gas clouds in interstellar space, are the birthplaces of stars. The glow of the nebula is due to the influence of several recently born, superhot stars embedded within the diaphanous clouds of gas and dust. Some 5 billion years ago, the region surrounding our soon-to-form Solar System probably looked like this.

1. From roughly a billion years after the big bang onward, protogalactic clouds that had different rotational characteristics have evolved to become different types of galaxies: spherical, disk-shaped, and irregular in shape. Complicating the issue is the likelihood that nearly all galaxies have collided with or passed near other galaxies during the billions of years they have been in existence. Collisions produce profound changes in galactic structure.

2. The universe is expanding, though not at smaller scales of distance. Our Solar System, Milky Way galaxy, and probably our Local Group of galaxies is not expanding. Only larger structures—clusters and superclusters of galaxies—appear to be pulling apart from one another. Also, the continuing expansion is known to be taking place at a slackening rate. Billions of years from now, the rate of expansion will be significantly less. We do not yet know if the universe will continue to expand forever or if it will someday reach a maximum state and then begin to contract by means of gravity. Accurate measurements of both the present rate of expansion and the total amount of visible and dark matter in the universe would, in principle, allow us to correctly distinguish between these two outcomes. Astronomers do not have, at present, precise measurements of either. We will take up this issue again in Chapter 17.

## Step 3: The Origin of the Solar System

   The Milky Way galaxy may have assumed much of its basic structure some 13 or 14 billion years ago, but our Solar System is a more recent arrival on the scene. *Radiometric dating* (Chapter 12) of mineral grains found on the surfaces of Earth and the moon and inside meteorites estab-

**PHOTO 3.16** This close-up photograph of a small section of the Orion nebula, obtained by the Hubble space telescope, shows disks of dusky and gaseous material around some of the smaller stars being formed. Much of the material in these disks will presumably congeal into planets at some future time.

lishes that the oldest solid materials of the Solar System have an age of about 4.6 billion years. This is what is taken to be the age of the Solar System—the time since the sun, the planets, and the smaller bodies of the Solar System congealed out of a cloud of interstellar gas and dust.

What did this cloud contain? By mass, it consisted of approximately 73% hydrogen, 25% helium, and about 2% other, heavier elements. The hydrogen and helium were already present, dating from the eras of pair production and nucleosynthesis that ended about a half hour after $t = 0$ during the big bang. The addition of the 2% heavier elements took place gradually as stars older than the sun lived, died, and, in some cases, ejected their newly created elements into interstellar space (as we shall learn in detail in Chapter 17). Almost all of the hydrogen and all of the helium were in gaseous form; the traces of heavier elements (joined with hydrogen in some cases) made up tiny grains of opaque material—the dust of an interstellar cloud.

The material that became the Solar System was once part of a large, relatively dense interstellar cloud, similar to many of those we see in interstellar space today (see Photos 3.15 and 3.16). The cloud fragmented into smaller clouds, one of which became the sun (and its accompanying planets). It is probable that for a time the sun was part of an **open cluster** of stars, containing dozens or hundreds of stars loosely bound to each other by gravity. Thousands of open clusters of a wide variety of ages are distributed within our galaxy's disk today.

Photo 3.17 shows a young cluster, probably much like the one our sun belonged to more than 4 billion years ago. Some clusters of this type are stable for billions of years, but others disintegrate over time, their member stars dispersing into the galactic disk. Our sun is probably an escapee of a now-defunct open cluster.

The gas-and-dust-cloud fragment that served as the progenitor of the Solar System is known as the *solar nebula*. Pictured in Figure 3.19 are the major steps of what is known as the solar nebula theory, or **condensation theory**—

**4.6 billion
years ago**

**PHOTO 3.17**   The Pleiades, named after the seven sisters of Greek myth, is an open star cluster about 230 LY distant. It can easily be seen by the naked eye as a small, dipper-shaped star cluster high overhead on autumn evenings. The Pleiades cluster is so young (a few million years old) that many of its stars are surrounded by what looks like a luminous blue haze. The haze is really starlight reflected off diffuse clouds of dust left behind after the cluster's stars were formed. Eventually, the strong radiation emanating from the brighter stars will blow the dust away. Our sun's stellar neighborhood looked much like this just after the Solar System's formation. The fainter members of the Pleiades we see here are comparable in brightness to the sun.

the modern explanation of how the diffuse material of the solar nebula was transformed into the sun and its family of smaller, orbiting bodies.

The condensation theory posits that the solar nebula at an early stage was contracting and yet spinning slowly at the same time. Gravity pulled most of its material toward the center, where it would accumulate and grow hotter to form the sun. A relatively small amount of leftover material settled into a thin, contracting disk surrounding the sun. Details of the sun's subsequent internal and external evolution will be traced in Chapter 17, so we will mostly ignore the sun here and focus on the formation of the planets and smaller bodies that formed from the leftover material in the disk.

The disk whirled faster as it contracted, like an ice-skater who spins faster by pulling in his or her arms. Material toward the inner part of the disk became quite hot, because it was being compressed as its material drifted closer to the forming sun and also because it was being heated by strong radiation that now streamed from the embryonic sun. Farther out in the disk, the material remained relatively cool.

At some point, about the time when the sun began its present phase as a stable star, the collapse of the disk halted; and the material in it began to cool off. Denser swirls existed in the disk, because there was turbulence in the solar nebula to begin with. Atoms and molecules began to stick to each other, especially within the denser swirls, forming chemical compounds. Close to the sun, where the temperature was (and still is) relatively high, only compounds containing heavy elements could condense, forming grains and larger lumps of metal and rock-forming minerals. Farther out from the sun, where it was much cooler, compounds of the lighter elements—water ice, methane, and ammonia—condensed into solid grains or crystals. Next, according to a leading variation of the condensation theory, most of the grains and small lumps gathered into clumps up to tens of kilometers in size called *planetesimals*. Some of these planetesimals became massive enough to attract neighboring planetesimals in a process called

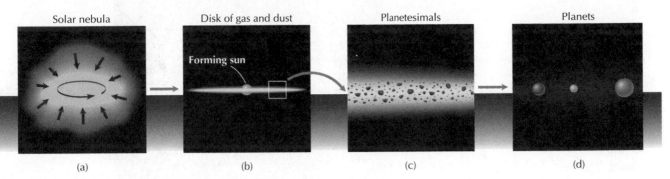

**FIGURE 3.19** In the condensation theory, the slowly rotating solar nebula contracts and flattens (a) to form the sun plus leftover material in the form of a disk surrounding the sun (b). The disk breaks up into planetesimals (c), and many of the planetesimals accrete to form planets (d).

*gravitational accretion*. The planetesimals that grew more quickly were more successful at attracting still more planetesimals from ever-wider regions in the disk.

In some of the final steps of accretion, the predecessors of today's planets—*protoplanets*—formed. The protoplanets close to the sun were relatively small and dense; they consisted of little else but the almost trivial 2% residue of heavy elements from the solar nebula. Upon cooling, they became the four terrestrial planets (Mercury, Venus, Earth, and Mars) more or less as we know them today.

In each of the terrestrial planets, the heaviest of the heavier elements, particularly iron and other metals, lie at the core; the lighter, rock-forming materials tend to dominate the upper layers: the mantle and crust. Two competing models try to explain this state of affairs: The *homogeneous accretion model* supposes that the terrestrial planets accreted from planetesimals more or less uniform in composition. Subsequent heating softened the homogeneous material of Earth so that it could "differentiate," or separate into different layers. Dense materials such as iron would sink to the core, while rock-forming materials, dominated by the lighter heavy elements, would float nearer the surface. The *heterogeneous accretion model* supposes that different elements and compounds would condense out of the solar nebula at different times. Metallic elements would condense and accrete first, forming the cores of the terrestrial planets; then the rock-forming compounds would condense and accrete on top of the cores. Perhaps both heterogeneous and homogeneous accretion played a role in the formation of the terrestrial planets.

The 2% residue of heavy elements that formed our planet is very important from a biological point of view, because the chemistry of life itself depends on carbon and other elements heavier than hydrogen and helium. Human life—indeed, any life on Earth—scientists think, could not have arisen and thrived without it.

**4 billion
years ago**

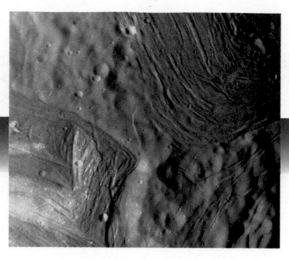

**PHOTO 3.18** Miranda, one of the smaller moons of the planet Uranus, exhibits a surface apparently fashioned by violent events in its past. Blocks of rock, 100 km or more across, are strangely juxtaposed against equally large blocks of ice or ice mixed with rock. Originally, Miranda's structure may have been simpler: a rock core overlain by an icy shell.

In contrast to the close-in terrestrial protoplanets, the far-out Jovian protoplanets picked up water, methane, ammonia, and other light compounds in addition to rocky and metallic materials. Four of them grew very large. Two of the four became massive enough to accrete large amounts of hydrogen and helium gas from the remaining solar nebula. These two became Jupiter and Saturn, both of which are big and massive, but of low density. The other two protoplanets, having less gravitational pull, attracted somewhat less hydrogen and helium. They became Uranus and Neptune, both of which are substantially smaller, less massive, and somewhat more dense than Jupiter and Saturn.

While the planets were still in their formative stages, the early sun's solar wind was intense enough to sweep away most of the remaining uncondensed gas of the solar nebula. The Solar System then consisted of the sun, the planets, and a substantial amount of debris. Some of the debris coalesced near individual planets to become moons (see Box 3.2 for an interesting description of Jupiter's four largest moons). A large part of the debris was either absorbed by the sun or left the Solar System because of gravitational interactions with one or more of the Jovian planets.

Some of the remaining debris continued to roam the interplanetary spaces. Over time, various pieces collided with the existing planets and moons, scarring them with impact craters. It is this record of impact we read today on the heavily cratered surfaces of many planets and moons.

Some of the more violent collisions with the larger planetesimals may have been responsible for several anomalies we observe today within the Solar System. Venus, for example, rotates very slowly in a retrograde (opposite) direction, quite possibly because it was knocked over or even temporarily disassembled during an early, catastrophic impact. Uranus was apparently knocked over on its side by a similar collision. Some astronomers think the bizarre surface of Uranus's moon Miranda (Photo 3.18)

**PHOTO 3.19**  The dark part of the circular feature shown here is Mare Orientale, a large, lava-filled basin on the moon. Like other lunar maria, this one originated from the impact of a large body that exploded on contact some 4 billion years ago. The shallow basin that resulted subsequently filled up with dark, dense, molten rock welling up from the moon's interior.

shows signs of having been fragmented by collision and reassembled by the subsequent gravitational attraction of the fragments. Earth's moon may have resulted from an oblique collision of a giant planetesimal or proto-planet with Earth. The collision would have torn off a large chunk of Earth's mantle (but little if any of its core) and flung it as hot vapor into orbit around Earth. The vapor, rich in rock-forming minerals, would have quickly cooled, solidified, and pulled together by gravitational attraction to form the moon.

Even if these scenarios are not entirely correct, there is unmistakable evidence of frequent bombardment on smaller scales. The moon's surface, for example, is pocked with millions of craters of various sizes and is dotted with several huge, lava-filled basins, many hundreds of kilometers across (see Photo 3.19). The smooth and dark appearance of these basins prompted Galileo to call them *maria*, a Latin word meaning "seas."

The cratering process on the moon seems to have taken place in two main episodes. For hundreds of millions of years after the lunar surface solidified (about $4\frac{1}{2}$ billion years ago), colliding planetesimals saturated the surface with impact craters ranging up to about 200 km across. Then, about 4 billion years ago, several very large planetesimals collided with the moon, blasting out basins that later filled with dark, dense, molten rock welling up from the moon's interior. The biggest of these impacts released so much energy that it nearly split the moon in two. Evidence of impact cratering is not unique to Earth's moon; it appears on the surface of every terrestrial planet and every other moon old enough and undisturbed enough to record the events of billions of years ago.

Earth, too, had its surface nearly pulverized by impact, but virtually all of its ancient craters have been erased through erosion processes (mainly by flowing water and wind) and by geologic processes that tend to alter or destroy large portions of Earth's crust. The moon has retained its record of

**4 billion
years ago**

**PHOTO 3.20** Arizona's Meteor Crater was produced by the impact of a solid metallic body on Earth's surface some 50,000 years ago. During the first half-billion years of Earth's existence, these events were commonplace.

cratering only because it has been airless, waterless, and geologically moribund for billions of years.

By about 1 billion years after the birth of the Solar System (3.6 billion years ago), most of the planetesimals had been swept up by the larger bodies of the Solar System or perturbed into orbits far larger than Pluto's orbit. The rocky remnants that remained in the inner Solar System (most orbiting the sun between Mars and Jupiter) are what we call *asteroids*. Many of the billions or perhaps trillions of small, icy, dusty planetesimals swarming amid the outer planets were ejected toward interstellar space, where they now drift slowly on eccentric orbits amid the huge Oort comet cloud. Relatively few of these comets ever chance to fall inward toward the sun and Earth, where they can sometimes be seen. Some of the larger, cometlike planetesimals may have settled into near-circular orbits somewhat farther out than Neptune. Pluto may be just one example of that group.

The period of intense bombardment in the Solar System may be long past, but impact events still occur occasionally. Earth and the other planets are like small targets in a shooting gallery of errant comets and asteroids. The remaining projectiles are far less numerous than they once were; but given enough time, major impacts on any given planet will still occur. The remains of dozens of impact craters have been found on Earth's surface, some the result of asteroid or comet collisions during the past 100 million years. The well-known Meteor Crater in northern Arizona (Photo 3.20) shows some of the consequences of a relatively small and recent collision. As has happened many times in its geologic history, Earth will one day be struck again by an impacting body large enough to cause catastrophic damage to our delicate and vulnerable biosphere.

A small number of astronomers are now looking for asteroids that might be on an immediate collision course with Earth. With enough advance warning, we might be able to acquire the necessary technology and resources to nudge any potentially threatening body out of harm's way.

## BOX 3.2
# The Galilean Satellites of Jupiter

We have seen that the Milky Way galaxy condensed out of a large, rotating cloud of primordial material early in the universe's history. Because the cloud was originally rotating, a large portion of our galaxy's mass is now confined to a thin disk made up of stars and diffuse gaseous and dusty material.

The Solar System originated from a much smaller, rotating gas and dust cloud within that galactic disk: the solar nebula. The rotation of the solar nebula is reflected in the motion of today's planets: They orbit the sun in essentially the same flat plane.

The same kind of thing has happened on even smaller scales within our Solar System. One example is Saturn's rings: The rings are made of swarms of independently moving, icy particles that have settled into an exceedingly thin disk above Saturn's equator. Another example is Jupiter and its four largest satellites (or moons).

Jupiter has at least 16 satellites, but only the 4 largest—Io, Europa, Ganymede, and Callisto—are visible in small telescopes (see Photo 3.21). They are called the *Galilean satellites* in honor of Galileo, who first studied them nearly 400 years ago. Galileo's careful observations and sketches showed how these satellites swing back and forth across Jupiter in the same straight line over a period of many nights. His interpretation was that all 4 satellites circle Jupiter in the same plane, with the innermost (Io) moving fastest and the outermost (Callisto) moving slowest. In this way, Jupiter seems to play the same central role among its neighboring moons as the sun does amid its retinue of planets.

**PHOTO 3.22**   Jupiter's four largest moons, as pictured by the *Voyager* spacecraft, are printed to scale here. They are Io (top left), Europa (top right), Ganymede (bottom left), and Callisto (bottom right).

During the early 1980s, photographs radioed back to Earth by the *Voyager 1* and *Voyager 2* spacecraft provided us with close-up views of Jupiter and several of its satellites. The Galilean satellites (Photo 3.22) were revealed to be four quite dissimilar worlds, each shaped by different environments, past and present.

Io, the innermost satellite, was revealed to be the most volcanically active body of the Solar System. Io is close to Jupiter, and Jupiter's large mass holds Io in a strong gravitational grip. As Io follows its eccentric orbit around the giant planet, that grip waxes and wanes, creating internal stresses and internal heat (we will learn more about these effects in connection with tidal forces in Box 5.3 in Chapter 5). Io has a density of 3.5 g/cm$^3$, not unlike our own moon, showing that it consists almost entirely of rock material. Io has no permanent features on its surface; all are quickly erased by the constant volcanism.

Europa's icy surface is smooth as a billiard ball, but its stripelike features give the illusion of some kind of topography. Apparently, liquid water has welled up from Europa's mantle and frozen inside a network of cracks. The cracks indicate that Europa has experienced considerable gravitational or other stresses in the not-too-distant past.   *(continued)*

**PHOTO 3.21**   Earth lies nearly in the plane of the orbits of Jupiter's Galilean satellites (the starlike images in this photograph), so that all four satellites appear to be arranged in a fairly straight line.

*(Box 3.2 continued)*   With a density of 3.0 g /cm³, Europa is believed to have a large, rocky core overlain by a thin, watery mantle, which is in turn surrounded by a thick ice crust.

Ganymede's peculiar grooved terrain suggests that its icy crust was greatly stressed, too—perhaps while cooling and shrinking. The grooves are obviously quite old (1–2 billion years) because there are many impact craters upon them. Ganymede's density of 1.9 g /cm³ indicates that it is a half-rock, half-ice world—solid rock on the inside, solid and dirty ice on the outside.

The outermost moon, Callisto, is saturated with impact craters. Its hard, icy surface is a battered survivor of early times (some 4 billion years ago) when the bombardment by planetesimals was intense. Like Ganymede's, Callisto's density of 1.8 g /cm³ indicates a composition of about half rock and half ice.

The Galilean satellites make up a well-ordered family of bodies formed at the same time as Jupiter. Jupiter itself accreted the lion's share of the planetesimals within its gravitational sphere of influence, but some material was left over in the form of a rotating disk. That disk fragmented, and the denser parts of it accreted to produce four moons. Later, Jupiter's immense gravity captured additional, smaller bodies, which were added as smaller moons.

At the same time that the Galilean satellites were forming, the embryonic Jupiter was becoming very warm because of the tremendous inflow of planetesimals and because of its contraction by means of gravity. It was, in fact, on its way to becoming a star, but it missed this stage because it had far too little mass. Consequently, temperatures inside Jupiter never climbed high enough to initiate hydrogen fusion—the event that allows a star, like our sun, to stabilize and maintain itself as an energy producer for billions of years. Jupiter had a hot flash of gravitational heating, but it quickly faded.

Jupiter's early heat created differences in temperature in the disk of leftover material that would become its moons, which led to a difference in composition in those moons, just as had happened with the planets of the Solar System. Heavy elements and compounds were the only materials that could condense close to the hot Jupiter. Far from the planet, where it was colder, lighter compounds such as water could condense as well.

Factors unique to the Jupiter system—specifically, Jupiter's gravity—have affected the subsequent evolution of the Galilean satellites. The outer two satellites today have radiated away virtually all the heat energy acquired during their formation; they are all but dead geologically. The inner two (especially Io) are affected greatly by heat-producing tidal interactions, which are much stronger near the planet than far from it.

**3.8 billion
years ago**

**PHOTO 3.23** These stromatolites at Shark Bay, Australia, consist of layers of sediment trapped by sticky mats of photo-synthesizing blue-green algae. The earliest stromatolite structures (older than 3 billion years) found today contain micro-fossils of organisms similar to today's bacteria and blue-green algae.

# Step 4: The Origin of Life on Earth

Earth continued to endure intense bombardment by planetesimals for perhaps 700 million years after its initial formation. Many of these planetesimals, like today's comets, were rich in frozen compounds such as water, carbon dioxide, methane, and ammonia. Colliding planetesimals like these could have contributed some or perhaps nearly all of Earth's water and the gases that made up its primitive atmosphere. The rest could have come from volcanic eruptions, which were probably frequent during the time after the bombardment. These eruptions would have released into the atmosphere large quantities of water vapor and other gases, as well as lava and ash.

It is not clear when Earth's surface solidified, but the discovery of solid mineral grains from Australia dated at 4.2 billion years of age indicates that at least some solid rocks had formed when Earth was no older than 400 million years. Intact chunks of crust dating from nearly 4.0 billion to 3.8 billion years ago still exist on some of today's continents, including areas of Minnesota, northern Canada, Greenland, and South Africa. According to the most recent scheme of classifying geologic time, this time in Earth's history represents the dawn of the Archean eon (see Figure 3.20). Preceding the Archean eon is the informally designated pre-Archean time, about which little is known. Following the Archean eon is the Proterozoic eon, which spans times from 2.5 billion to 570 million years ago. The entire span of geologic time before 570 million years ago has traditionally been called Precambrian time. The present eon is the Phanerozoic (see Figure 3.21).

The Archean and Proterozoic eons saw the development and growth of the *cratons,* or core units of crust, that represent the foundation units of our present continents. Today, the continents consist of the ancient cratons—which either underlie or lie exposed on parts of the continents—plus much additional material that has been added to the continental margins more

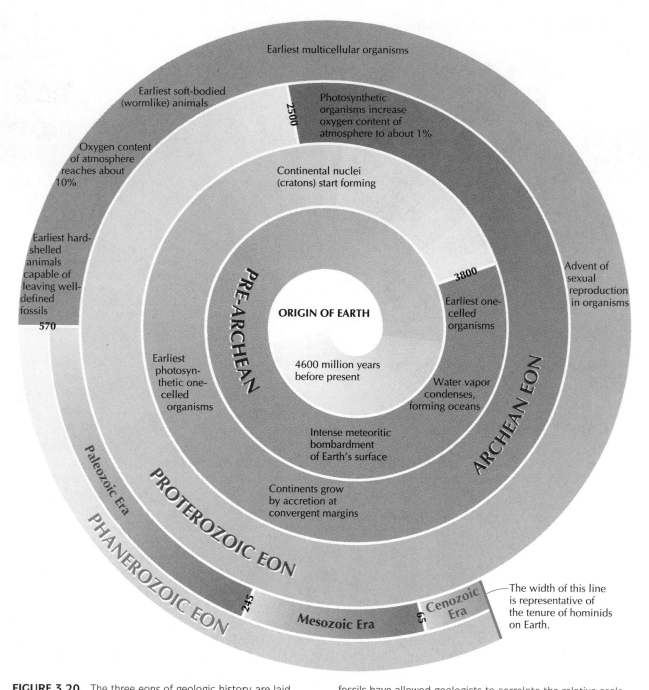

**FIGURE 3.20** The three eons of geologic history are laid out linearly on this spiral diagram. The most recent eon, the Phanerozoic, is divided into three eras, which are further divided into periods and epochs, as illustrated in further detail in Figure 3.21. The chronological order of the various subdivisions of time constitutes the *relative* geologic timescale. Modern methods of dating rock formations and fossils have allowed geologists to correlate the relative scale with an absolute scale of years. Absolute ages in this diagram are given in millions of years before the present (1000 million years = 1 billion years). *Note*: In many geology and natural science books, all geologic time before the Phanerozoic eon is referred to as Precambrian time.

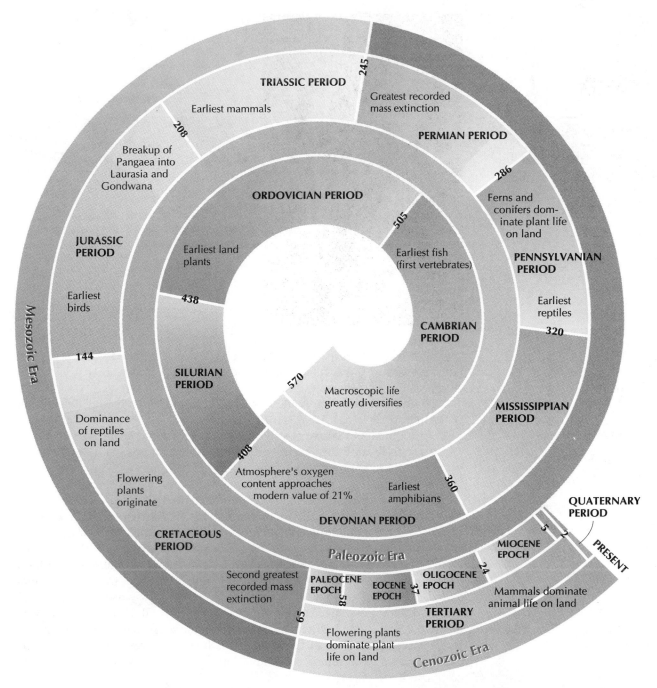

**FIGURE 3.21** The entire Phanerozoic eon, which constitutes the most recent 570 million years of geologic history, is laid out linearly along this spiral. The Pennsylvanian and Mississippian periods are collectively known as the Carboniferous period—a time of rampant growth of plants that later became compressed underground to form coal (a form of carbon). Not labeled on this diagram are the Pliocene epoch, 5 to 2 million years ago; the Pleistocene epoch, 2 million to 10,000 years ago; and the Holocene (or Recent) epoch, 10,000 years ago to the present. The divisions between the various eras, periods, and epochs are based on significant changes in the fossil record. One such division is the famous K-T (Cretaceous-Tertiary) boundary, which also divides the Mesozoic and Cenozoic eras. The mass extinction of all dinosaurs and many other forms of life at the K-T boundary paved the way for the ascendancy of mammals during the Cenozoic era.

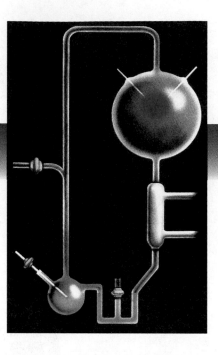

**PHOTO 3.24**  In this illustration of the Miller-Urey experiment, the apparatus consists of a "sea" of water (at bottom) under an "atmosphere" of hydrogen, methane, and ammonia (upper right). Electrodes produce sparks that simulate either lightning or ultraviolet solar radiation. In the experiment, a tarlike residue containing amino acids builds up at the bottom.

recently. The largest exposed part of the craton associated with the North American continent, called the Canadian Shield, sprawls across much of northeastern and eastern Canada.

No one knows what process started the initial birth of continents. One model supposes that the origin of cratons is related to some of the large meteorite impacts that were still occurring during the early Archean eon. Other models frame the origin and growth of continents solely in terms of plate tectonics, a process that may have been more vigorous in the past than it is today.

Life arose on Earth probably sometime during the early Archean eon. The first oceans, which filled the low areas of Earth's crust, may have boiled away several times as a consequence of some of the final, huge meteorite impacts. But as the bombardment lessened, Earth's surface assumed a more tranquil character. By the time Earth was about 1.1 billion years old, bacteria were thriving (as evidenced by the imprints of one-celled organisms geologists have found in 3.5-billion-year-old rocks). Evolutionary biologists have traced the "family tree" of all forms of life back to these bacteria, which themselves were remarkably sophisticated organisms. What is not clear at the moment is how these one-celled organisms evolved from the much smaller units of matter of which they are composed.

A famous experiment, first performed in the early 1950s by Stanley Miller and Harold Urey, showed how the simplest building blocks of life could have been constructed. In an attempt to re-create early conditions on Earth, Miller and Urey introduced electric sparks, simulating lightning, into flasks containing hydrogen, methane, ammonia, and water, simulating the early atmosphere and oceans (see Photo 3.24). After several days, the experiment yields a plethora of organic (carbon-rich) compounds, including amino acids, which are themselves the building blocks of proteins. Similar experiments have produced adenine and guanine, two important com-

**PHOTO 3.25** Giant tube worms and giant clams are among the exotic life-forms found around hydrothermal vents on the ocean floor. Some scientists think that the first self-replicating organisms evolved in a deep-ocean environment such as this.

ponents of DNA and RNA, which control cellular function and direct the process of self-replication that characterizes all life. Life, then, could have originated near the surface of the ocean in an atmosphere rich in gases that are not abundant in the atmosphere today.

Another idea, gaining in credibility, is that simple organic compounds were delivered intact to Earth by cometlike planetesimals and smaller debris still drifting about after the planets had formed. Some of the slower-moving debris, in particular, might have been gently caught by Earth's upper atmosphere and later sunk slowly to the surface. European and Soviet spacecraft flying by Comet Halley in 1986 measured it to be composed of one-third organic compounds. Complex compounds such as certain amino acids have even been identified in interstellar space. Clearly life's simplest building blocks existed outside Earth then as now, but did enough of them accumulate on Earth's surface to trigger reactions leading to self-replicating organisms?

Another possibility is that life began near *hydrothermal vents* on the seafloor. Marine geologists discovered these vents, where scalding-hot seawater emerges from cracks in Earth's crust, in 1977. Nearby, where the water is tepid, the scientists found bizarre forms of marine life (see Photo 3.25) that derive their energy not from the sun, like almost all other known life on Earth, but from bacteria that extract energy from chemicals expelled by the vents. Some believe that clay minerals deposited near these vents could have provided a suitable environment for chemical reactions that would have led to the first self-replicating, organized biological units, which may have resembled viruses. This hypothesis is attractive because hydrothermal vents, which were probably more abundant in the early Archean eon than now, could have provided a stable milieu for protracted chemical development. Earth's surface at the time was irradiated with strong ultraviolet rays from the sun that could have broken apart any complex, prebiological molecules.

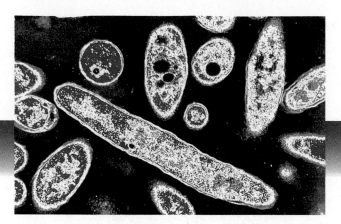

**PHOTO 3.26** Relatively simple organisms, such as these Legionella bacteria, once represented the pinnacle of complexity for biological evolution on Earth. Measured by the standards of resilience and longevity, however, some microorganisms might still be regarded as superior to the higher forms of life; they will almost certainly outlast humans. The adaptability of some earthly bacteria to extreme conditions suggests that similar life forms, if they ever arose on the planet Mars long ago, might remain there today, tucked away in warm, wet underground places.

The explanation of life's origin on Earth is one of science's greatest challenges. When and if an answer is found, it may shed light on the likelihood of life elsewhere in the universe. If life arose from nonliving, organic compounds in a relatively straightforward manner, then we would expect the same thing to happen wherever the right ingredients come together in a stable environment that includes liquid water.

The other planets of the Solar System are the easiest places to look for extraterrestrial life (life *outside* of Earth), but we find almost no liquid water there. The moon, we learned from firsthand exploration, is totally desiccated. Venus is far too hot; its meager share of water is present as a vapor trace in its thick, carbon dioxide–rich atmosphere. Mars has plenty of water, but it lies frozen on its surface and underground. However, at least some of this water was liquid for a time in the distant past, and some scientists believe that bacteria, or organisms of similar complexity, could have originated in warm, water-rich environments. At best, we might allow for the possibility that simple forms of life still exist on Mars today, having taken refuge near the throats of the Martian volcanoes. Aside from this scant possibility, extraterrestrial life of any kind within our Solar System seems very improbable.

If we could broaden our search for life to include other planets belonging to other stars, we might greatly improve our chances of finding it. As mentioned earlier, we have as yet obtained no direct evidence of such planets, although there is plenty of indirect evidence for the existence of planetary systems around other stars. If life exists on planets outside the Solar System, we could not hope to find direct evidence of its primitive origins, because we cannot travel such distances to study samples of it. However, intelligent life elsewhere (with a technological capability equal to or more advanced than ours) might eventually be detected in the form of radio waves or laser beam emissions radiated into space.

## The Rise of Plant and Animal Life

**500 million years ago**

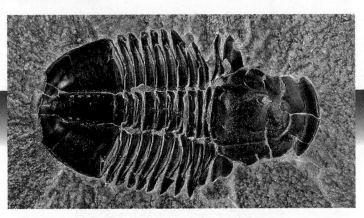

**PHOTO 3.27** Animals with shells, spines, and armor plates began to appear on Earth around 750 million years ago. This trilobite, whose hard parts made the impression we see in this fossil, lived in a warm shallow sea during the Cambrian period that ended about 500 million years ago.

Several teams of scientists are currently engaged in what is popularly known as SETI (*search for extraterrestrial intelligence*) programs. Several radio telescopes are being used by astronomers in full- or part-time systematic searches for radio waves encoded with intelligent signals. Those signals would come not from any stars themselves but from life-bearing planets that lie in orbit around them. Only recently has microelectronics technology progressed to the point where such a search can be done efficiently and (in the opinion of those who support SETI projects) with some reasonable chance of success. SETI, like other ambitious and speculative scientific ventures, holds the promise of much enlightenment if it is successful. We would learn, at the very least, that life—even intelligent life—is not unique to Earth. Most optimistically, if the signals received contained an encyclopedic amount of information, and we could decode it, human knowledge and civilization would be enriched in ways we cannot now imagine.

## Step 5: The Rise of Plant and Animal Life

Diversity and change are pandemic on Earth. Over geologic time, Earth has hosted a broad and ever-changing array of biological habitats. Some changes in the environment are regular and fast-paced, like the cycles of the days and the seasons. Others are slow, such as the global changes wrought by plate tectonics. Some are nearly instantaneous and catastrophic: landslides, floods, hurricanes, tornadoes, and rare meteorite impacts having global consequences.

**PHOTO 3.28** This lightning-damaged tree demonstrates the ability of an individual organism to accommodate change and survive. The pattern of its growth metaphorically suggests how life endures, despite the ravages of extinction.

Life has increased in diversity and in complexity for at least 3.5 billion years, not just in spite of these changes but probably because of these changes. In a stable environment, living organisms gather energy directly or indirectly from their environment, modify their surroundings to ensure their own preservation, and reproduce themselves. In a changing environment, however, they must do more if they are to survive: They must change themselves.

Individual organisms are capable of some adaptive change. For example, humans undergo physiological changes that can allow them to function better at high altitudes after a few days or weeks of acclimatization. These adaptations, however, cannot be directly passed on to offspring. But real and lasting changes in a line of organisms can and do take place over many generations: If offspring have some variability of traits (they almost always do), then those individuals with certain traits advantageous in a changed environment will thrive and reproduce, often at the expense of others. This process is known as *natural selection.*

After only a few generations, the traits of successful offspring, which are passed on to further descendants, can result in a population heavily skewed in favor of those favorable traits. After many generations, unfavorable traits can completely disappear from the population. Thus, incremental changes over long time periods can easily lead to organisms that have traits very different from their early ancestors'. This is the process of biological *evolution.* This is not to say that ancestral species inevitably die out and become replaced by new species. Modern biologists see the tree of evolution as a branching structure, with some twigs (species) and branches (genus, family, order, and so forth) surviving and others facing extinction.

The basic idea of evolution through natural selection, first proposed by Charles Darwin and Alfred Russell Wallace in the mid-1800s, has gained universal acceptance among scientists today. It inevitably assumes, however, that some mechanism exists to produce the variability that must

**PHOTO 3.29** The forces of natural selection are just as much in evidence today as they were in the distant past. One classic, contemporary example is the changing population balance of light and dark peppered moths of the same species in England. In the early 19th century, the light-gray form (camouflaged at right) was common and the dark-gray form was very uncommon. Half a century later, the darker moths were becoming common and the lighter moths were in decline. This dramatic change (which was reflected in the populations of other species of moths as well) was traced to the blackening of tree trunks by the severe air pollution common to the industrial regions of England at the time. The light-gray camouflage that suited the lighter moths on formerly light-gray tree trunks was now a disadvantage. The rare dark-gray moths, whose bodies and wings blended with the soot-covered trees, survived predation by birds and began to reproduce much faster. Since the enactment of strict pollution controls in England in 1952, the dark-colored variety is becoming rare again.

occur in offspring if biological change is to take place. That mechanism was found in the early 1950s, when James Watson, Francis Crick, and others decoded the structure and function of the DNA molecule.

Today, inheritable traits are seen as outcomes of the set of instructions contained in each organism's genetic code, which controls the organism's development. Genetic codes in DNA (or RNA in some simple organisms) can become scrambled because of errors in the copying process or because of external factors such as exposure to cosmic radiation. These genetic changes, called *mutations,* can result in traits that can either increase or decrease an organism's chances for survival. With natural selection at work, advantageous mutations are selected far more frequently than disadvantageous mutations in succeeding generations. That is, organisms inheriting advantageous mutations have a greater chance of surviving.

The term *survival of the fittest* is often used to describe how natural selection works. However, *fittest* does not always mean "strongest." In some cases, fitness for a changing environment can be met by organisms that evolve to become more complex. In other cases, particularly for microorganisms, fitness may be enhanced through symbiotic or cooperative relationships. The history of evolution, in fact, is just as much a story of cooperation as it is a story of competition. With these ideas in mind, we can begin to see how simple organisms, such as bacteria, could have evolved into multicellular organisms and how those organisms could have evolved into the higher organisms that inhabit Earth today.

The first one-celled organisms probably fed on (derived their energy from) organic molecules dissolved in Earth's primordial oceans. When these bacterialike organisms had depleted the ready supply of these molecules, the new environment hastened the evolution of new organisms that could manufacture their own food by way of an alternative path: *photosynthesis.* By absorbing sunlight and using it to manufacture their own "food," these one-celled, plantlike organisms thrived because they enjoyed

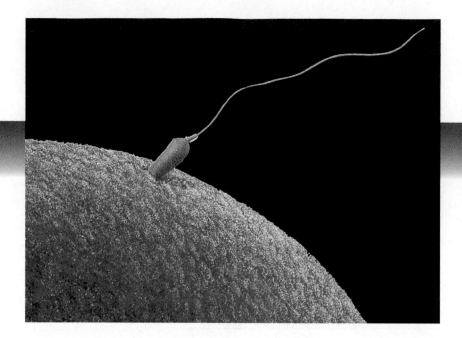

**PHOTO 3.30** Sexual reproduction originated among one-celled organisms. Today, this method is employed by most multicellular organisms including, of course, humans.

a selective advantage in the new, food-starved environment. (These evolutionary changes and many others are noted on the diagrams of geologic history presented in Figures 3.20 and 3.21.)

The advent of photosynthesis, perhaps 3 billion years ago, marked a turning point in Earth's chemistry as well. Free oxygen ($O_2$), which is a by-product of photosynthesis, began to build up in the atmosphere. The older lineages of *anaerobic* ("without oxygen") organisms either were poisoned by the gradually increasing oxygen, retreated to oxygen-free environments such as mud and sediment, or evolved mechanisms by which they could cope with the oxygen. According to one scenario, part of the latter group evolved through a symbiotic relationship in which an anaerobic host cell incorporated a smaller *aerobic* ("oxygen using") cell. This may have been the prototype of the first cells with cell nuclei. When these host cells divided, they passed along copies of their internal partners.

Perhaps 1.4 billion years ago, a new and more effective method of passing along genes evolved: sexual reproduction. This method of reproduction for cells sped up the pace of evolution because each offspring receives half its genetic makeup from each parent (but different offspring receive different genetic material from each parent). This scheme encourages more diversity among offspring, and therefore, more evolutionary pathways for any interbreeding organisms that are under environmental stress.

By about 1 billion years ago, cells were colonizing into multicellular units, and certain cells began to assume specialized functions in these colonies. This development gave rise to a wide variety of plants (photosynthesizing organisms) and animals (oxygen-using organisms) in the oceans. The earliest organisms with hard parts that could be preserved as fossils came on the scene about 570 million years ago, and from this time on (the Cambrian period), the development of life becomes much easier to read in ancient layers of rock.

**PHOTO 3.31** During the Pennsylvanian and Mississippian periods (360 to 286 million years ago) trees such as these and other vegetation covered vast areas of swampland at or near sea level. Periodic incursions of seawater covered the partly decayed remnants of these plants with thin layers of sediment. Over millions of years these buried sediments were converted by heat, pressure, and bacterial action to a residue rich in carbon. Today, this residue is mined as coal.

Up until about 400 million years ago, life may have remained largely restricted to the oceans, because Earth's land surface was exposed to lethal amounts of the ultraviolet light naturally radiated by the sun. At most, the land was coated in moist, protected places with a layer of photosynthetic bacteria underlain by anaerobic bacteria.

The oxygen buildup in the atmosphere was slow, but by about 400 million years ago, enough free oxygen existed at upper levels (at heights of 30–50 km in the stratosphere) for reactions to take place that converted some normal oxygen ($O_2$) to ozone ($O_3$). As it does today, stratospheric ozone effectively shielded Earth's surface from most of the ultraviolet radiation. With this turn of events, plant life quickly invaded the land. Those first plants, like today's mosses, were joined by ferns, conifers (cone-bearing trees), and a multitude of flowering plants.

Animals followed the plants onto land: first arthropods (insects and crustaceans), then amphibians (which probably evolved from fish). From amphibians evolved reptiles, including a group that became the dinosaurs. Eventually, some reptiles evolved to become birds and mammals. A branch of the mammals evolved to become primates, and one small branch of that family evolved to become today's human beings.

The emergence and proliferation of life as briefly traced in this section certainly played a central role in most of Earth's history. But other changes were occurring at the same time that shaped and reshaped the entire surface of the planet many times over.

By reading the sequential record of rock strata and fossils embedded in rock (see Box 3.3), geologists can construct possible scenarios for the history of any geographical region on Earth. These scenarios, however, were not meaningfully unified on a global scale until the advent of the theory of plate tectonics in the 1960s. Today, the plate tectonics theory, plus detailed measurements of the present movements of the continents, allows geosci-

## BOX 3.3
# A Geological Perspective of Time

Earth's crust is like an open book, telling of events stretching billions of years into the past. Unfortunately, the contents of the entire book of Earth's history are not easy to decipher. The following thought experiment will illustrate some of the difficulties and challenges faced by scientists who specialize in what is called *historical geology.*

Let us suppose you hold in your hands several copies of a history book you have never read. Imagine ripping every book apart at the seams, burning most of the early chapters, tearing and crumpling many of the remaining pages, and tossing everything about. You would end up with pieces of various sizes: some units might contain entire chapters, but many units would consist of fragments of pages or ashes. How would you go about reconstructing the information contained in the history book?

**PHOTO 3.32** The sedimentary layers exposed on the wall of the Grand Canyon in Arizona reveal an extraordinarily orderly sequence of geological events. Older rock formations lie at the bottom of the stack, near the area where the Colorado River continues to carve out the mile-deep canyon.

For centuries geologists have been faced with puzzles like this. Earth's 4.6-billion-year history is a record of ceaseless change. Very little of the present crust underlying the oceans was in existence even 200 million years ago, a consequence of plate tectonics. Most of the continental crust has been in existence for more than 200 million years; yet the older rocks that are part of it are not commonly exposed at the surface. Here and there, geologists find intact sequences of rock dating from much earlier times, and, of course, we can always dig down or drill into the crust to expose or obtain samples of older rock hidden underground. In some places, such as the Grand Canyon (Photo 3.32), there is enough information in plain view to fill several consecutive "chapters" of Earth's history. In most places, however, we are able to read but a single "page" or "paragraph," and that information is likely to pertain to relatively recent history.

The reconstruction of the contents of a real history book could proceed by two methods: (1) arranging paragraphs, pages, and chapters in an order that makes chronological and logical sense, and (2) arranging pages by page number. Method 1 is analogous to several techniques used by geologists to date the relative ages of rock formations. Relative geologic time is commonly divided into units called eons, eras, periods, and epochs (see Figures 3.20 and 3.21). Many of the arcane-sounding names of these units relate to certain geographic regions where rocks of given time periods were first studied. Rocks of Cambrian period, for instance, were first investigated in detail in Cambria (the Roman name for Wales, in the British Isles). Others have descriptive meanings. The Paleozoic, Mesozoic, and Cenozoic eras, for example, mean the times of "ancient," "middle," and "recent" life, respectively.

Method 2 is analogous to the absolute dating of particular rock formations. Today, this is primarily accomplished through the technique of *radiometric dating.* (Radiometric dating depends on measurements of the rate of radioactive decay of certain isotopes existing in trace amounts in rocks. We will discuss how it works in Chapter 12.) Absolute dating establishes an absolute geologic timescale in which the ages of rock formations can be expressed in years before the present.

We cannot determine all we want to know by either method applied individually. By relative dating alone, we may be able to determine that rock formation A preceded rock formation B, but we may not know how many years passed between the origin of A and the origin of B. Absolute-dating techniques, on the other hand, can be applied to some but not all rock formations. Reconstructing a history book would be easiest if we could pay attention to both the logical sequences of prose and the page numbers. Similarly, the sharpest focus on the geologic past is obtained if we can employ the techniques of both relative and absolute dating.

Much of the detective work involved in historical geology involves piecing together the relative timescale locally and globally. So let us take a closer look at some of the principles and methods behind these efforts.

## Principle of Uniform Change

Geologic change is an observational fact: Volcanoes explosively eject ash and gases, or they dribble molten rock down their flanks. When earthquakes occur, blocks of Earth's crust bump and grind against each other, often giving rise to mountainous topography. These constructive agents of geologic change are opposed by the destructive effects of weathering and erosion. The agents of erosion—rushing water, moving ice, and wind—tear away at elevated places and transport loosened materials, from tiny grains of silt to huge boulders, toward the sea.

Are these present changes irrelevant to the way Earth's surface has formed, or have they operated over an enormously long time, giving rise to Earth as we see it now? From the late 1700s to the mid-1800s, the former idea prevailed. Most geologists of that time accepted the notion of *catastrophism,* the shaping of Earth's crust by a series of recent (thousands of years in the past) global catastrophes culminating in a worldwide flood. Catastrophism was in close agreement with the account of creation as described in literal terms in the Bible, but it did not jibe with the flood of evidence geologists were gathering from places all over the world.

While catastrophism was still in vogue in the late 1700s, an alternative concept, *uniformitarianism,* was first introduced by the geologist James Hutton. Perhaps better referred to as the **principle of uniform change,** it supposed that, in Hutton's words, "the past history of our globe must be explained by what can be seen to be happening now." That is, relatively small, incremental changes, such as the changes produced by volcanoes, earthquakes, and erosion, are sufficient to explain how Earth came to be the way it is. Global catastrophes, according to this principle, are not required.

The principle of uniform change gradually gained acceptance in the 19th century, and today it is a guiding light (though not necessarily the only guiding light) in geology. We now think of uniform changes as generally occurring over geologic time and not human timescales. The modern concept of uniform change easily accounts for the sporadic, extraordinary events, or minicatastrophes (such as earthquakes, volcanoes, and floods) that happen regularly or irregularly on various parts of the earth. Over millions of years, incremental and sometimes violent geologic events have built entire mountain ranges and moved continents. The uniformity of geologic change can be compared to the way that the dozens of small explosions taking place every second in a car's engine can yield a smooth output of power.

Let us extend this car engine analogy a bit further: Just as you can "rev up" an engine, or alternately ease off on the gas pedal, large-scale geologic changes can proceed at ever-changing (continued)

**PHOTO 3.33** This gorge wall in Hamersley Range National Park, Australia, exhibits greatly folded strata.

(*Box 3.3 continued*)      rates. For example, volcanic activity on certain continents has waxed and waned over periods of tens of millions of years; glaciation has been more vigorous over the past 2 million years than it was over the previous 300 million years. We also have reason to believe that our planet's geologic "engine," driven by heat generated in Earth's interior, is aging and winding down. When considered over time spans of billions of years, uniform change is better interpreted as a slow decline of geologic activity.

Is there still a role for catastrophism in today's geological way of thinking? Yes, in the sense that uniform change over geologic history has been punctuated by events of extraordinary magnitude occurring every few million or tens of millions of years. Many of these events are believed to have been caused by giant meteorite impacts that have had global consequences. Some may be related to geologic episodes of frequent and widespread volcanism.

## Principle of Superposition

The **principle of superposition** simply says that in an undisturbed sequence of stratified (layered) rocks, each layer has been "superposed," or laid on top of a layer that is older than itself. In a stratified, undisturbed column of deposits, the lowermost layer is oldest and the uppermost layer is youngest (Figure 3.22).

Stratified deposits are most obviously seen in *sedimentary rocks,* which are commonly (but not exclusively) formed when particles of accumulated debris (called *sediment*) are subject to compaction by the weight of further layers of sediment or rock laid on top of them. Sedimentary strata can be found lying in horizontal layers, more or less as they were formed; but they can also be tilted, *folded* (bent), or *faulted* (broken) by the application of geologic forces (again, see Figure 3.22). In some unusual cases, stacks of layered sedimentary rock have been tilted vertically or overturned so that older layers lie on top of younger layers. Geologists look for subtle clues in the structure of the rocks to determine which side of the stack was originally up.

Superposition coupled with other geologic principles can give clues to the order in which geologic events have taken place in a given area. Consider, for example, the complex situation illustrated in Figure 3.23. By the principle of superposition, sedimentary layers A through G would have been deposited in a nearly horizontal stack, with A the oldest and G the youngest of just that group. Layers a through c were deposited after A through G, but there was a considerable gap in time, called an *unconformity,* during which layers A through G were tilted as a unit and partly eroded away. Because it overlies layers a through c, the lava flow on the surface is the youngest of all the formations shown.

Finally, let us assume that a large mass of *igneous rock* (solidified from a molten state), called

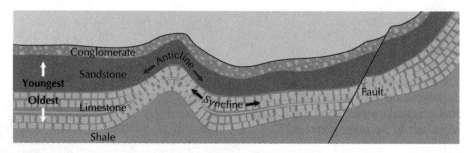

**FIGURE 3.22** According to the principle of superposition, stratified deposits, like these of sedimentary rock, are formed more or less horizontally, layer after layer, from the bottom up. On the left, the deposits lie in the same plane as they were formed. In the middle of the diagram, compression forces have folded the strata into an *anticline* (upward-point-ing fold) and a *syncline* (downward-pointing fold). On the right, the strata have been tilted and faulted (broken) by, perhaps, the core of a rising mountain range. Mountain building accompanied by severe faulting or folding can sometimes result in strata that are tilted vertically or even overturned.

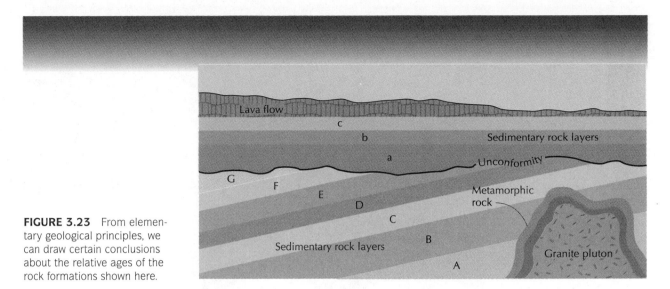

**FIGURE 3.23** From elementary geological principles, we can draw certain conclusions about the relative ages of the rock formations shown here.

a granite *pluton,* intrudes into the lower sedimentary stack. *Metamorphic rock* (altered by the heat and pressure of the intruding rock) lies adjacent to this pluton, showing that layers A through C were already intact as sedimentary units before the molten granite intruded into them, thereby altering them. The pluton must then be younger than layer C, but we do not know if it is as young as any of the layers above C. To determine the pluton's absolute age (the time since it solidified) precisely, we would need to obtain a sample of the rock in the pluton and analyze it by means of radiometric dating.

## Correlation

Figure 3.24 is a cross-sectional diagram illustrating the spatial relationships among many different rock formations in a given area. How can geologists map these formations? First, nature sometimes cooperates. In steep terrain, the erosive action of a river or a glacier can cut through solid rock and expose cross sections like those shown in the figure. The walls of the Grand Canyon beautifully exhibit the result of this carving process: Over the past several million years, the Colorado River has cut through a sequence of rock layers ranging in age from about 250 million years to nearly 2 billion years. Faulting—the shifting of one landmass with respect to another—can also expose cross sections.

Second, geologists can take an active role in probing what lies underground. By drilling many deep holes and analyzing the contents of the drill cores, they can construct detailed maps of the rock formations underground. For instance, a great deal is known about the sedimentary strata and bedrock underlying the Los Angeles Basin because of the tens of thousands of wells that have been drilled there in search of oil and water over the past century. Mine tunnels or excavations for open-pit mines accomplish the same thing.

Geologists can also gain some information about the presence and structure of underground formations by setting off small explosive charges just below the ground surface. By analyzing the vibrations, or *seismic waves,* that reflect back to the surface from many such blasts set off in different places, geologists can map boundaries that separate some of the formations.

Deciphering the geologic cross sections associated with local or regional areas is equivalent to assembling sequential pages belonging to our metaphorical history book. Remember that there are multiple copies of the book spread around, though many parts of each individual book have been destroyed. To learn more, we must try to correlate the sequences.

Correlation in geology is just this process. **Correlation** involves identifying similar geologic sequences over a broad geographic area and demonstrating their proper order in time.   *(continued)*

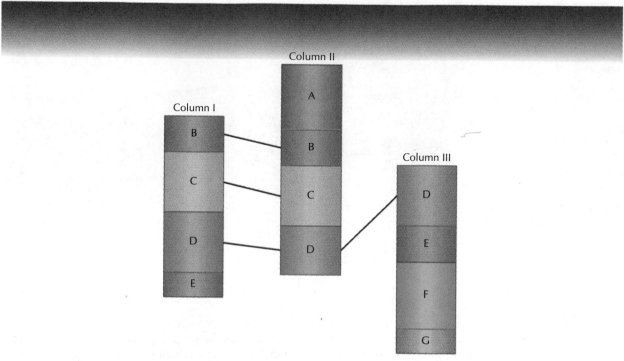

Column II

Column I

Column III

FIGURE 3.24  The process of correlation in geology strives to demonstrate equivalency among two or more rock sequences in different geographic areas. The relative positions of the sedimentary rock layers in columns I and II indicate that these columns were probably created in the same span of geologic time. Correlation between columns II and III can only rely on a similarity of fossils imbedded in layer D or some other indication characteristic of layer D only. How might columns I and III be correlated?

PHOTO 3.34  The fossilized bones of the fish *Diplomystus,* which flourished in the Eocene period, appears in this sedimentary rock formation from Wyoming.

*(Box 3.3 continued)*    When many sequences are correlated, a broader picture of geologic history emerges.

Fossils, which are the remains or traces of once-living creatures, greatly aid in the correlation process, especially in sedimentary rocks, where they are often abundant (see Photo 3.34). Fossils help because life has continually evolved throughout its nearly 4-billion-year-old history on Earth. Studies of fossils within layers of known relative or absolute age have shed light on the succession of various organisms; and a knowledge of that succession has, in turn, helped geologists make correlations among various sequences.

## Pangaea's Breakup

**200 million years ago**

**PHOTO 3.35** Did a huge asteroid or comet strike Earth at the close of the Cretaceous period, killing off the dinosaurs once and for all? Much astronomical and geological evidence gathered of late supports this hypothesis.

entists to extrapolate back in time and visualize how the continents and oceans were arranged many millions of years ago.

Paleogeographic maps, which are interpretations of the global geography at particular times in the geologic past, have been prepared for times as early as the Cambrian period (about 500 million years ago), but we cannot as yet place any great confidence in the earliest of them. Our conception of Earth's history becomes much clearer and more detailed when we approach the horizon of 200 million years ago, in what is called the Mesozoic era. A large body of evidence suggests that at around that time all of today's continents were melded together in a single huge landmass, dubbed *Pangaea* ("all lands").

## Step 6: Pangaea's Breakup

Pangaea, the Mesozoic supercontinent, was almost certainly the result of earlier separate continents that had drifted and converged. About 200 million years ago, Pangaea began to divide into two landmasses of about equal size: *Laurasia* in the north and *Gondwana* in the south. First Gondwana and then Laurasia fragmented to form smaller landmasses. These smaller landmasses—today's continents—continued to drift apart from one another, and they do so even today.

While today's continents were drifting apart and assuming their own identities in the mid–Mesozoic era, reptiles diversified and became the dominant form of life on land. The dinosaurs, the most successful group of reptiles, virtually ruled Earth for more than 140 million years. Just as mammals have diversified and adapted to a wide variety of environments today, dinosaurs were far from being homogeneous. They varied from individuals no larger than a rooster to the familiar giants with masses up to 100 metric tons (100,000 kg). Unlike all the known reptiles of today, which are cold-

**3 million
years ago**

**PHOTO 3.36** This collection of fossil skulls and fragments of jaw bones is from the oldest known group of hominids, the Australopithecinae, which inhabited Earth from 3.7 to 1.6 million years ago. These specimens were discovered in limestone caves in South Africa.

blooded, some dinosaurs may have been warm-blooded and therefore active under a wide variety of temperature conditions. Evidence also suggests that some species cared for their offspring after hatching, a trait associated with today's birds and mammals.

Mammals came on the scene during the early Mesozoic era, but they remained small, furry, warm-blooded creatures whose furtive ways of living kept them, for the most part, out of contact with the carnivorous reptiles. It is easy to marvel over and romanticize the dinosaurs, creatures whose exact appearance and behavior we can only make educated guesses about; but Earth today is inhabited by equally remarkable creatures. No known dinosaur, for instance, outranks the blue whale (which exists today) in mass and size. Nor could more than a hundred million years of reptile evolution at the top of the food chain produce anything as shrewd as the dominant mammal today, the human being.

The Mesozoic era opened, some 245 million years ago, with the greatest mass extinction of life ever recorded on Earth. That event paved the way for the rise of reptiles. The Mesozoic era closed, 65 million years ago, with the second greatest recorded extinction, which put an end to the reign of the dinosaurs and allowed mammals to proliferate and exploit a changed world. The history of life on Earth is rich in episodes of extinction like these, some or many of which seem to be associated with giant impacts on Earth's surface. For the extinction event of 65 million years ago, known as the K-T extinction after the symbolism for the Cretaceous (K) period and the Tertiary (T) period, the evidence for a giant impact seems especially convincing. We will examine this evidence further in Chapter 5.

`11:59:59 p.m.`

**200,000 years ago**

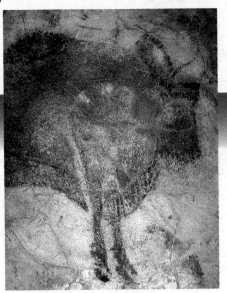

**PHOTO 3.37** Tens of thousands of years ago, humans began to translate ideas into artistic expression. This prehistoric cave painting of a bison testifies to the remarkable skills of an artist or artists who made use of natural dyes and other materials to create a vision of the world in two dimensions.

# Step 7: The Rise of the Hominids

Of all the different kinds of mammals that proliferated and diversified after the K-T extinction, one branch, the primates, gave rise to humans. Fossil evidence suggests that modern humans and their extinct close relatives, collectively known as hominids, probably evolved from a line of ape-like creatures similar to today's apes.

Fossil remnants of hominids extending back about 4 million years ago hint at several traits that characterize humans today: an upright posture with bipedal (two-legged) locomotion, a relatively large brain, hands and fingers capable of considerable manual dexterity, and the use and construction of tools. Evidence gathered so far indicates that these creatures probably originated in Africa and migrated to other continents from there. These brainy creatures eventually developed language, which allowed them to share complex information with each other and to transmit this information to succeeding generations.

While the various branches of the hominids were evolving, Earth's climate took a decided turn toward colder conditions. Starting about 2 million years ago, ice caps began to form in the Northern Hemisphere, and the latest *Ice Age* was underway. (Prior to this Ice Age, Earth had experienced at least four other major ice ages, stretching back in time to as early as 2.3 billion years before the present. Current geologic evidence suggests that Earth has been warm and essentially ice-free for 90% of the past 2.5 billion years.)

In the course of the latest Ice Age, 3-km-thick glaciers advanced and retreated several times, overtaking as much as 30% of Earth's land surface. During the most recent glaciation, which peaked about 18,000 years ago, so much water accumulated as ice on land that ocean levels sunk to as much as 100 meters below present sea level.

Note that the Ice Age was punctuated with at least 20 major *interglacials,* or warm spells, and numerous minor episodes of warming. From past cli-

**PHOTO 3.38** The longevity of cuneiform inscribed on clay may exceed that of any other form of symbolic expression. This cuneiform tablet is from Old Ugarit, Damascus, Syria.

mate patterns, many geologists think we are now in an unusually warm interglacial that began about 10,000 years ago and may well end some hundreds or thousands of years from now. Sooner or later, the ice-sheets will return, scratching across the landscape and crushing cities as far south as London and Philadelphia.

The most recent form of hominids, which includes all the human races of today, is called *Homo sapiens sapiens*. All humans were nomadic hunter-gatherers, subsisting on herds of animals and plants growing nearby, for the first three-fourths of their 40,000-year existence on Earth. Cro–Magnon cave paintings, 35,000 years old, found in France and Spain give eloquent testimony to the ability of these early people to render permanent records of the world around them. During one or more mild interglacial episodes, possibly as early as about 30,000 years ago, Asian hunter-gatherers migrated across a land bridge from Siberia to Alaska or traveled by boat to remote North or South American shores, thereby peopling both American continents.

During the onset of warmth following the last intense glaciation, humans began to take up an agricultural way of life. Starting as early as 10,000 years ago, hunting and gathering yielded to ways of life based on the care of planted vegetables, fruits, and grain crops, and the use of domesticated animals for tasks requiring brute strength. People began to settle into fixed communities with populations of hundreds or thousands. In densely populated regions, communities pooled their resources through trade and shared information with each other. Around 6000 years ago, the Sumerian culture of present-day Iraq developed the earliest known form of writing, called cuneiform. For the first time, knowledge could be codified in permanent symbols and disseminated easily through both space and time. Other cultures adopted their own written languages, thereby advancing to a historic (as opposed to prehistoric) stage.

The roots of mathematics and science developed simultaneously and often independently on several continents. Simple mathematics, including

**PHOTO 3.39** Stonehenge, in southwest England, and other similar stone monuments were built by Bronze Age peoples some 4000 years ago for keeping track of the changing positions of the sun and the moon and for keeping a calendar.

geometry, was employed in ancient India, the Near East, Egypt, and possibly China for the purpose of accounting, surveying, and calendar making as early as 5000 years ago. In the Americas the Maya, Incas, and other cultures developed remarkably accurate calendars based on cycles associated with the sun, the moon, and the planet Venus. The ancient Britons erected huge stone monuments, such as Stonehenge (see Photo 3.39), for keeping track of astronomical cycles. The Polynesians developed a form of celestial navigation that allowed them to island-hop over vast stretches of the Pacific Ocean. Some 5000 years ago, Egyptians discovered the relationship between the circumference of a circle and its diameter, the ratio known as pi ($\pi$).

Around 7000 years ago artisans in the Near East developed metallurgy, the process of extracting metal from ore and working the metal into tools and weapons. Metallurgic techniques were enhanced some 2000 years later with the discovery of metal alloys such as bronze. Even though the early artisans had no understanding of the theoretical underpinnings of their craft (what we would today call physics and chemistry), their accidental discoveries and trial-and-error experiments led to the same practical results we would strive for in today's industrial world. Knowledge of engineering and technology for practical matters (housing, water supply, food production) and for ceremonial matters (calendar making, the creation of objects of art) spread from one culture to another, or independently arose in cultures cut off from contact with the bulk of humanity. Without the benefit of underlying theory, but with plenty of practical experience, ancient craftsmen and workers erected monumental structures such as the Pyramids in Egypt and the Great Wall of China. Numerous lesser-known civil engineering accomplishments of antiquity (such as sophisticated irrigation, drainage, and road systems) dot nearly every continent.

The roots of today's scientific way of thinking sprang from the world of ancient Greece. The Greeks' leadership does not slight the remarkable

`11:59:59.99 a.m.`

**2000 years ago**

**PHOTO 3.40** This quadrant was fashioned from brass in Naples, Italy, in 1553. Quadrants of various designs were used as sighting devices to keep track of celestial movements and for navigation and surveying. The quadrant was eventually superceded by the sextant—a more accurate instrument.

mathematical and technological achievements of other cultures around the world, which at times paralleled or even led the advances made in Greece. The Greeks, however, directly contributed to what became modern science in two vital ways: First, the extensive literary and philosophical works of the ancient Greeks were widely translated and disseminated to later cultures. Second, the ancient Greeks originated the concept that the workings of the universe are comprehensible in terms of rational explanations. They suggested that natural processes, as opposed to supernatural processes, could explain the mechanisms we observe in our world. This concept lies at the heart of modern scientific inquiry.

By the fourth century B.C., Greek philosophers had conceived the idea that matter is made up of small, indivisible particles (atoms) held together by unknown forces; the concept that earthly materials are composed of varying amounts of the four elements *earth, air, fire,* and *water;* and the idea of a spherical Earth (as we learned in Box 2.2) surrounded by celestial bodies moving on circular paths. Although many of the early Greeks' concepts turned out to be not entirely correct, others were eventually vindicated.

# Step 8: The Roots of Modern Science

The modern scientific revolution started with Copernicus's revolutionary hypothesis that Earth and the other planets circle the sun (see Box 3.4). Galileo, however, is usually recognized as the father of the modern scientific method. He built and used telescopes (his first in 1609) as instruments of astronomical research, made and recorded his astronomical observations in a systematic manner, and used these observations to evaluate or test the competing geocentric and heliocentric models of the universe.

**PHOTO 3.41** By the close of the 19th century, the industrial revolution was in full swing in Europe and North America. This contemporary engraving shows steel being manufactured at Andrew Carnegie's Pittsburgh steel works in 1886.

Galileo's observations, which supported the *heliocentric* (sun-centered) conception, came into direct conflict with the authoritarian and powerful Roman Catholic Church, which had by then adopted as doctrine a *geocentric* (Earth-centered) model of all creation. Eventually, Galileo was forced by the Roman Inquisition to recant his views, probably more for political reasons than as a matter of disagreement on philosophical grounds. Galileo served a punishment of house arrest for the last ten years of his life.

Galileo's greatest contributions to science were in the field of physics, specifically in *mechanics,* the study of motion and force. These contributions paved the way for Isaac Newton's comprehensive theories, which for the first time, united all phenomena on Earth and in the heavens under a single set of laws.

After Newton's theoretical discoveries in the late 1600s, physics advanced rapidly in the fields of mechanics and *thermodynamics,* the study of heat energy. The science of chemistry evolved from earlier, unsuccessful attempts to change common materials into rare substances such as gold—a practice called *alchemy.* The chemists pursued the more practical, if less glamorous, ways of transforming matter and succeeded in producing a wide variety of useful substances. By the mid-1700s, widespread applications in physics and chemistry were kicking off the *industrial revolution* in Europe. That revolution is still spreading into the less developed countries of the world today.

The industrial revolution exploited what seemed to be practically limitless energy sources—initially coal and later oil and natural gas. The steam engine and other machines employed these new sources of energy for all kinds of practical tasks, including the manufacture of goods.

Progress in the fields of electricity and magnetism took place more slowly, until the development of electric batteries and generators in the first half of the 19th century. By the early 20th century, cities of the devel-

**PHOTO 3.42** The seeds of the information revolution were planted in the last century with the invention of the telegraph and telephone and with the establishment of the first communication networks, such as this early telephone exchange.

oped world were being electrified, a process that continues today in developing countries. At the same time, applications of chemistry, mechanics, and thermodynamics led to the dominance of the automobile in many parts of the industrialized world.

Today, nearly every facet of modern life is affected by past and present scientific and technological advances, and many of these advances are taking place at an accelerated rate. The developed nations have been thrust into an information age: an explosion of knowledge driven by powerful, new ways of gathering, storing, communicating, analyzing, and applying information. Television, computers, fax machines, and the like have radically altered the economy, culture, and politics of entire nations; they have profoundly affected, directly or indirectly, the lives of virtually everyone on Earth today.

Science, too, seems to be standing at a turning point. The costs of doing scientific research are rising, and taxpayers are increasingly unwilling to foot the bill for national efforts in basic science. There is more concern than ever that science is fragmenting into narrow, esoteric disciplines, each of which has developed its own worldview and speaks its own peculiar language, foreign to the public and other scientists alike. The information explosion has forced many specialists to devote a huge portion of their time to merely keeping up with their chosen specialty. Perhaps scientists should devote more time to synthesizing information from diverse fields. Perhaps humankind should guide its destiny by trying to understand the big picture, which is far more inclusive than the sum of all of its parts considered individually.

Scientists in some fields believe that their work is converging on answers to ultimate questions and that future advances will soon consist of mopping-up operations—devising new and clever ways to apply those answers. Thus, science for the sake of science should come to an end. Others cite the his-

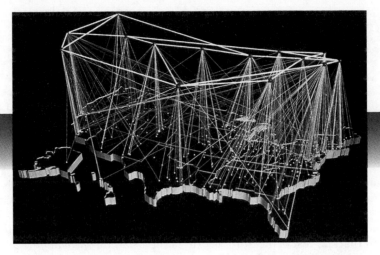

**PHOTO 3.43** This visualization of data traffic in a computer network serving scientists throughout the United States is color-coded to show frequency of use. Purple denotes low data traffic; white denotes high data traffic. The white lines at the top symbolize the backbone of the system that connects the main data centers. Each center serves many regional users. The rapid proliferation of computers and the spread of high-capacity data links, such as fiber-optic cables, to households, schools, and businesses in the 1990s is helping to fuel the current and forthcoming "information age."

torical record and point out that many answers to deep questions of the past have led to even deeper questions, and the answers found for those have led to even more profound questions.

Regardless of these philosophical issues, there is no denying the ability of science to solve, or at least ameliorate, the life-threatening crises that face us: the thinning of the ozone layer, the global climate change, the human population explosion, the decimation of species, and the spread of new diseases (such as AIDS) and old diseases (such as tuberculosis). In its most basic sense, science is knowledge and the desire to obtain knowledge. Without adequate knowledge of our complex and changing world, and without the resources and the will to apply that knowledge, there is little hope that our situation will improve.

What, then, lies in our future? Geologists can visualize a less active world, billions of years hence, when plate tectonics slows to a crawl and continents diminish in size. Astronomers can sketch scenarios of Earth suffering catastrophic meteorite hits in the short term (thousands or millions of years from now) or being rendered uninhabitable by steady, irreversible increases in the sun's luminosity over the long term (billions of years). Our immediate and serious concerns, however, center around the health of planet Earth, and the life that belongs to it, during the next century or so.

## BOX 3.4
# The Copernican Revolution

The ancient Greeks developed a model of the cosmos that featured a spherical, motionless earth at the center of everything else. This **geocentric model** (geo- means "earth"), reflected the common-sense view that we on Earth must be stationary because we do not feel that we are moving. This is one case where common sense fails us, for reasons we will learn in Chapter 5. Nonetheless, if we assume that Earth is indeed motionless, then we must come to the conclusion that the rest of the universe revolves around us. After all, the diurnal motion of all celestial bodies is an observational fact; and so are the peculiar, wandering motions of the sun, moon, and planets.

In the second century A.D., the Greek astronomer Ptolemy had refined the geocentric model to the point where it could be used to predict, with remarkable accuracy, future positions of the sun, the moon, and each of the known planets in the sky. Today, we still refer to his version of the geocentric model as the *Ptolemaic model.*

The Ptolemaic model pictured a universe that operates as an elegant clockwork, with perfectly spherical bodies moving on circular paths. Objects in the sky were thought of as being attached to crystalline (transparent), hollow spheres that turn steadily, with Earth at or near the center of each sphere. Stars and the Milky Way make up the outermost sphere, which is observed to rotate at the rate of once every 23 h, 56 min (as reckoned by our modern way of keeping mean solar time). The seven planets of antiquity, including the sun and the moon, were pictured as traveling on inner spheres, called *deferents,* each turning at a slightly different rate than that of the outer, starry sphere, so as to account for the fact that the planets "wander" with respect to the stars. The crystalline spheres of Ptolemy's geocentric model are graphically represented as circles in Figure 3.25 (top half).

Any geocentric model is necessarily complex because of the problem of *retrograde motion*—the observed reverse, or westward, motion that interrupts the usual *direct motion* (eastward drift) of the planets with respect to the fixed starry background. Retrograde motion in the planets occurs on a regular basis, roughly once every four months for Mercury, every one and a half years for Venus, every two years for Mars, and every one year plus a little for Jupiter and Saturn.

The geocentric model explains retrograde movements by means of *epicyclic motion.* Each planet moves on the rim of a small sphere or circle (called an *epicycle*), while the epicycle itself is attached to the larger crystalline sphere, or deferent. The combined motion of the epicycle turning at just the right rate and the epicycle itself moving around Earth gives rise to the periodic retrograde episodes we see from Earth.

Ptolemy's clever mechanism worked well in predicting where planets would be in the sky within his lifetime, but it was later shown to be inaccurate when applied to much longer time intervals. Ptolemy himself made no claim that his        *(continued)*

**FIGURE 3.25**  Ptolemy's geocentric model (top) features the sun and moon orbiting a stationary Earth on simple, circular paths. Each of the other five classical planets, or wanderers in the sky, revolves on the rim of a small, circular epicycle, which in turn moves along a large, circular orbit, called the deferent. The west-to-east movement of a planet's epicycle along its deferent is responsible for the planet's usual direct (west to east) observed motion relative to the background stars. At the same time, the motion of a planet swinging around the near side of its epicycle can surmount movement on the deferent and produce a temporary, east-to-west retrograde motion. An example of the combined motions along epicycle and deferent is given for the planet Mars. An observer on Earth, sighting along the thin lines shown, would see Mars temporarily reverse and then resume its direct motion as viewed against the background stars—the effect shown in the middle of the figure. The geocentric model also requires that the epicycles for Mercury and Venus always lie directly between Earth and the sun. This rule is consistent with the fact that we must always look in a direction close to the sun in the sky in order to see Mercury or Venus. (For simplicity in the geocentric diagram, the sphere of stars is shown as being stationary. Actually, the geocentric model requires that the sphere of stars and everything else in the heavens move relatively quickly east to west, to account for diurnal motion.)

Copernicus's heliocentric model (bottom) features the sun at the center of the known universe. The fixed, nonrevolving stars in the background and everything else that is seen in the earthly sky seem to move quickly east to west because of Earth's west-to-east rotation. The moon orbits Earth, but this is the only similarity between the two models. The six

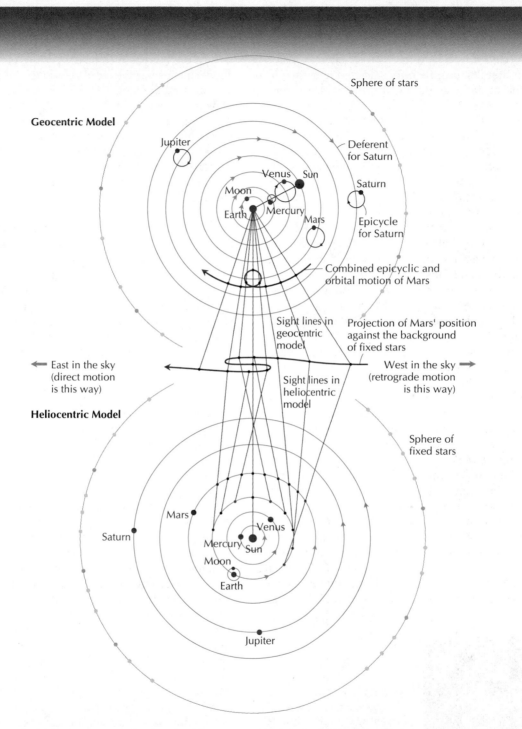

**Geocentric Model**

Sphere of stars

Jupiter

Venus   Sun

Moon

Mercury

Earth

Deferent for Saturn

Saturn

Epicycle for Saturn

Combined epicyclic and orbital motion of Mars

Mars

Sight lines in geocentric model

Projection of Mars' position against the background of fixed stars

← East in the sky (direct motion is this way)

West in the sky (retrograde motion is this way) →

**Heliocentric Model**

Sight lines in heliocentric model

Sphere of fixed stars

Mars

Saturn

Venus

Mercury   Sun

Moon

Earth

Jupiter

planets shown in the diagram revolve west to east around the sun and at speeds that decrease with increasing distance from the sun. As seen from Earth, all exhibit direct motion most of the time. Retrograde motion is explained as an illusion caused by Earth's circular movement around the sun. In the example given, again for the planet Mars, sight lines projected from Earth through Mars at various times show how

Mars can temporarily move retrograde with respect to the stars as seen from Earth. The retrograde motion occurs only when the faster-moving Earth is catching up to and passing the slower-moving Mars. The heliocentric diagram also shows why Mercury and Venus always appear near the sun: Both have orbits well inside Earth's orbit, and we must look more or less toward the sun in order to see them.

*(Box 3.4 continued)* model was necessarily correct in a literal sense; he only said that it was useful as a scheme for calculating future planetary positions in the sky.

After Ptolemy died, interest in astronomical investigation waned in most of the Western world, and the Ptolemaic model (virtually the last word on things astronomical) was accepted almost without question for the next 1400 years. Eventually, it was even worked into the religious dogma promulgated by those in authority in medieval Europe.

While scientific inquiry stalled in Europe, progress continued elsewhere. The Hindus invented our modern decimal number system and the concept of zero. Using these ideas, the Arabs developed trigonometry and furthered the techniques of algebra. The Arabs, who had access to many of the writings of the Greek astronomers, carried on important astronomical observations.

Interest in science was restored in Europe during the Renaissance, which was in full swing by the 16th century. Nicolaus Copernicus spearheaded the renewal of astronomy in 1542 with his hypothesis that the sun lies at the center of a system of planets, and that Earth turns on an axis and revolves around the sun. This **heliocentric model** (*helio-* means "sun"), or *Copernican model,* as it is often called, explains the observed motions of the planets as due partly to their actual motions and partly to illusory motion caused by the motion of Earth itself (see bottom half of Figure 3.25). One of these illusory motions is the relatively rapid east-to-west diurnal motion of all celestial bodies across the sky, a consequence of Earth's rotation. Another is the steady, eastward apparent motion of the sun along the ecliptic, this being the consequence of Earth's revolution around the sun.

**PHOTO 3.44** Astronomer Nicolaus Copernicus advanced the heliocentric (sun-centered) model of the Solar System in the 16th century. This model eventually replaced a much older and widely accepted model in which Earth occupied the center of the universe.

In the heliocentric model, all the planets (in the modern sense of the word) travel in the same direction around the sun. The farther out a planet is, the slower it moves and the longer it takes to circle the sun. Copernicus worked out the relative sizes of the orbits of the planets, placing them in correct order outward from the sun. Retrograde motion in the Copernican system was revealed as an illusion caused by the combined movements of Earth and the planet being observed. By taking into account Earth's cyclic movement around the sun in one year, Copernicus was able to calculate the time it takes for the other planets to circle the sun. Jupiter's period of revolution, for example, was calculated as the equivalent of about 12 Earth years.

In its original form, the Copernican model lacks the planets Uranus, Neptune, and Pluto, which were discovered after his death. It regards the stars as being points of light, of unknown physical character, at unknown but faraway distances. (No one had any real evidence, at that time, to suggest that stars could be like our own sun.) Thus, it seemed perfectly reasonable that the sun might be the center of the whole universe.

More importantly, Copernicus never completely abandoned the notion of perfectly circular orbits. He felt that small epicycles (not shown in the lower half of Figure 3.25), attached to the planetary orbits, would be necessary to explain certain observed irregularities in the motions of the planets. As we will see in Chapter 5, Johannes Kepler later explained these irregularities in terms of orbits that were elliptical in shape rather than circular. This explanation eliminated the need for epicycles.

Interestingly, Copernicus's model was not an entirely new idea. A minority of Greek philosophers had entertained the notion of a heliocentric cosmos some 1800 years earlier. But that notion was quickly rejected when they realized that none of the stars had any observable *parallax* (parallax was discussed in detail in Box 2.3). The philosophers thought that periodic changes in the positions of stars would have to occur if Earth really moved in a circle.

Stellar parallax does, in fact, exist, but it is so small that it was not detected until the year 1838. By then, telescope technology had advanced to the point where tiny, angular shifts of less than 1 arcsecond could be measured in the sky. The ancient Greeks, and also the astronomers of Copernicus's day, had no conception of the tremendous distances

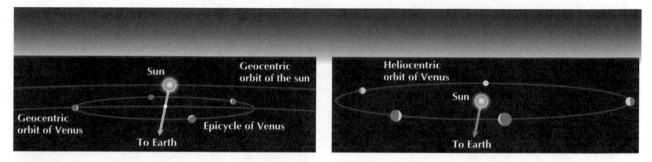

**FIGURE 3.26** In the geocentric model (left), the epicyclic path of Venus always lies between the sun and Earth. With the sun behind it all the time, Venus should invariably exhibit some kind of crescent phase. In the heliocentric model (right), Venus orbits the sun on a path smaller than Earth's orbit, which allows sunlight to fall on Venus at various angles and produces a full range of phases. Also, Venus should appear significantly larger at cresent phase, when it is close to Earth, and smaller at full phase when it lies at the far point of its orbit. After the invention of the telescope, astronomers could check which of these arrangements was correct. Galileo's early telescopic observations showed that Venus behaves exactly as the heliocentric model says it should.

that separate our Solar System from the other stars and, consequently, how small the effects of stellar parallax really are.

Despite the lack of any observable stellar parallax, the Copernican model gained widespread acceptance within a century. It was accepted partly because the philosophical elegance and simplicity of the model were appealing during the Renaissance, when many new and different ideas were being entertained.

More concrete reasons for accepting the Copernican model were provided by Galileo, more than half a century after the publication of Copernicus's hypothesis. Galileo turned his hand-built telescopes on the sky starting in 1609 and tried to interpret what he observed in terms of the competing geocentric and heliocentric models. In the movements of the four Galilean satellites around Jupiter, Galileo saw a reflection of the scheme Copernicus had proposed for the Solar System as a whole. Galileo's drawings of the satellites, sketched over many successive nights, showed that the inner satellites circle Jupiter faster than the outer satellites and that all four orbit Jupiter in the same plane. He also noted that Venus exhibits a complete range of phases (just as the moon does) at different times during its cycle. That set of observations proved that Venus circles the sun and not Earth (see Figure 3.26). These observations, and others of lesser significance, made by Galileo strongly supported the heliocentric model.

Today, there is no doubt which of the competing models is the correct one. Interplanetary spacecraft such as *Voyager 1* and *Voyager 2* have taken us on vicarious missions to most of the planets and have sent back pictures of our own spinning Earth in the context of the rest of the Solar System.

The conflict between the geocentric and heliocentric models is often cited as an example of the scientific method of inquiry (described in Chapter 1) at work. Obviously, the geocentric model has been relegated to the trash heap of discarded ideas. But the prevailing heliocentric model has hardly been frozen into the form envisioned by Copernicus or Kepler. With the help of Isaac Newton's laws of motion and gravity and Albert Einstein's relativity theory, the movements of the planets around the sun can now be modeled (mathematically and geometrically) with much greater accuracy than ever before.

Sometimes, even discarded models or theories can remain useful in some sense, even if they are incorrect as real-world modeling tools. For example, a vestige of the geocentric model survives today in the design of the planetarium projector—the device that projects spots of light mimicking the appearance of stars onto a planetarium's hemispherical dome. The globe used to project the star images turns on an axis, just like that of the star-studded, crystalline sphere imagined by the ancient Greeks. A number of separate mechanisms built into the projector are needed to control the images of the planets so as to reproduce their apparent, wandering movements in the sky over long periods of time. Each of these mechanisms features a wheel (or a gear) turning within a larger wheel.

**CHAPTER 3**

# Summary

Time is what a clock measures; it is related to cyclic events in nature. Time has an irreversible "arrow," or sense, that points from the past into the future. The arrow of time is related to the fact that entropy in any isolated part (or the whole) of the universe increases from the past to the future.

Space and time are united in a continuum called space-time. In our daily lives, we continually "travel" through time, whether or not we travel through space. Two or more events are unequivocally simultaneous only if they occur at the same point in space-time. The day is based on Earth's rotation. Sidereal time and mean solar time are based on the diurnal movements made by the stars and the sun across the sky, respectively. Standard time is the mean solar time kept by everyone within a broad zone of longitude. For convenience, Earth is divided into many time zones. The year is based on Earth's revolution around the sun. By a regular schedule of adjustments (leap years), our calendar keeps pace with the seasons. The seasons are caused by a tilt in Earth's axis of rotation, which causes more sunlight to fall on the Northern Hemisphere in June and less sunlight to fall on the same hemisphere in December. The ecliptic is the apparent path on the celestial sphere followed by the sun during one year's time. The moon and planets are confined to the zodiac, a band of sky centered on the ecliptic.

Equinoxes and solstices mark significant points on the sun's annual apparent journey along the ecliptic. They also mark the start of the astronomical seasons. The moon completes its cycle of phases in approximately one month's time.

When the sun, moon, and Earth (in that order) are arranged in a straight line, observers on part of the daylight side of Earth see a solar eclipse. When the sun, Earth, and moon are lined up, observers on the nighttime side of Earth see a lunar eclipse.

The big bang theory suggests that the universe started expanding from an extremely dense, hot state about 15 billion years ago. Galaxies of stars, including our own Milky Way galaxy, were probably forming within 1 billion years after the big bang.

The Solar System formed approximately 4.6 billion years ago, when part of an interstellar gas and dust cloud collapsed to make a central star and a surrounding disk of diffuse material. Earth and the other planets coalesced from many small and large chunks of material in that disk. The terrestrial planets, nearer the sun, accreted from the rare, heavier elements present in the disk. The Jovian planets, far from the sun, accumulated a large proportion of light elements in the disk, primarily hydrogen and helium.

Debris left over from the formation of the Solar System (asteroids and comets) has collided with Earth and the other planets frequently in the distant past, leaving impact craters. Impact events are much rarer today, since much of the debris has already been swept up.

From the principle of uniform change, the principle of superposition, correlation techniques, and other principles and methods of geology, Earth's past history can be deciphered through analysis of the current state of rocks exposed on its surface or sampled underground.

Oceans and continents existed on Earth as early as 4 billion years ago, and life probably arose in the oceans soon after. Life has increased in diversity and complexity up until the present time by the process of biological evolution. Evolution is possible because of genetic changes at the molecular level within cells and because of environmental changes.

Plant life began flourishing on land some 400 million years ago, and animals soon followed. Dinosaurs became the most successful land animals until 65 million years ago.

Earth's lithospheric plates have been interacting with each other for billions of years. Some 200 million years ago, tectonic processes began breaking up Pangaea, a landmass consisting of virtually all of today's continents. The continents continue to drift apart today.

A mass extinction, possibly caused by the impact of a large asteroid or comet, occurred on Earth some 65 million years ago. The extinction paved the way for the ascent of mammals, which formerly had been dominated by the dinosaurs. One evolutionary branch of the mammals eventually led toward present-day humans.

Over the past 2 million years, the global climate has fluctuated between cool, glacial (ice age) periods and warmer, interglacial periods. Human civilization started to advance rapidly near the end of the last glacial period, about 10,000 years ago.

The modern methods of science are descended from the musings of the ancient Greek philosophers and the observational and experimental techniques developed by Galileo during the Renaissance in Europe.

The heliocentric model of Copernicus placed the sun at the Solar System's center, in contrast to Ptolemy's earlier geocentric (Earth-centered) model. Galileo's telescopic observations supported the heliocentric model.

## CHAPTER 3
# Questions

## Multiple Choice

1. Time, according to Einstein, is relative in the sense that
   a) it flows only in one direction
   b) it seems to go by faster as we get older
   c) one observer may perceive two or more events as being simultaneous, while another observer may see the same events as being not simultaneous
   d) clocks are imperfect measuring devices

2. The day is based on one
   a) rotation of Earth
   b) revolution of Earth
   c) orbit of the moon around Earth
   d) orbit of the North Star around Earth

3. If you see the constellation Orion rising tonight at 8 P.M., you will see the same thing happen again
   a) at 8 P.M. 6 months from now
   b) at 8:08 P.M. 2 days from now
   c) at 7 P.M. 1 month from now
   d) 48 h of sidereal time later

4. How long does it take a star to cross the sky if it rises directly in the east and sets directly in the west?
   a) 6 h of sidereal time
   b) 12 h of sidereal time
   c) 6 h of normal (mean solar) time
   d) 18 h of normal (mean solar) time

5. If the year were exactly 367.2 days long, we would need to have a leap year every
   a) 2 years
   b) 4 years
   c) 5 years
   d) 1.95 years

6. The ecliptic is
   a) the plane of Earth's equator projected on the sky
   b) the straight line connecting the sun, Earth, and the moon
   c) the annual apparent path of the sun with reference to distant stars
   d) the plane of the moon's orbit

7. The equinoxes and solstices
   a) mark the midpoints of each of the four seasons
   b) come at intervals 3 months apart
   c) are different for different observers on Earth
   d) represent times when Earth is nearest to and farthest from the sun

8. The seasons are caused by
   a) the sun being closer to Earth in summer and farther away from it in winter
   b) the tilt of Earth's axis with respect to its orbit around the sun
   c) global air circulation patterns that shift frigid air from the poles toward the equator once a year
   d) the fact that Earth wobbles on its axis

9. Along the zodiac you would not expect to find
   a) the sun
   b) the moon
   c) the planets
   d) Polaris, the North Star

10. The moon's phases are caused by
    a) the changing distance between Earth and the moon
    b) the shadow of Earth falling on the moon
    c) sunlight reflected by Earth falling on the moon
    d) the changing angle of the sun's rays on the moon as viewed from Earth

11. The moon at full phase can never be seen on Earth
    a) at dawn
    b) at noon
    c) during evening twilight
    d) during lunar eclipse

12. Solar eclipses can occur only when the moon is at
    a) new phase
    b) first-quarter phase
    c) full phase
    d) last-quarter phase

13. In the universe as a whole, the general trend is toward
    a) expansion and heating
    b) expansion and cooling
    c) contraction and heating
    d) a steady state

14. The big bang occurred
    a) at the center of the universe
    b) at the edges of the universe
    c) everywhere at a single time
    d) everywhere, throughout all time

15. The cosmic background radiation is made up of
    a) matter that survived the wholesale annihilation of matter and antimatter during the first few seconds of the universe's existence
    b) photons emitted by the diffuse, expanding gas of the early universe
    c) photons emitted by the first generation of stars
    d) microwave static produced by radio and TV transmitters on Earth

16. Today, the universe expands because
    a) it is under the influence of a repelling force
    b) it is coasting outward under its own momentum
    c) things tend to get more disorganized with time
    d) matter continues to clump into galaxies

17. Dark matter is believed to be
    a) present only in the cores of stars
    b) spread over the entire volume of the universe
    c) accumulated on the surfaces of the planets as they circle the sun
    d) exerting a repulsive force that makes the universe expand

18. The Milky Way galaxy has a well-developed disk component because
    a) disks are the most stable large structures in the universe
    b) the globular clusters in our galaxy prevented stars from escaping
    c) the big bang caused the universe to expand preferentially in two dimensions, so that the whole universe is now shaped like a disk
    d) the protogalactic cloud our galaxy formed from was slowly spinning in some fashion

19. Globular clusters
    a) are among the oldest structures in our galaxy
    b) are among the youngest structures in our galaxy
    c) typically contain dozens or hundreds of stars
    d) stay within our galaxy's disk as they travel around its center

20. The age of the Solar System is approximately
    a) 20,000 years, a tiny fraction of the universe's age
    b) 570 million years, the same age as Earth
    c) 5 billion years, which is about a third the age of the universe
    d) 15 billion years, the same age as the universe

21. The most common elements present in the Solar System as a whole are
    a) hydrogen and helium
    b) oxygen and nitrogen
    c) hydrogen and oxygen
    d) iron and nickel

22. While the planets of our Solar System were accreting, they were
    a) cooling off
    b) becoming more uniform in composition
    c) developing the first forms of plant life, similar to mosses
    d) being heavily bombarded by planetesimals

23. There are few craters on Earth today (compared to what is on our nearby moon) because
    a) Earth has experienced few direct meteorite hits
    b) Earth's atmosphere has deflected almost all the errant comets and asteroids that have crossed our path
    c) Earth's gravitational pull is much less than that of the moon
    d) erosion and geologic (plate tectonic) activity on Earth's surface have erased them

24. Almost all the oxygen in today's atmosphere is a result of
    a) volcanic gases
    b) photosynthesis
    c) the decay of plant material
    d) meteorite impacts

25. Earth's biosphere is a suitable habitat for life mostly because
    a) it contains atmospheric oxygen
    b) it contains liquid water
    c) it is rich in iron and other metallic elements
    d) its atmosphere is rich in carbon dioxide

26. According to the principle of uniform change,
    a) plate tectonic activity on Earth is steadily increasing
    b) present changes on Earth are a key to past changes
    c) periods of glaciation alternate with interglacials
    d) life on Earth steadily evolves in complexity

27. According to the principle of superposition,
    a) older layers of rock generally overlie younger layers
    b) younger layers of rock generally overlie older layers
    c) sedimentary strata eventually are superposed, or squeezed together into a new form of rock
    d) various rock formations were precipitated out of Earth's original oceans such that old rocks lie inland and younger rocks lie near today's ocean shore

28. An unconformity in geologic strata
    a) results from the intrusion of igneous rock
    b) implies that the rock has been altered by heat and pressure
    c) could be a buried surface of erosion, indicating a gap in the geologic record
    d) means that sediments were laid down at a rate that was constantly changing

29. What significant event occurred about 65 million years ago at the Cretaceous-Tertiary (K-T) boundary?
    a) The larger dinosaurs were wiped out, leaving only smaller dinosaurs.
    b) Mammals originated from lesser forms of life.
    c) Plants gained a foothold on land.

d) Earth's biosphere experienced a major extinction event.

30. Life on land could not evolve into complex forms until
    a) anaerobic bacteria could thrive in the oceans
    b) Pangaea broke apart
    c) carbon dioxide enriched the atmosphere
    d) ozone in the stratosphere was able to block most of the ultraviolet rays in sunlight

31. During the Cenozoic era (after 65 million years ago),
    a) all reptiles became extinct
    b) mammals became the dominant form of animal life on land
    c) flowering plants originated
    d) marine creatures first developed hard parts (skeletons)

32. During the Ice Age of the Pleistocene period (about 2 million to 10,000 years before the present)
    a) ocean water levels were as much as 100 m below present sea level
    b) glaciers covered all of Earth's surface
    c) humans migrated from Australia to the Americas by way of Antarctica
    d) dinosaurs roamed the warmer, equatorial parts of Earth

33. Mathematics, science, and technology in ancient times developed in
    a) Europe and Asia only
    b) the Near East and Greece alone
    c) Egypt alone
    d) widely scattered places throughout the world

34. Galileo, whom many scientists regard as the originator of the modern scientific method, made important contributions in the fields of
    a) physics and astronomy
    b) alchemy and chemistry
    c) geology and meteorology
    d) chemistry and biology

35. Copernicus is known for his idea that
    a) Earth lies at the center of everything
    b) the moon circles Earth
    c) the sun lies at the center of a system of several planets, one of them being Earth
    d) the sun lies at the center of the Milky Way galaxy

# Questions

1. Refer to Figure 3.4. Can some stars in the sky be seen during *both* April and October?

2. Pictured in Photo 3.45 is phenomenon poetically called "the old moon in the new moon's arms." Why, during the moon's waxing crescent or waning crescent phase, can we see its dark hemisphere faintly illuminated? (*Hint*: Use Figure 3.12 to imagine what an observer standing on the moon's dark side would see in the sky overhead.)

**PHOTO 3.45** This long-exposure photograph of the crescent moon reveals a faint glow, sometimes called "earthshine," on the moon's nighttime side.

3. On a crescent moon, the moon's cusps (horns) always point away from wherever the sun is. Why?

4. Briefly review some of the contemporary ideas dealing with how primitive living organisms developed from nonliving molecules.

5. Why is it safe to look at a *totally* eclipsed sun but not at an eclipse in which the sun is, say, 99% covered by the moon?

6. What are some of the ways that a geologist could map a geological structure on and below the ground in a local area?

7. What is the purpose of correlation in geology?

8. How has the mechanism of evolution produced diversity and change during the history of life on planet Earth?

9. What supporting evidence for the heliocentric model did Galileo provide with the help of his observations through telescopes?

10. Compare and contrast the geocentric and heliocentric models of the cosmos.

# Problems

1. Box 3.1 states that when laser light is shortened to a pulse that lasts 3 fs, it is reduced to a length of only about 1 $\mu$m. Confirm this statement.

2. How old are you in "heartbeats"? Don't forget to take your resting pulse, which you can measure in beats per minute.

3. What time resolution (shutter speed or electronic flash duration) would you need in order to photograph a bullet 1 cm long moving at a speed of 500 m/s? Assume that the bullet appears as a blurred image 2 cm long on your photograph.

4. Find your latitude on a map, and then determine the maximum altitude (in degrees) of the sun at your location on the dates of summer solstice, winter solstice, and either equinox.

5. Relative to the sun, Earth turns 15° in longitude per hour of time. At the latitude of most of the United States, this corresponds to a distance of about 1200 km. If there is a town A, 240 km west of town B, and both towns are at the same latitude, approximately how much earlier or later does the sun rise at town A than at town B?

6. The modern way of doing science (the scientific method) was pioneered by Galileo nearly four centuries ago. Assuming that Galileo lived exactly 400 years ago and that the big bang occurred 15 billion years ago, how many seconds before 12 midnight did Galileo live as registered by a clock that could compress the entire history of the universe into 24 h?

# Questions for Thought

1. Would time exist if there were no recurring events or changes of any kind with which to measure it?

2. If you lived on Earth centuries ago, before the invention of mechanical clocks, how could you roughly keep track of relatively short intervals of passing time, such as seconds, minutes, or hours?

3. A growing human being becomes more and more "organized" from gestation to adulthood. Does the existence of life itself violate the principle that entropy tends to increase with time?

4. Why is it possible, in one's travel through space-time (as shown in Figure 3.2), to move horizontally across the graph but impossible to move only vertically on the graph?

5. Time obviously plays an important role in historical geology and astronomy. Does it also play an important role in physics and chemistry?

6. In China, everyone follows the standard time kept in Beijing, the capital, even though China sprawls across four theoretical time zones. What are some of the advantages and disadvantages of such a time system?

7. If you were assigned the task of inventing a calendar for all to use on Earth, how would you subdivide the year into smaller units? You need not call these units months, weeks, or days.

8. Many people think that Earth has seasons because the sun is closer in the summer and farther away in winter. The sun's distance does, in fact, change very slightly during the course of the year. Earth is closest to the sun in January and farthest from the sun in July. How do you think this changing distance affects the seasons in Earth's Northern Hemisphere? In Earth's Southern Hemisphere?

9. The tropics of Cancer and Capricorn on Earth are at latitudes of $23\frac{1}{2}°$ N and $23\frac{1}{2}°$ S, respectively. The Arctic and Antarctic circles on Earth are at latitudes of $66\frac{1}{2}°$ N and $66\frac{1}{2}°$ S, respectively. ($90°$ minus $23\frac{1}{2}°$ equals $66\frac{1}{2}°$.) Why do you think these particular latitude parallels on Earth are significant enough to have been given special names?

10. During a total solar eclipse, the round silhouette of the moon appears to be totally black, but is it really black? (*Hint*: Review Question 2 in the Questions section.)

11. How do you think it is possible for a rock or chunk of metal the size of a school building colliding with solid land to excavate a crater nearly 1 mi across (see Photo 3.20)?

12. What practical or cultural changes do you think will occur in our world if we discover (perhaps by the reception of radio signals from intelligent life elsewhere) that we are not alone in the universe?

13. Is the industrial revolution still going on today? Do you think there is an information revolution today that is supplanting the industrial revolution?

## Answers to Multiple-Choice Questions

| | | | | | |
|---|---|---|---|---|---|
| 1. c | 2. a | 3. d | 4. b | 5. c | 6. c |
| 7. b | 8. b | 9. d | 10. d | 11. b | 12. a |
| 13. b | 14. c | 15. b | 16. b | 17. b | 18. d |
| 19. a | 20. c | 21. a | 22. d | 23. d | 24. b |
| 25. b | 26. b | 27. b | 28. c | 29. d | 30. d |
| 31. b | 32. a | 33. d | 34. a | 35. c | |

# CHAPTER 4
# The Universe of Matter

*All ordinary matter is composed of atoms of simple substances, called elements, which number about 100. Most elements are metals, and nearly all are solid at ordinary temperatures. About 20 elements are nonmetals such as sulfur (the yellowish powder) and bromine (the brown liquid). Several nonmetallic elements are gases such as the nitrogen and the oxygen in the air we breathe.*

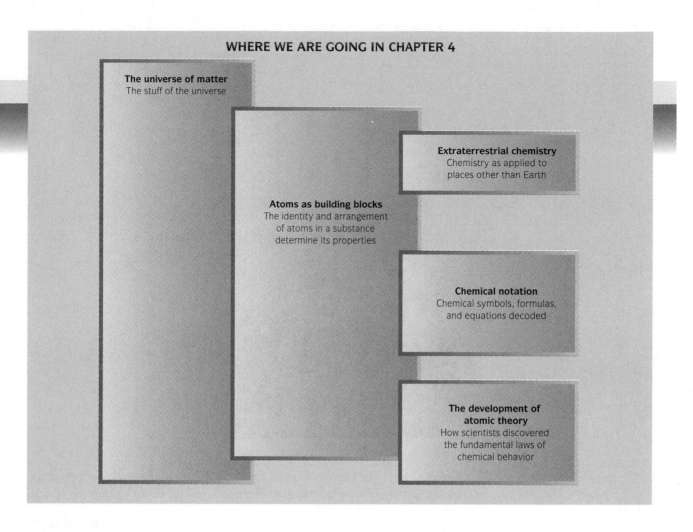

## WHERE WE ARE GOING IN CHAPTER 4

**The universe of matter**
The stuff of the universe

**Atoms as building blocks**
The identity and arrangement
of atoms in a substance
determine its properties

**Extraterrestrial chemistry**
Chemistry as applied to
places other than Earth

**Chemical notation**
Chemical symbols, formulas,
and equations decoded

**The development of
atomic theory**
How scientists discovered
the fundamental laws of
chemical behavior

Of the ingredients that make up our universe—time, space, matter, and energy—matter is perhaps the most concrete. We can see solid or liquid matter in bulk, touch it, and (especially in the case of food) often experience it through all our senses. Gases, such as air, constitute a more circumspect kind of matter, eluding some or all of the senses.

In some ways, though, the notion of matter is not so concrete, especially when we try to look at the universe on the subatomic scale. In the macroscopic world, we usually think of matter as being localized (there is a place where a pebble *is* and places where it *is not*). On the subatomic scale, the concept of precisely localized matter is invalid. Tiny particles of matter such as electrons can never be pinned down to a specific place, and yet somehow they manage to occupy space.

To work our way out of this conundrum and restore clarity to the concept of matter, we will merely state an operational definition of matter: *Matter has mass and occupies space (or volume)*. Both mass and volume are easily measurable in the macroscopic world. This definition will suffice for

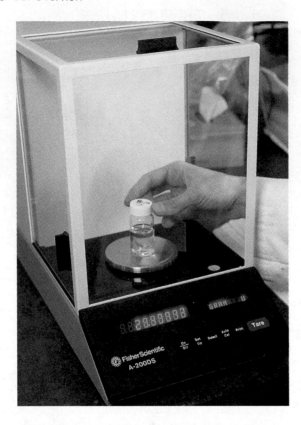

**PHOTO 4.1** This modern digital electronic balance can measure masses to a precision of 0.0001 g.

the purposes of this chapter, which is concerned primarily with *chemistry*, a science that deals with transformations of matter and relies heavily on measurements of mass and volume.

Much of modern chemistry was advanced by careful measurements of the masses of substances undergoing chemical reactions. Measurements of a body's mass can be made either by weighing the body or by testing for its *inertia*, or the tendency for the body to resist any change in its motion. In chemistry, weight is usually the easier of the two to measure, so chemists make frequent use of precision scales or *analytical balances* (see Photo 4.1) in the laboratory. Although an analytical balance directly responds to the weight (downward force on a mass owing to gravity) of the load being measured, it can easily be calibrated to read in mass units such as grams or kilograms.

Here is a summary of our general approach to chemistry in this chapter: Atoms are the building blocks out of which all bulk matter is made. Atoms of a single kind or atoms of different kinds combined are responsible for the great diversity of the matter we observe in the universe. When bonds between atoms are made or broken and atoms rearrange themselves in some way, there is a chemical reaction going on.

Later in this book, we will explore the importance of energy in chemical reactions (Chapter 6) and the nature of the chemical bonds that hold atoms together (Chapter 9). In Chapter 15, we will survey many of the chemical elements and discover how they can be organized by means of the periodic table of the elements.

## Atoms as Building Blocks

Like all scientists in their respective fields, chemists have succeeded in extracting order from a body of knowledge that, initially, seemed chaotic. The first step in this process is, of course, to classify things. Centuries ago, hundreds of different kinds of matter (called *pure substances*, but often known as simply *substances*) were known, each with a unique set of properties. Today, the number of known substances exceeds 9 million, and hundreds of thousands more are being discovered or synthesized every year. The task of organizing all this matter is made easier by recognizing that there is a common denominator associated with it all: **atoms.**

In explaining the macroscopic behavior of a substance, chemists must know not only what kinds of atoms make up a substance but also how those atoms relate to one another. Do the atoms of a substance move freely and independently, or are their motions constrained in some way? Are atoms of a substance "lone wolves," or do they bond together in pairs, in trios, or in large clusters? Are the members of a cluster of atoms identical or different?

Fortunately for chemists, atoms are limited in kind. The various kinds of atoms are the tiny building blocks that make up macroscopic quantities of matter. Although much of the science of chemistry is concerned with macroscopic behavior, its more fundamental task is to discover everything possible about the characteristics of different kinds of atoms, the ways in which atoms combine to form more complex structures, and the ways in which atoms can be rearranged into different structures.

Since each substance has several properties associated with it, and since atoms can combine in several ways to form matter, there are many useful ways to classify matter. Some schemes are keyed to a single property; for example, any substance can be classed as a conductor, a semiconductor, or a nonconductor (or insulator) of electricity. Other very useful classification schemes subdivide matter according to broader sets of behavior. Here, we will examine two broad classifications:

1. the division of matter into four principal physical states or *phases:* solid, liquid, gas, and plasma;
2. the organization of matter into elements, compounds, and mixtures.

Later, in Chapter 15, we will explore the grandest, most complex, and most meaningful classification scheme yet devised for matter: the periodic table.

## Phases of Matter

Matter in the universe comes in many forms, including the four common states or phases known as *solid, liquid, gas,* and *plasma.* Matter in the solid, liquid, and gaseous phases dominates Earth and the other planets. When any kind of matter is heated to a sufficiently high temperature, it becomes plasma. Plasmas prevail in extremely high-temperature environments, such as inside the sun and the stars. The common phases of matter can be distinguished from each other in several ways; some of them are illustrated in Figure 4.1.

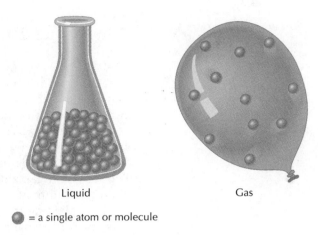

Solid                          Liquid                          Gas

● = a single atom or molecule

**FIGURE 4.1** Atoms or molecules are closely packed in solids and liquids. In gases (and also plasmas), the rapidly moving atoms or molecules are farther apart.

*Solid matter* is rigid and maintains a definite shape. It strongly resists being compressed, and to a lesser extent, it resists being pulled apart. Solids expand and contract only slightly with temperature changes.

The other three common phases of matter are classed as *fluids*, which means that they can easily change shape and flow.

*Liquid matter,* like solid matter, resists compression and expands or contracts only slightly with alterations in temperature. Liquids, however, have little resistance to being pulled apart, which means that they are virtually formless except when occupying a container. On Earth, gravity pulls on a confined liquid so that its top surface becomes flat and horizontal while the rest conforms to the container's shape. Most substances are less dense in their liquid state than they are in a solid state.

*Gaseous matter* has indefinite volume and an infinitely variable and very broad range of density. A gas expands so as to uniformly fill any container it is introduced into. Unlike solids and liquids, gases are readily compressible. Changes in the volume of a gas are accompanied by changes in temperature or pressure or both.

*Plasma* (which is not the same as the plasma component of blood) is a hot, ionized form of gas that has certain electromagnetic properties not possessed by ordinary gases. Plasma is relatively unimportant in the field of chemistry.

Since we already know from Chapter 2 that matter consists of tiny particles (either atoms or *molecules,* which are clusters of atoms), let us see how these particles are arranged for matter in various phases. A good analogy is that of students in a classroom. During a lecture, the students sit in closely spaced seats; they may fidget a bit, but generally they do not move relative to each other. Similarly, the particles of a solid can be thought of as occupying fixed, closely spaced positions. The particles themselves may vibrate in place, but overall they remain at rest relative to each other.

When the classroom lecture is over, the students head for the exits and flow out the doors. They remain close to each other, perhaps jockeying for position to save time in exiting. Similarly, the particles of a liquid lie close to each other, but they are not necessarily stuck together.

Finally, after leaving the classroom and going outside, the students disperse across the campus in a somewhat random fashion. Similarly, matter in a gaseous state consists of comparatively widely spaced atoms or mole-

**PHOTO 4.2** In the laboratory (and sometimes in nature), water can exist in solid, liquid, and gaseous phases simultaneously.

cules moving freely and randomly. There is no limit to how far these particles can travel, so a gas expands indefinitely unless it is confined. When confined to a fixed volume, a gas completely fills its container; and the many gas particles colliding with the inside container walls produce an outward pressure (gas pressure). In Figure 4.1 the balloon containing gas particles remains inflated because of this pressure.

# A Chemical Division of Matter

In a common classification scheme used by chemists (Figure 4.2), matter is first divided into pure substances and mixtures. A **pure substance** is made up of individual particles or arrays of particles with a definite (not variable) composition. All pure substances have some definite physical properties, such as density, hardness, melting point, boiling point, thermal conductivity, and electrical conductivity. Pure substances are further divided into

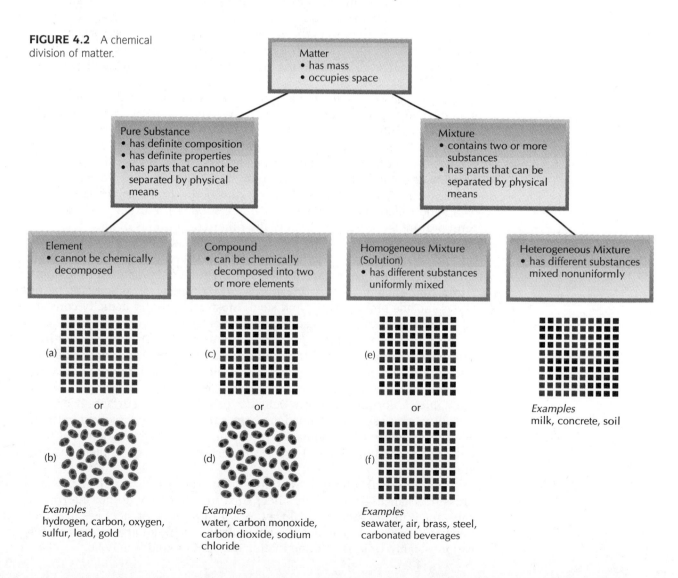

**FIGURE 4.2** A chemical division of matter.

*elements* and *compounds*, as we shall further explain after the following discussion of mixtures.

**Mixtures** consist of two or more pure substances that can be separated by physical means. Salt and pepper mixed together constitute a *heterogeneous mixture;* if you had the patience, you could physically separate its parts with a microscope and tweezers. The same thing can be said for beach sand, dirt, or concrete, all of which consist of clearly differentiated parts. In contrast, a *homogeneous mixture,* which is the same thing as a **solution,** consists of components so uniformly mixed that you could not separate them mechanically, as with tweezers. Other physical means, however, could be tried. They might include heating, cooling, or changes in pressure in order to produce phase changes (freezing, melting, evaporation, or condensation) that would separate the solution's components.

Seawater is a good example of a *liquid solution* whose components can be separated by evaporation. Seawater consists of pure water plus a variety of dissolved salts, primarily sodium chloride (table salt). In commercial salt evaporation operations such as those along the south shores of San Francisco Bay (see Photo 4.3), brine (salty water) is isolated in shallow evaporation ponds, where most of the water is driven away by the sun's heat. The concentrated brine is then moved through a series of other ponds where different kinds of salt crystallize out in turn as more and more water is removed by evaporation.

Air is a *gaseous solution*—a homogeneous mixture of the gases nitrogen, oxygen, argon, water vapor, and more. Air can be separated into components by chilling it. Each gas will condense out of the remaining air in turn as the temperature is reduced. Water vapor is the first to condense. In nature, cold air tends to be dry because most of its water vapor has already condensed out as liquid water (in clouds, rain, or dew) or as ice crystals (in clouds or frost). Small traces of carbon dioxide in the air will condense out as a solid called dry ice at about $-78\,°C$ ($-108\,°F$). Further reductions of temperature are needed to liquefy the other constituents of air.

Alloys are *solid solutions* of two or more metals (or sometimes metals and nonmetals) mixed together as liquids and then cooled. Alloys can be separated by physical means by subjecting the melted material to temperatures high enough to evaporate at least one component. Some examples of alloys include bronze, brass, and steel. Most gold jewelry actually consists of alloys of gold, silver, and sometimes other metals.

Soft drinks are sugar solutions with a "twist." During the bottling process, carbon dioxide gas is forced into (dissolved in) the sugar solution under great pressure. When you remove the bottle cap, thereby reducing the pressure, the carbon dioxide separates from the liquid components of the solution, forming tiny gas bubbles that rise to the surface. Beer and sparkling wines such as champagne have carbon dioxide dissolved in them as a consequence of fermentation under pressure. Fermentation itself is a naturally occurring chemical process that converts sugars in plant material to alcohol.

The **physical changes** we have mentioned—freezing, boiling, and pressure changes—profoundly affect mixtures, but they do not alter pure substances in any fundamental way. Water (a pure substance) is still water down to the very last molecule, whether it is ice, liquid water, or water vapor (steam). Gold is gold, no matter how hot or cold it may be.

**PHOTO 4.3**   As water evaporates from the brine contained in these ponds along the shore of San Francisco Bay, various salts crystallize out on the bottom.

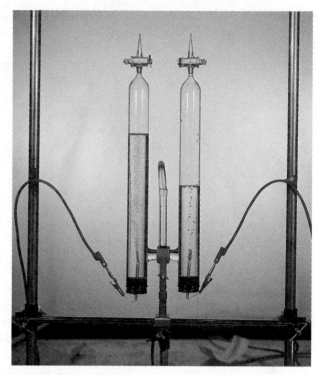

**PHOTO 4.4**   As water is electrolyzed, hydrogen gas builds up in the tube on the right and oxygen gas builds up in the tube on the left.

The pure substances themselves (refer again to Figure 4.2) can be divided into two groups: (1) **compounds,** those that can be chemically decomposed, and (2) **elements,** those that cannot. Chemical decomposition includes such processes as burning, fermentation, food spoilage, corrosion, and electrolysis.

Chemical decomposition often results in the transformation of a compound into two or more new, pure substances. Each of the new substances has physical properties that make them distinct from each other and distinct from the compound they were derived from. For instance, let us consider the decomposition of water, which can be accomplished by adding electric energy in a process called *electrolysis* (Photo 4.4). The products of this reaction are oxygen and hydrogen, which are different from each other and from water itself. Both products in this particular reaction are elements, because they cannot be decomposed any further by chemical means. The reverse reaction can also take place: Hydrogen and oxygen can combine to form water. Processes like this, which involve changes in the identity (the set of properties) of matter, are called **chemical changes** or **chemical reactions.**

Another example of chemical change occurs when the paraffin wax of a burning candle seemingly disappears. Careful measurements show that oxygen (in the air) is consumed as well and that water vapor and carbon dioxide are given off. All the gases involved in this particular reaction are invisible (colorless and transparent); yet it is possible to trap them as the wax is burning and measure their masses. Measurements show that whenever a total of, say, 1 g of paraffin and oxygen reacts within a candle flame, then a total of 1 g of water vapor and carbon dioxide (and sometimes other products, such as soot) is produced. Clearly, there has been no disappearance of anything, just a transformation of matter—a chemical change. In this particular reaction, the products—water vapor and carbon dioxide—are *not* elements, but rather, compounds. Both can be decomposed into elements: Carbon dioxide when broken down yields carbon and oxygen, and water when broken down yields hydrogen and oxygen.

As we learned in Chapter 2 when exploring the size realm of atoms, elements are substances consisting exclusively of atoms having the same number of protons. *The number of protons in the nucleus of an atom determines the electron structure in the outer part of the atom, which in turn controls the chemical behavior of that atom.* In addition, *isotopes* of a given element can exist; isotopes are atoms with the same number of protons but different numbers of neutrons in their nuclei. Generally, different isotopes of the same element have little to do with that element's chemical behavior. About 90 naturally occurring elements have been detected on Earth. Some 20 other elements, all unstable, have been artificially synthesized or detected during radioactive decay processes.

At room temperature and sea level pressure, most elements are solid; and their individual atoms are arranged in some kind of regular geometric (crystalline) pattern, as suggested in Figure 4.2(a). Only two elements, mercury and bromine, are liquid at the room temperature of about 20 °C (68 °F). Their atoms lie close together but as a whole possess little or no geometric structure.

Eleven elements are gaseous at room temperature. Some elemental gases, like helium and neon, are *monatomic,* which means that their atoms are independent of one another. Others, such as hydrogen, oxygen, and nitrogen, are *diatomic;* which means that their atoms tend to join into two-atom molecules, as illustrated in Figure 4.2(b). The element oxygen is typically diatomic in the atmosphere, but it can also exist in a rare triatomic (three-atom) form called *ozone.* Some elements can exist as hollow structures of many atoms each. The newly discovered carbon fullerene molecules discussed in Chapter 9 include hollow, spherical cages of 60 carbon atoms and other similar configurations.

Since compounds can be decomposed by chemical means, they always consist of two or more elements. Like the elements, compounds can assume different phases and forms. They can exist in crystalline forms as lattice works of different kinds of atoms, arranged in a regular and periodic fashion, as suggested in Figure 4.2(c); or they can be collections of independent molecules, as in Figure 4.2(d). Water in its ice form is an example of the former, and carbon monoxide gas is an example of the latter. Solid sodium chloride has a crystalline structure similar to the one shown in Figure 4.2(c), with alternating sodium ions and chlorine ions locked in place. The particles are ions rather than atoms, because of the particular nature of the bonds existing between them (a subject we will explore further in Chapter 9).

Keep in mind that it is possible to have a given quantity of any pure substance (element or compound) in more than one phase at the same time. A glass of ice water, for example, contains two different arrangements of water molecules (crystalline and liquid); yet all parts of the ice-and-water mix are the same compound.

Confusion may arise when we compare compounds and solutions. Both are homogeneous on the molecular scale of size. How can we tell the difference? This question can be resolved by way of two experimental tests:

1. Subject a sample of the material in question to temperature and/or pressure changes and see what happens. Such physical changes can separate the constituents of a solution (as we have already discussed in connection with seawater, air, and alloys) but not of a compound. A compound, with definite physical properties, will not separate into two or more new substances as it melts, boils, condenses, or freezes.

2. Separate (by any method available) the constituents of many samples of a given compound or solution. If the ratios of the masses of those constituents are always constant, then the material is a compound. If the ratios change from sample to sample, then the material must be a solution. Saltwater, for example, is a solution because the ratio between the sodium chloride and water is infinitely variable: You can add any number of tiny table salt grains (up to a certain limit) to a glass of water to form solutions of various concentrations. This situation is suggested in Figures 4.2(e) and (f); both diagrams show two different kinds of intimately mixed particles, but the two kinds of particles need not have a definite ratio between their number (or mass). On the other hand, when various samples of water are decomposed, the amount of oxygen produced *always* has eight times the mass of the amount of hydrogen produced, and the ratio of hydrogen to oxygen atoms produced is *always* two to one. Water, therefore, is a compound.

# Chemical Notation

Chemistry has its own set of symbols and notations. **Chemical symbols** are one- or two-letter abbreviations for each of the more than 100 known elements. Listed in Table 4.1 are the names and symbols of 8 elements having the simplest structures, starting with hydrogen, the simplest possible. In the pages and chapters ahead you will gradually be introduced to more

**TABLE 4.1**
**The Eight Simplest Elements**

| Atomic Number $Z$ (number of protons) | Mass Number $A$* (number of protons plus neutrons) | Chemical Symbol | Name of Atom/Element |
|:---:|:---:|:---:|:---:|
| 1 | 1 | H | Hydrogen |
| 2 | 4 | He | Helium |
| 3 | 7 | Li | Lithium |
| 4 | 9 | Be | Beryllium |
| 5 | 11 | B | Boron |
| 6 | 12 | C | Carbon |
| 7 | 14 | N | Nitrogen |
| 8 | 16 | O | Oxygen |

*The mass number given refers to the most common isotope of each element.

of the elements and their symbols. In Chapter 15, all known elements will be presented and organized in the form of the *periodic table of the elements.*

Sometimes, when chemists write a chemical symbol, they must add additional information about the nuclear composition of a particular atom, or that of every atom of a pure isotope of an element. In such cases, the following notation is often employed:

$$^{A}_{Z}X$$

where X is the chemical symbol of any element, $Z$ is the **atomic number** (number of protons), and $A$ is the **mass number** (total number of nuclear particles). Note that the number of neutrons is $A - Z$, since there are no other nuclear particles besides protons and neutrons. Let us look at some examples of this notation.

$^{1}_{1}H$      This is ordinary hydrogen, with one particle (a proton) in its nucleus.

$^{2}_{1}H$      This is deuterium (heavy hydrogen), which has one proton and one neutron—two nuclear particles in all.

$^{12}_{6}C$      This is the most common isotope of carbon (carbon-12).

$^{14}_{6}C$      This is a rare, heavier, radioactive isotope of carbon (carbon-14).

$^{238}_{92}U$      This is a uranium isotope with 92 protons, and 146 neutrons, for a total of 238 particles in the nucleus. It is also called uranium-238 (U-238).

**Chemical formulas** are useful in describing the way atoms bond to form molecules and compounds. Here are several examples.

$O_2$      This is diatomic oxygen, the common form of oxygen that makes up about 1/5 of the mass of Earth's atmosphere.

$O_3$      This is ozone, a pollutant at ground level but an important trace gas present in the upper atmosphere.

$H_2O$      This is water, with molecules consisting of two atoms of hydrogen and one atom of oxygen.

$H_2O_2$      This is hydrogen peroxide, a compound made of molecules having a total of four atoms.

$CO$      This is carbon monoxide, a poisonous gas with molecules of one carbon atom paired with one oxygen atom.

$CO_2$      This is carbon dioxide, having two O atoms for every C atom; it is an important trace gas in the lower atmosphere.

$NH_3$      This is ammonia, a cluster of three hydrogen atoms and a single nitrogen atom; it is a gas at room temperature unless dissolved in water to form an ammonia solution.

$CH_4$      This is methane (natural gas), the simplest of a class of compounds known as hydrocarbons.

$NaCl$      This is sodium chloride (table salt), a crystalline compound consisting of a lattice of vast numbers of sodium (Na) and chlorine (Cl) ions. In this case the formula expresses the 1:1 ratio of Na and Cl.

Note from these examples that a subscript numeral (for instance, the 2 in $O_2$) always refers to the chemical symbol that precedes it. By common custom, the subscript numeral 1 is not written. Sometimes, parentheses are used to set one group of atoms apart from another atom or group of atoms. For example, the compound calcium nitrate is written as $Ca(NO_3)_2$. In this compound, two nitrate ($NO_3$) "groups" are linked to one calcium (Ca) atom.

**Chemical equations** describe transformations of matter by tracking the rearrangement of atoms during a chemical reaction. For example,

$$H_2O \rightarrow H_2 + O_2$$

This formula describes the decomposition of water into diatomic hydrogen gas and diatomic oxygen gas. $H_2O$ is the *reactant* and $H_2$ and $O_2$ are the *products*.

The formula as written may be adequate qualitatively (descriptively) but is not accurate in a quantitative sense, because it does not correctly describe the relative numbers of atoms that participate in the reaction. We would say that this chemical equation is *unbalanced*. Balancing it is quite easy, requiring just two steps. We need at least two O atoms on the left side, because we already have two O atoms on the right, so our formula becomes

$$2H_2O \rightarrow H_2 + O_2$$

The number 2 we wrote in front of the reactant in the formula, called a *coefficient*, multiplies the entire $H_2O$ unit. (If no coefficient appears in front of a reactant or product, then the coefficient is automatically 1.) We now have two water molecules, containing a total of four H's and two O's, on the reactant side; and we have two H's and two O's on the product side. We need twice as many H's on the product side to balance the equation. The second step yields

$$2H_2O \rightarrow 2H_2 + O_2 \qquad \text{(balanced chemical equation)}$$

Please note that chemical equations are not equivalent to algebraic (mathematical) equations. The arrow symbol is not the same as an equal sign; rather, it is a symbol of transformation. In chemical equations you *can* unilaterally multiply any of the reactants or products by a coefficient in an effort to balance it. In an algebraic equation, you cannot do this. All chemical equations presented in the text of this book will be written in their balanced form.

Here is another example of a chemical reaction; this one is a description of what goes on when you ignite a gas burner in your kitchen. The burning of methane, which is the principal component of household natural gas, can be expressed as

$$CH_4 + 2O_2 \rightarrow CO_2 + 2H_2O$$

The formula says that methane is oxidized (combines with oxygen), yielding carbon dioxide and water vapor. Quantitatively, each molecule of methane combines with two diatomic molecules of oxygen, and the result is one molecule of carbon dioxide and two molecules of water.

Sometimes, chemical formulas include additional notations such as a parenthetical (s) or (l) or (g) after each reactant and product, denoting solid, liquid, or gas. (All reactants and products in the equation describing methane burning are gaseous just before and after the reaction.)

All chemical reactions are reversible, and many take place simultaneously in both directions. In pure water, the reactions

$$H_2O \rightarrow H^+ + OH^- \qquad \text{and} \qquad H^+ + OH^- \rightarrow H_2O$$

take place simultaneously, although the second reaction is greatly favored. The very rare $H^+$ and $OH^-$ products of the first reaction are really ions, the "dissociated" fragments of a water molecule. The $OH^-$ ion (hydroxide ion) carries an extra electron, so it has a charge of $-1$. The $H^+$ ion (hydrogen ion) lacks an electron, so its charge is $+1$. (Ion charges of $+1$ and $-1$ are

## BOX 4.1
# Extraterrestrial Chemistry

It is a central axiom of physical science that the fundamental laws of nature, including those of chemistry, are universal. That is, they are not dependent on either place or time. As we look out into space (and also back into time), we find all sorts of patterns of regularity that fit perfectly with this idea.

Despite the sameness of the *laws* that govern chemistry throughout the universe, the results of "extraterrestrial" chemistry (chemistry taking place outside Earth) are on the whole quite different from the results of "terrestrial" chemistry (chemistry on Earth). There are at least two reasons for this difference.

1. The universe at large consists primarily of the elements hydrogen and helium, along with about 100 other elements present in varying small or trace amounts. Helium itself is chemically inert; so hydrogen is the keystone element in extraterrestrial chemistry. Earth, on the other hand, is largely composed of elements heavier than hydrogen and helium. Hydrogen is but one of several elements that play important roles in terrestrial chemistry.

2. The extreme physical conditions of most extraterrestrial environments discourage chemical bonding. High temperatures tear apart molecules into their constituent atoms, and low-temperature–low-pressure conditions may make it unlikely for atoms to find each other and bond in the first place. On or near the surface of our planet, the prevailing conditions of temperature and pressure promote an astounding variety of chemical reactions, both biological and nonbiological. These just-right conditions have allowed liquid water—and life as we know it—to exist.

Compared with terrestrial chemistry, most of the extraterrestrial chemistry we know about is comparatively simple and/or slow-paced. The surfaces of many stars, the sun included, are just cool enough ($\approx 3000$ to $6000$ K) for some atoms to bond in the form of simple molecules. Spectral analysis of light from the sun's surface reveals that a small but significant amount of the compound titanium oxide (TiO) exists there. At the higher temperatures characteristic of the interior of the sun and the other stars, no atoms can bond and no chemical reactions can take place. Deep down in the cores of stars, where temperatures exceed $10^7$ K, nuclear reactions, not chemical reactions, occur.

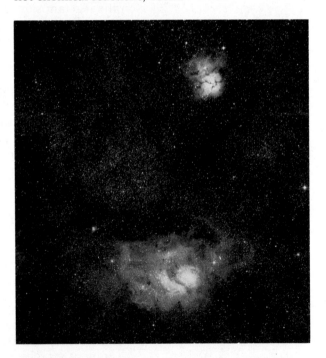

**PHOTO 4.5**   The Lagoon (red) and Trifid (blue) nebulas lie in a gas- and dust-rich region about 3500 LY closer to the center of our galaxy than our Solar System does. The hydrogen gas of the Lagoon nebula glows by virtue of many hot, luminous stars embedded in the nebula. The cooler, dust-rich Trifid nebula reflects the blue light of adjacent stars. The entire region surrounding both nebulas is relatively rich in interstellar molecules.

often represented simply by superscript plus and minus symbols.) We could also write the two equations as

$$H_2O \rightleftharpoons H^+ + OH^-$$

where the double arrow means that both reactions take place simultaneously.

Interstellar space is strewn with cold, wispy material consisting of hydrogen and helium gas and minor amounts of dust. The disk of our galaxy harbors many giant molecular clouds, each having 100,000 or more solar masses of gas and dust thinly spread over a volume many tens of light-years wide (see Photo 4.5). Matter in these clouds is far more dense than matter elsewhere in interstellar space but far less dense than the best vacuum we have achieved in Earth laboratories. Atoms collide infrequently in interstellar clouds; but when they do, they have a reasonably good chance of sticking together and forming stable molecules. So far, astronomers have discovered about 100 fairly simple kinds of interstellar molecules, including several already mentioned in this chapter: $H_2$, $H_2O$, CO, $NH_3$, and $CH_4$. One interesting molecule found in trace amounts in the giant molecular clouds is ethyl alcohol, $C_2H_5OH$. In total, there is enough alcohol in some clouds to make about $10^{28}$ bottles of hard liquor! These and other molecules have slowly built up over millions of years through collisions among atoms and molecules.

With the help of telescopes, flyby spacecraft missions, and direct sampling by robotic spacecraft, we know a fair amount about the surface chemistry of the planets and the debris of our solar system. In 1976 the United States landed two robotic spacecraft, *Viking 1* and *Viking 2,* on Mars (see Photo 4.6). Among other tasks, the *Vikings* searched for subtle chemical signatures that might have been left behind in the Martian soils by microorganisms. No signs of life were found, but the Vikings' experiments revealed exotic forms of nonbiological chemical activity. Solar ultraviolet light shining on the surface of Mars breaks down molecules made of carbon, sterilizes the soil, and produces compounds that are highly reactive in the presence of water and other substances.

A small armada of spacecraft sailed by Comet Halley during its most recent visit to the inner Solar System in 1986. The volatile (easily evaporated) sub-

**PHOTO 4.6**   Here, the robotic arm of the *Viking* spacecraft is seen digging a trench. On board the spacecraft, three separate experiments tested for the presence of past or present biochemical activity in the Martian soil samples.

stances that make up the comet were found to be largely water and various compounds of hydrogen, oxygen, carbon, nitrogen, and sulfur. Some of the compounds detected were *hydrocarbons,* the stuff of fuels such as natural gas and gasoline.

# The Historical Development of Atomic Theory

John Dalton (Photo 4.7), an English schoolmaster and experimental chemist, was the first person to posit a comprehensive theory of how atoms function as the building blocks of matter. In a series of papers published during the period 1803–1810, Dalton borrowed from earlier concepts about the atom and contributed his own ideas as well. We will now state the postulates of Dalton's atomic theory, comment on their origin and rationale, and discuss their validity in light of what we now know about the structure and behavior of atoms.

**PHOTO 4.7** John Dalton (1766–1844), a champion of the atomic theory of matter.

**Elements are composed of tiny, indivisible particles (atoms).** At the time of the ancient Greeks, two models of the nature of matter were recognized. One model considered matter to be continuous and subject to being divided into arbitrarily smaller and smaller pieces. The other model held that matter is composed of discrete particles of invisibly small, but finite, size.

The influential Greek philosopher Aristotle held the majority view that matter has a continuous nature and that it consists of four elements (*earth, air, fire,* and *water*), which possess various admixtures of four fundamental qualities (*hot, cold, moist,* and *dry*). Thus, in theory, any substance could be transformed into any other by tinkering with its qualities. For example, wet clay (moist earth) is transformed into pottery (dry earth) because the moist quality is replaced by dry when the clay is fired. These erroneous ideas, which prevailed for many centuries, led the *alchemists* (the predecessors of chemists) of the Middle Ages on fruitless quests to transform common metals into valuable ones such as gold.

The minority view, advanced by Democritus and several other ancient Greek philosophers, held that matter consists of a finite number of invisibly small, discrete particles that can combine with each other in various ways to produce a wide range of substances. Democritus coined the term *atom,* which means "indivisible."

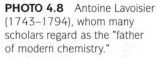

**PHOTO 4.8** Antoine Lavoisier (1743–1794), whom many scholars regard as the "father of modern chemistry."

In 1661, the English experimentalist Robert Boyle clarified the notion of an element. Elements, he proposed, are substances that cannot be decomposed into simpler substances. A century later, experiments by the French chemist Antoine Lavoisier (Photo 4.8) demonstrated that nonelements (compounds) can be broken down into elements and also that elements can combine chemically to form compounds.

Today, we know that atoms (as Democritus envisioned them) exist, but they are not truly indivisible. Atoms *can* be disassembled, but not by chemical means. The radioactive decay of certain isotopes is one example of atom deconstruction. Another example is the fission process used in conventional nuclear reactors, which literally splits apart the nuclei of heavy atoms. Thus, elements *can* be transformed into other elements, as the alchemists once believed. Ironically the "base" metal lead, and not gold, is a common end product of element-transforming nuclear processes.

**Atoms are indestructible.** Much insight was gained into chemical transformations when Lavoisier, in 1774, developed a sensitive balance

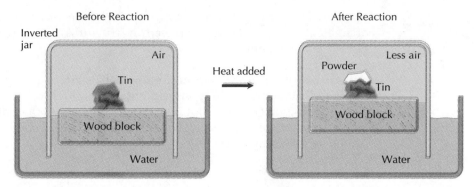

**FIGURE 4.3** Lavoisier heated tin metal inside a sealed jar by means of focusing the sun's rays on it with a magnifying glass. The tin reacted with something in the air that was trapped inside the jar, and a white powder replaced some of the tin. The reaction proceeded until about 20% of the volume of the air was missing, whereupon it stopped, even though some tin remained. In an earlier experiment with tin heated in a sealed container of air, Lavoisier had carefully weighed the container and contents before and after the reaction and found the total weights to be exactly the same. Soon after completing these experiments, Lavoisier learned that a new, reactive gas had been isolated by the chemist Joseph Priestly in England. The new gas, Lavoisier discovered for himself, reacted with tin and formed the same white powder as in his previous experiments with tin and air. On the basis of all his results, Lavoisier inferred that the new gas, which he called oxygen, comprises about 20% of the air. The other 80% of the air, he reasoned, consists of a gas or gases that will not combine with tin under the particular conditions of his experiments. Today, we know that the white powder is the compound tin oxide and that it consists of tin and oxygen atoms.

## Law of Conservation of Mass

There is no detectable change in the total mass during a chemical reaction.

capable of measuring the mass of 1/100 of a drop of water. In a series of chemical experiments designed to take place inside sealed containers (see Figure 4.3), and with the help of his state-of-the-art balance, Lavoisier showed that the mass of the reactants about to undergo a chemical reaction is exactly the same as the mass of the products after the reaction takes place. This behavior is summarized in what we now call the law of conservation of mass: There is no detectable change in the total mass during a chemical reaction.

The word *detectable* is needed in the modern statement of mass conservation because of Einstein's discovery in 1905 that energy and mass are equivalent. We now know that every chemical reaction is accompanied by either a release or an absorption of energy, and that the outward or inward flow of energy is equivalent to the disappearance or appearance of a tiny amount of mass during the reaction. Most *chemical reactions* involve a mass change of about one part in a billion—hardly detectable, even with current instruments. (*Nuclear reactions* release or absorb millions or billions of times more energy than chemical reactions and result in detectable changes of mass.)

In the modern view, atoms participating in a chemical transformation are neither created nor destroyed but they may gain or lose extremely slight amounts of mass when the bonds between them are made or broken.

**Atoms of a given element are identical in character.** Unlike pebbles of different size, shape, or mass that make up a quantity of gravel, Dalton felt that atoms of an element must be identical in all their characteristics. Today, we know that the picture is a bit more complicated: Atoms of an element can have different masses by having more or fewer neutrons, but that quality does not affect their chemical behavior. Chemical behavior is controlled by the patterns made by electrons in the outer parts of an atom, and those patterns (electron structures) are identical for atoms of the same element.

**Atoms of different elements are different in character.** Dalton believed that atoms of different elements have different masses and sizes. This proposition is true, but today we emphasize that differences in electron structure are responsible for variations in chemical behavior. Atoms of different elements are different in size as a consequence of different electron structure.

**Chemical compounds are formed when atoms of different elements join together to make identical units.** Dalton called these identical units "compound atoms," and we know them today as *molecules*. The mass of any molecule is equal to the total mass of the atoms that compose it.

Dalton prepared a chart of known elements and compounds that presented the structure of 17 different molecules, including carbon monoxide, carbon dioxide, and water. He erred with water, attributing to it the formula HO rather than $H_2O$. Later experiments and observations by Gay-Lussac and Avogadro (described below) led to the right ratio of mass for atoms of hydrogen and of oxygen and a knowledge of the correct ratio of hydrogen and oxygen atoms in the water molecule.

**The different kinds of atoms in a compound are present in simple numerical ratios (1:1, 1:2, 3:1, 3:2 and so on).** This postulate was supported by experiments conducted in the late 1700s by the French chemist Joseph Proust and others who enunciated what became known as the law of definite proportions: A compound always contains two or more elements combined in a definite proportion by mass. As viewed through the prism of Dalton's atomic theory, the law of definite proportions is extended to ratios of atoms as well, and those ratios happen to be relatively simple ones, at least for common compounds. (You do not find compounds consisting of, say, 17 atoms of element A and 43 atoms of element B.)

A classic example of the law of definite proportions is the chemical reaction between hydrogen and oxygen that produces water. A typical hydrogen-oxygen experiment is diagramed in two ways in Figure 4.4. As shown on the left side of that figure, a mixture of 16 g of oxygen gas and 4 g of hydrogen gas, when ignited by heat or a spark, produces 18 g of water vapor, 2 g of leftover hydrogen gas, and no oxygen gas. All the oxygen gas is gone; so it must have been incorporated into the water vapor. Since 2 g of hydrogen gas remain in excess, the missing 2 g must have been incorporated into the water vapor as well. The excess hydrogen is present after the reaction because (according to the law of definite proportions) the compound water contains hydrogen and oxygen combined in a definite proportion. That proportion in this and all similar experiments happens to be 16 g of oxygen for every 2 g of hydrogen. Reduced to lowest terms, this ratio is 8 to 1. From these facts, Dalton concluded that oxygen atoms are 8 times more massive than hydrogen atoms. He was mistaken about the ratio of *atoms* only because he assumed that water molecules are made of one hydrogen and one oxygen atom each (HO).

**Law of Definite Proportions**
A compound always contains two or more elements combined in a definite proportion by mass.

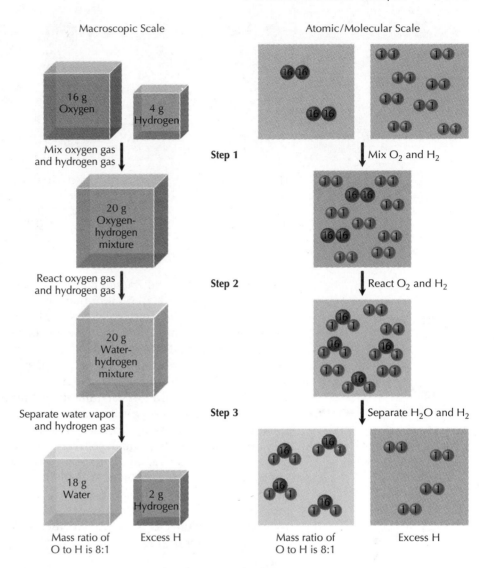

**FIGURE 4.4**  These steps outline the transformation of given quantities of oxygen and hydrogen into water and excess hydrogen. In Step 1, $O_2$ and $H_2$ physically combine to form a gaseous solution (mixture). In Step 2, $O_2$ and $H_2$ chemically react to form $H_2O$ and excess $H_2$. In Step 3, $H_2O$ and the excess $H_2$ are physically separated. The circled numbers on the right half of this illustration refer to the mass number of each atom (O = 16 and H = 1), which is a measure of the mass of each atom.

We now know that each water molecule has two H atoms and one O atom: $H_2O$. Thus O atoms must really be 16 times more massive than H atoms. The modern visualization of what goes on in the experiment is shown on the right side of Figure 4.4. At room temperature, both H and O are diatomic, and their molecules ($H_2$ and $O_2$) can combine in any ratio of mass (or number of atoms) to form a mixture. The molecules quickly intermix to form a gaseous solution. No chemical change takes place until something ignites the mixture, causing a very rapid (explosive) chemical reaction. The reaction involves the disassembly of the $O_2$ and $H_2$ molecules and their reassembly into $H_2O$ molecules. In our example, excess H atoms remain after the reaction, and they recombine into $H_2$ molecules again. The product of the reaction, $H_2O$ molecules, plus the excess $H_2$ molecules mixed in, can be separated by physical means, such as chilling the mixture until all the water condenses.

Dalton would not have been mistaken about the composition of the water molecule had he been receptive to the experimental results of one of

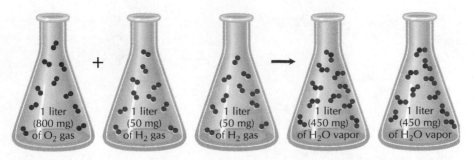

**FIGURE 4.5** One volume of $O_2$ chemically combines with two volumes of $H_2$ to form two volumes of $H_2O$, provided all volumes shown have a common temperature and pressure. The same number of molecules is present in each equal volume, as stated by Avogadro's law. The law works for gases because the molecules in a gas are relatively far away from one another. Since we know that one volume of $O_2$ has half as many molecules as two volumes of $H_2$, and we can experimentally determine that the $O_2$ volume has 8 times (800 mg/100 mg = 8) the mass of the two $H_2$ volumes, we can conclude that each O atom has 16 times the mass of each H atom.

---

### Avogadro's Law
Under identical conditions of temperature and pressure, equal volumes of gases of any kind contain the same number of molecules.

his contemporaries, the French chemist Joseph Gay-Lussac. In 1809, Gay-Lussac discovered by experiment that two volumes of hydrogen gas had to combine with one volume of oxygen gas to produce two volumes of water vapor, provided that all volumes are measured at the same temperature and pressure (see Figure 4.5). If Dalton's conception of water as single hydrogen and oxygen atoms being paired had been correct, then oxygen atoms would have had to split in two in order to join with hydrogen to form water. This violates Dalton's statement that atoms are indestructible.

In 1811, the Italian physicist Amadeo Avogadro put forward an assertion, now known as Avogadro's law, that under identical conditions of temperature and pressure, equal volumes of gases of any kind contain the same number of molecules. This law works for gases, in which the molecules are widely spaced, but not for liquids or solids. Avogadro correctly pointed out that Gay-Lussac's results make sense if hydrogen and oxygen are both diatomic in the gas phase. Also, there have to be twice as many hydrogen molecules as oxygen molecules in the volumes that combine to form water vapor. In modern notation,

$$2H_2 + O_2 \rightarrow 2H_2O$$

Avogadro's law lets us know the relative masses of the molecules in volumes of gas at the same temperature and pressure. The masses of gas volumes measured in experiments like that of Figure 4.5 indicate that oxygen molecules are 16 times as massive as hydrogen molecules.

**Atoms of two or more elements combine in different ratios to produce more than one compound.** Dalton discovered that in many instances, two elements can combine in different proportions to form different compounds. Dalton also found that the different proportions of the same elements in different compounds are always in simple ratios to one another. For example, carbon (C) and oxygen (O) can combine in one reaction to form carbon monoxide (CO) and in another reaction to form carbon dioxide ($CO_2$). The mass ratio of C and O in CO turns out to be 3 to 4; in $CO_2$ the mass ratio is 3 to 8. The amount of O participating in the second case versus the first case is twice as much: a 2 to 1 ratio. This result is consistent with the existence of C and O atoms that can combine in at least two different ways.

Today, we can use the mass number of atoms as a fairly accurate guide to their masses. By referring to the mass numbers of C and O in Table 4.1, you can see why C and O atoms combined in a 1:1 ratio and a 1:2 ratio have mass ratios of 3:4 and 3:8 respectively.

A further example illustrates what is said in Dalton's statement about elements combining. Chemical reactions between nitrogen (N) and oxygen can yield many compounds. One gram of N reacts with 0.571, 1.142, 1.713, 2.284, and 2.855 grams of O to form five of those compounds. Dividing each of these numbers by 0.571 yields the whole numbers 1, 2, 3, 4, and 5.

Dalton's statement about the way elements combine is formally called the law of multiple proportions. It supports the view that compounds must be made of more than one homogeneous kind of atom and that homogeneous kinds of atoms can combine in various ratios.

Some parallels can be drawn between the development of atomic theory and the development of the heliocentric (Copernican) model of the cosmos discussed in Box 3.4. Both of these essentially correct ideas were tossed aside at the time of the ancient Greeks in favor of popular alternatives that ultimately turned out to be erroneous.

The heliocentric model of Copernicus included some inaccuracies that were largely eliminated by later modifications. In a like manner, many errors and inaccuracies in Dalton's theory were eventually corrected. Even today, the nature of atomic liaisons is not fully elucidated, nor have the dynamics of the Solar System been modeled so perfectly that we can predict with confidence where the planets will be millions of years hence. Still, the basic character of Dalton's conception of the world of the small has triumphed, as has the notion that the sun lies at the center of the Solar System.

**Law of Multiple Proportions**
Atoms of two or more elements may combine in different ratios to produce more than one compound.

## CHAPTER 4
# Summary

Matter, a fundamental part of our universe, has mass and occupies space. Matter can be broadly classified by its physical characteristics into solids, liquids, gases, and plasmas.

Matter can be broadly classified by its chemical characteristics into pure substances and mixtures. Pure substances can be further divided into compounds and elements. Compounds can be decomposed by chemical means into elements. Elements cannot be decomposed by any chemical process.

Atoms are a fundamental building block of matter. An element is composed of similar atoms, and the characteristics of those atoms determine the prop-

erties of the element. A compound is made up of more than one kind of atom combined in a definite proportion. The properties of compounds depend both on the kinds of atoms present and on the way that those atoms are arranged.

The laws of chemistry are universal, but the results of those laws as applied to environments existing on and outside our planet are often very different. Terrestrial chemistry is extraordinarily rich, because many elements exist on Earth and because advantageous ranges of temperature and pressure allow atoms of those elements to combine.

**CHAPTER 4**
# Questions

## Multiple Choice

1. The ratio by mass of carbon to oxygen in 100 g of CO (carbon monoxide) is 3 to 4. The ratio in 200 g of the same compound is
   a) 3 to 4
   b) 6 to 4
   c) 6 to 1
   d) 200 to 1

2. Methane and oxygen are in a sealed container. The total mass of the container and the gases within it is 453 g. The gases are ignited so that some of the methane and all of the oxygen are consumed. The total mass is then
   a) much less than 453 g
   b) a few percent less than 453 g
   c) 453 g
   d) a few percent more than 453 g

3. Molecules
   a) must consist of two or more atoms of the same kind
   b) must consist of two or more atoms of different kinds
   c) are aggregates of two or more atoms of any kind
   d) are never present in solids

4. How many neutrons exist in an atom of the isotope given the symbol $^{16}_{8}O$?
   a) 4
   b) 8
   c) 16
   d) 24

5. How many atoms are present in a sugar molecule represented by the chemical formula $C_{12}H_{22}O_{11}$?
   a) 11
   b) 12
   c) 22
   d) 45

6. Which of the following chemical equations is unbalanced?
   a) $S + O_2 \rightarrow SO_2$
   b) $2H_2S + 3O_2 \rightarrow 2H_2O + 2SO_2$
   c) $SO_2 + H_2O \rightarrow H_2SO_4$
   d) $C_3H_8 + 5O_2 \rightarrow 3CO_2 + 4H_2O$

7. Which of the following is a homogeneous mixture?
   a) nitrogen gas
   b) seawater
   c) salt
   d) wood

8. Which of the following is a heterogeneous mixture?
   a) air
   b) water
   c) steel
   d) salad dressing

9. If a pure substance cannot be chemically decomposed, then it is
   a) an element
   b) a compound
   c) a solution
   d) a homogeneous mixture

10. Pure water contains
    a) only $H_2O$ molecules
    b) $H_2O$ molecules and a tiny fraction of $H^+$ and $OH^-$ ions
    c) relatively few $H_2O$ molecules and a large fraction of $H^+$ and $OH^-$ ions
    d) $H_2O$ molecules, $H_2$ molecules, and $O_2$ molecules

11. The chemical elements include
    a) earth, air, fire, and water
    b) simple substances that are always gaseous at room temperature
    c) more than 100 distinct substances, each of which consists only of atoms having the same number of protons
    d) more than 100 distinct substances, each of which consists only of atoms having the same number of nuclear particles

12. Atoms of an element
    a) never link up with each other
    b) are always identical to each other in every respect
    c) sometimes link together in geometric patterns
    d) cannot be disassembled by any means

13. Atoms of two different elements
    a) cannot join to form a compound
    b) will always join and form only one compound
    c) must join when brought together and form at least one compound

d) may or may not join to form one or more compounds, depending on how they interact with each other

14. The science of chemistry has nothing to do with
    a) the structure of the atom
    b) the structures of molecules
    c) a forest fire
    d) the motion of an apple falling from a tree

## Questions

1. How could you show that water is a compound and *not* an element or a homogeneous mixture of simpler components?

2. In what ways are solid, liquid, and gaseous matter the same? In what ways are they different?

3. What is it that is "balanced" in a balanced chemical equation?

4. How and why are the results of chemical processes generally different on Earth than in the universe at large?

5. In the law of definite proportions, what quantities are related in fixed proportions?

6. Why doesn't Avogadro's law work for liquids or solids?

## Problems

1. If three volumes of hydrogen gas ($H_2$) chemically combine with one volume of nitrogen gas ($N_2$), how many volumes of ammonia gas ($NH_3$) result? Assume that all volumes are at the same temperature and pressure and that Avogadro's law holds.

2. Write a chemical equation for the reaction taking place in Problem 1.

3. How many grams of hydrogen gas ($H_2$) need to combine with 28 g of nitrogen gas ($N_2$) so that both are consumed to form ammonia gas ($NH_3$)? (*Hint:* Refer to Table 4.1.)

## Questions for Thought

1. Is honey a solution or a compound? Can you think of an experiment you could perform in the kitchen that would answer this question?

2. If the world were to consist of only four elements—earth, air, fire, and water—which elements do you think would make up humans or any other organism?

## Answers to Multiple-Choice Questions

1. a    2. c    3. c    4. b    5. d    6. c
7. b    8. d    9. a    10. b    11. c    12. c
13. d    14. d

Parts II and III (Chapters 5–12) of this book are structured on the premise that the physical world is governed by a relatively small number of physical laws and principles. Nearly all of these laws and principles find their simplest expressions within the science of physics. Physics, therefore, serves as the backbone of the next eight chapters. In greater or lesser detail, various "applications" of fundamental physics will be explored within the disciplines of chemistry, earth science, and astronomy. For example, certain fundamental principles governing the behavior of gases and the distribution of heat apply directly to the study of weather and climate on Earth (the science of meteorology). The subject of meteorology, then, appears in Chapter 7, which deals with heat and its effects on matter.

Part II (the next four chapters) traces much of the early development of physics, starting from the ideas of the ancient Greek philosopher-scientists. In Chapter 5 we will learn how earlier notions of motion and gravity were corrected or refined by Galileo Galilei and Isaac Newton. Chapter 5 is rich with examples taken from astronomy.

In Chapter 6 we will explore the concept of energy, particularly as it pertains to the energy of moving bodies or of bodies having the potential to move. A central concept of Chapter 6 is the notion that energy cannot be created or destroyed, but it can be readily transformed. Transformations of energy account for the fact that our universe is dynamic and ever changing, both in the biological and the nonbiological senses. Chapter 6 includes some chemistry, since all chemical reactions involve transformations of energy.

Heat, a specific type of energy, is the focus of Chapter 7. Heat and matter are inextricably tied: Heat flowing into or out of a body produces internal change within the body. Chapter 8 takes up the subject of waves. All waves carry energy and have certain common characteristics and behaviors. These generalities will be illustrated through applications from physics, astronomy, and the earth sciences. For example, the detailed understanding of the properties of seismic waves has allowed geologists to probe the various layers of Earth's interior and map its structure in considerable detail.

# PART II
# The Mechanical Universe

# CHAPTER 5
# Motion and Gravity

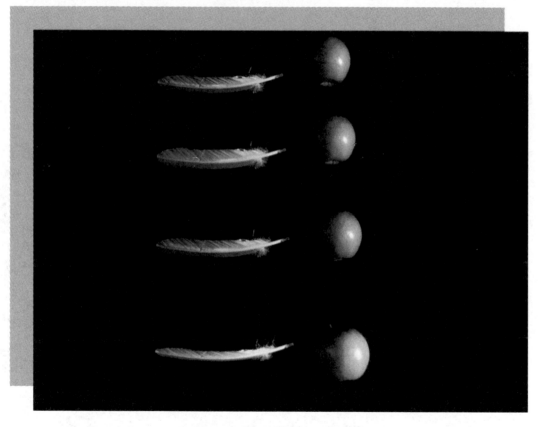

*Gravity acts with equal visible effect on an apple and a feather when both objects are allowed to fall within a vacuum chamber. The absence of air allows us to isolate the effects of Earth's gravitational pull on any body. The science of physics attempts to break down the complexity of everyday events, such as apples and feathers falling through air, into more fundamental phenomena that are easy to analyze.*

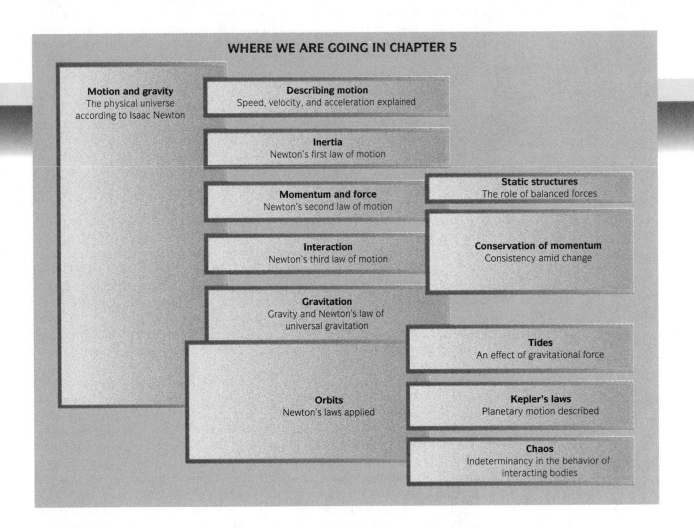

## WHERE WE ARE GOING IN CHAPTER 5

**Motion and gravity**
The physical universe according to Isaac Newton

**Describing motion**
Speed, velocity, and acceleration explained

**Inertia**
Newton's first law of motion

**Momentum and force**
Newton's second law of motion

**Static structures**
The role of balanced forces

**Interaction**
Newton's third law of motion

**Conservation of momentum**
Consistency amid change

**Gravitation**
Gravity and Newton's law of universal gravitation

**Tides**
An effect of gravitational force

**Orbits**
Newton's laws applied

**Kepler's laws**
Planetary motion described

**Chaos**
Indeterminacy in the behavior of interacting bodies

In the fourth century B.C., Aristotle conceived an explanation for motion so simple and so rational that it would guide human thought on the subject for two millennia to come. Motion on Earth, asserted Aristotle, is divided into two kinds: "natural" motion and "violent" (forced) motion. Natural motion was explained as the tendency of various earthly objects to seek their proper resting places. Thus, stones quickly fall to the ground (by virtue of "gravity") because stones are a part of Earth. Smoke rises (by virtue of "levity") and comes to rest with the air because of smoke's resemblance to air. Leaves, feathers, and other lightweight objects fall slowly and gently, because their somewhat airy composition opposes their tendency to fall by gravity. In other words, the rate at which a body rises or falls depends on its composition.

Once a body lies in its proper place, Aristotle supposed, it should stay at rest unless acted on by agents capable of causing "violent" motion. These agents include winds, flowing water, and the pushing or pulling forces exerted by humans and animals.

Today, we recognize that Aristotle's "gravity" is the direct outcome of one of the four fundamental forces—specifically, the gravitational interaction between a falling body and Earth itself. "Levity" (or "buoyancy," in modern usage) may seem to be a fundamental force exerted in an upward direction, but it is not. It is an indirect consequence of gravity, as we will learn in Chapter 7 in connection with pressure.

Aristotle's philosophical picture as outlined here is flawed in another, very important sense: It cannot be applied universally. It fails to explain the motion of heavenly bodies. If Earth is truly stationary, then by virtue of what we see, everything in the sky must be in perpetual motion. But how can the heavenly bodies keep moving without any apparent agents of violent motion? Aristotle resolved this dilemma simply by saying that the heavens and Earth are separate realms: Earthly bodies come to rest because it is natural for them to do so. Heavenly bodies persist in motion because it is their nature to do so.

By the 16th and 17th centuries, the old notions about motion were falling apart. Nicolaus Copernicus's heliocentric model of the Solar System cast serious doubt on the idea that Earth must lie at rest at the center of the universe. Galileo Galilei overthrew the idea that undisturbed bodies in motion naturally come to rest. He showed that whenever any external agent (such as friction) acting on a moving body is diminished, the more that body tends to maintain a given state of motion. Galileo also showed that gravity acts with equal effect in changing the motion of bodies subject to its downward pull. This disproved Aristotle's notion that more massive bodies must always fall faster.

A crucial breakthrough in the understanding of motion and gravity came when Isaac Newton introduced his *three laws of motion* and his *law of universal gravitation.* Together, these four laws unified the disparate worlds of Earth and heavens. They brought a sense of order and predictability to the universe and to all of its interacting parts. Newton's laws and their applications—known as Newtonian mechanics, or simply *mechanics*—are the primary focus of this chapter. Historically, Newtonian mechanics has been regarded as being philosophically *deterministic,* which means that a precise knowledge of the present state of any system of interacting parts would, in principle, allow a physicist to precisely work out both the past history and the future outcome of the system.

Elegantly simple and comprehensive as they are, Newton's laws have hidden flaws. They rest on certain assumptions now known to be incorrect. Today, Newton's laws are regarded as being approximate descriptions of nature—though very close approximations for most circumstances and for most practical applications in the macroscopic world. A more general, relativistic formulation of mechanics, based on Einstein's theories rather than Newton's laws, must be invoked if we are to make meaningful sense out of what goes on near the speed of light or close to a black hole. Quantum mechanics must take over if we are to explain the often strange behavior of atoms. Apart from relativity and quantum mechanics, the new study of *chaos* (in the scientific and mathematical sense) is shedding new light on a wide range of perplexing behavior. We will touch upon the scientific notion of chaos in this chapter and, in so doing, learn that the falling of a leaf, the rising of a smoke plume, and even the toilings of planets and stars are not so deterministic after all.

## Describing Motion

Motion involves a change in the position of a body in space coupled with a change in time. Motion can be described in terms of *speed, velocity,* and *acceleration,* each of which has a distinct meaning in physics.

## Speed

**Speed** is a measure of how "fast" a body is moving. Specifically, it is defined as the distance traveled by a body divided by the time required for the body to travel that distance. Symbolically,

$$v = d/t$$                                      Equation 5.1a

where $v$ means speed, $d$ means distance, and $t$ means time. (The letter $v$ is used to represent speed because, as we shall soon see, speed is closely related to velocity).

Equation 5.1a can be written in two alternative forms by solving for distance and for time:

$$d = vt$$                                       Equation 5.1b

$$t = d/v$$                                      Equation 5.1c

Equation 5.1b should already be familiar. We used it in Box 2.3 in connection with range finding by means of sound echoes and radar.

Speed can be measured in any combination of units of distance and time. Speedometers in automobiles, for example, are calibrated in units of both miles per hour (mi/h) and kilometers per hour (km/h). When the distance covered is short, speed measurements may be given in units of feet per second (ft/s) or meters per second. Meters per second (m/s) is the SI unit for speed. It is a derived SI unit, since it incorporates the SI base units for distance and time. Typical speeds of many moving things (in meters per second and kilometers per second) are given in Table 5.1.

| TABLE 5.1 Typical Speeds | Motion | Approximate Speed (m/s) | Approximate Speed (km/s) |
|---|---|---|---|
| | Light (in vacuum) | 300,000,000 | 300,000 |
| | Sun around our galaxy | 250,000 | 250 |
| | Earth around the sun | 29,600 | 29.6 |
| | Moon around Earth | 1,000 | 1.00 |
| | Commercial jet airliner | 300 | 0.30 |
| | Automobile (at 67 mi/h) | 30 | 0.03 |
| | Runner sprinting (maximum) | 10 | 0.01 |
| | Person walking | 1.5 | |
| | Snail crawling | 0.01 | |
| | Sperm swimming | 0.00005 | |

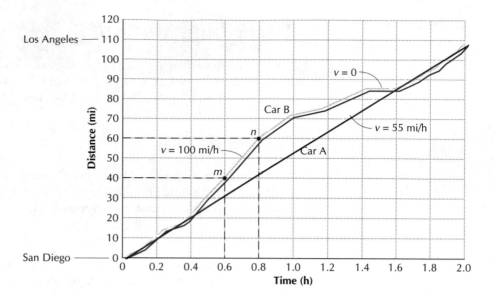

**FIGURE 5.1** A graphical representation of two different car trips from San Diego to Los Angeles. Car A moves at a constant speed of 55 mi/h and averages 55 mi/h over the whole trip. Car B's instantaneous speed varies between 0 and 100 mi/h, yet car B also averages 55 mi/h over the whole trip.

Since speed can change with time, it is important to distinguish between instantaneous speed and average speed. *Instantaneous speed* (the speed at any given instant of time) is distance divided by time for an infinitesimally small time interval. *Average speed* is distance divided by time for some arbitrary and usually not-so-small interval of time. The following example illustrates the difference between the two.

Late at night, a person can drive from San Diego to Los Angeles, a distance of 110 mi, at a constant 55 mi/h. By applying Equation 5.1c, we find that the time of the trip is

$$t = d/v$$
$$= 110 \text{ mi}/(55 \text{ mi/h})$$
$$= 2 \text{ h}$$

This result corresponds to the trip made by car A in the distance-versus-time graph in Figure 5.1.

If the same trip is made during the day, the driver will almost certainly get stuck in slow or stopped traffic along the way. Still, it may be possible (though not legal) to make it in 2 h. Car B, whose journey is also graphed in Figure 5.1, did this by speeding in certain areas where traffic was light. Owing to severe traffic congestion near Los Angeles, car B sat at an utter standstill ($v = 0$) for about 10 min. Earlier, its maximum speed occurred between points *m* and *n*, 40 and 60 mi from San Diego. The straight line between those points on the graph indicates that car B's speed was constant through that stretch. Its speed there was

$$v = d/t$$
$$= (60 \text{ mi} - 40 \text{ mi})/(0.8 \text{ h} - 0.6 \text{ h})$$
$$= 20 \text{ mi}/0.2 \text{ h}$$
$$= 100 \text{ mi/h}$$

which is well above the posted 55-mi/h speed limit on the road between the two cities.

For *both* trips, the average speed was 55 mi/h over the 2-h interval. The late-night driver's instantaneous speed always equaled the average speed; the daytime driver's instantaneous speeds were at different times less than, equal to, and greater than the average speed.

## Frame of Reference

We must know more than a body's speed if we are to determine where the body is going. We need to specify a *frame of reference* (or *reference frame*) in which the speed measurement is being made. For most everyday motion, we subconsciously choose Earth's surface as our frame of reference. Ambiguities arise when we speak of someone or something moving inside something else that is moving. For instance, the speed of a person walking up the aisle of a train moving at 60 mi/h along train tracks could be described as 2 mi/h relative to seated passengers on the train (that is, within a reference frame fixed to the train) or as 62 mi/h relative to the ground (that is, within a reference frame fixed to the rails on which the train moves).

One's freedom to choose a reference frame brings into question the meaning of the words *at rest* or *stationary*. For example, the car on the treadmill in Figure 5.2 is moving at either zero speed or 20 mi/h, depending on the views of two different observers. If we choose our reference frame to be fixed on the sun, then we humans are moving through that reference frame at almost 67,000 mi/h—the orbital speed of Earth around the sun. Our perspective of being in motion changes in a similarly dramatic way if we picture our Solar System speeding though a reference frame fixed on the Milky Way galaxy. Even broader perspectives reveal our galaxy moving relative to other galaxies, and other galaxies moving relative to us. Physicists believe that *all* motion is relative and that there is no such thing as absolute rest. Consequently, an object's speed has concrete meaning only when a reference frame is specified or clearly implied.

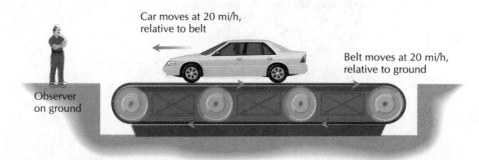

Car moves at 20 mi/h, relative to belt

Belt moves at 20 mi/h, relative to ground

Observer on ground

**FIGURE 5.2**   All motion is relative. Relative to a spider clinging to the top of the treadmill belt, the car has a speed of 20 mi/h. Relative to the "stationary" observer at left, the car has a speed of zero. Is anything truly stationary, or at rest, in this picture?

# Velocity

A body's speed within a reference frame only partially describes its motion. A more complete description includes the body's *direction of motion* as well. When we state *both* a body's speed and direction of motion, we are specifying its **velocity.** If we say that a person moves at 1 m/s (a slow walking pace), we specify the person's speed only. If we say that a person walks 1 m/s south (or west, or southwest), we specify the person's velocity.

Two objects with the same speed do not necessarily have the same velocity. A car moving at 60 mi/h *north* on Interstate 35 differs in velocity from another car moving at 60 mi/h *south* on the same road, though both cars share the same speed relative to the ground. Two baseballs, one moving *up* at 10 m/s and the other moving *down* at 10 m/s, differ in velocity as well.

Notice in the examples just given how the stated direction of motion (north and south, up and down) is related to an implied reference frame. Just as speed must be related to a known reference frame, so too must the direction of a body's motion be related to a known reference direction or directions.

It is often crucial to distinguish between instantaneous velocity and constant velocity. *Instantaneous velocity* is distance divided by time for an infinitesimally small time interval, plus information about the direction of motion during that same infinitesimally small time interval. If a body's instantaneous velocity does not change over time, then its velocity is constant. *Constant velocity* means constant speed and constant direction during a given (and not infinitesimally small) time interval. A body having a constant velocity is said to be in *uniform motion.*

Is it possible for a body to have a constant speed and yet not be moving uniformly? Try walking or running around an oval racetrack at a constant speed and ask yourself that question (see Figure 5.3). Your velocity is changing if you slow down or speed up while going in a straight line, if you

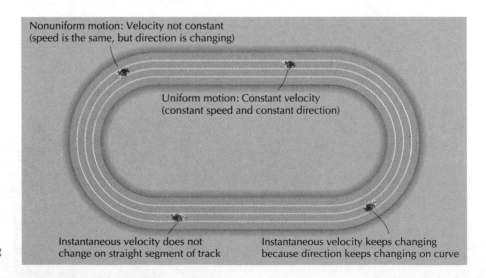

**FIGURE 5.3**    A runner maintaining a constant speed is in uniform motion (moving at constant velocity) only when moving in a straight line.

Nonuniform motion: Velocity not constant (speed is the same, but direction is changing)

Uniform motion: Constant velocity (constant speed and constant direction)

Instantaneous velocity does not change on straight segment of track

Instantaneous velocity keeps changing because direction keeps changing on curve

change your direction while keeping the same speed, or if you change your speed and your direction at the same time.

## Vectors and Scalars

Velocity is an example of a vector quantity. Speed is an example of a scalar quantity. A **vector quantity,** or **vector,** expresses both *magnitude* ("how much") and *direction*. A **scalar quantity,** or **scalar,** expresses magnitude only. Length (without any indication of direction), area, volume, mass, density, time, and temperature are all scalars because no direction is associated with them.

*Direction,* in the context of our discussion, means direction in space. It may be tempting to think that temperature is a vector rather than a scalar because it goes "up" and "down." But temperature goes up and down only in a metaphorical sense. Temperature increases and decreases do cause mercury columns in standard thermometers to rise and fall, but temperature itself heads nowhere. Time is another example of a scalar quantity with purely metaphorical directions (past and future) ascribed to it.

We will represent vectors by boldface type in equations appearing in this book: for example, **v**, which means velocity (a vector), can be distinguished from $v$, which means speed (a scalar). We will soon make our acquaintance with another vector, *force,* which is denoted in type by the bold letter **F**. When force is symbolized by $F$, we are referring only to the magnitude of the force and not to its direction. On diagrams, vectors often appear as arrows whose lengths are proportional to the magnitudes of the quantities being represented (see Figure 5.4).

Differences between scalars and vectors become apparent when we add two or more of each. If we combine (add) two or more masses, they always add normally: 2 kg of water plus 3 kg of water always make a total of 5 kg of water. Vector quantities, on the other hand, add in a way that takes their direction into account. For example, the top of the treadmill belt in Figure 5.2 is moving to the right at 20 mi/h, while the car on it is moving to the left (with respect to the belt) at 20 mi/h. The two velocities do not add to 40 mi/h; instead, the velocity is zero relative to the observer on the ground. If, however, we reversed the direction of the moving car on the

**FIGURE 5.4** Arrows can be used to represent vectors on a flat plane, such as this page. Some velocity vectors are illustrated here. By common convention, the length of each arrow drawn is proportional to the magnitude of the quantity being represented.

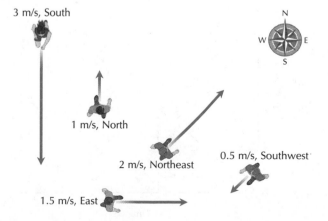

3 m/s, South

1 m/s, North

2 m/s, Northeast

0.5 m/s, Southwest

1.5 m/s, East

belt, the car would move at a velocity of 40 mi/h away from the observer on the ground.

# Acceleration

Velocity expresses all we need to know about a body's motion during an infinitesimally small interval of time. However, we can go one step further and describe how a body's motion *changes with time,* which is what physicists call acceleration. **Acceleration** is a change of velocity.

You may have the idea that acceleration means only an increase in speed. You should abandon this notion! A body is accelerating if its speed is changing (increasing *or* decreasing), if its direction of motion is changing, or if both its speed and direction of motion are changing. (Remember that velocity specifies both speed and direction. If either or both of these things are changing, then there is acceleration.)

Like velocity, acceleration is a vector quantity. To get a feeling for what acceleration means, try the following exercise, which is best performed outdoors or in a gymnasium: From rest, start walking in a straight line so that after 1 s your speed is 1 m/s (a slow walk). Gradually and smoothly, increase your pace so that after 2 s you are moving at 2 m/s (a brisk walk). At 3 s your speed should be 3 m/s (jogging pace). At 4 s, if you can still keep up, you should be moving at 4 m/s (a fast run). These increases in speed with time are graphed in Figure 5.5. A world-class sprinter would, by continuing in this manner, reach 10 m/s (about 22 mi/h) after 10 s.

For as long as you engage in this exercise, your speed continually changes. But something else does remain constant: the ratio between the change in your speed and the time during which the change took place. This ratio is the magnitude of your acceleration ($a$), which is given by

$$a = \Delta v / \Delta t$$    Equation 5.2

where $\Delta v$ (read "delta v") is the change in speed over some time interval $\Delta t$ (read "delta t"). Strictly speaking, this equation refers to the average acceleration over the time interval in question—though for this exercise it does not matter, since the acceleration is assumed to be constant. There has

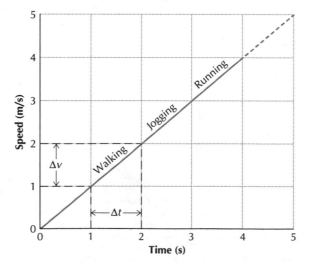

**FIGURE 5.5**  If a person's speed uniformly increases by 1 m/s for every 1 s of elapsed time, then we say that his or her acceleration is 1 m/s per second, or 1 m/s².

been a $\Delta v$ of 1 m/s for every $\Delta t$ of 1 s. Substituting into Equation 5.2, we get

$a = \Delta v/\Delta t$
$\quad = (1\ m/s)/1\ s$
$\quad = 1\ m/s^2$

The unit $m/s^2$ can be read as "meter(s) per second squared," but its meaning is "meter(s) per second per second."

Note that we need only two of the four fundamental properties—distance and time (not mass or charge)—to express the magnitude of acceleration. Acceleration is the ratio between speed and time, and speed itself is derived from distance and time. In other words, we express acceleration by means of distance used once and time used twice.

Now, let us consider the direction of acceleration. Three cases are shown in Figure 5.6. In every case, the magnitude of the acceleration is the same, 1 m/s². In case (a), which is the exercise we described earlier, the acceleration and velocity vectors point in the same direction. In case (b), the runner is slowing down. His acceleration is in a direction opposite to the direction in which he is moving. The magnitude of his acceleration is still 1 m/s² because his speed is changing by 1 m/s with every second of time. In case (c), a woman is moving around a small circular track. Her speed remains constant, but her direction of motion is steadily changing.

Case (c) requires more explanation. Not only is the runner's direction of motion changing all the time, but also the direction of her acceleration is changing. As long as she maintains the same speed and does not deviate from the circular path, her acceleration vector keeps turning so that it always points inward toward the center of that path. Bodies moving on circular arcs at constant speed are said to be undergoing *centripetal* ("center-seeking") *acceleration.*

The magnitude, *a,* of the centripetal acceleration is given by

$$a = v^2/r \qquad\qquad \text{Equation 5.3}$$

where $r$ is the radius of the circular path and $v$ is the speed. The equation says that $a$ and $r$ are inversely related: Increasing the radius decreases the acceleration magnitude, and decreasing the radius increases the acceleration magnitude. For this reason, it is more difficult to make a sharp turn (a turn with a smaller radius of curvature) than a gradual turn. Note also that the $v$ term in Equation 5.3 is squared. This means that a runner or a car moving through the same curve at *twice* the speed experiences *four* times the acceleration ($2^2 = 4$).

In the case of the woman in Figure 5.6, values of $v = 3$ m/s and $r = 9$ m produce a centripetal acceleration having a magnitude of

$a = v^2/r$
$\quad = (3\ m/s)^2/9\ m$
$\quad = (9\ m^2/s^2)/9\ m$
$\quad = 1\ m/s^2$

Other combinations of speed and radius could have yielded the same 1 m/s² acceleration: $v = 1$ m/s and $r = 1$ m, $v = 2$ m/s and $r = 4$ m, $v = 16$ m/s and $r = 4$ m, and many more.

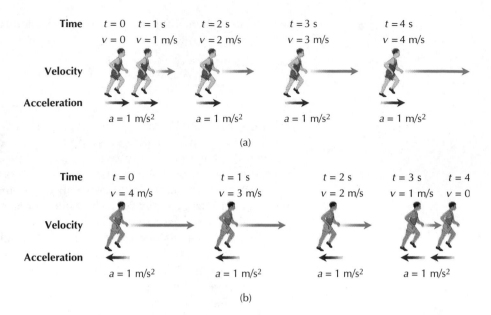

(a)

(b)

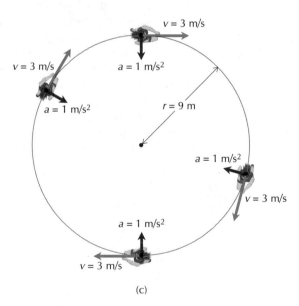

**FIGURE 5.6** Acceleration involves a change in speed only, as in cases (a) and (b); a change in direction only, as in case (c); or a change in both speed and direction. In case (c), the direction of the runner's velocity is changing uniformly because she is moving at a constant speed around a circle. Her acceleration vector is changing in such a way that it always points inward toward the center of the circular path. The runners in (a) and (b) are undergoing *linear acceleration,* and the runner in (c) is undergoing *centripetal acceleration.*

(c)

You can readily feel accelerations (or changes of motion) as long as their magnitudes are not too small. Step on the gas pedal of your car and you have the sensation of the seat pushing you forward. Hit the brakes hard and your tightly fastened seat belt "pushes" you back toward the seat. Turn the steering wheel right and you feel a push to the right. Turn the steering wheel left and you are forced to the left. When your car glides along a smooth, flat road in a straight line at constant speed, you feel very little sensation of movement, because there is very little change in your velocity (the magnitude of your acceleration is practically zero). On a road filled with potholes, every jerk up and down or back and forth lets you know that your velocity is changing in a significant way.

# Inertia

It was Galileo who introduced the idea that any "material" body (a body with mass), whether or not it moves, tends to resist a change in its motion. This resistance to change in motion is referred to as **inertia.** Through a series of carefully thought-out experiments, Galileo came to the conclusions that a body at rest tends to remain at rest, and that once a body is moving, nothing at all is required to *keep* it moving.

Galileo's idea of inertia was a clear philosophical break from the earlier Aristotelian tradition, which decreed that some constantly applied force would be needed to keep an earthly body moving. On the contrary, Galileo saw forces such as friction and gravity as opposing the tendency of a moving body to persist in a given state of motion.

Only under somewhat contrived circumstances do we see any everyday examples of unrestrained inertia in moving bodies. A bowling ball rolling down a flat, hard surface and a hockey puck sliding on an ice rink approximate Galileo's vision of what the world would be like without friction and gravity. Gravity is certainly present in these instances; however, it cannot act to change the motion of a body that remains on a flat surface.

Measurements and common experience alike show that the more mass a body has, the more inertia (or resistance to change of motion) it has. If you kick a soccer ball and then a bowling ball, it is obvious which of the two has more resistance to change of motion. We do not know *why* inertia exists. However, we do know that a body's inertia is proportional to the quantity of matter contained in it.

In 1642, the same year that Galileo died, Isaac Newton was born. Newton, who devoted much of his life to creative mental pursuits, paid close attention to the work of his predecessors, including Galileo. Newton carried on where Galileo had left off, conceiving his three laws of motion and the law of gravitation while in his early twenties. Together, these four laws con-

**PHOTO 5.1** Galileo in Venice, demonstrating the satellites of Jupiter. Galileo's contributions to the emerging science of physics were perhaps more significant than his better-known contributions to astronomy.

stitute a unified theoretical framework describing motion and gravity throughout the universe.

# Newton's First Law

Newton's first law of motion, which is also called the law of inertia, is a rephrasing of Galileo's concept of inertia. Using the terminology introduced earlier, we shall state Newton's first law as follows: A body remains at rest or continues in uniform motion except when compelled to change its motion by forces acting upon it.

In the first law, Newton does not describe what these forces are or how they act; however, we already know that they include agents such as friction and gravity. The first law also contains an important hidden assumption. It applies only to *inertial reference frames*—frames of reference that are either at rest or moving with constant velocity. In other words, inertial reference frames are those that are not accelerating. Thus, the first law does not apply to how a body might move inside a train, with respect to the train, whenever the train is increasing or decreasing its speed, changing its direction of motion, or both.

What are the consequences of inertia according to the first law? When a car hits a brick wall, any front-seat passenger not wearing a seat belt simply continues in uniform motion toward the dashboard and front window. (Hopefully, an inflated air bag intercedes at that point to slow the passenger's motion gradually.) Similarly, when a car with bald tires rounding a curve hits oil-soaked, or ice-coated, flat pavement, friction between the tires and road nearly disappears and the car moves uniformly—not necessarily in the direction the driver wishes.

More abstractly, let us imagine that in the moment after a home run is hit, all forces on the moving baseball—gravity, air resistance, and others—were to suddenly disappear: The baseball sails out of the ballpark in a straight line and maintains a constant speed forever. (To be truthful, the idea of anything sailing through outer space in perfectly uniform motion is unrealistic. Beyond Earth lie the moon, the sun, other planets, and everything else in the universe. Each of these bodies tugs on the baseball with some very tiny but nonzero force.)

# Momentum and Force

We have already stated that a greater inertia (resistance to change in motion) is associated with bodies of greater mass. Similarly, it is everyone's common experience that greater weight is associated with greater mass. From all experimental evidence gathered so far, it appears that these two properties of mass—inertia and weight—are exactly proportional to each other when measured in the same gravitational environment.

Whether we think of a body's mass in terms of its weight or in terms of its inertia, mass alone insufficiently describes the properties of a body in motion. Similarly, velocity alone does not embody all the properties a moving body has at some instant, even though it describes perfectly well where

**PHOTO 5.2** The gaseous debris of the Crab nebula, the remnants of a supernova whose light first reached Earth in the year 1054, continue to expand at a speed of about 1500 km/s. At the outer rim of the nebula, where gravity and other forces are exceedingly weak, the now very thinly spread gas is virtually in uniform motion, in accordance with Newton's first law.

and how that body is moving through space. Newton made use of *both* concepts, mass and velocity, in his explanation of how a body responds in its motion when acted on by a force (an agent capable of changing a body's motion).

## Momentum

When we take into account both a body's mass and its motion at some instant, we are referring to what is called *momentum,* or "quantity of motion." A body's **linear momentum** (**p**), or momentum in the direction it is moving, is defined as its mass multiplied by its velocity:

$$\mathbf{p} = m\mathbf{v}$$                                                         Equation 5.4

Note that **p** is a vector; it has both a magnitude and a direction associated with it. The **p** and **v** vectors for a given body in motion point in the same direction. The SI units for the magnitude of momentum are kg·m/s. For example, a 1-kg mass moving at 1 m/s speed has a momentum of 1 kg·m/s. A 5-kg mass moving at 7 m/s has a momentum of 35 kg·m/s.

Since velocity is relative—it can be measured with respect to any reference frame—so, too, is momentum relative. The pencil resting in your hand has zero momentum relative to you, but it has a rather significant 100 kg·m/s or so of momentum relative to the sun (as a consequence of Earth's motion around the sun).

The utility of linear momentum becomes apparent when you consider how it feels to be beaned on the head with a fast-moving Ping-Pong ball and a similarly fast-moving golf ball. Both objects rebound (change their velocity) in about the same manner, which makes it clear that velocity alone does not say much about either object's quantity of motion. Obviously, the golf ball's greater mass (and therefore greater momentum) imparts a far greater wallop. The momenta of various bodies listed in Table 5.2 will give you some idea of the quantity of motion carried by various earthly and celestial bodies.

**TABLE 5.2**
**Typical Masses and Momenta**

| Body | Mass (kg) | Body in Motion | Momentum of Moving Body (kg·m/s) |
|---|---|---|---|
| Sun | $2 \times 10^{30}$ | Sun around the galaxy | $5 \times 10^{35}$ |
| Earth | $6 \times 10^{24}$ | Earth around the sun | $2 \times 10^{29}$ |
| Moon | $7 \times 10^{22}$ | Moon around Earth | $7 \times 10^{25}$ |
| Space shuttle | $3 \times 10^{4}$ | Shuttle in low Earth orbit | $2 \times 10^{8}$ |
| Car | 1000 | Car at legal speed limit | $3 \times 10^{4}$ |
| Human being | 60 | Human at record sprint | 600 |
| Tennis ball | 0.06 | Tennis ball after hard serve | 2 |
| Spitball | 0.001 | Spitball flung at 5 m/s | 0.005 |

# Force

According to Newton's first law, a body will not change its motion unless compelled to by a force or forces acting upon it. But what is force? Force is the push or the pull you can exert with your muscles. Forces are responsible for holding together the tiny particles of a solid. A force called *friction* retards objects that are sliding, spinning, rolling, or in some other way moving against or through something. Force is the downward pull of gravity and also the gravitational attraction between Earth and other celestial bodies. Generally speaking, **forces** are agents of change. All forces, and all changes, are traceable to the four fundamental forces of nature: the strong, weak, electromagnetic, and gravitational forces.

Force ($\mathbf{F}$) is a vector, which is intuitively obvious if you consider how you are able to exert pushes or pulls in various directions with your muscles. Like all vectors, forces add in a way that takes their direction into account. In some instances, two or more forces acting upon a body may produce no apparent change on the body. If you try to squeeze together the covers of this book with your hands or knees by applying equal forces in opposite directions, the net force (vector sum of the forces) on the book is zero and the book goes nowhere. We say that the book is in *equilibrium,* which means that its motion remains unchanged. If both hands push in the same direction on the book, the book accelerates and is not in equilibrium. An entire branch of mechanical engineering called *statics* deals with forces in equilibrium (see Box 5.1).

A force or forces acting on a body might not result in any observable change, but change is always there. If you lean against a brick wall, the wall flexes imperceptibly and the molecules inside the bricks undergo very slight shifts in position. The change is reversible in this instance, because the wall will return to its earlier state when you stop leaning on it.

When dealing with bodies capable of some change in motion, we are concerned with forces capable of changing motion. These forces are called *external forces,* because they act on a body from the outside. External forces are exemplified by pushing or pulling on a car from the outside to change

**PHOTO 5.3**   This balanced rock is not moving because the vector sum of the forces on it is zero—it is in *equilibrium.*

the car's motion. We will not be concerned in this chapter with the *internal forces* that act from within a body, such as the electromagnetic forces that bind the atoms of a solid body. These forces do not change the motion of the body as a whole. Internal forces are exemplified by pushing or pulling on the front seat of a car while sitting in the back seat in order to speed it up or slow it down. Such efforts are as fruitless as trying to raise yourself by pulling upward on your clothing.

## Force and Momentum

Newton's second law of motion, in its original form, relates force and momentum. It describes how an external force, acting on an unrestrained body, changes that body's momentum: The rate of change of a body's momentum is equal to the net external force impressed upon the body; the change of momentum is made in the same direction as the applied net force. In the language of mathematics,

$$\mathbf{F} = \Delta\mathbf{p}/\Delta t \qquad\qquad\qquad \text{Equation 5.5}$$

The expression $\Delta\mathbf{p}/\Delta t$ means the time rate of change of momentum. The net force vector $\mathbf{F}$ and the change of momentum vector $\Delta\mathbf{p}$ point in the same direction. The larger the net force applied to a body, the faster the momentum changes. Conversely, the smaller the force applied, the more slowly the momentum changes. For intervals of time $\Delta t$ that are not infinitesimally small, $\mathbf{F}$ is interpreted to be the *average* net force over the time interval.

**Newton's Second Law of Motion**
The rate of change of a body's momentum is equal to the net external force impressed upon the body; the change of momentum is made in the same direction as the applied force.

Notice that the second law does not say that force is related to any particular value of momentum but that it causes a *change of momentum*. Momentum itself (and velocity, too) is relative. Momentum (a vector) may increase or decrease in magnitude as a result of an applied force. Or it may remain the same in magnitude yet change in direction. For example, a car's momentum increases when the car is accelerating forward and decreases when the car is braking. When the car makes a turn at constant speed (and constant momentum magnitude), the car's momentum changes in direction only.

Using Equation 5.5, we can work out the units for force (in the SI). We divide the SI units for momentum (kg·m/s) by the SI unit for time (s); the result is kg·m/s². For simplicity, this ungainly collection of mass, length, and time units is given a name: *newton* (abbreviated N), named in honor of Isaac Newton. One newton is not much force; it is roughly equivalent to the force (weight) exerted by a small apple resting in the palm of your hand.

## BOX 5.1

# The Design of Static Structures

A building is an excellent example of a static structure, an object that remains rigid and unmoving even as it is subjected to external forces such as gravity. In the context of Newton's second law, $\mathbf{F} = m\mathbf{a}$ (see Equation 5.6a on page 200), the engineer or architect wants the vector forces on each part of a building to sum to zero, so that each part undergoes zero acceleration and the whole building stays put and does not collapse.

Each brick inside the wall of a building, for example, experiences a downward pull due to its own weight, plus the downward force applied to it by the weight of all the bricks, mortar, and other material above it. At the same time, the brick experiences an upward force applied to it by the brick or mortar underneath it. When all the forces sum to zero, the brick is in equilibrium. A brick may feel a great deal of *compression,* but it does not collapse in upon itself because bricks (and any other solid building material) greatly resist being compressed.

Building materials are usually less successful at resisting tension, which is any force that tends to pull them apart. Tension is the opposite of compression. In a diving board under stress (see Figure 5.7), there is tension along the upper surface and compression along the lower surface. If too much weight is brought to bear on the board, the top surface of the board will crack and break first.

**PHOTO 5.4** The Parthenon, an ancient Greek temple, features closely spaced stone columns that support the brittle beams that lie above.

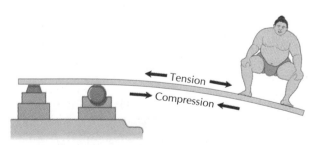

**FIGURE 5.7** Diving boards made of fiberglass and epoxy are flexible yet resistant to considerable tension. Too much tension, however, will cause the board to break apart at the top first.

Equation 5.5 tells us that a strong average force applied over a short time interval can accomplish the same change of momentum as a weak average force applied over a longer time. For example, consider the force required to stop a car moving at typical highway speeds. With normal braking, it takes about 10 s to bring the car to a halt. If the car crashes into a brick wall, the same thing is accomplished in about 0.1 s. Let us find out how much average force must be applied in each case if the car has (from Table 5.2) an initial momentum of 30,000 kg·m/s and a final momentum of zero. We are assuming no change of direction, so we are concerned only with the scalar form of Equation 5.5. For the braking car,

$$F = \Delta p / \Delta t$$
$$= (30{,}000 \text{ kg·m/s})/10 \text{ s}$$
$$= 3000 \text{ kg·m/s}^2$$
$$= 3000 \text{ N}$$

**PHOTO 5.5** Highway overpasses like this one in California are supported mostly as a result of compressive forces acting within the supporting arch.

Stone and brick are very good at resisting compression but are poor at resisting tension. When these materials are used in structures, they are arranged in ways that minimize tension. For example, the horizontal beams of stone in ancient structures like the Parthenon have many closely spaced supporting columns supporting them in order to minimize the tension that arises on the lower sides of the beams as they sag between the columns. The excellent ability of steel to handle both compression and tension makes it better for uses such as horizontal beams. Concrete works well, too, as long as it is poured around a cage of reinforcing steel bars.

Arches are a more elegant solution to the problem of minimizing tension. Arches can be seen in the supports for ancient Roman aqueducts, in the windows of old brick buildings, and in bridges. The combined weight of each part of the arch and the load above produces mostly compression and very little tension. Rotate an arch around a vertical axis and you have a dome, which can cover a large area without the need for internal columns. Domes on a small scale can be built with material as unsubstantial as snow.

**PHOTO 5.6** In this igloo shelter made of blocks of compressed snow, the downward pull of gravity on the upper blocks is channeled toward the lower blocks that lie upon the snow-covered ground. The interlocking ice crystals on the edges of the blocks will ensure that the structure remains rigid until the spring thaw.

For the crashing car,

$F = \Delta p / \Delta t$

$\qquad = (30{,}000 \text{ kg·m/s})/0.1 \text{ s}$

$\qquad = 300{,}000 \text{ kg·m/s}^2$

$\qquad = 300{,}000 \text{ N}$

The momentum change ($\Delta p$) for the car is the same, but the average forces and the times are different. The crashing car experiences a very large average force applied over a short time. As is typical in collisions, the magnitude of that force quickly increases to a maximum and quickly decreases thereafter. The braking car experiences an average force 100 times smaller, which is applied over a time interval 100 times longer. The driver of the braking car can choose to apply steady braking force or to pump the brakes, thereby applying intermittent braking forces. Either way, the average force is the same as long as the car slows to a stop in 10 s.

**Problem 5.1**
You push straight downward on a book that is resting on a table, applying 100 N of force for 10 s. By how much does the momentum of the book change during the 10-s interval?

**Solution**
Assuming that the table does not give way, the book does not change its momentum. Why? Because when you push on the book with 100 N of downward force, the table pushes up on the book with a force equal to 100 N plus the weight of the book itself. The net force on the book remains zero. If the net force were not zero, the momentum of the book would change, and the book would start moving.

# Force and Acceleration

Newton's second law as expressed in Equation 5.5 is valid over an extremely broad range of circumstances. We will now introduce an alternative (but limited) form of the law that is more commonly applied to everyday phenomena involving force and motion.

Recall from Equation 5.4 that $\mathbf{p} = m\mathbf{v}$. With this equation in mind, Equation 5.5 can be written as

$$\mathbf{F} = \Delta\mathbf{p}/\Delta t = \Delta(m\mathbf{v})/\Delta t$$

If, moreover, we are dealing with a body of unchanging mass $m$, then momentum can change only by virtue of a change of velocity $\mathbf{v}$. With that assumption, we can therefore rewrite the above equation as

$$\mathbf{F} = m(\Delta\mathbf{v})/\Delta t$$

Since, from Equation 5.2, we know that $\mathbf{a} = \Delta\mathbf{v}/\Delta t$, we can substitute $\mathbf{a}$ for $\Delta\mathbf{v}/\Delta t$, and rewrite the equation as

$$\mathbf{F} = m\mathbf{a} \qquad\qquad\qquad \text{Equation 5.6a}$$

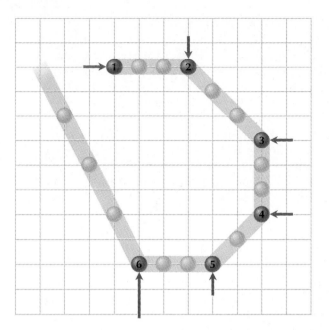

**FIGURE 5.8**   In this simulation of a body moving on a flat, frictionless surface, changes of direction or speed come about as the result of blows (forces) delivered over the same, very short time interval. The larger arrow at position 6 represents a blow having twice the force of the others. At position 1, an object initially at rest accelerates to the right only during the brief time of the blow, and it continues moving at constant velocity after the blow to position 2. There, it receives a downward blow and accelerates in the same direction as the blow, downward. The object's *velocity,* however, is not directly downward since the object was already moving to the right when the blow took place at position 2. At position 3, the left-directed blow cancels the object's motion to the right. After the stronger blow at position 6, the object continues moving with constant velocity in accordance with Newton's first law.

Equation 5.6a is Newton's second law of motion as stated in terms of the familar concepts of force, mass, and acceleration. Both **F** and **a** are vectors, and they point in the same direction—which means simply that a body will accelerate in the direction of the net force that acts upon it. The second law says that for a body of constant mass, *the larger the net external force, the larger the acceleration:* $\mathbf{F} \propto \mathbf{a}$. It also says that for a body undergoing the same acceleration, *the larger the mass, the larger the net force needed:* $\mathbf{F} \propto m$.

The Equation $\mathbf{F} = m\mathbf{a}$ is also useful in the following rearranged form:

$$\mathbf{a} = \mathbf{F}/m \qquad\qquad\qquad \text{Equation 5.6b}$$

When a net force is applied to a body, the body accelerates in the same direction as that net force. Notice we did not say that a body's *velocity* must be parallel to the force applied. Velocity and force vectors are parallel only if the body is stationary to begin with or if the force is applied in the same direction as the body's velocity vector. Figure 5.8 illustrates the effects of acceleration on a body moving without friction when various short-duration forces are applied to it.

The equal sign in $\mathbf{F} = m\mathbf{a}$ is valid if we use consistent units, such as those of the SI: N for force, kg for mass, and $m/s^2$ for acceleration. One newton of force can be thought of as the force required to give a 1-kg mass an acceleration of 1 $m/s^2$. One newton of force is equivalent to 0.225 pound of force, or roughly 1/4 pound, which is the weight of a small apple.

**Problem 5.2**
(a) What force (magnitude only) is needed to give a 0.17-kg hockey puck an acceleration of 100 $m/s^2$?
(b) What happens to the puck when the force is no longer applied?

*(continued)*

*(Problem 5.2 continued)*

**Solution**

(a) $F = ma = (0.17 \text{ kg})(100 \text{ m/s}^2) = 17 \text{ N}$. This is a rather feeble force to be applied during the game of ice hockey. If applied over a very brief time by means of a blow with a hockey stick, the puck would undergo a relatively small change of velocity.

(b) Assuming that the puck is moving on a perfectly level, frictionless surface, its velocity remains constant. A real hockey puck on a flat surface of ice would experience sliding friction—a force having a direction opposite to that of the moving puck—and would slow down.

**Problem 5.3**

A 60-g tennis ball moving at 30 m/s is struck by the net of a tennis racquet. After a collision that lasts 0.01 s, the ball is springing back in the opposite direction with a speed of 20 m/s.

(a) What is the momentum of the ball before and after being struck with the racquet?

(b) What is the change of momentum of the ball?

(c) What average force was applied to the ball while it was in contact with the racquet?

(d) What was the average acceleration of the ball while it was in contact with the racquet?

**Solution**

(a) $p_{before} = mv_{before}$ 　　　　　 $p_{after} = mv_{after}$

　　　　$= (0.06 \text{ kg})(30 \text{ m/s})$ 　　　　　$= (0.06 \text{ kg})(-20 \text{ m/s})$

　　　　$= 1.8 \text{ kg·m/s}$ 　　　　　　　$= -1.2 \text{ kg·m/s}$

(We use a minus sign for the velocity and momentum after collision to indicate that the ball's direction has been reversed.)

(b) $\Delta p = (1.8 \text{ kg·m/s}) - (-1.2 \text{ kg·m/s})$

　　　$= 3.0 \text{ kg·m/s}$

(c) $F = \Delta p / \Delta t$

　　　$= (3.0 \text{ kg·m/s})/0.01 \text{ s}$

　　　$= 300 \text{ N}$

This average collision force of about 68 pounds is typical of a fast game of tennis.

(d) $a = F/m$

　　　$= 300 \text{ N}/0.06 \text{ kg}$

　　　$= 5000 \text{ m/s}^2$

This acceleration is obviously large, but it is experienced very briefly. If the same acceleration could be maintained for 1 s instead of 0.01 s, the tennis ball would depart from the racquet at nearly 5 km/s.

Newton's second law as expressed in Equations 5.6a and 5.6b meets with difficulty when the speed of the body (relative to some reference frame) being accelerated is extremely high. As the body approaches the speed of light, the mass of the body changes significantly, in accordance with Einstein's special theory of relativity. Changes of momentum, however, remain proportional to whatever force is applied (as stated in Equation 5.5), no matter what reference frame is chosen for measuring speed.

# Interaction

In many elegant ways, the natural world is full of symmetry. Visual symmetry (the geometrical symmetry that we can see with our eyes) exists in the internal design and often in the external shape of crystals. Abstract symmetry, such as the way that positive and negative electrical charges can electrically balance each other, pervades the worlds of the small and the large.

## Action and Reaction

**Newton's Third Law of Motion**
Whenever one body exerts a force on a second body, the second body exerts a force of equal magnitude and opposite direction on the first body.

Newton's third law is a simple statement of symmetry, which is often paraphrased as "for every action, there is a reaction." Explicitly, Newton's third law of motion says: Whenever one body exerts a force on a second body, the second body exerts a force of equal magnitude and opposite direction on the first body. This law means that forces are really *interactions*. No force that is exerted on a body can exist by itself. Any one force must be accompanied by a twin force exerted in the opposite direction.

You cannot exert a force on something without that something exerting an oppositely directed force on you. If you kick a brick wall, your foot will hurt. The force you exert on the wall results in an equal and opposite force exerted by the wall on your foot. When you drive a nail with a hammer, the action of the hammer on the nail is mirrored by the action of the nail on the hammer in bringing the hammer to rest.

When a bat strikes a ball, the ball momentarily accelerates in one direction, while the bat momentarily accelerates in the opposite direction (Figure 5.9). The acceleration of the bat is felt as a recoil during the brief moment of contact. It is clear that the two objects, bat and ball, do not have equal accelerations. Why? Because their masses are unequal. By Newton's second law, the same force exerted on bat and ball should result in a greater acceleration for the ball, because its mass is less. The moving bat accelerates (in this case it slows down while in contact with the ball), but not by as much.

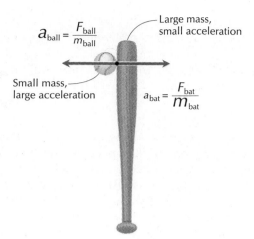

$$a_{ball} = \frac{F_{ball}}{m_{ball}}$$

Large mass, small acceleration

Small mass, large acceleration

$$a_{bat} = \frac{F_{bat}}{m_{bat}}$$

**FIGURE 5.9** When a ball is struck by a bat, the magnitudes of the forces experienced by the bat ($F_{bat}$) and the ball ($F_{ball}$) are equal at any given instant. These two forces are symbolized here by oppositely directed arrows of the same length. By Newton's second law (Equation 5.6b), however, $a_{ball}$ is greater than $a_{bat}$ because $m_{ball}$ is less than $m_{bat}$.

Note that each application of Newton's third law applies to forces *that act on two different interacting bodies.* As we saw earlier, it is possible for two or more forces to act on a single body. Two equal but opposite forces acting on the same body are *not* an action-reaction pair.

To clarify this notion, imagine a book lying on a table. There are two forces acting on this book: the downward force of gravity and the upward force the table exerts on the book that keeps the book from falling. Because these two forces are equal and opposite, the net force on the book is zero and the book does not accelerate. Newton's third law does not apply to these forces, because we are considering forces exerted on *one* body—the book alone. The third law applies to equal and opposite forces on *two* bodies that are interacting with each other. Here are two examples of action-reaction pairs for a book resting on a table: (1) The book presses down on the table and the table pushes up on the book. (2) Earth's entire mass attracts the book with a downward force of a certain magnitude (which is why the book would fall if the table were not there) and, in turn, the book attracts Earth with a force of the same magnitude.

What happens when a book is released from an elevated position and allowed to fall? According to Newton's third law, there are two consequences: The book accelerates downward toward Earth and, at the same time, *Earth accelerates upward* to meet the book. This really does happen! However, Earth's acceleration is far too small to be noticed or measured because Earth's mass (and therefore its inertia) is so large.

The third law's significance is often overlooked in common experience, though its effects can be noticed. When a cannon is fired and the cannonball accelerates through the barrel, the cannon receives an equal and opposite push, which is why it jumps back. Any mechanism that throws a projectile—a gun firing a bullet, a taut bow releasing an arrow—experiences a similar recoil. When you walk, your feet push against the ground and the ground in turn pushes against your feet. The ground's reaction force on your feet propels you forward.

A jet airplane operates by forcing air and exhaust gases through its engines, front to back. As the exhaust is pushed toward the rear of the

**PHOTO 5.7** The force propelling this space shuttle upward against gravity is provided by the gases expelled downward.

plane, the plane is pushed forward by the exhaust. Spacefaring rockets carry propellents that need no influx of air (or oxygen) from the outside, so they, unlike jet planes, can operate perfectly well above the atmosphere. The exhaust gases resulting from the combustion of the propellents are accelerated to very high speeds and expelled from rear nozzles. The accelerated gases, in turn, push back and accelerate the rocket and its payload. In order to receive a forward thrust, a rocket need not push against anything except what it expels.

## Conservation of Momentum

In the previous chapter (dealing mostly with chemistry), we introduced the *law of conservation of mass*—the idea that the mass of the products of a chemical reaction equals the mass of the reactants. The general term *conservation* refers to a characteristic or property of something that remains the same after some kind of change. There are several conservation laws associated with physics, and all are extremely important.

# Conservation of Linear Momentum

All the experiments that have been performed on moving bodies (on both the microscopic and macroscopic scale) have led physicists to conclude that in the absence of a net external force, the linear momentum of a body, or of a system of bodies, is always conserved. This statement is known as conservation of linear momentum. Some examples of this law follow.

When billiard balls roll short distances over a smooth, flat surface, there is very little friction and gravity does not act to change motion. If we could eliminate *all* friction, a ball of given mass would maintain the same linear momentum, $\mathbf{p} = m\mathbf{v}$, in accordance with both the law of inertia and the law of conservation of momentum. Let us assume these conditions are in place. What happens when one moving billiard ball A collides head-on with a stationary ball B of equal mass? Experiments show, and Figure 5.10(a) suggests, that ball A suddenly stops on contact and ball B starts moving at the speed ball A had just before the collision. Before the collision, all the momentum in the system of ball A and ball B resided in the motion of ball A. After the collision, all the momentum resides in the motion of ball B. What remains the same in this system before, during, and after the collision is the total momentum; *linear momentum is conserved.*

In Figure 5.10(b), one moving hockey puck with a very sticky rim collides with an identical but stationary hockey puck on a flat, icy surface. Measurements show that the two pucks, which stick together after colliding, move at half the speed of the puck that was moving before the collision took place. An accounting of the total momentum ($m\mathbf{v}$) of the two pucks after the collision shows that it equals the momentum carried by the one moving puck before the collision.

If collisions take place head-on, as in Figures 5.10(a) and 5.10(b), there is no change of direction associated with the momentum vector. Oblique collisions would certainly introduce some complications, but measurements would still show that the vector sum of the momentum before and after the collision is the same.

There is an obvious qualitative difference between the type of collision shown in Figure 5.10(a) and the type shown in Figure 5.10(b). The idealized *elastic* collision shown in Figure 5.10(a), which is similar to the bounce of a golf ball on a concrete sidewalk, differs from the *inelastic* collison shown in Figure 5.10(b), which is rather like a water balloon hitting the ground. This qualitative difference is related to energy, a subject we will take up in the next chapter. Regardless of this difference, linear momentum is always conserved.

In Figure 5.10(c), the cherry bomb suspended on a string can be thought of as being a group of many small particles. After all, it *will be* many small, shredded pieces as soon as it explodes. Before the explosion takes place, the velocity of each particle in the cherry bomb is zero, and the sum of the momenta (note that *momenta* is the plural of *momentum*) of all the particles is zero. After the collision, each particle has its own momentum relative to where the cherry bomb *was,* but the momenta of particles moving in any one direction are canceled by the momenta of particles moving in the opposite direction. The vector sum of the momenta of all the particles is zero. Mind you, this example assumes that gravity does not play

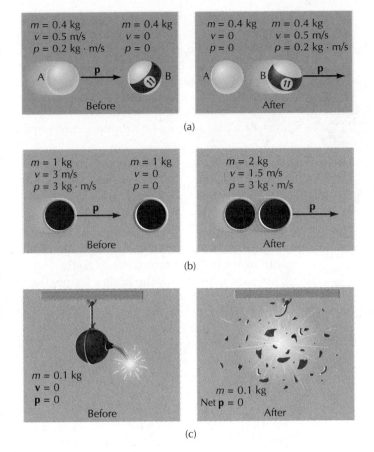

**FIGURE 5.10** Three examples of conservation of linear momentum (**p**). (a) Before *and* after the collision, the total momentum is 0.2 kg·m/s to the right. Both the magnitude and the direction of the momentum remain the same. (b) Before *and* after the collision, the total momentum is 3 kg·m/s to the right. (c) Before the explosion, the momentum of the cherry bomb was zero because its velocity was zero. After the explosion, the *vector sum* of the momenta of the debris is zero.

a role as an external force. That is, it explains how an exploding bomb would behave if it were detonated in a weightless environment.

In all three examples, the total momentum of each system remains constant, although the momentum is distributed differently before and after the collision or explosion. A whole host of more complicated experiments have shown that linear momentum is always conserved in collisions and explosions, whenever the effects of external forces (gravity, air resistance, and so forth) are eliminated.

## Conservation of Angular Momentum

The simple examples of straight-line (linear) motion we have introduced thus far are relatively easy to analyze, but they are not necessarily typical of much of the motion taking place around us and at every scale. Molecules undergo both *translational* (point-to-point) and *rotational* (spin) motion as they move about chaotically within a body. Even subatomic particles of vanishingly small size are known to behave as if they spin. In the macroscopic world, moving fluids feature vortices, ranging from the miniature whirlpools set up in a stirred cup of coffee to the powerful currents of moist air whirling about in a tornado. In the larger universe, virtually all bodies—from moons and planets to galaxies—rotate and revolve. Thus,

*angular motion,* or motion around a point or an axis, is important to consider if we are to understand a great many aspects of our universe.

In the same way that linear momentum describes the "quantity of motion" of a body headed in some direction, **angular momentum (l)**, describes the "quantity of angular motion" of a body moving sideways or obliquely past a given point or axis. Angular momentum is easy to define when we consider the special case of a particle of mass $m$ moving in a circle of radius $r$ at speed $v$ around a central point. The magnitude of $l$ is then

$$l = mvr$$

Equation 5.7

**FIGURE 5.11**   The right-hand rule arbitrarily establishes the sense of direction for angular momentum. When the fingers of your right hand curl around the top in the same direction that the top rotates, your right thumb points in the direction of **l**. If a left-hand rule were adopted instead, then **l** would point down.

The angular momentum of a planet (approximated as a "particle") *revolving* around the sun in a circular orbit would thus be the product of the planet's mass, its speed in orbit, and the distance between the planet and the sun.

The angular momentum associated with a solid body's *rotation* is the sum of the angular momenta of all the individual particles that it contains. This sum depends on the body's shape, the distribution of mass within the body, and the location and orientation of the axis around which the body spins. The job of computing this sum is made easier with the help of calculus, especially if the object is symmetrical.

A full description of angular momentum **l**, which is a vector, must include its direction. By definition, this direction is taken to be along the axis of rotation or revolution around which the body moves. To distinguish between the two opposite directions along an axis, we may use the *right-hand rule* (see Figure 5.11). When the right-hand rule is applied to Earth's rotation (counterclockwise when viewed from above the North Pole), we find that **l** points out of the North Pole and toward the north celestial pole.

Experiments show that some agent is needed to cause a body to spin faster or to cause a body to slow its spin. This agent, which is a force applied in some manner sideways to the body's axis of rotation, is called a *torque.* When you grasp a steering wheel and apply forces to turn the wheel, you are applying torques to the shaft that runs down the steering column. Torque can be loosely thought of as the result of a force that tends to produce rotation.

In everyday life, spinning toy tops and freely spinning bicycle wheels experience frictional torques that slow their rate of spin. However, in the absence of a net external torque, the angular momentum of a body, or of a system of bodies, is always conserved. The truth of this statement, which is known as conservation of angular momentum, has been demonstrated in a wide variety of experiments. Since angular momentum is a vector, its direction as well as its magnitude is conserved.

A spinning bicycle wheel would turn forever, thereby maintaining its angular momentum, if its hub had frictionless bearings and if it were spinning in a vacuum with no air resistance. Earth, too, has been spinning on its axis for billions of years by virtue of its tendency to maintain its angular momentum. However, it has always been subject to weak torques caused by its gravitational interactions with the sun and the moon. For over four billion years, these torques have acted as a brake to slow our planet's rotation rate from about once every 10 h to once every 24 h (as we shall see in Box 5.3).

## Conservation of Angular Momentum

In the absence of a net external torque, the angular momentum of a body, or of a system of bodies, is always conserved.

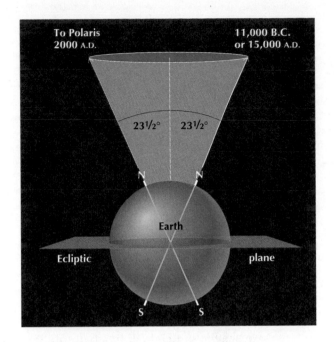

**FIGURE 5.12** Earth's axis completes one precessional cycle in about 26,000 years by tracing out a cone in space. Owing to the precession effect alone, the axis would maintain the same $23\frac{1}{2}°$ tilt, but other factors cause small variations in the tilt angle over long periods of time.

Because the direction associated with a body's angular momentum tends to stay fixed, it is difficult to reorient the axis of a spinning body, such as the axle of a spinning bicycle wheel or the spinning disk inside a gyroscope. (Perhaps your instructor has a demonstration bicycle wheel that will allow you to experience what this is like. Grasp the spinning wheel by the axle and try to change its orientation. What you feel may astonish you.)

A toy top's angular momentum keeps it spinning and contributes to its stability; but once its angular momentum is decreased by friction, gravity pulls it over. Of course, there is something else going on, too: If a spinning top's axis is not at a perfect right angle to a flat surface, it precesses (wobbles). Whenever a spinning top is tipped, Earth's gravity, which pulls in a direction that is not parallel to the top's axis, changes the direction of each particle of matter in the top. The combined motion of all the particles, which are rigidly held in place inside the top, contribute to an overall wobbly motion. This wobbly motion persists for as long as the top spins.

A spinning, wobbly top has an analog in the rotation of Earth. Because of the $23\frac{1}{2}°$ tilt of Earth's axis, the combined gravitational pull of the moon and the sun on Earth (which does not always act along a line parallel to Earth's equatorial plane) causes Earth's axis to slowly precess. Unlike a top, which might execute one complete wobble in 10 or 20 turns, Earth must turn about 9.5 million times in order to complete one precessional cycle. Thus the cycle lasts about 26,000 years (see Figure 5.12).

Probably the most interesting aspect of conservation of angular momentum comes into play in nonrigid bodies (including fluids), which may change in shape or in size. Since the magnitude of angular momentum ($l = mvr$) of a particle of mass $m$ moving around an axis is conserved, any increase in $r$, the particle's distance from the axis, must be accompanied by a corresponding decrease in $v$, the speed of the particle. The opposite is also true: If $r$ decreases by a factor of, say, 2, then $v$ must increase by

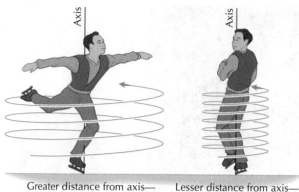

Greater distance from axis—
slower spin

Lesser distance from axis—
faster spin

**FIGURE 5.13**   The angular momentum of a spinning figure skater depends upon the rate at which he spins and the distribution of mass in his body with respect to his axis of rotation. Because angular momentum is conserved, any movement of mass outward from the axis produces a slower rate of spin, and any movement of mass inward produces a faster rate of spin.

a factor of 2. The quantity $mvr$ must remain constant. We can see this effect in action (Figure 5.13) when a figure skater (of constant mass $m$) pirouetting on the tip of one blade draws his arms and one leg inward toward his body (thereby decreasing $r$ and increasing $v$) or extends them outward (thereby increasing $r$ and decreasing $v$).

We see the same effect in a bathtub drain. The streams of water spiraling inward circulate ever faster the closer they get to the drain. Something similar happens in a hurricane, where streams of air speed up and circulate faster as they move inward toward the hurricane's eye.

The Solar System evolved from a spinning disk of gas and dust (solar nebula), which conserved its angular momentum by rotating faster as it contracted. At first, our forming sun (the protosun) at the center of the disk gained much angular momentum from the falling material spiraling in. Later, since there was friction (probably aided by magnetic forces) between the protosun and the diffuse material of the protoplanetary disk, angular momentum stored in the fast-spinning protosun was transferred back to the material that would soon coalesce into the planets. The sun's rotation has been slowed to a crawl as a consequence; today it completes one rotation in just under a month. Despite the sun's enormous mass, its angular momentum now amounts to a paltry 1% of the Solar System's total. Although the *distribution* of angular momentum in the Solar System underwent radical changes during the Solar System's formative stages, the magnitude and direction of the total angular momentum have remained constant.

# Gravitation

Newton's laws of motion describe how things move (or don't move, as the case may be) in response to applied forces or a lack of them. They do not, however, address the fundamental nature of any force in particular. We now know that all forces operating in the physical world are derived from just four fundamental forces or interactions. In the 16th and 17th centuries, only one of these four—gravity—could be described with any degree of confidence. Newton showed that the same gravity that everyone was familiar with on Earth extends beyond Earth. While doing so, he deduced the mathematical formula that quantitatively describes how gravity works—not just

**PHOTO 5.8**    As the Flying Indians of Veracruz "unwind" from the top of the pole, the speed of their motion around the pole tends to decrease as the radius of their "orbital" paths around the pole increases. Friction and gravity (which produce external torques) play roles in controlling the rate of the descent, so angular momentum is not strictly conserved as the Indians descend on their spiral paths.

for some particular environment (such as at Earth's surface), but for all circumstances anywere in the universe. Consequently, his formula is sometimes called the *law of universal gravitation*.

# How Gravity Affects Motion

Newton believed, as Galileo did before him, that when air resistance is eliminated (or at least greatly minimized), all bodies falling toward Earth's surface have the same constant acceleration. According to legend, Galileo dramatically demonstrated this in front of numerous witnesses when he dropped various dense objects of different weights from the top of the Leaning Tower at Pisa in Italy and showed that their times of fall were virtually identical. Obviously, Galileo was clever enough not to drop a feather during the same demonstration. He knew that the effect of air resistance on a feather would nearly overwhelm any tendency of the feather to fall.

When we repeat this kind of experiment today, we find that over relatively small distances of, say, 50 m or less, various dense bodies hit the ground at essentially the same time. A low-density object such as a Ping-Pong ball or a feather takes longer to fall in air, because air resistance (the frictional force of the air moving upward relative to the object) increases quickly in intensity until it is equal to the object's weight (the downward gravitational force on it). When the two forces, air resistance and weight, become equal in magnitude, the falling object accelerates no further. We then say it has reached its *terminal speed* (Figure 5.14).

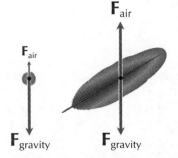

**FIGURE 5.14**    Two bodies having the same mass, a pea and a feather, are falling in the air at about 1 m/s speed. The feather has already reached its terminal speed (maximum speed of falling in air) of about 1 m/s, because the upward force of the air pushing against the large area of the feather (the air resistance) is equal to the force of gravity on the feather (its weight). The pea continues to pick up speed because its weight is greater than the air resistance on it; meanwhile, the pea experiences more air resistance as it falls faster. The pea attains terminal speed when the air resistance on it becomes as large as its weight.

**PHOTO 5.9** Sky divers accelerate downward after jumping from a plane, yet they quickly (after about 10 s) reach terminal speed as the force of the air moving against them becomes equal to their weight. The spread-eagled, belly-down orientation of these sky divers produces a terminal speed of about 125 mi/h.

There is no air resistance when bodies fall in a vacuum (a space with nothing in it). Under vacuum conditions, experiments show that all free-falling masses accelerate at exactly the same rate. This rate, which is called the acceleration of gravity at Earth's surface (**g**), is approximately 9.8 m/s$^2$ downward. For several reasons, **g** varies slightly at different places on Earth's surface. For simplicity, we shall assume from now on that the magnitude of the acceleration of gravity ($g$) is 10 m/s$^2$. This rounded-off value makes calculations easier, though we do sacrifice a little accuracy. Remember that for the following discussion we are assuming a negligible air resistance.

If $g \approx 10$ m/s$^2$, then a freely falling body released from a stationary position at time $t = 0$ attains a speed of about 10 m/s at $t = 1$ s; 20 m/s at $t = 2$ s; 30 m/s at $t = 3$ s; and so on. Now let us determine the distance $d$ moved by the body, at 1-s intervals, as it falls. For this calculation we again make use of the formula $d = vt$. Since the speed ($v$) of the falling body is changing, we need to use the *average* speed over the time interval.

At $t = 1$ s, the falling body has attained a speed of 10 m/s. This particular speed, however, was not maintained over the whole time interval between $t = 0$ and $t = 1$ s. It was merely the speed at the end of the interval. The average speed is the average of the initial speed (zero) and the final speed (10 m/s). Using the average speed, we get $d = vt = (5$ m/s$) (1$ s$) = 5$ m. The body falls 5 m in the first second.

At $t = 2$ s, the falling body has attained a speed of 20 m/s. By then, its average speed (since it was released 2 s earlier) is 10 m/s, so $d = vt = (10$ m/s$) (2$ s$) = 20$ m. The body has fallen a total of 20 m. Applying this process once more, the body at $t = 3$ s has an average speed of 15 m/s over the 3 s, and so has fallen a total of 45 m. Table 5.3 summarizes and extends these results.

## What Is Weight?

Experiments confirm that *all* bodies, regardless of mass or weight, falling side by side in a vacuum experience the same acceleration. Why? First, let us state explicitly what weight is: The **weight** of a body is the gravitational force exerted on it.

| TABLE 5.3 | Time ($t$) | Speed ($v$) | Total Distance ($d$) |
|---|---|---|---|
| Time, Speed, | 0 | 0 | 0 |
| and Distance | 1 s | 10 m/s | 5 m |
| for Objects | 2 s | 20 m/s | 20 m |
| Accelerated | 3 s | 30 m/s | 45 m |
| at 10 m/s² | 4 s | 40 m/s | 80 m |
| | 5 s | 50 m/s | 125 m |

*Note:* The trends shown in this table can be generalized by the formulas $v = at$ and $d = gt^2/2$.

Since weight is a force, it can be calculated by means of Newton's second law, $F = ma$. When calculating the weight $F_g$ of a given mass $m$, we substitute $g$ for $a$, since $g$ is the acceleration that applies to masses that are allowed to fall freely near Earth's surface. The magnitude of the weight ($F_g$) is then

$$F_g = mg$$ 

Equation 5.8

The direction of the weight is, of course, downward. Since weight is proportional to mass, a body with (say) twice the mass of another has twice the weight. It also has twice the inertia, or resistance to change in motion. In other words, twice as much gravitational force is available to accelerate an object that is twice as hard to accelerate because of its doubled inertia. The result is that the acceleration is the same (see Figure 5.15).

The interplay between the two characteristics of mass—inertia and weight—explains why the various objects Galileo released from the top of the tower arrived on the ground at the same time.

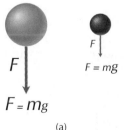

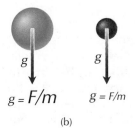

**Problem 5.4**

If a woman has a mass of 60 kg, approximately how much does she weigh (in SI units)?

**Solution**

The implicit assumption here is that the woman is on Earth, where $g \approx 10$ m/s². So

$$F_g = mg$$
$$= (60 \text{ kg})(10 \text{ m/s}^2) \approx 600 \text{ N}$$

This weight of 600 N is equivalent to 135 pounds.

**FIGURE 5.15**   (a) The weight of a more massive body is greater than the weight of a less massive body, because the gravitational force $F$ is greater. (b) The acceleration $g$ is the same for both, because the more massive body has more inertia and therefore a proportionally greater tendency to resist the greater pull of gravity.

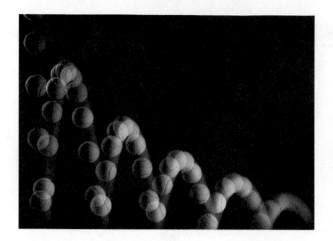

**PHOTO 5.10**   This strobe shot of a bouncing ball illustrates several things. Each trajectory between bounces is parabolic, which is consistent with gravity accelerating the ball in the vertical direction only. The ball loses some of its vertical speed, and also a little of its horizontal momentum, immediately after each bounce, because each bounce involves a collision that is not quite elastic. The ball would repeat exactly the same trajectory between bounces if the bounces were perfectly elastic and there was no air resistance.

# Projectile Motion

We will now consider what happens when a body or a projectile is thrown horizontally, such as a bullet fired from a level gun. Once the bullet leaves the gun barrel and is no longer being accelerated by the explosive charge that got it moving, the bullet has a tendency to move at constant velocity, in accordance with Newton's first law. (For simplicity, we will ignore the retarding force of air resistance, which is actually considerable for a speeding bullet.)

Earth's gravity does not affect the bullet's speed in the horizontal direction, because gravitational force does not act in that direction. Therefore, the bullet continues to cover equal amounts of horizontal distance in equal periods of time. The bullet does, however, accelerate downward as it is pulled by Earth's gravity, in accordance with Newton's second law. What is the magnitude of this acceleration? It is the same as that of a freely falling body, approximately 10 m/s$^2$. No matter what *horizontal* speed the bullet has, the *vertical* speed (downward) is 10 m/s after 1 s, 20 m/s after 2 s, and so on. Also, the bullet falls from its projected straight-line path a total of 5 m after 1 s, 20 m after 2 s, and so on, just as our summary of distance fallen in Table 5.3 indicates. More examples of the independence of horizontal and vertical motion are illustrated in Figure 5.16.

# Newton's Formulation of Gravity

It is said that Newton began thinking intensely about gravity after watching a ripe apple break free from a branch and accelerate toward the ground. Newton's thoughts were certainly not restricted to this rather mundane occurrence. Galileo had already demonstrated that all massive bodies fall at the same constant acceleration toward Earth. Also, by this time Johannes Kepler had developed three laws of planetary motion (Box 5.2), which successfully described how (if not why) the planets moved about the sun in Copernican, as opposed to Ptolemaic, orbits. Kepler and several

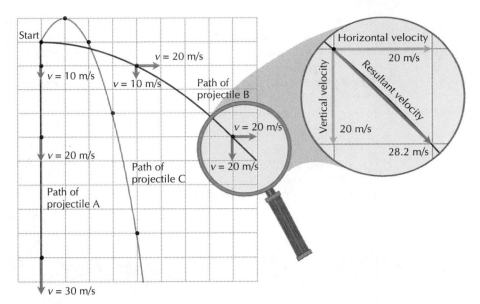

**FIGURE 5.16** Successive positions for three projectiles are shown at 1-s intervals. Projectile A is released from a stationary starting position. After 1 s it has dropped 5 m and moves at 10 m/s. After 2 s it has dropped 20 m and moves at 20 m/s. Projectile B starts with a velocity of 20 m/s, directed horizontally to the right. After 1 s it has dropped 5 m and moves at 10 m/s downward and at 20 m/s to the right. After 2 s it has dropped 20 m and moves at 20 m/s downward, like projectile A. Projectile B, however, maintains its horizontal speed, even as it picks up vertical speed due to the accelera- tion of gravity. The resultant, or combined velocity of projectile B after 2 s is greater than that of either its horizontal or its vertical components, as we see in the enlargement. Projectile C is thrown from the starting point at 10 m/s upward (and incidentally sideways at 5 m/s), so that it reaches the top of its parabolic trajectory in 1 s. How fast is it moving vertically and horizontally in the successive positions shown? (*Note:* This graphical simulation assumes that $g = 10$ m/s$^2$ and that air friction is negligible.)

other astronomers and scholars had been guessing that gravity is an attraction between celestial bodies that gets weaker with distance. Newton, however, envisioned that the same "gravity" that attracts an apple to the ground is responsible for keeping the moon in its orbit around Earth and for keeping Earth and the other planets in their orbits around the sun.

Putting all the pieces together, Newton came to the following conclusion: There exists an attractive (gravitational) interaction between any two material bodies in the universe, and the magnitude of the force on each body is directly proportional to the product of the masses of the bodies and inversely proportional to the square of the distance between their centers.

Today, Newton's law of gravitation in symbolic form looks like this:

$$F_g = Gm_1m_2/R^2 \hspace{3cm} \text{Equation 5.9}$$

There are four variables in this equation: $F_g$ is the attractive force of gravity on each of the two interacting bodies; $m_1$ and $m_2$ are the masses of these two bodies; and $R$ is the distance between the centers of the two bodies. The word *center* as used here refers to *center of gravity,* which is the same as the geometric center of a spherically symmetric body like a planet or a star. The symbol $G$, called the universal constant of gravitation, is a constant of proportionality. It expresses the inherent strength of the gravitational force; that is, it tells us how strong gravitational interactions are in a

### Newton's Law of Gravitation

There exists an attractive (gravitational) interaction between any two material bodies in the universe, and the magnitude of the force on each body is directly proportional to the product of the masses of the bodies and inversely proportional to the square of the distance between their centers.

general sense. In SI units, $G$ has the experimentally determined value $6.67 \times 10^{-11}$ N·m$^2$/kg$^2$. As far as we know, the value of the universal constant $G$ is the same everywhere and remains constant over time.

Newton's law of gravitation says that the attractive force due to gravity between any two bodies is greater whenever either or both of the masses of those bodies is greater, and whenever the distance between the bodies is less. Of course, bodies may interact by forces other than gravity. Magnets can both attract and repel each other by a force that has nothing to do with gravity.

Since $R$ appears in the denominator of the expression on the right of Equation 5.9, and because $R$ itself is squared, the gravitational force between any two bodies *decreases* rapidly as the distance between them *increases*.

---

## BOX 5.2
# Kepler's Laws of Planetary Motion

Johannes Kepler's three *laws of planetary motion* reconciled Copernicus's fundamental notion of a heliocentric universe (see Box 3.4) with the reality of observational facts. As a gifted young mathematician, Kepler had the good fortune of serving under the astronomer Tycho Brahe, whose systematic observations of the sun, moon, and planets in the late 16th century constituted the most complete and accurate set of data on the Solar System at the time.

Kepler eventually acquired Tycho's carefully kept records and used them to show that the Copernican model was flawed in that it presumed *circular* orbits. In trying to superimpose curves of various shapes onto the known orbit of Mars, Kepler discovered that the closed curve called an *ellipse* seemed to fit best. After generalizing the idea of elliptical orbits for all the planets, Kepler went on to discover further relationships. In summary, Kepler's three laws describe how (not necessarily why) the planets move around the sun in a geometrical sense and how the sizes and periods of the orbits of the various planets are related to each other in a mathematical sense.

**Kepler's first law** states that *each planet moves around the sun in an orbit that is an ellipse, with the sun at one focus of the ellipse.* Ellipses can be of various shapes (eccentricities) and can be visualized in a number of ways. Tilt a flat disk so that you see it obliquely, and its edge will appear as an ellipse. The greater the tilt, the flatter (the more "eccentric") the

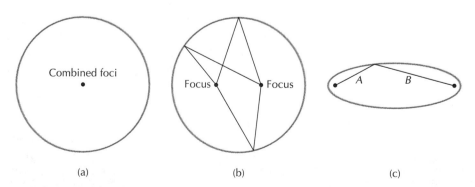

(a)　　　　　(b)　　　　　(c)

**FIGURE 5.17**  Planets can follow a variety of circular or elliptical orbits in accordance with Kepler's first law. The sun lies at the center of any circular orbit (a) and occupies one focus of any noncircular orbit, (b) or (c). Most of the planetary orbits in our Solar System are similar to circles, with foci very close together. Pluto has a somewhat eccentric orbit that resembles the ellipse in (b). Comets typically follow orbits as eccentric as the ellipse shown in (c). For any ellipse, the sum of the segments $A$ and $B$ to any point on the elliptical curve is constant.

Conversely, the gravitational force rapidly increases as the distance between the two bodies decreases. The law of gravitation is an example of what is called an *inverse square law,* because the quantity in question, the force of gravity, varies in proportion to the inverse square of the distance. (We will explore other inverse square laws that deal with other phenomena in future chapters.) To understand how the law works in a comparative sense, study the examples given in Figures 5.19 and 5.20.

Newton had to overcome some serious mathematical difficulties in order to prove his law of gravitation. One difficulty was deciding how a spherical body attracts a particle (for example, how the entire mass of Earth attracts an apple). The answer is that a spherical body gravitationally attracts everything as if all of its mass resided at its center. For this reason,

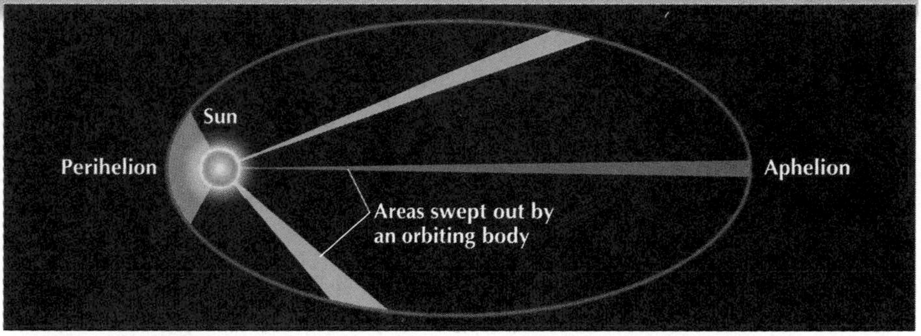

**FIGURE 5.18**   Kepler's second law states that the line joining a planet (or a comet) and the sun sweeps out equal areas (four of which are shown in this diagram) in equal intervals of time. Consequently, the orbiting body must move faster when near its closest point to the sun (called *perihelion*) and slower when near its farthest point from the sun (called *aphelion*).

ellipse. If you slice a cone perpendicular to its axis, you get a circle; slice it obliquely, but not steeper than either side of the cone, and you get an ellipse (see Figure 5.23).

Ellipses can also be generated by finding all the points on a plane at which the sum of the distances to two fixed points (called *foci,* which is the plural of *focus*) is constant. The farther apart the foci, for the same fixed sum of distances, the more eccentric, or elongated in shape, the ellipse is (see Figure 5.17). If the foci are superimposed, the result is a circle, which can be regarded as a special case of an ellipse. (Newton's law of gravitation permits circular orbits, though no planet in our Solar System happens to have one.) The planets are indeed in elliptical orbits, some more eccentric than others. Comets, which orbit the sun just as planets do, typically have very eccentric orbits.

Kepler's first law states that the sun occupies one focus of the elliptical orbit of a planet. The other focus is "empty." The notion of empty focus is a good example of the lack of perfect compatibility between a mathematical or geometrical model and

reality, a situation that arises often in science. A mathematical ellipse has two foci, but the sun, obviously, can only be at one place inside an eccentric orbit.

**Kepler's second law** says that *the line joining a planet and the sun sweeps out equal areas of space in equal intervals of time.* This relationship, as illustrated in Figure 5.18, implies that planets travel faster when near the sun and slower when far from the sun—which is exactly what they do. Comets, in their highly eccentric orbits, move very rapidly when they are close to the sun and very slowly when at the far reaches of their orbits. For this reason, a typical comet with a period of decades or centuries will become conspicuous in our sky (when near the sun) for no more than a few weeks.

Kepler's first and second laws describe the behavior of *individual* planets or of other bodies orbiting the sun. **Kepler's third law** can be used to compare the orbit of one planet with the orbit of another. The characteristics being compared are the period $P$ (the time it takes a planet to complete one orbit), and the average   *(continued)*

the distance $R$ in the law of gravitation refers to the distance between the centers of the bodies. Another difficulty was accounting for the fact that the orbits of the planets are not perfectly circular but elliptical (Box 5.2). In order to solve problems like these, Newton conceived the mathematical techniques we now call differential and integral calculus. (A rival mathematician, Gottfried Leibniz, working independently, invented the same techniques at about the same time.) Interestingly, this remarkable intellectual feat was, for Newton, not an end in itself but a stepping-stone that allowed him to test his hypotheses about motion and gravity.

A crucial test for the law of gravitation was checking to see whether it could explain the behavior of the moon in its orbit. In Newton's day, the size of that orbit had been determined (by means of parallax measurements); its period was known (by simple observation); and therefore, its

| TABLE 5.4 Comparison of $P$ and $R$ for Several Planets | Planet | $P$ (Earth years) | $R$ (AU) | $P^2$ | $R^3$ |
|---|---|---|---|---|---|
| | Earth | 1.000 | 1.000 | 1.000 | 1.000 |
| | Mars | 1.881 | 1.524 | 3.537 | 3.537 |
| | Jupiter | 11.862 | 5.203 | 140.7 | 140.8 |
| | Saturn | 29.456 | 9.534 | 867.7 | 867.9 |

*(Box 5.2 continued)*    "radius" $R$ of the orbit. Kepler's third law states that *the squares of the periods are proportional to the cubes of the average radii.* As long as we use astronomical units (AU) to express the average radius and years (that is, Earth years) to express the period, Kepler's third law can be written simply as

$$P^2 = R^3$$

Table 5.4 lists the measured values of $P$ and $R$ for several of the planets and compares the values of $P^2$ and $R^3$ for those planets. We now know that the values of $P^2$ and $R^3$ for each planet do not precisely match, except for Earth. Considering the accuracy of planetary data in Kepler's day, however, Kepler was justified in considering the third law to be exact.

All of Kepler's laws, it turns out, merely approximate the behavior of the real Solar System. Once Newton had devised his more general laws of motion and accounted for the movements of all heavenly bodies in terms of the law of gravity, he was able to derive each of Kepler's laws using his own laws alone. Kepler's laws are approximate because, in the context of Newton's laws, they

account for the behavior of planets acted upon by the large gravitational pull of the sun alone, and they ignore the smaller gravitational interactions among the planets themselves.

Newton's laws are deeper than Kepler's laws for another reason. Kepler's laws merely describe *how* the planets move and not *why* they move. Kepler did not know what keeps the planets going and why they follow the orbital paths they do. Newton's laws go further. They explain why the planets move (in a mechanical sense and not a philosophical sense) in terms of general concepts like inertia, gravity, and momentum.

Kepler's laws are appealing in their simplicity and also in their applicability to systems both larger and smaller than the Solar System. They can be used to describe the circular or elliptical orbits of artificial satellites around Earth, to model the behavior of the multiple moons of Jupiter or Saturn, and to describe the orbital motions of binary stars.

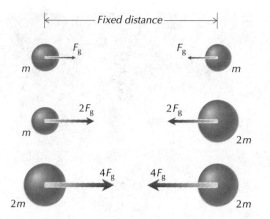

**FIGURE 5.19** The arrows depict the comparative gravitational forces on bodies of different masses at a constant center-to-center distance. Note that gravitational force is really an interaction: Each body is pulled toward the other. According to the law of gravitation, the force doubles if the mass of one body doubles $(2 \times 1 = 2)$ or quadruples if the masses of both bodies double $(2 \times 2 = 4)$.

orbital speed was known. Just like a car rounding a curve, the moon must experience some centripetal force—and an accompanying centripetal acceleration—that keeps it on its curved path. The magnitude of that acceleration, computed by means of Equation 5.3, turns out to be approximately 0.00272 m/s². This feeble acceleration toward Earth's center is all that is needed to keep the moon from escaping. Freely falling bodies near Earth's surface, like ripe apples, accelerate at the much greater 9.8 m/s².

The distance to the moon is known to be very nearly equivalent to 60 Earth radii. So according to the $1/R^2$ term in Newton's law of gravitation, the centripetal acceleration experienced by the moon in its orbit should be $1/60^2 = 1/3600$ as much as the acceleration experienced by an apple falling out of a tree. Sure enough, the ratio of the observed accelerations turns out to be $0.00272/9.8 = 1/3600$. These relationships are illustrated diagrammatically in Figure 5.21.

Newton's law of gravitation was vindicated by this test and also by many more applications of the law during the succeeding centuries. The astronomer Edmund Halley used it to successfully predict, many years in advance, the return of a bright comet that would later bear Halley's name. In the mid-1800s, the law of gravitation was used to predict the existence of a previously unknown outer planet that was thought to be gravitationally *perturbing,* or subtly affecting, the orbital paths of the known planets. The predictions led to the discovery of the planet Neptune. Even today, the

**FIGURE 5.20** The arrows depict the comparative gravitational forces on bodies of equal mass at different center-to-center distances. If the separation of the two bodies doubles, the force is $1/2^2 = 1/4$ as strong. If the separation of the two bodies triples, the force is $1/3^2 = 1/9$ as strong. At four times the distance, the force is 1/16 as strong, and so forth.

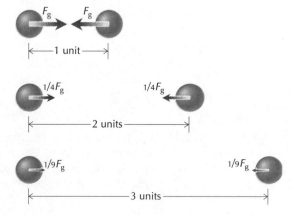

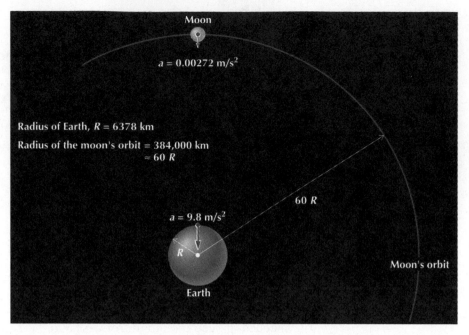

**FIGURE 5.21** Bodies falling near Earth's surface accelerate at 9.8 m/s² toward Earth's center. In order to follow the curved path of its orbit, the moon must have an acceleration of approximately 0.00272 m/s² toward Earth's center. The ratio of these accelerations is approximately 3600. The ratio of the radius of the moon's orbit to the radius of Earth is 60. Newton's law of gravitation predicts that the acceleration (and therefore the force on any given body) due to Earth's gravitational pull should be $(60)^2 = 3600$ times weaker at the moon than on Earth's surface—which it is.

flight trajectories of our interplanetary spacecraft are planned in meticulous detail by applying the law of gravitation.

Let us do a couple of calculations to demonstrate how the law of gravitation operates in our immediate surroundings.

**Problem 5.5**

What is the gravitational force between two 100 kg persons who are standing 1 m apart? (Assume that the 1 m refers to the distance between their centers.)

**Solution**

$$F_g = \frac{Gm_1m_2}{R^2}$$

$$= \frac{(6.67 \times 10^{-11} \text{ N·m}^2/\text{kg}^2)(10^2 \text{ kg})(10^2 \text{ kg})}{(1 \text{ m})^2}$$

$$= 6.67 \times 10^{-7} \text{ N}$$

This is an extremely small force, less than a millionth of a newton. We never notice gravitational forces between individual people because they are so very weak, mainly by virtue of the fact that gravity is an inherently weak force.

**Problem 5.6**

What is the gravitational force between Earth and a 100-kg person standing on it? The known mass of Earth is $5.98 \times 10^{24}$ kg. The radius of Earth, which is essentially the distance between the 100-kg person and Earth's center, is $6.38 \times 10^6$ m.

**Solution**

$$F_g = \frac{Gm_1m_2}{R^2}$$

$$= \frac{(6.67 \times 10^{-11} \text{ N·m}^2/\text{kg}^2)(10^2 \text{ kg})(5.98 \times 10^{24} \text{ kg})}{(6.38 \times 10^6 \text{ m})^2}$$

$$= \frac{3.99 \times 10^{16} \text{ N·m}^2}{4.07 \times 10^{13} \text{ m}^2}$$

$$= 980 \text{ N}$$

This force of 980 N is the weight of the 100-kg person, which is equivalent to about 220 lb. This force is considerable, mainly by virtue of Earth's enormous mass. (Note that the person's weight can also be computed by $F_g = mg$, where $g = 9.8$ m/s$^2$.)

## Gravity as a Force Field

Newton's law of gravitation describes the attractive force between any two objects that have mass. In a somewhat more abstract way, we can think of gravitational attractions as being the result of interacting *gravitational fields*—that is, fields of gravitational influence existing in space around every material body. Thus, we could say that the moon travels in a curved path around Earth because it is responding to Earth's gravitational field. The idea of a field can be extended to other forces as well. We will discuss electric and magnetic fields in Chapters 9 and 10.

When force fields are represented by imaginary lines in space (or lines drawn on paper), they become visual metaphors that can help us understand some of the details of interaction. Modern physicists may argue that force fields really do not exist—that all interactions are really due to exchanges of particlelike entities ("gravitons" in the case of gravity)—but the utility of force fields is unquestioned.

Point masses or spherical bodies have simple gravitational fields around them as long as the influences of other masses are negligible. In Figure 5.22, the gravitational field around Earth has a simple pattern if we ignore the influences of other bodies. The nearest such body capable of significantly disrupting the field would be the moon. All the field lines in Figure 5.22 point inward, suggesting that any body introduced into the field would be pulled in the direction of the field, which is toward Earth. Any small mass $m$ placed within the field will experience a force parallel to the direction of the field at that point, and the magnitude of the force will depend on the density (or spacing) of the field lines at the point. When the lines are drawn in three-dimensional space, instead of the two-dimensional space depicted in Figure 5.22, their density, or closeness, varies in proportion to the inverse square of their distance from Earth's center. Thus, the density of the lines gives us a good feel for how the field becomes weaker with increasing distance from Earth.

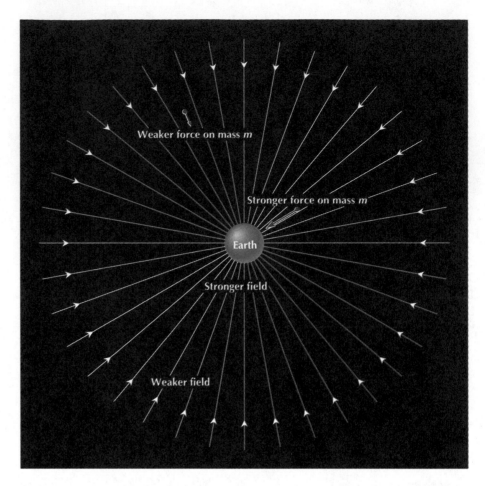

**FIGURE 5.22** In this map of the gravitational field around Earth, the imaginary lines of force, or field lines, shown indicate both the magnitude and the direction of the field. Any number of lines can be drawn to represent the field, but they must be drawn such that the stronger the field, the closer together the lines are. A small mass *m* placed at any spot in the field will experience a force parallel to the direction of the field (toward the center of Earth). The strength of the gravitational force on mass *m* is proportional to the density of the field lines where mass *m* is placed.

Any body moving *across* the field lines associated with a single large mass will have its path deflected in the direction of the field. If the body is moving perpendicular to the lines at just the right speed, its path will be a circular arc. If it is not moving perpendicular to the field lines, or if it is moving faster or slower at a given point than the speed required for a circular arc, it will still move along some kind of *conic section* (the boundary of a cross section of a cone cut by a plane). The four conic section types—the circle, the ellipse, the parabola, and the hyperbola—are illustrated in Figure 5.23.

Gravitational fields extend into space for an infinite distance around any particle or body of larger mass, though the strength of the field around any mass falls off inversely with the square of the distance. Mass exists nearly everywhere in the universe, so the gravitational field at any one point in space is the vector sum of the enormous number of fields present at that point. On Earth and for a considerable distance above it, Earth's contribution to the overall gravitational field overwhelms every other contribution.

The gravitational field on and not far above Earth's surface is quite uniform, because the distribution of Earth's mass is very nearly symmetrical. Nevertheless, small irregularities do exist, in part because of underground deposits of materials that are either more dense or less dense than average. Geologists searching for dense metallic ores or low-density pockets of

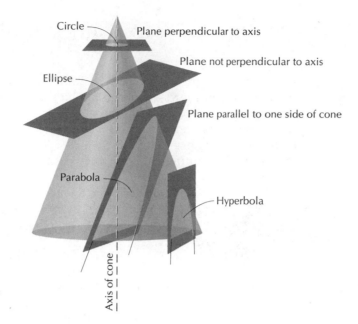

**FIGURE 5.23** Circles and ellipses are closed conic sections produced by cutting a hollow cone with a flat plane. Parabolas and hyperbolas are open conic sections produced by cutting a hollow cone at an angle parallel to or steeper than the angle of the cone's sides, respectively. Kepler's laws refer to closed orbits that are either circular or elliptical. Newton's laws, however, allow for orbits of both the open and closed type.

petroleum or natural gas have been able to exploit these very small irregularities. Instruments called *gravimeters,* usually consisting of an ultrasensitive spring balance, are taken to various spots in a given area to measure the gravitational field. An analysis of anomalies in the field often indicates geological structures of commercial interest (see Figure 5.24).

A similar technique has been used by space scientists to map dense deposits within the moon. By carefully tracking the paths of spacecraft orbiting the moon and noting irregularities in them, scientists located a number of mass concentrations, or *mascons,* on and below the lunar surface. They are associated with the lunar *maria,* which appear as large, flat, circular basins filled with dark-colored rock (see Photo 3.15, Chapter 3). The basins were originally produced by giant meteorite impacts around 4 billion years ago. The basins were subsequently flooded by dense lava that rose from the moon's interior.

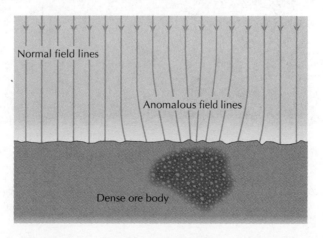

**FIGURE 5.24** Irregularities in the gravitational field at Earth's surface can be measured in an effort to locate concentrations of dense ore underground.

# Orbits: Newton's Laws Applied

Newton's laws of motion and the law of universal gravitation brought order to the universe, or at least order to humankind's perception of it. With the help of Newton's laws—and also Kepler's laws (Box 5.2), which are derivations of Newton's laws—astronomers had available a powerful means of modeling orbits and predicting the future positions and speeds of any two bodies interacting gravitationally. We will now look at two examples of orbital motion, simplified so as to consider the motions of only two bodies at a time.

## BOX 5.3

# Tides: A Complex Gravitational Interaction

      Tides on Earth are an interesting phenomenon affected by several entirely predictable astronomical factors. Chief among them is the position of the moon relative to Earth. The moon can be pictured as "raising" two tidal bulges on the watery parts of Earth: one facing the moon and the other facing away from the moon (Figure 5.25). In the normal pattern of the tides, the cycle of rising and falling ocean levels occurs twice each day, because Earth turns (rotates) underneath the tidal bulges, but the tidal bulges stay fixed relative to the moon.

It would seem as though the moon should gravitationally draw the water in Earth's oceans toward itself and thereby produce a single tidal bulge. Why are there two tidal bulges? Newton's laws can help us answer this question. Refer to Figure 5.26 as we explain.

In accordance with Newton's third law, Earth and moon mutually attract each other and revolve around the barycenter (center of mass, see page 228) between them. Earth's orbit due to the moon's pull is relatively small (because Earth's mass is large). For illustrative clarity in Figure 5.26, Earth's orbit has been greatly exaggerated in size, and the barycenter of the Earth-moon system has been moved far to the left.

The law of gravitation tells us that the side of Earth nearest the moon is pulled more strongly

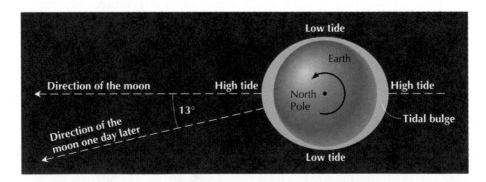

**FIGURE 5.25** Earth's two tidal bulges, which are never more than a meter high over the deep parts of the ocean, stay oriented along a line pointing to the moon. Earth's rotation carries points on its surface, such as coastlines, into and out of the bulges, causing two high tides and two low tides per day. After one full day, however, the moon has moved along its orbit (through an arc of about 13°), and Earth must turn for a longer time (an average of 50 min) to catch up to the new positions of the tidal bulges. Thus the tidal cycles on any given coastline occur an average of 50 min later on any given day then they did on the previous day.

# Earth and Moon

The idealized model of the Earth-moon system sketched in Figure 5.27 exemplifies all three of Newton's laws of motion plus the law of gravity. We assume, for the sake of simplicity, that both bodies move in circular (not elliptical) orbits and that both are isolated from all other gravitational influence, such as the pull of the sun or other planets.

At any given instant, the moon is moving tangent to its orbit at a velocity **v**. Newton's *first law* says that the moon would continue at this velocity forever in the absence of external forces. The moon, however, departs

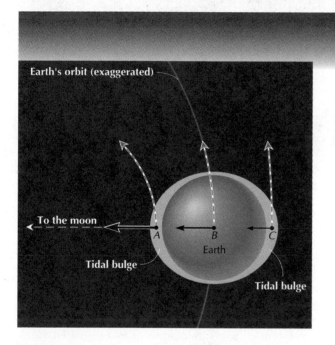

**FIGURE 5.26** As Earth moves "around" the moon, different parts of Earth experience different centripetal forces (solid arrows) toward the moon. Point *A*, closest to the moon, experiences the greatest centripetal force; point *C* experiences the smallest centripetal force. The result is a tidal, or stretching, force that tends to elongate the shape of Earth along a line pointing to the moon. Points *A*, *B*, and *C* have a weak tendency to follow the divergent paths shown by the dashed arrows, but this does not happen because of Earth's strong self-gravitation.

**PHOTO 5.11** Low tide and high tide on the shore of the Bay of Fundy, on the Canadian east coast, can differ in height by more than 50 ft. The long, narrow shape of the bay is responsible for amplifying the tidal differences.

toward the moon than the side of Earth farthest from the moon. We say that the moon impresses a *differential gravitational force,* or a *tidal force,* across the breadth of Earth. This differential force is extremely small in comparison to Earth's own gravitation, which has molded Earth into a sphere.

As indicated by solid arrows in Figure 5.26, point *A* is pulled more strongly toward the moon than point *B*, and point *B* is pulled more strongly

toward the moon than point *C*. The result is that Earth becomes stretched into the shape of an egg— but only very slightly, a matter of a few centimeters at most. Water, being a fluid, responds   *(continued)*

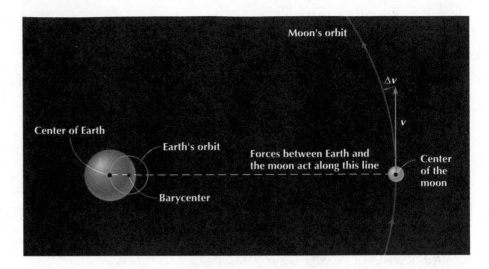

**FIGURE 5.27** In this idealized, not-to-scale model of the moon's orbit around Earth, the moon departs from a straight-line path because of the centripetal force impressed on it by Earth's gravity. At the same time, Earth gets pulled around the moon by the centripetal force impressed on it by the moon's gravity. Earth's orbit is 81 times smaller than the moon's, because Earth is 81 times more massive than the moon.

*(Box 5.3 continued)*    more readily to the subtle differential force of the moon's gravity than does Earth's more rigid crust and interior. Relative to Earth's interior, the water tends to flow toward and accumulate on two opposite sides of the globe, one facing the moon, the other facing away from the moon. The bulges are no more than about a meter high in the open ocean; but just as the bow of a ship pushes up water in front of it, the bulges become higher when colliding with continental coastlines.

While Earth is following its orbit "around" the moon, point *A,* which is experiencing more centripetal force, has a tendency to bend more sharply toward the moon than point *B.* Point *C* tends to bend less sharply toward the moon than point *B.* If these three points were acted on by a tidal force strong enough to overcome their attraction for each other, they would follow the divergent paths indicated by the dashed arrows in Figure 5.26. Earth would break apart if this were the case.

The tidal picture becomes more complex when we consider the sun's gravitational influence on Earth. The sun is more massive than the moon (which makes its tidal influence stronger), and it is also farther away (which makes its tidal influence weaker). These factors conspire to produce a solar tidal force on Earth that is approximately half as large as the lunar tidal force. The sun, therefore, does not control the tides, but it does reinforce and diminish the effect of the lunar tidal force. Reinforcement occurs twice each month, when the

moon's phase is new or full and all three bodies lie on a straight line (recall the lunar phase diagram in Chapter 3, Figure 3.12). Reinforcement in this manner produces what are called *spring tides*: a maximized range between the daily high- and low-tide levels.

Diminished tidal effects occur whenever the moon is at first-quarter or third-quarter phase—again, two times per month. At these times, the solar and lunar tidal forces (which are both vectors) add so that the total force is less than the lunar tidal force alone. The diminished tides are called *neap tides*: a minimized range between the daily high- and low-tide levels.

Because of the moon's (and to a lesser extent the sun's) tidal actions on Earth, Earth's rotation is continually slowed by *tidal friction,* or *tidal braking.* Most of this friction is a result of the world's coastlines bumping into the tidal bulges. The corresponding reaction to this action is a westward push—opposite to Earth's west-to-east rotation. As a result, the time required for Earth to make one complete rotation is increasing, and thus, our days are lengthening by a tiny but measurable amount, about 25 billionths of a second daily.

Newton's third law tells us there is reciprocity in *all* interactions. Although the moon's overall attraction toward Earth is equal to Earth's overall attraction toward the moon, Earth's differential gravitational force (tidal force) across the width of the moon is much greater than the differential force

from a straight-line path because there is an external force on it: Earth's gravitational pull on the moon's mass. That pull, whose magnitude can be calculated from Newton's *law of gravitation,* acts as a centripetal force on the moon. This centripetal force gives rise to a centripetal acceleration in accordance with Newton's *second law* (Equation 5.6b).

The moon departs from its projected, straight-line path by acquiring a speed $\Delta v$ toward Earth after a small interval of time $\Delta t$. Its acceleration during that interval of time ($a = \Delta v / \Delta t$) is 0.00272 m/s$^2$, as we saw in Figure 5.21. The moon's perpetual "sideways" acceleration toward Earth forces the moon to follow a closed path around Earth.

Newton's *third law* tells us that if Earth is pulling on the moon, then the moon must be pulling on Earth with a force of the same magnitude but opposite direction. Thus, Earth, too, moves and accelerates, but not by very much. Because Earth's mass is 81 times greater than the moon's mass, Earth's acceleration is 81 times smaller than the moon's acceleration; and its orbit "around" the moon is 81 times smaller. Strictly speaking, both Earth

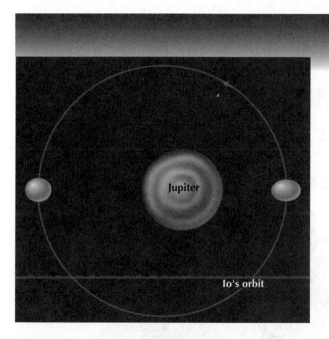

**FIGURE 5.28**   As Io moves alternately closer to and farther from Jupiter along its elliptical path, the tidal force varies accordingly, squeezing and unsqueezing the hapless moon. Io orbits in Jupiter's equatorial plane which, as shown here, lies in the plane of the page.

across Earth. Although it was once spinning quite rapidly, the moon has slowed to a virtual standstill by tidal braking over a relatively short period, perhaps hundreds of millions of years. Today, the periods of the moon's rotation on its axis and revolution around Earth are the same. The moon is tidally locked, which is to say that one side of the moon (the "near" side) always faces us as the moon revolves around us.

The near side happens to contain most of the maria present on the moon, along with the mascons, or mass concentrations, associated with them. The moon's far side has few maria. It is not hard to explain why most of the maria face us. Just before the moon became tidally locked, it was spinning in a wobbly fashion like a loaded die. The moon's denser hemisphere, which was attracted toward Earth with greater force, ended up facing Earth— just as a loaded die is likely to come to rest with its heavy side down.

The tidal braking that took place on the moon before it became tidally locked was a result of the moon's having to continually change shape as it spun. Earth's tidal force elongates the body of the moon so that its long axis must always point toward Earth, even when it is spinning. The continual warping of the moon generated heat in the same way that a lump of modeling clay gets warmer when you squeeze it repeatedly with your hands. Once generated, this heat (a form of energy) was radiated away into space, but not without drawing away energy stored in the moon's rotation.

A similar, but different, kind of tidal heating occurs within Io, the innermost of Jupiter's Galilean satellites (see Figure 5.28). Io is slightly farther from Jupiter than our moon is from us, but Jupiter itself is 318 times as massive as Earth.   *(continued)*

and the moon orbit around a point called the *barycenter* (or center of mass), which lies along the line between the centers of the two bodies. The Earth-moon barycenter is 81 times closer to Earth's center than to the moon's center.

## Artificial Satellites

The paths of artificial satellites—as well as our own natural satellite, the moon—are controlled by Earth's gravity. Newton foresaw the existence of today's artificial satellites when he imagined what would happen to a cannonball fired at a sufficient height above Earth's atmosphere so that air resistance could be ignored. No matter what the initial speed of the cannonball, it drops about 5 m from its projected, straight-line path in the first second, 20 m in the next second, and so on (recall Table 5.3). If the cannonball is fired fast enough, roughly 8 km/s, it maintains whatever altitude it has, because the ground falls away underneath the ball at the same rate

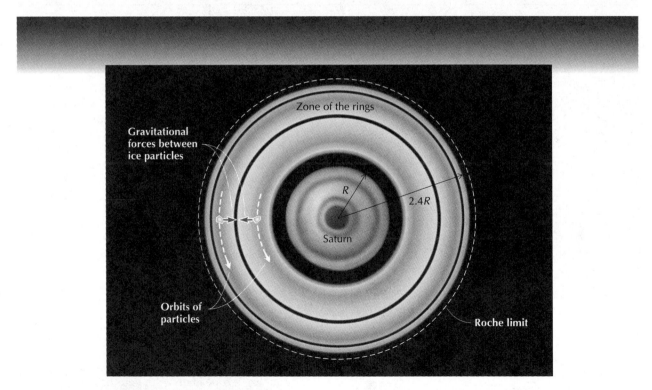

**FIGURE 5.29** Saturn's ring system consists of individual particles orbiting Saturn in a flat plane, spread out to a distance nearly as large as the Roche limit, approximately 2.4 times the radius of Saturn. Within the Roche limit, the gravitational attraction between any two ring particles cannot overcome the tendency for the particles to move apart owing to their different centripetal accelerations and different orbital speeds. Outside the Roche limit, the gravitational force between individual particles is greater than the tendency for the particles to move apart. *Voyager 1* and *Voyager 2* discovered "shepherd moons" lurking near the Roche limit which effectively inhibit outward movement by the ring particles. The *Voyagers* also discovered much unexpected fine structure in the rings, which was probably caused in part by gravitational interactions between ring particles of different sizes. The rings themselves are probably temporary: When the particles collide, they lose some of their rotational energy and tend to fall inward toward Saturn.

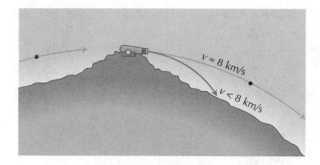

**FIGURE 5.30** Newton believed that if a cannon could be mounted on a mountain high enough to poke above Earth's atmosphere, a cannonball fired horizontally from it at a sufficiently fast speed (about 8 km/s) would orbit Earth. Such a cannonball would circle once in about 90 min.

that the ball falls. This situation is due to the curvature of Earth (see Figure 5.30).

In the late 1950s, both the United States and the former Soviet Union developed the necessary rocket boosters to place satellites into the kind of orbit Newton had envisioned. Once the boosters do their work, satellite payloads can cruise along in "low" orbits, 200 km or so above Earth's sur-

**PHOTO 5.12** Saturn and its ring system as photographed by *Voyager*. Three of Saturn's satellites—Tethys, Dione, and Rhea—are visible, as well as some of the dusky radial features known as "spokes" across the rings. The spokes may consist of sheets of dust suspended above the plane of the icy rings.

*(Box 5.3 continued)*   The tidal force on Io is immense. Io spun down very quickly after its formation, and today Io is tidally locked with respect to Jupiter. Still, there is an enormous amount of heating within Io because it revolves in an elliptical orbit. Io is alternately squeezed and unsqueezed in a period of just 1.8 days, which explains why it is the most volcanically active moon in the Solar System.

The existence of Saturn's rings is another example of tidal forces at work in the Solar System. The innumerable ice particles and boulders that make up the rings may have resulted from the breakup of an icy moon that strayed too close to Saturn some tens of millions of years ago. Regardless of how the material got there in the first place, it will never coalesce into a larger body or bodies as long as it remains within the *Roche limit* associated with Saturn. Inside the Roche limit (Figure 5.29), the tidal force on any two particles exceeds the mutual gravitational attraction of the two particles for each other, and the particles tend to drift apart. Outside the Roche limit, self-gravitation overcomes the disruptive tidal force, and any two nearby particles traveling at a similar speed will drift toward each other.

Tidal interactions can even occur between stars. In close binary stars—stars that orbit each other at close range—tidal forces may severely warp the shapes of the stars and may result in the transfer of mass from one star to another. Mass transfer interactions between stars are responsible for some of the rare, eruptive events we call *novas* and for some of the rare cataclysmic stellar explosions we call *supernovas*.

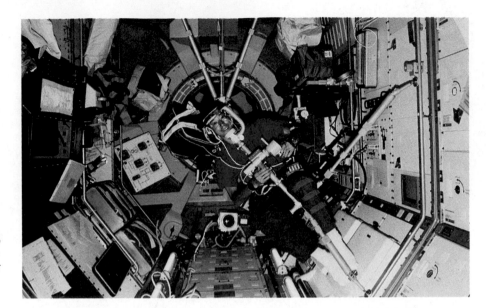

**PHOTO 5.13**  This astronaut floating free inside the Spacelab orbiting satellite exerts no force on the walls of the spacecraft or anything else. He is artificially weightless.

face, without any further propulsion. (Satellites in such low orbits are dragged down after months or years owing to the small but significant amount of air resistance existing even at these rarefied heights in the atmosphere. Satellites moving in higher orbits, where the atmosphere is thinner still, remain in orbit much longer.)

An astronaut inside or outside any spacecraft orbiting Earth experiences *artificial weightlessness*. Earth's gravity is still present in outer space, though it does become weaker with increasing height. The astronaut has "weight" while in orbit in the sense that weight is due to the gravitational attraction of Earth on the astronaut's mass. Even so, the astronaut is falling (accelerating) toward the Earth's center at exactly the same rate as his or her spacecraft. This result agrees with Galileo's discovery that free-falling bodies moving side by side accelerate in the same way. The sensation of artificial weightlessness is not hard to imagine. Astronauts in orbit continually experience what thrill seekers feel for a few seconds when bungee jumping (Photo 5.14). It takes a while to get used to it.

All bodies moving freely in a gravitational field are in a state of artificial weightlessness. The laws of Newton tell us that the paths these bodies follow, when close to a spherically symmetric body like Earth or the sun, are conic sections: circles, ellipses, parabolas, or hyperbolas. The shape of a body's orbit is determined by the body's velocity at any given point in the gravitational field. Figure 5.31 shows the effects of Earth's gravitational field on several different satellites.

The various conic section orbits are viewed from another perspective in Figure 5.32. To stay in a circular orbit, a satellite orbiting close to Earth must move at about 8 km/s and parallel to the ground, as we have seen during our discussion of Newton's imaginary cannon. If, however, the satellite is given a boost of forward propulsion at some point in its orbit (labeled *perigee* in Figure 5.32), it starts to follow an elliptical orbit. Further increases in speed due to further propulsion lead to longer and more eccentric elliptical orbits. At one focus of each of these elliptical orbits lies Earth's center, in accordance with Kepler's first law of planetary motion and New-

ton's more general laws of motion and gravitation. Such satellites move fastest at perigee and slowest at apogee (recall Kepler's second law).

If just enough speed (about 11 km/s) is imparted to a low-orbiting Earth satellite, it departs from Earth on a parabolic orbit; if the speed is greater still, then the orbit is hyperbolic. These bodies will never return, because Earth's gravity will never succeed in overcoming their outward momentum. The critical value of speed (11.2 km/s at Earth's surface)—just enough to ensure an open-ended orbit—is known as the *escape speed*.

The *Voyager 1* and *Voyager 2* spacecraft, which were propelled away from Earth's orbit in the early 1970s to start a grand tour of the outer planets, were given sufficient speed to escape from not just Earth but also the

**FIGURE 5.31** Satellite A has sufficient speed to keep it in a circular orbit. Satellite B is moving somewhat faster than the speed required to keep it in a circular orbit. It follows an elliptical orbit. Satellite C is moving so fast that its inertia overcomes the effect of Earth's gravitational field. It departs on a hyperbolic orbit (or a parabolic orbit if the satellite's speed is barely sufficient to overcome gravity) and will not return. The slow-moving satellite D is drawn inward by the field along the elliptical path shown and collides with Earth. Newton showed that the paths of bodies moving under the influence of a single force that varies inversely as the square of the distance must be conic sections.

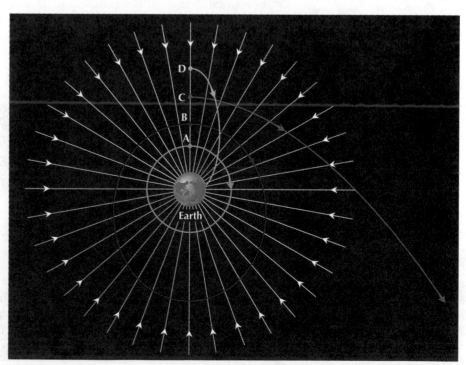

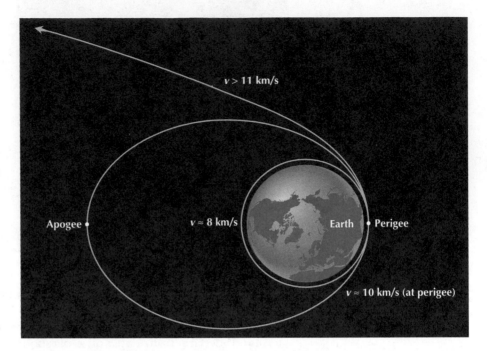

**FIGURE 5.32** A satellite must travel at about 8 km/s to stay in a low, circular orbit around Earth. Additional speed gained through propulsion at perigee produces an elongation of the orbit. The faster the speed at perigee (up to about 11 km/s), the larger and more eccentric the elliptical orbit will be. Satellites moving faster than the escape velocity of 11 km/s at perigee assume open-ended orbits.

sun. They are now beyond the orbit of Pluto, heading outward, never to return to our Solar System.

In general, the more massive and the more dense a planet is, the greater is the acceleration of gravity at the planet's surface and the greater is its escape speed. For example, the acceleration of gravity at Jupiter's "surface" (Jupiter's surface is taken to be the level of its cloud tops) is about 2.5 times greater than the acceleration of gravity on Earth's surface. The escape speed for Jupiter is about 60 km/s. The moon's escape speed is about a fifth that of Earth's, and its surface gravity is about a sixth that of Earth's. (See Table 5.5 for more examples of surface gravity and escape speeds.)

**PHOTO 5.15** The *Galileo* spacecraft, headed for Jupiter, took this photograph of the asteroid Ida on August 28, 1993. Ida, some 56 km in length, was revealed to have its own tiny satellite, only 1.5 km across. The size and shape of the satellite's orbit has not yet been determined.

**TABLE 5.5**
Acceleration
of Gravity and
Escape Speed for
Various Bodies

| Body | Acceleration of Gravity* (in g units) | Escape Speed* (km/s) |
|---|---|---|
| Earth's moon | 0.17 | 2.4 |
| Mars | 0.38 | 5.0 |
| Earth | 1.00 | 11.2 |
| Jupiter | 2.53 | 60 |
| Sun | 275 | 620 |
| A black hole† (of 1 solar mass) | $15 \times 10^{12}$ (at the event horizon) | 300,000 (at the event horizon) |

\* These quantities refer to values *at* the surface of the body. The values are less for any height above the surface.

† A black hole's gravitational pull is sufficient to trap all light emitted within its spherical "event horizon." Inside the event horizon, the escape speed is greater than the speed of light. Outside the event horizon, the escape speed is less, and light moving outward can escape.

---

**BOX 5.4**

# Chaos and the Solar System

   For a time, Newton's laws seemed to suggest that the universe runs like a perfect machine: If we could measure with infinite precision the positions and motions of every body in a given system (or, for that matter, in the whole universe), we would then be able to predict, with perfect accuracy, the positions and motions of those same bodies at any time in the future. Scientists once thought they could approach this theoretical ideal simply by improving their measurement techniques and by acquiring the ability to make the immense number of calculations needed to determine how each part of a system influences every other part.

Today we are wiser. First, we know from quantum mechanics (Chapter 14) that it is impossible, even in principle, to make perfectly accurate measurements of anything. Second, many scientists have come to the conclusion that in systems of three or more interacting bodies, a tiny change in the motion of one body may cause profound, and truly *unpredictable,* changes in the motions of the other bodies in the system.

The unpredictable behavior associated with interacting bodies has become known as chaos. In the physical sciences, **chaos** refers to situations in which the behavior of interacting bodies depends sensitively on the initial conditions and on the environment in which the interactions are taking place. Chaos seems to govern phenomena   *(continued)*

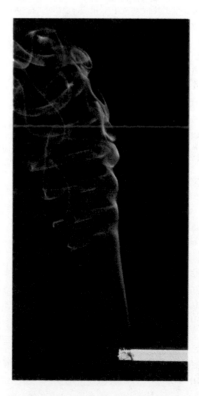

**PHOTO 5.16** The initially smooth and predictable flow of smoke from the tip of a burning cigarette quickly becomes turbulent and chaotic. Predictability breaks down for particles of smoke within the turbulent zone.

Planets with larger escape speeds tend to retain their atmospheres, and planets with smaller escape speeds tend to lose their atmospheres. In Earth's atmosphere, the average speed of the gas molecules is well under the escape speed of about 11 km/s. At any moment, only a tiny fraction of the gas molecules are moving faster than this. They may escape, but only if they lie near the top of the atmosphere, where they might chance to move upward without colliding with other molecules. Temperature also plays a role in the ability of a planet to retain an atmosphere, because higher temperatures mean faster-moving molecules. Earth is massive enough,

*(Box 5.4 continued)*    as diverse as planetary and stellar motions, the motions of atoms or molecules in a gas, patterns of weather and climate on Earth, the swinging of a pendulum buffeted by a capricious breeze, the fluttering of a falling leaf, smoke rising from a cigarette, and the flow of water in a stream. For all of these phenomena, future outcomes for individual particles or units in the system may be fairly predictable over some relativity short interval of time, but that predictability breaks down for longer intervals of time.

Paradoxically, order can be found in many chaotic systems—not in the behavior of individual units of a system but, rather, in the system as a whole. A case in point is a leaf falling from a twig on a quiet day: The leaf starts its journey downward essentially in free fall. For a fraction of a second it moves downward in a predictable way. As the leaf picks up speed, its acceleration slows, and the forces of the randomly colliding air molecules on its underside cause the leaf to start fluttering. We cannot predict exactly how the leaf will flutter or exactly what path it will take to reach the ground, but we can certainly say that the leaf will fall to the ground and that it will flutter as it does so.

A snowflake develops in a cloud in a manner that depends sensitively on conditions such as temperature and pressure. All snowflakes have an underlying six-sided pattern of development (see Photo 5.17), yet no two snowflakes in a cloud are likely to share exactly the same history. Each snowflake moves on its own, impossible-to-predict path through the cloud, experiences different temperatures and pressures, and therefore develops in a unique way. We cannot predict exactly how a snowflake will grow, but we can certainly say that it will reflect the underlying hexagonal structure of water in the solid phase.

**PHOTO 5.17**    Water molecules tend to arrange themselves into hexagonal units when condensing or freezing into ice, which is why every snowflake has a six-sided structure. The hexagonal units themselves can combine in an enormous variety of ways, depending on the external environment.

Until recently, the details of chaotic processes eluded those who tried to study them, because it was impossible or impractical to perform the trillions of calculations necessary to model systems of many interacting parts. Today, supercomputers churn away at such tasks; even so, computer runs of many days are sometimes required to obtain meaningful results.

In recent years, astronomers and mathematicians have been uncovering evidence of chaotic dynamics in the Solar System. Far from resembling a clock with a repetitious and predictable mechanism, the Solar System may actually be inherently unstable. When researchers run several parallel computer simulations of future planetary motion—all differing slightly in their initial conditions—the results turn out to be broadly divergent.

dense enough, and cool enough to have held onto an atmosphere for billions of years.

The moon, however, is not. The moon's original atmosphere consisted of molecules moving with an average speed not much less than the moon's escape speed of 2.4 km/s. The faster molecules escaped, leaving the slower-moving ones behind. Heat from the sun kept stirring up the remaining molecules so that the faster ones continued to escape. The moon lost its atmosphere during a cosmic blink of an eye, and it remains an airless world today.

One study, which considered only the interactions among the sun and the nine major planets, showed that if Mars were nudged just a millimeter from its present path, even far-off Pluto would eventually have its orbit changed in a significant way. There would be little variation at first, but after several million years, the changes in Pluto's orbit would start to grow larger exponentially.

Similar studies indicate that chaos may differ from complete randomness in the sense that changes in a given planet's orbit tend to occur in spurts of a few percent at infrequent intervals. The interactions of the moving bodies of the Solar System tend to reinforce an overall order, though an orderly state is not guaranteed indefinitely.

Other computer simulations have linked changes in the tilt of Mars's axis of rotation to chaotic wobbles in the planet's orbit around the sun. Over a period of tens of millions of years, these studies show, the tilt of Mars's axis varies in an irregular and unpredictable fashion between angles of about 10° and 50° (today it is 25°). These changes, if real, would cause extreme fluctuations in the climate of Mars. Photographic evidence gathered by spacecraft (see Photo 5.18) indicate that Mars has had past climates very different from what it has now.

Earth's axial tilt (now $23\frac{1}{2}°$) varies over a relatively small range, 22° to 28°, so our climate is relatively stable (ice ages notwithstanding). Earth's axial tilt stays within narrow bounds only because our massive moon lies nearby; the moon's role in producing precession in Earth's rotation promotes stability. Mars has no large satellite to perform the same role; it has only two tiny satellites (probably captured asteroids) of trivial mass.

The study of chaos and its underlying order is still in its infancy, yet applications of chaos theory seem to cut across a broad spectrum of the sciences.

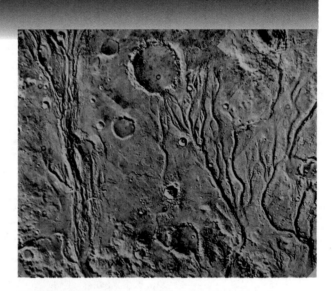

**PHOTO 5.18**  Dry stream channels on Mars hint of an earlier time when Mars had a warmer climate, a thicker atmosphere, and water flowing across its surface.

Rising and falling populations of biological species seem to exhibit chaotic behavior. Superimposed on the orderly beating of a human heart are irregularities of rhythm that many researchers take to be the mark of chaos. Economists are trying to find order and predictability within the fluctuating, chaotic cycles of national or world economy. (Quite naturally, some of them are hoping to read the stock market well enough to make a fortune in timely investments!) We will encounter chaos again in this book in connection with weather (Chapter 7), which is, as everyone knows, maddeningly unpredictable.

# Summary

Motion can be described in terms of speed, velocity, and acceleration. Speed is the distance a body travels divided by the time it takes to do so. A body's velocity is expressed by indicating its speed and direction of motion. Both speed and velocity are relative; they must be expressed with respect to a defined or implied frame of reference. Acceleration is a measure of the time rate of change of the velocity. Acceleration can involve a change in speed, a change in direction of motion, or both.

Velocity and acceleration are examples of vectors, quantities that have both magnitude and direction. Speed is a scalar quantity, one that has magnitude only.

Newton's first law of motion suggests that inertia, the tendency of a body to resist any change in its motion, exists for all bodies. In the absence of a net external force, a body remains at rest or in uniform motion.

A body's momentum (linear momentum) is defined as its mass multiplied by its velocity. Force can be thought of as an agent that can change the momentum of a body. One formulation of Newton's second law of motion relates force, momentum, and time: When acting on a body, a net force produces a change of momentum; the greater the force, the faster the momentum changes with time. Another formulation of Newton's second law relates force, mass, and acceleration: Force equals mass times acceleration.

Newton's third law of motion describes forces as interactions: When one body exerts a force on a sec-

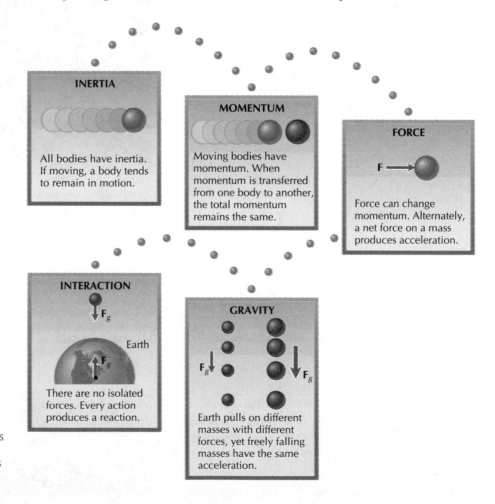

**FIGURE 5.33** The core ideas of this chapter and Newtonian physics are summarized in this graphic.

**INERTIA**

All bodies have inertia. If moving, a body tends to remain in motion.

**MOMENTUM**

Moving bodies have momentum. When momentum is transferred from one body to another, the total momentum remains the same.

**FORCE**

F →

Force can change momentum. Alternately, a net force on a mass produces acceleration.

**INTERACTION**

$F_g$

Earth

$F_g$

There are no isolated forces. Every action produces a reaction.

**GRAVITY**

$F_g$     $F_g$

Earth pulls on different masses with different forces, yet freely falling masses have the same acceleration.

ond body, the second body always exerts an equal and opposite force on the first body.

The linear momentum of a body or a system of bodies, isolated from net external forces, is conserved (remains the same). Angular momentum is a quantity associated with a mass or a system of masses moving around some axis. The angular momentum of a body or a system of bodies, isolated from external influences, is conserved as well.

Newton's law of gravitation mathematically describes how the force of gravity between two bodies depends on the masses of the two bodies and the distance between them. The law of gravitation is an inverse square law, which means that the gravitational attraction between any two bodies decreases with the square of the distance between the bodies.

The force of gravity on Earth's surface is relatively large primarily because Earth's mass is large. Weight is simply a force—the force of Earth's gravity (or some other planet's gravity, if applicable) on a given mass.

Kepler's laws of planetary motion describe how the planets move in their elliptical orbits around the sun. Each of Kepler's laws can be derived from Newton's more general laws of motion and gravity.

Newton's laws of motion and gravitation can be used to describe the orbit of any body around another. Such orbits are either closed circles or ellipses, or open-ended parabolic or hyperbolic curves. Bodies moving on open-ended curves are traveling at escape speed or faster.

Tides on Earth are the result of a differential force across the width of our planet caused by the gravity of the moon and the sun. Tidal interactions have profoundly affected other bodies in the Solar System besides Earth.

In many-body systems such as the Solar System, very small changes in one part of the system may eventually give rise to profound changes elsewhere. This behavior—sensitive dependence on initial conditions or environment—is the hallmark of chaos. Order may be found in chaotic systems when the behavior of the system as a whole is considered.

## CHAPTER 5
# Questions

## Multiple Choice

1. A body traveling at constant speed in a circle has
   a) constant velocity
   b) constant acceleration magnitude
   c) no net force applied to it
   d) a net force that keeps it from stopping

2. Isaac Newton
   a) was the first person to use a telescope for studying the stars
   b) invented a peculiar geocentric model of the Solar System
   c) unified the concepts of gravity on Earth and gravity in the heavens
   d) was the first to discover that planets travel in elliptical orbits

3. Which of the following quantities is a vector?
   a) mass
   b) volume
   c) time
   d) momentum

4. A car travels from New York to Philadelphia (100 km) at an average speed of 100 km/h, instantly turns around, and travels back to New York at an average speed of 100 km/h. Its average speed over the whole trip is
   a) zero
   b) 50 km/h
   c) 100 km/h
   d) 200 km/h

5. If a body is being accelerated,
   a) it must be moving in the direction of the net force applied to it
   b) it cannot be stationary
   c) a net force must be acting on it
   d) it cannot be changing its direction

6. Baseball X is thrown horizontally and baseball Y is thrown upward. Once they leave the pitcher's hand,
   a) ball X has a greater acceleration
   b) ball Y has the greater acceleration
   c) they both have the same acceleration
   d) neither has any acceleration

7. It is always correct to say that a body's inertia is proportional to its
   a) mass
   b) weight
   c) force
   d) acceleration

8. With a quick pull, you can withdraw a sheet of paper from under a stack of 20 or 30 coins. This is a good illustration of Newton's
   a) first law
   b) second law
   c) third law
   d) law of gravitation

9. When a body is in equilibrium,
   a) there must be no force acting on the body
   b) there must be fewer than two external forces acting on the body
   c) the vector sum of two or more external forces acting on the body must be zero
   d) the body must be stationary

10. Which is a correct statement about Newton's second law of motion?
    a) The greater the net external force on a body of given mass, the greater the momentum of the body.
    b) The greater the net external force on a body of given mass, the faster the body moves.
    c) The greater the net external force on a body of given mass, the more the body accelerates.
    d) Forces always give rise to changes in momentum, or, alternately, acceleration.

11. For a body free to move in any direction,
    a) a net external force produces a change in the body's momentum
    b) a net external force produces a change in the body's velocity
    c) a net external force makes the body accelerate
    d) all of the above are correct

12. According to Newton's third law of motion,
    a) single forces cannot exist on a single object
    b) for every force there is an equal and opposite force, and each force acts on a different object

c) action and reaction forces are equal in magnitude and direction
   d) every object in the universe is in equilibrium

13. Momentum is conserved during
    a) elastic collisions only
    b) inelastic collisions only
    c) head-on collisions only
    d) all collisions

14. When an ice-skater pulls in his arms, his
    a) angular momentum increases
    b) angular momentum decreases
    c) spin rate increases
    d) spin rate decreases

15. If Earth's polar caps were to melt so that ocean levels would rise,
    a) Earth would spin faster
    b) Earth would spin slower
    c) the length of our year would increase
    d) the length of our year would decrease

16. When you start pedaling a bicycle, what force directly (not indirectly) causes you to accelerate in the forward direction?
    a) the force of your feet on the pedals
    b) the force of the sprocket on the chain
    c) the force of the rear wheel on the pavement
    d) the force of the pavement on the rear wheel

17. All objects freely falling in the same gravitational environment accelerate at the same rate. This is equivalent to saying that
    a) mass and weight are the same thing
    b) inertia and weight are proportional to each other if measured in the same gravitational environment
    c) the acceleration of gravity $g$ is the same everywhere
    d) air friction always applies equally to all objects

18. A baseball thrown with a velocity of 20 m/s straight upward will return to the pitcher in about
    a) 1 s
    b) 2 s
    c) 4 s
    d) 8 s

19. A baseball thrown with a velocity of 20 m/s straight upward will reach a height of about
    a) 5 m
    b) 10 m
    c) 15 m
    d) 20 m

20. The gravitational force between two people holding hands is very weak primarily because
    a) gravitational forces are inherently weak
    b) the force between the people is canceled by their combined weight
    c) electromagnetic forces between two people cancel any gravitational force
    d) gravity does not operate over distances as small as a meter

21. If Earth were 5 times farther from the sun than it is now, the sun's gravitational force on Earth would be
    a) 5 times stronger
    b) 1/5 as strong
    c) 1/10 as strong
    d) 1/25 as strong

22. Kepler's second law implies that
    a) all planets must travel in circular orbits
    b) a planet must move fastest when it is farthest from the sun
    c) a planet must move fastest when it is closest to the sun
    d) the more massive the planet, the longer its period

23. Kepler modified
    a) the geocentric model
    b) the heliocentric model
    c) Newton's laws
    d) Galileo's ideas about inertia

24. Kepler's laws
    a) explain the nature of the force that keeps the planets moving around the sun
    b) can be used to derive all of Newton's laws
    c) are applicable to both the geocentric and heliocentric models
    d) are applicable to other systems of moving bodies besides the Solar System

25. Without gravity, the moon would
    a) fall toward Earth
    b) move in a direction directly opposite Earth
    c) travel in a straight line tangent to its present orbit
    d) suddenly stop in its orbit

26. An astronaut inside an orbiting satellite feels weightless because
    a) she wears a space suit
    b) both the satellite and the astronaut are accelerating toward Earth at the same rate
    c) Earth's gravitational pull is zero in space
    d) the moon's gravitational attraction for the satellite balances Earth's gravitational attraction

27. The usual tidal pattern on ocean shores is
    a) a high tide on one day and a low tide on the next day
    b) one high tide and one low tide daily
    c) two high tides and two low tides daily
    d) three high tides and three low tides daily

28. An effect caused by Earth's tides is a
    a) speeding up of Earth's rotation
    b) slowing down of Earth's rotation
    c) shortening of the year
    d) lengthening of the year

29. Newton's laws are
    a) a perfect description of how moving bodies behave
    b) considered obsolete and seldom used anymore
    c) useful as very close approximate descriptions of the actual behavior of moving bodies under most circumstances
    d) described by none of the above

# Questions

1. How do instantaneous speed and average speed differ from each other?

2. How do speed and velocity differ from each other?

3. What is inertia?

4. One form of Newton's second law of motion relates force to momentum. Another form of the same law relates force to acceleration. Why, then, are momentum and acceleration not the same thing?

5. Venus spins east to west on its axis, that is, clockwise if the planet is viewed from above its north pole. By applying the right-hand rule, determine the direction of Venus's angular momentum.

6. Why must the Earth turn about 9.5 million times in order to complete one precessional cycle of 26,000 years?

7. What is weight? Why is the weight of something not an absolute quantity?

8. What is the difference between tension and compression?

9. What is meant by *chaos* in the scientific or mathematical sense of the word?

# Problems

1. From the data listed in Table 5.1, determine how far you would travel in one year if moving continuously at the typical speeds of (a) a snail, (b) a walking human, (c) a cruising automobile, (d) a jet aircraft in flight, and (e) light.

2. How long would it take to reach the sun (150,000,000 km away) if you could fly straight toward it at the cruising speed of a jet airliner (300 m/s)?

3. A car starts from a speed of 5 m/s and undergoes a constant acceleration of 2 m/s$^2$ afterward. How much time is needed for the car to reach a speed of 15 m/s?

4. If a 900-kg car moving at 20 m/s is slowed to a stop in 0.1 s as it crashes into a brick wall, what is the average force applied to the car by the brick wall during the time of impact? How much average force does the car impart to the brick wall during the impact? (Use the formula $F = \Delta p/\Delta t$.)

5. How much force is required to give a 900-kg car an acceleration of 3 m/s$^2$? What will the direction of the acceleration be? (Assume that friction is negligible, and that the car is rolling on a flat surface.)

6. After 6 s, what is the speed of a freely falling body released from rest? How far has the body fallen by this time? (Assume $g \approx 10$ m/s$^2$ and there is negligible air resistance.)

7. From your mass in kilograms, figure your weight in newtons. (Your mass in kilograms can be calculated by dividing your weight in pounds by 2.2.)

8. Using your mass in kilograms and your weight in newtons, calculate the mass of Earth using Newton's law of gravitation (Equation 5.9). In this equation, $F_g$ is your weight, $m_1$ is the mass of Earth, $m_2$ is your mass, and $d$ is the distance between you and Earth's center, $6.38 \times 10^6$ m.

9. If there existed a planet in the Solar System with an average orbital radius of 4 AU, what would its period be, in years, according to Kepler's third law (Box 5.2)?

10. What is the size of Earth's orbit around the barycenter between Earth and the moon? Make use of the facts that Earth's mass is 81 times greater than the moon's mass and that the average distance between Earth and the moon is 384,000 km.

# Questions for Thought

1. What do you think causes the apparent "levity" of rising smoke or hot air, or the tendency of a column of mercury to rise in a barometer? Is levity a real force on a par with gravity as Aristotle thought? (*Note:* Full explanations of these phenomena will appear in Chapter 7. For now, try to understand how gravity and levity differ.)

2. Of his intellectual achievements, Newton said that he had seen so far because he had "stood on the shoulders of giants." Who were some of the giants he was referring to?

3. Do you think Newton's first law and the concept of inertia are related to the idea that motion is relative?

4. If an elevator in a tall building is accelerating downward at 9.8 m/s$^2$ because its cable has been severed, the people inside are "weightless" until the elevator crashes in the basement. If a car on flat ground is accelerating at 9.8 m/s$^2$, the passengers inside do not feel weightless. Why is there a difference?

5. Why is a bicycle relatively stable when it is moving and unstable when it is stationary?

6. Do you think the universe as a whole could have an overall (net) angular momentum? How could we tell if it did?

7. List as many examples as you can of symmetry in nature. The symmetries can be either geometrical or abstract.

8. If all of humanity were to face east and start walking, how would this affect Earth's rotation?

9. Earth bulges a little at its equator, so that its equatorial radius is slightly larger than its polar radius. Earth also spins, so that a person at the equator moves in a circle once a day, but a person at the pole does not. How do these two factors affect a person's weight at the equator and a person's weight at either pole?

10. How would the design of large static structures in space (such as the space station proposed by the U.S. government) differ from the design of static structures on Earth?

11. We commonly define *straight up* as being parallel to the direction of a string or rope holding a stationary, hanging mass. Is this necessarily the same direction as a line perpendicular to Earth's flat surface?

12. What would the gravitational field be like at the center of Earth? How much would you weigh if you could inhabit a hollow space at the center of Earth?

13. List as many natural or artificial phenomena as you can that exhibit chaotic (in the scientific sense) behavior. Do you think there is some kind of order superimposed on the disorderly behavior of every chaotic system? Do you think any entirely orderly systems exist? Do you think any entirely disorderly systems exist?

## Answers to Multiple-Choice Questions

| | | | | | |
|---|---|---|---|---|---|
| 1. b | 2. c | 3. d | 4. c | 5. c | 6. c |
| 7. a | 8. a | 9. c | 10. c | 11. d | 12. b |
| 13. d | 14. c | 15. b | 16. d | 17. b | 18. c |
| 19. d | 20. a | 21. d | 22. c | 23. b | 24. d |
| 25. c | 26. b | 27. c | 28. b | 29. c | |

# CHAPTER 6
# Energy

*All changes are accompanied by transformations of energy. When fireworks ignite, energy is not created and then released. Rather, energy stored in the unexploded fireworks—chemical potential energy— is transformed into light, heat, and sound.*

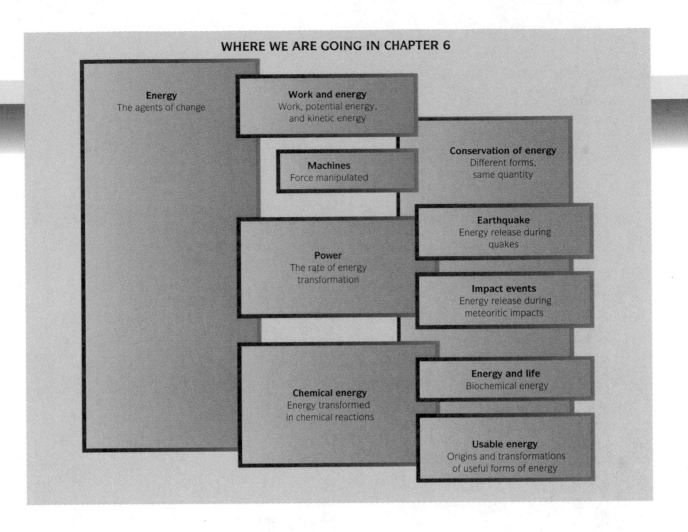

Attempting to define the word *energy* in a simple way is as futile as explaining what *art* means in ten words or less. Like art, which exists in many forms and communicates many kinds of messages, energy exists in many different forms and applies to a wide range of phenomena. Although both energy and art are abstract and conceptually "slippery," energy, as opposed to art, is a quantitative concept. In all its many guises, energy has at its root a particular quantity that can be precisely measured and expressed in definite units. There is unity in energy in that all forms of energy can be measured in the same units. There is diversity in energy in that energy comes in many forms.

Six common forms of energy you have probably heard about are *mechanical energy, heat energy, chemical energy, electric energy, radiant energy, and nuclear energy.* These particular forms of energy and their applications will be introduced individually in various parts of this chapter and in succeeding chapters in this book. For your convenience they are summarized in Table 6.1.

| | Form of Energy | Description | Examples |
|---|---|---|---|
| **TABLE 6.1** **Common Forms of Energy** | Mechanical energy | The external kinetic and potential energy of moving bodies or of bodies having the potential to move | A moving car Water behind a dam A stretched rubber band Machinery in operation |
| | Thermal energy (heat) | The internal energy contained in matter by virtue of the kinetic and potential energies of the atoms or molecules that make up matter. | All material bodies |
| | Chemical energy | The energy associated with chemical reactions; primarily, the potential energy of electrons in the bonds between atoms | Energy contained in fuels such as gasoline and coal Food energy |
| | Electric energy | The energy associated with the position and motion of electric charges | Electricity in a power line or in an electrical circuit |
| | Radiant energy (or electromagnetic energy) | The energy of oscillating electric and magnetic fields moving at the speed of light | Sunlight Radiant heat from a fire X-rays Radio waves |
| | Nuclear energy | The energy associated with nuclear reactions; primarily, the potential energy of particles inside atomic nuclei | The sun's energy source A nuclear power plant A nuclear bomb |

*Note:* These are not fundamental types but, rather, common forms that may involve one or more of the broad categories of energy (potential, kinetic, and rest) in the context of interactions governed by one or more of the four basic forces (strong and weak nuclear, electromagnetic, and gravitational).

We shall begin our investigation of energy in this chapter by relating energy to work, an easily understood concept. One definition of energy, in fact, is "the ability to do work." We shall then examine in detail two broad categories of mechanical energy, called *potential energy* and *kinetic energy*.

All changes in our world, whether they occur by natural or artificial means, are accompanied by transformations of energy. These changes can be as monstrous as the detonation of a nuclear bomb or as minute as an insect beating its wings. Naturally occurring energy transformations have fueled the evolution of our universe during the last 15 billion years or so and have led to the existence of life on at least one planet: Earth.

Through the remainder of this book, you will discover how the concept of energy applies universally across all the physical sciences. In the latter part of this chapter, and in several succeeding chapters, some of the consequences associated with the use of "practical" energy will be examined. Humankind's harnessing of energy for modern applications has made life much more comfortable for many people but, at the same time, problematical for the health of our planet.

## Work and Energy

Energy implies action or activity—or at least the capacity for these things. Unlike matter, which we usually think of as concrete and tangible, energy is intangible. We feel the sun's radiant heat energy on a hot summer day, but we cannot touch the thing we experience. The hanging weights of an old-fashioned clock possess the energy needed to drive the clock mechanism, but we cannot sense in a direct way what that energy is.

Despite its abstract nature, energy is an extremely useful concept in science. To begin to understand energy, let us start with a working definition: **Energy** is the capacity to do work.

## Work

We must, of course, define work. The **work** done on a body is the force on the body multiplied by the distance through which the force acts. A simplified formula can be written for work:

$$W = Fd$$

Equation 6.1

where $W$ is work, $F$ is the force applied, and $d$ is the distance through which the force acts. The variable $F$ in the equation refers to a constant or average force acting on a body through a given distance. The variable $d$ refers to the distance traveled by the body *in the direction of the applied force,* while the force is being applied. A hockey puck sliding along a flat surface has no work done on it by gravity because it never moves *in* the direction of the applied force, which is downward.

It is entirely possible for an applied force to produce no movement at all. In this case, no work is done on the body by the applied force. A weight lifter does work on a barbell while lifting it, but no work is done on the barbell when it is held motionless overhead (Photo 6.1). More examples of work or the absence of work are shown in Figure 6.1.

Work can be quantified, since both force and distance are measurable quantities. If 1 newton (N) of force acts through 1 meter (m) of distance, we say that 1 newton-meter (N·m) of work is done. For simplicity, the newton-meter is called a *joule.* The joule (J) is the SI unit for work. Since energy can be thought of as stored work, or the capacity to do work, energy has the same SI unit: the joule. The joule is a derived unit, consisting of one application of mass (in kilograms), two applications of length (in meters), and two applications of time (in seconds):

$$1 \text{ J} = 1 \text{ N·m} = 1 \text{ (kg·m/s}^2)(\text{m}) = 1 \text{ kg·m}^2/\text{s}^2$$

As is customary in the metric system, large quantities of work or energy may be expressed in units of thousands of joules (kilojoules, kJ) and mil-

**PHOTO 6.1** No work is done on a barbell when it is held motionless overhead. However, work (against gravity) was done on the barbell in order to raise it.

lions of joules (megajoules, MJ). The enormous range of energy, in joules, associated with various phenomena is diagramed in Figure 6.2.

You can "do work" on a body in a variety of ways. In Figure 6.3(a), work is done on a box by sliding it along the floor. In this instance, the work being done overcomes a resistance—a frictional force. The particular force needed to keep the box sliding would depend on the weight of the box, the smoothness of the floor, and other factors. In Figure 6.3(b), where friction

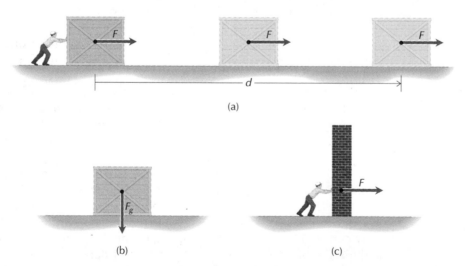

(a)

(b)                                    (c)

**FIGURE 6.1** The work done on a body equals the force on the body times the distance through which the force acts. In (a), a worker slides a crate along a flat surface through a distance d. A horizontal force F is applied through the distance d to overcome the force of sliding friction between the floor and the bottom of the crate. The work done on the crate is equal to F times d. In (b), the downward force of gravity $F_g$ on a crate, which is the weight of the crate, does not result in any downward movement, so the work done on the crate is zero. (What would happen if the crate fell through a hole in the floor?) In (c), a horizontal force applied to a brick wall results in no movement, so no work is done on the wall.

is considered negligible, work is done on a box by overcoming its inertia. The box accelerates. In Figure 6.3(c), the downward force of gravity is doing work on a mass because that mass is moving in the direction of the applied force.

**FIGURE 6.2**  In the SI, the common currency of energy is the unit called the joule (J). The energy, in joules, of some common and uncommon phenomena are given on this logarithmic bar. Many of the values indicated are estimates.

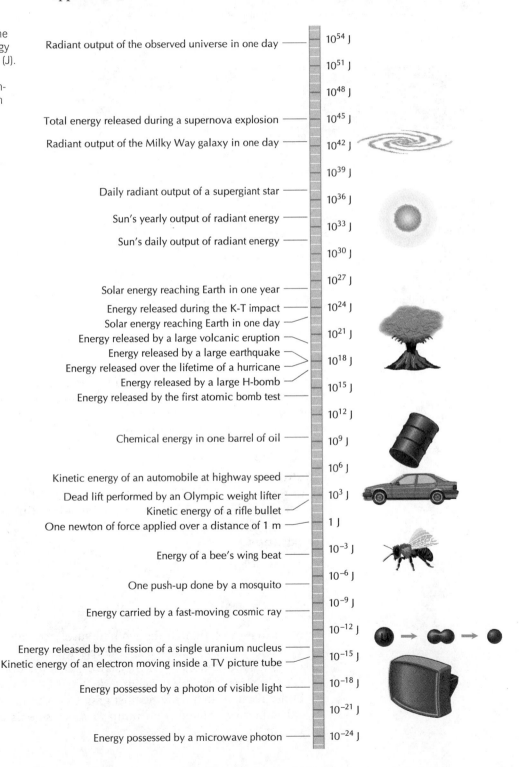

|  |  |
|---|---|
| Radiant output of the observed universe in one day | $10^{54}$ J |
| | $10^{51}$ J |
| | $10^{48}$ J |
| Total energy released during a supernova explosion | $10^{45}$ J |
| Radiant output of the Milky Way galaxy in one day | $10^{42}$ J |
| | $10^{39}$ J |
| Daily radiant output of a supergiant star | $10^{36}$ J |
| Sun's yearly output of radiant energy | $10^{33}$ J |
| Sun's daily output of radiant energy | $10^{30}$ J |
| Solar energy reaching Earth in one year | $10^{27}$ J |
| Energy released during the K-T impact | $10^{24}$ J |
| Solar energy reaching Earth in one day | |
| Energy released by a large volcanic eruption | $10^{21}$ J |
| Energy released by a large earthquake | $10^{18}$ J |
| Energy released over the lifetime of a hurricane | |
| Energy released by a large H-bomb | $10^{15}$ J |
| Energy released by the first atomic bomb test | |
| | $10^{12}$ J |
| Chemical energy in one barrel of oil | $10^{9}$ J |
| | $10^{6}$ J |
| Kinetic energy of an automobile at highway speed | |
| Dead lift performed by an Olympic weight lifter | $10^{3}$ J |
| Kinetic energy of a rifle bullet | |
| One newton of force applied over a distance of 1 m | 1 J |
| Energy of a bee's wing beat | $10^{-3}$ J |
| One push-up done by a mosquito | $10^{-6}$ J |
| Energy carried by a fast-moving cosmic ray | $10^{-9}$ J |
| | $10^{-12}$ J |
| Energy released by the fission of a single uranium nucleus | $10^{-15}$ J |
| Kinetic energy of an electron moving inside a TV picture tube | |
| Energy possessed by a photon of visible light | $10^{-18}$ J |
| | $10^{-21}$ J |
| Energy possessed by a microwave photon | $10^{-24}$ J |

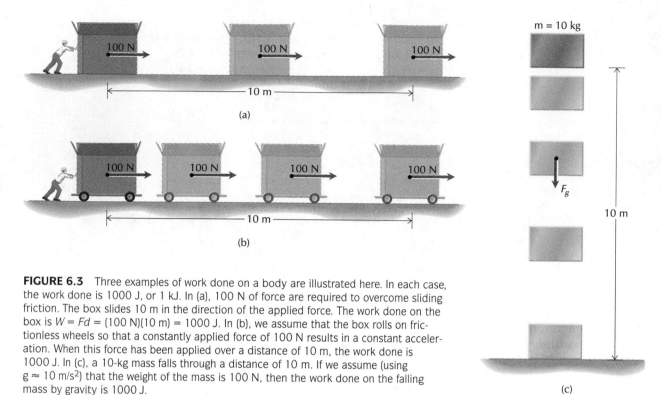

**FIGURE 6.3**   Three examples of work done on a body are illustrated here. In each case, the work done is 1000 J, or 1 kJ. In (a), 100 N of force are required to overcome sliding friction. The box slides 10 m in the direction of the applied force. The work done on the box is $W = Fd = (100 \text{ N})(10 \text{ m}) = 1000$ J. In (b), we assume that the box rolls on frictionless wheels so that a constantly applied force of 100 N results in a constant acceleration. When this force has been applied over a distance of 10 m, the work done is 1000 J. In (c), a 10-kg mass falls through a distance of 10 m. If we assume (using $g \approx 10 \text{ m/s}^2$) that the weight of the mass is 100 N, then the work done on the falling mass by gravity is 1000 J.

Yet another way to do work on something is to lift it. If you lift something, you do work against gravity. You also do work against gravity whenever you increase the height of your own body by climbing steps or by other means involving your own physical effort.

**Problem 6.1**

A woman who weighs 600 N climbs a flight of stairs 5 m high. How much work does she do in raising herself to this height?

**Solution**

$$W = Fd$$
$$= (600 \text{ N})(5 \text{ m})$$
$$= 3000 \text{ J}$$

Note that even though the woman surely traveled more than 5 m to reach the top of the stairs (the exact distance would depend on the pitch, or steepness, of the staircase), she applied a force vertically, *against* gravity, through a distance of 5 m. She would have done just as much work against gravity if she had climbed a vertical ladder or walked up a ramp to raise herself to the same 5-m height.

**PHOTO 6.2** By virtue of its elevated position, water stored at a high place (above the waterwheel) has potential energy. This water has the potential to do work, and it does so when it races downward and pushes on the waterwheel's blades.

## Potential Energy

Work is required to increase the height of any body. The greater the mass of the body, and the greater the height through which the mass is raised, the greater the work required to raise it. In an elevated position, the mass possesses something intangible but real: potential energy. **Potential energy** is the energy posessed by a body by virtue of its position in a force field. Potential energy is a useful concept, because when a body possesses potential energy, it is capable of doing work.

Potential energy was just defined in a general sense. The more restrictive *gravitational potential energy* (which we shall label *GPE*) refers to the energy possessed by a body by virtue of its position (or height) in a gravitational force field, such as Earth's gravitational field. The greater the height to which a body is raised, the more *GPE* it has. Other kinds of potential energy exist for other kinds of force fields (electric or magnetic fields, for example), but we will restrict our immediate discussion to *GPE*.

Potential energy can be thought of as stored work. Work is required to raise a mass. The raised mass can do work on something else if and when it falls again. Falling things that have been harnessed to do work include the slowly falling mass that keeps a weight-driven clock moving, a pile driver's massive ram that falls upon a piling and hammers it into the ground, and the water falling from a dam that can be harnessed to push an old-fashioned waterwheel around (Photo 6.2) or spin a modern turbine.

On or near Earth's surface (where $g$ can be considered constant), an object's *GPE* is simply the work required to raise the object to a given height. Since work is force times distance (height in this case), and the weight of a mass is the force one must work against to raise it,

$$GPE = (\text{weight})(\text{height}) = F_g h = mgh$$

Equation 6.2

**249**

**Problem 6.2**

After you raise a textbook like this one (with a mass of about 1.5 kg and a weight of about 15 N) to a height of 1 m above a tabletop, how much *GPE* does the book have relative to the height of the tabletop?

**Solution**

$$GPE = F_g h$$
$$= (15 \text{ N})(1 \text{ m})$$
$$= 15 \text{ J}$$

*Note:* An implicit assumption being made here is that the book starts from rest and is brought to rest at the 1-m height above the table. You would have done more work on the book if it were still moving upward at the 1-m height.

The book in its elevated position has acquired 15 J of *GPE* because you had to do 15 J of work on the book in order to raise it. Similarly, in Problem 6.1, the woman at the top of the stairs would have a *GPE* of 3000 J, because she performed 3000 J of work to raise herself to that level. In cases like these, work done on a body against gravity results in a corresponding increase of *GPE* stored in that body.

Note that gravitational potential energy, or any other kind of potential energy, is relative. Thus for *GPE*, a reference level or height must be specified or implied. In our example of the textbook lifted above the tabletop, the implied reference level was the tabletop. The same book would have more *GPE* if its height were measured relative to the level of the floor. *GPE*, or any other potential energy for that matter, can sometimes be negative. For example, an object that has fallen through a hole in the floor would have negative *GPE* relative to its former position on the floor.

# Kinetic Energy

When we release an object and let it fall freely, its *GPE* decreases, but something else increases: kinetic energy (*KE*). **Kinetic energy** is the energy possessed by a body by virtue of its motion. The greater the mass of the body and the faster it moves, the greater is its kinetic energy. The formula for kinetic energy is

$$KE = \tfrac{1}{2} mv^2 \qquad\qquad \text{Equation 6.3}$$

**Problem 6.3**

A 70-kg track star attains a speed of 7 m/s. What is his *KE* at that speed?

**Solution**

$$KE = (1/2) \, mv^2$$
$$= (1/2)(70 \text{ kg})(7 \text{ m/s})^2$$
$$= (1/2)(70 \text{ kg})(49 \text{ m}^2/\text{s}^2)$$
$$= 1715 \text{ J}$$

In order to accelerate to 7 m/s, the track star did work on the ground with every stride by applying a force, through a distance, with his feet. By Newton's third law, the ground applied the same force, in the opposite direction (forward), through the same distance. The work done on the track star by the ground engendered an increase in his *KE*, from 0 to 1715 J. The athlete has *KE* by virtue of his motion.

Like potential energy, kinetic energy is relative. That is, *KE* depends on speed, and speed itself must be measured with respect to a reference frame that is either specified or implied. In Problem 6.3, the moving track star has a *KE* of 1715 J relative to the ground he is running on.

An important property of *KE* is that it is proportional to the *square* of a body's speed. Each doubling of speed produces a fourfold increase in *KE*; each tripling of speed produces a ninefold increase in *KE*; and so on. For this reason, it is much harder (actually, four times harder) to stop a car if it is moving only twice as fast. Similarly, head-on car collisions at around 50 mi/h are serious but often survivable; the same collisions at 70 mi/h are usually fatal.

The **mechanical energy** of a system of interacting bodies is defined as the sum of the kinetic energies and gravitational potential energies possessed by all the different parts of the system. We will now see how this particular energy is conserved (or remains the same) as long as the system is not acted on by forces other than those associated with the system itself. Indeed, we will see how this principle can be broadened to include *all* forms of energy.

## Conservation of Energy

We have learned from Equation 6.3 that any increase or decrease in the speed of a body corresponds to a gain or loss of *KE*. Similarly, by Equation 6.2, we know that any change in the height of a body results in a change in its *GPE*. Let us investigate a case in which these two qualities are related to each other.

## Conservation of Mechanical Energy

Imagine a 1-kg book suspended so that it can fall freely 20 m to the floor, as pictured in Figure 6.4. We will choose the reference level for measuring *GPE* as the floor, so that the book's *GPE* is zero when it is on the

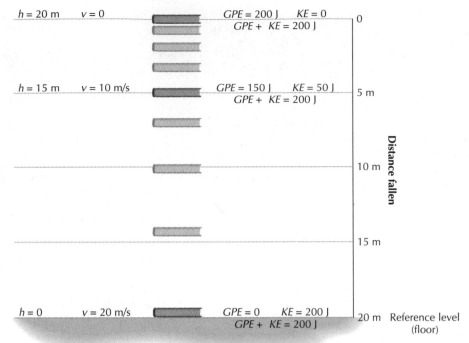

**FIGURE 6.4**  A 1-kg book is released from rest at a height (*h*) of 20 m relative to some reference level. After 1 s, the book has fallen 5 m to a height of 15 m above the reference level, and its speed (*v*) is 10 m/s. After 2 s, the book has fallen a distance of 20 m, where *h* is zero and *v* is 20 m/s. Substituting values of *v* and *h* into the formulas for *GPE* and *KE* yields the values given.

floor. By Equation 6.2 (and for $g \approx 10$ m/s$^2$), the *GPE* of the book at a height of 20 m is 200 J. Just before the book is dropped, its *KE* is zero. The sum of these two energies, *GPE* and *KE*, is 200 J.

After being released, the book loses *GPE* and gains *KE*. When the book is a fourth of the way to the floor, its *GPE* is reduced by a fourth. When halfway to the floor, its *GPE* is down to half. When the book reaches the floor, all its *GPE* is gone.

How much *KE* does the book gain on its way to the floor? To determine that, we can start with the data for freely falling bodies given in Table 5.3 in Chapter 5. We learn that it takes 2 s for an object to fall 20 m, by which time it is moving at a speed of 20 m/s. Substituting this value of speed into Equation 6.3, we find that the *KE* is 200 J. Again, the sum of the two energies, *GPE* + *KE*, is 200 J. Calculations done for any height show that the sum of *GPE* and *KE* remains a constant 200 J as the book falls from its initial height of 20 m.

All freely falling bodies transform *GPE* into *KE* as they descend. This is but one example of energy being transformed in such a way that the total mechanical energy within the system being considered remains constant. Strict conservation of mechanical energy, however, depends on the absence of external forces other than gravity. When an object falls through air or any other retarding medium, a small amount of the *KE* carried by the falling book is lost owing to air friction. Also, when the book arrives on the floor, its downward motion ceases and all its *KE* is lost. The book's mechanical energy (*GPE* + *KE*) suddenly becomes zero as the upward force exerted by the floor brings the book to a halt. Mechanical energy is decidedly *not* conserved when the book hits the floor because an external force then acts on the system.

# Energy Conservation

Whenever we analyze transformations of energy in an all-inclusive sense (not just by focusing on one type of energy, such as mechanical), we find that the total amount of energy in an isolated system remains constant. This is thought to hold for the whole universe, as well as for any part of the universe we can consider as being isolated. This idea is summarized as the law of conservation of energy, one of the most important generalizations in science: Energy cannot be created or destroyed; it can only be changed from one form to another.

It is instructive to ask how the falling book pictured in Figure 6.4 acquired its initial 200 J of energy. One possible answer would be that someone raised the book to its 20-m height. That process would have required 200 J of work to be done on the book. Work is an energy-transforming process, and in this instance the work was transformed into *GPE*. But just where did the energy required to raise the book by human effort come from in the first place? We will answer this by tracking down the immediate and ultimate sources of "muscle energy."

The energy used by muscles to expand and contract and do work is derived from the *chemical potential energy* contained in food (see Box 6.4). This type of potential energy is associated with the positions of charged particles—electrons—in atoms and molecules. When atoms or molecules are rearranged during chemical or biochemical reactions, energy is either taken in or released. Digestion in humans breaks down food into simple molecules, such as glucose (sugar). Glucose molecules, which are rich in chemical potential energy, can combine with oxygen molecules, thereby releasing energy. Some of this released energy causes muscle cells to contract, in order to move the hand that raises the book. The rest of the energy is released in the form of heat; muscle exercise always generates heat.

Let us go further by asking where the energy stored in food comes from. Directly or indirectly, food possesses the energy of stored sunlight. In plants, the sun's radiant (electromagnetic) energy is trapped by means of the complex chemical reaction called *photosynthesis*. The sun derives *its* energy ultimately from the *nuclear energy* stored in the nuclei of hydrogen atoms in the core of the sun. When hydrogen nuclei fuse (or join), to form helium nuclei, nuclear energy is released largely in the form of photons, which are individual units of electromagnetic energy. This electromagnetic, (or radiant), energy eventually works its way out to the surface of the sun and escapes into space. Earth, which lies in the space surrounding the sun, intercepts a tiny fraction of the sun's radiant energy.

Where did the nuclear energy in the sun come from? All we can say with reasonable certainty is that from very early on during the big bang, the major building blocks of matter in the universe—the lightweight hydrogen and helium nuclei—already possessed nuclear energy. Lightweight nuclei have the potential to join with others, and in so doing, they release a large amount of energy.

Staying with our example of the falling book, let us consider what happens to the energy possessed by the book immediately *after* it falls and comes to rest on the floor: The 200 J of mechanical energy the book had while falling suddenly disappear. What happens to this energy? Some of it is transformed into sound waves: the noise of the book hitting the floor.

Most of it becomes what is loosely called heat and more precisely called *thermal energy* (which is the subject of the next chapter). Essentially, the impact stirs up atoms and molecules in the book and on the floor, causing them to move or vibrate faster. The book and the floor are now slightly warmer; they both have more thermal energy (an internal form of energy) than they did before.

In our discussion of momentum in Chapter 5 we introduced the notion of elastic and inelastic collisions. During collisions, momentum is always conserved, but *kinetic energy is not necessarily conserved.* An **elastic collision** is one in which kinetic energy *is* conserved. If the book in our example could somehow (unrealistically) collide elastically with the floor, it would bounce back to its former height, and there would be no increase in thermal energy. Our case of the book hitting the floor and hardly bouncing at all is an example of an **inelastic collision,** a collision in which kinetic energy is *not* conserved. The kinetic energy of the book suddenly drops to zero and thermal energy is generated in its place. Remember, however, that the *total energy,* including kinetic and thermal, is conserved during any kind of collision.

If the book has exactly 200 J of *KE* as it collides inelastically with the floor, then 200 J of thermal and acoustic (sound) energy appear after the collision. The thermal and acoustic energy never disappear; they just spread out into the surrounding environment. The thermal energy soon becomes so dispersed as to be undetectable. The sound waves reflect from or become absorbed by surrounding objects. Objects that absorb sound are heated, immeasurably so in this case. By the law of conservation of energy, 200 J of energy are dispersed into a larger system that contains the book, the floor, and much more.

In mechanical and biological systems, thermal energy is a complicating aspect we cannot afford to ignore. It is an inevitable by-product of the work done by humans and machines alike. Take, for example, the stair climber in Problem 6.1. Although she does 3000 J of work in raising herself to a particular height, her metabolic (biochemical) expenditure of energy is greater. Her muscles produce *waste heat* (a release of excess thermal energy) as well as propel her up the stairs. In a similar manner, all machines (see Box 6.1) perform with imperfect efficiency, because they suffer from friction, and the work done by friction invariably releases thermal energy.

Let us look at four more examples of energy transformation: a mass vibrating horizontally on a spring, a swinging pendulum, the up-and-down motion of a roller coaster, and a comet traveling in an elliptical orbit around the sun. All four examples illustrate transformations of mechanical energy in systems of moving bodies that are idealized to some extent. For the most part, transformations into thermal energy are ignored.

## A Vibrating Mass

A mass vibrating horizontally while fastened to a stiff spring, as illustrated in Figure 6.5, exhibits a repetitive motion of the type called *simple harmonic motion.* The spring has a tendency to resist being either stretched or compressed, and that resistance is greater when the spring is stretched or compressed to a greater degree. When attached to the spring, the mass

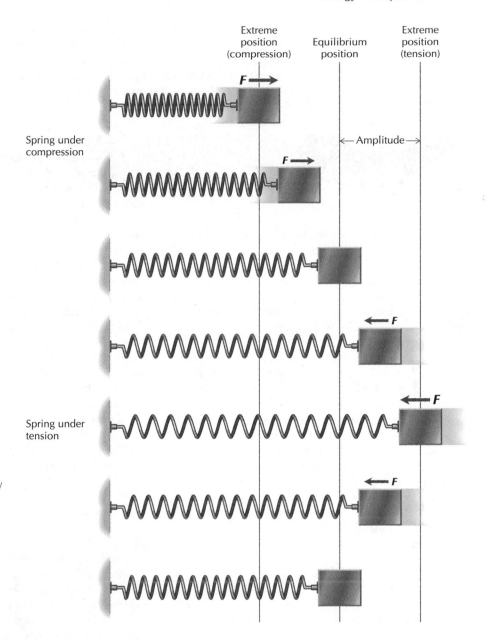

Extreme position (compression)    Equilibrium position    Extreme position (tension)

Spring under compression

←—Amplitude—→

Spring under tension

**FIGURE 6.5** An idealized mechanical system consisting of a mass oscillating horizontally on a spring experiences a continuing transformation between elastic potential energy (stored within the spring) and kinetic energy (possessed by the mass whenever it moves). The amplitude depends on the amount of energy initially added to the system and on the characteristics of the spring.

has a certain *equilibrium* position at which it will remain at rest if initially at rest. Whether moving or not, the mass at equilibrium has no force acting on it (except gravity, which is being ignored here because the mass can move only horizontally).

Let us see what happens when we use an external force to push the mass to either the left or the right of the equilibrium position and then release it. The work required to move the mass is stored as *elastic potential energy* in the spring. (This elastic potential energy arises because, on a microscopic scale, the atoms in the solid metal, which are linked to each other by forces of electromagnetic origin, resist being forced either closer together or farther apart.) The farther we push the spring, the more work we do and the more *PE* (elastic potential energy) is stored. Once it is released, the mass accelerates toward the equilibrium position but (because of its inertia) overshoots it. The mass ends up oscillating between two extreme positions. At either extreme position, the mass is momentarily at rest, the *KE* of the mass is zero, and all of the system's mechanical

## BOX 6.1
# Machines

A machine is any device that can change the magnitude or direction of a force. Machines can be as simple as a crowbar used as a lever or a jack used to raise a car, or as complex as the drivetrain of an automobile.

It is commonly thought that machines do work—or at least make work easier to do. Machines make it possible for us to *do* certain kinds of work, but they do not save work. Just as there is no "free lunch" in the monetary world, neither can you get something for nothing in the physical world. In accordance with the law of conservation of energy, the work output of a machine can never be greater than the work required to operate it.

The simplest analysis of any machine ignores the very real losses of energy associated with friction and considers only mechanical energy—potential and kinetic energy on a macroscopic scale. For such idealized machines, work input ($W_{input}$) equals work output ($W_{output}$). Since work is force times distance:

$$W_{input} = W_{output}$$
$$(F_{input})(d_{input}) = (F_{output})(d_{output}) \quad \text{Equation 6.5}$$

Consider what is perhaps the simplest machine of all, a lever (see Figure 6.6). Ideally, the work required to depress the long end of the lever equals the work delivered to raise the load on the short end. For a suitably chosen fulcrum position, a relatively small force exerted through a relatively large distance on the long end produces a large force through a small distance on the short end. A ratio of 4 to 1 in distance, as shown in Figure 6.6, produces a *mechanical advantage,* or force-multiplying factor, of 4. The mechanical advantage of a lever can be readily adjusted by choosing different fulcrum positions. Given the proper lever and fulcrum, as asserted by the ancient Greek philosopher Archimedes, one could move the entire Earth.

Another simple machine is the pulley shown in Figure 6.7, whose sole function is to change the direction of a force. Its mechanical advantage is 1.

Yet another simple machine is a jack used to raise a car. An idealized car jack with a mechanical advantage of 100 multiplies the force used to push the handle by 100, but it also requires that you push the handle through a distance 100 times greater than the height the car is raised. Like most machines, a car jack changes the direction of a force as well as its magnitude. Some car jacks require repeated downward pushes on a handle, whereas others (of the screw type) require that you rotate a crossbar to drive the screw. In each case, the applied force is directed upward.

For certain purposes, a machine can increase the output distance at the expense of a reduction in out-

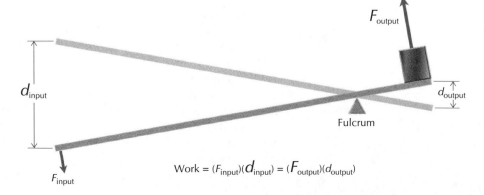

**FIGURE 6.6**   A lever is a simple machine that can apply a large force over a small distance at the output end in response to a smaller force applied over a larger distance at the input end. Ideally, if the ratio of input and output distances is 4 to 1, as shown here, the force on the output end is 4 times greater.

$d_{input}$

$F_{output}$

$d_{output}$

Fulcrum

$F_{input}$

Work $= (F_{input})(d_{input}) = (F_{output})(d_{output})$

energy is stored as *PE* in the spring. At the equilibrium position, the *PE* is momentarily zero and the energy is all *KE*. How much *KE*? Measurements show that the *KE* of the mass when passing the equilibrium position equals the *PE* stored in the spring when the mass lies at either of the extreme positions. The constant interchange between *KE* and *PE* keeps the mass oscil-

**FIGURE 6.7** A simple pulley changes the direction of a force, not the magnitude. Its mechanical advantage is 1. More complicated pulley systems can have a mechanical advantage greater than 1.

put force, as a bobbin winder does when it wraps thread or yarn on a spool (Figure 6.8). A bicycle in high gear is another example of a machine with a distance-multiplying ability but a mechanical advantage less than 1.

Idealized machines—those without friction—conserve mechanical energy. Real machines do not. For real machines, work output equals work input *minus* work lost owing to friction (the lost work is manifested as heat, or thermal energy.) Real machines, then, are never regarded as being perfectly efficient. Efficiency in machines is defined as

$$\text{efficiency} = W_{\text{output}}/W_{\text{input}} \qquad \text{Equation 6.6}$$

In some machines, such as electric motors, the work input is better understood as an input of energy (electric energy in this case). Work input in the formula also stands for input energy. The efficiencies of most simple machines tend to be high, since few moving parts are involved. An exception is a rusty car jack, for instance, which might have an efficiency of something like 0.5, or 50%. The efficiencies of complex machines run the gamut: The best electric motors approach an efficiency of around 95%. The automobile—regarded as an integrated unit that converts fuel energy into motion—is a very inefficient machine. Waste heat vented by the engine, friction in the moving parts of the engine and drivetrain, and the energy demands of

accessories, such as air-conditioning units, all contribute to lowering an automobile's efficiency to an abysmal 10–15%.

For centuries, people have dreamed of making "perpetual-motion machines": machines that run forever without any added input of energy. If we could build a machine with an efficiency of 100%, then we could keep it running indefinitely simply by feeding all of its output back into its input. No one has succeeded (or ever will succeed) in doing this, because every machine has some friction in its moving parts. The most far-fetched perpetual-motion machines are those that could somehow create energy—that is, be more than 100% efficient. Such a machine would violate the law of conservation of energy.

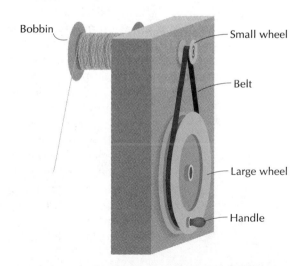

**FIGURE 6.8** This hand-driven bobbin winder uses a belt wrapped around two unequally sized wheels to spin a bobbin. When the handle is turned through a small arc, thread is spun rapidly onto the bobbin. Very little force is needed to wrap the thread and speed is of the essence, so this machine was designed to magnify distance instead of force. The mechanical advantage is less than 1.

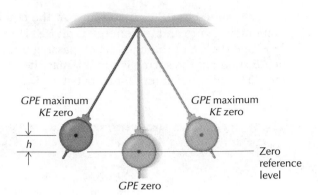

**FIGURE 6.9** Gravitational potential energy and kinetic energy are conserved in a frictionless pendulum. If the maximum height *h* of the pendulum bob is measured relative to $h = 0$ at the center position of the bob, then $GPE = mgh$.

Since the bob is momentarily motionless at the left and right positions, $KE = 0$ there. Kinetic energy is greatest at the center position and is equal to the $GPE$ of the bob at the left and right positions.

lating. An idealized system of this sort would oscillate forever; but real springs generate heat as they flex, thus removing mechanical energy from the system. We say that the oscillation in a real spring is *damped*: Its *amplitude,* or range of vibration, gets smaller and smaller.

## A Pendulum

An idealized pendulum (Figure 6.9), swinging back and forth with a relatively small amplitude, also exhibits simple harmonic motion. The downward pull of gravity continually tries to return an oscillating pendulum bob (hanging mass) to its equilibrium, or vertical, position. In this case, there is a continual interplay between *KE* and *GPE*. Kinetic energy is maximized at the bottom of the pendulum's swing, because the bob is moving fastest there. Gravitational *PE* is maximized at the extreme positions of the pendulum's swing, where the bob reaches its maximum height. In real pendulums used in clocks, mechanical energy is gradually converted into waste heat by friction, so some additional energy, usually in the form of little nudges delivered at some point in each oscillation, must be constantly fed into the system. The input of energy can come from gradually falling weights (as in a weight-driven clock) or by repeated magnetic impulses (as in certain battery-operated pendulum clocks). In windup watches, the tension of a coiled spring stores the energy needed to keep a tiny balance wheel oscillating back and forth.

## A Roller Coaster

As implied by their name, roller coasters spend most of their time in motion simply coasting. When being towed to the highest "hill" on the track, the coaster steadily gains *GPE* from an external source. At the top of the hill, gravity takes over and the coaster is on its own, its speed and path determined by the twists and turns of the track and by the pull of gravity (see Figure 6.10). As it falls, the coaster's initial *GPE* is transformed into *KE*, but that *KE* is partially transformed back into *GPE* as the coaster climbs the

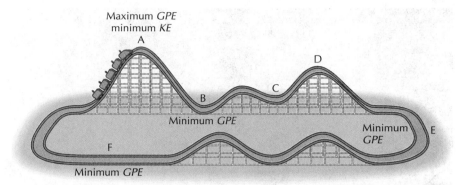

**FIGURE 6.10** This simplified roller coaster acquires its maximum *GPE* at point *A* by being towed up the hill. At point *A*, the coaster is allowed to fall, accelerating from a small initial speed. In the absence of any friction, the coaster would complete the circuit and return to point *A* at the same speed at which it left. Its total mechanical energy, *GPE + KE*, would then be conserved everywhere along the circuit; and the speed of the coaster would depend strictly on its height. The higher the coaster is at any given moment, the greater its *GPE*, the less its *KE*, and the less its speed. The lower the coaster is at any moment, the less its *GPE*, the greater its *KE*, and the greater its speed.

In a realistic roller coaster, frictional forces do the work of slowly braking the moving cars. Mechanical energy is not conserved; instead, it is gradually converted into heat. The *GPE* lost by the cars in going from point *A* to point *B* equals the *KE* gained at point *B* plus the heat energy lost owing to friction. The height of point *D* must not be so great that the sum of the cars' *KE* at point *B* plus frictional losses between points *B* and *D* exceeds their *GPE* at point *D*; otherwise, the cars will fail to make the summit and roll backward toward point *C* or *B*. As the cars pass points *B, E,* and *F,* which are all at the same height, the *GPE* is the same; but the *KE*, and therefore the speed, progressively declines. For a real roller coaster, brakes are applied at point *F* to reduce the residual *KE* to zero and bring the cars to a halt for passenger unloading and loading.

lesser hills ahead. Were it not for air resistance and friction in the coaster's wheels, the coaster could climb a hill equal in height to the one from which it was released, and the coaster's mechanical energy would be conserved.

### A Comet's Movement

The conservation of energy principle holds for the orbits of comets and other bodies with noncircular orbits. A comet's distance from the sun changes greatly as it revolves, and so does its speed (see Figure 6.11). The closer a comet is to the sun, the smaller is its *GPE*, the larger is its *KE*, and the faster it moves. Because friction can be considered negligible for bodies moving through interplanetary space, a comet's total mechanical energy, *GPE + KE*, remains constant. The relationship between a comet's speed and its distance from the sun can also be understood in terms of the comet's angular momentum ($l = mvr$), which is conserved. The smaller the orbital distance ($r$), the greater the speed ($v$).

## Rest Energy

Besides the two broad categories of energy discussed so far, potential energy and kinetic energy, there is a third, all-inclusive category: rest energy. **Rest energy** (or *mass energy*, as it is sometimes called) is the energy equivalent of mass. As a consequence of Einstein's postulates for the special theory of relativity (Chapter 13), the energy ($E$) a body possesses *by*

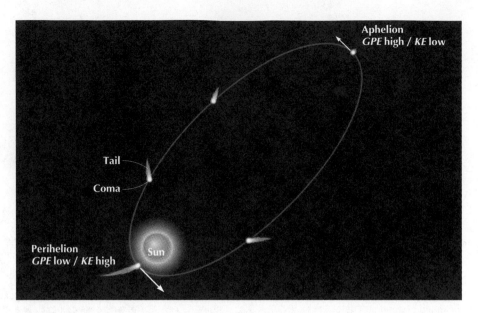

**FIGURE 6.11**   An idealized comet subject only to the gravitational pull of the sun conserves its total mechanical energy, *GPE* + *KE*. On its way "in" toward the sun, the comet gains speed (and *KE*) at the expense of *GPE*. On its way "out," the comet loses *KE* at the same rate it regains *GPE*.

The analysis of a real comet's orbit is much more complicated than this. If a comet passes near a planet or other massive body besides the sun, that body will exert a significant attractive tug on the passing comet, thereby altering its path.

A comet passing near the sun may also be subject to *nongravitational forces.* The hard-to-predict nongravitational forces include the reaction force a comet receives when heated gases suddenly erupt from its surface, and the steady reaction forces that result from material streaming away from the comet's *coma,* or head. The ejected material is pushed away from the coma by the sun's radiant energy and its solar wind. For this reason, a comet's tail always points approximately away from the sun.

*virtue of the existence of its mass* is equal to its mass (*m*) multiplied by the speed of light (*c*) squared:

$$E = mc^2$$

Equation 6.4

Rest energy is an internal form of energy distinct from the thermal energy we mentioned earlier. Even the tiniest bit of matter has a large amount of energy locked up within it. To get some idea of the equivalency of mass and rest energy, we'll calculate the energy equivalent of a 1-kg mass at rest using Equation 6.4: Square the speed of light and multiply it by 1 kg. The result is $E \approx 10^{17}$ J. This amount of energy could propel a million-ton payload from Earth to the moon!

For certain practical reasons, we cannot take a kilogram of dirt or any other substance and expect to easily extract much of its rest energy. If we could easily do that, then matter in the universe would not be very stable. Still, every reaction, whether chemical or nuclear, involves the appearance or disappearance of some mass. Energy-producing chemical reactions, such as the burning of gasoline, involve so slight a transformation of rest energy into other forms that the mass missing after the reaction is negligibly small (remember the qualification mentioned in connection with the law of conservation of mass in Chapter 4). Nuclear reactions, however, do involve measurable losses of mass. For example, in the sun's core more than 600 million metric tons of hydrogen are converted through nuclear fusion to helium every second. The helium produced has 0.7% (or 1/143) less mass than the original hydrogen.

On the subatomic scale, in a process called *pair annihilation,* the complete conversion of mass into energy takes place whenever a particle of matter and a corresponding particle of antimatter meet and annihilate to form two gamma ray photons of high energy. Conversely, the pure energy of gamma rays can be converted into mass in a process called *pair production.* These wholesale interconversions between mass and energy occur when subatomic particles collide at high speeds during particle accelerator experiments, and they commonly occurred during the earliest moments of the big bang.

## Power

Energy (or work, which transforms energy) and power are often confused. The difference between energy and power is like the difference between a worker's total payment for a particular job (say, $150) and his or her rate of pay (say, $10 per hour). Energy is a fixed quantity, independent of time. **Power** is the rate at which work is done, or, equivalently, the rate at which energy is transformed:

$$P = W/t = E/t$$                                    Equation 6.7

Since the SI units for work and time are, respectively, the joule and the second, power is measured in joules per second. By definition, 1 J/s equals 1 watt (W). The watt is the SI unit for power, but you will often see, for larger quantities of power, the units kilowatts (kW) and megawatts (MW). (Watts, kilowatts, and megawatts are, of course, the familiar units used in electrical distribution systems. They refer to the rate at which electric energy is either produced or expended, as we will see in Chapter 10.)

---

**Problem 6.4**
A woman who weighs 600 N climbs a flight of stairs 5 m high in exactly 10 s. What was her average power output during the 10 s?

**Solution**
First, calculate the work. From Problem 6.1, $W = 3000$ J. Then use Equation 6.7:

$P = W/t$

$\quad = (3000 \text{ J})/(10 \text{ s})$

$\quad = 300 \text{ W}$

(Note that our symbol for the variable "work" is an italic *W,* and the unit watt is abbreviated as a roman W.)

---

An older unit for work is the horsepower (hp), which was devised over two centuries ago by measuring the power output of a workhorse. The horsepower unit is now defined as being equal to 746 watts—approximately 3/4 kilowatt. A champion workhorse might deliver this amount of power, or 746 J of work per second, for a period of several hours. Trained athletes can develop around 1/4 hp, or nearly 200 W, for long-duration events such as the marathon footrace. Humans can very briefly develop as much as 7 hp (5 kW) for athletic feats like weight lifting and shot putting.

We should emphasize again that power and energy are different but related quantities. The energy released by a single, loud handclap is about the same as the energy you release when rubbing your hands together for a few seconds, though it may seem as if the former process is much more

---

## BOX 6.2
# Earthquake!

Large earthquakes are the least predictable and perhaps the most feared of Earth's natural upheavals. Although we now have a clear understanding of where and why earthquakes occur, we cannot reliably predict when they will occur—at least, not in the short term. At best, the prediction of future earthquakes is similar to a meteorologist's long-term forecast of a coming year's rainfall or average temperature for a given region.

The driving force behind earthquakes is well understood in terms of the theory of plate tectonics. (The subject of plate tectonics was introduced in Window M of the cosmic zoom in Chapter 2 and will be explained in detail in Chapter 16.) The slow, jostling movements of Earth's lithospheric plates stress (apply forces to) the various rock formations underground. Depending on the kind of stress applied to them, rocks can experience compression, tension, or both. Elastic potential energy is stored in these rocks in much the same way that a spring (like the one in Figure 6.5) stores energy whenever it is either compressed or stretched. If stressed enough, rigid masses of rock pressed against each other can slip and start sliding so as to lose their elastic potential energy. These movements typically take place along one or more linear fractures within Earth's crust or upper mantle. These fractures, called *faults,* range in length from less than a kilometer to hundreds of kilometers. Lengthy faults can have many strands, or parallel branches; if many strands are present, they are collectively known as a *fault zone.* Along what is probably the world's most famous fault zone, California's San Andreas Fault, ground movements have profoundly affected the topography, and they occasionally lay waste to cities and their infrastructures (see Photo 6.3).

Once sudden movement starts along a fault, elastic potential energy is converted into kinetic energy of the moving rock masses. When the movement stops, much of the kinetic energy is converted into seismic waves (*seismic* means "relating to an earthquake") that radiate in all directions, much like the circular ripples that radiate from a fallen stone that loses *KE* as it hits the surface of a pond. Some seismic waves propagate outward along Earth's surface, causing the ground to heave and quake. Other seismic waves pass through Earth's interior, reflecting and bending as they encounter different layers inside. We will return to the study of these seismic waves in Chapter 8.

If rock formations along an active fault easily crumble or slip past each other, then the accumulated strain on the rocks is released often, triggering earthquakes that are both small and relatively frequent. However, if rocks on opposing sides of a fault tend to stick to each other and resist sliding, then the movements are likely to be infrequent and violent.

The intensity of an earthquake, as it is felt at a given site, depends on several factors; among them are the amount of slip (displacement) along the fault, the length of the rupture, the distance of the site from the rupture, and the type of rock or soil traversed by the seismic waves. When intensity measurements of a quake made from many sites are pooled and analyzed, geoscientists can estimate the total amount of energy released by the quake event.

A common measurement of a quake's overall energy release—often quoted in news reports today—is its *moment magnitude,* or simply *magnitude.* Until the 1980s, ratings of earthquake magnitude referred to the scale devised by the seismologist Charles Richter in 1931. That scale, though, which relied on ground movements as recorded by earlier instruments, does not accurately express the energy released by the biggest quakes. The new, more accu-

energetic. Similarly, Earth's normal daily share of solar energy far exceeds the energy released by the largest earthquakes and violent volcanic eruptions (see Figure 6.2). The quantity called power introduces the element of time: The less time it takes for a given amount of energy to be released, the more powerful the event. Earthquakes (see Box 6.2) and the impact of large pieces of celestial debris on Earth's surface (see Box 6.3) are examples of phenomena that are short-lived and extremely powerful.

rate scale of moment magnitude, which is based on the amount of displacement along a fault and the length of the rupture, has been defined so that it closely agrees with the old Richter scale for all but the very largest quakes.

Earthquake magnitude scales are logarithmic so as to encompass an enormous range. Generally, each successive integer magnitude on the scale represents a tenfold increase in the amplitude, or range of vibration, recorded at any given fixed site. The magnitude scales indirectly measure total energy as well. Each successive integer magnitude represents a factor of about 30 in energy release. Thus, a magnitude 7.0 event releases about 30 times more energy than a magnitude 6.0 event, and a magnitude 8.0 event releases some 900 times ($30 \times 30 = 900$) more energy then a magnitude 6.0 event. Magnitudes range from less than 2 (insensible to humans, even at close range) to more than 8, which is considered a "great" earthquake.

The total energy released by a great quake (about $10^{18}$ J) is impressive, yet it is almost insignificant compared to Earth's daily receipt of solar energy, a quantity some 100,000 times larger. Remember, though, that such a quake releases nearly all its energy during a period lasting no more than a few minutes. The *rate* of energy release (power) is large, and it is concentrated for the most part on a small fraction of Earth's surface.

Moment magnitudes are based on both the displacement and the length of rupture, but these two factors are themselves somewhat related: The longer the rupture, the greater the displacement tends to be. For example, San Francisco's great earthquake of 1906, moment magnitude 7.7, ruptured the San Andreas Fault along a distance of 440 km and produced offsets of up to 6 m. In contrast, the much weaker 1989 Loma Prieta quake, moment magnitude 7.0, in the same region, was produced by

**PHOTO 6.3**   The 1994 Northridge earthquake extensively damaged parts of the Los Angeles metropolitan region and severed several important transportation links. Although its 6.7 magnitude was not as large as that of other quakes in California, the energy of the quake was intensely focused on a region with a large population density.

a movement of about 1 m along a roughly 50-km-long stretch of the San Andreas Fault. The Loma Prieta quake fell short of being a great quake, but it still managed to claim 62 lives and cause $6 billion in property damage.

Major movements along certain sections of the San Andreas Fault are expected within the lifetime of most readers of this textbook. The last big lurch along the southern section of the San Andreas Fault occurred in 1857, with displacements of up to 5 m. Strain has been accumulating ever since. Southern Californians await the inevitable "big one"—the major earthquake that many geologists guess will occur on the southern San Andreas Fault within 30 years.

# Chemical Energy

According to modern atomic theory, chemical reactions involve rearrangements among atoms, along with the making or breaking of *chemical bonds* (the nature of bonds will be discussed in Chapter 9). Apart from any knowledge of what these bonds really are, several important aspects of chemical reactions may be understood solely in terms of energy and its conservation. Every chemical reaction is accompanied by a transformation of energy, and all transformations of energy are consistent with the law of conservation of energy.

# Exothermic and Endothermic Reactions

When hydrogen and oxygen combine to form water vapor in an explosive flash, much energy is released in an obvious way. The chemical equation for this reaction can be written

$$2H_2 + O_2 \rightarrow 2H_2O + E$$

$$E = 13.6 \text{ kJ/g of } H_2O = 142 \text{ kJ/g of } H_2$$

For every gram of water produced, 13.6 kJ (13,600 J) of energy are released. Alternately, for every gram of hydrogen that participates in the reaction, 142 kJ are produced. As chemical reactions go, this amount of energy is large. Any reaction that liberates energy is known as an **exothermic reaction.** (The prefix *exo* means "outside"; *thermic* refers to thermal energy, the primary form of energy release in chemical reactions.)

a

b

**PHOTO 6.4**   (a) The reaction of elemental potassium (potassium metal) and water evolves hydrogen gas, which ignites. The entire reaction is strongly exothermic. (b) The reaction between the chemicals barium hydroxide and ammonium thiocyanate is endothermic. When mixed at room temperature, these reactants yield a product that is very cold.

The reverse of the reaction is the decomposition of water. Water molecules can be decomposed, or broken apart, by heating water vapor to very high temperatures or by forcing an electric current through the water, in the process called *electrolysis*. This reverse reaction is written

$$2H_2O + E \rightarrow 2H_2 + O_2$$
$$E = 13.6 \text{ kJ/g of } H_2O = 142 \text{ kJ/g of } H_2$$

This and other reactions that absorb energy are called **endothermic reactions** (the prefix *endo* means "inside"). The law of conservation of ener-

---

## BOX 6.3
# Impact Events

 Every day, Earth sweeps up about 100 tons of debris from interplanetary space. The vast majority of this debris, particles ranging in size from dust motes to grains of sand, immediately disintegrates as it rains down upon the upper atmosphere.

A particle no bigger than a grain of sand can produce a rather bright, fiery-looking trail in the night sky—a *meteor* or "shooting star"—as it plunges through the atmosphere at tens of kilometers per second (Photo 6.5). A solid particle the size of a marble can produce a fireball nearly as bright as the moon. A meteor as big as a baseball, if hard and dense and moving relatively slowly, may survive its fiery plunge and reach Earth's surface. Once on the ground, such bodies are known as *meteorites*.

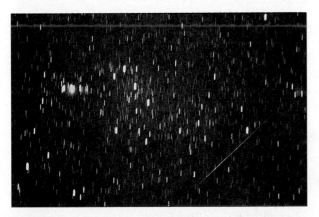

**PHOTO 6.5**  Small bits of space debris are constantly swept up by Earth's atmosphere. The meteor that caused the visible streak in this photograph was about as bright as the planet Jupiter in the sky and remained luminous for less than a second.

Larger bodies come in contact with Earth much less frequently than smaller bodies. Once a year, on the average, a body with a diameter of around 1 m strikes Earth or disintegrates in the atmosphere. Bodies of about 10-m size may come our way approximately every century, and bodies of 100-m size may come every 10,000 years or so. (See Figure 6.12.)

Dense objects, such as the iron-rich bodies that make up the small fraction of asteroids that cross Earth's orbit, may plunge through Earth's atmosphere without too much difficulty and strike the ground. To get some idea of the energy released by a relatively large impact of this sort, let us calculate the kinetic energy of a metallic body, roughly 100 m across (about the length of a football field) with an assumed mass of 4 billion kilograms, moving at a typical speed of 50 km/s relative to Earth. From Equation 6.3, we get

$$KE = (1/2)mv^2$$
$$= (1/2)(4 \times 10^9 \text{ kg})(5 \times 10^4 \text{ m/s})^2$$
$$= 5 \times 10^{18} \text{ J}$$

This kinetic energy is about 100 times more than the energy released during the explosion of a large nuclear bomb. If an object having this much *KE* collides with the ground, the impacting body is instantly vaporized. The resulting explosion excavates a crater more than a kilometer across and 100 to 200 m deep. The excavated debris, along with the vaporized remnants of the impacting body, is flung upward and outward. It then falls, coating the ground for a radius of many kilometers around the crater. Similar "splash" (or ejecta) patterns can be seen around the impact craters observed on many of the moons and planets throughout the Solar System.

The most recent known event of this magnitude was the collision that produced the   *(continued)*

gy states—and experiments confirm—that the energy liberated in the water-producing, exothermic reaction equals the energy absorbed by the water-destroying endothermic reaction. This relationship holds for any reversible reaction. The energy supplied in the reaction's endothermic direction equals the energy liberated in its exothermic direction, and vice versa.

It may seem a bit mysterious that energy can be absorbed during an endothermic reaction, be stored by the products of that reaction, and then be released when the products participate as reactants in the reverse exothermic reaction. By way of analogy, we may understand these steps as being similar to the sequence of lifting a book, holding it, and then letting it fall. In

*(Box 6.3 continued)*    Meteor Crater in Arizona (refer to Photo 3.20 in Chapter 3) an estimated 50,000 years ago. Because little geologic evidence is left behind by similar-sized bodies striking the ocean, Earth has probably swept up other bodies of similar size in the time since the Arizona crater was formed.

Bodies less dense than the iron-rich asteroids such as stony asteroids and comets, which consist of frozen gases and dust, have a better chance of exploding in midair before reaching the ground. Recent studies show that a mysterious blast over Tunguska, Siberia, in 1908 can be attributed to the sudden breakup of a 60-m-diameter stony asteroid at a height of about 8 km. The sudden deceleration caused by air friction at that height would have smashed the asteroid into numerous small pieces. Those pieces, in turn, would have slowed at an even greater rate, causing enough heat to produce an explosive fireball. Whatever the exact cause, the midair explosion was powerful enough to flatten hundreds of square kilometers of forest and produce pressure waves that traveled twice around the plan-

**PHOTO 6.6** This iron meteorite, discovered in Imperial County, California, has been sliced open and etched with acid to show its structural pattern. The large crystal structure shown indicates that the parent body of the meteorite (probably one of the larger asteroids) was once molten inside, slowly cooled, and later broke apart during a collision or collisions with other asteroids.

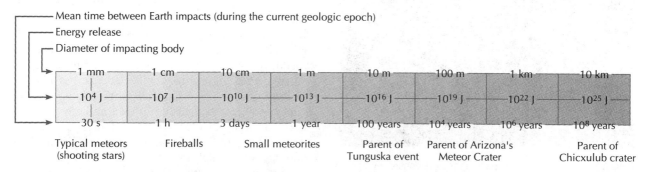

| | | | | | | | |
|---|---|---|---|---|---|---|---|
| 1 mm | 1 cm | 10 cm | 1 m | 10 m | 100 m | 1 km | 10 km |
| $10^4$ J | $10^7$ J | $10^{10}$ J | $10^{13}$ J | $10^{16}$ J | $10^{19}$ J | $10^{22}$ J | $10^{25}$ J |
| 30 s | 1 h | 3 days | 1 year | 100 years | $10^4$ years | $10^6$ years | $10^8$ years |

Mean time between Earth impacts (during the current geologic epoch)
Energy release
Diameter of impacting body

Typical meteors (shooting stars)    Fireballs    Small meteorites    Parent of Tunguska event    Parent of Arizona's Meteor Crater    Parent of Chicxulub crater

**FIGURE 6.12**   The larger the size of a meteoritic body entering Earth's atmosphere, the greater the release of kinetic energy. Few bodies smaller than 1 m in diameter survive intact during their fiery plunge through the retarding atmosphere. The rare impacts of bodies 10 km in diameter or larger may cause the extinction of the majority of Earth's biological species.

both cases, potential energy (*PE*) is gained, stored, and released. Gravitational potential energy is stored in the book owing to its position in a force field, namely Earth's gravitational force field. *Chemical potential energy* is stored in the particular molecule or molecules formed by an endothermic reaction, and that stored *PE* has the potential of being released.

Chemical potential energy is the *PE* that negatively charged electrons have with respect to their position in an electric force field set up by the positively charged nuclei of atoms. For example, after one $O_2$ molecule and two $H_2$ molecules combine exothermically to make two $H_2O$ molecules, the electrons in the $H_2O$ molecules move in new patterns around the H and O

et. Several nomadic herdsmen were killed or rendered unconscious for hours or days as a result of the blast, and one man was knocked off his chair at a trading post about 70 km away from the impact site.

Roving asteroids or comets of around 10-km diameter may strike Earth about once every 100 million years. In doing so, they catastrophically damage the entire biosphere by lofting huge quantities of sunlight-absorbing dust into Earth's atmosphere and by burning much of the world's vegetation to ashes. The last event of this magnitude probably took place 65 million years ago. This was the K-T (Cretaceous-Tertiary) extinction event of 65 million years ago, which triggered the final demise of the dinosaurs and ultimately led to the ascent of the mammals (as we outlined in Chapter 3).

Evidence for a catastrophic K-T impact comes in several forms:

1. A rich deposit of the element iridium occurs in 65-million-year-old sediments from around the world. This element is rare in Earth's crust but is much more abundant in comets and asteroids. The iridium could be the fallout from a giant impact of a single large comet or asteroid or possibly of two or more large meteoritic bodies over a small interval of geologic time.

2. Slivers of "shocked quartz," which can only be formed by a shock wave comparable to that of a nuclear explosion, are found in the same sediments.

3. Lumps of "tektite glass" are also found imbedded in the sediments. They could have formed when a spray of molten-rock droplets that were ejected from the impact crater cooled and solidified into bits of glass.

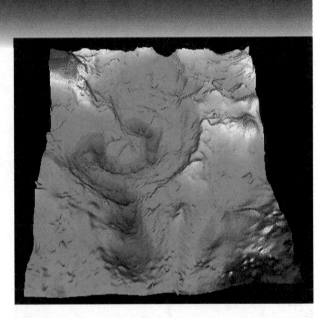

**PHOTO 6.7** This false-color, computer representation of the buried Chicxulub crater in Mexico is based on gravity measurements made by orbiting spacecraft.

4. A crater has been found of just the right age and of a sufficient size to account for the ejecta found worldwide. This is Mexico's Chicxulub crater (Photo 6.7), a basin now buried by sediments on the coastline of the Yucatán Peninsula. Gravity measurements indicate that the feature has a diameter of at least 180 km, which makes it Earth's largest known crater. The impacting body responsible for this crater is thought to have been at least 10 km across.

Astronomers estimate that more than 1000 asteroids large enough to pose a serious threat to Earth's biosphere cross or pass near Earth's orbit. Thus far, only about 120 of them *(continued)*

nuclei. The new patterns involve less potential energy. The lost potential energy becomes kinetic energy on a molecular scale (thermal energy, or heat), and some of the thermal energy is converted into the light and the percussive "boom" that accompanies an explosive reaction.

Most of the chemical reactions we tend to notice are exothermic. They include the *combustion* (rapid, oxygen-combining, or "burning") reactions associated with flammable substances like wood, gasoline, and gunpowder. A good deal less obvious are most endothermic reactions, such as the process of *photosynthesis* (Box 6.4), which traps radiant energy from the sun and converts it to potential energy in plants. The functioning of Earth's biosphere relies on a complex web of exothermic and endothermic reactions that continually recycle both matter and energy.

## Activation Energy

Many chemical reactions are really net reactions, encompassing two or more distinct steps. These steps can include both endothermic and exothermic processes. Let us look more closely at the seemingly simple combustion reaction between hydrogen and oxygen that forms water. The

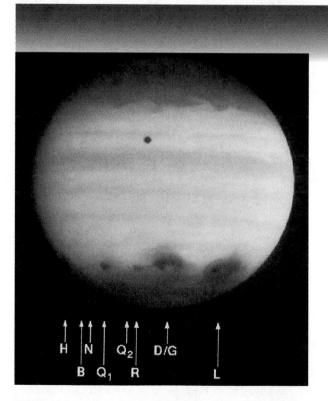

**PHOTO 6.8** These ultraviolet images taken by the Hubble space telescope show several of the sites in Jupiter's atmosphere hit by fragments of Comet Shoemaker-Levy 9. The sites are labeled in the order of the fragments of the comet seen before the impacts took place.

*(Box 6.3 continued)* have been found. Almost all the rest could be located within a decade by means of a purposeful search using a battery of small to midsize electronically controlled telescopes. By tracking these asteroids, astronomers could determine well in advance when any of them would strike Earth. With plenty of warning time, we could presumably muster the technology and effort to divert any threatening projectile out of a collision path. The urgency of this task was underscored in July 1994 when a long string of kilometer-size fragments of the shattered Comet Shoemaker-Levy 9 struck Jupiter (Photo 6.8) and released the energy equivalent of tens of thousands of nuclear bombs.

Protecting Earth from the vastly greater number of smaller bodies, tens of meters in diameter (like that of Tunguska impactor), is much more problematical and probably not worth the trouble. Such bodies strike Earth with a frequency of once every few hundred years, but the vast majority explode over the ocean or sparsely populated areas. Each of these events affects a limited area on Earth and releases less energy than that of a large volcanic eruption.

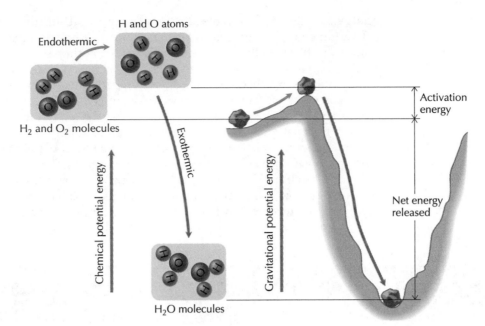

**FIGURE 6.13**  The activation energy required to break the chemical bonds in $H_2$ and $O_2$ molecules is greatly surpassed by the release of energy when H and O atoms bond in a new way (shown as simplified here) to form water, $H_2O$. The net release of energy, as heat and light, equals the net loss in chemical potential energy. In an analogous way, a stone that must be rolled upward over a canyon rim in order to fall to the bottom releases a large amount of gravitational potential energy at the expense of a small amount of activation energy (or work, in this case). Both water molecules and a stone resting in a deep chasm are relatively stable. Much energy is needed to decompose water into hydrogen and oxygen, and much energy is needed to raise the stone back to the level of the rim.

$2H_2 + O_2 \rightarrow 2H_2O$ reaction does not occur spontaneously at room temperature. Hydrogen and oxygen mixed together will not combine and release energy unless a concentrated amount of thermal energy (usually, a small spark or a lighted match suffices) triggers the start of the reaction.

In the high-temperature region of the spark or flame, molecules are moving rapidly. When they collide, they do so with enough force to separate the atoms of the $H_2$ and the $O_2$ molecules. The following is the first of one (though not the only) possible series of steps leading to the formation of water: $H_2 \rightarrow H + H$, and $O_2 \rightarrow O + O$. This first, entirely endothermic step involves the absorption of *activation energy*—the energy needed to break the bonds of the $H_2$ and the $O_2$ molecules. The activation energy is supplied by the kinetic energy of the fast-moving molecules. Once broken apart, the individual H and O atoms have more potential energy than they had before.

The momentarily unfettered H and O atoms can recombine into their original state as $H_2$ and $O_2$ and thereby lose the potential energy they just gained, or they can recombine into a new and even more stable configuration, $H_2O$. They prefer the latter, $H + H + O \rightarrow H_2O$, which involves a greater loss of potential energy (see Figure 6.13). The net loss in potential energy is transformed into a net gain of thermal energy. The net energy liberated by most combustion reactions is relatively large compared to the activation energy required.

Once molecules start reacting somewhere in a combustible substance, the thermal energy released may be large enough to supply activation energy to other nearby molecules. They react and, in turn, supply more activation energy to molecules farther afield. The reaction may spread like wildfire until virtually all of one reactant or another is used up. Self-sustaining combustion may spread relatively slowly (as in the burning of leaves, paper, or wood) or rapidly enough to cause an explosion (as in hydrogen combustion when plenty of oxygen is mixed with the hydrogen).

In some cases, the atoms or molecules of a substance already have enough thermal energy at room temperature to supply the necessary activation energy to start certain reactions. Pure sodium metal and pure chlorine gas, for example, very rapidly combine with one another to form the compound sodium chloride (NaCl). This reaction happens spontaneously, even at very low temperatures. Luckily for us, most combustion reactions favor temperatures much higher than room temperature. If they did not,

## BOX 6.4
# Energy and Life

Every living thing depends on a steady supply of useful energy. Organisms such as phytoplankton (algae that support the food chain in the oceans) and green plants of all kinds transform the sun's radiant energy into chemical potential energy. Inside the green foliage of plants, carbon dioxide from the atmosphere and water drawn from the plant's roots combine in an endothermic reaction called *photosynthesis*. The products of this reaction are glucose (a form of sugar), which remains inside the plant, and oxygen gas, which is released into the atmosphere. Photosynthesis consists of two sets of complex reactions, but its net effect can be summarized by the following statement:

$$6CO_2 + 6H_2O + \quad E \quad \rightarrow C_6H_{12}O_6 + \quad 6O_2$$
$$\underset{\text{(carbon dioxide)}}{} + \text{(water)} + \text{(energy)} \rightarrow \text{(glucose)} + \text{(oxygen)}$$

Further biochemical reactions inside plant tissue may combine glucose molecules into carbohydrates such as starch and cellulose. Molecules of starch store energy for further use by the plant. Cellulose is used for structure: The wood fibers of a tree and the woody stems of other plants are essentially made of cellulose.

Animals make their metabolic living by eating plants, by eating animals that have eaten plants, or by eating animals that have eaten animals that have eaten plants. Thus the flow of energy that originated in the sun eventually gets distributed to a wide variety of organisms.

Do animals, which are higher on the food chain than plants, have anything to give in return? Of course! Recycling has been the norm in the natural world for billions of years. Take, for example, the metabolic process that powers animals, called the *oxidation of glucose*. In an animal's digestive tract, complex carbohydrate molecules such as starch are disassembled into simpler glucose units. Next, enzymes inside the cells break up glucose molecules into simpler components. These components react with oxygen molecules to form carbon dioxide and water, thereby releasing metabolic energy. For an average human at rest, metabolic energy is generated—mostly in the form of heat—at the rate of about 100 W. (When you think about how much heat escapes from a standard 100-W lightbulb, it's easy to

**PHOTO 6.9**   The green color of these leaves indicates the presence of chlorophyll, a substance that makes photosynthesis possible.

much of the world's biomass and fuels would have long ago gone up in flames by spontaneously combining with the oxygen of the air.

## Chemical Energy and Stability

Water is chemically stable, because a relatively large amount of energy is required to break water molecules apart. Using the visual analogy of Figure 6.13, we can see that much energy must be supplied to decompose water—just as much work is required to raise a stone, fallen to the floor of a canyon, back to the canyon's rim.

In general, chemical reactions that produce a great deal of energy also produce stable products. Conversely, reactions that either release relative-

understand why a small, closed room filled with people can get stiflingly hot in a short time.) The oxidation of glucose can be summarized as

$$C_6H_{12}O_6 +\ \ 6O_2\ \rightarrow 6CO_2 + 6H_2O +\ \ \ E$$

(glucose) + (oxygen) → (carbon dioxide) + (water) + (energy)

This exothermic process is the reverse of photosynthesis. Photosynthesis releases oxygen to the air, a gas needed by animals to keep their metabolism going. And the metabolism of animals gives back carbon dioxide to the atmosphere, a gas that plants need (see Figure 6.14).

The recurring role of nitrogen in the biosphere is another example of natural recycling. The element nitrogen is an essential component of proteins, which all plants and animals need. But neither plants nor animals can make direct use of the gaseous nitrogen that is abundant in our atmosphere. Instead, plants obtain nitrogen from nitrogen compounds in the soil and pass it on to animals when plants are consumed.

Nitrogen is steadily depleted in the soil as plants grow. In undisturbed ecosystems, the nitrogen is quickly replenished, primarily by the decay of animal waste and of dead animals and plants. In disturbed ecosystems, including most of the world's agricultural zones, the nitrogen cycle is often broken by overzealous harvesting, erosion of topsoil, and the practice of dumping nitrogen-rich sewage into lakes and oceans rather than returning it to the land. Nitrogen-rich fertilizers can and do save the day, but at the expense of much extra energy needed to manufacture, transport, and dispense them.

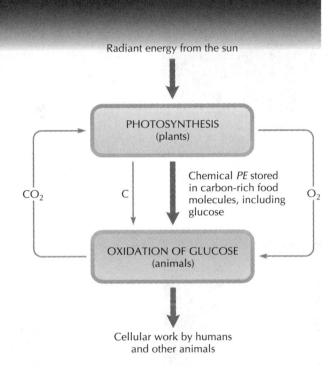

**FIGURE 6.14** Humans and other animals participate in the recycling of oxygen, carbon, and carbon compounds through the biosphere. Although matter can be completely recycled through an idealized system such as the one shown here, the flow of energy is primarily one way. It originates in the sun, passes through plants and animals, and is ultimately dispersed as thermal energy, or waste heat, as a result of cellular work.

ly little energy (weakly exothermic) or else absorb energy (endothermic) tend to result in products that are unstable or easily decomposed. The "fireworks" that occur during the chemical combination of sodium and chlorine are an indication that its product, sodium chloride, is stable. It remains stable even at very high temperatures. Conversely, chlorine and sodium are themselves very unstable. A great deal of energy must be supplied to sodium chloride in order to decompose it into its constituent atoms.

# Chemical Reaction Rates

In physics, energy and power have different definitions. Similarly, in chemistry, the *total energy* released by a reaction and the *rate of energy release* during the same reaction differ. For example, a large amount of hydrogen can be explosively combined with the oxygen in the air. Alternatively, the same mass of hydrogen can be allowed to flow from a supply tank through a small orifice where it will nurture a small flame that can burn for hours.

The rates of most chemical reactions are sensitive to four principal factors: the *concentration* and the *surface area* of the reactants, the *temperature* at which the reaction takes place, and the participation of agents called *catalysts*.

a

## Concentration

Reaction rates are generally proportional to the concentration of each reacting substance. A candle, for example, will burn about five times faster in pure (100%) oxygen than it will in the air, which has an oxygen content of 21%. Concentrated acids are more effective than diluted acids when they are used as "reagents," or reacting agents. The more molecules per volume that can participate in a reaction, the more frequently the molecules interact and the faster the reaction proceeds.

## Surface Area

When reactions occur on the surface of a solid or a liquid, the amount of surface area involved has an effect on the speed of the reaction. The more finely sliced or divided the mass of a given reactant, the more surface area it has and the faster it reacts. Take, for example, the combustion of wood and wood products. A crumpled-up piece of paper, with its large surface area, burns easily. Many small, thin pieces of wood (kindling) will burn more slowly than paper but faster than a single, fat chunk of wood which has relatively little surface area exposed to the air. For this reason, paper and kindling are useful in starting a campfire made of large logs.

The surface area effect comes into play during the combustion of gasoline in the cylinders of a car's engine. Gasoline is sprayed as a fine mist into each cylinder along with air, just before a spark plug ignites the mixture. The smaller the droplets of gasoline, the greater the area of contact between the gasoline and the oxygen of the air and the faster the reaction. Similarly, many catastrophic explosions have occurred as a result of coal dust, cotton dust, grain dust, sawdust, or even metal dust igniting inside mines, grain elevators, and other enclosures.

b

**PHOTO 6.10**   A candle burning in (a) air and (b) pure oxygen.

## Temperature

For nearly all reactions, an increase in temperature results in a more rapid reaction rate. A hot campfire burns faster than a cooler one, all other factors being equal. Higher temperatures are associated with faster-moving molecules, and faster molecules have a greater chance of reacting when they undergo collisions. A handy rule of thumb, called the *10 °C rule,* can be invoked to estimate the rate of a chemical reaction near room temperature. It states that for every 10 °C (18 °F) increase in temperature, the reaction rate roughly doubles. The rule is valid for most, but not all, chemical reactions. The 10° rule helps explain why food spoils or becomes stale so rapidly when left in the sun, and why food is preserved longer when refrigerated or frozen.

## Catalysts

Chemicals that either speed up or slow down the rate of a reaction are known as **catalysts.** A catalyst participates in the reaction it affects, yet it remains unchanged after the reaction is over. In this sense, a catalyst can be compared to an usher in a dark theater. The usher greatly shortens the time it takes a partron to find an assigned seat, and afterward, the usher is available to help other patrons find their seats. Some catalysts, especially certain metals, participate in reactions by providing a surface on which reactants can be concentrated and activated. Other catalysts act by briefly joining with a reactant to form an unstable compound that may trigger a reaction that would otherwise not easily occur. Many chemical processes in living things—for example, digestion—would take place at a very sluggish rate without the help of *enzymes,* which are biochemical catalysts. Thousands of different enzymes in the human body regulate the complex web of biochemical reactions that keep us alive.

# Chemical Fuels

Chemical fuels (as opposed to nuclear fuels) are substances that can be utilized for the controlled release of chemical potential energy. Ideal fuels are strongly exothermic when burned in the oxygen of the air, cheap and abundant in nature (or at least easy to synthesize from a raw state), safe to store and transport, and nonpolluting when burned. Of the panoply of fuels available today, none satisfies all these desirable traits.

For tens of thousands of years, firewood has been used for cooking and heating. It remains an important, though relatively feeble, source of chemical energy in many of the world's most poorly developed countries. If managed properly, firewood is a renewable resource. Unfortunately, it is being harvested far too rapidly in the regions where it is needed most; its use will surely diminish with time. Wood combustion also emits into the atmosphere an assortment of polluting by-products, including soot, which consists of tiny carbon particles.

*Fossil fuels* supported the industrial revolution and remain the fuels of choice for meeting most of the world's energy needs today. The fossil fuels (coal, oil, and natural gas) are so named because they were formed from the remains of buried organisms that lived millions to hundreds of millions of years ago.

Coal, historically the first fossil fuel to be exploited for energy on a large scale, consists of the carbon-rich remains of tropical plants that lived and died around 300 million years ago. If coal were pure carbon, its reaction with oxygen at high temperature could be written as

$$C + O_2 \rightarrow CO_2 + E$$

$$E = 9 \text{ kJ/g of } CO_2 = 33 \text{ kJ/g of C}$$

The 33 kJ of thermal energy (heat) per gram of carbon produced by burning pure carbon compare somewhat unfavorably with the heat derived from hydrogen combustion, 142 kJ per gram of molecular hydrogen ($H_2$), but compare favorably with the yield of most firewood, about 15 kJ per gram of wood. Molecular hydrogen does not exist abundantly in nature, and it is difficult to store safely because of its tendency to explode when in contact with air. Coal, on the other hand, is abundant and relatively easy to mine in many parts of the world, is not very flammable when stored properly as solid chunks, and is easy to transport, especially by rail. Coal was the world's chief source of useful energy prior to the 1940s. Its use has since been surpassed by that of oil (petroleum). Today, coal meets about 25% of the demand for energy in the United States. Most of it is burned to generate electricity at power plants.

Earth's coal reserves might last another 300 years, assuming the rate of coal consumption does not grow with time. Coal combustion has serious drawbacks, however. All coal contains some sulfur and other impurities, and some of these impurities are carcinogenic (cancer causing). The common bituminous (soft) coal has a sulfur content averaging about 3%. The unwanted substances are difficult or impossible to extract during the mining stage. When the coal is fired, the sulfur in it reacts with oxygen to form sulfur dioxide: $S + O_2 \rightarrow SO_2$. The sulfur dioxide, propelled upward through the stacks of coal-burning power plants, mixes with moisture in the atmosphere and is eventually converted to sulfuric acid, which collects in clouds. Sooner or later, near or far, the acidified moisture falls as acid rain. Acid rain, which may also consist of nitric acid and other acids, has been implicated in massive die-offs of vegetation and the destruction of freshwater fish and aquatic life in various parts of the world.

Carbon dioxide, a product of the burning of coal and other fossil fuels, was long considered harmless. Carbon dioxide is not a polluting substance in the usual sense. However, its rapidly increasing concentration in the atmosphere is causing great concern among many scientists and environmentalists. Carbon dioxide and certain other gaseous products of modern civilization trap solar radiation by means of the *greenhouse effect*. Theoretically, an excess of solar radiation trapped in the atmosphere by the greenhouse effect may lead to an increase in the average temperature of Earth's surface and atmosphere, or *global warming*. We will continue our discussion of the greenhouse effect and global warming in the next chapter.

Currently, most of the world's practical energy is derived from oil. Crude oil, or petroleum, is thought to have been formed from the remains of marine organisms buried by sediments and then subjected to heat and pressure for millions of years. As we shall see in Chapter 9, the polyglot

**PHOTO 6.11** A large amount of chemical potential energy is stored in this pile of coal. Unfortunately, the mining of coal and the release of its energy involve serious environmental drawbacks.

mixtures of *hydrocarbons* (carbon- and hydrogen-containing compounds) in petroleum can be refined into the more uniform mixtures we know as the liquid fuels gasoline, diesel fuel, kerosene, and heating oil, plus motor oil and paraffin wax. Petroleum also plays a vital role in the synthesis of plastics and many other modern products.

Oil is currently abundant in many parts of the world, though its total reserves may not last another century at the present rate of consumption. Whether in crude or refined form, oil is relatively easy to transport, store, and use, which accounts for its worldwide popularity, especially as fuel for transportation. Currently, the United States relies on oil for about 40% of its energy needs, a figure that is increasing even as its domestic sources become depleted. The United States now obtains much of its supply from overseas. The cost of this imported oil, which includes the political and military efforts that help keep oil flowing in from regions like the Middle East, should be recognized as the high cost that it really is.

The ideal, or complete, combustion of a fuel like gasoline produces only carbon dioxide ($CO_2$) and water vapor ($H_2O$). The following chemical equations describe the complete combustion of two components of gasoline, heptane ($C_7H_{16}$) and octane ($C_8H_{18}$):

$$C_7H_{16} + 11O_2 \rightarrow 7CO_2 + 8H_2O$$

$$2C_8H_{18} + 25O_2 \rightarrow 16CO_2 + 18H_2O$$

Most grades of gasoline have heats of combustion of about 50 kJ/g, which compares favorably with that of coal. Unfortunately, the complete combustion of gasoline is seldom attained in practice, even in well-tuned automobile engines. Some pollution is inevitable, as in the following reactions, which are representative of incomplete combustion.

$$C_7H_{16} + 9O_2 \rightarrow 4CO_2 + 8H_2O + 2CO + C$$

$$2C_8H_{18} + 20O_2 \rightarrow 11CO_2 + 15H_2O + 3CO + C_2H_6$$

For incomplete combustion reactions such as these, the products may include poisonous carbon monoxide (CO) gas, carbon (C) in the form of soot, and hydrocarbons such as ethane ($C_2H_6$).

Another difficulty arises as a result of the very high temperature existing in the cylinders while the gasoline is burning. Under high-temperature conditions, nitrogen and oxygen (the two main constituents of the air in the cylinders) have a tendency to combine into nitrogen compounds such as nitric oxide (NO) and nitrogen dioxide ($NO_2$). Once expelled through the exhaust pipe, these compounds may interact with atmospheric moisture to form a dilute nitric acid solution, which eventually falls as acid rain. Moreover, in the presence of strong sunlight, nitrogen dioxide can decompose into nitric oxide and atomic oxygen (O). The atomic oxygen reacts with normal oxygen ($O_2$) to form ozone ($O_3$), which is an irritating pollutant when present in the lower atmosphere. Some of the ozone produced may, in turn, react with hydrocarbon pollutants (such as ethane) in the air. The resulting mixture of noxious substances is known as *photochemical smog*, the type of air pollution common in automobile-oriented cities like Los Angeles.

Photochemical smog is distinct from another variety of smog, or "smoke fog," that occurs in areas polluted by the uncontrolled burning of coal. Severe air pollution of this sort was common in England during much of the industrial revolution and continues to occur today in several of the industrialized, impoverished countries of the former Soviet Union.

Over the past two or three decades, significant strides have been made in the United States and other developed countries toward limiting pollution from cars and other sources. Newer antipollution technologies include computer-controlled fuel injectors and catalytic converters in cars and "scrubbers" that can chemically remove sulfur compounds from the smokestacks of coal-burning electric power plants. These technologies are quite costly, and they will have to be applied more widely and with greater stringency if our present quality of life is to be maintained in the face of continued or accelerated worldwide use of coal and oil fuel.

*Natural gas* (cooking and heating gas) is close to being an ideal fossil fuel. It normally burns clean and produces nothing but carbon dioxide and water. Natural gas is a mixture of the lightest and simplest hydrocarbons, primarily methane ($CH_4$). Methane has a high heat of combustion:

$$CH_4 + 2O_2 \rightarrow CO_2 + 2H_2O + E$$

$$E = 56 \text{ kJ/g of } CH_4$$

*Natural gas* exists in underground cavities in association with petroleum deposits. An inevitable by-product of oil drilling, it can be collected at oil wells and piped wherever it is needed. When compressed and chilled, it can be shipped in tankers in the form of liquid natural gas. Unfortunately, a great deal of natural gas is simply vented at the wellheads and burned off (Photo 6.12). This is a colossal waste of useful energy dictated by the sometimes unfavorable economics involved in its recovery, storage, and transportation from remote sources. Natural gas is a superb but dwindling resource that could be nearly depleted 50–100 years from now. It currently fulfills about 20% of the U.S. energy demand.

Less common fuels, which are not fossil fuels, include ethyl alcohol (ethanol) and methyl alcohol (methanol). Ethanol, the more common of

the two, can be derived from agricultural sources. It is the result of the fermentation of sugar in potatoes, sugar cane, and grains. Some premium gasolines sold on the market contain about 10% alcohol.

Hydrogen shows much promise as a practical fuel of the future. It has an extraordinarily high heat of combustion (142 kJ/g), and its combustion produces no pollution, not even the pseudopollutant carbon dioxide. As we have seen, the sole reaction product of hydrogen combustion is water. The technology exists for storing and transporting hydrogen (at some risk, of course), and conventional automobile engines can be modified to handle it. Unfortunately, pure hydrogen does not exist in any appreciable quantity on or below Earth's surface. Hydrogen can be chemically extracted from water—but, of course, that takes a lot of energy. Scientists are currently searching for ways to economically "split" water molecules by utilizing the pure energy of sunlight and thereby obtain hydrogen in an environmentally benign way.

## Usable Energy

Whenever we flip a light switch, accelerate a car from a stoplight, or eat store-bought food, we become consumers of "usable" or "practical" energy. In various guises, usable energy is transmitted over power lines, consumed in vehicles, and released as dietary calories in the food we eat. The modern way of life, as known by many throughout the developed parts of the world, cannot be maintained without the continual and prodigious transformation of usable energy into forms such as heat, light, and mechanical energy. In the United States, the total rate of usable energy consumption (or power) per capita is about 10 kW. This is an astonishing 10 kJ of energy consumed every second, or 846 MJ per day, per person—enough to illuminate one hundred 100-W lightbulbs every hour of the day and night.

Many familiar forms of energy are actually intermediaries between some fundamental source of energy and what that energy eventually becomes after one or more transformations. Fossil fuels, for example, possess the energy of sunlight that fell upon Earth long ago. They can be trans-

Nuclear fusion reactions inside the sun convert nuclear energy into radiant energy. The energy moves out to the sun's surface and escapes into space, primarily in the form of visible light, infrared, and ultraviolet; some of this energy is intercepted by Earth.

Solar energy can be used directly for heat.

Spontaneous nuclear reactions (natural radioactivity) inside Earth convert nuclear potential energy into heat (geothermal energy) that

Nuclear fission reactions can proceed on Earth by artificial means (inside nuclear reactors). This produces heat that

Solar heating of water to the boiling point

can be used directly for heat.

Solar energy heats Earth's continents, oceans, and atmosphere.

Solar energy is absorbed by biological material and stored as chemical potential energy.

Solar energy can be directly converted to electricity by photovoltaic cells.

can turn water into pressurized steam that can expand and push against the blades of a turbine, which produces mechanical energy.

Uneven heating causes winds in the atmosphere that can be harnessed by windmills to provide

Biomass (i.e., firewood) can be burned to produce

Biomass undergoes chemical changes underground and eventually becomes coal, oil, and/or natural gas (fossil fuels).

heat and light, primarily for small-scale, household/subsistence uses.

Solar energy evaporates water, which condenses in clouds, falls as rain, and flows downhill, giving up gravitational potential energy. This release of hydro power

Through chemical reactions, biomass can be converted into a fuel or a fuel additive, like ethanol. When burned inside an engine, derivatives of biomass yield

mechanical energy, for transportation (by land, sea, and air), and for industrial purposes.

Mechanical energy can spin an electric generator, producing electricity.

mechanical energy, which is used to spin electric generators, which produce electricity.

Fossil fuels are burned to release their chemical potential energy to provide

mechanical energy, primarily for agricultural purposes (pumping water).

spins hydroelectric generators, thereby converting mechanical energy into electricity.

heat, for household and industrial purposes.

historically has yielded mechanical energy, primarily for industrial purposes (through waterwheels).

heat.

light.

By means of various devices electricity can be converted into

information that can be encoded and decoded by electronic circuits and transmitted by electromagnetic waves.

mechanical energy, which is used to spin electric generators that produce electricity.

mechanical energy, primarily for household and industrial uses.

chemical potential energy, stored in batteries and industrial chemicals, for later use.

**PHOTO 6.13**   This is one of three solar collector arrays operated by Luz International in California's Mojave Desert. The solar energy is used to convert water into steam, the steam drives a turbine, and the turbine drives an electric generator. Together, the three arrays cover 1000 acres and provide approximately 300 MW of electric power—some 90% of the world's grid-connected solar energy production. The production of practical energy by means of collecting the sun's rays has not yet been widely applied.

ported, stored, and burned at any time to produce practical energy on demand. Household electricity is the intermediate form of energy that passes between an electric generating station and the electric or electronic devices used in the home. Electric generators do not produce energy; they merely convert mechanical energy from some external source into electric energy. The external source might be the energy of falling water (as in a hydroelectric plant) or the heat, derived from some combustible source, used to drive a steam turbine that in turn drives a generator (as in a coal- or oil-fired power plant).

To explore in depth the origins of usable energy, we must follow energy transformation chains that may stretch across great distances and far back in time. As diagramed in Figure 6.15, nearly all forms of usable energy in the world today can be traced to nuclear energy released from the core of the sun. A secondary, much smaller source is the nuclear energy released from the world's conventional nuclear power plants. A third, still less important source is *geothermal energy,* the thermal or heat energy present inside Earth as a result of its natural radioactivity. We come to the surprising conclusion that all of these ultimate sources involve nuclear reactions of one sort or another!

A minuscule fraction of the world's usable energy (not diagramed in Figure 6.15) is derived from a fundamentally different source: the gravitational (tidal) interaction between the moon and Earth. There are only two

**FIGURE 6.15 (left)**   Energy is transformed into usable forms in many ways; most of them are diagramed here. As suggested by the width of the arrows, nuclear fusion reactions inside the sun are the source of most of the energy we use today. Some usable energy comes from electric power plants employing nuclear fission, and a tiny fraction comes from geothermal sources. Also as indicated by the arrows, the bulk of our present supply of energy comes through the intermediary of fossil fuels.

**PHOTO 6.14** Electricity generated at nuclear power plants, like this one at Three-Mile Island, Pennsylvania, contribute about 8% of the energy used in the United States.

**PHOTO 6.15** The ancient technology of extracting power from the wind has been put to use in three regions of California where the winds are fairly dependable and often very strong. Modern wind turbines such as these have blades specially shaped to take maximum advantage of winds of various speeds. An electric generator is attached to the shaft of each rotating set of blades.

electric generating stations on Earth that have the capability of harnessing the power of the tides. Each is positioned at the narrow mouth of a bay to take advantage of ocean tides surging in and out.

Currently, fossil fuels serve as the most robust links between energy sources and energy uses. All fossil fuels are considered to be nonrenewable. Once extracted from Earth, they will never be replaced during any time short enough to benefit the world in the foreseeable future. As fossil fuels become increasingly scarce, humans will have to rely more and more on today's lesser-used energy resources and, perhaps, on entirely new sources of usable energy. One such new source might be controlled nuclear fusion, which is a purely experimental technology at present.

The consumer's choice of particular kinds of practical energy depends upon many factors, including availability, cost, convenience, and applicability. Electricity, in particular, can be transmitted almost instantaneously over thousands of miles, but this advantage is somewhat offset by the difficulty of storing electric energy. Most electricity is consumed as soon as it is generated, so the supply of electricity must be continually adjusted to match the demand. Today, most users of electricity in North America are connected to a power grid that distributes energy from hundreds of power plants to a vast number of users. By pooling the ever-changing demands made by different geographical areas at different times, the overall supply of and demand for electricity are balanced out to a great extent.

**PHOTO 6.16** Most forms of usable energy can be traced back to the sun.

# Summary

Work is the product of force and the distance through which the force acts. Energy is defined as the capacity to do work. Energy takes many forms, and it can be converted from one form to another.

Two different forms of energy are potential energy and kinetic energy. Potential energy is the energy a body has by virtue of its position in a force field. Kinetic energy is the energy a body has by virtue of its motion.

Whenever any kind of change is taking place, energy is being transformed. When work is done, energy is being transformed. Energy cannot be created or destroyed; it can only be changed from one form to another.

All matter contains rest energy, which is proportional to mass. The rest energy stored in matter is enormous. Rest energy can be converted to other forms; therefore, all energy-producing reactions result in a (usually tiny) loss of mass.

Machines are devices that change the magnitude or direction of a force. A machine can never produce more work or energy than it requires.

Power is the rate at which work is done, or the rate at which energy is expended or transformed.

Earth movements, which take place in response to stress applied to rock masses along faults, produce earthquakes. The energy released during the rupture of a fault depends on both the length of the rupture and the total displacement along opposite sides of the fault.

Earth remains susceptible to catastrophic meteorite impacts. The larger the impacting body, the greater the energy released on impact and the rarer the event.

All chemical reactions involve the release or the absorption of energy. Reactions that liberate energy are known as exothermic; reactions that absorb energy are known as endothermic. In general, the more exothermic a reaction is, the more stable the products of that reaction are. The rate of a given reaction depends on the concentration and the surface area of the reactants, the temperature, and the presence or absence of catalysts.

Much of the world's energy demand is currently being met by burning chemical fuels, primarily fossil fuels—coal, oil, and natural gas. Nearly all usable energy, whether derived from fossil fuels or other sources, can trace its origin back to the sun. A small but significant fraction comes from nuclear fission reactions in nuclear power plants. Biochemical reactions in Earth's biosphere are almost entirely driven by an input of energy from the sun, starting with the process of photosynthesis.

# CHAPTER 6
# Questions

## Multiple Choice

1. Work is best described as a process of
   a) energy increase
   b) energy decrease
   c) energy disappearance
   d) energy transformation

2. A 1.5-kg book is held 2 m above the floor for 5 s. The work done on it is
   a) 0 J
   b) 3 J
   c) 15 J
   d) 98 J

3. Estimate (using $g \approx 10$ m/s$^2$) how much work is required to raise a 1.5-kg book from the floor to a height of 2 m.
   a) 3 J
   b) 15 J
   c) 20 J
   d) 30 J

4. Estimate (using $g \approx 10$ m/s$^2$) how much *GPE* a 1.5-kg book has at a height of 2 m.
   a) 3 J
   b) 15 J
   c) 20 J
   d) 30 J

5. How much kinetic energy does a 60-kg person have when running at 4-m/s speed?
   a) 120 J
   b) 240 J
   c) 480 J
   d) 960 J

6. An object falling freely (without friction) toward Earth is
   a) gaining *GPE* at the expense of *KE*
   b) gaining *KE* at the expense of *GPE*
   c) releasing heat at the expense of *KE*
   d) releasing heat at the expense of *GPE*

7. A falling sky diver who has reached terminal speed (constant downward velocity) in the atmosphere is
   a) gaining *GPE* at the expense of *KE*
   b) gaining *KE* at the expense of *GPE*
   c) releasing heat at the expense of *KE*
   d) releasing heat at the expense of *GPE*

8. When the speed of an object is doubled, its
   a) *KE* is doubled
   b) *GPE* is doubled
   c) rest energy is doubled
   d) momentum is doubled

9. According to the law of conservation of energy, energy can be
   a) created but not destroyed
   b) destroyed but not created
   c) both created and destroyed
   d) neither created nor destroyed

10. A pendulum's *GPE* is greatest
    a) at the bottom of its swing
    b) at the top (either side) of its swing
    c) when its kinetic energy is greatest
    d) when its total energy is zero

11. A machine cannot
    a) convert work into heat
    b) change the magnitude and/or direction of a force
    c) deliver more work than the work supplied to it
    d) deliver less work than the work supplied to it

12. About how much more energy is released by a magnitude 7 earthquake than by a magnitude 5 earthquake?
    a) 2 times
    b) 60 times
    c) 100 times
    d) 900 times

13. Alaska's 1964 Good Friday earthquake was one of the most powerful on record; many times more energy was released than during the 1906 San Francisco earthquake. The rupture along the fault responsible for the Alaska quake was about how long? (*Hint:* See the discussion of moment magnitude in Box 6.2.)
    a) 50 km
    b) 100 km
    c) 200 km
    d) 800 km

14. The profound biological extinction at the boundary between the Cretaceous and Tertiary (K-T) periods is thought to have been caused primarily by
    a) a massive meteorite impact

b) a sudden rise in sea level

c) the breakup of the supercontinent of Pangaea

d) a vigorous episode of volcanism lasting hundreds of thousands of years

15. All chemical reactions
   a) involve a catalyst
   b) are either exothermic or endothermic
   c) produce products that have a more complex structure than the reactants do
   d) produce products that have a less complex structure than the reactants do

16. If a given reaction is exothermic, the reverse reaction
   a) is exothermic
   b) is endothermic
   c) may involve no energy change
   d) is any of the above, depending on the reaction

17. The combustion of ordinary fuels (wood, coal, oil, etc.) involves
   a) a rearrangement of protons in the nucleus of an atom
   b) a rearrangement of neutrons in the nucleus of an atom
   c) a rearrangement of the electrons in their orbits with respect to atoms or molecules
   d) the disappearance of a small number of atoms

18. When sodium metal and chlorine gas react with each other, a great deal of energy is given off. The product of this strongly exothermic reaction is
   a) stable
   b) unstable
   c) stable or unstable, depending on how fast the reaction took place
   d) stable or unstable, depending on how concentrated the reactants were

19. The most damaging by-product associated with coal combustion is
   a) carbon monoxide (CO) gas
   b) dihydrogen oxide ($H_2O$) vapor
   c) sulfur dioxide ($SO_2$) gas
   d) nitrogen dioxide ($NO_2$) gas

20. Which of the following fuels is least polluting?
   a) coal
   b) gasoline
   c) wood
   d) natural gas

21. Photosynthesis produces

a) glucose
b) carbon dioxide
c) water
d) energy

22. An enzyme is a
   a) type of plant
   b) carbohydrate
   c) biological catalyst
   d) a hydrocarbon fuel

23. Energy consumption per person in the United States today occurs at a rate of about
   a) 100 W
   b) 1 kW
   c) 10 kW
   d) 100 kW

24. In the United States and in the world today, the three most commonly consumed sources of energy are
   a) coal, nuclear fission, oil
   b) coal, natural gas, hydropower
   c) coal, oil, natural gas
   d) nuclear fission, hydropower, oil

25. Among the fuels made of or derived from biological material, which type has the greatest known reserves?
   a) coal
   b) oil (petroleum)
   c) natural gas
   d) wood

## Questions

1. Explain how there is both unity and diversity in the concept of energy.

2. How are energy and work related?

3. What is mechanical energy concerned with?

4. How do the concepts of energy and power differ from each other?

5. A bungee jumper climbs to the top of a 60-m tower and jumps off. The elastic cord tied to her ankles starts slowing her descent at a height of 30 m, and in the end she is suspended upside down at a height of 15 m. What energy transformations has her body experienced while she moves, and in what order?

6. In your own words, what is simple harmonic motion? Give some examples.

7. By what means can the rate of a chemical reaction be changed?

8. Explain why many chemical reactions do not occur spontaneously.

## Problems

1. A person must perform 20 J of work in order to push open a heavy door, by its handle, through an arc of 90°. How much work is required to push the same door through an arc of 90°, with hands placed on the middle of the door, halfway between the handle and the hinges? (*Note:* The person pushes in a direction perpendicular to the surface of the door in both cases.)

2. Find a nearby hill, mountain, or staircase of known height. Climb to the top of it, and record your time. Using your mass in kilograms or your weight in newtons, calculate how much work you performed against gravity during the climb. Also, calculate your average power, or rate of work, during the climb.

3. If you could transform all of your *KE* while you are running into *GPE*, to what height would your body be raised? (This kind of energy transformation is accomplished in part by a pole vaulter, who converts much of his initial *KE* into the upward movement that carries him over the bar.) To start this problem, you must know your mass in kilograms, and estimate your running speed in meters per second.

4. Calculate the rest energy associated with your body's mass. Where does this energy fall in the range of energies given in Figure 6.2?

5. Using a rusty jack, you raise the front end of a car by pumping the handle 100 times through a distance of 1 m. The car's front end, which weighs 5000 N, is raised to a height of 0.5 m. If the jack has an efficiency of 20%, how much average force is required for each stroke of the jack handle?

6. Approximately how many times faster will cottage cheese spoil if heated to 42 °C in the sun, instead of being stored in a refrigerator kept at 2 °C?

## Questions for Thought

1. In Figure 6.3(b), let us assume that the box and the wheels under the box have a combined mass of 100 kg. If we apply Newton's second law, $a = F/m$, it would seem that a horizontal force of 100 N applied to this mass would result in an acceleration of 1 m/s$^2$ to the right. Experiment proves otherwise. The acceleration is somewhat less than 1 m/s$^2$. Why?

2. A jet airplane on a transcontinental flight climbs to a high altitude, levels off, and finally descends toward its destination. Beginning with the ultimate source of the plane's energy (which you can assume to be energy from the sun), diagram how energy flows toward, into, and out of the plane before and during its flight. Start off by consulting Figure 6.15.

3. How do the oars on a boat and the claw on a clawhammer act like simple machines?

4. If a planet moves around the sun in a perfectly circular orbit, does the sun's gravity do any work on the moving planet? How does your answer relate to the fact that both *GPE* and *KE* are constant for that planet as it goes around the sun?

5. The typical meteor visible in the dark night sky is caused by the sudden heating and disintegration, at a height of about 80 km, of a particle no bigger than a grain of sand. Why do you think it is possible to see such a tiny particle burning up at such a great distance away?

6. Assume that every 1 million years, on average, a large asteroid or comet strikes Earth, causing global aftereffects that would kill (if it happened today) a significant fraction of the world's human population of about 5 billion. For an average life span of 50 years, then, each person on Earth has one chance in 20,000 of being alive during such a collision. If scientists could identify, track, and ultimately divert nearly all of the potentially threatening large asteroids and comets that come our way, the odds against such a catastrophe would greatly improve. A search program costing about $50 million could identify and track about 90% of the objects on Earth-intersecting trajectories. This cost works out to

approximately 1¢ for every living person on Earth. Do you think this is cheap enough "insurance" to help guard against such rare, catastrophic impact events?

7. Fire is a chemical reaction. Everyone knows that water quenches most fires. Why is water so effective?

8. Identify the sources of energy, including particular fuels, that you use for transportation, household heating, household electricity, and other needs. You may need to contact your local electrical utility to find out what fuels or processes are being used to generate the electricity used by your household.

## Answers to Multiple-Choice Questions

| | | | | |
|---|---|---|---|---|
| 1. d | 2. a | 3. d | 4. d | 5. c |
| 6. b | 7. d | 8. d | 9. d | 10. b |
| 11. c | 12. d | 13. d | 14. a | 15. b |
| 16. b | 17. c | 18. a | 19. c | 20. d |
| 21. a | 22. c | 23. c | 24. c | 25. a |

# Heat

*This infrared thermogram gives a graphic picture of the flow of thermal (heat) energy outward from the windows, walls, and roof of this building. Far more thermal energy exists in the environment outside the building, but heat still flows outward because the temperature inside the building is higher than the temperature outside.*

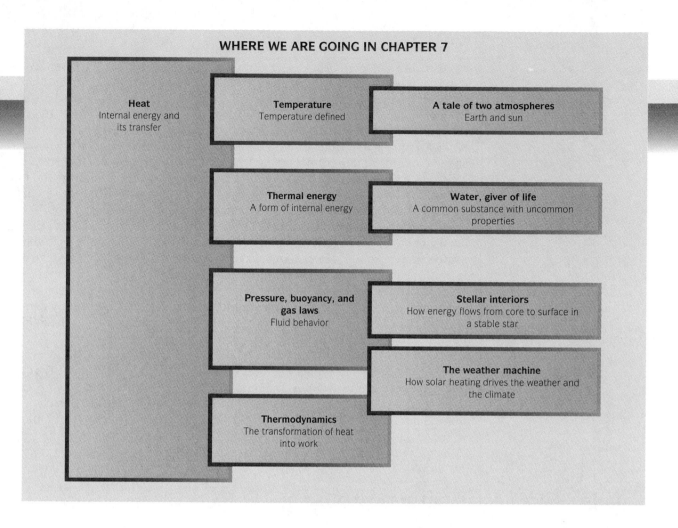

**WHERE WE ARE GOING IN CHAPTER 7**

**Heat**
Internal energy and its transfer

**Temperature**
Temperature defined

**A tale of two atmospheres**
Earth and sun

**Thermal energy**
A form of internal energy

**Water, giver of life**
A common substance with uncommon properties

**Pressure, buoyancy, and gas laws**
Fluid behavior

**Stellar interiors**
How energy flows from core to surface in a stable star

**The weather machine**
How solar heating drives the weather and the climate

**Thermodynamics**
The transformation of heat into work

Generations of scientists struggled with the notion of heat before its true nature became apparent. On one level, heat (or the lack of it) is related to the sensations we vaguely describe as "hot" and "cold." Temperature, a property associated with heat, became a concrete idea only after the invention of the first thermometers in the late 16th and early 17th centuries. By the middle of the 19th century, the atomic theory envisioned earlier by John Dalton and others was being successfully invoked to explain the nature of temperature, the nature of heat, and the ability of substances to expand, contract, or transform into different phases in response to changes of temperature or heat.

We already know that heat is a form of energy. Like all forms of energy, heat is capable of causing change. Conversely, heat results because of change. Heat can be generated through work against friction (rub your hands together rapidly and forcefully, and you will feel heat). We also speak of *radiant heat,* which is really electromagnetic radiation. The warm sensation we get from the sun's rays or the hot flames and coals of a campfire are consequences of changes in the source of that warmth.

Because the word *heat* has so many connotations in ordinary language, many physicists prefer to think of heat solely as energy in transit. This energy, however, is not the organized mechanical energy associated with the position (potential energy) or the motion (kinetic energy) of a body of macroscopic size. Rather, heat is related to the random motions of tiny particles (atoms or molecules) that vibrate, rotate, and move about at random speeds and in random directions within a body. These motions constitute a certain internal energy or *thermal* (heat) *energy,* which all material bodies have. A body can lose or gain thermal energy by means of heat (energy in transit) transferred to or from another body or another place.

Our study of heat and thermal energy in this chapter includes both theoretical and practical considerations. We will distinguish between heat and temperature. We will learn about ways that thermal energy can be transferred as heat. We will learn what physical changes bulk matter undergoes when thermal energy is added or removed. We will study the connections among temperature, pressure, and volume in gases. And we will study conversions between heat energy and mechanical energy, a branch of thermal physics known as *thermodynamics.*

By applying some of the principles of thermal physics, we will investigate the properties of the gaseous outer shells (the atmospheres) of Earth and the sun, trace the modes of energy transfer in the sun, and delve into the mechanisms of weather on Earth. Much of what is learned in this chapter will also be applied to the study of geologic processes in Chapter 16 and to the evolution of stars in Chapter 17.

# Temperature

Ours is a world of temperature opposites. We draw water from hot and cold faucets. We feel hot in summer and cold in winter. Our senses tell us that heat flows into us to make us warm and that we absorb cold (the opposite of heat) on a frigid day. Our senses mislead us in several ways.

Cold and hot are physiologically distinct sensations, but thermometers reveal that both are associated with one and the same property. The hotter (on the temperature scale) a patch of your skin becomes, the more thermal energy it contains and the "hotter" the sensation you feel. The cooler the same patch of skin, the less thermal energy it contains and the "cooler" the sensation. When you touch a hot skillet, thermal energy (or heat, which is thermal energy in transit) rapidly flows from the skillet into your skin. Molecules in your skin begin vibrating more rapidly, and that gives rise to a hot sensation. When you hold a cube of ice in your hand, heat from your skin rapidly flows outward toward the ice. Molecules in your skin slow down and you feel a corresponding cold sensation. There is always some "body heat" moving outward toward your skin from within, but not enough to replenish the heat lost to the ice—until after you let go of the ice cube. In a real sense, there is no such thing as "cold."

The sense of touch is often a poor "instrument" for measuring temperature. For example, as long as the air inside a sauna remains dry, it can be brought to a temperature equal to the boiling point of water without causing any great discomfort for the people inside (for some minutes, anyway).

On the other hand, a person's brief immersion in a hot tub filled with boiling water would give lethal burns. The transfer of heat from hot water to skin when the two are in contact is far more rapid than the transfer of heat from hot air to skin. Why? Primarily because water is much more dense than air. The jiggling molecules of hot water are hundreds of times more densely packed than the jiggling molecules of hot air, and vastly more of them can interact with the molecules of the skin in the same period of time.

There are plenty of places in our universe where very high temperatures exist in regions that contain very little thermal energy. As we shall see in Box 7.1, the exceedingly thin, uppermost zone of Earth's atmosphere has a temperature of thousands of degrees, and the sun's outer atmosphere (corona) is millions of degrees hot.

# What Is Temperature?

The concept of temperature is distinct from the concepts of both thermal energy and heat. Operationally, we can sidestep any real notion of what temperature is and simply say that temperature is what a thermometer measures. Most thermometers exploit the fact that virtually all matter (such as the mercury used in a conventional thermometer) expands and contracts with changing temperature. In a more fundamental sense, however, temperature is related to the microscopic behavior of moving atoms and molecules.

Matter in any phase consists of particles that move about, vibrate, and rotate in some way. In gases, translational (place-to-place) motions are much more important than vibrational or rotational motions. A close-up snapshot view of the moving molecules inside a closed container of pure oxygen, Figure 7.1(a), would reveal a great range of molecular speeds. Yet there is a certain *average* speed associated with all the molecules in the sample and a certain average *KE*, as well. [Remember that *KE* is related to speed: $KE = (1/2) mv^2$.] The faster these oxygen molecules move, on the average, the higher the temperature of the gas.

Mixtures or solutions of gases, such as air, Figure 7.1(b), consist of different kinds of molecules and atoms, with different masses. The lighter particles will, on the average, be moving faster than the heavier particles.

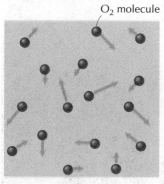

(a) A pure gas

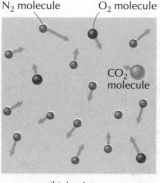

(b) A mixture

**FIGURE 7.1** The individual molecules of a gas have random speeds and directions. Together, the molecules have a certain temperature associated with their motion. In (a), a gas made of the same molecules, temperature is related to both the average molecular speed and the average molecular *KE*. In (b), a mixture of gases such as air, the heavier molecules, on the average, move slower than the lighter molecules, but they still have the same average *KE*. Temperature is still related to the average *KE* of all the molecules.

BOX 7.1

# A Tale of Two Atmospheres

  Earth's atmosphere consists of several thin (relative to Earth's size) layers of gas enveloping its rocky and watery surface. The sun's "atmosphere," by analogy, consists of three distinct layers surrounding the sun's gaseous, but opaque, interior. Earth's atmosphere is a two-way "window." Solar energy streams inward toward the surface, and an equal amount of energy is radiated back into space from the heated continents, the oceans, and the air itself. In the sun's atmosphere, by contrast, prodigious amounts of energy pass outward into space and virtually no energy flows inward. These facts alone illustrate a fundamental difference between a star and a terrestrial planet such as Earth. A star is a powerful and self-reliant emitter of energy. A terrestrial planet reflects some of the energy coming to it from its parent star and passively absorbs the remainder. Let us further compare the anatomy of the two very different atmospheres of Earth and the sun.

## Earth's Atmosphere

Earth's atmosphere consists almost entirely of nitrogen and oxygen gas, with small amounts of water vapor and argon, plus even smaller traces of carbon dioxide, methane, and other gases. Nearly all of the atmosphere's consituents are linked in some way to Earth's crust, oceans, and biosphere. Water vapor is rapidly recycled in the *hydrologic* (water) *cycle* through processes such as precipitation (rain and snow), runoff, and evaporation. Oxygen, nitrogen, carbon dioxide, and methane all participate in slow but grand cycles involving living and nonliving matter. Our atmosphere also serves as an insulating blanket, trapping enough solar energy to maintain conditions conducive to life, and it functions as a protective shield against certain life-destroying forms of solar radiation.

The denser parts of the atmosphere (those dense enough to breathe) are astonishingly thin compared to Earth's overall size—thinner, comparatively, than the eggshell on an egg. Both density and atmospheric pressure decrease rapidly with increasing height. At an elevation of 3 km (nearly 10,000 ft), about 30% of the mass of the atmosphere lies below us. The air at this altitude, some 30% thinner than the air at sea level, rushes in and out of our lungs with noticeable ease as we breathe. We pant under exertion, since each breath also contains about 30% less oxygen.

At an elevation of 5 km (16,400 ft), half the mass of the atmosphere lies below us. This is very near the upper limit of permanent human habita-

**PHOTO 7.1**   Most of the atmosphere's "weather" takes place in the troposphere, the lowest 10 km or so.

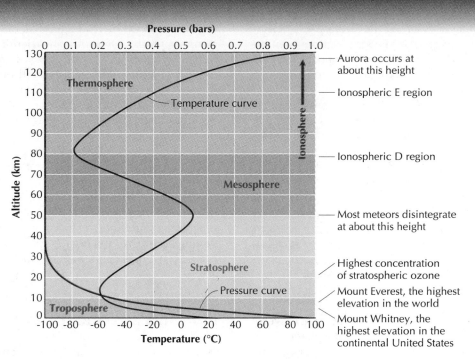

**FIGURE 7.2** Temperature characteristics can be used to define four major zones in Earth's atmosphere: troposphere, stratosphere, mesosphere, and thermosphere. Temperature increases with height in the stratosphere and the thermosphere because significant amounts of high-energy solar radiation are absorbed there. About 3/4 of the mass of the entire atmosphere is confined to the roughly 10-km-thick troposphere, where nearly all weather phenomena occur. Only 1/1000 of the atmosphere lies at altitudes above the stratosphere. Pressure in this graph is given in units of bars. The value of 1 bar is very nearly equal to sea level pressure.

tion. (Mountain climbers have reached the summit of Mount Everest, at an elevation of about 8.8 km, or 29,000 feet, without the use of oxygen; but in doing so, they have risked permanent damage to body and brain.) At an altitude of 50 km, the density and pressure are only a thousandth of their sea level values. At 150 km, density and pressure are about a billionth of what they are at sea level. The rapid decline in pressure with altitude is plotted in Figure 7.2.

Experience tells us that the higher we go, the colder it gets. Temperature is normally highest on or near the ground and decreases with height, because relatively little solar energy is directly absorbed by air, which is mostly transparent. Earth's land and oceans, on the other hand, are good absorbers of solar energy. They heat up in response to the influx of solar energy and, in turn, give up thermal energy to the surrounding air. Thus, the lower atmosphere is heated mostly from the ground up, not from the top down.

At higher levels in the atmosphere, the relation between temperature and elevation becomes more complex. The zigzagging curve of Figure 7.2 reveals how temperature alternately decreases and increases with elevation through the atmosphere's four major concentric "spheres," or shell-like zones.

In the lowest zone, the *troposphere,* the average temperature generally decreases up to about 10 km in altitude, a figure representing the average thickness of the troposphere (the thickness varies somewhat in latitude, because Earth's rotation makes the entire atmosphere bulge slightly at the equator). The troposphere contains significant quantities of water vapor, water in the form of clouds and rain, and suspended dust. Air in the troposphere is almost constantly in motion, moving parallel to Earth's surface in what are called *winds,* and, moving vertically in what are called *updrafts* and *downdrafts.*

Above approximately 10 km, temperature starts to increase with height in the zone    *(continued)*

(*Box 7.1 continued*)    called the *stratosphere* (*strato* means "flat" or "layered"). The stratosphere is much drier, thinner, and clearer than the troposphere, and there is relatively little mixing of air from one elevation to another.

Ozone is largely responsible for the unexpected warming that takes place in the stratosphere. Ozone ($O_3$) is produced in small but significant concentrations at altitudes of around 20 km through chemical reactions involving the absorption of solar ultraviolet (UV) radiation. The absorbed UV is transformed into thermal energy.

Near 50 km the temperature reaches a surprisingly balmy 10 °C or so—not much cooler than room temperature. It declines again with increasing height in the third major zone, the *mesosphere* (*meso* means "middle"). Pressure in the mesosphere, bottom to top, drops from about 0.1% to about 0.001% of sea level pressure.

At around 80 km, the temperature trend reverses again in the outermost major atmospheric zone, the *thermosphere.* The relatively few gas atoms existing in the thermosphere are exposed to the full brunt of solar radiation. Powerful solar X-ray and ultraviolet photons strip electrons from these atoms, so that a significant fraction of them become ionized. As in the stratosphere that lies below, some of the incoming solar energy is absorbed and transformed into thermal energy. This explains the warming trend of the thermosphere.

The thermosphere itself encompasses four distinct ion-rich layers, collectively known as the *ionosphere.* The ionosphere has practical value in that it acts as a mirror for certain frequencies of radio waves transmitted from Earth's surface. When beamed obliquely upward, these waves can reflect and return to the surface at points hundreds of kilometers away. The wave frequencies used by shortwave radio broadcasting stations are capable of bouncing off both the ionosphere and the ocean. When transmitted properly, they can zigzag their way from continent to continent or around the world.

The thermosphere extends from 80 km to an indistinct outer limit of about 500 km, where the temperature approaches a sizzling 2000 °C. Despite this high temperature, relatively little thermal energy is present because of the extremely low density of atoms residing in the thermosphere.

The majority of the satellites in low-Earth orbits pass partially or wholly through the thermosphere, where the effects of air drag are small but ultimately consequential. Over time, as friction slows them, these satellites spiral inward toward the denser layers of the atmosphere. As they fall on steeper and steeper descent paths, the thickening air rushing past heats them to incandescence. Most reentering satellites disintegrate completely before reaching the ground.

Beyond about 500 km lies an exceedingly tenuous region known as the *exosphere.* As the prefix *exo* implies, the exosphere provides an avenue of escape for the very few atoms (mostly hydrogen and helium) that chance to drift that high. With little chance of colliding with another atom, an exospheric atom moving faster than the escape speed (see Chapter 5) of about 11 km/s can slip away into interplanetary space.

## The Sun's Atmosphere

Astronomers have divided the solar atmosphere into three zones—photosphere, chromosphere, and corona—on the basis of distinguishable physical characteristics (see Figure 7.3). Many other stars are thought to possess atmospheres with the same basic structure.

The visible surface of the sun, which appears as a blazing disk to our eyes, is actually the innermost and densest of the sun's three layers, the photosphere ("sphere of light"). The photosphere consists of a 200-km-thick shell of glowing, semitransparent gases with an average temperature of 5800 K. At this temperature, gases emit strongly over the entire visible-light portion of the electromagnetic spectrum, producing what we perceive as white light. Below the photosphere lies the sun's interior, where temperature increases all the way to the core. In this zone the sun's gaseous material (plasma) is almost entirely ionized and effectively opaque.

On many occasions, parts of the photosphere are dotted by darker, cooler (4000–5000 K) patches called *sunspots.* Luminous bodies always emit less light at lower temperatures and more light at higher temperatures. We view sunspots juxtaposed against parts of the photosphere that are 1000–2000 K hotter (see Photo 7.2).

The sun's middle atmospheric layer, the *chromosphere* ("sphere of color") averages about 10,000 km in thickness and consists of gases much less dense than those in the photosphere. Temperature increases outward through the chromosphere, up to about 100,000

**PHOTO 7.2**  This high-resolution photograph of a small part of the sun shows a cooler, dark sunspot dotting the normal, somewhat hotter photosphere. The grainy appearance of the photosphere (called *solar granulation*) is caused by slightly hotter and brighter columns of gas moving upward while cooler gas is sinking.

K at its upper edge. Even so, relatively little thermal energy resides in the chromosphere because the matter there is so tenuous.

The chromosphere, which emits only about a thousandth as much light as the photosphere, can be seen most clearly during the beginning and ending stages of totality of a total solar eclipse. For a few seconds, just as the moon's dark disk is beginning to completely cover (or about to uncover) the bright photosphere, a sliver of glowing chromosphere becomes visible against the dark background sky. Its distinctively rosy tint led astronomers to give it a name meaning "color." Close-up observations show that the outer edge of the chromosphere is turbulent, with many small spikelike jets, called *spicules,* that rise and fall on time scales of a few minutes.

Temperature rises still further in the *corona,* the outermost layer of the sun's atmosphere, which is visible during a total solar eclipse. This faint (about a millionth as bright as the photosphere) white halo of plasma, heated to a temperature of 1 or 2 million kelvins, extends indefinitely outward from the outer boundary of the chromosphere. Because the density of the corona is so low (averaging far less than a billionth of the density of the air we breathe), the thermal energy contained in it is very small.

After decades of study, astronomers have been unable to agree on the cause of the *(continued)*

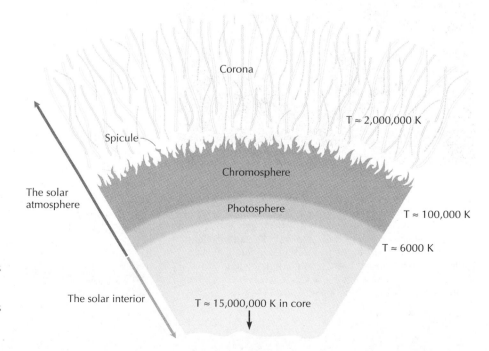

**FIGURE 7.3**  The three-part solar atmosphere (not to scale, as illustrated here) consists of a photosphere, having a thickness of only about 0.03% of the sun's radius; a thicker chromosphere, about 1.5% of the sun's radius; and a corona that extends outward indefinitely.

Corona

$T \approx 2{,}000{,}000$ K

Spicule

Chromosphere

The solar atmosphere

Photosphere

$T \approx 100{,}000$ K

$T \approx 6000$ K

The solar interior

$T \approx 15{,}000{,}000$ K in core

*(Box 7.1 continued)*    extremely high coronal temperature, though it may have something to do with the sun's magnetic field. Magnetic fields are implicated in the formation of sunspots and in the fountainlike solar prominences (Photo 7.3) that often extend above sunspots into the coronal region.

Although virtually all of the radiant energy coming to Earth from the sun originates in the photosphere, smaller amounts of high-energy radiation, especially X-rays, are sent our way from within the glowing plasma of the corona. The corona also acts as an "exosphere" for the sun. Through it, the sun continually loses small amounts of matter in the

**PHOTO 7.4**   An intense aurora borealis (northern lights) display.

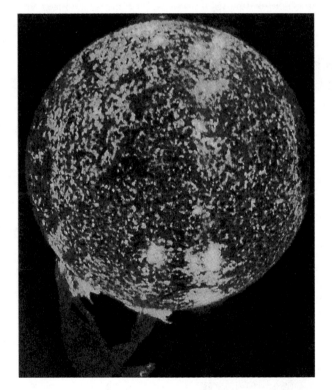

**PHOTO 7.3**   This false-color X-ray image of the sun, obtained by a telescope borne aloft on a rocket, shows a looping solar prominence rising far into the coronal region. Prominences consist of plasma (mostly hydrogen) constrained to follow the lines of force associated with the sun's magnetic field. The reddish tints of the features shown here are used for illustrative purposes. There is no color associated with X-rays, though prominences do appear pinkish in color when they are seen by the eye during an eclipse.

form of high-speed ions and charged subatomic particles. Thus, a wind—a *solar wind*—of charged particles blows outward past the planets of the Solar System toward interstellar space. Violent disturbances on the sun, such as *solar flares,* can fling bursts of particles outward and increase the intensity of the solar wind. If strong enough, the solar wind can visibly interact with atoms in Earth's thermosphere (typically at altitudes of 100–200 km), causing the *aurora,* or northern and southern lights. From regions near the North and South poles, these displays appear as dancing sheets and rays of colored light in the night sky (Photo 7.4).

Nevertheless, in any given sample of a gaseous mixture kept at the same temperature, there is a common average *KE* associated with all the different types of particles. Temperature is a measure of that average *KE*.

In liquid and solid substances, the particles have less freedom of motion; each one vibrates about an average position that changes little or not at all relative to the average positions of other, neighboring particles. Nonetheless, the same characteristic holds: there still exists an average *KE* associated with whatever motions the particles are undergoing. Temperature is related to this average *KE*. The **temperature** of a substance is a measure of the average kinetic energy possessed by its particles. The greater the average kinetic energy of the particles, the higher the temperature of the substance, and vice versa.

With few exceptions, materials expand when their temperature is raised and contract when their temperature is lowered. In the familiar liquid-in-glass-type thermometers, the liquid (either mercury or a colored alcohol) inside a tiny bore in the glass expands outward to fill the bore as the temperature increases, and it contracts inward as the temperature decreases. The glass also expands and contracts, but not as much as the liquid. Gases, in particular, are very sensitive to temperature changes. Air expands in volume by several percentages when subjected to the typical increase in outside temperature from night to day. A crude thermometer can be easily fashioned from a sealed quantity of gas that is allowed to expand when heated and contract when cooled. You may check this effect for yourself by chilling an inflated balloon, placing it in a freezer. When you remove the shrunken balloon from the freezer, it expands to its normal, inflated size.

**PHOTO 7.5** This expansion joint in the middle of a concrete bridge allows the deck of the bridge to expand in length with higher temperatures.

**PHOTO 7.6** Liquid nitrogen (a very cold fluid) is being poured on the balloon at right. As a result, the air molecules inside the balloon slow down and push outward with less force. The balloon shrinks.

Three temperature scales are in common use in the world today. The familiar *Fahrenheit scale* was invented by Daniel Gabriel Fahrenheit nearly 300 years ago. Zero on Fahrenheit's scale (labeled 0 °F) was chosen to be the lowest temperature easily attainable in the laboratories of his day—that of a mixture of water, ice, and salt. The normal internal temperature of a human body is just under 100 °F. Water freezes at 32 °F and boils at 212 °F under normal conditions at sea level. The Fahrenheit scale remains in common use in the United States, though most other countries have switched to the Celsius scale, the common measure of temperature in the metric system.

The *Celsius* (or centigrade) *scale,* in wide use in most of the world today, was invented in 1741 by Anders Celsius. The *centi-* prefix in *centigrade* refers to the 100 equal intervals, or degrees, spanning the range between the freezing point of water (0 °C) and the boiling point of water at sea level (100 °C). The Celsius scale has proven quite useful in modern science, because water is used so often in physical experiments, especially in the laboratory work associated with chemical and biological research.

A third temperature scale, the *Kelvin scale* (after William Thomson Kelvin, or Lord Kelvin, a pioneer in the study of thermodynamics), was defined to take advantage of the fact that a minimum temperature exists: **absolute zero.** From a simplistic (classical, not quantum mechanical) point of view, absolute zero is the equivalent of zero thermal energy within a body. If you could somehow extract all *KE* from the moving, vibrating, or rotating particles within a body, absolute zero would be attained. According to the theory of quantum mechanics, absolute zero is an extrapolation that cannot be attained. In practice, however, physicists can produce temperatures of less than a millionth of a degree above absolute zero in the laboratory.

The Kelvin, or absolute, scale begins, logically enough, with absolute zero (labeled 0 K) and increases in intervals of magnitude equivalent to those of the Celsius scale. Absolute zero is approximately −273 °C, so the Kelvin temperatures for the freezing and boiling points of water are approximately 273 K and 373 K, respectively. (By modern convention, the degree

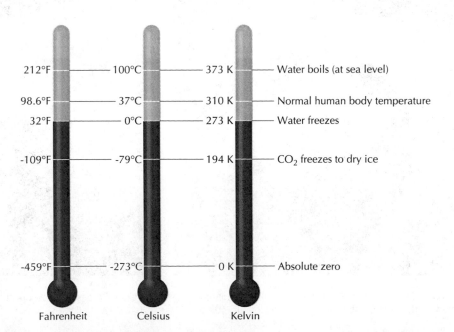

**FIGURE 7.4** Comparison of the three common temperature scales. The Celsius and Kelvin scales are closely related: To get Kelvin units from Celsius units, simply add 273. Conversion formulas for the Fahrenheit scale are given in Appendix C.

| | | | |
|---|---|---|---|
| 212°F | 100°C | 373 K | Water boils (at sea level) |
| 98.6°F | 37°C | 310 K | Normal human body temperature |
| 32°F | 0°C | 273 K | Water freezes |
| -109°F | -79°C | 194 K | $CO_2$ freezes to dry ice |
| -459°F | -273°C | 0 K | Absolute zero |

Fahrenheit          Celsius          Kelvin

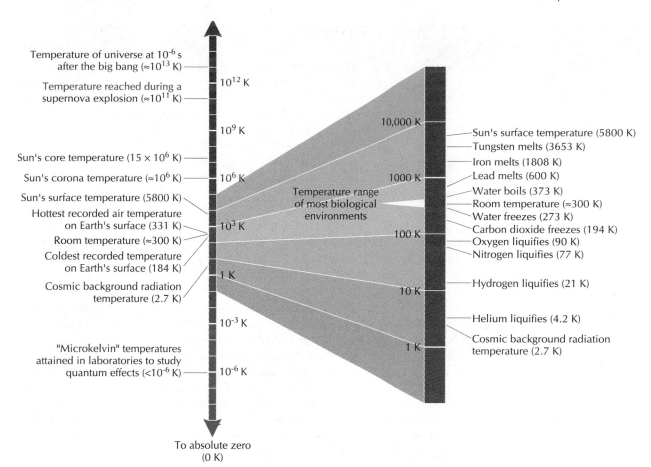

Temperature of universe at $10^{-6}$ s after the big bang ($\approx 10^{13}$ K)

Temperature reached during a supernova explosion ($\approx 10^{11}$ K)

$10^{12}$ K

$10^9$ K

Sun's core temperature ($15 \times 10^6$ K)

Sun's corona temperature ($\approx 10^6$ K)

$10^6$ K

Sun's surface temperature (5800 K)

Hottest recorded air temperature on Earth's surface (331 K)

Room temperature ($\approx 300$ K)

$10^3$ K

Coldest recorded temperature on Earth's surface (184 K)

Cosmic background radiation temperature (2.7 K)

1 K

$10^{-3}$ K

"Microkelvin" temperatures attained in laboratories to study quantum effects ($<10^{-6}$ K)

$10^{-6}$ K

To absolute zero (0 K)

10,000 K

Temperature range of most biological environments

1000 K

100 K

10 K

1 K

Sun's surface temperature (5800 K)
Tungsten melts (3653 K)
Iron melts (1808 K)
Lead melts (600 K)
Water boils (373 K)
Room temperature ($\approx 300$ K)
Water freezes (273 K)
Carbon dioxide freezes (194 K)
Oxygen liquifies (90 K)
Nitrogen liquifies (77 K)

Hydrogen liquifies (21 K)

Helium liquifies (4.2 K)

Cosmic background radiation temperature (2.7 K)

**FIGURE 7.5** An enormous range of temperatures, calibrated in kelvin units, is depicted on this logarithmic scale. From the fixed, minimum value of temperature, absolute zero (0 K), the scale increases without end. Biological organisms thrive within a narrow range of temperature—far less than a single power of ten on the Kelvin scale. A good point of reference for humans is room temperature, which is typically a little less than the rounded-off figure of 300 K (300 K = 27 °C = 81 °F).

symbol is omitted whenever Kelvin units are stated in print. The figure 373 K is read as "373 kelvin" or "373 kelvins.") Room temperature, the typical temperature of the indoor environment, is a little under 300 K (300 K is the temperature of a warm day, about 27 °C, or 81 °F). Kelvin temperatures are especially useful in physics and are often useful in chemistry, because any change in the Kelvin temperature of a quantity of gas is accompanied by a proportional change in either the volume or the pressure of the gas (we will explore these relationships, called the *gas laws,* later in this chapter). Astronomers prefer the Kelvin scale, because stars are made of gas and many stellar properties depend in a straightforward way on Kelvin temperature.

All three temperature scales are graphed for comparison in Figure 7.4. A minimum value (absolute zero) exists on each scale, but otherwise, the scales are open-ended. An enormous range of temperatures exists in the physical world, though in our daily lives we experience only a tiny fraction of that range. Several representative temperatures are graphed logarithmically on a scale calibrated in Kelvin units in Figure 7.5.

## Thermal Energy

When two bodies of different temperature are in contact with one another, heat spontaneously "flows" from the higher-temperature body to the lower-temperature body. We know this is fact from common experience as well as from laboratory experiments. If you touch a hot fireplace poker, heat flows in the direction of your hand. If you touch a cold doorknob, heat flows out of your hand and into the metal of the doorknob. Your hand gets cooler and the metal gets warmer. Heat, as we noted before, is really thermal energy being transferred from one place to another. By the law of conservation of energy, the thermal energy gained by the warming body is equal to the thermal energy lost by the cooling body. It is also clear that we can "generate" heat by doing work against friction. When we rub our bare hands together vigorously, it warms the skin. Our skin gains thermal energy at the cost of mechanical energy, or work. But what exactly is thermal energy?

In a microscopic sense, **thermal energy** is the sum of the internal potential and kinetic energies of the atoms and molecules of a body. Add up these energies for all the particles of a body and you have the measure of thermal energy for that body. In gases, most of the thermal energy is stored in the translational *KE* of the freely moving particles. In liquids and solids, a considerable fraction of the thermal energy is stored in rotational and vibrational movements and as *PE* associated with the particles' attracting or repelling each other by means of electromagnetic forces.

In the macroscopic sense, it is easy to think of thermal energy as a kind of invisible fluid that bodies can take in, store, and pour out. This fluid metaphor (which is an old one) gives us insight into the way thermal energy is gained and lost by bodies. However, there is no substance to this hypothetical fluid; it is energy.

The thermal energy contained in a body in a given phase (solid, liquid, gas) depends on more than just temperature. It obviously depends on the quantity of matter, or mass, of the body. One kilogram of water at 10 °C contains twice as much thermal energy as 0.5 kg of water at the same temperature, simply because there are twice as many water molecules in the 1-kg quantity. The thermal energy stored in a body also depends on the body's particular "capacity" to store heat. That capacity depends on a property of the body called its *specific heat capacity,* which we will discuss in the next few pages.

## Measuring Thermal Energy

Like any other kind of energy, thermal energy can be measured in the SI unit of joules (or its multiples, such as kilojoules and megajoules). Quite commonly, however, you will see thermal energy, or "heat," expressed in units of calories, which are directly related to the thermal characteristics of water. By definition, 1 *calorie* (cal) equals the amount of heat required to raise the temperature of 1 gram (g) of water by 1 °C (or, equivalently, 1 K). Thus, for example, 10 cal would raise 1 g of water by 10 °C, and 10 cal would raise 10 g of water by 1 °C.

A common multiple of the calorie is the kilocalorie (1 kcal = 1000 cal). If you add 1 kcal of heat to 1 kg (1000 g) of water, the water's temperature increases by 1 °C. On a typical kitchen gas stove, this is equivalent to running a burner on high for about 5 s underneath a pot filled with 1 liter of water (1 liter of water has a mass of 1 kg and its temperature rises by about 1° in those 5 s).

One kilocalorie can also be written as 1 Calorie, with an uppercase C (Cal). This *Calorie,* which is really the kilocalorie referred to by scientists, is the familiar unit of food energy used by dietitians. If a particular hamburger patty is said to contain 200 Cal, then it should, if burned completely, yield 200 kcal of thermal energy. When consumed and digested, the hamburger's molecules contain 200 kcal of chemical potential energy available for various uses within the cells of the body.

Since energy is energy no matter what its form, the various measures of a given quantity of energy must be equivalent to each other. (A half-dollar is worth the same whether it is expressed as $0.50 or 50¢.) For the units of calories and joules,

$$1 \text{ cal} = 4.186 \text{ J}$$
$$1 \text{ kcal} = 4.186 \text{ kJ}$$

This equivalence, which is called the *mechanical equivalent of heat,* was first arrived at through experimental work done by James Prescott Joule in the 1840s. Joule's apparatus (like that of Figure 7.6) consisted of weight-driven paddle wheels turning inside a container of water. The known loss of mechanical energy was computed by the loss of *GPE* as the weight fell, and the corresponding gain of thermal energy in the known mass of water was determined by the temperature increase of the water.

## Specific Heat Capacity

When thermal energy spontaneously flows out of a higher-temperature body and into a lower-temperature body, and both bodies are isolated from all others, the process continues until both bodies have reached some common "equilibrium" temperature—we say they are in *thermal equilibrium*. The equilibrium temperature depends on the initial temperatures of the bodies, on the masses of the bodies, and on a property called specific heat capacity (for examples, see Figure 7.7).

**Specific heat capacity** refers to the ability of a substance to store thermal energy. Explicitly, it is the ratio between the net heat added to or removed from a substance, per unit mass, and the resulting temperature change this heat exchange produces in the substance. Specific heat capacity is frequently expressed in units of calories per gram per degree Celsius (cal/g per °C) simply because its numerical value for water on that scale is unity (recall that, by definition, 1 cal added per gram of water raises water's temperature by 1 °C). The specific heat capacities of several substances are

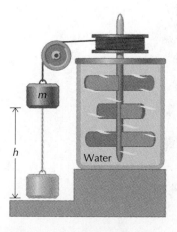

**FIGURE 7.6** The mechanical equivalent of heat was first determined by James Joule through experiments as straightforward as this one: As a weight of known mass falls, it turns immersed paddles, which heat a known mass of water by friction. There is a complete conversion from mechanical to thermal energy because the water stays put and does not circulate after the paddles stop moving. The thermal energy added to the water (in calories) can be determined by measuring the temperature increase of the water. The work done by the paddles (in joules) is the same as the work done by the falling mass (*mgh*). The result shows that it takes 4.186 J of mechanical energy to increase the thermal energy of the water by 1 cal.

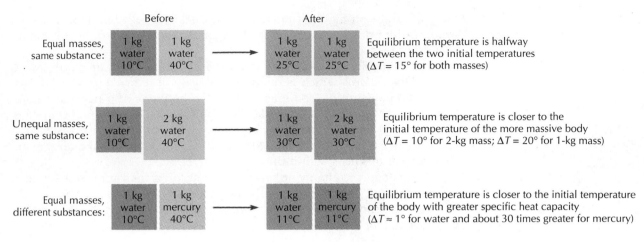

Before                                    After

Equal masses, same substance:
| 1 kg water 10°C | 1 kg water 40°C | → | 1 kg water 25°C | 1 kg water 25°C |

Equilibrium temperature is halfway between the two initial temperatures ($\Delta T = 15°$ for both masses)

Unequal masses, same substance:
| 1 kg water 10°C | 2 kg water 40°C | → | 1 kg water 30°C | 2 kg water 30°C |

Equilibrium temperature is closer to the initial temperature of the more massive body ($\Delta T = 10°$ for 2-kg mass; $\Delta T = 20°$ for 1-kg mass)

Equal masses, different substances:
| 1 kg water 10°C | 1 kg mercury 40°C | → | 1 kg water 11°C | 1 kg mercury 11°C |

Equilibrium temperature is closer to the initial temperature of the body with greater specific heat capacity ($\Delta T \approx 1°$ for water and about 30 times greater for mercury)

**FIGURE 7.7** When two isolated bodies come in contact with each other, heat spontaneously flows from the warmer to the cooler body. The amount by which a body's temperature ($T$) rises or falls depends on the body's mass, on the specific heat capacity of its material, and on the amount of thermal energy transferred in or out. The specific heat capacity of mercury is only about a thirtieth as great as that of water. Therefore, when equal masses of water and mercury experience the same change of thermal energy, the temperature change ($\Delta T$) of the mercury is about 30 times greater than the temperature change of the water.

given in Table 7.1. Specific heat capacity depends weakly on temperature and pressure, and it may be quite different for substances existing in different phases (solid, liquid, gas). Note in the table that ice, water, and steam have different specific heat capacities.

Compared with most substances, water (especially in liquid form) has an unusually large specific heat capacity. Thus, much heat must be added to raise water's temperature, and much heat is released whenever water's temperature decreases. This is a fortunate circumstance for human beings and other mammals who must maintain a nearly constant internal temperature. Our water-rich bodies possess (by analogy with mass) a certain "thermal inertia," or resistance to change in temperature, that helps us cope with hot and cold environments.

| TABLE 7.1 Specific Heat Capacities of Various Substances | Substance | Specific Heat Capacity (cal/g per °C) |
|---|---|---|
| | Water | 1.00 |
| | Ethyl alcohol | 0.60 |
| | Ice | 0.50 |
| | Steam | 0.48 |
| | Aluminum | 0.22 |
| | Glass (typical) | 0.19 |
| | Air | 0.17 |
| | Iron | 0.11 |
| | Copper | 0.094 |
| | Mercury | 0.033 |

*Note:* Values given are for room temperature, except for ice and steam.

The element mercury, on the other hand, has a specific heat capacity of 0.033 cal/g per °C—about a thirtieth of that of water—in its liquid phase. Its temperature responds very quickly to the influx or outgo of heat, which is one reason it is commonly used in liquid-in-glass thermometers.

# Thermal Energy Transfer

Thermal energy is transferred from place to place as heat by three different mechanisms: conduction, convection, and radiation. All three can be easily demonstrated with a candle flame—if you care to perform the simple exercises described next and pictured in Figure 7.8.

1. While holding one end of a steel nail with your thumb and finger, place the other end into the top (hottest) part of a candle flame. Within several seconds, you will begin to feel heat slowly seeping through the nail. Thermal energy is being transferred by *conduction*.
2. Place your hand a safe distance above the flame and feel the currents of heated air (mixed with combustion products) moving upward. These currents carry thermal energy by *convection*.
3. Place a sensitive part of your skin, say the back of your finger, alongside the candle flame. The heat you feel emanating from the flame is *radiation*.

Heat transfer plays an important role in virtually all energy-transforming systems: in stars, in planets, and in the heat-driven machinery (heat engines) that we depend on so much in modern life. We will examine these three mechanisms from a microscopic point of view.

## Conduction

What we are discussing here is really *thermal* conduction, which is distinct from but usually related to *electrical* conduction. **Thermal conduc-**

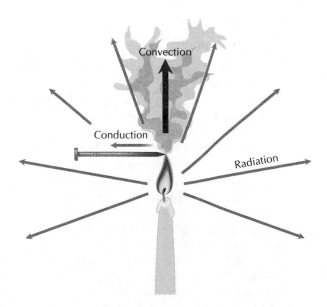

**FIGURE 7.8** Thermal energy is transferred by means of three mechanisms. Conduction occurs most readily in solids and liquids, though it is more important in solids. Convection occurs in fluids: liquids, gases, or plasmas. Radiation travels at the speed of light and requires no medium.

**tion** is the transfer of random kinetic energy within a substance without any net motion of the constituent particles of the substance.

Conduction is important in solids because the particles (atoms or molecules) of a solid cannot move from place to place, but they do vibrate. If one part of a solid is heated to a higher temperature (like one end of a nail in a flame), the increased vibrations of the particles at the hot spot interact with neighboring particles and cause them to vibrate more vigorously. Those particles, in turn, bump against their neighbors and agitate them. The thermal energy associated with the vibrating particles diffuses outward from the hot spot, typically at a slow rate. In metals, conduction is aided by the presence of an abundance of free electrons, which move about through the lattice of the metal atoms. When these free electrons are moving from place to place, they also constitute an electric current; hence, metals are good conductors of both heat and electricity.

Metals are classified as good thermal *conductors,* though some metals are better conductors than others. Copper-bottomed pots and pans are popular because they conduct heat more easily than pots made from most other metals. Materials that tend to inhibit the passage of heat, such as wood and fiberglass, are classified as thermal *insulators.* The fiberglass insulating mats placed inside the walls of buildings help to keep thermal energy inside from escaping during winter weather and help to keep it from invading from the outside during summer weather. Air is a good insulator as well. Double-pane windows are extraordinarily effective as thermal insulators because of the air trapped between the panes. Down, the fine, feathery insulation that grows close to the skin of ducks and geese, traps pockets of air and helps these animals maintain their body temperature. Down (or synthetic fibrous materials with similar properties) is commonly used in cold-weather outer clothing and in sleeping bags.

## Convection

**Convection** is the transfer of thermal energy by large-scale movements within a fluid (liquid or gas). Convective movement commonly takes place in fluids whenever there is a difference in temperature across the fluid and the fluid is acted on by gravity. Virtually all fluids expand and become less dense with increasing temperature. For reasons to be detailed later in this chapter, the warmer, expanded regions of a fluid rise against gravity; and the cooler, denser regions of the fluid sink. You can observe convection in action when rapidly heating water in a transparent container on a stove. As the water warms (and expands slightly) along the bottom and sides of the container, it rises. Arriving at the surface, the water cools and quickly sinks toward the bottom again. You see what are called *convection currents* in the water.

Convection is the basis of most weather phenomena. Warm regions of air in the atmosphere expand and rise (as hot air does above a candle flame), and cool regions of air sink. Winds arise when air rushes in to replace regions of air that have risen.

Convection can also be "forced" in the sense of a forced-air heating system inside a building. Hot air from a furnace is blown by means of an electric fan through various rooms, where it circulates and warms, by contact, the contents of the rooms.

**PHOTO 7.7** These campers are warmed primarily by radiant (mostly infrared) energy from the flames and the hot coals of the campfire.

## Radiation

**Radiation** is the transfer of energy by means of electromagnetic waves. As mentioned earlier in this book, electromagnetic waves can also be thought of as streams of particlelike entities called photons. By either way of thinking, radiation of this sort, or *radiant energy,* travels at the speed of light when traversing empty space. Unlike the other two mechanisms, which involve the transfer of heat through some kind of material substance, radiant energy travels best when it interacts with nothing at all.

All bodies emit radiant energy of various kinds (radio, infrared, visible light, ultraviolet, X-rays, and gamma rays). Bodies at temperatures of between a few hundred kelvins and a few thousand kelvins emit a large fraction of their energy in the infrared. As our skin absorbs energy from any source of infrared—a candle flame, the sun, or even another human at close range—the molecules in the skin begin to vibrate faster. This agitates nerve endings, and the message sent to the brain is translated as a feeling of radiant heat.

Some materials are relatively transparent to radiation of particular kinds, and others are opaque. The sun, having an average surface temperature of 5800 K, radiates the bulk of its energy in three ranges: visible, infrared, and ultraviolet. That radiation quickly crosses the 150-million-kilometer-wide nearly empty gap between the sun and Earth, but much of it is absorbed or reflected when passing through Earth's atmosphere. Ignoring the effects of clouds, the atmosphere is mostly transparent to visible light, somewhat opaque to infrared, and mostly opaque to ultraviolet.

## Phase Changes and Heat

Among the three ordinary physical phases of matter—solid, liquid, and gas—there are but two fundamental phase changes or phase transitions. As we go up the temperature scale, *melting* and then *evaporation* (or *vaporization*) occur for most substances. As we go down the temperature scale, the opposite occurs: *condensation* and then *freezing* (or *solidification*). Some substances, such as carbon dioxide ($CO_2$), skip a phase unless under considerable pressure. At normal atmospheric pressure $CO_2$ cannot exist as a liquid; it either *sublimates* from a solid called "dry ice" into a gas, or it con-

**PHOTO 7.8** As dry ice (frozen carbon dioxide) sublimates, its vapor tends to sink, because it is denser than air. The white mist consists of water droplets condensed from the air by the cold $CO_2$ vapor.

denses from a gas directly to a solid state. Water ice does this to a small extent: Snowbanks can slowly sublimate at high altitudes. At low enough temperatures, water vapor can condense as ice (snowflakes) in a cloud and as frost on a solid surface.

Phase changes always involve a significant transfer of thermal energy. Melting and evaporation, in particular, require that energy be absorbed. When a solid melts (or else sublimates), the absorbed thermal energy releases atoms or molecules from their rigid positions. If atoms or molecules are in a liquid phase, they must acquire more energy in order to evaporate and become gas atoms or molecules.

Surprisingly, all phase changes take place at constant temperature. You may not realize that the water dripping off an ice cube held in your hand is at the same temperature as the ice itself. Both remain at 0 °C as long as water and ice are in close contact with each other. To understand this phenomenon, let us follow the temperature and phase changes of water over a wide range of temperature. Refer to Figure 7.9 as we proceed.

We will start with a quantity of ice at −20 °C. A measureable amount of heat is allowed to flow into the ice from some outside source. The ice warms in response to the influx of heat at a rate that depends on its specific heat (0.50 cal/g per °C). Only when the ice reaches 0 °C does melting begin. During melting, the thermal energy being added does not increase the translational *KE* of the water molecules; rather, it is transformed into the work that is required to release water molecules in the ice from their rigid positions. As long as ice and water are in contact with each other (as in any well-mixed ice-water bath), the temperature remains a constant 0 °C. Any extra *KE* acquired by a liquid water molecule very quickly gets transformed into work that further melts the ice.

The melting of the ice requires 80 cal for every gram of ice; this value is the **latent heat of fusion** for water. The term *latent heat* refers to the energy required for the phase change to take place, and *fusion* in this context refers to melting. When 1 g of ice melts, 80 cal of heat are absorbed into what becomes 1 g of water. That influx of heat, stored as thermal energy, has the potential of being released during the opposite of melting, which is

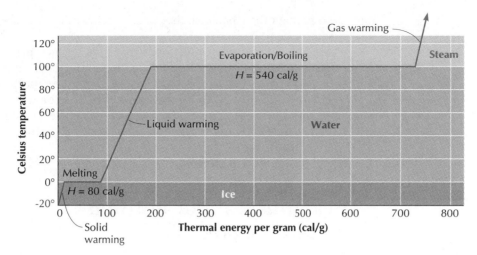

**FIGURE 7.9**  Phase changes take place at a constant temperature, as shown by the level parts of the graph. First, ice absorbs 0.5 cal/g during each 1 °C increase in temperature up to 0 °C. Then, a rather substantial 80 cal/g are needed to melt the ice at no increase in temperature. Once in the liquid phase, water absorbs 1 cal/g for every 1 °C increase in temperature; this is a total of 100 cal/g for the 100 °C increase to the boiling point. Next, a very substantial 540 cal/g must be absorbed by the water in order to change it into steam. Finally, the steam absorbs approximately 0.5 cal/g for every increase of 1° over 100 °C.

freezing. When water freezes, 80 cal/g of heat are released to the surroundings. Your freezer has to work extra hard when you make ice cubes, because it must get rid of the extra heat generated by the water-to-ice phase change.

As soon as all the ice in an ice-and-water mixture is melted, the temperature can increase again as heat is added. As we have seen, the rate of increase is 1 °C for every calorie added to each gram of water. (This rate is half as fast as the corresponding rate of temperature increase for ice.) At 100 °C, boiling, or rapid evaporation, commences. If the water and the vapor above it are kept in contact at the same pressure, the temperature remains constant as long as there remains some water in the water-and-vapor mixture.

Because water molecules in the liquid phase are very "sticky"—they cling to each other through electromagnetic forces—it requires a large amount of energy to pull them apart so that they can travel freely as a gas. The amount of energy needed, 540 cal/g, is called the **latent heat of vaporization** for water. When water boils, each gram of water evaporated requires 540 cal. Conversely, whenever water vapor condenses, the same 540 cal/g of thermal energy are released. The comparatively high heat of vaporization of water is the reason people are burned much more severely by steam at 100 °C than by water at 100 °C. The steam condenses on the skin, instantly giving up its substantial heat of vaporization. Incidentally, the visible "steam" that billows from a whistling teapot is not really steam, or water vapor, at all. By then, almost all the steam has condensed back into tiny water droplets. You can put your hand in it without being burned severely. Look at (but don't touch!) the invisible gap between the small hole at the end of the teapot's spout and the visible column of "steam." A narrow stream of water vapor at 100° passes through there.

Once all the water is boiled to steam, the steam can increase in temperature as more heat is added. The rate of temperature increase is about the same as that of ice, because the specific heats of ice and steam are near-ly the same (recall Table 7.1).

Water has the ability to snuff out most fires for two principal reasons. By coating a substance on the fire, it excludes the oxygen that is needed to keep the combustion reaction going. Furthermore, the rapid vaporization of water in hot flames soaks up a great deal of thermal energy and chills the surroundings. Lower temperatures mean slower reaction rates.

Just the opposite process—the condensation of water vapor to liquid water and the consequent release of much thermal energy—feeds hurricanes and other tropical storms.

# Evaporation and Boiling: A Closer Look

A liquid need not boil for evaporation to take place. A puddle of water on a sidewalk readily evaporates. Its rate of evaporation increases as the sun comes out and more heat flows into it. This common kind of evaporation takes place only on the surface of the liquid. *Boiling,* on the other hand, is a rapid evaporation that takes place inside a liquid as well as on its surface. The bubbles you see rising through a volume of boiling water are not air; they are evaporated water (water vapor alone). From a microscopic perspective, let us see what is going on during the processes of evaporation and boiling.

Any liquid substance contains particles (atoms or molecules) moving at random velocities. Inside the liquid, each particle is caught in the web of the attractive forces between it and the particles adjacent to it. In addition, there is usually an outside force involved, the force of air pressure, that tries to squeeze the particles together.

On the surface of the liquid, some of the particles may be moving fast enough and in the right direction (up) to break free and escape (see Figure 7.10). Once they do, they become the particles of a gas. If the temperature of the liquid is raised by adding thermal energy, the average speed of the particles increases, and the particles escape at a faster rate. If enough thermal energy is added to raise the temperature of the liquid to the boiling point, the particles become so agitated that they can break free to form small bubbles of gas *within* the liquid. These gas bubbles quickly rise to the surface and escape. Boiling is just a rapid form of evaporation.

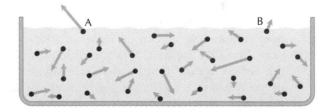

**FIGURE 7.10**  Molecules below the surface of a liquid cooler than its boiling point have little chance of breaking free of their neighbors because their velocities, indicated here by arrows, are too small to overcome the "stickiness" of intermolecular forces. On the surface, however, molecule A escapes because it is moving fast enough and in the right direction. The slower-moving molecule B fails to escape. The warmer the liquid, the faster the molecules move and the faster they escape from the surface.

If you increase the pressure on a liquid, its boiling temperature goes up because there is a greater outside force squeezing the molecules together and preventing their escape. Kitchen pressure cookers, which allow a buildup of steam inside, are designed to operate at pressures greater than the pressure of the atmosphere outside. That allows the water to boil at a higher temperature and cook food faster.

Lowering the pressure reduces the boiling point. At higher altitudes, where atmospheric pressure is less, it takes longer to boil an egg because the boiling temperature is reduced somewhat, and less heat is transferred at that reduced temperature. At 5% of the sea level pressure, the boiling point is lowered to approximately room temperature. In a vacuum chamber, liquid water almost instantly flashes into water vapor: The forces between the water molecules are simply not strong enough to hold the jiggling molecules together. Freeze-dried coffee and other food products are made by subjecting the original product to near-vacuum conditions. The evaporated water is pumped away, leaving behind a dehydrated product. Vacuum sealing of the product prevents hydration from moisture in the air, ensuring a long shelf life.

## Evaporation and Condensation: A Closer Look

The process of evaporation, whether slow or rapid, tends to cool the remaining liquid. The faster-moving particles (those with greater *KE*) preferentially escape, leaving behind particles that are moving slower. For this reason, the evaporation of water or sweat from skin or clothing cools the body.

Different liquids evaporate at different rates. Alcohol evaporates faster than water at the same temperature because the attractive forces between alcohol's molecules (intermolecular forces) are weaker (see Figure 7.11). Consequently, rubbing alcohol or antiseptic liquids containing alcohol chill the skin more rapidly than water does.

During condensation (the opposite of evaporation) fast-moving gas molecules lose speed and *KE* by colliding with and sticking to the slower-moving molecules of either a liquid or a solid. These collisions increase the

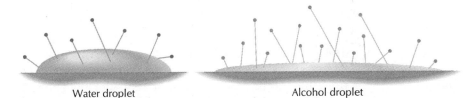

Water droplet                    Alcohol droplet

**FIGURE 7.11** Alcohol evaporates faster than water at the same temperature because the attractive forces between its molecules are weaker. In either case, the faster-moving molecules escape first, leaving behind molecules whose average speed and temperature are lower. Liquid alcohol chills faster than water because it loses molecules faster. The strong intermolecular forces associated with water also give it a pronounced tendency to ball up into spherical droplets in the absence of other forces. On a flat surface such as a smooth tabletop, gravity tends to squash a water droplet, as shown. An equal volume of alcohol speads farther on a flat surface because its intermolecular forces are weaker. This is yet another reason why alcohol spilled on a flat surface evaporates faster than the same volume of spilled water: The greater surface area of the alcohol allows more molecules to escape.

*KE* of the recipient molecules, which raises the temperature of the solid or liquid. The warm, clammy feeling you get on a humid day or in a steamy shower room results from fast-moving water molecules giving up their *KE* as they condense on your skin. In warm air saturated with moisture (100% relative humidity), there may be no evaporative cooling at all but plenty of warming by condensation—a miserable combination.

# Pressure, Buoyancy, and Gas Laws

Nearly all the large bodies in the universe are composed of some kind of fluid. Stars are great spheres of plasma—hot, ionized gas. The Jovian planets are not as much gas planets as they are liquid planets, at least inside. The gases in their interiors are compressed to a density near that of water. Earth's atmosphere is fluid, and so is much of Earth's interior. The outer core is liquid metal, and the mantle consists largely of rocky material capable of deforming and flowing at a slow rate.

Convection is the primary means by which thermal energy is transported by or through fluids. Convection thrives on temperature differences, and temperature differences are endemic to nearly every planet and star in the universe. If the range of motion of a heat-carrying fluid is confined in some way, convection currents arise and persist for as long as a temperature imbalance exists. We can see and measure these convection currents in the water circulating about the oceans, in the winds of the atmosphere, and in the seething plasma of the solar surface.

To understand the details of convective processes, which are so common and important throughout the universe, we must explore the concept of pressure and a behavior associated with fluids known as *buoyancy*.

## Pressure

Force (a vector) was introduced in Chapter 5. Pressure is related to force. When a force *F* acts on an area *A* perpendicular to it, the **pressure** (*P*) is the ratio between the force and the area:

$$P = F/A$$
Equation 7.1

If 1 pound (lb) of force is applied to a surface of area 1 square inch (in.$^2$), we say that the pressure is 1 pound per square inch (psi). The SI unit for pressure, the pascal (Pa), is much smaller than this. It is defined as a force of 1 N applied over a surface of area 1 m$^2$.

How much pressure is a pascal? Imagine the ground-up pulp of a small apple weighing 1 N being spread evenly over a card table, roughly 1 m$^2$. That is a pascal, and it is not much pressure. An American dollar bill resting flat on a table also exerts roughly 1 Pa of pressure; it would take a whole tabletop of bills laid side by side to equal 1 N of force.

Because the pascal unit is so small, the multiples of $10^3$ Pa (1 kilopascal = 1 kPa) and $10^5$ Pa (1 bar = 100 kPa) are often used. The maximum tire pressure ratings stamped on the sidewalls of automobile tires are given in both kPa and psi units; typically, these values are 240 kPa and 35 psi.

One bar of pressure is very close to the value of the average atmospheric pressure at sea level on Earth, so it and the millibar (1 mb = $10^{-3}$ bar = 100 Pa), are commonly used in meteorology. One bar is also equivalent to approximately 15 psi.

The following problem will clarify the difference between pressure and force.

---

## BOX 7.2
## Water, Giver of Life

    When Earth's daylight hemisphere is viewed from space, the planet seems to float like a luminous blue and white bubble in the blackness that surrounds it. The color, unique among the planets of our Solar System, is due largely to water. Liquid water covers about 70% of Earth's surface, mostly in a global ocean whose parts (Pacific, Atlantic, Indian, and so on) are arbitrarily divided. A permanent cap of ice covers Antarctica, and glaciers and snowfields are found on every other continent except Australia. A veneer of ice waxes and wanes with the seasons over the northernmost and southernmost parts of the ocean. Everywhere, water vapor cycles into clouds—white veils that twirl and drift over about half the land and sea at any one time. The total volume of water on Earth (including ice, 2% of the total) is a staggering 1.4 billion cubic kilometers. If Earth's crust were reshaped to form a smooth ball, water would cover it to a depth of about 2.7 km.

Water is so common on Earth that it is easily taken for granted. Yet compared with other substances, water is highly unusual in many of its physical and chemical characteristics. And it is certainly true that life as we know it depends on water.

As noted earlier, water has an extraordinarily high specific heat capacity and stores heat well. Thus, the temperature of any large volume of water (the water in a deep lake, for example) varies only slightly in response to short-term changes in the weather. Ocean temperatures in many parts of the world may vary by only a few Celsius degrees, even as the solar energy seasonally waxes and wanes by a factor of 3 or 4. Ocean currents that distribute heat are partly responsible. So is the enormous heat capacity of so large a mass of water. The thermal stability of the oceans moderates the world climate as a whole and gives some areas, such as coastal Southern California and the Mediterranean region, remarkably even-tempered climates.

The warm ocean currents that loop northward and southward from tropical waters gradually lose their heat as they move toward the colder polar climes. They return to the tropics as cold currents, ready to absorb solar energy as they once more pass through tropical latitudes. These ocean currents transport by convection about 20% of the excess solar energy bestowed on the tropical regions to the energy-deficient polar regions.

A similar kind of energy transfer is accomplished by the movements of water vapor and other gases in the atmosphere. They are responsible for about 80% of the heat transported from the equator to the poles. Water plays an important role in both global and local atmospheric processes because of its large latent heat of vaporization. The vast amount of solar energy required to evaporate water from Earth's oceans and lakes is later released as heat into the atmosphere when the evaporated water condenses back to liquid droplets and forms clouds.

Nearly all liquid substances expand with increasing temperature and contract with decreasing temperature. Water is exceptional in two ways: First, water expands when it freezes, which is why ice cubes and icebergs float on liquid water. By assuming an open, hexagonal *(continued)*

**Problem 7.1**

(a) A woman weighing 600 N wears flat-bottomed shoes. She stands on a flat floor with a total of 300 cm$^2$ (0.03 m$^2$) of shoe area in contact with the floor. How much pressure do her shoe bottoms exert on the floor?

(b) The same woman wearing spiked-heel shoes rocks back and momentarily balances all her weight on a single spiked heel. If the bottom of that spiked heel has an area of 1 cm$^2$ (0.0001 m$^2$), how much pressure does she apply to the floor under that heel?

**Solution**

(a) $P = F/A$

$\quad\quad = (600\ \text{N})/(0.03\ \text{m}^2)$

$\quad\quad = 20\ \text{kPa}$

*(Box 7.2 continued)*    molecular structure (see Figure 7.12), water molecules in the solid phase occupy about 9% more volume than they do when randomly arranged in the liquid phase.

Second, although water expands with higher temperature over most of its range as a liquid, it actually contracts slightly when being warmed from 0 °C to about 4 °C (see Figure 7.13). This unusual behavior is extremely important for large bodies of water subjected to subfreezing temperatures. If water were like most liquids, water chilled to 0 °C would invariably sink; and lakes and polar oceans would freeze rapidly from the bottom up in winter, killing all but the most hardy organisms living in it. Instead, water at 4 °C (its densest state) sinks, and water at less than 4° floats, where it can turn into floating ice if it is subjected to further cooling. The

**PHOTO 7.9**   A puddle of water that froze overnight, showing fractures in it due to expansion during the phase change from liquid to solid.

floating ice and the cold water underneath it act like an insulating blanket that helps keep thermal energy stored in the water near the bottom from escaping quickly. Deep lakes are slow to freeze over and seldom freeze solid, even though they may be subjected to months of subfreezing air temperatures at their surfaces.

We also tend to take for granted that water exists in two and often all three of its common phases on our home planet. Where else in the Solar System could you touch water and ice in a thawing lake and simultaneously breathe water vapor in the air

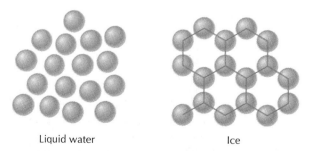

Liquid water                              Ice

**FIGURE 7.12**   When water freezes, it expands in volume by approximately 9%. During the phase transition, the molecules of water assume a hexagonal pattern, with extra space inside the hexagons.

This is equivalent to about 3 psi.

(b) $P = F/A$

$$= (600 \text{ N})/(0.0001 \text{ m}^2)$$

$$= 6000 \text{ kPa}$$

The value of 6000 kPa is equivalent to about 900 psi—more than enough pressure to deeply pit a linoleum floor. The force applied to the floor is the same 600 N in both cases, but the pressure, and also the area of the floor affected, is very different.

Chemists, geologists, and other scientists who study the properties of materials under great pressure use a viselike instrument called a diamond anvil, which works a bit like the narrow ends of two spiked heels pressing against each other with tremendous force. The mineral diamond is used in

around you? The three-phase existence of water on Earth not only promotes the thermal feedback systems that keep our global temperature within fairly narrow boundaries, but it also has increased the number of ecological niches that have encouraged the great diversity of life.

Several chemical and physical properties of water are related to the electrical characteristics of its molecules, a subject to be covered in detail in Chapter 9. Because of their electrically sticky nature, water molecules readily cling to each other and to the molecules of many other substances. Water can dissolve a great variety of substances, and it acts as an *electrolyte,* or electrical conductor, when dissolved ions are present in it. Because of these properties, liquid water facilitates the chemical reactions associated with life on the cellular level.

Whether the bulk of Earth's water arrived on its surface through the ejections of early volcanoes or by impacting comets is uncertain. However, scientists are quite certain that our planet, rich as it is in surface water, is unusual. Had Earth formed too close or too far away from the sun in the Solar System, Earth would probably have had its water boiled away, as on Venus, or permanently frozen, as on Mars. Earth's mass, size, rotation rate, axial tilt with respect to the sun, and the shape of its orbit were, and still are, just right for the conditions that have allowed liquid water to exist and for life to thrive here for billions of years.

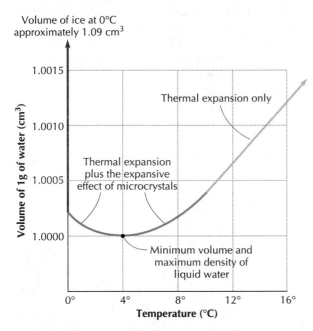

**FIGURE 7.13** Normally, both solids and liquids expand with higher temperature, because greater thermal activity pushes their atoms or molecules farther apart from each other. When water is chilled below about 10 °C, however, microscopic ice crystals start to form. In water between 4 °C and 0 °C, the microcrystal formation overwhelms the effect of thermal expansion. When the phase transition from water to ice is completed, the volume expands by about 9%.

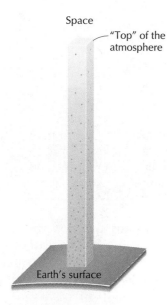

Space

"Top" of the atmosphere

Earth's surface

**FIGURE 7.14** A column of air 1 m² in cross section stretching from sea level to outer space contains about 10,000 kg, or about 100,000 N, of air. Most of the air is concentrated close to Earth's surface. A weight of 100,000 N applied over a surface area of 1 m² gives a pressure of 100,000 Pa, which is equivalent to 1 bar. A similar column having a cross-sectional area of 1 in² contains approximately 15 lb of air.

the anvil, both for its hardness and for its transparency. Researchers can look through the diamond and see what is happening to the sample material as it is compressed. Experimenters using diamond anvils have produced pressures greater than those at Earth's center—greater than 360 million kilopascals or 3.6 million bars.

Pressure increases with depth in the atmosphere and in bodies of water. The weight of a column of air 1 m² in cross-sectional area stretching upward from sea level to the top of the atmosphere is approximately 100,000 N. Thus, atmospheric pressure at sea level averages about 100,000 Pa, or 1 bar (see Figure 7.14). A similar column of air 1 in.² in cross-sectional area weighs approximately 15 lb.

Note that the force exerted by the atmosphere on the surface of a human body at sea level is huge. An average-sized person with 2000 in.² of skin area is being acted upon by a total force of 30,000 lb, or 15 tons! Why, then, do we not all collapse inward? We stay whole because our bodies are permeated with fluids that push outward with 15 lb of force for every square inch of surface. Normally, inward pressure equals outward pressure for every part of our bodies.

Pressure increases with depth much faster in liquids than in air, because liquids are denser. In water there is 1 bar of additional pressure for approximately each 10-m (about 33 ft) increase in depth. Saltwater is slightly more dense than freshwater; therefore, pressure increases a little faster with depth in the ocean than in freshwater lakes or swimming pools.

Since mercury (the liquid metal) is very dense, pressure in it increases very rapidly. There is one additional bar of pressure for every 0.76 m (30 in.) of depth in mercury. Thus, mercury is the fluid of choice for precision barometers (see Figure 7.15), which are capable of measuring very small changes in air pressure. A barometer using water would necessarily be awkward: Sea level air pressure raises a column of water about 10 m, so a water barometer would need to stand about as high as a three-story building.

Pressure is a scalar quantity within fluids, so there is no particular direction associated with it. The same magnitude of pressure applies in all directions at any given depth. You can sense this when you dive to the bottom of a pool: You feel the increased pressure of the water on your

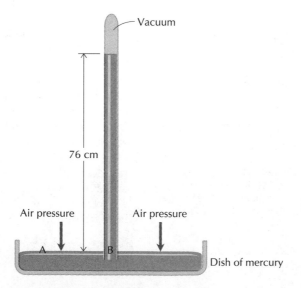

Vacuum

76 cm

Air pressure          Air pressure

A          B

Dish of mercury

**FIGURE 7.15** A mercury barometer. At sea level, the pressure of the atmosphere, approximately 1 bar, pushes down on the surface of liquid mercury in a dish and forces the mercury up an evacuated tube. One bar of air pressure pushes the mercury up 76 cm, because any 76-cm-high column of mercury exerts 1 bar of pressure on the surface below it. Points *A* and *B* are at the same level and at the same pressure—1 bar.

eardrums, but that sensation is the same (at a given depth) no matter how you turn your head. The direction-independent quality of pressure helps explain the phenomenon of buoyancy.

# Buoyancy

The ancient Greek philosophers ascribed to the hot gases or smoke of a fire a quality they called "levity," or the inherent tendency of certain bodies to rise. Today, "levity" can be explained as buoyancy (an apparent upward force on a body in a fluid) overcoming gravity. Buoyancy itself, however, derives from the gravitational pull on a fluid.

A body of any size or composition, if placed in a fluid, experiences a certain force—a *buoyant force*—that acts in the upward direction. This is true regardless of whether the body tends to rise in the fluid, like a cork in water, or to sink, like a rock in water. To see why this buoyant force exists, imagine a cube composed of any substance immersed in a fluid such as water (see Figure 7.16). The pressure exerted on any one of the four vertical faces of the cube is equalized by the pressure on the cube's opposite face. The pressure exerted on the top and bottom faces, however, is unequal, because pressure increases with depth. The buoyant force on the cube is really the difference between two forces that arise from the pressure inside the fluid, a stronger one exerted upward from the bottom and a weaker one exerted downward from the top.

How strong is this buoyant force? The ancient Greek mathematician Archimedes was the first to discover an important generalization about it. Archimedes' principle states that the magnitude of the buoyant force on a body always equals the weight of the fluid displaced by that body. That is, the weight of the fluid that has been *replaced* by the volume of the body is just as strong as the upward, buoyant force on the body.

However, the buoyant force alone does not tell us whether a body will rise or sink in a fluid. We must also consider the weight ($W = mg$) of a given object, which is the same whether or not the body is in a fluid. Three cases are illustrated in Figure 7.17:

1. If a body is *less dense* than the fluid it is immersed in, the upward buoyant force on it is *greater than* the (downward) weight of the body. The body will rise to the surface until only a part of its volume displaces the fluid. When the weight of the displaced fluid equals the weight of the body, the body is in equilibrium, and we say that it floats. For example, wood, which is less dense than water, floats in water. Iron, which is only about 60% as dense as mercury, floats in a mercury bath.

**Archimedes' Principle**
The magnitude of the buoyant force on a body always equals the weight of the fluid displaced by that body.

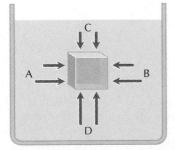

**FIGURE 7.16** A body immersed in a fluid experiences inward-pointing forces, due to pressure, that increase with depth. On the cube shown, the forces (*A* and *B*) acting on the left and right sides are equal and opposite. The top and bottom forces (*C* and *D*) are not equal, because the surfaces acted on by those forces are at different depths. A buoyant, or upward, net force ($D - C$) results. With further analysis, it can be shown that a buoyant force exists on any object displacing a fluid, no matter what its shape or orientation.

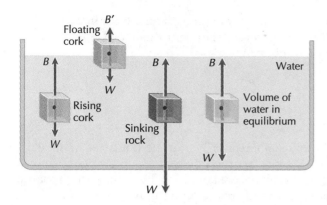

**FIGURE 7.17**   The same buoyant force (*B*) exists on all objects of the same volume in the same fluid. This buoyant force is equal in magnitude to the weight of the fluid displaced by the object. An object (like a cork) that is less dense than say, water, will tend to rise when totally immersed, because its weight (*W*) is less than the buoyant force. Once floating, the cork displaces just enough water so that the *B'* and *W* forces cancel each other. An object (like a rock) that is more dense than water will tend to sink, because its weight is greater than the buoyant force on it. Any volume of water within water will neither rise nor sink, because the *B* and *W* forces on it are equalized.

2. If a body is *more dense* that the fluid it is in, the upward buoyant force on it is *less than* the weight of the body. It sinks. Note, however, that the *net* downward force on the sinking or sunken body is less than the body's weight alone. Hence, it is easier to lift rocks when they are immersed in water.

3. Any given volume of a fluid submerged inside *the same* fluid (at the same temperature) will neither go up nor go down. The buoyant force and weight associated with the fluid in that volume are equal and opposite.

Buoyant forces occur just as readily in gases as in liquids. Your own body is buoyant to some degree, because it displaces air. However, that displaced air has a weight of only about 0.1% the weight of your body; therefore, you do not feel particularly buoyant when immersed in air. Jump into a pool, however, and you will find that you will either barely float (if your lungs are inflated) or have a slight tendency to sink (if you exhale completely). The tissues of the human body have an average density slightly greater than that of water, but inflating the lungs decreases the overall density.

## Gas Laws

In an effort to model the behavior of gases in as simple a way as possible, physicists and chemists have invented the notion of an *ideal gas.* In

**PHOTO 7.10**   In the dense waters of Israel's Dead Sea, which has a dissolved salt content of approximately 25%, people are buoyed up to a surprising extent.

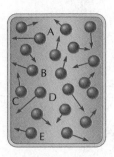

**FIGURE 7.18** The pressure exerted on the walls of any container of gas depends on the average *KE* of the atoms or molecules and on their density. Only those molecules (lettered) moving toward and colliding with the left wall will contribute to an outward pressure on that wall.

such a gas, the particles (which are idealized as being point particles with virtually no size and structure) are pictured as darting about freely except when they collide with other gas particles or with the walls of a container (if any) that confines the gas. All collisions between particles are assumed to be elastic: The particles rebound with perfect elasticity, which means that the total *KE* they have after collision equals what they had before collision. Most real gases closely resemble the theoretical ideal, though their behavior can be quite different from ideal under conditions of extreme pressure or low temperature.

In an ideal gas, macroscopic properties such as temperature and pressure are linked directly to simple particle behavior on a microscopic scale. Temperature, as we have seen, is proportional to the average *KE* of the moving particles. Pressure in any fluid is the effect of a myriad of molecules or atoms striking a surface, such as the walls of a container filled with gas (see Figure 7.18).

The pressure on the walls of a container of fixed volume filled with gas can be increased in two ways: (1) by increasing the temperature of the gas, which increases the force with which gas particles collide with the wall; and (2) by increasing the density of the gas (adding more gas to the container), which increases the number of particles striking the wall during any given interval of time.

There are a number of ways you can manipulate a fixed amount (mass) of gas to change its pressure, volume, and temperature. If you increase or decrease the temperature of a gas while keeping its volume constant, pressure increases or decreases accordingly, because you are changing the force with which each particle rebounds from the fixed walls of the container. You can check this effect for yourself by measuring a tire's "cold" pressure before driving, and then its "hot" pressure after it has been subjected to lots of friction during a high-speed drive. As long as the tire remains inflated so that its volume remains essentially constant, pressure increases with temperature.

If you can keep a gas at a constant temperature, any increase in pressure is accompanied by a proportional decrease in volume, and any decrease in pressure is accompanied by a proportional increase in volume. Thus, a sealed, empty plastic soft-drink bottle in a constant-temperature environment (inside your car, for example) expands when you travel to higher altitudes, where the air pressure is less, and collapses when you come back down.

If you can keep a gas at a constant pressure, temperature affects the volume of the gas. With an increase in temperature, the gas expands; with a decrease in temperature, the gas contracts. This is how hot-air balloons operate (Photo 7.12). Because the air masses inside and just outside the balloon envelope are in contact, they are under essentially the same pressure. (If they were not, air would rush into or out of the balloon.) When hot air is pumped into the balloon, the hot-air molecules do work by pushing outward on the envelope. The air inside occupies more volume and its buoyant force increases. With enough buoyancy, the hot air inside the balloon can overcome the weight of the balloon envelope and the gondola, and the balloon will rise.

Let us rephrase the three relationships involving temperature (*T*), pressure (*P*), and volume (*V*) we have just described, this time in a symbolic way that is consistent with the theoretical behavior of ideal gases and

**PHOTO 7.11** This sealed metal can has been collapsed by an increase in air pressure on the outside.

**PHOTO 7.12** Hot-air balloons are "lighter than air" because the heated air inside expands and becomes less dense than the air outside—and therefore more buoyant.

with the results of controlled experiments performed on real gases that approximate the ideal.

| | | |
|---|---|---|
| For constant $V$, | $P \propto T$ | (Gay-Lussac's law) |
| For constant $T$, | $P \propto 1/V$ | (Boyle's law) |
| For constant $P$, | $V \propto T$ | (Charles's law) |

In the first and third relations, $T$ necessarily refers to *Kelvin* temperature. The Kelvin scale must be used because any change in the average $KE$ of gas particles—and hence the ability of those particles to exert pressure or spread apart to increase volume—is directly proportional to their Kelvin temperature. All three relationships, above, which are really special cases, can be combined in a single, more general law, the **ideal gas law,** which says that

$$PV/T = \text{constant} \qquad\qquad \text{Equation 7.2}$$

The constant in the ideal gas law is a number that depends on the amount of the gas in question. The law says that the particular combination of quantities $P$ times $V$ divided by $T$ does not change for any sample of a gas.

In the real world, the applications of the ideal gas law are often complex, because all three variables can change at the same time. For example, when a gas is quickly compressed inside a cylinder through the action of a piston, there is a simultaneous reduction in volume, an increase in temperature, and an increase in pressure. It is difficult to predict how the changes in $P$ and $T$ will be apportioned. But if the gas is compressed very slowly, so that any thermal energy generated by the compression is readily conducted away by the walls of the cylinder, the situation is much simpler: The temperature remains virtually constant, so (in accordance with Boyle's law) any decrease in volume is accompanied by a proportional increase in pressure.

**Problem 7.2**
If the air in a hot-air balloon is warmed from 27 °C to 30 °C at constant pressure, what change occurs in the volume of the interior of the balloon?

**Solution**

When pressure is constant in the ideal gas law (Equation 7.2), the law becomes identical to Charles's law: $V \propto T$ for constant $P$. Remember that $T$ refers to Kelvin temperature; so our balloon warms from 300 K to 303 K. This is a 1% increase in absolute temperature, which leads, by Charles's law, to a 1% increase in volume. (Small changes like these in temperature and volume are all that a hot-air balloon pilot needs for control over the balloon's altitude when the weather is calm.)

The interlocking effects of pressure, temperature, and volume explain what is behind many of the changes occurring in our atmosphere (as we shall soon see). They also apply to the bulk of the visible matter in the universe—the gases (or plasma) in stars. Gas pressure keeps every "normal" star in the universe from collapsing under its own weight (see Box 7.3).

## BOX 7.3
# Stellar Interiors

About 90% of all stars (the sun included) shining in the universe today are classified as *main sequence stars* by astronomers. Main sequence stars are alike in several fundamental ways, though individually they may vary greatly in mass, size, temperature, and especially *luminosity* (a measure of brightness, or energy output). An important common characteristic of all main sequence stars is their stability over long periods of time. In the case of the sun, this stability has allowed life on Earth to persist and flourish for billions of years.

Main sequence stars are stable because they are in a state of overall balance internally. This balance takes two forms: *hydrostatic equilibrium* and *thermal equilibrium*. Hydrostatic equilibrium means that at any depth below the star's surface, the outward net force due to gas pressure balances the weight (inward force) of all the layers of gas above that layer (see Figure 7.19). Hydrostatic equilibrium generally applies to Earth's atmosphere as well. The tendency of air to expand outward is balanced by the inward force of Earth's gravity. Thermal equilibrium in a star means that the star's surface radiates just as much energy as the energy it receives from below. Thermal equilibrium ensures that a star's temperature remains stable.

Although we cannot literally "see" very deeply into a star, stellar interiors can be analyzed with the help of equations formulated on the basis of the known laws of physics. A theoretical model of the sun's interior, for example, might start with the known characteristics of the solar photosphere: its temperature, pressure, density, and so forth. (Astronomers can "see" the photosphere with telescopes and measure its characteristics by means of spectroscopic measurements, as detailed in Chapter 11.) When hydrostatic and thermal equilibrium are assumed, the equations can be used in a straightforward way to calculate the changing values of temperature, pressure, density, and other parameters at greater and greater depths, all the way to the center of the sun.    *(continued)*

**FIGURE 7.19** If a star is in hydrostatic equilibrium, then the pressure in each layer of the star balances the weight (due to gravity) of the material lying above that layer. As symbolized by the arrows, both pressure and weight steadily increase with depth.

# The Weather Machine

Now that we have been exposed to some of the connections between matter and energy (particularly the notions of pressure and buoyancy, and the gas laws), we can examine how these connections relate to the branch of the earth sciences called *meteorology*—the study of weather and climate. *Weather* refers to short-term conditions of temperature, humidity, wind speed, and so forth, at a given locality. *Climate* means generalized or averaged weather for a given region.

The word *machine* suggests an arrangement of parts designed to produce an orderly outcome from an input of energy. In a broad sense, the intricate systems and subsystems governing weather and climate behave as a machine does. In this section, we will explore the generalities and some of the specifics of the metaphorical machine that keeps Earth's atmosphere and oceans churning with vitality. What drives this machine is energy from the sun.

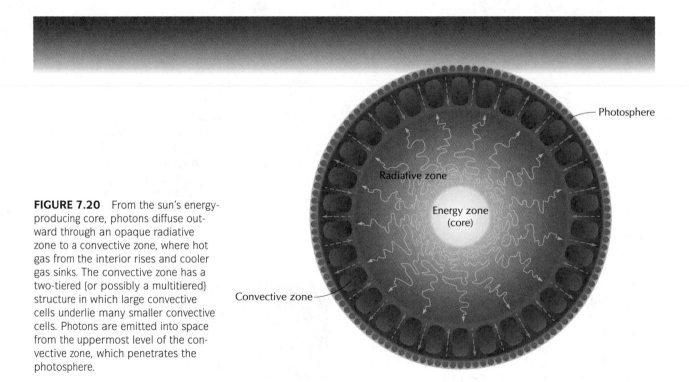

**FIGURE 7.20**  From the sun's energy-producing core, photons diffuse outward through an opaque radiative zone to a convective zone, where hot gas from the interior rises and cooler gas sinks. The convective zone has a two-tiered (or possibly a multitiered) structure in which large convective cells underlie many smaller convective cells. Photons are emitted into space from the uppermost level of the convective zone, which penetrates the photosphere.

*(Box 7.3 continued)*  In practice, detailed models of a star's interior require millions or billions of calculations, which is a task appropriate for high-speed computers. For the sun, the theoretical models predict a central temperature of 15 million kelvins and a central density of about 150 g/cm³ (150 times the density of water). These results are not unexpected, since the hydrogen fusion reactions that power the sun and all other main sequence stars require a minimum temperature of 10 million kelvins at a suitably high density and pressure.

The theoretical models also suggest how the energy flowing out of a star's core eventually reaches the surface. One such model for the sun is shown in Figure 7.20. The radiant energy produced in the core begins its journey outward in the form of gamma rays

# Energy In, Energy Out

The sun accounts for more than 99.9% of the energy input at Earth's surface. (Other inputs, such as heat moving upward from Earth's interior and tidal interactions, make up the small remainder.) Meteorologists often make two simplifying assumptions about the influx of solar energy: First, the sun's radiant energy output is considered constant. Second, Earth is assumed to be in thermal equilibrium. That is, the solar energy received by Earth is equal to the amount of energy Earth gives back to space.

These assumptions have weaknesses. Generally, the sun is considered to be a stable, nonvariable star (in contrast to the small minority of stars whose light output varies conspicuously over time). Precise measurements of the sun's luminosity, obtained from orbiting spacecraft, show that the sun brightens and dims slightly (by much less than 1%) in step with the well-known 11-year cycle of solar activity. During the peak of each cycle, phenomena such as sunspots, prominences, and solar flares are more fre-

(high-energy photons). Because the plasma inside stars is relatively opaque, these gamma rays do not hurry straight out to the surface at the speed of light. Rather, in the manner shown in Figure 7.21, they constantly interact with the matter around them and slowly diffuse outward. In the sun, this process of radiative diffusion takes place primarily in what is called the radiative zone, which extends out from the center to about 70% of the sun's radius.

Above the radiative zone is a convective zone, where convection currents move streams of plasma upward to the photosphere. By then, the plasma has cooled to thousands instead of millions of kelvins. There it can radiate photons into space and thereby cool, contract, and sink back into the solar interior. The simplified model pictured in Figure 7.20 shows a thin layer of small convective cells overlying a thicker layer of larger convective cells in the convection zone. The small cells are comparable in size to Earth, and their upwelling tops are visible as the solar granulation seen earlier in Photo 7.2.

Only when the sun's hot, glowing material actually reaches the photosphere can the photons within it finally escape and travel through space, without further hindrance, at the speed of light. In the case of the sun's interior, the entire process of radiative diffusion and convective transfer takes nearly 1 million years.

Theoretical modeling of the interiors of stars having different masses than the sun's indicates great differences in the way energy moves outward

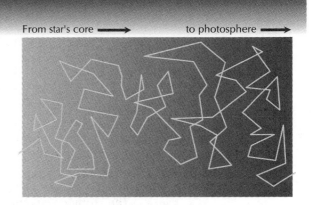

From star's core ⟶            to photosphere ⟶

**FIGURE 7.21**  Energy seeps outward very slowly through the radiative zones of stars, because photons are continually being absorbed and reemitted—somewhat like a baton being passed from runner to runner. (Near the sun's core, the average distance between "handoffs" is less than 1 cm.) Although millions of handoffs take place each second, the directions of the emitted photons are random, so the path of escape toward the star's surface is circuitous and very long in both distance and time.

from the core to the surface. Main sequence stars of around 1/10 solar mass seem to be entirely convective. Main sequence stars several times more massive than the sun apparently have small convective inner regions and thick radiative envelopes.

quent and intense; and the sun's luminosity is maximized. Data gathered by astronomers over the past four centuries show that the 11-year cycles have persisted since at least the early 1700s. Before that, there was very little solar activity during the years 1645–1715, a period called the *Maunder minimum*.

Further insight into past solar activity has been gained by studying traces of radioactive carbon trapped in the annual growth rings of ancient trees. Radioactive carbon is continually formed in the upper atmosphere by means of cosmic ray collisions, and more of it is produced when the sun is relatively inactive. Some of this radioactive carbon joins with oxygen to become carbon dioxide and cycles into the lower atmosphere, where it is taken in by trees. The record of radioactive carbon present in tree ring sequences is thought to mirror the sun's past activity and therefore its luminosity.

The analysis of tree rings points to not one but several instances of reduced solar luminosity. These episodes last for decades and occur every two to four centuries. These long-lasting minima have been correlated to significant changes in the global climate. For example, the Maunder minimum coincided with the so-called Little Ice Age in Europe and a drought in the American Southwest. Other minima, corresponding to unusually cold spells in different parts of the world, occurred in the years 1410–1530 and 1280–1340.

By extrapolating into the future on the basis of the past, we can expect the sun's luminosity to wane again sometime in the next century or two. This would tend to bring on global cooling along with regional changes of climate; some regions might become significantly drier, for example. Remember, however, that solar luminosity is just one of many physical factors that govern the complicated mechanism of world climate.

The second assumption—energy in equals energy out—is valid only on a global scale, averaged over at least one year. During the day, more energy falls on a given part of Earth than can be radiated back into space, but this situation reverses as night falls. Similarly, more solar energy falls on the tropics than the poles, but winds and ocean currents are always busy transporting the excess thermal energy from places of warmth to places of cold.

Other factors, both natural and artificial, are capable of creating imbalances in Earth's energy equation. Consequently, fluctuations in the average global temperature from year to year are to be expected. Major volcanic eruptions play a role. For example, the volcanic aerosols (tiny particles of ash and sulfur compounds) blasted into the upper atmosphere by the 1815 eruption of the Indonesian volcano Mount Tambora were apparently responsible for the following "year without a summer" experienced in Europe and elsewhere. Over the weeks and months after the eruption, the aerosols, floating about in the stratosphere, had spread over much of the world and were filtering out significant amounts of sunlight. The 1991 eruption of Mount Pinatubo in the Phillipines triggered similar but less dramatic effects.

As an example of an artificial (and largely unknown) influence on climate, we need only consider the dramatic increase in the concentration of greenhouse gases (carbon dioxide, methane, and other gaseous products of industrial civilization) in our atmosphere over the past century. Figure 7.22

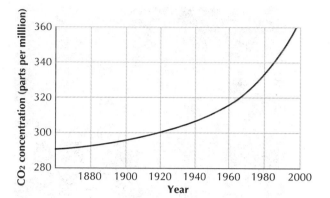

**FIGURE 7.22** The carbon dioxide content in the atmosphere has been rising at an ever-increasing rate for more than a century.

plots the measured (and earlier estimated) increases in the carbon dioxide content of our atmosphere since the late 1800s. Greenhouse gases are known for their ability to trap radiant energy close to Earth's surface. It is generally, but not universally, believed that increasing concentrations of greenhouse gases will lead to future global warming, the so-called **greenhouse effect.** Some say that the expected global warming either may not occur or may not be very dramatic. For instance, increases in cloud cover (owing to more heat trapped near the ground) might produce enough cooling to cancel the warming trend.

If we ignore the weaknesses of the two assumptions we have been discussing, then a simplified picture of the flow of energy in and out of Earth emerges (see Figure 7.23). The energy input, called incoming solar radiation (*insolation,* for short), is distributed by several pathways through the atmosphere, land surface, and oceans. Some of the solar energy—which consists almost entirely of ultraviolet, visible, and infrared radiation—is returned to space essentially unchanged through reflection or scattering. However, most of the radiation is absorbed by the land, ocean, and atmosphere; then it is reemitted purely as infrared radiation, or radiant heat.

**FIGURE 7.23** Averaged over time for the whole Earth, the total solar energy going in (insolation) equals the energy going out. About 30% of the radiant energy coming in from the sun is reflected or scattered back to space. About 70% is absorbed by Earth's atmosphere and surface. The energy of this 70% share is transformed into infrared radiation—thus driving the weather machine—and ultimately returns to space.

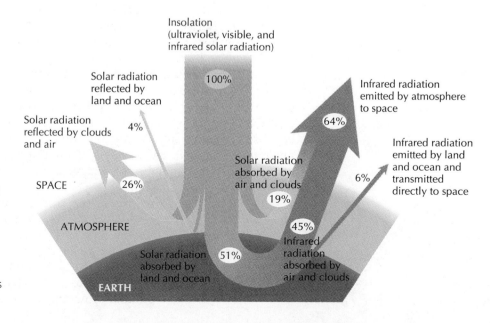

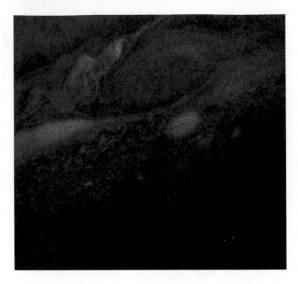

**PHOTO 7.13**   Jupiter's Great Red Spot, a gigantic spinning vortex somewhat analogous to a hurricane on Earth, derives its energy from heat moving upward from Jupiter's interior. Jupiter is still contracting, continuing the process that led to its formation over 4 billion years ago. It releases thermal energy as it does so. The sun may drive atmospheric circulation on Earth but not necessarily on other planets.

The sum of the different energies returned to space is 100%, which is equal to the insolation.

Energy has been conserved in this process, but what has been transformed? Essentially, higher-frequency radiation (mostly visible light) has been transformed into lower-frequency radiation (infrared). The transformation is accompanied by work that is done. This is manifested largely as movements of air and water—the "cogs" and the "wheels" of the metaphorical weather machine. We will examine these cogs and wheels first on a global scale and then on much smaller regional or local scales.

## Global Air and Water Circulation

When solar energy is absorbed by the ground, the oceans, and the atmosphere, it is converted into thermal energy. As noted earlier, this thermal energy may be distributed to other places by means of conduction, convection, and radiation.

Convection is the primary mechanism by which heat energy moves from place to place on Earth. Let us view this process on a global scale. We can think of global convection as starting in the tropics at or near Earth's equator. The air is relatively warm there, no matter what the season, because the sun's rays at midday always strike at a near-perpendicular angle to the surface. The air is humid in the tropics because warm air can easily hold large amounts of water vapor evaporated from the tropical oceans.

As the tropical air warms in the sunlight, it expands, becomes more buoyant, and starts rising. As it rises, it cools and releases some of its moisture in the form of rain. By the time this air reaches the upper troposphere, it has cooled and shrunk so that it can rise no further. It can, however, be pushed to the side (north and south) by air rising beneath. The air migrating north and south eventually must come back down. Ideally, when it does, it flows along the ground, returning to the equator in order to replace

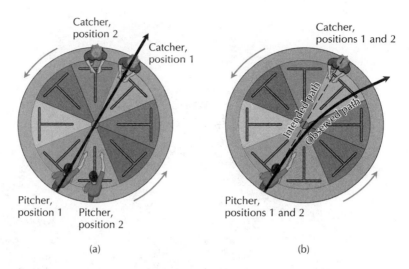

**FIGURE 7.24** Coriolis effect. (a) As seen by an observer looking down on a rotating merry-go-round, a ball pitched toward a catcher at position 1 appears to travel in a straight line. The catcher, however, misses the ball because of her motion to position 2 on the far rim. (b) As seen by the pitcher, himself rotating with the merry-go-round, the ball appears to veer to the right in a clockwise arc.

the air that is rising there. This idealized, closed circuit of moving air constitutes a convection current.

Climatological models show that if Earth did not rotate, the rising air would flow straight toward the polar regions, descend there, and then flow back to the equator, forming just two huge convective cells, one in each hemisphere. Earth's rotation, however, gives rise to an effect called the *Coriolis effect* (or *Coriolis force*), which profoundly alters this simple global pattern. How the Coriolis effect works is illustrated in Figure 7.24. By analogy with the merry-go-round pictured there, imagine looking down on Earth's Northern Hemisphere. You see Earth turning counterclockwise, but people on Earth are not particularly aware of their rotational motion. However, by firing a projectile (say a long-range missile) far enough, a person could tell that Earth is turning to the left, because the missile would veer to the right, relative to the ground (and any observers on the ground) underneath it.

The circulation of air in the atmosphere is affected in the same way. The Coriolis effect causes moving air to be deflected to the right, or clockwise (when looking down), in the Northern Hemisphere. In the Southern Hemisphere, the deflection is opposite: to the left, or counterclockwise. These deflections are slight and of trivial importance over a short range of distance, but they are very significant when air travels over distances of hundreds or thousands of kilometers.

The Coriolis effect alters global air circulation by deflecting moving air sideways to form three primary convective cells in each hemisphere instead of one (see Figure 7.25). Thus, rising tropical air moves poleward and descends at north and south latitudes of about 30°. It then curls west, near the surface, and returns to the tropics. The air warms as it descends at latitudes of about 30°, producing typically warm and dry weather. The majority of the world's deserts lie along belts centered at 30° latitudes. Other factors besides global air circulation are involved in producing extremely arid climatic regimes, as we will soon see.

At the midlatitudes (roughly 30° to 60°), the circulation reverses: Air moves toward the equator at upper levels and toward the poles near the surface. At latitudes of about 60°, rising air gives up its moisture in the form of rain and snow, producing the cool, wet climates so prevalent there.

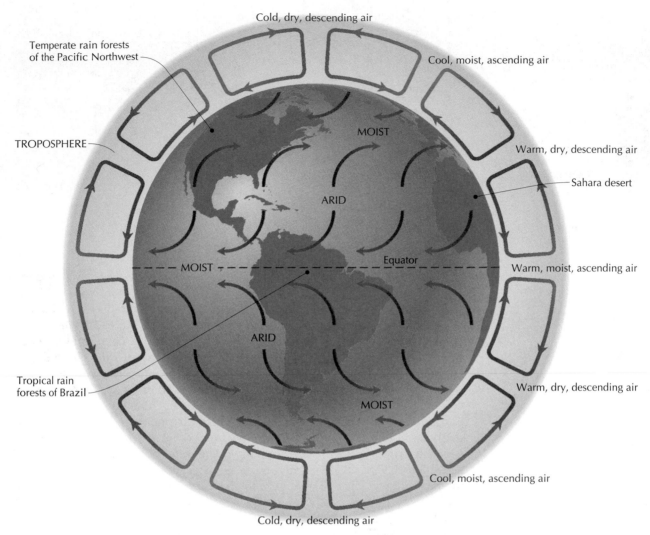

**FIGURE 7.25** In this idealized representation of global air circulation, air warmed at the tropics rises and flows north and south toward the poles. Because of the Coriolis effect, these air currents are deflected to the right in the Northern Hemisphere and to the left in the Southern Hemisphere. As a consequence, three separate systems of convective cells, stratified according to latitude, form in each hemisphere. These cell systems, seen in cross section at the edge of this illustration, help distribute heat energy from the tropics toward the poles and determine the prevailing winds at various points on the surface. In reality, the general patterns shown here are greatly perturbed by regional geography (the distribution of continents and oceans and the influence of mountain ranges and other land features) and by the changing seasons.

At arctic and antarctic latitudes, the circulation is reversed again. At or near the poles, cold, descending air produces what are essentially frozen deserts. Snowfalls are rare at the poles, but each one incrementally adds to the burden of accumulated ice that seldom melts.

On or near the boundaries between the three major convective cells, broad rivers of moving air develop in the uppermost troposphere and lower stratosphere. They contain narrow cores of fast-moving air, called *jet streams,* that influence the weather beneath them. Over North America, the jet streams follow ever-shifting, broadly curving, west-to-east paths that are linked to the motion of storm systems.

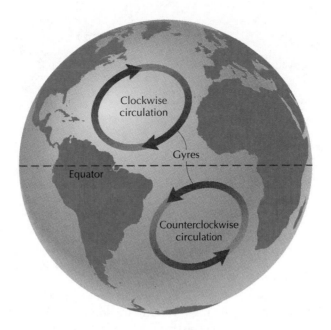

**FIGURE 7.26** Convection in the ocean and in the overlying atmosphere, coupled with the Coriolis effect, organizes the ocean's surface currents. Here are two of the largest circular currents, or gyres. Notice how these two gyres, in opposing hemispheres, circulate in opposite directions.

The Coriolis force also helps organize the surface movements of water in the world's oceans into several great *gyres,* or circular currents, that move around the peripheries of the ocean basins. Two great gyres are shown in Figure 7.26. Gyres are driven in part by convection currents within the oceans, caused by the expansion of warm water and the contraction of cool water; but mostly they are a consequence of prevailing ocean winds blowing across the ocean surface. The gyres, and the many smaller currents associated with them, greatly affect the local coastal climate. For example, much of the southeast coast of North America is warmer than it otherwise would be because of the northward flow of warm water in the Gulf Stream offshore. This flow is tremendous: about 300 times the average flow of the Amazon, the world's mightiest river. In contrast, the northern California coast, despite its rather low latitude of about 40°N, experiences a predominantly cool and foggy climate. The cool California current runs south just offshore, carrying with it cold water from the North Pacific Ocean.

# Local and Regional Air Circulation

Convection occurs wherever there are differences in temperature. Sometimes, temperature differences arise because of local geographic factors, such as land existing next to ocean or mountains existing next to plains. Along many coastlines (see Figure 7.27), a *sea breeze* commonly blows in off the ocean in the daytime, and a *land breeze* blows from the land out to sea at night. The greater the temperature difference between land and sea, the stronger the breeze. In a similar manner, the flow of heated air up along a sun-warmed mountain slope in the daytime is often countered by a cold breeze that moves down the same slope toward the lowlands when night falls. Reversible wind patterns like these persist as long as some stronger regional effect does not interfere.

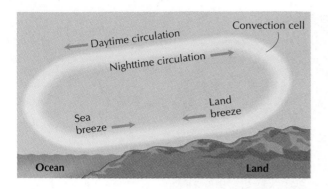

Daytime circulation

Nighttime circulation

Convection cell

Sea breeze

Land breeze

Ocean                                              Land

**FIGURE 7.27** Local convection currents, such as these along a seacoast, are caused by temperature differences. The ocean temperature responds very sluggishly to changes in sunlight, but the terrestrial temperature changes rapidly and over a broad range. In the daytime, the hot, sun-bathed land gives up heat to the overlying air, which then rises. The air cools as it expands when rising. Much of it flows toward the ocean, where it descends to a surface that is cooler than the land. The air returning to the land to complete the convection cycle consititutes a cooling *sea breeze.* At night, when and if the land temperature falls below the ocean temperature, the cycle is reversed and a *land breeze* blows from the land out to the sea.

On a regional or continental scale, any *air mass,* or large parcel of air, that lies above a given region for a few days or more tends to take on the characteristics of the land or the sea beneath it. Because different regions of Earth are heated unequally and convection is frequently occurring in or around those regions, large air masses never stay in one place indefinitely. They are pushed this way and that. If an air mass from one region invades another, the region being invaded may experience a significant change in weather, often lasting for several days or more. The interior United States is particularly subject to invasion by several kinds of air masses (see Figure 7.28), and its weather is notoriously changeable.

Convective movements in the atmosphere involve updrafts and downdrafts as well as winds (horizontal movements of air). Descending air produces areas of higher pressure near Earth's surface, while ascending air produces areas of lower pressure. These areas, each of which may influence the weather over a span of hundreds of kilometers, are known as *highs* and *lows* (abbreviated as H and L on regional or world weather maps). Highs tend to dominate in the 30°-latitude belts and at the poles because that is where air descends and compresses. Lows tend to dominate in the tropics and the 60°-latitude belts because air moving upward leaves behind a slight partial vacuum. The specific pattern made by highs and lows around the globe at any given time is influenced by several other factors, such as geography (including the arrangement of the continents) and the changing seasons.

Highs feature descending air that moves outward along arcing paths determined by the Coriolis force—to the right in the Northern Hemisphere and to the left in the Southern Hemisphere. Such descending whirlpools of air are known as *anticyclones.* This air warms while it compresses, giving rise to generally fair weather.

Since air always moves from a region of higher pressure to a region of lower pressure, air spun off by highs tends to converge on lows. While approaching an intense low, air currents from nearby highs are sucked inward, thereby bending right (in the Northern Hemisphere) owing to the Coriolis force. As these currents continue inward, they begin to spiral in a counterclockwise sense as they converge on the low. The converging air finds escape by spiraling upward at or near the center of the low (see Figure 7.29). Such ascending whirlpools of air are given the generic name *cyclones.* The rising air cools and may cause, depending on the intensity of the low, cloudy weather and rain or snow.

In the tropics, intense lows can evolve into the gigantic spiraling storms known in various parts of the world as tropical cyclones, typhoons, and

**FIGURE 7.28** Four main types of air masses, invading from several directions, influence the weather over the contiguous United States. The tropical air masses tend to push north more often in summer, and the polar air masses sweep south more frequently in winter.

**PHOTO 7.14** A tropical cyclone or hurricane, like this one in the Indian Ocean pictured from space, may begin near the equator as a smaller cyclonic disturbance. If conditions are just right, the storm will intensify as water evaporated over the warm ocean surface condenses around the small, central core (eye of the hurricane) and gives up its latent heat. The heat released feeds a spiraling updraft with winds that may exceed 100 mi/h. Smaller vortices within the most intense updrafts just outside the eye can contribute to peak wind speeds in areas swept by the central part of the hurricane. For example, a vortex spinning at 20 mi/h and circling the eye at 130 mi/h will produce ground speed winds on opposite sides of the vortex of 110 and 150 mi/h. Hurricanes diminish in intensity and eventually dissipate when they move over land, in part because they become starved for the moist, tropical air they feed on and in part because they encounter frictional forces when interacting with the uneven terrain.

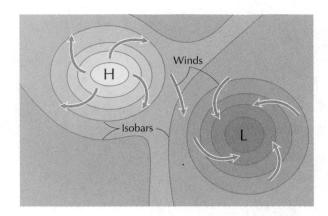

**FIGURE 7.29**  In the Northern Hemisphere, as shown here, air spirals clockwise and descends from the areas of peak high pressure. Air converging on areas of peak low pressure is sucked in so rapidly that it circulates in a direction opposite to that of the Coriolis effect diversion. Air converging on a low finds escape by rising in or near its center. The isobars shown here are lines of equal pressure. The closer the isobars are, the greater the pressure difference is across a given distance, and the more intense the winds are across them.

hurricanes. Hurricane Andrew, one of the most damaging storms in world history, flattened large parts of southern Florida in 1992 with winds in excess of 200 mi/h (Photo 7.15). Hurricanes can spin off tornadoes, which are much smaller than hurricanes but have winds that can exceed 300 mi/h. Tornadoes, however, are more often spawned by clashes between fast-moving cool and warm air masses. These clashes take place frequently over the interior United States during the spring and early summer.

If the flow of air from highs toward lows succeeded in equalizing air pressure, the weather machine would sputter and die. This, of course, does not happen because convection, driven by the sun, never stops. Some highs and lows remain stationary for weeks, but typically, they move from place to place, swelling or shrinking in intensity. The patterns associated with this activity are chaotic (in the scientific or mathematical sense described in Box 5.4). Computer models that mimic weather patterns indicate that any small change in the initial conditions (temperature, pressure, wind speed, and so forth) at any one place can profoundly affect the weather

**PHOTO 7.15**  On August 24, 1992, Hurricane Andrew moved across a densely populated region of southern Florida, damaging or destroying about 72,000 homes and rendering homeless about 200,000 people. At $20 billion in damage, Andrew was the costliest storm in U.S. history. Owing to a well-executed evacuation plan, no more than 20 people lost their lives as a direct result of the storm.

**PHOTO 7.16** The winds around this waterspout (a tornado over water) on Lake Okeechobee, Florida, were strong enough to spawn a spectacular display of lightning.

somewhere else at a future time. Some scientists half jokingly refer to this as the "butterfly effect," the idea that a butterfly flapping its wings in a particular way in, say, Peoria, Illinois, could determine whether or not it rains weeks later at some spot half a world away.

If weather is chaotic in a mathematical sense, then meteorologists have no real hope of ever predicting it, with reasonable certainty, more than a few days ahead. On the other hand, chaos theory may provide the necessary key for understanding certain semiregular patterns of global climate change. A case in point is a series of climatic circumstances and events called *El Niño*. One sign of the onset of the El Niño cycle, which recurs every few years, is the abnormal warming of the waters of the eastern Pacific Ocean, especially along the equatorial coast of South America. The increase in ocean temperature there over seasonal norms can be as much as 10 °C (18 °F). The El Niño cycle has been linked to changes in the locations of highs and lows over much of the Western Hemisphere. These changes, in turn, influence local weather over a large part of Earth.

El Niños of the early 1980s and the early 1990s seemed to be responsible for droughts in normally wet areas and for heavy rainfall in normally dry areas. Hurricanes such as Andrew and the great floods along the Mississippi and other rivers of the midwestern United States in 1993 may have been the indirect result of global shifts in highs and lows triggered by the long-lived 1992–1993 El Niño. Any direct inference, however, would be premature, since another event capable of temporarily altering world climate—the 1991 eruption of Mount Pinatubo in the Phillipines—occurred at nearly the same time.

Although imperfectly understood at present, El Niños are thought to represent a direct link between cyclical changes in the ocean and cyclical changes in the atmosphere. A fuller understanding of this linkage will hinge upon climatological data to be gathered over the next several years or decades.

# The Hydrologic Cycle

Nature's way of circulating water from the ocean to the air, to the land, and back to the ocean is called the *hydrologic cycle,* or water cycle. Virtually every aspect of weather and climate is connected to this cycle, because the lower atmosphere is laden with varying degrees of water vapor at all times. We shall first look at the weather-related components of the hydrologic cycle; then we will map the whole cycle in a graphic form.

Water gets into the atmosphere by evaporation and stays there until it falls, usually somewhere else, as precipitation: rain, snow, and much less importantly, dew and frost. The measure of water vapor content in the air is called *humidity.* It can be expressed in two ways: absolute and relative humidity. *Absolute humidity* is the mass of water vapor contained in a given volume of air. Its importance is suggested in Figure 7.30, which shows that warmer air has a greater ability to take up and hold water vapor. When air at a given temperature contains the maximum amount of water vapor it can hold at that temperature, we say that the air is *saturated.* Less water vapor will saturate cool air; therefore, if warm, saturated air is cooled, some of its water vapor will condense into liquid water.

*Relative humidity,* commonly expressed as a percentage, is defined as the amount of water vapor in air of a given temperature divided by the quantity of water vapor required for saturation of that air. If air having a fixed amount of water vapor is warmed, its relative humidity decreases; if it is cooled, the relative humidity increases. If air is cooled to the *dew point*—the temperature at which it becomes saturated with water vapor—then its relative humidity is 100%. Further decreases in temperature may cause the excess water vapor to condense into tiny droplets, which are the substance of clouds and fog. Relative humidity is often quoted in weather reports because it adds another dimension to the "feel" of the air. Low humidity imparts not only dryness but also a feeling of extra coolness, because the body's moisture evaporates quickly from the skin. On humid days, the air feels warm and sticky, because body moisture tends to cling to the skin and evaporates slowly or not at all.

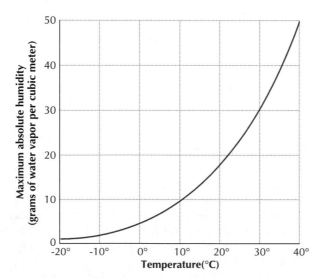

**FIGURE 7.30**   The higher the temperature, the more water vapor a cubic meter of air can contain. Any point on the curve represents a relative humidity of 100%.

Condensation of water vapor can take place when air is chilled below the dew point, but it will not take place without a surface for the vapor to condense on. On Earth, this technical point is moot, because condensation surfaces exist everywhere. The outside of a mug filled with a cold beverage chills the air immediately next to it and beads of water form on its surface if the air around it is humid enough. As blades of grass on a cool or cold night radiate away heat, dewdrops or frost crystals may form on them. Even the most pristine and transparent air is peppered with a surprisingly dense assortment of microscopic dust particles, called *condensation nuclei,* on whose surfaces water vapor can condense. When condensation occurs in the open air, a cloud or a fog is born (fog is merely a cloud on or close to the ground).

Clouds vary in size, shape, and altitude (see Photo 7.18), yet the formation of any cloud requires three basic things: (1) Water vapor must be present in the air in sufficient quantity, (2) the air must be cooled to the dew point, and (3) condensation nuclei must be present. In all regions except for deserts, where water vapor is scarce, the critical factor is cooling. Here are several mechanisms by which the necessary cooling can take place.

1. A moist mass of air moving horizontally is forced upward by a mountain range. Under such conditions, the air expands rapidly and cools by about 1 °C for each 100 m of altitude gain. If the temperature falls to the dew point, condensation takes place and clouds form. Further cooling and condensation may lead to precipitation. This process, called *orographic lifting* is common along mountain ranges (like those along the west coast of North America) that lie in the path of moist ocean winds. Orographic lifting sometimes produces dramatic cloud caps that may obscure the tops of prominent mountains.

When orographic lifting is combined with consistent prevailing winds, large inequities in rainfall may result. The Hawaiian Islands are well known for this effect: The storms that move in off the ocean tend to give up their moisture along the northeast (windward) sides of the islands, leav-

**PHOTO 7.17**  Inland from the Southern California coast, terrain known as desert "badlands" (left) lies in the rain shadow of mountains (right) that receive ten times as much precipitation.

**PHOTO 7.18** The scientific names of cloud types are based on their shape, altitude, or other characteristics. (a) Alto-cumulus clouds (a mackerel sky) over Finland (*alto* = "high"; *cumulus* = "lumpy"). (b) Cirrus clouds (or mare's tails) over Finland (*cirrus* = "wispy"). Cirrus or *cirro*-type clouds are always at very high altitude. (c) The turbulent undersides of nimbus clouds (*nimbus* = "rain"). (d) Cirrostratus clouds and a 22° ice crystal halo around the moon (*stratus* = "flat" or "layered"). The ice crystals in these and other cirro-type clouds refract light and may produce interesting optical

ing little to fall on the southwest (lee) sides. In a similar way, the coastal mountains of western North America cast a "rain shadow" on the arid lands that lie in their lee (see Figure 7.31). It is no accident that Death Valley, one

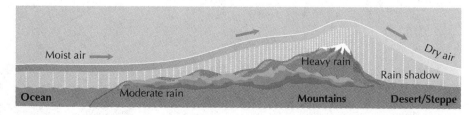

**FIGURE 7.31** In this meteorological scenario, typical of the behavior of winter storms along the West Coast of the United States, moist air moving in from the ocean is wrung nearly dry as it ascends over a mountain barrier. Desert or steppe (semiarid) lands lie in the rain shadow of the mountains.

e

f

g

h

effects. (e) Cumulonimbus (thunderhead) clouds. These clouds, which may tower 6 mi high, unleash the heaviest rains. (f) Coastal stratus clouds at Santa Catalina Island, California. On or near the ground, these clouds are experi-enced as fog. (g) Cumulus clouds over Utah. These clouds form at moderate altitudes on ascending columns of air. (h) Inside a cumulus cloud near Mount Whitney in California's Sierra Nevada.

of the lowest, hottest, and driest places on Earth, lies in the rain shadow of California's towering Sierra Nevada range and other lesser ranges.

2. Moist air heated by contact with the warm ground rises because of its buoyancy, cools, and later sinks. This process sets up convection currents that, driven by the hot sun, can be vigorous enough to build towering cumulonimbus clouds and unleash thunderstorms. Throughout much of the interior and eastern United States, this scenario is played again and again during the summer as maritime tropical air moves north and is stirred up by solar-driven convection.

3. Large air masses can collide along a boundary called a *front* (see Figure 7.32) and interact in ways that depend on their relative movement and respective temperatures. A *warm front* arises when the leading edge of a moving air mass is warmer than the air mass it overtakes. Because it is more buoyant, the warm air rises over the mass of cold air along a wedge-shaped boundary that typically moves slowly. Depending on the tempera-

Air rises, expands, cools; rainfall often results

Inrushing warmer, moist air "ramps up" retreating wedge; cooling and precipitation may occur

Advancing wedge of cold air

Cold front

Warm front

**FIGURE 7.32**  Both cold fronts (left) and warm fronts (right) involve an interaction between air masses of different temperature.

ture difference, air masses that clash in this way can produce anything from a sequence of increasingly dense clouds to light but steady precipitation. A *cold front* develops when a moving mass of cooler air wedges under and displaces a relatively static, warmer air mass. The relatively fast moving and steep boundary of a cold front forces the displaced warm air to rise quickly, which, in the case of a large temperature difference, can unleash violent thunderstorms. Cold and warm fronts, as well as two other types—occluded and stationary fronts—are commonly found in the interior parts of the United States, where air masses frequently collide.

4. Moist marine air blowing onto a chilled land surface (particularly at night) can reach the dew point and so form low stratus clouds or fog. In this situation, warmer, drier air originating from the land's interior often lies above. The result is a *temperature inversion* (Figure 7.33), a reversal of the normal trend of decreasing temperature with increasing height in the troposphere. Temperature inversions may prevent polluted air, generated on the ground, from rising very far. The effect of temperature inversions, exacerbated by the presence of nearby mountains that prevent the escape of stagnant air, contributes greatly to the air pollution problems of Los Angeles and other western cities.

5. On clear, cold nights with little or no wind, moist air that settles into valleys and basins can be chilled greatly after the ground has radiated away

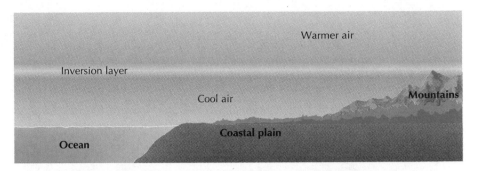

**FIGURE 7.33** Temperature inversions frequently occur on coastlines where a relatively cool ocean lies close to the source of warm air somewhere inland. At night, moisture in the cool marine air moving inland may condense to form a low-lying bank of stratus clouds or fog. During the day, the clouds usually burn off (evaporate), but the temperature inversion may persist. If so, the cooler air underneath may stagnate.

much of the heat it acquired the previous day. If chilled below the dew point, a layer of ground-hugging "radiation fog" may form.

Whenever condensation takes place in a cloud of liquid water droplets, the individual small droplets of water coalesce and grow into larger droplets that may become heavy enough to fall. *Sleet* (which is often mistaken for hail) consists of small, pelletlike chunks of ice, or raindrops that freeze before reaching the ground. In clouds cold enough to be made of ice crystals, water vapor molecules can be accreted (captured) by the existing ice crystals, and the crystals may grow into elaborate six-sided structures, or snowflakes.

Inside a turbulent cumulonimbus (thunderhead) cloud, ice crystals are carried along by convection currents moving upward and downward. Some may make several trips up and down, all the while accreting more water and ice to their surfaces. Eventually, they become heavy enough to fall out of the cloud as hail. Pea-size hail is common, but hailstones the size of a grapefruit have fallen over parts of the American Midwest—a testimony to the enormous force of turbulent winds within some clouds. When sliced

**PHOTO 7.19** The cool, dense marine layer (filled by stratus clouds) over the Los Angeles Basin is trapped by a strong temperature inversion.

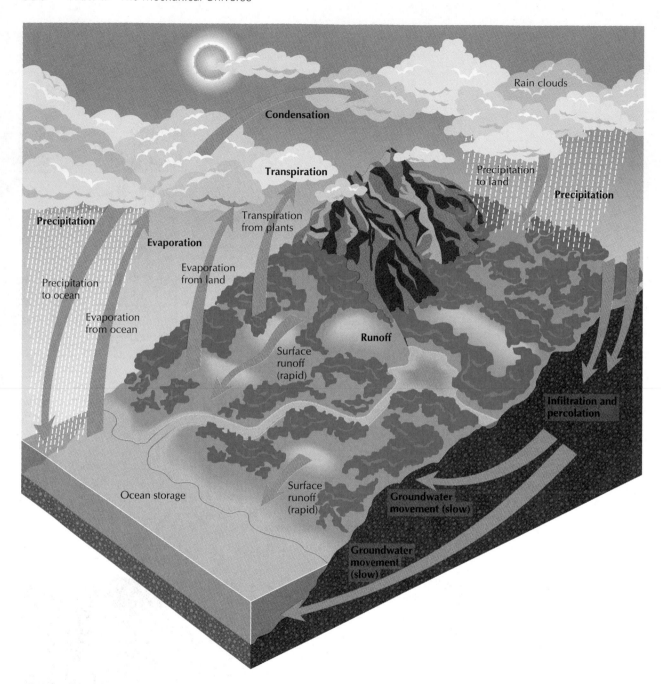

**FIGURE 7.34** Earth's hydrologic cycle involves water in all its phases: solid, liquid, and gas.

open, these large hailstones show an onionlike structure, each layer representing a separate trip into the frozen upper part of the cloud.

Since condensation in clouds is aided by the presence of condensation nuclei, clouds have been "seeded" in some arid regions in an attempt to wring out extra precipitation. Typically, this modern form of rainmaking involves aircraft flying over storm clouds and releasing tiny silver iodide crystals, whose structure mimics that of ice. It is difficult to evaluate the success of these efforts, since every cloud is unique and it is impossible to gauge the amount of rainfall that would have fallen from a cloud had it not been seeded.

Cloud formation, precipitation, and the transportation of moisture by means of winds and moving air masses are the meteorological components of the hydrologic cycle. Movements of water on land—on the surface and to a lesser extent underground—complete the cycle. Figure 7.34 illustrates the grand, global scheme of the hydrologic cycle.

## Thermodynamics

Thermodynamics is the study of energy transformations involving heat, especially the conversion of heat energy into work by devices called *heat engines*. Examples of heat engines include the internal combustion engines that power most cars and trucks, jet engines of aircraft, old-fashioned steam engines, and steam turbines used in most modern electric power stations.

Thermodynamics also concerns itself with the transfer of heat from one place to another so as to create and maintain a lower temperature within a higher-temperature environment. This process requires an input of mechanical energy accomplished by machines called *heat pumps*. Heat pumps are heat engines running in reverse; they include refrigerators, freezers, and air conditioners.

## Heat Engines

A heat engine can be as simple as the piston-and-cylinder arrangement shown in Figure 7.35. A source of heat increases the temperature of an enclosed quantity of gas. The faster-moving molecules of the gas exert a greater force against the inside surface of the piston and cause it to move upward. If the piston is used to lift a load, the raised load gains GPE. Alternatively, the piston rod could be linked to a machine capable of doing useful work. Either way, there is a conversion of thermal energy into work.

The engine in Figure 7.35, however, is of little practical value. Once the molecules of the gas have done all the work they can by pushing on the piston (or once the piston pops out of the cylinder) the engine quits working. To keep the heat-to-work conversion process going, a practical piston-type engine must have some means of returning the piston to its original position so that it may be once again pushed outward by the expansion of a heated gas. This can be accomplished by expelling the expanded gas (by means of a valve in the bottom of the cylinder), thereby allowing the piston to fall, and by introducing a new and cooler quantity of gas into the enclosed space of the cylinder. (We cannot keep the old, expanded gas sealed in the cylinder, because we would have to supply as much work in compressing it as the work our engine already delivered by pushing the piston upward.) Once the cooler gas is sealed in, further additions of heat will push the piston upward again.

Now we have an engine capable of completing a cycle. As long as we continue to add heat from an external source and also remove some heat (by getting rid of the exhaust gases that can do no further work), our engine can continue to go through repetitive cycles. In common engines of

**FIGURE 7.35** This simplified heat engine converts heat energy to mechanical energy by means of heating an enclosed quantity of gas.

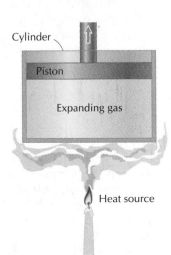

this sort, the back-and-forth (linear) motion of the piston rod is converted to rotational motion by means of a crankshaft mechanism. This same mechanism controls the timing of the valves that open and close to facilitate the gas exchange.

*Internal combustion engines* (either gasoline or diesel) operate in the basic manner just described, except that the heat input is supplied by the ignition and combustion of fuel and oxygen *inside,* rather than outside, the engine's cylinder(s). Typical car engines have four to eight individual piston-cylinder units synchronized in a manner that smoothes out the power output of the whole assembly.

Regardless of their particular design, all heat engines capable of running continuously follow a repetitive three-step process: (1) Heat flows into the engine from a high-temperature source, (2) the engine does work, and (3) heat is expelled from the engine at a lower temperature.

# The Laws of Thermodynamics

**First Law of Thermodynamics**
Energy is conserved in all of its transformations within an isolated system.

**Second Law of Thermodynamics**
Within an isolated system, some (but never all) heat from a source can be converted into work or mechanical energy.

The science of thermodynamics, a major division of physics, rests on two fundamental laws that have never been disproved. The first law of thermodynamics states that energy is conserved in all of its transformations within an isolated system. The second law of thermodynamics states that within an isolated system, some (but never all) heat from a source can be converted into work or mechanical energy.

The first law is, of course, just the law of conservation of energy to which we have already been introduced. The second law puts limits on a particular kind of energy conversion. It says that you cannot take heat from a source and convert *all* of it to mechanical energy; during the conversion *some* heat must be wasted. *Wasted* in this context means that it cannot be utilized to produce mechanical energy. (In some cases, though, the waste heat may have some practical value. For example, an automobile heater—a very useful accessory when the weather is cold—extracts waste heat generated by the engine before it is released to the surrounding environment through the tailpipe.

Heat engines are expressly designed to convert thermal energy arising from some source into work. But in practice, they are capable only of converting *some* of that thermal energy into work. The unused portion of the thermal energy is discarded (or vented) as waste heat. By the first law, the distribution of energy can be summarized as

heat input  =  work output  +  waste heat output          Equation 7.3

This relationship is expressed diagrammatically in Figure 7.36.

The second law says that it is impossible to extract all of the thermal energy from a source and convert it into mechanical energy. The reason that a heat engine will never be perfectly efficient at what it is designed to do can be understood by looking at the nature of thermal energy on the molecular level within, for instance, a simple piston-type engine. When heat is introduced into the cylinder, the average *KE* possessed by the randomly moving gas molecules inside increases, and each of those gas molecules is capable of exerting a stronger force on the walls of the cylin-

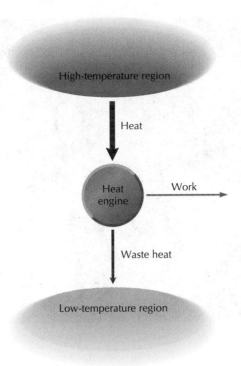

**FIGURE 7.36** A heat engine extracts heat from a high-temperature source and converts some of it into mechanical energy. The heat that escapes conversion into mechanical energy is distributed to a lower-temperature environment by means of convection, conduction, and radiation. A car engine, for example, releases much heat to the environment in the form of hot gases sent through the tailpipe, heat conducted through the engine block and other parts, and the net outward flow of radiation into the surrounding environment.

der. Not all of the randomly moving gas molecules can participate in doing work, however. Only those moving in the right direction—against the piston—can do work (see Figure 7.37). The rest bounce off the immovable walls of the cylinder and do no work (remember that any force acting through no distance yields no work). If the movements of the gas molecules could somehow be spontaneously organized so that they all pushed on the piston and nothing else, then we could entertain the notion of perfect thermodynamic efficiency. This, however, is impossible. Thermal energy is a random form of energy; there is no particular direction associated with the movements of gas molecules.

The efficiency of a heat engine can be defined as the ratio between the engine's work output and its heat input. The first law of thermodynamics, which applies to all energy-transforming devices, says that the efficiency of *any* energy-transforming device can never be greater than a ratio of 1, or 100% (in other words, you cannot get something for nothing). The second law puts further limits on the conversion of thermal energy into work or mechanical energy: It says that no heat engine can ever be 100% efficient. A formula, which may be derived from the second law, can be used to specify the ideal or maximum efficiency of any heat engine. Such an ideal engine would be frictionless and would suffer heat losses only through convection. The maximum efficiency depends on only two things: $T_{hot}$, the hottest Kelvin temperature attained by the gas inside a heat engine, and $T_{cold}$, the Kelvin temperature of the cooler environment to which waste heat is discarded.

**FIGURE 7.37** Relatively few of the gas molecules moving about randomly in this cylinder have the necessary motion to propel the piston upward and thereby do work.

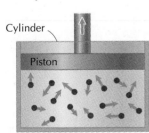

$$\text{Maximum theoretical efficiency} = \frac{T_{hot} - T_{cold}}{T_{hot}}$$

Equation 7.4

**PHOTO 7.20**   This steam turbine rotor (left) inside a power station has been removed for maintenance. Steam turbine rotors may have thousands of individual fan blades.

Maximizing $T_{hot}$ and minimizing $T_{cold}$ increases the maximum possible engine efficiency. In practical situations, it is often easier to increase $T_{hot}$, since $T_{cold}$ is usually the ambient temperature of whatever surrounds the engine, typically the atmosphere or water drawn from a nearby source.

Large steam turbines (see Photo 7.20) of the sort used to drive electric generators in power stations have efficiencies that are comparatively high. A simplified version of such a turbine appears in Figure 7.38. Superheated steam (steam under pressure), originating from a boiler fired by some source of heat (a furnace or a nuclear reactor), is piped through an enclosure with several fan units connected to the same shaft. The steam cools from a temperature of $T_{hot}$ to a temperature of $T_{cold}$ as water vapor molecules move past the fans and transfer some of their *KE* to the blades. In a typical power station, steam at about 740 K is fed into the turbine and

**FIGURE 7.38**   In this greatly simplified steam turbine, hot steam from a boiler moves past rotor blades and turns them. Steam is expelled at a lower temperature and circulates through a condenser back to the boiler. A large commercial steam turbine contains thousands of rotor blades attached to several rotors on the same shaft. It also has several sets of stationary blades that direct the flow of steam across the moving blades in the most effective manner.

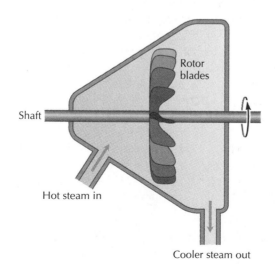

exhausted into a partial vacuum at about 370 K, just under the boiling point of water. These temperature values, substituted into Equation 7.4, give a maximum efficiency of 0.5, or 50%. Friction in the moving parts of the turbine and other factors usually reduces the actual efficiency to less than 40%.

Internal combustion engines do rather poorly on the efficiency scale, in part because of the several steps they must go through to convert heat to work, and in part because they contain many moving parts and linkages subject to friction. The gasoline engine of a car is about 25% efficient in practice, and further losses of energy owing to air resistance and friction in the car's drivetrain and wheels reduce the car's overall efficiency to a meager 10–15%.

Can a heat engine be built with a maximum theoretical efficiency of 100%? The answer is no. We would need an environment with a temperature ($T_{cold}$) of 0 K into which we could expel waste heat. The coldest parts of our universe have an ambient temperature of about 2.7 K, and no part of the universe can spontaneously get colder than that. There are also limits, imposed by quantum-mechanical considerations, on the minimum amount of heat energy (and therefore temperature) any collection of particles can have. An absolute-zero environment can never be attained, even in principle.

## Heat Pumps

The second law of thermodynamics has several alternative expressions, one of which is: *Heat flows spontaneously from a hotter body to a colder body, never the other way around.* In other words, the random darting of fast-moving atoms or molecules in a higher-temperature body will cause the slower-moving atoms or molecules of a lower-temperature body to speed up as long as heat can freely flow between the two bodies.

Can we reverse the flow of heat by forcing heat to flow from a colder region to a hotter region? Yes, we can but the process requires work. Heat pumps such as refrigerators, freezers, and air conditioners do not *produce* cold; they remove heat energy from one region (the inside of a sealed enclosure or the inside of your house) and deposit it into a surrounding, warmer region. Thus, an operating refrigerator warms up the kitchen, and an air conditioner exhausts hot air to the outside of a house. Thermodynamically speaking, a heat pump is the reverse of a heat engine (see Figure 7.39). Heat is forced to flow from a low-temperature region to a high-temperature region by applied work.

The mechanisms of heat pumps usually involve a fluid that changes phase as it circulates between the colder and warmer regions. Evaporation of the fluid inside the colder region *absorbs* heat, and condensation of the resulting gas outside the colder region gives up heat to the warmer surroundings. These phase changes will not occur unless the gas is compressed and condensed while it is circulating within the warmer region. The work needed to force heat out of a colder region and into a warmer region is done by a compressor, usually powered by an electric motor, which requires an influx of energy from some external source.

Heat pump mechanisms can also be used to force heat *into* an enclosed space. As long as the temperature outside is not too cold, a heat pump can

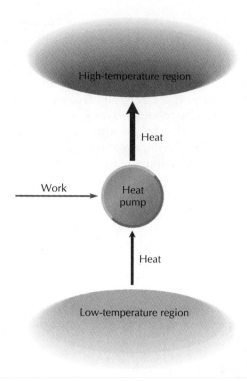

**FIGURE 7.39**  A heat pump forces heat to flow from a low-temperature region to a high-temperature region. This process requires an input of mechanical energy. The heat added to the high-temperature region is the sum of the heat removed from the low-temperature region and the heat generated by the expenditure of mechanical energy inside the pump. Thus, heat pumps are net producers of heat.

be reversed so that it pumps heat from the outside to the inside of a house. This method of heating can be practical in regions where winters are not too severe.

# Entropy and the Second Law

We introduced the concept of entropy in Chapter 2 through examples of order and disorder. Entropy is a measure of disorder, so disorderly systems are said to have comparatively more entropy and orderly systems are said to have comparatively less entropy. A system consisting of particles that move at random speeds and in random directions has no recognizable internal order at all. We would say that system is in a state of maximum entropy. The random kind of energy we call thermal energy always has a great deal of entropy associated with it.

Mechanical energy (gravitational potential energy and the kinetic energy of moving macroscopic bodies), on the other hand, has a lot less entropy associated with it. It can be harnessed easily and efficiently to perform various mechanical tasks: A weight poised to fall can apply force in a single direction and thus do work in an efficient manner as it falls. The moving parts of a machine can apply forces in definite, nonrandom directions.

The transformation of mechanical energy into work nearly always involves friction, and friction produces heat that quickly becomes so dispersed that it is essentially unavailable for further use. Even if we could gather up that dispersed energy, by the second law we could only partially transform it back into mechanical energy. What we see in this instance is a steady increase in entropy and the degradation of a more organized form of energy (mechanical energy) into a less organized and more random form of energy (heat). This tendency toward the degradation of energy with time is universal.

Yet another form of the second law, also known as the law of entropy, summarizes these notions: The entropy of an isolated system never decreases. In practice, for all but the simplest isolated systems, disorder increases with time. This is true not just for heat engines and machines with friction in their moving parts but also for all systems that are truly isolated.

There is another way to look at energy transformations in light of the second law. From a practical standpoint, mechanical energy is considered to be of "high quality" because it is very useful in its native form (*KE* and *GPE*) and it can easily and efficiently be transformed into alternative forms, such as electric energy and heat. Other forms of high-quality energy include chemical potential energy (such as the energy locked up in fuels), electric energy, and the radiant energy that comes from the sun. All are easily transformed to other kinds of energy, including heat, if it is wanted. Thermal energy (heat) is not nearly as versatile or useful. It is considered to be of low quality because it disperses quickly and because it is not as easy to convert heat into mechanical energy, electric energy, or various kinds of potential energy.

Ecologists and environmental scientists (scientists who study the web of connections between the living and nonliving parts of our environment) consider the sun's radiant energy to be of high quality. Sunlight is abundant, frequently available, and easily transformed into other forms through biological and nonbiological processes. Framed in this perspective, Earth and its biosphere can be thought of as constituting a unified system in which high-quality solar energy is steadily transformed into low-quality thermal energy. The system is extremely complex, however, and the overall transformation of energy is accomplished through many pathways of change. These changes promote the many physical and chemical cycles on Earth that make possible the phenomenon of life.

By means of convection, solar energy drives the hydrologic cycle that keeps water circulating between Earth's surface and its atmosphere. This physical cycle includes a small subcycle involving the biosphere; water, after all, is cycled through all living organisms. Solar energy that is converted to chemical energy in plant tissues supports the food chain upon which virtually all organisms rely, and the organisms themselves play a vital role in circulating oxygen and carbon dioxide through the atmosphere. Carbon, nitrogen, phosphorus, and other elements essential to life are also cycled through the biosphere, primarily by processes driven by the influx of solar energy. The end result of these and other solar-driven processes (diagrammed in Figure 7.40) is the production of waste heat, or low-quality energy. This waste heat becomes widely dispersed in the environment, and its energy content is ultimately radiated into space.

The law of entropy is cited by some as proof that there must be some supernatural process directing the phenomenon of life. After all, how can a living being, an exquisitely organized system in itself, arise spontaneously out of disorder? The flaw in this analysis is that it ignores the fact that any organism—and indeed Earth itself—is not an isolated system. When a system is the recipient of abundant energy from the outside, conditions of

**Law of Entropy**
The entropy of an isolated system never decreases.

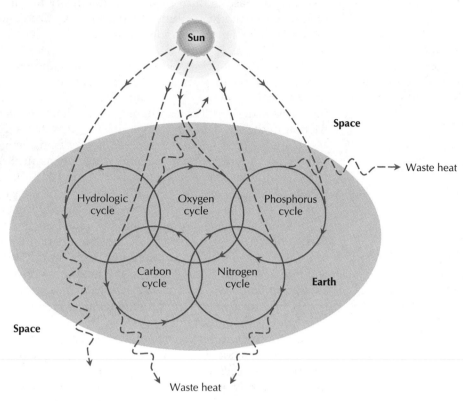

**FIGURE 7.40**   The flow of energy (dashed-line paths) through Earth's crust, oceans, atmosphere, and biosphere starts with high-quality energy from the sun and ends with low-quality heat energy, which is released into the environment and ultimately radiated into space. Diagramed here are several of the many cycles (accompanied by the transforma-tion of solar energy into waste heat) in which chemicals vital for life are circulated through living and nonliving materials. The diagramed circles overlap to suggest interdependencies among the cycles: certain functions of each cycle depend on the functions of other cycles.

decreasing entropy may occur, at least temporarily. You as a living creature have developed the capability of concentrating, hoarding, and utilizing high-quality biochemical potential energy. (In a similar way, a heat pump can temporarily reverse the normal flow of heat from place to place, as long as the pump is turned on, or "alive" in a sense.) An influx of high-quality energy allowed you to gather and assemble the raw materials that now make up your body. Beyond what is needed for growth and development, large amounts of high-quality energy are needed just to maintain your body, whose metabolism and various activities produce heat. After you die, the raw materials of your body will become dispersed again. From an energy standpoint, your whole life will have been spent taking in high-quality energy from the sun and transforming it into processes that sooner or later produce heat. Your life cycle in no way violates the law of entropy for the Earth-sun system, because in the long run, high-quality energy from the sun will have been converted, through you, into heat.

There may be no escaping the "grim reaper" of increasing entropy, even if you think you can immortalize yourself by bequeathing wisdom to your successors. That wisdom, whether stored in printed form, on magnetic tape, or on CD media, will inevitably decay unless it is carefully maintained by the influx of still more high-quality energy. However, after an enormous period of time (billions of years, at least) the high-quality energy possessed by our sun will be virtually exhausted, and it may be impossible to gather enough dispersed heat energy to do any useful work or maintain anything at all.

## CHAPTER 7
# Summary

Temperature is a measure of the average kinetic energy possessed by the atoms or the molecules of a body. Nearly all substances expand in volume when there is an increase in temperature.

Temperature is commonly measured on one of three scales: Fahrenheit, Celsius, and Kelvin. The Kelvin scale of temperature is fundamental because it is calibrated so that zero corresponds to absolute zero, the lowest temperature possible.

Thermal energy (often referred to as heat energy) is the total energy a body has by virtue of the internal kinetic and internal potential energies of its particles. Thermal energy is commonly measured (in the metric system) in units of joules or calories. One calorie is equivalent to approximately 4.2 J. Also, 1 kcal = 4.2 kJ of energy.

The amount of thermal energy stored within a body of uniform composition and phase depends on its mass, its temperature, and its specific heat capacity. Liquid water has a specific heat capacity of 1 cal/g per °C, which means that 1 cal of heat added to 1 g of water will raise its temperature by 1 °C.

Heat can be thought of as thermal energy in transit. Heat will spontaneously flow from a higher-temperature region to a lower-temperature region, never the reverse.

All phase changes occur at constant temperature. Heat is absorbed by the atoms or molecules of melting solids and evaporating liquids. Heat is given off by the atoms or molecules of condensing gases and solidifying liquids. The process of boiling is a form of rapid evaporation.

Heat flows from place to place by one of three mechanisms: Conduction is a relatively slow transfer of heat that takes place primarily within dense materials, especially metals. Convection is the transfer of heat by means of a moving fluid. Radiation is the transfer of heat by electromagnetic waves or, equivalently, photons.

Pressure and density decline steadily with altitude in Earth's atmosphere, yet the temperature rises and falls with altitude in a complex way. Most weather (clouds, rain, and so on) takes place in the troposphere, the lowermost 10 km or so of the atmosphere. In the next layer, the stratosphere, ozone absorbs most of the ultraviolet solar radiation.

The temperature of the sun's outermost layers (its atmosphere) increases with height. The lowest, coolest, and densest layer, the photosphere, emits the vast majority of the energy we see as visible sunlight. Energy generated in the sun's core moves outward through its interior by a slow process of radiative diffusion and convection.

Pressure, a scalar, is defined as force divided by area. Its SI unit is the pascal (Pa), which is $1 \text{ N/m}^2$. Pressure increases with depth in fluids.

Objects immersed in fluids always experience a buoyant (upward) force that is equal in magnitude to the weight of the fluid displaced by the object. This fact is known as Archimedes' principle. The net force on an immersed object (which takes into account the object's weight) determines whether the object moves up or down in the fluid in which it is immersed.

In gases there is an interrelationship among pressure, volume, and temperature. For most gases, $PV/T$ = a constant value, a relationship known as the ideal gas law. When any one of the variables, $P$, $V$, or $T$, is held constant, there emerges a simple relationship between the other two variables.

Weather and climate are manifestations of a global "engine," powered by the sun, that moves heat and moisture from place to place on small and large scales. The weather machine runs mainly by virtue of convection currents that course through the atmosphere and oceans.

The Coriolis effect, brought about by Earth's rotation, helps to organize convective currents in the lower atmosphere into spinning whirlpools. Areas of high pressure (highs) are characterized by air spiraling outward and downward and are associated with fair and warm weather. Areas of low pressure (lows) feature air that spirals upward, cools, and produces unstable weather. On a global scale, convection and the Coriolis effect in the atmosphere produce alternating bands of wet and dry climates that differ according to latitude.

Evaporation, condensation, and precipitation are the important atmospheric components of the global hydrologic cycle. Water is cycled through clouds,

which can form and give up their moisture in a variety of ways.

Heat engines are devices that convert some heat energy into work. They extract a given quantity of heat energy from a high-temperature source and exhaust a smaller quantity of heat energy to a low-temperature environment. The reverse of a heat engine, a heat pump, may use mechanical energy to move heat energy from a low-temperature environment to a high-temperature environment.

The first law of thermodynamics says that in any kind of energy conversion, the same amount of total energy exists after the conversion as before. The second law of thermodynamics says that heat energy can never be totally converted to mechanical energy; some heat energy must be exhausted or wasted.

# CHAPTER 7
# Questions

## Multiple Choice

1. Heat is
   a) measured in Fahrenheit, Celsius, or kelvin units
   b) always proportional to the average speed of atoms or molecules in a body
   c) energy in transit from one body or place to another
   d) a type of force

2. If two objects have the same temperature, then the
   a) heat stored in one object must equal the heat stored in the other
   b) heat stored per gram in one object must equal the heat stored per gram in the other
   c) average kinetic energy of the molecules of both objects is the same
   d) average internal kinetic plus internal potential energy of the molecules of both objects is the same

3. The Celsius (C) and Kelvin (K) temperature scales are related to each other in the following way:
   a) K = C + 273
   b) C = K + 273
   c) K = (5/9)C + 32
   d) K = C (they are identical)

4. If a researcher wants to double the average kinetic energy of the molecules in a gas at 300 K, the temperature of the gas must be changed to
   a) 150 K
   b) 400 K
   c) 500 K
   d) 600 K

5. Room temperature on the Kelvin scale is roughly
   a) 72 K
   b) 273 K
   c) 300 K
   d) 400 K

6. Absolute zero can be thought of as the
   a) coldest laboratory temperature that could be reached when the Kelvin scale was invented
   b) freezing point of water at sea level pressure
   c) point at which liquid helium boils at sea level pressure
   d) extrapolated point on the temperature scale at which all molecular motion ceases

7. One kilocalorie (1 kcal) is the amount of energy required to
   a) raise the temperature of 1 g of water 1 °C
   b) raise the temperature of 1 kg of water 1 °C
   c) cause 1 kg of ice to melt
   d) cause 1 kg of water vapor to condense

8. When water vapor is condensing or liquid water is freezing,
   a) the temperature of the water is decreasing
   b) the temperature of the water is staying the same
   c) energy supplied by the surrounding environment is being absorbed into the water
   d) the pressure on the water must be increasing

9. Heat from a candle flame or from glowing coals in the fireplace is most slowly transferred outward by the mechanism of
   a) conduction
   b) convection
   c) radiation
   d) emission of photons

10. The bubbles rising inside an open pot of boiling water at sea level are
    a) air dragged under the surface by rapid convection
    b) dissolved oxygen in the water being released
    c) water vapor below a temperature of 100 °C
    d) water vapor at or very near a temperature of 100 °C

11. When heat flows into a body and its temperature increases, the increase does not depend on the body's
    a) mass
    b) shape
    c) chemical composition
    d) physical phase (solid, liquid, gas)

12. When condensation (of a gas) takes place,
    a) heat energy is absorbed
    b) heat energy is given off
    c) the gas always becomes a liquid
    d) the gas always becomes a solid

13. Changes of phase (such as solid to liquid) take place at constant
    a) temperature
    b) pressure
    c) heat
    d) volume

14. While pressure steadily declines with increasing altitude in Earth's atmosphere, temperature
    a) increases steadily with increasing altitude
    b) decreases steadily with increasing altitude
    c) increases and decreases several times with increasing altitude
    d) remains approximately constant at room temperature over the entire range of altitude

15. The troposphere is heated primarily by
    a) direct absorption of the sun's visible light
    b) infrared radiation radiated by the land and the oceans
    c) heat released by the condensation of water vapor in clouds
    d) geothermal energy released from Earth's interior

16. The stratosphere
    a) blocks the solar wind
    b) absorbs solar ultraviolet radiation
    c) reflects radio waves
    d) absorbs hydrogen and helium

17. The ionosphere consists of
    a) heavy concentrations of ozone
    b) ions that block most solar ultraviolet
    c) ions that may reflect radio waves
    d) a haze of tiny water droplets

18. The solar temperature is lowest in the
    a) interior
    b) photosphere
    c) chromosphere
    d) corona

19. The sun's core temperature is approximately
    a) 15,000,000 K
    b) 100,000 K
    c) 6000 K
    d) 3000 K

20. Energy moves outward from the sun's core to its surface primarily by means of
    a) conduction and convection
    b) conduction and radiation
    c) radiation and convection
    d) radiation alone

21. The vast majority of the sun's visible light comes to us directly from the
    a) core
    b) photosphere
    c) chromosphere
    d) corona

22. Auroras (northern or southern lights) are
    a) caused by an interaction between the solar wind and Earth's upper atmosphere
    b) the glowing traces left in Earth's upper atmosphere after the passage of a large meteor
    c) lightning discharges above the tropospheric clouds
    d) plumes of volcanic dust floating in the stratosphere and caught in the sun's rays at dusk

23. In a glass of water just barely above 0 °C,
    a) slightly warmer water tends to rise
    b) slightly cooler water tends to sink
    c) tiny ice crystals are present
    d) energy must be added so that water molecules can link together and form ice

24. A mercury barometer works because
    a) the vacuum at the closed top of the glass tube exerts an upward force on the mercury inside the tube
    b) the pressure in the air surrounding the barometer forces the mercury up the tube
    c) surface tension (the effect of attraction among a liquid's atoms or molecules) pulls the mercury up the tube
    d) an electric current in the mercury produces a repulsive magnetic force

25. If a gas is kept within a container of fixed volume,
    a) molecules in it move in an orderly way
    b) molecules in it move at a definite speed that depends on temperature
    c) molecules in it continually give up energy when they collide with one another, which keeps the temperature from falling
    d) there is no uniformity of speed or direction associated with the molecules at any time

26. A bar of soap sinks in a bathtub. The buoyant force on the soap is
    a) zero
    b) less than zero
    c) equal to its weight
    d) less than its weight

27. The gases that make up the majority (by mass) of Earth's atmosphere are
    a) oxygen and water vapor
    b) nitrogen and oxygen
    c) hydrogen and helium
    d) oxygen alone

28. The chief source of atmospheric heat is
    a) incoming solar radiation (insolation)
    b) tidal friction
    c) ultraviolet radiation absorbed by ozone in the stratosphere
    d) geothermal energy

29. Heat is transported from the equator to the poles mainly by
    a) winds
    b) ocean currents
    c) conduction of heat through air and water
    d) infrared radiation trapped under cloud cover

30. Winds are propelled by
    a) convection currents in the atmosphere
    b) downdrafts caused by Earth's gravity
    c) clouds that expand when they form
    d) Earth's rotation

31. Because of the Coriolis effect, Northern Hemisphere winds have a tendency to
    a) be deflected to the right
    b) be deflected to the left
    c) spiral inward
    d) curve upward

32. When maritime tropical air sweeps north into the central United States, that region experiences
    a) clear and dry weather
    b) warm, humid, cloudy, and often wet weather
    c) cold, dry winds
    d) snow

33. The boundary of an advancing mass of warm air that passes over and displaces a cooler air mass is called a(n)
    a) warm front
    b) cold front
    c) cyclone
    d) El Niño

34. When saturated air is cooled,
    a) it becomes better able to take up water vapor
    b) the relative humidity goes down
    c) the water vapor in it condenses on dust particles
    d) no further water vapor can condense

35. Generally speaking, precipitation is scarce
    a) in the tropics (near latitude 0°)
    b) only at latitudes of about 60° in both hemispheres
    c) only at latitudes of about 30° in both hemispheres
    d) at latitudes of about 30° in both hemispheres plus near the North and South poles

36. Cyclones center around
    a) regions of low pressure
    b) regions of high pressure
    c) points in between regions of intense high and intense low pressure
    d) low-pressure regions that never stray from the midlatitudes

37. The work output of an ideal heat engine
    a) equals its heat input
    b) equals the difference between its heat intake and heat exhaust
    c) depends only on its input temperature
    d) depends only on its exhaust temperature

38. A refrigerator
    a) creates cold
    b) changes heat to cold
    c) causes heat to disappear
    d) removes heat from one place and transports it elsewhere

39. The law of entropy is violated by
    a) the condensation of water vapor, which spontaneously produces heat
    b) human beings, which can gather up energy and transform it into organized outcomes
    c) the sun, which spontaneously generates prodigious amounts of radiant energy
    d) no physical system

# Questions

1. How do the concepts of temperature and heat (or thermal energy) differ?

2. Although the SI unit for thermal energy is the joule, many scientists find an older, metric-based unit—the calorie or its multiple, the kilocalorie—to be more useful for calculations. Why?

3. Why does heat flow between a warm object and a cooler object when the two are in contact with each other?

4. Why is a spray of water effective in putting out most fires?

5. Does the fact that gasoline is *volatile* (it evaporates quickly) tell you anything about the forces that exist between the molecules that make it up?

6. Why is it normally impossible to see (without special instruments) the sun's chromosphere and corona?

7. How can the sun's corona be extremely hot on the basis of temperature and yet not be a source of much heat?

8. Summarize the unusual chemical and physical properties of water. How does the existence of abundant water on Earth promote equable conditions for most forms of life?

9. How does pressure differ from force?

10. What is buoyancy? Under what conditions will an object rise or sink when immersed in a fluid?

11. List and briefly describe the three mechanisms by which thermal energy is transmitted as heat.

12. By what mechanism(s) does energy from the core of the sun reach its surface? In what form does solar energy move from the surface of the sun to Earth? How is solar energy distributed to Earth's atmosphere and surface? What becomes of the solar energy absorbed by Earth?

13. Why do some coastal areas experience an alternating onshore and offshore (sea and land) breeze on a daily schedule?

14. Why is stable (generally fair) weather associated with anticyclones, and unstable (cloudy or stormy) weather associated with cyclones?

15. What three circumstances are necessary for the formation of clouds?

# Problems

1. A 100-m-long steel bridge stands at a place where the temperature never drops below −10 °C nor rises above 40 °C. When the bridge is at the minimum temperature, how wide should the expansion gap at one end of the bridge be? Assume that the whole bridge expands 0.01 % in length for every 10 °C increase in temperature.

2. Starting with 1 kg (1 liter) of water at 25 °C, how long does it take to raise that water to the boiling point, provided that your stove burner set on "high" delivers 1 kcal of heat every 5 s? Assume that there are no heat losses to the environment, and ignore the fact that the pot containing the water must be heated along with the water. Remember that it takes 1 kcal of energy to raise the temperature of 1 kg of water by 1 °C.

3. How much heat, in joules, is required to accomplish the task described in Problem 2? If the pot is placed on a turned-on electric stove burner

emitting a constant amount of heat, how many watts of electricity are required to keep the burner on. (Remember from Chapter 6 that 1 W = 1 J/s.)

4. Ten grams of ice at 0 °C are placed in 80 g of water at 10 °C. The mixture is stirred well and the ice melts. What is the temperature of the water then?

5. A strong breeze exerts a force of 20,000 N on a square-rigged sail 4 m high and 5 m wide. How much pressure is exerted on the sail?

6. Assuming that pressure increases by 1 bar for every 10-m increase of depth in the ocean, what is the approximate water pressure at the bottom of the 11-km deep Mariana Trench, the deepest point in the ocean? Compare this pressure with the pressure at the center of Earth.

7. A container of fixed volume holds 1 kg of air at atmospheric pressure (assume it is 1 bar). If an additional kilogram of air is pumped into the container and there is no change in temperature, what is the new pressure? What is the justification for your answer?

# Questions for Thought

1. Some people think it is quite a demonstration of courage or of mind over matter for a person to scoot barefoot several paces over a bed of hot coals (of wood), without burning the soles of the feet. From a sober scientific point of view, how would you explain this ability? (Hint: There is more than one physical principle or aspect involved. One aspect often overlooked is the fact that fire walkers tend to have sweaty feet, a result of anxiety.)

2. Why can't liquid water exist on the surface of Mars, despite the fact that some areas on or near its equator warm almost to room temperature at Martian noon?

3. If astronauts did not wear protective space suits during EVAs (extra-vehicular activities, or space walks) in space, their blood would boil. Why? Also, why is the bubblelike glass pane on the

helmet of a space suit coated with a reflective film?

4. In the steamy climate of Hong Kong, it feels more pleasant to be outside in the early morning when the temperature is rising than in the early evening when the temperature is falling, even though the thermometer reads the same at both times. Why? (Assume that direct sunlight plays no part; you stay in the shade.)

5. Where does "outer space" begin with respect to Earth? Does it begin at the top of one of our atmosphere's major zones, or is its meaning relative and arbitrary?

6. Thermos bottles are designed to keep hot liquids hot and cold liquids cold. In the highest-quality thermos bottles, called Dewar flasks (Figure 7.41), a double-walled enclosure surrounds the liquid, and there is a vacuum in the space between the two walls. Both walls are coated with reflective metal on both sides, and the lid is tightly sealed. How does such a bottle inhibit heat transfer by means of conduction, convection, and radiation?

**FIGURE 7.41** A Dewar flask.

7. Think of several examples of how you feel heat or cold, and explain each of these sensations in terms of the mechanisms of conduction, convection, and/or radiation.

8. How can the sun's interior be blindingly bright and opaque at the same time?

9. If a pressure gauge indicates a reading of 30 psi for a given tire, what is the total pressure in the air inside the tire? (Hint: Remember that the

atmospheric pressure *outside* the tire is approximately 15 psi.)

10. Charles's law says that for an ideal gas kept under constant pressure, volume and Kelvin temperature are directly proportional. If you vary the temperature of a gas kept at constant pressure and measure changes in its volume, what insight do you think you would gain concerning the concept of absolute-zero temperature?

11. How can the principle of hydrostatic equilibrium be applied to an inflated balloon?

## Answers to Multiple-Choice Questions

| | | | | | |
|---|---|---|---|---|---|
| 1. c | 2. c | 3. a | 4. d | 5. c | 6. d |
| 7. b | 8. b | 9. a | 10. d | 11. b | 12. b |
| 13. a | 14. c | 15. b | 16. b | 17. c | 18. b |
| 19. a | 20. c | 21. b | 22. a | 23. c | 24. b |
| 25. d | 26. d | 27. b | 28. a | 29. a | 30. a |
| 31. a | 32. b | 33. a | 34. c | 35. d | 36. a |
| 37. b | 38. d | 39. d | | | |

# Waves

*Sound waves, like all waves, carry energy. The sound waves impinging on this wine glass, moments before this photograph was taken, were of just the right frequency to cause the glass to vibrate, like a bell, at its natural frequency. The sound and the vibrations are so intense that the glass shatters.*

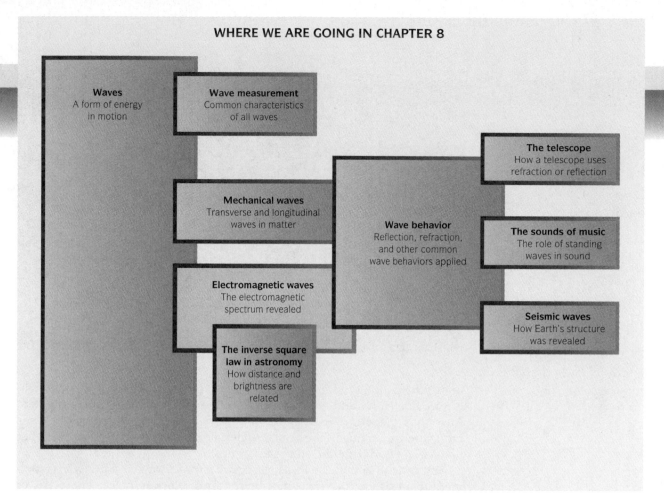

Imagine a tennis ball falling toward the smooth surface of a pond. Immediately before it strikes the water, the ball has kinetic energy by virtue of its motion. There is a small splash, and the tennis ball bobs up and down a couple of times before settling on the pond's surface. The ball is now stationary; its kinetic energy is zero. Where has the energy gone?

The answer is found in the waves that spread in ever-widening circles from the point of impact. The peaks and valleys of the waves certainly travel outward, but does the water itself move outward from the point of impact? If it did, then water would move away from the point of impact and accumulate at the edges of the pond. Clearly, it does not.

Unlike the energy carried by a body moving from one place to another (a falling tennis ball, for example), water waves embody a kind of energy moving *through* and not *with* the water. A pollen grain floating on the pond surface is pushed forward and up as the crest (topmost part) of a wave approaches. It then falls backward and down as the crest passes by and a trough (the low spot between crests) approaches. The water molecules

**PHOTO 8.1** Like ocean waves, ripples on a sand dune are caused by the dragging force of the wind. Unlike most waves, which can propagate through particles in a medium without displacing a particle's average position, sand particles on the surface of a dune are transported from place to place as the ripples creep across the surface.

under the pollen grain participate in this same motion. They follow roughly circular paths and return to their original positions after a complete wave has passed. When many successive waves pass by, each water molecule on or near the surface undergoes a cyclical motion, somewhat like a planet orbiting the sun many times.

By generalizing these ideas, we can say that a wave is a periodic (repetitive) disturbance that transmits energy from place to place. There are, of course, a great variety of waves. Some waves, such as water waves and sound waves, require some kind of *medium* (physical substance) for their transmission. Other waves, such as light waves, have no trouble propagating through a space completely devoid of a medium—though, to be sure, something is being disturbed in that empty space.

Light waves in a vacuum move at the maximum possible speed, the speed of light. Other waves may move millions or billions of times slower. Waves differ in their interactions with matter. Light waves, for example, readily interact with the tissues of the human body; we say that humans are largely *opaque* to light waves. Air molecules do not respond to radio waves passing by; we say that air is *transparent* to radio waves.

Using our senses alone, we are made aware of some waves as they pass by and are unaware of others. We feel the energy (and the sickening effect) of passing water waves when aboard a ship on a storm-tossed sea. We hear sound waves with our ears, see light waves with our eyes, and feel waves of infrared radiation (radiant heat) with the nerve endings beneath our skin. We do not, however, normally feel or otherwise sense the radio wave signals emanating from broadcast antennas, cellular telephones, and other electronic devices. These waves, like all waves, are capable of carrying *information* as well as energy.

The study of waves in this chapter will emphasize the similarities rather than the differences among waves. All waves share certain common behaviors, primarily *reflection, refraction, diffraction, interference,* and the *Doppler effect.* Once we understand how one type of wave reflects or refracts, we can extend the concept of reflection or refraction to other types of waves.

The casual study of waves enriches our knowledge of many of the ordinary things we can see, hear, and feel. On a deeper level, scientists have acquired an enormous body of knowledge about things in the universe we cannot see or reach simply by measuring and analyzing waves. Geologists, as we shall see in this chapter, have discovered virtually all we know about Earth's interior structure solely by means of analyzing seismic waves generated by earthquakes. Astronomers were totally ignorant of the physical nature of the stars and the universe at large until *spectroscopy,* the method by which light waves are separated and analyzed, was applied to starlight in the mid–nineteenth century. The topic of spectroscopy will be taken up in Chapter 11, "Light and the Atom," after we have explored several important ideas concerning the production of light.

## Wave Measurement

All waves have certain common characteristics, though the measure of these characteristics will vary with the type of wave. To visualize these common characteristics, we will utilize the image of water waves—ripples on the surface of a pond or ocean swells (see Figure 8.1). Light, sound, and other wavelike phenomena do not look like water waves, but they do have qualities that are analogous to the moving crests and troughs of water waves.

By following the moving crests or troughs, researchers can measure the *speed* of the moving series of waves. Small ripples in a pond travel at less than 1 m/s, but ocean swells typically move at 10 m/s or faster. Winds are the primary cause of the waves that develop in open bodies of water. The force of friction between the moving air and the water surface raises small wavelets, which can combine to form larger waves.

Sound, which travels at about 343 m/s through air at room temperature, consists of an alternating series of *compressions* (denser regions under greater pressure) and *rarefactions* (less dense regions under less pressure). These features are analogous to the crests and troughs of water waves. Sound waves readily travel through water (at about 1500 m/s, or 1.5 km/s)

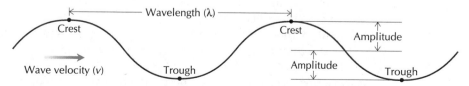

**FIGURE 8.1**   Wave anatomy consists of the following characteristics: velocity, wavelength, and amplitude. Frequency or period may be specified instead of wavelength. Although wave features such as crests and troughs move steadily from left to right in this diagram, the *material* through which a wave is passing experiences no net movement.

and other liquids and through solids. The speed of sound through steel is approximately 5 km/s; you can hear the sound of a far-off train by pressing your ear to the track, though this exercise is best left to your imagination.

Light travels at the incredible rate of 300,000 km/s in empty space. It is instructive to compare the speeds of light and sound in units of meters per second: 300 million m/s for light, and about 343 m/s for sound in air. Light outraces sound by a factor of almost a million. When lightning strikes nearby, you see the light after a negligibly small interval of time. The sound of thunder, which is produced by the sudden heating and expansion of the air around the lightning bolt, takes about 3 s to travel each kilometer of distance and about 5 s to travel each mile of distance. By counting the seconds between the lightning and the thunder, you can estimate your distance from a lightning bolt. If the thunder and lightning seem simultaneous, you are too close for comfort.

A more complete description of the motion of a wave would include its direction as well as its speed. This is, of course, wave *velocity*. Directional information is particularly important when describing the behavior of light as it reflects from smooth surfaces or passes through lenses. When dealing with these issues, we often speak of light *rays*: imaginary lines that trace the paths of the energy carried by light waves.

The **wavelength** (abbreviated by the Greek letter $\lambda$, lambda, in formulas) of a repeating series of water waves is simply the distance between two successive crests or between two successive troughs. The wavelength of sound is the distance between adjacent compressions or rarefactions in the traveling series. Light, imagined as a wave, consists of a series of wavelike "ripples" in space. Each ripple represents a complete cycle associated with changing electric and magnetic fields. The wavelength of light represents the distance between adjacent ripples.

The **frequency** ($f$) of any type of wave is defined as the number of waves that pass a given point in 1 s. If, for example, 8 successive ripples reach the edge of a pond in 1 s, we would say that their frequency is 8 waves per second, or 8 Hz. The abbreviation Hz refers to the unit *hertz*, which means cycles per second. In standard SI units, the hertz is equivalent to 1/s (the reciprocal of 1 s). When frequencies are measured in thousands, millions, or billions of hertz, the units kilohertz (kHz), megahertz (MHz), and gigahertz (GHz) are used, respectively. Audible sound frequencies range from about 20 Hz to about 20 kHz (for people with excellent hearing). What our ears detect as the pitches of sounds—the "highs" and the "lows" on the musical scale—are directly related to the frequencies of those sounds. Sound frequencies higher than about 20 kHz are known as ultrasound. Dogs can hear ultrasound frequencies up to 50 kHz, and bats can both vocalize and detect chirplike squeaks with frequencies of up to 120 kHz. Some typical radio wave frequencies are those of the commercial AM (540 to 1600 kHz) and FM (88 to 108 MHz) radio broadcasts. Visible light has a frequency of somewhat less than $10^{15}$ Hz, or $10^6$ GHz.

The **period** ($T$) of a series of waves is the time interval between the passage of successive waves. The period of a wave is simply the reciprocal of the frequency, and vice versa. For example, the period of waves having a frequency of 8 Hz is 1/8 s, or 0.125 s. The period of the ocean swells breaking upon the Southern California shoreline averages about 10 s, which means that the frequency of these waves is about 0.1 Hz.

The wave characteristics speed ($v$), wavelength ($\lambda$), and frequency ($f$) are related in the following manner, a statement called the *wave equation:*

$$v = \lambda f \qquad\qquad \text{Equation 8.1}$$

From this equation it is apparent that when the wave speed is constant (which is true for most kinds of waves under most circumstances), relatively long wavelengths are linked to relatively low frequencies. Conversely, shorter wavelengths imply higher frequencies.

### Problem 8.1
What is the wavelength of the note A ($f = 440$ Hz) above middle C on the musical scale? (Assume $v = 343$ m/s.)

### Solution
Solving for $\lambda$ in Equation 8.1, we get $\lambda = v/f$.

$$\lambda = v/f$$
$$= (343 \text{ m/s})/(440 \text{ 1/s})$$
$$= 0.78 \text{ m}$$

### Problem 8.2
Simple FM antennas are made with metal rods or wire elements that are about 1.5 m long. These are half-wave antennas constructed to respond to radio wavelengths twice that long: around 3 m. What is the approximate frequency of the FM radio broadcasts these antennas are tuned for? (*Note:* Light and all other electromagnetic waves move more slowly as they make their way through matter. The slowdown in air is very slight, so we can still use $v = 300{,}000$ km/s $= 3 \times 10^8$ m/s.)

### Solution
Solving for $f$ in Equation 8.1 and substituting the symbol $c$ (the speed of light) for $v$, we get $f = c/\lambda$.

$$f = c/\lambda$$
$$= (3 \times 10^8 \text{ m/s})/(3 \text{ m})$$
$$= 10^8 \text{ 1/s}$$
$$= 100 \text{ MHz}$$

This frequency is near the middle of the FM broadcast band, which spans 88–108 Mhz.

Only one more general characteristic of waves needs to be introduced: amplitude ($A$). For water waves, **amplitude** is defined as half the height of a wave from crest to trough. For sound waves of a given frequency, amplitude is related to the intensity, or loudness, of the sound as heard by a given listener. For light waves of a given frequency, amplitude is related to the intensity, or brightness, of the light as it is seen by an observer. The intensities of sound and light are proportional to the square of both the amplitude and the

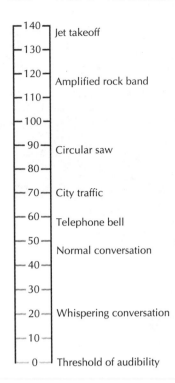

**FIGURE 8.2**   The decibel (dB) scale is used to describe the intensity (loudness) of various sounds. The scale is logarithmic, so that each increase of 10 dB represents a tenfold increase in sound intensity. The situations noted refer to sounds heard at close range. If a sound becomes stronger by 10 dB, our ears perceive the change to be considerably less than a factor of 10 louder. Because of this curious "compressed" response to changes in sound levels, persons with normal hearing can discern sounds as faint as 0 dB and tolerate (but with some pain, perhaps) sounds of over 120 dB, a range of more than $10^{12}$ (a trillion). Chronic exposure to sounds of 90 dB or greater leads to permanent hearing damage.

frequency of the waves. The more severe the disturbance (the greater the amplitude) produced by a given wave, the greater the energy carried by that wave. The more waves per second (the higher the frequency), the greater the flow of energy.

Because sound and light intensities span such a large range, they are often specified with the help of logarithmically configured scales. Sound intensities are measured in units of *decibels* (see Figure 8.2). The **decibel scale** is defined so that any change of 10 decibels (dB) represents a tenfold change in sound intensity.

Astronomers use a somewhat different logarithmic scale of **apparent magnitudes** to express the observed brightnesses of celestial bodies. On this scale a difference of five magnitudes is equivalent to a factor of 100 in brightness. Vega, a bright star of the summer sky, shines at apparent magnitude 0.0. Polaris, the North Star, is of approximately magnitude 2. The faintest stars visible to the naked eye on clear, dark nights are of approximately magnitude 6. The modern apparent magnitude scale is derived from a similar scale used since ancient Greek times. In antiquity, the brightest stars were said to be of first magnitude, the second-brightest stars were classed as second magnitude, and so on.

## Mechanical Waves

We know that a wave is a periodic disturbance in "something" that carries energy from place to place. That something can be a medium (some kind of matter) or it can be empty space. If a medium is required, the waves traveling through it are categorized as **mechanical waves.** They include waves on a liquid surface, waves passing though a taut string or wire, sound waves, and seismic waves. Mechanical waves will not propagate through a vacuum (perfectly empty space).

The simplest kinds of mechanical waves can be analyzed with the help of an object moving in *simple harmonic motion*—for example, the vibrating mass on a spring pictured in Figure 6.5 (Chapter 6). If a mass (or a person's hand) vibrating up and down rhythmically shakes the loose end of a cord stretched horizontally, as in Figure 8.3, a series of waves, similar in appearance to water waves, moves down the cord. These waves are categorized as **transverse waves** because the medium through which they travel (the cord) oscillates in a direction transverse (across, or perpendicular) to the direction the waves are moving. The more tightly the cord is stretched , the faster the waves move. The vibrating strings or wires of a guitar, violin, or

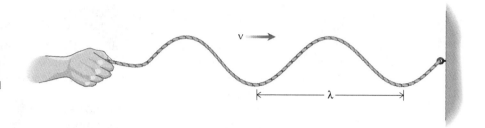

**FIGURE 8.3** Transverse waves will travel down a stretched cord when one end is shaken. The more taut the cord, the faster the waves travel.

piano have transverse waves passing through them. The waves bounce back and forth so quickly that it is impossible to follow the motion of the waves with the eye.

If a mass (or a person's hand) vibrating back and forth rhythmically shakes a wire Slinky (pictured in Figure 8.4) so as to alternately compress it and expand it, alternate compressions and rarefactions can be seen moving down the slinky. These waves are categorized as **longitudinal waves;** they oscillate longitudinally (parallel) to the direction of the moving waves. The stiffer the wire is in the slinky, the faster the waves move. Sound waves are longitudinal. They can be created by the paper cone of a loudspeaker that is alternately pushing forward against a medium (usually air) and pulling back. Each forward pulse compresses the air and each backward pulse rarefies (makes thinner) the air next to the cone. The rate at which the compressions and rarefactions propagate outward depends on the interactions between nearby air molecules. The higher the temperature of the air, the more active the air molecules are, the more quickly they interact with each other, and the faster the speed of the sound is.

Transverse waves will travel only through solids. Transverse motion requires that particles in the medium moving back and forth drag adjacent particles with them. In fluids, the particles will merely slide by one another, and the wave dies out. Longitudinal waves can travel through any medium, solid or fluid. When any one particle in a medium pushes forward in the direction of the traveling wave, it will push on or otherwise interact strongly with the particles ahead of it. Like falling dominoes, the wave progresses forward.

Waves on a surface or a boundary between two fluids (water and air, for example) are of a hybrid type. As we have already seen, particles near the surface of a water wave move both longitudinally (forward and back) and transversely (up and down) in a coordinated way. These particles execute roughly circular paths as successive waves pass.

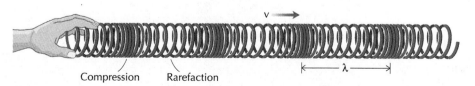

Compression    Rarefaction

**FIGURE 8.4** Alternate compressions and rarefactions can be seen moving through a wire Slinky when it is pushed and pulled at one end in a rhythmic manner. The moving compressions and rarefactions constitute a longitudinal wave moving in a solid medium.

# Electromagnetic Waves

The idea that certain waves can pass through nothing at all, with no medium to support them, became widely accepted in the early 20th century. These "nonmechanical" waves include electromagnetic waves and gravitational waves.

**Gravitational waves,** theoretically, are produced in empty space by moving masses; they are so incredibly weak that none have been detected thus far. Increasingly sophisticated experiments are being designed to test for their existence. **Electromagnetic waves** are essentially time-varying, coupled electric and magnetic fields moving through space at the speed of light (more details are given in Chapters 10 and 11). They are associated with moving electric charges. These waves are embodied in light and its "cousins"—radio waves, infrared radiation, ultraviolet radiation, X-rays, and gamma rays. Electromagnetic waves are enormously more energetic than gravitational waves, a fact that becomes more apparent when we compare the inherent strengths of electromagnetic and gravitational interactions. From Table 2.2 in Chapter 2 we see that electromagnetic interactions are stronger by a factor of about $10^{38}$.

It is an oversimplification to think that light and the other electromagnetic radiations are purely waves, but that is exactly the approach we will take in this chapter whenever we discuss them. In Chapter 11 we will explore the other side of light's duplicitous personality—that light is composed of particlelike entities called *photons*.

From the water wave metaphor, electromagnetic waves can be thought of as having "crests" and "troughs" that move past given points in space. As the crests and troughs pass, the electric and magnetic fields existing at those points oscillate back and forth. For now, we need only know that the fields oscillate transversely with respect to the direction the waves are moving. Electromagnetic waves are transverse waves.

When electromagnetic waves emanate from a small source (or *point source*), the intensity of the waves decreases in proportion to the square of the distance from the source. When an observer walks directly away from an illuminated lightbulb, every doubling of the distance makes the light appear 4 times dimmer. Tripling the distance makes the light appear 9

**FIGURE 8.5** The radii of circles intersected by the circular beam shown ($r$, $2r$, $3r$, and so on) increase at the same rate as the distance the beam travels ($R$, $2R$, $3R$, and so on). The areas of the circles penetrated by the beam, however, increase with the square of the radius $r$. Therefore, the beam's intensity decreases in proportion to the square of the distance. The total energy moving outward in the beam remains the same, but it becomes more diluted as the beam spreads.

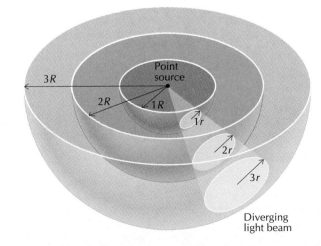

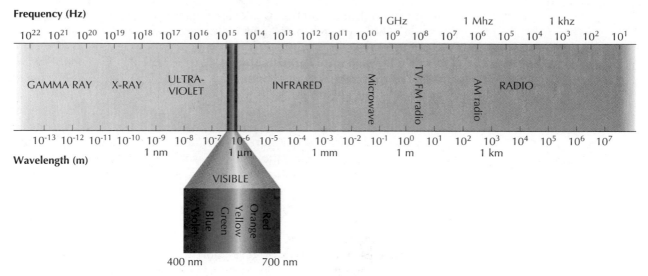

**FIGURE 8.6**   The electromagnetic spectrum includes six major parts with ill-defined boundaries: radio, infrared, visible, ultraviolet, X-ray, and gamma ray. The visible part of the spectrum spans a relatively small range of frequencies and wavelengths.

times dimmer; quadrupling the distance makes it appear 16 times dimmer; and so on. We have already seen this same inverse square behavior in connection with Newton's law of gravitation. For light, this behavior can easily be understood with the help of a geometrical diagram (see Figure 8.5). Inverse square behavior is not unique to light; it applies to any type of energy that radiates outward. Sound, for example, decreases rapidly in intensity as sound waves move outward from a small source such as a loudspeaker horn. The inverse square law for light is particularly important in astronomy, where it is employed to determine the distances to stars (see Box 8.1).

The entire range of electromagnetic frequencies or wavelengths is known as the *electromagnetic spectrum* (see Figure 8.6). The spectrum is theoretically open-ended; there is no definite minimum or maximum frequency or wavelength possible. However, electromagnetic radiation outside the range plotted in Figure 8.6 is neither easy to produce nor easy to detect. The electromagnetic spectrum has been traditionally divided into six major parts; each part is discussed in the following sections.

## Radio Waves

**Radio waves** encompass the lowest frequencies and longest wavelengths of the electromagnetic spectrum. The word *radio* comes from the word *radiate,* which means "to move outward along a radius." All radio waves radiate without hindrance through empty space and the lower atmosphere, and some can even penetrate Earth's oceans and interior. Radio waves can be produced by means of electrons surging back and forth in a wire, or a transmitting antenna. Another wire of similar length and orientation can be used as a receiving antenna. Electrons will surge back and forth in the receiving antenna in response to the signal radiated from the transmitting antenna, and that weak signal can be amplified by a receiving device: a radio, a television set, or a cellular phone.

Radio waves may be used to transmit sound, but they themselves are not sound. Rather, sound (and other information such as the video part of a television transmission) can be encoded into the transmitted radio signal and decoded by the receiving device. One of the simplest methods of encoding, or "modulating," information is *amplitude modulation,* a way of varying the amplitude of a high-frequency radio wave so that it defines a lower-frequency wave (see Figure 8.7). The lower-frequency wave produced can change in both frequency (over a range that corresponds to the sound frequencies) and amplitude. Another method is *frequency modulation,* in which the frequency of the radio wave varies with time but the amplitude does not. These methods lend their initials to the standard AM and FM radio broadcast bands. More sophisticated modulation schemes encode information for television broadcasts.

**Microwaves** (radio waves of very short wavelength) have frequencies of greater than about 1 GHz and wavelengths of less than about 1 m. Much of the information traveling across the Unites States by telephone line is at some point encoded in microwave transmissions that pass between dish-shaped antennas mounted on tall buildings or towers. The dish antennas are effective in focusing the microwaves into a narrow beam. Since

## BOX 8.1

# The Inverse Square Law and Astronomy

Of the many tools used by astronomers to estimate very great distances, the inverse square law of light propagation is the most powerful and wide ranging. Every astronomer knows that the apparent brightness of a star is proportional to the inverse square of its distance: Doubling a given star's distance makes it appear $(1/2)^2 = 1/4$ as bright. Increasing a given star's distance by a factor of 10 makes it appear $(1/10)^2 = 1/100$ as bright.

If all stars emitted exactly the same amount of energy, distance estimates would be simple. For instance, a star appearing 1/4 as bright as another would be twice as far away, and a star appearing 1/100 as bright would be 10 times farther away. We could use the sun as a standard for brightness comparisons, since we already know its distance. In the real world, however, the energy outputs of various stars are vastly different. We must also know the energy output of the star if we wish to determine its distance.

When applying the inverse square law, astronomers make use of a formula with three variables:

*apparent magnitude* (a measure of apparent brightness), *luminosity* (a measure of energy output, or intrinsic brightness), and *distance.* If the value of any two of these variables is known, the third is determined by a simple calculation.

The apparent magnitude of any visible star can be measured without much difficulty. A *photometer* (sensitive light meter) attached to a telescope will do the job. In order to calculate the star's distance, we need to know the star's luminosity, which is not always easy to measure. Quite often, educated guesswork is involved. All stars have characteristic spectra (we shall see some examples in Chapter 11), and certain subtle features in those spectra reveal much about a star's luminosity.

As a simplified example of this process, suppose we observe two stars having identical spectra, and one of the stars is five magnitudes (a factor of 100) brighter in appearance than the other. Because their spectra are identical, we can reasonably assume that both stars are virtually identical in their attributes, including luminosity. Applying the inverse square law, we conclude that the brighter star is 10 times closer. This gives us relative distances, but how can we find the distance to either star in absolute terms?

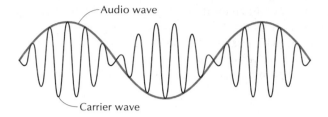

Audio wave

Carrier wave

**FIGURE 8.7** In amplitude modulation (AM), a carrier wave oscillating at a typical frequency of around 1 MHz produces an audio wave of much lower frequency by waxing and waning in amplitude. An AM radio receiver filters out the high-frequency oscillations of the carrier wave and sends the remaining audio wave to a loudspeaker or headphones.

microwaves pass easily through all layers of the atmosphere, they are the radio frequency of choice for satellite and spacecraft communication. Microwaves are also utilized in *radar* (*ra*dio *d*etecting *a*nd *r*anging) systems, which transmit and receive microwaves in short-duration bursts.

Microwave ovens exploit the fact that water molecules are good absorbers of microwave energy. When water molecules in food or a beverage absorb the microwaves emitted inside the oven, they vibrate and rotate faster. The kinetic energy they have acquired spreads to other molecules nearby, and the thermal energy content (and temperature) of the food or beverage increases.

Astronomers search the microwave band of the spectrum for signatures of many astrophysical processes. By gathering microwaves from the sky

That information must come by independent means. For example, if the nearer star's distance had previously been measured by trigonometric parallax (recall Box 2.3), we would then know the distance to both stars in absolute terms.

In practice, the inverse square law, coupled with several other, fundamentally different ways of determining distance, has allowed astronomers to construct reasonably accurate spatial maps of our galaxy and much of its contents. The study of one particular class of stars called *Cepheid variables* has yielded enormous gains in our understanding of intergalactic distances. All Cepheid variables pulsate (expand and contract) in a particular rhythmic manner, and they wax and wane in luminosity as they do so. Studies of many nearby Cepheids of known distance and luminosity have revealed an important correlation: The longer the period of pulsation of a Cepheid, the greater its mean luminosity.

Knowing this rule, we can calculate the distance (by applying the inverse square law) of any arbitrarily distant Cepheid simply by measuring its period, thereby getting an estimate of mean luminosity, and measuring its mean apparent magnitude. The longer-period Cepheids are among the most lumi-

nous stars in a galaxy; they shine like lighthouse beacons and are visible as distinct points of light across millions of light-years of space. The identification and measurement of Cepheid variables in the Andromeda galaxy in the 1920s gave astronomers conclusive proof of its enormous distance (about 2 million light-years by current estimate). One of the ongoing tasks of the Hubble space telescope is to identify Cepheids in galaxies tens of millions of light-years away, so that astronomers can more accurately determine the distances to nearby clusters of galaxies. This knowledge has a direct bearing on estimates of the size and age of the universe.

Note that any application of the inverse square law over large distances rests on the implicit assumption that the laws of physics are universal and that stars of a given type (such as Cepheid variables) are fundamentally the same everywhere. There is no reason, as yet, to suspect that this is not the case. If these basic assumptions are ever proved wrong, though, it will be difficult or impossible to develop any coherent conception of the universe.

with radio telescopes, they have measured and mapped the distribution of hydrogen gas in the interstellar space around us and within other galaxies. Precise measurements by the COBE satellite (p. 111) of microwave energy coming from the entire sky have allowed astronomers to detect tiny irregularities in the otherwise smooth cosmic background radiation that bathes the entire universe.

## Infrared Radiation

**Infrared** (abbreviated IR) **radiation** refers to radiation beyond the red end of the visible spectrum. Infrared radiation is absorbed by most substances, causing them to heat up. The nerves in our skin are quite sensitive to heating, so IR radiation on the skin brings on the sensation of radiant heat. Any object having an ambient temperature of a few hundred kelvins will emit rather large amounts of IR radiation; the body heat emanating from a warm human body is an obvious example.

Earth's atmosphere is only partially transparent to IR energy from the sun and other celestial sources, because it contains water vapor, carbon dioxide, methane, and other IR-absorbing gases. The water vapor content of the atmosphere decreases quickly with altitude, which is why the sun often shines with an unaccustomed fierceness at high elevations. On a sunny day atop Mount Everest, a climber can overheat in the sun's rays and yet freeze in a shadow. To minimize or avoid atmospheric absorption, astronomers site infrared astronomical telescopes on high mountaintops, send them aloft in high-altitude balloons, or launch them into Earth orbit.

## Visible Light

**Visible light** spans the relatively small range of radiant energy that the human eye responds to. With the help of three different types of cone (color-discriminating) cells in the retina, we discern six major bands of color within the visible spectrum: red, orange, yellow, green, blue, and violet. People who are color-blind are deficient in one or two of the cone cell types; they see a more limited set of spectral colors. Mixing colors from various parts of the spectrum can produce thousands of hues. Color and the various ways of producing it are subjects we will explore in further depth in Chapter 11.

Any radiating body at a temperature of several thousand kelvins emits visible light copiously. A good example is the sun, which has a surface temperature of 5800 K. It is no mystery why our eyes have evolved to be light-sensitive, rather than infrared- or ultraviolet-sensitive organs. First, the sun is brightest by far in the wavelengths of visible light. Second, the sun's radiation must pass through Earth's atmosphere, which is largely transparent to visible light, opaque to most ultraviolet light, and only semitransparent to infrared light.

We learned earlier that radio waves can be produced by electrons surging back and forth in a wire. Visible light also derives from changes made by electrons, but on a smaller and more energetic atomic scale. When an electron makes a transition in an atom from one energy level to another, a small burst of light (a photon) is emitted. This aspect of light is explored in Chapter 11.

# Ultraviolet Radiation

**Ultraviolet** (abbreviated UV) **radiation** refers to the radiation beyond the violet end of the visible spectrum. Although humans cannot see UV radiation, delayed physiological reactions are likely to follow any long or intense exposure to it. For most people, UV exposure provokes the buildup of melanin, a dark-colored pigment, in the skin. Sunburn occurs when the skin is overexposed to solar UV or tanning lamps. Sunburn is an indication of damage to skin cells, and severe or repeated damage can trigger cancerous growths. Curiously, many substances that are transparent to visible light are opaque to most UV radiation. Ordinary glass is such a substance. It is difficult to get a suntan in the sunlight passing through a glass window.

Only the longest wavelengths of UV, near the color violet on the visible spectrum, can penetrate the ozone layer in the stratosphere. Without any protection from the ozone layer, we would be exposed to the sun's short-wavelength UV, which is injurious to biological tissue. This shorter UV (along with X-rays and gamma rays) is known as **ionizing radiation,** because it has the ability to ionize molecules and atoms. Ionizing radiation can also consist of streams of fast-moving atomic particles. Ionizing radiation harms cells to some extent, even when it is very weak in intensity, because each ionizing photon, or fast-moving particle, has the capability of disrupting a cell's molecular structure. In medicine, *radiation therapy* is sometimes used in the treatment of cancer. It involves irradiating malignant (cancerous) tissue with beams of ionizing radiation intense enough to kill cancer cells and yet minimally damage normal cells nearby.

# X-rays

**X-rays** acquired their rather cryptic name during experiments a century ago that demonstrated the existence of a new, highly penetrating type of radiation. That radiation, given the name X for its then-unknown character, could pass through visibly opaque matter and expose photographic plates. In time, these unknown rays were identified as a range of electromagnetic radiation somewhat higher in frequency than UV.

X-rays can be artificially produced by bombarding a metal target, inside an evacuated tube, with high-speed electrons. The resulting collisions can eject the inner-orbiting electrons of the metal atoms. X-rays are emitted when outer-orbiting electrons rush in to fill the vacant inner orbits. Several astrophysical processes produce copious quantities of X-rays. The sun's superhot corona emits X-rays by virtue of its extremely high temperature.

The higher the frequency of X-rays, the greater their ability to penetrate matter. "Soft" (lower-frequency, longer-wavelength) X-rays are used for medical and dental X-ray photographs of tissue and bone. "Hard" (higher-frequency, shorter-wavelength) X-rays are used in industry to probe for defects in metallic structures. The typical medical or dental X-ray photograph is really a *shadowgraph,* a pattern of light and dark areas corresponding to the shadows of bone and soft tissue cast by a point source of X-rays.

## Gamma Rays

The highly penetrating **gamma rays** occupy the short-wavelength, high-frequency extreme of the electromagnetic spectrum. Gamma radiation is produced in many ways. As we have seen earlier, the core region of the sun is flooded with gamma rays resulting from the nuclear fusion reactions that power the sun. On Earth, gamma rays are produced at random during the radioactive decay of certain naturally occurring isotopes. This is part of the natural background of ionizing radiation everyone living on Earth is exposed to.

*Cosmic rays* are another important source of ionizing radiation on Earth. Raining down on Earth's atmosphere at very nearly the speed of light, cosmic rays consist of ions (mostly protons) probably originating from distant supernovas in our galaxy or possibly outside our galaxy. When a cosmic ray particle collides with atoms or molecules in the upper atmosphere, a "secondary shower" of gamma rays and other subatomic particles is produced. Many of these fragments can penetrate through the atmosphere and reach the ground.

## Wave Behavior

To "see" a rose is to see waves of light energy coming from the direction of the rose. To "hear" a bird's song is to hear waves of sound coming from the bird's throat. To "feel" an earthquake is to feel waves of ground motion passing underfoot. What is not so obvious about these waves is that all have undergone one or more changes of some kind before they arrive. Everything we see and hear, and part of what we feel, is governed to some extent by wave behavior.

The wave behavior of *reflection* often plays a major role in what we perceive. If objects did not reflect some light, then everything we see that is not self-luminous would appear black. We see the blur of waves bouncing back and forth along a plucked guitar string, and we hear the sound waves generated by the string's rhythmic motion. We hear the guitar's echo throughout the room.

The wave behavior called *refraction* is more subtle. Light rays are refracted (bent) when passing through a glass of water, a prism, a magnifying lens, and even through air. Eyeglasses refract light so as to compensate for certain limitations in human vision. Sound may refract: A foghorn clearly heard from many miles away has probably had its sound waves bent and focused as they passed through distinct layers of air. Usually subtle as well are the effects of *diffraction, interference,* and the *Doppler effect* in various kinds of waves.

We do not have space in this book to delve deeply into theoretical explanations of the various wave behaviors. Rather, we shall focus primarily on the macroscopic consequences of each behavior. If we understand how one type of wave behaves while it is refracting, then we will be able to understand, by analogy, how other kinds of waves refract, too. Let us start with the wave behavior that is simplest to explain, reflection.

## Reflection

Light bounces off a mirror. Sound reflects from a wall and returns as an echo. The crest of an ocean swell strikes a cliff face and its reflection ripples back out to sea. In every case, the energy carried by the wave changes direction in a simple way as the wave strikes a flat surface. Experiment shows that the angles made with respect to the surface by the incoming (*incident*) energy and the outgoing (*reflected*) energy are the same. Using the language of *optics,* which is the study of light's directional and wavelike behavior, we can state the law of reflection: The angle of incidence equals the angle of reflection.

Figure 8.8 shows how light rays, which represent the direction light energy travels, reflect from a smooth, flat surface. This behavior is not unique to waves; it also occurs for bodies rebounding from a smooth surface during an elastic collision. These kinds of reflections are seen when molecules of a gas strike the walls of a car engine's cylinders or when a hard rubber ball bounces off the interior walls of a handball or racquetball court.

Many things influence whether or not a given wave will reflect at a boundary. Mechanical waves moving through a fluid medium such as air or water (or a solid but flexible medium such as a tensioned guitar string) tend to reflect when they encounter anything that is rigid. Thus sound reflects off walls; ocean waves reflect when they strike a seawall; waves travel back and forth many times through a plucked string anchored on both ends. With the exception of some X-rays and gamma rays, electromagnetic radiation readily reflects from metal surfaces. These surfaces need not be "solid" (as opposed to porous). A wire screen with metal wires 1 m apart has no trouble reflecting radio waves of $\lambda = 1$ m or greater. Radio waves less than 1 m long would slip right through the screen. For this reason, microwaves ($\lambda \approx 1$ cm) generated inside a microwave oven fail to travel through the small holes in the metal door plate. Waves of visible light, which are thousands of times shorter in wavelength than microwaves, pass unhindered through the holes, so you can view what is happening inside.

**Law of Reflection**
The angle of incidence equals the angle of reflection.

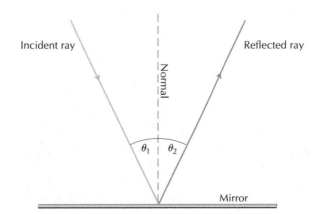

**FIGURE 8.8**   According to the law of reflection, the angle of incidence ($\theta_1$) of a ray of light equals the angle of reflection ($\theta_2$). By convention, both angles are measured with respect to the *normal* to the surface. Here, *normal* means "perpendicular." The incident ray, the normal, and the reflected ray always lie in the same plane. How would a ray incident at 0° reflect?

**PHOTO 8.2** The reflection of this city skyline is somewhat diffused by the wind-rippled surface of the water.

Optical mirrors are like reflecting screens, too. The metal coating applied to glass to form a mirror need be only as thick as several layers of atoms. The lattice structure of the metal atoms in the reflecting coating is opaque to visible light but allows the much shorter waves of X-rays and gamma rays through.

The manner in which waves reflect can be broadly categorized as *specular* or *diffuse,* according to the smoothness of the reflecting surface (see Figure 8.9). A **specular** (or ordinary) **reflection** occurs for surfaces that are smooth when compared to the wavelength of the waves being reflected. We describe the microscopically smooth surfaces of mirrors and window panes as being shiny, because they reflect light in an organized manner. Rough surfaces produce **diffuse reflection** in which the reflected waves or rays scatter haphazardly. White paper exemplifies diffuse reflection. The ink-

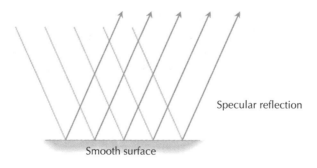

Specular reflection

Smooth surface

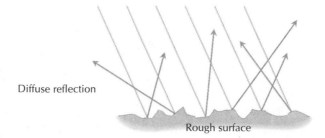

Diffuse reflection

Rough surface

**FIGURE 8.9** Smooth (shiny) surfaces produce specular reflections. Rough surfaces produce diffuse reflections.

**PHOTO 8.3** Specular reflections can occur on a calm water surface. Notice how the virtual image in the water is upside down. The "twist" in the image can be understood with the help of Figure 8.10.

free parts of the page you are reading reflect about as much light as a typical household mirror, but the tiny interwoven fibers that make up the paper constitute a very uneven surface at the microscopic level.

*Images* are associated with specular reflections of light. When you stand in front of a flat mirror, some of the background light incident on various parts of your body diffuses toward the mirror. According to the law of reflection, you must look in specific directions "into" the mirror to see light approaching the mirror from different parts of your body (see Figure 8.10).

**FIGURE 8.10** Your self-image as seen in a flat mirror is virtual because no light originates or collects behind the mirror. The virtual image seen in the mirror is twisted: The image of your right hand in the mirror appears to be on the right. This hand would correspond to the left hand of a real person facing you.

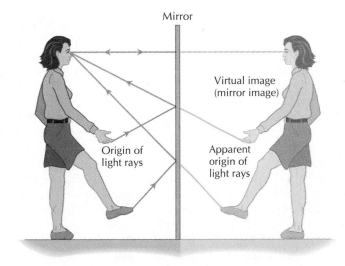

An image produced by the sight lines of light entering your eyes appears in front of you, seemingly behind the mirror. This type of image is called a **virtual image,** because no light rays come from the place where the image appears to be. Virtual images can also be seen in convex mirrors (mirrors that curve outward), such as those commonly used for automotive rearview mirrors. Virtual images appearing in convex mirrors are smaller than those seen in flat-mirror images because incident light rays tend to diverge as they strike various parts of the convex surface.

Another kind of image, called a **real image,** is characterized by the convergence of light rays upon a point or a plane. A camera lens produces real images at the back of an ordinary camera, which is where the light-recording film is placed. Slide and movie projectors project real images as well. Although lenses frequently do the job of organizing light to produce real images, concave mirrors (mirrors that curve inward) are capable of doing the same thing (see Figure 8.11). When light from a common direction (say, the incident parallel rays of light from a star or other distant point source)

## BOX 8.2
# The Telescope

Since Galileo's time, the telescope has been the quintessential tool of astronomical exploration. Today, astronomers crave the use of ever-larger telescopes for two reasons: first and foremost, for their greater light-gathering ability, and second, for their ability to deliver sharper images (or their *resolution*).

Telescopes come in essentially two types: the *refractor telescope* and the *reflector telescope.* The fundamental workings of each type, as illustrated in Figure 8.12, are quite simple. In the **refractor telescope,** a large lens called the *objective* gathers light from a distant source (or sources) and brings the light to a focus at the *focal plane* of the telescope. For simplicity, rays of light from only a single star passing through the objective are shown in the figure. The objective lens is of a *convex* or *converging* type. Light rays entering and exiting the lens at angles other than along the normal to the surface are refracted inward. The central ray shown in the figure, along the normal, continues straight through. (In practice, to minimize the tendency of light to disperse into various colors when refracting through a single lens, the objective of a high-quality telescope is a compound lens having two or three closely spaced lens elements.) Once past the focal plane, the light diverges and is intercepted by a small con-

vex lens (or compound lens), the *eyepiece,* which refracts the light into a parallel beam small enough to enter the pupil of the human eye.

The magnifying ability of a telescope used visually is purely a matter of convenience. The eyepiece merely magnifies the image on the focal plane. By interchanging one eyepiece for another, the viewer can magnify that image as much or as little as desired. By comparison with a telescope's light-gathering ability and resolving power, magnification is relatively unimportant for astronomical telescopes. Even at very high magnifications, stars still appear as points of light in a telescope because they are so far away.

In the **reflector telescope,** a large, concave *primary mirror* reflects and converges rays of starlight toward a point in front of the mirror. In the *Cassegrain reflector telescope,* the most common of several reflector telescope designs, the converging beam is intercepted by a small, slightly convex *secondary mirror.* Reflected again, the light passes through a hole in the primary mirror and converges at the focal plane behind the primary. An eyepiece may be placed behind the focal plane for viewing purposes. Or as in all telescopes used for research, various instruments may be placed at the focal plane for the purpose of photography, *spectroscopy* (the analysis of light spectra), or *photometry* (the measurement of light intensity). Nearly all the instruments used with modern astronomical telescopes

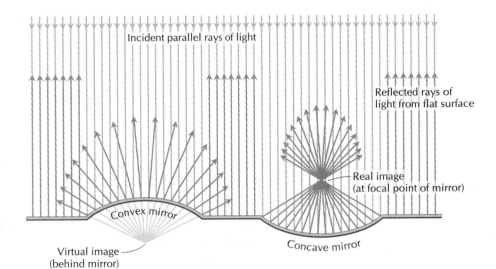

**FIGURE 8.11** A convex mirror diverges incident parallel rays of light. A concave mirror converges incident parallel rays of light to a point. Flat mirrors neither diverge nor converge the incident light.

are electronic. For example, the old-fashioned cameras using photographic film or plates have largely been replaced by supersensitive CCD (*charge-c*oupled *d*evice) electronic cameras.

A telescope's light-gathering ability is proportional to the area (and hence the square of the diameter) of its objective lens or primary mirror. Thus, modest increases in diameter yield big gains in the amount of light collected. The more light collected, the greater is the ability of the telescope to detect faint, faraway objects; and the faster the light of any celestial object, bright or dim, can be analyzed. For a variety of reasons, reflector telescopes are much cheaper to construct than refractor    *(continued)*

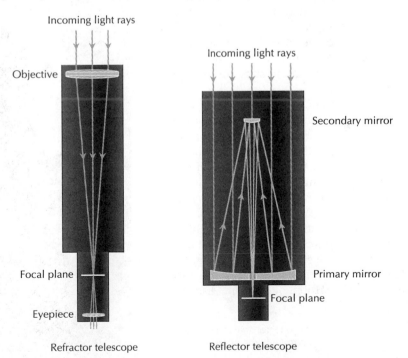

**FIGURE 8.12** Refractor telescopes gather and focus starlight by means of a converging objective lens. Reflector telescopes accomplish the same thing by means of a concave primary mirror.

is reflected by a concave mirror, it converges to a *focal point,* where a real, pointlike image of the star lies. If light is incident on the concave mirror from different directions (say, from the moon, which is an "extended" source of light) the rays from each particular direction end up converging to different points in front of the mirror. Those points lie on a plane: the *focal plane* of the mirror. That is where the real image of the moon lies.

Most large telescopes use a concave mirror to gather and focus light coming from celestial sources (see Box 8.2). The concave metal surface of a common satellite dish antenna functions in much the same way; it collects radio energy beamed toward Earth from satellites hovering above the equator in geostationary orbits. Similarly, our outer ear consists of relatively rigid cartilage having a cuplike form. The cup shape reflects and captures sound waves efficiently.

The paths traveled by the energy carried by waves are reversible. Thus, for example, light from a small source can be placed at the focal point of a concave mirror and projected outward in a narrow beam. This is how a flashlight works. The concave reflector behind the lightbulb gathers light rays and projects them into a parallel beam. Reverse the rays of light hitting the concave mirror in Figure 8.11 to see how this works. Similarly, any dish antenna can be used to transmit highly directional radio energy as well as receive radio energy originating from a single direction.

*(Box 8.2 continued)*    telescopes in the larger sizes. This is why reflector telescopes dominate at major observatories. The current record holder is the 10-m (400-in.) Keck reflector telescope located on Mauna Kea near the top of Hawaii's Big Island. The telescope's unusual segmented primary mirror consists of a tilelike array of 36 separate concave mirrors, each of which reflects incoming starlight to a common focal plane.

A telescope's theoretical resolving power, which is limited by the diffraction of light, improves in proportion to the diameter of the objective or primary mirror. In practice, large Earth-based telescopes fail to attain their maximum resolution when directed toward the sky because light from space must first pass through Earth's turbulent atmosphere. The Hubble space telescope orbiting Earth is the first large instrument to operate in an environment free of the optical degradation engendered by the atmosphere. Despite its modest 2.5-m diameter, it regularly produces far sharper images than any Earth-based telescope. The Hubble telescope has been

**PHOTO 8.4**   This 1-m-diameter reflector telescope at Mount Laguna Observatory, California, has a photometer attached at its Cassegrain focus.

# Refraction

**Refraction** is the bending of a wave by virtue of a change in its speed. The process can be visualized in the ocean swells approaching a beach (Figure 8.13). As these swells move from deep water to the shallows near the shore, they begin to drag against the sea bottom and slow down. (At some point near the shore, a wave breaks as its crest slides forward much faster than the water underneath can move.) When a wave crest is moving "straight in" toward the shore, all parts of it slow down simultaneously, and there is no change in direction. However, when a wave crest approaches the shore at some oblique angle (see the bottom of Figure 8.13), some parts of the wave reach the shallows earlier and begin dragging sooner. The direction of the wave changes so as to approach the shoreline in a more straight-in fashion. In the language of optics, we say that these waves bend toward the normal of the shoreline. This is the main reason that, when standing on a beach, you never see waves moving parallel or nearly parallel to the shore.

Light can behave in a similar way because it may change speed when crossing the boundary between one transparent substance and another. In glass, for example, the interaction of photons with atoms in the glass causes a series of delays that add up to a substantial drop in the effective speed of the light. (When moving between atoms through empty space, however,

designed to detect a broad range of wavelengths, including ultraviolet wavelengths inaccessible to telescopes on the ground.

In recent decades, astronomy has benefited greatly from new information gathered in parts of the electromagnetic spectrum other than the visible. Alternative "telescopes," such as radio telescopes, have been developed to collect nonvisible wavelengths.

The concave surface of a radio telescope dish antenna gathers and focuses radio waves in the same way that the primary mirror of a reflector telescope manipulates light. The dishes are made as large as costs will allow to maximize both resolution and *sensitivity* (the analog of light-gathering ability). Further improvements, especially in resolution, can be made by sending the signals of several radio telescopes to a central computer, which correlates the signal information and generates a detailed radio "image," or map. These maps are like photographs in the sense that they show the intensity of radio energy coming from various directions in the sky. Working at maximum resolution, the Very Large Array (VLA), the most powerful compound radio telescope of this type, can produce radio wave images that are about as sharp as the visible-light images routinely obtained by large, Earth-based optical telescopes. The VLA, however, may require hours of observing time and computation to accomplish this sharpness.

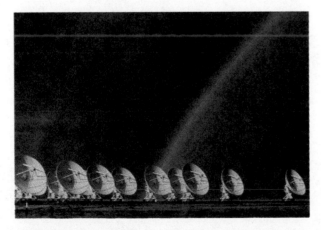

**PHOTO 8.5** The Very Large Array (VLA) radio telescope in New Mexico consists of 27 movable dish antennas. When the antennas are spread farther apart, resolution is improved. When the antennas are moved closer together, the area of sky viewed by the telescope is broadened.

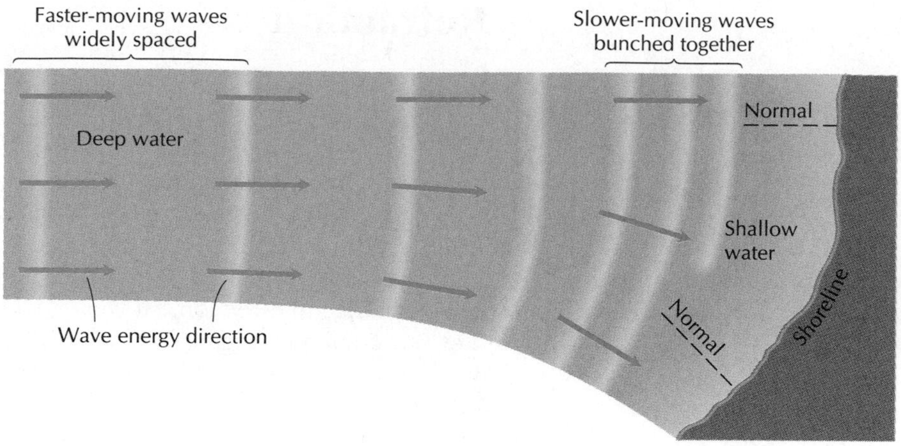

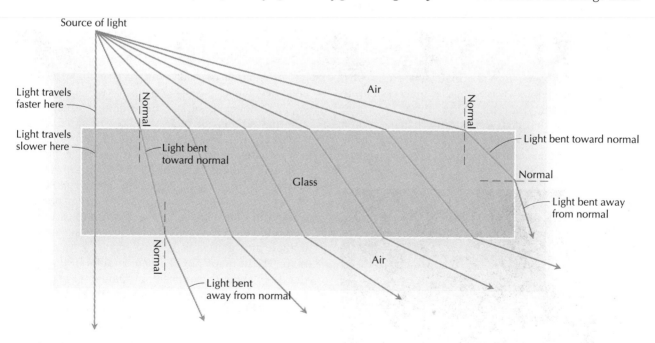

**FIGURE 8.13**   Waves approaching the shore parallel to the normal of the shoreline (top) slow down but do not change direction. Waves approaching at some angle relative to the normal of the shoreline (bottom) slow and also tend to bend in the direction of the normal.

photons travel at exactly $c$, the speed of light through a vacuum.) In air, the slowdown of light is very slight. If a ray of light passes from air to glass and back to air again, it will experience changes in speed—and in direction, too, as long as the ray strikes a surface obliquely (see Figure 8.14).

Refraction takes place in the human eye when the lens in the front of the eyeball (Figure 8.15) gathers light rays from the outside and brings them

**FIGURE 8.14**   Light bends toward the normal when entering a medium in which its average speed decreases and away from the normal when it passes into medium in which its average speed increases. [In air, light travels very slightly slower than $c$; in glass, light moves at approximately $(2/3)c$.] The greater the angle of incidence, the more light is bent or refracted. After passing through the top and bottom parallel faces of this glass block, the light rays assume their original direction. Notice the light ray on the far right; the corner of the block behaves as a prism, bending the light so that it exits in a very different direction. White light disperses (separates) into its component colors whenever it undergoes refraction, but this relatively small effect is ignored in this diagram. Also ignored here is the fact that some incident light can *reflect* off the surfaces shown. For this reason, you can see faint reflections in a window and on the smooth surface of a pond.

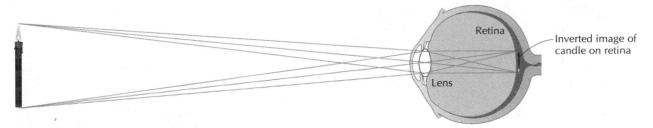

**FIGURE 8.15** In this simplified diagram of the human eye, the convex-shaped lens in front gathers light from afar and brings it to a focus on the retina. Notice how the image of the candle projected on the retina is upside down, or *inverted*. When processed through the neurons of the brain, the mental image of the candle appears upright.

**PHOTO 8.6** A pencil appears bent when partially immersed in a glass of water because light rays coming from the submerged parts of the pencil refract (away from the normal to the surface) when they pass into the air.

to a focus on the retina on the back side of the eyeball. Light-receptive cells on the retina's surface send tiny electrical signals through a network of nerves to the brain, and we "see" images as a result.

Water, clear plastic, and other transparent solid and liquid substances are all capable of refracting rays of light to a substantial degree. Air's ability to slow down and refract light is slight, but quite noticeable under a variety of circumstances. The sun's distorted appearance at sunset is a consequence of *atmospheric refraction*: the dragging and bending downward of light waves as they enter denser and denser parts of the atmosphere. When the sun appears to be just touching the horizon, it already has set; its light has traveled a curved path to reach your eyes. Light coming from the sun's bottom edge passes through lower and thicker parts of the atmosphere and is bent more than the light coming from the sun's top edge. The result is a sun image that appears squashed. Refraction in the air is also responsible for mirages, which are a consequence of light traveling along an upward-bending, rather than a downward-bending, path (Figure 8.16).

Sound waves refract through the atmosphere when they encounter air of a different temperature. Recall that sound travels faster in warmer air and slower in cooler air. Normally, cooler air overlies warmer air, so sound projected horizontally bends upward, as light does in the mirage effect. Thus, very loud sounds, such as thunder, are usually heard no more than a few miles away. When warm air overlies colder air (the condition known as a temperature inversion), sound waves refract downward instead and are audible over a much greater distance. Voices may be heard a mile or more across a cold lake if a layer of calm, chilled air rests on the water sur-

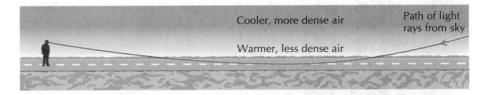

**FIGURE 8.16** Mirages commonly appear on long, straight roadways that bake in the sun. Light travels faster through the warmer air lying just above the hot pavement and slower through the cooler air higher up. The observer at left sees light from the sky coming in from a distant part of the road surface. The illusion is that there is a puddle of water in the roadway reflecting the sky light.

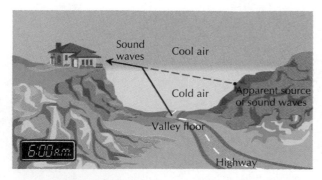

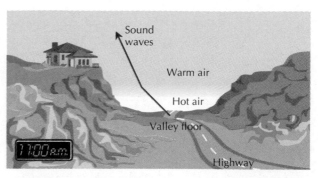

**FIGURE 8.17**   The author's home lies on the rim of a valley susceptible to temperature inversions. On calm nights, cold air from the surrounding mountains sinks into the valley while warmer air lies above. By early morning, the sounds of traffic on a busy highway a mile away on the valley floor boom in loud and clear. These sounds seem to come from the far side of the valley. By late morning, the hot Southern California sun has created a pool of hot air in the valley and sounds of the highway are barely audible.

face. Sound refraction effects can also be quite startling near valleys subject to nighttime cold-air drainage (see Figure 8.17).

Light may behave in a fascinating way inside glass surrounded by air or inside any other medium surrounded by a less dense medium. As illustrated in Figure 8.18, light can undergo *total reflection* (or *total internal reflection*) when it strikes the inner surface of the glass at a large angle to the normal. When shaped into a long, thin cylinder with smooth walls, transparent materials can act like "light pipes," or **optical fibers,** to transmit light from place to place. Once light is beamed into either blunt end of an optical fiber, it ricochets along the inside walls until it exits the far end. Today, more and more information is transmitted through fiber optic cable systems that contain bundles of optical fibers. Each tiny fiber carries light from a laser that can send billions of pulses per second down the fiber at nearly the speed of light. Encoded in the "off" and "on" states of the laser beam are the basic units of information used by every computer and digital device: zeros and ones.

In a similar though not identical manner, sound can be channeled over great distances in the deep ocean. The speed of sound in seawater increases with both temperature and pressure. The tropical and temperate oceans of the world contain a *thermocline* (a boundary of sudden temperature change), at a depth of around 1 km, that separates the upper, warm layer of water from the nearly freezing water underneath. Sound travels considerably faster in the warm water above the thermocline because of the large temperature difference. Below the thermocline, sound travels faster with increasing depth solely because the pressure increases. The net result is a minimum-velocity "channeling" layer within the thermocline. When moving along the channeling layer, upward-slanting sound waves are refracted downward through the warmer water above, and downward-slanting sound waves are refracted upward through the highly pressurized water below. Thus, any sound generated in the channeling layer propagates outward with extraordinary efficiency. During field tests, the foghornlike sounds generated by underwater loudspeakers suspended in the channeling layer have been detected by underwater microphones halfway around the world. Whales use the channeling layer to communicate with each other in the

**PHOTO 8.7**   Light emerging from the ends of transparent optical fibers.

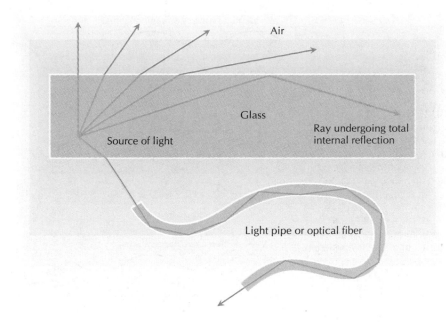

**FIGURE 8.18** When light strikes the inside of a glass surface at a large enough angle relative to the normal, it cannot refract into the air outside. Instead, it undergoes total reflection inside the glass. Light pipes and optical fibers made of flexible, transparent materials exploit this property. After entering a fiber at a blunt end, a ray of light reflects inside many times before it exits at another blunt end. Very little energy is lost during each internal reflection.

open ocean. The low-frequency sounds a whale emits may be audible to other whales hundreds of kilometers away.

## Interference

Beachgoers and surfers know that waves often come in "sets," two or three large waves followed by several smaller waves. This is a common, though quite complex, example of **interference,** a phenomenon unique to waves.

It would be absurd to think that two particles or chunks of matter on a collision course could pass through one another; yet waves of any size can do exactly that. When two or more waves pass a given point in space at a given instant of time, they *interfere,* or combine, to produce a height (intensity) at that point that is the sum of the heights of the individual waves passing through the point.

Figure 8.19(a) illustrates the situation called **constructive interference** between two waves. Crest meets crest and trough meets trough, and the two waves reinforce each other throughout. **Destructive interference** is illustrated for the same waves in Figure 8.19(b). Crest meets trough, trough meets crest, and the result is a cancellation of wave energy wherever the two waves are combining. When two parallel waves of different character interfere, as in Figure 8.19(c), the result is a complex wave that defies simple description.

Simple *interference patterns,* consisting of pathways of constructive interference and destructive interference, occur whenever waves of the same frequency originating from two point sources interfere (see Figure 8.20). When projected along a straight line more or less parallel to the outward-moving waves, the interference pattern forms a series of *interference fringes*: alternating spots of wave reinforcement and wave cancellation.

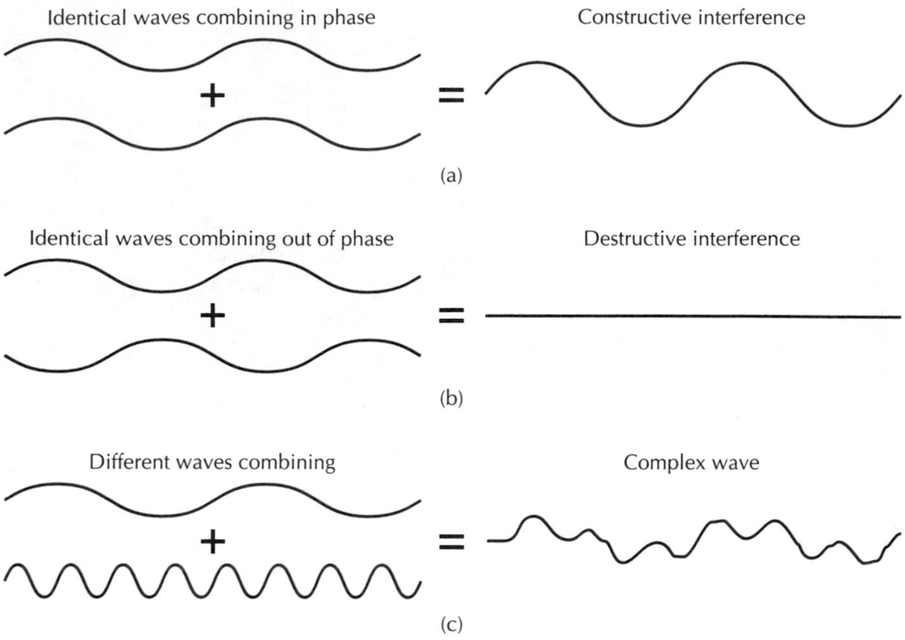

Identical waves combining in phase

Constructive interference

+ =

(a)

**FIGURE 8.19** (a) If two identical, parallel waves combine with one another in phase (in step with each other), the result is a single wave of the same wavelength and speed, but larger amplitude. (b) If the same two waves combine out of phase, the result is a cancellation of wave amplitude. (c) When parallel waves of different wavelength and amplitude combine, they form an interference pattern (and a wave) that is complex. This interference pattern is neither perfectly constructive nor destructive.

Identical waves combining out of phase

Destructive interference

+ =

(b)

Different waves combining

Complex wave

+ =

(c)

When interference fringes were first detected some two centuries ago in light projected from two closely spaced sources, the evidence was taken to be conclusive proof that light travels in the form of waves. Today, the same type of experiment can be easily performed with a laser beam. When the beam is shined through two closely spaced pinholes or narrow slits, each hole or slit acts as a separate source of waves of the same frequency. (The waves diverge, or diffract, outward to some degree from each hole or slit. Diffraction will be discussed shortly.) When projected across a room

**FIGURE 8.20** In this two-source interference pattern, wave crest meets wave crest and wave trough meets wave trough along the line labeled "constructive interference." Wave crests and troughs cancel along the line labeled "destructive interference." At some distance from the two sources, the interfering waves give rise to interference fringes consisting of alternating areas of large and small wave amplitude.

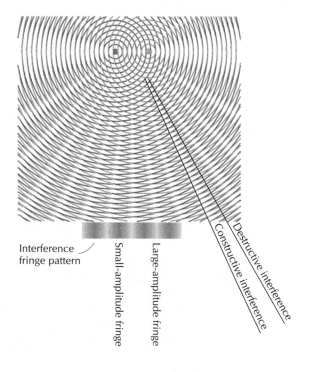

Interference fringe pattern

Small-amplitude fringe

Large-amplitude fringe

Destructive interference

Constructive interference

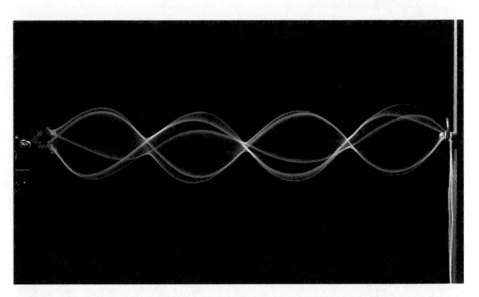

**PHOTO 8.8**  Transverse standing waves can be produced in a string fixed at one end and driven by a constant-frequency mechanical oscillator at the other end. The distance between adjacent nodes (points of minimum vibration) or antinodes (points of maximum vibration) is equivalent to one-half the wavelength of the waves passing back and forth through the string. If the string is pulled tighter, the wave speed increases, the wavelength increases, and the number of standing wave "loops" that fit across the length of the string decreases. Decreasing the string tension allows more loops to fit across the string.

onto a screen, light from the two beams combines to form a pattern of regularly spaced light and dark interference fringes.

Another form of interference pattern is seen in **standing waves,** waves that apparently stand still. Standing wave patterns develop when waves reflect back and forth between two fixed points and interfere constructively and destructively at fixed points in between. A variety of standing wave patterns can be generated in vibrating strings (see Photo 8.8). One of the

**PHOTO 8.9**  The Tacoma Narrows Bridge at the moment of initial collapse. The stalled car belonged to a reporter who crawled off the bridge to safety. Notice how the deck of the bridge at right is twisted along the centerline.

more spectacular and unusual examples of a standing wave was seen in the original Tacoma Narrows suspension bridge in Washington State. Not long after its completion in 1940, high winds pushing sideways across the deck caused it to start oscillating violently in a tortional (twisting) mode. "Galloping Gertie" eventually collapsed in the continuing gale. Today, the rebuilt Tacoma Narrows Bridge and others of its kind are fitted with stiffening beams to prevent such disasters.

Standing waves on a smaller scale are everywhere: in the vibrating strings of a guitar, in the heaving membrane of a drum, and in the vibrating air columns trapped in flutes, organ pipes, or open bottles "singing" in the wind. Without standing waves, it would be difficult to make any kind of music (see Box 8.3). In addition, lasers employ standing waves of light, as we shall learn in Chapter 11.

## BOX 8.3
# The Sounds of Music

Classical music enthusiasts and hard rock music fans might argue forever about the distinction between music and "noise," but physicists can at least point to an objective standard of sound refinement. The pinnacle of purest sound, to a physicist, is the sound made by a tuning fork.

After being struck with a rubber hammer, the two tines of a tuning fork oscillate back and forth in simple harmonic motion. When moving against the air, each tine triggers a series of longitudinal waves of steadily increasing and then steadily decreasing pressure. The shorter or the stiffer the tines are, the higher is the frequency of vibration and the higher is the *pitch* of the sound produced. If sound waves from a tuning fork are picked up by microphone and the resulting electrical signal is fed into an oscilloscope (Photo 8.10), the waveform displayed on the screen is a *sine wave* resembling an ideal water wave. A sound wave represented by a sine wave corresponds to a *pure tone*. To the ear, it sounds rather flat and uninteresting.

The "quality" of sound (what musicians call *timbre*) improves when different frequencies, related in some way to each other, combine to form more complex sound waves. Most musical instruments produce sounds that would be described as rich and distinctive, because their vibrating elements oscillate in a complex but organized way.

When mechanical energy is added to the taut string of a guitar or violin by plucking it or stroking it with a bow, most of the energy is converted into a dominant standing wave. When a string oscillates in its simplest mode of vibration, a *node* (a point of minimum vibration) lies at each end of the string, and an *antinode* (a point of maximum vibration) appears in the middle. The wavelength of the wave bouncing back and forth is fixed by the length of the string; it is twice the length of the string. The frequency of the standing wave (and the frequency of the sound waves produced by the displacement of the string) depends on the velocity of the wave moving back and forth along the string, and that in turn depends on the string's tension, thickness, and other properties. The frequency of this dominant

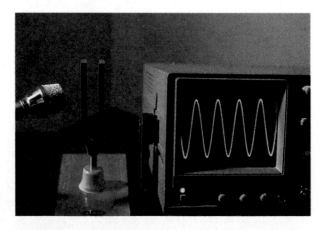

**PHOTO 8.10**   When converted to an electrical signal and displayed on an oscilloscope screen, the pure tone of a tuning fork appears as a sine wave.

# Diffraction

**Diffraction** refers to the bending of waves around an edge or an obstacle, or the spreading of waves as they pass through some kind of gap or opening. Unlike refracting waves, which bend when passing into a different medium, diffracting waves stay in the same medium as they change direction. Figure 8.22 shows how ocean waves are diffracted around an obstacle in their path. Only a fraction of the wave energy bends into the "shadow zone" behind the obstacle.

The longer the wavelength of a wave in comparison to the size of an obstacle, the more easily it is diffracted. Diffraction effects are relatively minor for short-wavelength waves striking large obstacles or passing through large openings.

---

standing wave is called the *fundamental frequency.* It may be increased by tightening the string, shortening the string, or both.

Superimposed on any dominant wave in the string are other waves, called *harmonics* or *overtones,* with frequencies that are multiples of the fundamental frequency (see Figure 8.21). These harmonics are generally small in amplitude, but their

contribution to the displacement of the string affects the timbre of the sound that is generated.

The emphasis of one harmonic over another in a vibrating string depends partly on the shape and composition of the instrument, particularly the "sounding" part—the belly of a violin or guitar, the sounding board of a piano. The sounding part acts to amplify certain frequencies through    *(continued)*

**FIGURE 8.21**   A vibrating string fixed at both ends supports a variety of superimposed standing waves. The dominant wave has the lowest frequency, the fundamental frequency ($f$). The higher-frequency harmonics are multiples of the fundamental frequency. Note that all the waves shown here, plus waves of even higher frequency (and generally lesser amplitude) not shown here, occur on the string at the same time.

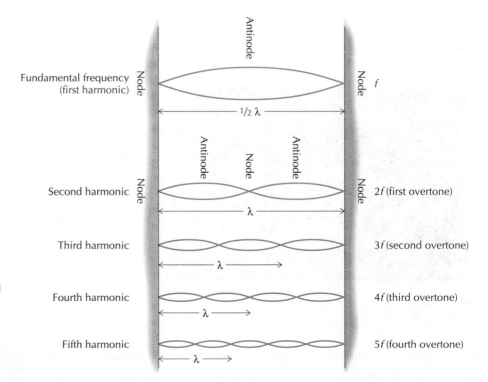

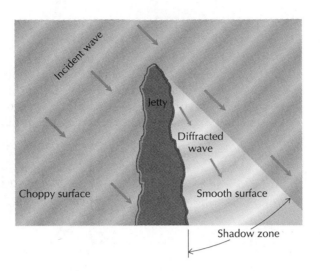

**FIGURE 8.22** Ocean swells striking the edge of an obstacle (such as a jetty) diffract, or bend, around the obstacle to some degree. Longer wavelengths diffract more easily than shorter wavelengths around an obstacle of the same size. If the ocean surface is ruffled with short-wavelength choppy waves from local winds, then those choppy waves are cut off from the "shadow zone" behind the jetty.

*(Box 8.3 continued)* a process called **resonance.** Any rigid body has a natural tendency to vibrate at one or more characteristic frequencies (called *natural frequencies*) when struck. When exposed to vibrations or sound waves having the same frequency (or frequencies), that body will resonate (resound loudly).

In a wind instrument, the vibrations created by air moving across the opening of a tube set up a vibrating air column whose fundamental frequency depends on the length of the tube. The air column is really a standing wave (often with harmonics) consisting of sound of a given frequency bouncing back and forth along the length of the tube. The longer the tube, the lower is the fundamental frequency of the sound produced. A flute or a clarinet has holes along the side that can be plugged to effectively shorten the length of the tube and increase the pitch of the sound. Pipe organs have separate pipes of different lengths, one for each note.

Certain mixtures of sound frequencies are considered pleasing, at least to those steeped in Western musical tradition. The combination of a tone and its first harmonic, which differ by a factor of 2, is usually considered "harmonious" or agreeable. These two frequencies are said to be an *octave* apart, because there are eight notes on the musical scale between them. Other pleasing combinations of two or more tones (*chords*), have simple frequency ratios, such as 1:4, 2:3, and 2:3:4. Frequency ratios that are less simple, such as 8:9 and 15:16, seem less harmonious and more discordant.

In an orchestra or in a band, many instruments and voices may add to produce an enormously complex waveform. That waveform is analyzed by the listener's brain and mentally interpreted as having many layers of harmony, plus tempo and other patterns that somehow seem meaningful. Perhaps it is one's toleration of discordant frequencies, irregular tempos, and other quirks in various forms of music that sets the boundary between what a person considers to be musical sound and simply noise.

**PHOTO 8.11** A large pipe organ, like this one in the Mormon Tabernacle at Salt Lake City, Utah, may have thousands of pipes ranging from 18 cm to 9 m in length. Some of the pipes produce the relatively pure tones of a simple whistle; others with vibrating reeds produce tones of a more complex timbre.

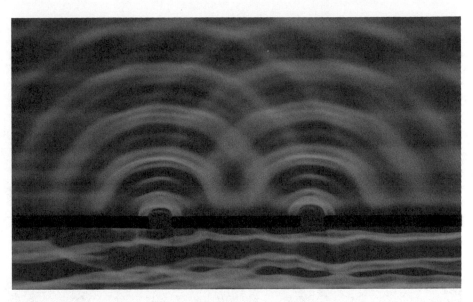

**PHOTO 8.12** When plane waves moving across a ripple tank strike a barrier and pass through two small openings, each opening acts as a separate source of waves of the same frequency. The bending of the waves around the corners of the opening (diffraction) is pronounced, because their wavelength is large compared with the size of the opening. Beyond the openings, the waves cross each other and form a two-source interference pattern.

The diffraction of sound waves allows you to hear loud sounds around the corner of a building, even if there are no other structures nearby to reflect those sounds. Rumbling sounds (having low frequency and long wavelength), such as those of heavy equipment, diffract better than sounds of higher pitch.

Even light, with its tiny wavelengths, can diffract. The effect is quite noticeable when rays of light pass through a pinhole: Most of the light continues straight through, but some spreads to the side. Diffraction is an important limiting factor in the ability of a telescope to resolve images clearly. Generally, the larger the diameter of the telescope lens or mirror, the less pronounced are the effects of diffraction and the greater is the resolving power of the telescope.

**PHOTO 8.13** The apparent "aura" around the silhouette of this hiker is partly the result of sunlight diffracting around strands of his hair and small fibers in the outermost parts of his clothing.

# The Doppler Effect

The **Doppler effect** is an apparent change in wave frequency owing to relative motion between wave source and observer. If an observer moves toward a stationary source of waves, the number of waves passing the observer per second (that is, the frequency) increases. Conversely, an observer moving away from the wave source notices a frequency decrease. If, under different circumstances, a source of waves moves toward a stationary observer; the waves it emits become bunched together in between the source and the observer, and the observer notices that they are of higher frequency. If the source moves away, the waves become spread out; and the stationary observer notices that they are of lower frequency. In all these cases, the observer notices a *Doppler shift* in the frequency of the waves.

## BOX 8.4
## Seismic Waves

Nearly all knowledge of Earth's interior comes from the analysis of the *seismic waves* (earthquake waves) that periodically shake our planet. These waves cannot be detected during the time they are speeding throught the layers of rock deep underground; they make their presence known only when they emerge on Earth's surface, both near and far from their source.

All waves are affected by the medium they travel through or the boundaries they encounter. Seismic waves are no exception; they refract and reflect inside Earth, telling cryptic tales to *seismologists* (geologists who study seismic phenomena). The decoding of seismic waves is an excellent example of how common wave behaviors such as refraction and reflection can be exploited to gain new knowledge.

When rocks move along a geologic fault, seismic waves move outward from the site of the rupture, spreading far and wide in all directions. **Surface waves** propagate along the ground, sometimes wiggling transversely like a snake and sometimes rippling along like water waves. Surface waves are responsible for most of the damage to buildings during major earthquakes.

Seismic waves traveling along the ocean surface are known as **tsunamis,** or tidal waves. They may be produced by a sudden change in the depth of the seafloor during an earthquake or by other mechanisms, such as volcanic eruptions and massive landslides occurring along a coastline. In the open ocean, a powerful tsunami may hardly be noticed as it passes a ship because its amplitude is small (a meter or so) and its wavelength is very large (100 km or more). Tsunamis move extremely fast (over 700 km/h) in deep water; however, they slow down greatly when arriving in the shallow water around an island or a continent. The wave amplitude grows as the faster-moving water from behind piles up, and the wave breaks with tremendous force. Large tsunamis may be 10 m or more in cresting height, and some may wash many kilometers inland by traveling over low coastal terrain. Over 150 tsunamis have struck the Japanese islands in recorded history, including one in 1896 that killed over 27,000 people.

Much of the wave energy generated by earthquakes travels through Earth's interior in the form of seismic **body waves.** When a major earthquake strikes, its concussion is felt by humans as far away as 1000 km and by *seismographs* or *seismometers* (seismic wave recording devices) throughout the world (see Photo 8.14). After a 1960 earthquake in Chile, one of the most powerful in recorded history, seismic stations were kept busy for a month recording the event's fading reverberations as they rattled around inside Earth! Our planet shook like a bell with a natural frequency of about 0.0008 Hz (each complete vibration cycle took about 20 min).

Body waves are segregated into two types: P waves and S waves. **P waves** (*p*rimary or *p*ressure waves) are longitudinal; they propagate like low-frequency sound waves through both the solid and liquid parts of Earth's interior. The alternate pulses

In sound, Doppler shifts correspond to changes in pitch (see Figure 8.23). The high-pitched whine of a low-flying jet plane as it approaches an observer changes to a low rumble as it recedes. A car with its horn blaring seems to change its tune as it whizzes by in an oncoming lane.

The Doppler effect does not apply to observers or sources of sound moving sideways with respect to each other. Doppler shifts only apply when there is some motion "in the line of sight," that is, along a line joining the source and the observer. If the motion is "across the line of sight," or perpendicular to the observer's line of sight at some moment, there is no change of distance and therefore no Doppler shift.

The Doppler effect has myriad uses in modern technology and research. Doppler radar units are used by police officers to measure traffic speeds. The unit works by emitting a burst of microwaves of precisely known frequency toward an approaching or receding vehicle. The reflected

of P wave compressions and rarefactions travel at about 6 km/s near the surface. **S waves** (secondary or shear waves) are transverse; they wiggle back and forth laterally, and their shearing motion prevents

**PHOTO 8.14**  Back-and-forth ground movements are translated into the motion of a pen across a slowly turning drum in an instrument called a *seismograph*. The widest excursion of the pen (maximum amplitude) on the drum during a given quake is taken to be the basis of the now-obsolete Richter scale of earthquake magnitude.

them from propagating through liquid material. The wiggling S waves travel slower than P waves, at about 3.5 km/s near the surface. Both waves travel faster through Earth's deeper layers and therefore are subject to refraction. Seismologists have no trouble distinguishing between the two body wave types on a seismogram of a distant quake: Both are generated simultaneously, but the P waves arrive more quickly.

Some of the earliest studies of seismic body waves showed that they suddenly speed up at a depth of approximately 5 km below the oceans and roughly 35 km below the surface level of the continents. This boundary is taken to be the division between Earth's crust and the mantle below, which consists of denser rocks.

The detailed study of the behavior of P and S waves, as they have been recorded for decades by a worldwide network of seismic stations, has been the key that continues to unlock the secrets of Earth's deep structure. Some clues were apparent very early with the discovery of certain "shadow zones" of body waves on parts of Earth opposed to the source of the waves (see Figure 8.24). The large S wave shadow zone is due to S wave absorption in a core region that is presumably liquid. The smaller P wave shadow (a circular band on Earth's surface) is due to P waves being separated by refraction at the core-mantle boundary.

By analyzing the speeds and directions of the waves recorded, geoscientists have come to the following conclusions: Earth's 2900-km-thick mantle is underlain by a much denser core   *(continued)*

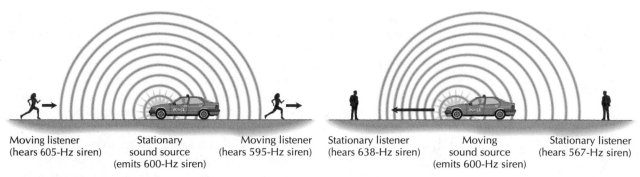

Moving listener
(hears 605-Hz siren)

Stationary
sound source
(emits 600-Hz siren)

Moving listener
(hears 595-Hz siren)

Stationary listener
(hears 638-Hz siren)

Moving
sound source
(emits 600-Hz siren)

Stationary listener
(hears 567-Hz siren)

**FIGURE 8.23** Whenever the distance between a listener and a sound source is decreasing, the pitch of the sound waves heard by the listener is higher. This situation occurs for both listeners on the left side of each diagram. Whenever the distance is increasing, the pitch heard by the listener is lower. This occurs for both listeners on the right side of each diagram. For the woman moving at running pace, it is difficult to detect the slight change in pitch as she passes the police car. But for the observer on the right in the right diagram, there is a noticeable drop in the pitch of the siren as the car moves by at 45 mi/h.

burst is intercepted by the same unit. Electronic circuitry measures the Doppler shift in the frequency of the returning waves and computes the speed. Using more sophisticated Doppler radar equipment, meteorologists can follow, in real time, the precise motions of storm systems. In medicine, Doppler-shifted, high-frequency sound waves are used to monitor the flow of blood through arteries. The sound waves bounce off red blood cells that move no faster than a few centimeters per second.

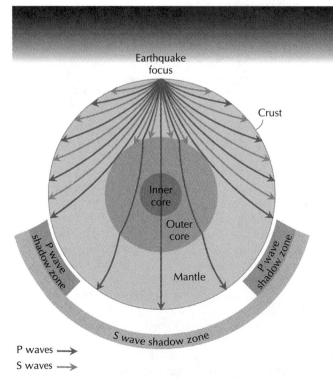

Earthquake focus

Crust

Inner core

Outer core

Mantle

P wave shadow zone

P wave shadow zone

S wave shadow zone

P waves ⟶
S waves ⟶

**FIGURE 8.24** The analysis of body waves generated by large earthquakes has yielded several important conclusions about Earth's interior structure. The waves of any single earthquake can be analyzed by hundreds of seismic stations around the world. Both P waves and S waves speed up at greater depths in the mantle owing to increasing density. Thus they curve (refract) upward within the mantle. S waves damp out quickly in the outer core because it is liquid. This leaves a broad S wave shadow zone (where no direct S waves are received) on parts of Earth opposed to the source of the waves. P waves, which can travel through fluid as well as rigid media, propagate freely in the inner core; they refract sharply at the core-mantle boundary if they are not incident along the normal. P wave shadow zones lie in areas bounded by the arriving waves that pass just above the core-mantle boundary and those that barely nick the boundary and refract through the outer core. P waves passing straight through Earth's center are unbent, but they speed up when passing through the inner part of the core. For this reason, the inner core is believed to be solid, or at least more rigid than the outer core. In addition to the wave paths shown here, there are many weaker, reflected body waves whose times of arrival and directions can be analyzed with sensitive arrays of seismometers.

**PHOTO 8.15**   The Doppler effect is visually apparent in these water waves, produced in a ripple tank by a moving, vibrating source.

Using the Doppler shifts evident in the light (and other electromagnetic radiations) coming from stars and patches of diffuse matter in space, astronomers have mapped out internal motions within the Milky Way galaxy and other galaxies. They have also measured the relative motions of galaxies themselves. One of the most startling revelations in all of science—the expansion of the universe—has come from studies of the Doppler-shifted light from distant galaxies. This discovery will be discussed in detail in Chapter 17.

*(Box 8.4 continued)*   region having a radius of 3400 km. The core's volume is less than a fifth of Earth's volume, but it accounts for about a third of Earth's mass. At an average density of about 11 g/cm$^3$, the core must be composed of compressed metal, primarily iron, which is abundant among the metals found in the Solar System as a whole. The large density of the core makes sense when one compares the average density of Earth (5.5 g/cm$^3$) with the average density of its outer layers (crust, about 2.8 g/cm$^3$; mantle, about 4.5 g/cm$^3$).

The core is in a more or less liquid (although very dense and compressed) state on the outside—the outer core—but has a solid or semisolid inner part. At Earth's very center, the iron (and smaller amounts of other metals) are compressed to a density of about 13 g/cm$^3$. The central temperature, perhaps 5000 K, is far above the melting point of iron at low pressure, but the enormous pressure near the center (3 million bars or more) squeezes the metal atoms together into a rigid arrangement.

Today, multiple arrays of seismometers are deployed in various parts of the world. One array in Montana covers 200 km and contains over 500 individual instruments. Similar in their basic operation to the VLA radio telescope (Box 8.2), these seismic arrays are able to resolve the direction and timing of seismic waves to a far greater degree than single-instrument seismic installations. They are also sensitive enough to pick up and analyze body waves *reflected* from various internal boundaries. This data, for example, has allowed geoscientists to map irregularities in the core-mantle boundary, which is not smooth but quite turbulent. These studies may lead to a better understanding of the forces that move tectonic plates and the episodes of massive volcanism Earth has experienced many times in its geologic past.

# CHAPTER 8
# Summary

Waves are periodic disturbances traveling through either a physical medium or empty space. All waves can be characterized by their velocity, wavelength, frequency, and amplitude.

Mechanical waves require a medium for travel. They include sound waves, water waves, waves in vibrating strings, and seismic waves. Mechanical waves can be longitudinal, transverse, or a combination of the two.

Electromagnetic waves oscillate transversely. They pass freely through empty space and through certain kinds of media. Electromagnetic waves (and all other waves) emanating from a point source diminish in intensity in proportion to the square of the distance from the source.

The electromagnetic spectrum of frequencies or wavelengths includes the major categories of radio waves, infrared radiation, visible light, ultraviolet radiation, X-rays, and gamma rays. Listed in that order, these radiations have higher frequencies and shorter wavelengths.

For waves that reflect, the angle of incidence is equal to the angle of reflection. Most large telescopes employ a concave surface to reflect light (or other electromagnetic waves) to a focus.

Waves may refract, or bend, when they change speed. Refraction often happens when waves cross the boundary between one medium and another. Some telescopes and many other optical devices (including the eye) employ convex lenses to gather and focus light. Waves can also change direction by diffracting around a corner or obstacle.

Waves can pass through each other and interfere, or combine. The combined wave is the sum of all the waves it is composed of. A standing wave may be produced when a wave reflects back and forth and interferes with itself. Standing waves in strings and other media may contain harmonic frequencies. The interpretation and appreciation of musical sound depends in part on combinations of various sound frequencies and their harmonics. Waves change in their apparent frequency (Doppler effect) when there is any relative motion in the line of sight between the source of the waves and the observer of the waves.

Earthquakes trigger seismic waves, which are propagated through Earth and along its surface in a variety of ways. The detailed analysis of seismic body waves has revealed much about Earth's interior structure.

# CHAPTER 8
# Questions

## Multiple Choice

1. All waves transport
   a) electric charge
   b) mass
   c) sound
   d) energy

2. Transverse mechanical waves propagate only through
   a) crystalline solids
   b) solid and semisolid materials
   c) liquids
   d) liquids and gases

3. In transverse waves, particles of the medium move
   a) in circles
   b) in elliptical paths
   c) parallel to the wave velocity
   d) perpendicular to the wave velocity

4. Water waves are
   a) longitudinal
   b) transverse
   c) longitudinal and transverse
   d) electromagnetic

5. An audible sound wave could have a frequency of
   a) 0.1 Hz

b) 100 Hz
c) 1 MHz
d) 100 MHz

6. How much louder is the sound of an amplified rock band (115 dB) than the sound of normal conversation (45 dB)?
   a) 70 times
   b) 128 times
   c) 10,000 times
   d) 10,000,000 times

7. If thunder follows lightning by 12 s, how far away did the lightning strike?
   a) 2 km
   b) 3 km
   c) 4 km
   d) 8 km

8. Sound is
   a) a longitudinal wave
   b) a transverse wave
   c) a partly longitudinal and partly transverse wave
   d) an electromagnetic wave

9. Sound waves travel fastest in
   a) a vacuum
   b) gases
   c) liquids
   d) solids

10. The higher a sound wave's frequency, the
    a) faster its speed
    b) higher its pitch
    c) louder it sounds
    d) longer its wavelength

11. Higher amplitudes are associated with
    a) lower frequency
    b) faster speed
    c) slower speed
    d) more energy

12. Radio waves travel
    a) much slower than light
    b) slightly slower than light
    c) the same speed as light
    d) faster than light

13. Visible light has wavelengths closest to
    a) 1 m
    b) 1 mm
    c) 1 $\mu$m
    d) 1 nm

14. The property of light waves that leads to the phenomenon of color is
    a) velocity
    b) amplitude
    c) intensity
    d) wavelength

15. Infrared radiation
    a) has a shorter wavelength than red light
    b) moves through Earth's atmosphere without hindrance
    c) has a higher frequency than red light
    d) has a lower frequency than red light

16. Ionizing radiation includes
    a) some ultraviolet
    b) X-rays
    c) gamma rays
    d) all of the above

17. When fast-moving electrons strike a metal plate, the plate emits
    a) protons
    b) atoms
    c) X-rays
    d) radio waves

18. When a wave changes direction because of a change in speed, it is
    a) reflected
    b) refracted
    c) diffracted
    d) absorbed

19. Echoes are due to the
    a) reflection of sound waves
    b) refraction of sound waves
    c) diffraction of sound waves
    d) interference of sound waves

20. When parallel light reflects from a concave mirror,
    a) the rays of light diverge from the mirror
    b) a real image is produced in front of the mirror
    c) a virtual image is produced in front of the mirror
    d) a real image appears in front and a virtual image appears behind the mirror

21. The image formed by one's reflection in a flat mirror is
    a) real
    b) virtual
    c) inverted
    d) magnified

22. Laser light travels long distances within a glass optical fiber by means of
    a) interference
    b) total internal reflection
    c) bending owing to refraction inside the glass
    d) diffraction

23. A wave can change speed and yet not be refracted (bent) only when the incident ray
    a) is incident at 0° to the normal of the boundary surface
    b) is incident at 45° to the normal of the boundary surface
    c) is incident at nearly 90° to the normal of the boundary surface
    d) consists of light rays

24. Bending by diffraction occurs when a wave
    a) changes speed
    b) moves from one medium to another
    c) passes near an obstacle or through a gap
    d) spreads out from a central source

25. When the frequency of external waves corresponds to an object's natural frequency, a response arises which is called
    a) resonance
    b) harmony
    c) diffraction
    d) interference

26. Which frequency is not a harmonic of the fundamental frequency $f$?
    a) $(1/2) f$
    b) $2 f$
    c) $3 f$
    d) $4 f$

27. The Doppler effect always produces a change of
    a) frequency
    b) intensity
    c) amplitude
    d) direction

28. Police radar units operate on the principles behind
    a) reflection and refraction
    b) reflection and diffraction
    c) diffraction and the Doppler effect
    d) reflection and the Doppler effect

29. High-quality astronomical telescopes are optimized for
    a) magnification
    b) magnification and resolution
    c) light-gathering ability and resolution
    d) none of the above

30. Which of the following correctly describes the behavior of seismic body waves?
    a) P waves are absorbed in Earth's core; S waves are not.
    b) Both P waves and S waves are absorbed in Earth's core.
    c) Both P waves and S waves refract through Earth's mantle.
    d) Neither P waves nor S waves will reflect at any boundary inside Earth.

# Questions

1. How are wavelength and frequency related for waves of a particular constant speed?

2. How do frequency and wavelength vary for the different types of waves or radiations in the electromagnetic spectrum?

3. How do transverse and longitudinal waves differ?

4. What does the word *modulation* mean in connection with waves?

5. Which electromagnetic waves pass most easily through Earth's atmosphere?

6. Why is ionizing radiation (such as X-rays or gamma rays) harmful to biological tissue?

7. How do specular reflection and diffuse reflection differ from each other? How are they the same?

8. How does refraction explain how mirages work?

9. To measure the velocity of speeding cars, police officers always aim their radar units toward oncoming or receding cars, never toward the side of a car moving by. Why?

10. What methods have earth scientists used to "see" Earth's interior and thereby map its interior structure?

# Problems

1. Water waves kicked up by the passage of a boat have a speed of 2 m/s. What is their wavelength

if they cause a nearby buoy to bob up and down every 4 s? (Remember to covert the period of the waves to frequency before using Equation 8.1.)

2. An anchored ship is rocked by waves with crests 50 m apart moving at 10 m/s. What are the frequency and period of these waves?

3. The sun and the full moon have approximate apparent magnitudes of −27 and −12, respectively. How many times brighter does the sun appear to us than the full moon?

4. Two identical stars, A and B, differ by ten magnitudes on the apparent magnitude scale. What is the ratio of their distance from us?

## Questions for Thought

1. Sometimes, a nearly instantaneous flash of lightning is followed by the sound of rolling thunder, which continues for several seconds. How can the flash of lightning take place in a very brief time, yet the sounds originating from it, traveling at nearly constant speed, keep arriving over a much longer time period?

2. Make any observation of any kind of wave—for example, ocean waves viewed from the end of a pier or the small water waves you can produce on a pool's surface. Try to estimate their wavelength (in meters), frequency (in Hertz), and speed (in meters per second). Are the values you estimated for these quantities in keeping with the wave equation (Equation 8.1)?

3. When you drive at night through rain, the pavement ahead of you appears much blacker than it normally does. At the same time, there is extra glare in your eyes whenever an oncoming car approaches. Both of these effects make driving on a busy, undivided road on a wet night a nerve-racking experience. How can these two effects be explained in terms of the specular and diffuse reflections of headlights beamed toward the road surface?

4. Why does it take a long time to heat up dehydrated food in a microwave oven?

5. If a body creating some kind of disturbance travels through a medium as fast as (or faster than) waves in that medium can travel, what sort of wave is produced? This situation occurs when a ship is moving faster in water than the water waves it kicks up can travel, and when an airplane is traveling at a supersonic speed (faster than sound through the air). Use the right half of Figure 8.23 or use Photo 8.15 to imagine what the waves coming from the fast-moving body would look like. How would water waves like these affect a rowboat near the passing ship? How do people on the ground perceive a passing supersonic aircraft?

6. Using Figure 8.10 as a guide, try to determine the minimum height of a wall mirror in which you could see yourself standing straight up at full length. Does your answer depend on your distance from the mirror?

7. Box 8.1 stated that by knowing the apparent magnitude and the luminosity of a star, we can quickly calculate the distance. In practice, we often see stars through a haze of interstellar dust. How would the presence of dust affect the result of the distance calculation? Do you think there is any way to correct for the effect of intervening dust?

8. Much of what we perceive comes to us in the form of waves. What other carriers of information can we detect?

9. Mysterious "booms" shake the Southern California coastal area several times a year. They often occur during clear weather in autumn, when temperature inversions are common over the region. The booms, strong enough to break windows in some cases, often vary in intensity from place to place. The booms are thought to coincide with supersonic military aircraft that maneuver in secret 50–150 km offshore. Develop a plausible scenario explaining why such aircraft might be responsible for booms that can be heard so far away.

## Answers to Multiple-Choice Questions

| | | | | | |
|---|---|---|---|---|---|
| 1. d | 2. b | 3. d | 4. c | 5. b | 6. d |
| 7. c | 8. a | 9. d | 10. b | 11. d | 12. c |
| 13. c | 14. d | 15. d | 16. d | 17. c | 18. b |
| 19. a | 20. b | 21. b | 22. b | 23. a | 24. c |
| 25. a | 26. a | 27. a | 28. d | 29. c | 30. c |

Macroscopic objects derive their properties from the behavior of the atoms of which they are made. The behavior of atoms and of the small particles inside them—especially electrons—provides a key for understanding the origins of electricity, magnetism, chemical reactions, and light. These are the primary topics we will focus on in Part III of this book, which consists of the next four chapters. As in Part II, we will again introduce a small number of basic physical laws, principles, and ideas, and then we will see how they apply in the various physical science disciplines.

Chapter 9 introduces electric charge and Coulomb's law, which governs interactions between charged bodies from the atomic scale on up. All of chemistry has at its roots interactions between tiny charged bodies: electrons patrolling the outer precincts of atoms and whole atoms or molecules attracting or repelling each other.

Chapter 10 investigates streams of moving charge (electricity) and a phenomenon that arises when charges move—magnetism. We will look at some of the many practical and practically indispensable modern devices made possible by the interaction of electricity and magnetism. We will learn that electricity and magnetism, coupled together, also give rise to electromagnetic radiation, including light.

Chapter 11 explores the quantum nature of light and its connection with the small, discrete jumps made by electrons as they orbit an atom. We will delve somewhat deeply into the nature of color and the spectrum, and we will detail the uses of spectral analysis in physics, chemistry, and astronomy.

Last, in Chapter 12, we will explore the energy-rich world of the inner atom, the ultimate source of most of the energy pulsing through the universe today. Despite its checkered history in war and in peace, the harnessing of nuclear energy on Earth entices us with the promise of virtually unlimited amounts of usable energy at our fingertips.

# PART III
# The World of the Atom

# Electric Charge

*Like charges repel. By touching these negatively charged spheres, the girl acquires an overall negative charge so that each strand of her hair pushes away from all other strands. Repulsive electrostatic forces such as these, along with the attractive electrostatic forces that occur between bodies having unlike charges, explain much of the behavior that takes place among atoms and molecules.*

## WHERE WE ARE GOING IN CHAPTER 9

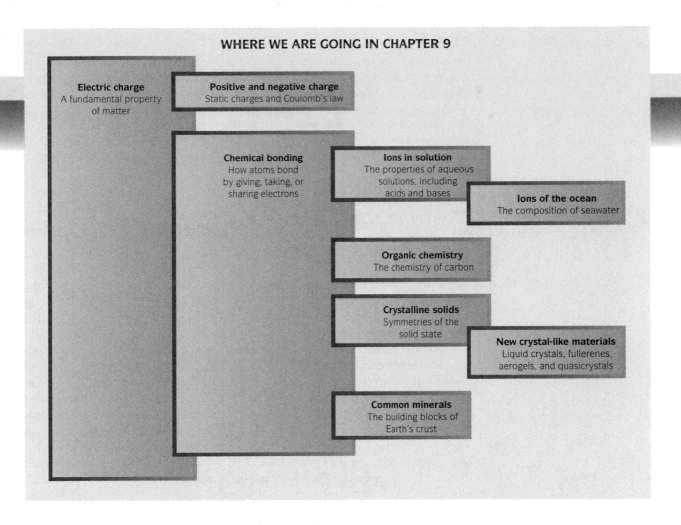

**Electric charge**
A fundamental property of matter

**Positive and negative charge**
Static charges and Coulomb's law

**Chemical bonding**
How atoms bond by giving, taking, or sharing electrons

**Ions in solution**
The properties of aqueous solutions, including acids and bases

**Ions of the ocean**
The composition of seawater

**Organic chemistry**
The chemistry of carbon

**Crystalline solids**
Symmetries of the solid state

**New crystal-like materials**
Liquid crystals, fullerenes, aerogels, and quasicrystals

**Common minerals**
The building blocks of Earth's crust

A strange kind of unease comes over you as every hair on your head stands on end. Out of the darkened sky comes a blinding flash, followed by a gut-shaking, earsplitting boom. Suddenly, it is over. With pounding heart and hyperventilating breath, you notice the air is filled with the acrid scent of ozone mixed with smoke from a scorched and shattered tree nearby.

Although humans have gained much control over the basic form of energy embodied in a lightning bolt, electricity's raw, random, and uninhibited display in nature's arena still has the power to terrify and impress nearly everyone—even if it results in no harm.

Our investigation of electrical matters in the next two chapters starts with a qualitative discussion of what electric charge is. Only two kinds of charge exist: "positive" and "negative" charge. Early scientists discovered that there is always some force, attractive or repulsive, existing between charged bodies. *Static electricity* generally refers to phenomena associated with charged bodies that are not moving relative to each other. Much of fundamental chemistry on the atomic or molecular level can be understood in terms of electrostatic interactions. Thus, chemistry, including the chemical composition of Earth's minerals, is emphasized in this chapter. Movements of electric charge (called *electric currents* or simply *electricity*), along with magnetism, are an inevitable spin-off of the study of electric charge. The static aspects of electricity will be dealt with mostly in this chapter; the dynamic aspects of electricity and magnetism will be dealt with in the following chapter (Chapter 10).

# Positive and Negative Charge

Some basic experimental facts of electricity on a macroscopic scale can be easily demonstrated with simple equipment and simple procedures. If a glass (or plastic) rod is rubbed vigorously with silk, it becomes electrified, or capable of attracting small bits of matter such as straw, hair, or facial tissue. If a rod made of hard rubber (or the resin *amber*) is rubbed with fur, it becomes electrified and will attract small bits of matter as well.

For thousands of years, people have known that friction between certain materials could create electrostatic effects, or static electricity. Because these effects are minor and tend to disappear quickly, especially when the air is humid, little progress was made in understanding them until the time of the Renaissance. By then, experiments indicated that many different materials could be electrified or charged and, furthermore, that charged bodies could be classified into two and only two categories. The first category includes glass and porcelain. The second category includes amber, hard rubber, and silk. The two categories can be distinguished from each other by the following behavior: If a body electrified with either kind of charge is brought close to another body electrified in the same way, the two bodies repel one another. If a body electrified with one kind of charge is brought close to another body electrified with the other kind of charge, the two bodies attract each other (see Figure 9.1). This behavior can be stated as like charges repel each other; unlike charges attract each other.

> Like charges repel each other; unlike charges attract each other.

Benjamin Franklin suggested names for the two kinds of charge: *positive* (that possessed by an electrified glass rod) and *negative* (that possessed by an electrified hard-rubber rod). This was a fortunate and enduring choice of terms, because the two different categories are opposites in the same way that positive and negative numbers are on opposing sides of zero on a number line. We now know that matter in bulk normally has a net charge of zero. The process of rubbing one material against another does not create electric charge; it merely causes a separation of charge. When an initially uncharged piece of silk is rubbed against an initially uncharged glass rod, the rod acquires a certain quantity of positive charge while the silk acquires the same quantity of negative charge. Taken together, the charges balance each other, just as any positive number combined with (added to) its counterpart on the negative side of the number line sums to zero. So after being rubbed against each other, a silk cloth and a glass rod attract each other (Figure 9.2). If allowed to come together, they will gradually discharge and become electrically neutral.

The two kinds of charge identified through macroscopic experiments have at their origin the electrical nature of matter on the atomic scale. As

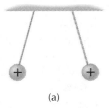
(a)

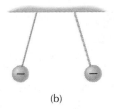
(b)

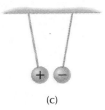
(c)

**FIGURE 9.1** This simple laboratory experiment demonstrates how charged bodies interact with each other. (a) When each of two pith balls or lightweight plastic balls are charged by contact with an object having a large positive charge (say, a glass rod that has been rubbed against silk), both balls become positively charged and repel each other. (b) If the two balls are charged by contact with a negatively charged object (such as a hard-rubber rod that has been rubbed against fur), both become negatively charged and repel each other. (c) When the balls are electrified by contact with oppositely charged objects, they attract each other.

we saw in Chapter 2, electrically positive protons and electrically neutral neutrons inhabit the atomic nucleus, while the relatively lightweight and negatively charged electrons move about the nucleus in, for the most part, certain definite patterns. All three of these so-called elementary particles have a fixed rest mass and electric charge.

$$\text{Proton mass} \quad = 1.673 \times 10^{-27} \, \text{kg} \quad \text{Proton charge} \quad = +1.6 \times 10^{-19} \, \text{C}$$
$$\text{Neutron mass} = 1.675 \times 10^{-27} \, \text{kg} \quad \text{Neutron charge} = 0$$
$$\text{Electron mass} = 9.11 \times 10^{-31} \, \text{kg} \quad \text{Electron charge} = -1.6 \times 10^{-19} \, \text{C}$$

Note that the mass of either the proton or the neutron is more than three powers of ten (about 1800 times) greater than the mass of the electron. The charges carried by the proton and the electron, however, are equal in magnitude, if not in sign. The SI unit of electric charge given here, the *coulomb* (abbreviated C), is a rather large amount of charge. The amount of charge you can put on a party balloon by rubbing it on your clothing is not much more than a nanocoulomb (a billionth of a coulomb).

The electron charge of $-1.6 \times 10^{-19}$ C, and its positive counterpart, the proton charge, are certainly tiny; but they are the basic units of charge in nature. (More fundamental, of course, are the quarks. They have either 1/3 or 2/3 of the electron or proton charge, and they are irretrievably bound up together in trios inside the nuclear particles. Quarks may have existed as separate entities at an early stage of the big bang.) All charged bodies, then, have charge quantities that are multiples of the electron or the proton charge. In the atomic realm, the "unit" charges of the electron and proton are often labeled as $-1$ and $+1$, respectively. Also, in the atomic realm, the masses of both the proton and the neutron are taken to be approximately

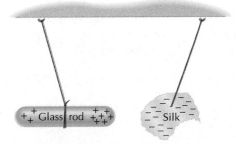

**FIGURE 9.2** When a glass rod and a silk cloth are rubbed together, each acquires the same amount, but opposite type, of charge. If brought together again, the opposite charges neutralize each other. Numerous experiments of this nature show that electric charge is conserved; that is, charges can be separated, but the total amount of charge in the universe remains constant.

one "mass unit," even though the mass of the neutron is very slightly greater. The electron, which is about 1/1800 the mass of either the proton or the neutron, is considered to have approximately zero mass units.

A rough modern explanation of the simple electrostatic experiment depicted in Figure 9.2 is this: Electrons, being of very low mass and forming the outer parts of atoms, have varying degrees of mobility in various materials. Rubbing silk and glass together transfers electrons (typically about $10^{10}$ of them) to the silk from the glass. This gives the silk a net negative charge of about a billionth of a coulomb. The rod, lacking some electrons, is left with a net positive charge of about a billionth of a coulomb. This tiny imbalance of charge is enough to produce a noticeable attractive force between the two bodies when they are brought near each other.

If the two charged bodies are surrounded by air, they will gradually discharge by either losing or gaining electrons to or from ions or molecules in the air. The discharge is sped up by the presence of water vapor in the air. (If your professor fails to successfully demonstrate electrostatic effects through simple experiments in the open air, blame the weather, not your professor!) Electrostatic demonstrations work much better when charged bodies interact in a vacuum.

## Coulomb's Law

Somewhat mysteriously, a charged body can attract things that are *uncharged* (neither positive nor negative), as well as things that are oppositely charged. A comb charged with static electricity will attract dust particles and bits of paper. A balloon rubbed against clothing may stick to a wall or the ceiling of a room. What explains this apparent loophole in the rule that opposites attract and likes repel?

The answer comes only after studying the way that the strength of the electrostatic attraction or repulsion varies with distance. Careful measurements performed with charged bodies under controlled conditions show that the magnitude of the electrostatic force ($F_e$) between two charged bodies depends on the charges ($q_1$ and $q_2$) on the bodies and varies with the inverse square of the distance ($R$) between them. If the magnitude of the charge on *either* body is doubled, then the force between the two bodies doubles. If the magnitude of the charge on *both* bodies is doubled, then the force becomes four times as great. If the charges remain the same but the distance between the two bodies is doubled, the force becomes one-fourth as great; if the distance is tripled, the force becomes one-ninth as great; and so on. This behavior is amazingly reminiscent of Newton's law of gravitation. Interactions between electric charges, in fact, follow an inverse square law behavior, just as the gravitational force obeys an inverse square law when masses are involved. Summarizing the behavior outlined above for charges, we have **Coulomb's law:**

$$F_e = Kq_1q_2/R^2$$

Equation 9.1

It is named in honor of the 18th-century physicist Charles Coulomb, who helped discover it. The constant $K$ (which is analogous to the universal con-

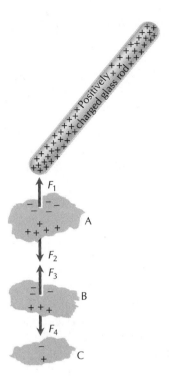

**FIGURE 9.3**  A large positive charge in the glass rod induces a separation of charge in tissue paper A. The force of attraction ($F_1$) on the upper side of A is stronger than the force of repulsion ($F_2$) on the bottom side. The positively charged bottom side of A, when near tissue paper B, induces a separation of charge in B, giving rise to unequal forces $F_3$ and $F_4$ that cause a net attraction. Paper B, in turn, attracts paper C, but with a net force weaker than that between papers A and B. A chain of paper bits stuck together will remain intact until a significant number of electrons flow into the glass. This happens slowly, since there is very little contact, on a microscopic scale, between the fibrous paper and the smooth glass or between any two pieces of paper.

stant of gravitation $G$ in Newton's gravitational law) has an experimentally determined value of $9 \times 10^9$ N·m²/C².

The dependence of electrostatic force on distance in Coulomb's law explains why any charged body (say, a positively charged glass rod) can attract and pick up bits of uncharged matter. Those bits of matter can even form chains, like the bits of tissue paper illustrated in Figure 9.3. When the glass rod is brought near a piece of paper, mobile electrons on the paper's surface are attracted toward it (opposites attract). The side of the paper nearest the glass becomes negatively charged as the electrons migrate to it. The side of the paper farthest from the glass, lacking the electrons that have been pulled toward the glass, becomes positively charged. (When considered as a whole, the piece of paper is *electrically polarized*; opposite electric charges lie at opposite ends of the paper.) The glass rod simultaneously attracts and repels the paper, but attraction wins out, since the distance between areas of unlike charge is less than the distance between areas of like charge.

Coulomb's law not only plugs the apparent loophole in the "likes repel, unlikes attract" law but also is an accurate description of how charges interact with each other on any spatial scale, even very tiny ones. Take, for example, the *electrostatic force* responsible for holding together an ordinary hydrogen ($^1_1$H) atom. The magnitude of the force between the proton and the electron in this atom can be computed by the known charges of the electron and the proton and by the known average distance between the two particles, which is about $5.3 \times 10^{-11}$ m (see Figure 9.4). For comparison, we will also compute the *gravitational force* between these same particles. From Chapter 2, recall that the electromagnetic (electric) and gravitational forces are the only two of the four fundamental forces operative

**FIGURE 9.4**  In this simplified model of the hydrogen atom, a single electron orbits a single proton at the distance given.

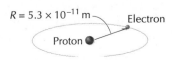

$R = 5.3 \times 10^{-11}$ m — Electron

Proton

outside the realm of the atomic nucleus. The electric force (Equation 9.1) is

$$F_e = Kq_1q_2/R^2$$
$$= \frac{(9 \times 10^9 \text{ N·m}^2/\text{C}^2)(-1.6 \times 10^{-19} \text{ C})(+1.6 \times 10^{-19} \text{ C})}{(5.3 \times 10^{-11} \text{ m})^2}$$
$$= -8.2 \times 10^{-8} \text{ N}$$

(In electrostatics, a minus sign in front of the force magnitude means attraction; a positive value indicates repulsion.) The gravitational force (Equation 5.9) is

$$F_g = Gm_1m_2/R^2$$
$$= \frac{(6.7 \times 10^{-11} \text{ N·m}^2/\text{kg}^2)(1.76 \times 10^{-27} \text{ kg})(9.1 \times 10^{-31} \text{ kg})}{(5.3 \times 10^{-11} \text{ m})^2}$$
$$= 3.8 \times 10^{-47} \text{ N}$$

The magnitude of the electric force (roughly, a ten-millionth of a newton, according to our calculation) keeping the electron in its orbit may seem small, but it is large when we consider how tiny a hydrogen atom is. The gravitational force between the proton and the electron is far smaller—more than $10^{39}$ times smaller, in fact. In the realm of the hydrogen atom, and indeed in all atoms, electric forces vastly overwhelm gravitational forces for two reasons: (1) There is a definite separation of charge within an atom, positive charge residing in the nucleus and negative charge in the orbiting electrons; and (2) the electric force is intrinsically much stronger than the gravitational force.

Electric forces lose their importance when we look at the world on larger and larger scales. Larger chunks of matter tend to be electrically neutral, with an even and well-mixed proportion of protons and electrons. On the larger scales, attractive and repulsive electric forces tend to cancel each other out. Gravitational forces, on the other hand, assume more importance on the larger scales because gravity forces are always attractive and affect all masses, whether or not they have any net charge.

These thoughts can lead us to a summary of fundamental similarities and differences between the laws governing electric and gravitational forces.

- Both the gravitational force and the electric force are subject to inverse square laws. The force in each case gets weaker with distance.
- The constant $G$ in Newton's law of gravitation refers to interactions between masses. It is a small number, indicating that gravity is relatively weak in a general sense. The constant $K$ in Coulomb's law refers to interactions between charges. It is a large number, indicating that electrical interactions are relatively strong in a general sense.
- Through the gravitational force, bodies having mass attract, never repel, other bodies. Through the electric force, bodies having a net charge can either attract or repel other bodies.

# Conductors and Insulators

In some materials—glass, rubber, silk, hair, wood, and paper, to name a few—charge moves about only with difficulty. Nevertheless, as we saw earlier, rubbing some of these materials together can result in a slight transfer of electrons. Materials like these that greatly resist the movement of electrons (the flow of electricity) are classified as electric **insulators.** In other materials, electrons can move easily, thereby carrying charge expeditiously. These materials, including all metals, are classified as electric **conductors.**

The particulars of moving electric charge (*electric current*) will be explored in detail in Chapter 10. For now, you should know that the atoms of insulators hold onto their electrons very tightly and that every insulator is a nonmetal substance. Conductors, on the other hand, consist of at least some charged particles that are free to move from place to place. In metals, the freely moving charged particles are electrons that are not tightly bound to their parent atoms. In liquid solutions that can conduct electricity (*electrolyte solutions*), ions are the carriers of electric charge.

A broader look at the electrical properties of materials reveals two other categories: semiconductors and superconductors. **Semiconductors,** the most familiar being the elements silicon and germanium, normally conduct electricity with an ability somewhere between that of conductors and insulators. When altered slightly in composition and sandwiched together, semiconductors can be fashioned into devices called diodes and transistors. These devices act like gates or valves that can control the flow of electricity. The development of the transistor, in particular, gave birth to the microelectronics industry, which continues to revolutionize the way we live.

When some materials are cooled to temperatures near absolute zero, they behave as perfect conductors, or **superconductors,** having no resistance at all to the flow of electrons. Superconducting materials already have many practical applications, but their operation only under conditions of very low temperature renders them impractical for a wide range of other uses. Chemists and physicists are continuing their quest to find and fabricate materials that behave as superconductors at temperatures closer to room temperature.

# Chemical Bonding

Electric forces provide the glue by which atoms, ions, or molecules bind to each other in solid and liquid substances. When any two atoms (or clusters of atoms) are linked together, the vector sum of all the electric forces, which attract and repel, is equal to zero. The two particles maintain a certain optimum, or equilibrium, distance from each other; and they resist being either pulled apart or forced closer together. This behavior is called *chemical bonding* (see Figure 9.5).

Nature employs several strategies for bonding. In this chapter, we highlight the two simplest: ionic and covalent bonding.

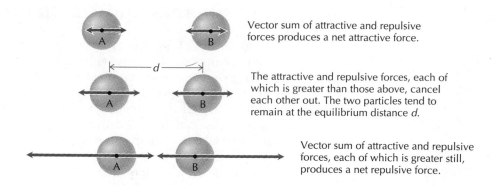

**FIGURE 9.5**   In this generalized case of chemical bonding, two particles (A and B) resist being pulled apart and resist being forced closer together. Typically, the atoms in molecules or molecules linked together are not completely stationary; instead, they vibrate and may rotate.

Vector sum of attractive and repulsive forces produces a net attractive force.

The attractive and repulsive forces, each of which is greater than those above, cancel each other out. The two particles tend to remain at the equilibrium distance $d$.

Vector sum of attractive and repulsive forces, each of which is greater still, produces a net repulsive force.

# Ionic Bonding

**Ionic bonds** occur when electrons are lost by atoms of one element and gained by atoms of another element. The electron shifts are explained by the quantum mechanical model of the atom and reflected in the organization of the periodic table of the elements, topics to be covered in Chapter 15. For now, we only need to know that there are certain atoms that tend to either lose or gain electrons. Those that lose electrons during chemical reactions are **metals,** and those that gain electrons are **nonmetals.** The shift of electrons may also involve *atom groups,* or clusters of atoms that maintain their identity when chemical bonds are made or broken.

The atoms of metals have one or more electrons loosely bound to their atomic nuclei. *A metal atom can combine chemically by losing one or more of its electrons to a nonmetal atom (or atom group).* After the loss takes place, the metal atom becomes a positively charged ion (because it has lost one or more of its negatively charged electrons).

The atoms of nonmetals tend to gain one or more extra electrons. This produces a negatively charged ion. *A nonmetal atom (or atom group) can combine chemically by gaining one or more electrons from metal atoms.* (Nonmetal atoms can also *share* electrons with each other. This covalent type of bonding is explained in the next section).

When one or more electrons are transferred from a metal atom to a nonmetal atom or atom group, the ions so formed are oppositely charged and have a tendency to attract each other and cling together, as shown in Figure 9.6. The ions assume positions at an equilibrium distance apart from each other as if they were connected by a rigid girder. To visualize this situation, think of a round stick plugged into two wooden hubs of a Tinkertoy set. The ionic bond is particularly strong because there is always a significant separation of charge (one or more electrons' worth of charge) between the two ions that bond. Ionic compounds are typically hard solids that resist both compression and tension. The bond, however, is not as resistant to sideways forces. Just as Tinkertoy structures can be easily destroyed by sideways forces that twist or break the sticks, ionic compounds tend to be brittle.

The combination of sodium (a soft metal) and chlorine (a nonmetal gas) illustrates several features common in ionic bonding. A chlorine atom will react (combine) with almost anything that will give it the single electron it "desires." Each sodium atom has one loosely bound electron, which

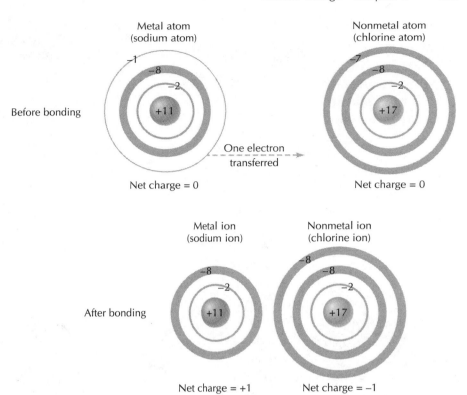

**FIGURE 9.6**  The electrons bound to atoms tend to group themselves within *shells,* schematically represented here by gray circles. In this example of ionic bonding, the outermost electron in a sodium atom is captured by the outer shell of a chlorine atom when the two atoms interact. Once the electron transfer has occurred, the two atoms, having become oppositely charged ions, strongly attract each other. The two will draw close, but not too close because of the mutual repulsion between the electrons in the outer shells of both ions. In practice, when many ions of each kind are present, they arrange themselves in a crystal lattice (a geometric pattern) such as the one shown in Figure 9.7.

can easily be donated to some other atom that desires it. Sodium and chlorine are a good match for ionic bonding. The two substances unite in a fiery, exothermic reaction that produces the ionic compound sodium chloride (NaCl), which is common table salt.

The reaction involves many sodium and chlorine atoms coming together to form a regular pattern made of sodium and chloride ions (see Figure 9.7). This pattern, a *crystal lattice,* represents a state of minimum potential energy. There is more potential energy associated with sodium and chlo-

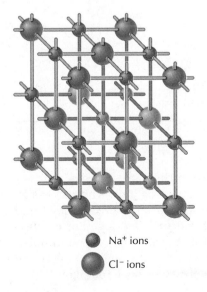

- Na⁺ ions
- Cl⁻ ions

**FIGURE 9.7**  The ions that make up a sodium chloride crystal arrange themselves in a lattice that has a "face-centered" cubic structure, as shown here. Crystals of other compounds or of solid elements have other types of crystal structure.

---

**TABLE 9.1**
Valences of Some Common Elements and Atom Groups

| Element or Atom Group | Symbol with Valence* |
|---|---|
| Hydrogen | $H^+$ |
| Lithium | $Li^+$ |
| Sodium | $Na^+$ |
| Potassium | $K^+$ |
| Magnesium | $Mg^{2+}$ |
| Calcium | $Ca^{2+}$ |
| Barium | $Ba^{2+}$ |
| Iron | $Fe^{2+}$, $Fe^{3+}$ |
| Copper | $Cu^+$, $Cu^{2+}$ |
| Silver | $Ag^+$ |
| Zinc | $Zn^{2+}$ |
| Mercury | $Hg_2^{2+}$, $Hg^{2+}$ |
| Aluminum | $Al^{3+}$ |
| Tin | $Sn^{2+}$, $Sn^{4+}$ |
| Lead | $Pb^{2+}$, $Pb^{4+}$ |
| Fluorine | $F^-$ |
| Chlorine | $Cl^-$ |
| Bromine | $Br^-$ |
| Iodine | $I^-$ |
| Oxygen | $O^{2-}$ |
| Sulfur | $S^{2-}$ |
| Nitrogen | $N^{3-}$ |
| Phosphorus | $P^{3-}$ |
| Ammonium (group) | $NH_4^+$ |
| Chlorate (group) | $ClO_3^-$ |
| Hydroxide (group) | $OH^-$ |
| Nitrate (group) | $NO_3^-$ |
| Cyanide (group) | $CN^-$ |
| Sulfate (group) | $SO_4^{2-}$ |
| Carbonate (group) | $CO_3^{2-}$ |
| Silicate (group) | $SiO_3^{2-}$ |
| Phosphate (group) | $PO_4^{3-}$ |

*The valences $^+$ and $^-$ refer to $1^+$ and $1^-$, respectively.

rine atoms when they are segregated from each other than when they are joined in the lattice as ions. The process is similar to a bunch of marbles being thrown into a large mixing bowl with steep sides. Very quickly all marbles will end up piled together on the bottom, where their collective gravitational potential energy is minimized. When sodium and chlorine react, they do so quickly, and the loss of potential energy manifests itself in a sudden release of heat and light.

Many different kinds of atoms and atom groups can participate in ionic bonding. But how can we tell when and how atoms or atom groups will bond ionically? Table 9.1 can help. It is a list of the *valences,* or unit charges, assumed by various atoms and atom groups when they participate in ionic bonding. A +1 valence for a metal atom indicates that it can readily lose one electron and therefore become an ion with a unit positive charge (+1). A +2 valence indicates a metal atom's tendency to lose two electrons; and so on. Some metals, like iron (Fe) have more than one valence. Iron can participate in ionic bonding by giving up either two or three of its electrons. Nonmetals have valences such as −1 and −2, which indicates that they are receptors of electrons and that they become ions with negative charges. (The valences listed in Table 9.1 can be determined from experiment. They can also be derived by knowing how and where electrons orbit in various atoms, as we shall see in Chapter 15.)

The formulas of ionic compounds, which indicate the ratios among various atoms of elements in them, are easily determined by knowing valences. For example, Na and Cl combine in a one-to-one ratio to become NaCl, because each Na atom gives one electron to one chlorine atom. When calcium (Ca), with a valence of +2, and fluorine (F), with a valence of −1, combine, the result is the compound $CaF_2$. That is, two F atoms are needed to accept the two electrons a single Ca atom tends to give away. Still another example is CaO (calcium oxide, commonly known as lime). Each Ca atom donates two electrons, and each O atom accepts two. The ratio between Ca and O atoms, reduced to lowest terms, is one to one.

## Covalent Bonding

The most common and versatile method of combining atoms is **covalent bonding,** which involves the sharing of one or more pairs of electrons. The covalent substances so formed can be as simple as common hydrogen or oxygen gas, with one or more pairs of electrons shared between two identical atoms, or as complex as some biological molecules, which can contain millions of atoms and involve millions of shared pairs of electrons. Covalent compounds are usually composed of nonmetal atoms, though some metal atoms can participate, too.

As shown in Figure 9.8, each hydrogen atom in a hydrogen molecule ($H_2$) has one electron to share, and there is a total of two shared electrons between the two atoms. Oxygen atoms have a tendency to share two electrons each. For nitrogen atoms it is three electrons each, and for carbon atoms it is four electrons each. This behavior can be symbolized in two ways: electron dot notation or dash notation. Each dot represents a single electron and each dash represents a shared pair of electrons (one each con-

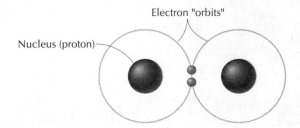

Electron "orbits"

Nucleus (proton)

**FIGURE 9.8**  Two hydrogen atoms have less chemical potential energy when they are covalently bonded, as seen in this simplified model, than when they are separate and not bonded. (A more accurate picture of the space occupied by the electrons in the $H_2$ molecule is given in Chapter 15.) The two electrons are shared in such a way that they spend more time in between the two nuclei than elsewhere. This is an energetically favorable configuration and it binds the nuclei together.

tributed by the atoms involved in the bond). Here are some examples.

| | | |
|---|---|---|
| H:H | H—H | hydrogen molecule ($H_2$) |
| O::O | O=O | oxygen molecule ($O_2$) |
| N:::N | N≡N | nitrogen molecule ($N_2$) |
| O::C::O | O=C=O | carbon dioxide ($CO_2$) |

The electrons represented here by the dots are only the electrons being shared, not necessarily the total number of electrons orbiting the atomic nuclei. Each dash represents a single covalent bond, so we say that the atoms in a molecule of $H_2$ have a single bond attached to them. The atoms in $O_2$ have a double bond, and the atoms in $N_2$ have a triple bond. (After our discussion of electron shell structures and the periodic table in Chapter 15, you will have a better understanding of why each oxygen atom shares two of its electrons, each nitrogen atom shares three of its electrons, and so on.)

Dot notation can be used to symbolize how electrons are *unequally* shared in the covalent bonds between some atoms. For example, molecules of the covalent compounds hydrogen chloride (hydrochloric acid) and water can be symbolized as

H :Cl   (HCl)

H :O
  ··      ($H_2O$)
  H

In both cases, the shared electrons (which are negatively charged) spend more time on one side of the molecule than on the other. As individual units, these molecules act as if they were electrically polarized (that is, each one has a positively charged end and a negatively charged end). They behave in much the same way that the small pieces of electrically polarized paper in Figure 9.3 do. Molecules of this sort are named *polar molecules*. The substances they comprise are called *polar covalent compounds*.

Polar molecules are, in a sense, slightly "sticky." The polar molecules of a liquid like water attract and cling to each other, and they attract and cling to other polar molecules as well. This feature explains several behaviors—*cohesion, surface tension, adhesion,* and *capillary action*—that characterize many liquids. **Cohesion** is the tendency for like molecules, such

**PHOTO 9.1** A water strider's legs dimple the surface of a pond but do not break through. Here is surface tension made visible.

as those of water, to attract each other. **Surface tension,** a consequence of cohesion, is the tendency of polar liquids (especially) to pull toward each other and collectively assume a shape that has maximum volume and at the same time minimum surface area (in other words, a sphere). Cohesion and surface tension are responsible for producing the roughly spherical shape of small dewdrops and raindrops and for the ability of small insects such as water striders (Photo 9.1) to skate across the surface of a pond.

**Adhesion** is the tendency of unlike molecules to stick to each other. A familiar example is the *meniscus,* or turned-up edge of a water surface when in contact with the walls of a glass container. The adhesive forces between water and glass molecules are stronger than the cohesive forces between the water molecules themselves, so molecules on the water surface "climb" upward along the inside of the glass. **Capillary action,** the ability of water or other polar liquids to pull themselves through small tubes, occurs when the bore of the tube is so small that the surface adhesion is stronger than any other force acting on the liquid. Capillary action is exploited by trees, which defy gravity by pulling up mineral-laden water through tiny tubes in their trunks and branches, and by animals, whose circulatory systems include the tiny blood vessels (capillaries) that deliver oxygen-laden blood to every part of the organism.

A special kind of chemical bond called a **hydrogen bond** often exists between a hydrogen atom in a polar molecule and the negatively charged portion of another polar molecule. A hydrogen atom is left with a considerable positive charge when it shares its single electron with atoms such as oxygen, fluorine, and nitrogen in a polar molecule. It is therefore attracted to the negatively charged parts (often oxygen, fluorine, or nitrogen atoms) of other polar molecules nearby. Hydrogen bonding is responsible for the weak mutual attraction of molecules of water in its liquid state and for the stronger attraction of water molecules for each other in the solid state (ice). Hydrogen bonding with nitrogen atoms plays an important role in biochemistry; for example, when DNA molecules split apart into two strands and later reassemble themselves, hydrogen bonds are broken and restored.

Hydrogen bonds are not the same as the covalent bonds that bind atoms together into molecules. Instead, they are bonds *between molecules* that arise as a result of polar covalent bonds existing between the atoms of those molecules (see example in Figure 9.9). Hydrogen bonds are relatively weak: about 5 or 10% as strong as a typical covalent bond.

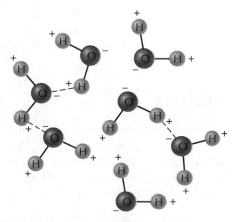

**FIGURE 9.9** The polar covalent bonds (solid lines) between atoms in water molecules result in a slight separation of charge. In liquid water, schematically shown here, some molecules are hydrogen-bonded (dashed lines) to others by the attractions between H and O atoms.

## Ions in Solution

A **solution** is a homogeneous (uniform) mixture of different elements or compounds. Solutions can form from the mixing of two liquids, of two gases, of a liquid and a gas, or of a solid and a liquid. For convenience, we use the terms **solvent** and **solute** to denote the larger and the smaller quantities of substances that, when mixed together, make a solution. We say, for example, that sugar (a solute) dissolves in water (a solvent).

## Like Dissolves Like

When substances are combined, the electrical properties of their atoms or molecules determine whether they will mix to become a solution or mix in some heterogeneous way. If substances made of nonpolar molecules—say, gasoline and motor oil—are physically brought together, they freely intermix, because there are no large intermolecular forces between their molecules [see Figure 9.10(a)] and because the molecules are moving about randomly owing to ordinary thermal motion. Vegetable oils and fats are also nonpolar; they can be dissolved in petroleum derivatives such as gasoline.

In contrast, a nonpolar substance like oil and a polar substance like water will not freely intermix when stirred together. The cohesive forces between water molecules cause them to clump together and thereby squeeze out any nonpolar oil molecules nearby [see Figure 9.10(b)]. Since oil is less dense than water, the oil molecules will migrate toward the top of a container, and the water molecules will sink.

When the crystalline grains of a polar solute, such as table salt, are sprinkled into water, the polar molecules of water surround and pull apart ions on the surface of each salt crystal, thereby dissolving, or dissociating, the crystal (see Figure 9.11). Nonpolar solvents, such as gasoline or oil, cannot dissolve salt, because their nonpolar molecules barely interact with the ions of the crystals.

These behaviors can be generalized in the rule "like dissolves like": Solutes and solvents with similar electrical characteristics can be blended into solutions. Solutes and solvents having different electrical characteristics tend to not mix with each other.

Soaps and cleansers, when dissolved in water, make solutions with the ability to dissolve a wide variety of substances. Soap molecules are polar at one end and nonpolar at the other. The nonpolar ends surround and dissolve dirt and grease, which are composed largely of nonpolar molecules. The polar ends of soap molecules readily cling to water molecules. The soapy solution, with its load of suspended dirt and grease, easily rinses away.

## Solubility

Some solutes and solvents easily intermix in any proportion. For example, any amount of water can be added to ethylene glycol (an automobile radiator coolant and antifreeze agent), or vice versa, to make a solution. The amount, or *concentration,* of a solid, liquid, or gaseous solute in a given amount of liquid solvent, however, may be limited. If an arbitrarily

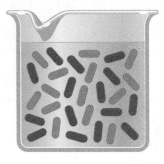

(a)

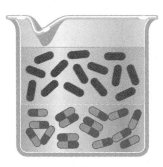

(b)

Oil
Gasoline
Water

**FIGURE 9.10**  (a) Nonpolar compounds such as oil and gasoline easily intermix. (b) The polar molecules of water will not mix homogeneously with the nonpolar oil molecules.

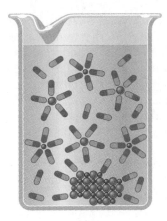

• Na$^+$ ions
• Cl$^-$ ions
⊜⊕ Water molecule

**FIGURE 9.11** Ionic crystals such as sodium chloride are subject to attack by polar solvents such as water in the manner shown schematically here. The negative ends of water molecules pull away the positive ions, while the positive ends of water molecules pull away the negative ions. While doing the work of dissolving, or dissociating, the crystal, the water molecules lose some of their thermal energy; and the temperature of the solution decreases. Dissociation, in this case and in most other cases, is an endothermic process.

large quantity of table salt is stirred into a glass of water, for example, an excess of salt piles up on the bottom of the glass. Such a solution is *saturated*; it contains the maximum amount of solute possible. An examination of Figure 9.11 shows why unlimited amounts of salt cannot dissolve in water: There are simply not enough unattached water molecules around to exert the necessary electric forces.

At 20 °C, a maximum of 36 g of NaCl will dissolve in 100 g of $H_2O$. This concentration, which represents a saturated solution at that temperature, can be expressed as the **solubility** of NaCl in $H_2O$ at 20 °C. The solubility of solid compounds in water is nearly always temperature-dependent, sometimes strongly so. Figure 9.12 compares the very different solubility characteristics of two common salts, NaCl and $KNO_3$ (potassium nitrate).

Normally, when a saturated solution is cooled, the solute responsible for the saturation crystallizes, or *precipitates* out of solution. Sometimes, however, if the cooling is slow and the solution is not subject to any mechanical disturbance, a solute may remain in solution even when its solubility is exceeded. The result, known as a *supersaturated solution,* is analogous to an unstable equilibrium in mechanics. The slightest jarring of a supersaturated solution can bring on an instant crystallization of the excess solute.

The physical properties of liquid solutions change as their concentrations change. With increasing concentration, freezing points usually go down and boiling points usually go up. For example, seawater, with a salt content of about 3.5%, freezes at −1.9 °C and boils at 100.3 °C. The concentration of salt in the water-ice bath of a household ice-cream maker is typically large enough to depress the temperature of the bath by several degrees. In a similar vein, road surfaces in some parts of the United States are salted during cold, wet spells so as to prevent the formation of ice at temperatures under 0 °C. The ethylene glycol, $C_2H_4(OH)_2$, antifreeze solutions used in automobile cooling systems remain liquid at well under 0 °C and at well over 100 °C. These solutions will not freeze inside a parked car on wintry nights nor easily evaporate and escape from a car whose engine is running hot.

**FIGURE 9.12** The solubility of sodium chloride in water depends only slightly on temperature. Potassium nitrate is relatively insoluble at low temperature and much more soluble at higher temperatures.

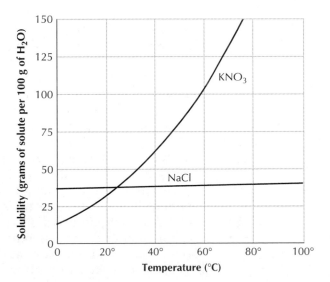

The solubility of gases in liquids depends on several factors. Unlike solid solutes, a gas becomes *less* soluble in a liquid as the temperature increases. In addition, a gas becomes more soluble when it is under greater pressure and often less soluble when subjected to mechanical jarring. All three of these relationships are apparent in the behavior of carbonated soft drinks, beer, and "sparkling" alcoholic beverages such as champagne. In unopened containers, these beverages contain carbon dioxide gas dissolved under considerable pressure. When pressure is relieved by opening the container, the carbon dioxide bubbles out of solution, because it becomes less soluble. Carbonated beverages "fizz" faster on hot days than on cold days; and, of course, we all know how soft drinks or champagne can spurt out of a bottle that has been shaken or jarred in some way.

Oxygen is another gas that readily dissolves in water, especially cold water. Fish extract this dissolved oxygen by means of their gills. Some fish, such as trout, thrive on oxygen-rich cold water and languish or die in water that is warm and oxygen-poor.

This fact about oxygen is related to the reason large power plants sited along riverbanks often have huge cooling towers. The cold river water is used to absorb the waste heat generated by a power plant's turbines, and then it returns to the river at a higher temperature. For environmental reasons, primarily to protect aquatic life in the river, it is necessary to limit the temperature of the discharged water. This step is accomplished by circulating the water through cooling towers, where much of its thermal energy is released into the atmosphere.

## Aqueous Solutions

Water is the most ubiquitous and versatile polar solvent on Earth, so it is not surprising that aqueous (water) solutions play important roles in chemistry and biochemistry. When ionic compounds (as well as some covalent compounds containing hydrogen) dissolve in water, they release ions that freely intermix with the water molecules around them—a process known as *dissociation.* Substances that can dissociate in water are known as **electrolytes,** and the resulting solutions are known as *electrolyte solutions.* Other aqueous solutions contain molecules that do not dissociate into ions; these molecules belong to *nonelectrolytes,* sugar and alcohol being two examples.

An electrolyte solution can conduct electricity by virtue of the free movement of its dissolved ions (charged particles), just as a metal can conduct electricity by virtue of its free electrons. Nonelectrolyte solutions, on the other hand, are essentially electrical insulators. Pure (or *deionized*) water is a weak electrolyte, capable of conducting only very small currents.

We have already seen how ions of an element can behave differently from atoms of the same element. Sodium and chlorine are good examples: When locked together as ions in a crystal lattice, they lose their great ability to react with other elements or compounds. When ions are in solution, as in aqueous solutions, their physical properties are quite different.

Here is a typical example of the difference. The chloride ion ($Cl^-$) in solution is colorless; has a mild, pleasant taste; does not react with most metals or hydrogen; and combines with silver ions ($Ag^+$) to form the insol-

**PHOTO 9.2** Each type of ion contributes its properties to all solutions that contain it. The color properties of four different metal ions can be seen in these four nitrate solutions. Cobaltous nitrate, $Co(NO_3)_2$, containing cobalt ($Co^{2+}$) ions, appears red. Cupric nitrate, $Cu(NO_3)_2$, containing copper ($Cu^{2+}$) ions, appears blue. Nickel nitrate, $Ni(NO_3)_2$, containing nickel ($Ni^{2+}$) ions, appears green. Zinc nitrate, $Zn(NO_3)_2$, containing zinc ($Zn^{2+}$) ions, is colorless.

uble compound silver chloride (AgCl) whenever silver ions are present. In contrast, chlorine gas, which consists of diatomic molecules ($Cl_2$), has a greenish color; has an irritating taste and odor; reacts with both metals and hydrogen; and does not react with silver ions. Chlorine gas is poisonous to all living things in somewhat small concentrations; chlorine ions in solution are generally not.

Modern chemical theory suggests that each type of ion in solution has a distinct set of properties, and that the overall properties of any solution depend on the proportions of the different kinds of ions dissolved in that solution. In other words, each type of ion contributes its properties to all solutions that contain it.

The taste, color, and other properties of, say, a sodium chloride solution are a blend of the properties of both sodium and chloride ions. Another solution might contain copper ions ($Cu^{2+}$). These ions reflect blue more easily than any other color of visible light, so any solution containing copper ions will have a tendency to appear blue.

The properties of two ions in particular, the hydrogen ion ($H^+$) and the hydroxide ion ($OH^-$), merit special attention. Aqueous solutions containing significant amounts of these ions can be categorized as acidic and basic, respectively.

# Acids and Bases

Acids and bases play important roles in many industrial and biological processes. Many of the familiar materials and foods we encounter nearly every day are either acidic or basic. Aspirin, vitamin C, vinegar, and citrus fruits are acids. Soaps, cleansers, baking soda, and milk of magnesia are bases. Here are the simplest definitions of acid and base: An **acid** is a compound that increases the concentration of hydrogen ions ($H^+$) when dissolved in water. A **base** is a compound that increases the concentration of hydroxide ions ($OH^-$) when dissolved in water.

Acid solutions have a characteristic sour taste and can change blue litmus paper to red. Strong acids can cause painful burns on the skin and can easily decompose a variety of materials. Weak acids—such as the citric acid of citrus fruits, the acetic acid in vinegar, and the lactic acid in yogurt—may add a pleasant, tart taste to food. (Of course, you should never use your tongue to taste acids or any other chemicals other than those in food.)

Acids can be classified as strong or weak depending on the ability of their molecules to dissociate in the presence of water. In strong acids, such as hydrochloric acid (HCl) and sulfuric acid ($H_2SO_4$), hydrogen atoms or ions are easily dissociated from their parent molecules. In weak acids, such as acetic acid ($HC_2H_3O_2$), the hydrogens are not as easily dissociated. (The formula for acetic acid may be written as shown here to indicate that only one hydrogen in an acetic acid molecule dissociates in water; the other three hydrogens do not.) Furthermore, any acid (or any base, for that matter) can be diluted with any amount of water to make a solution that is less concentrated and, therefore, effectively less strong than it is in concentrated form.

A few substances containing no hydrogen at all can react with water to increase the concentration of hydrogen ions. A well-known example is carbon dioxide ($CO_2$) gas. When dissolved in water, it reacts with some of the water molecules to form hydrogen ions and hydrogen carbonate ions ($HCO_3^-$):

$$CO_2 + H_2O \rightarrow H^+ + HCO_3^-$$

Some hydrogen carbonate ions can further dissociate into hydrogen ions and carbonate ions ($CO_3^{2-}$):

$$HCO_3^- \rightarrow H^+ + CO_3^{2-}$$

The resulting acid solution, containing what is often referred to as *carbonic acid,* is very weak. Nonetheless, the carbonic acid in a freshly opened bottle of soda water (carbonated water) has a tartness strong enough to taste.

Basic solutions have a characteristic bitter taste, turn red litmus paper to blue, and have a slippery feel. Strong bases also cause burns on the skin and can decompose many different materials. Sodium hydroxide (NaOH), used in oven cleaning solutions, is a strong base because it dissociates freely in water, giving up a large number of hydroxide ions.

Soluble weak bases are rare, but some chemicals lacking the OH group can nonetheless combine with water to form basic solutions. One example is ammonia gas ($NH_3$), which, when dissolved in water, yields ammonium ions ($NH_4^+$) and hydroxide ions. The resulting ammonia solution is used widely in household cleansers.

The term *alkali* is used as a synonym for *base,* and solutions arising from alkali substances dissolved in water are also known as alkaline solutions. An important alkali is potash, originally an extract of wood ash containing potassium carbonate ($K_2CO_3$). Today, the term *potash* refers mostly to potassium chloride (KCl), a substance that is mined and mainly used to make fertilizer.

Interestingly, pure water has both acidic and basic properties, although both are weakly expressed. There is a very slight tendency for water molecules to dissociate in the presence of other water molecules. At any given instant, about 2 out of every 10 million ($10^7$) water molecules in a

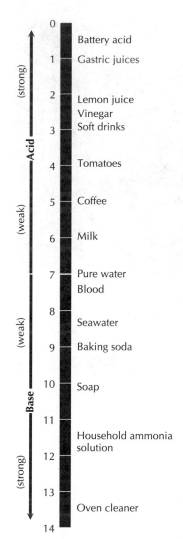

**FIGURE 9.13** Typical pH values for a variety of common substances. Pure water, at pH 7, is a poor electrolyte. Strong acids and strong bases are good electrolytes, since they contain large numbers of H$^+$ and OH$^-$ ions, respectively.

given sample of pure water are dissociated into H$^+$ and OH$^-$ ions. With respect to acids and bases, water is considered to be neutral. In acidic solutions, H$^+$ ions outnumber OH$^-$ ions, and in basic solutions OH$^-$ ions outnumber H$^+$ ions.

The *pH scale* was invented to quantify how acidic or basic a solution is. The scale is logarithmic and defined so that each integer difference represents a tenfold increase in hydrogen ions as pH gets smaller, and a tenfold decrease in hydrogen ions as pH gets larger. The scale is calibrated so that pure water has a pH value of exactly 7. The pH scale graphically presented in Figure 9.13 includes some typical values for a variety of common substances.

The symmetry between the acid and the base sides of the pH scale is mirrored whenever acids and bases are mixed. What occurs is called *neutralization:* Acids neutralize bases and bases neutralize acids. When, for example, hydrochloric acid is mixed with an equal measure of sodium hydroxide solution (of the same strength), the following equation describes what happens:

$$H^+ + Cl^- + Na^+ + OH^- \rightarrow Na^+ + Cl^- + H_2O$$

Two of the four different kinds of ions present in the two reactants, namely, H$^+$ and OH$^-$, recombine to form water in the product. The Na$^+$ and Cl$^-$ ions remain in solution. This solution is dissolved sodium chloride—in other words, saltwater.

Since, in the reaction, Na$^+$ and Cl$^-$ ions appear on both sides, we can cancel them and write the following net reaction:

$$H^+ + OH^- \rightarrow H_2O$$

As we see, neutralization is really a reaction between hydrogen and hydroxide ions to form water. This reaction is exothermic, and greatly so for strong acid-base combinations like concentrated hydrochloric acid and concentrated sodium hydroxide. For this reason, strong acids and bases should never be quickly mixed together; the neutralization process is safer when either or both of the solutions have been diluted.

When a neutralized solution is evaporated to dryness, the ions in it precipitate out to form a solid that chemists call a *salt*. Sodium chloride is only one example of this important category of compounds. In many foods, the salts potassium chloride (KCl) and sodium chloride (NaCl) supply the K$^+$ and Na$^+$ ions needed for the proper functioning of the nervous system. (These ions, along with many others such as calcium, magnesium, and iron, are known as "minerals" in the nutritional sense of the word. The word *mineral* is also used in a geologic sense. It refers to the crystalline grains that make up rocks and the various chemical elements and ions dissolved in seawater; see Box 9.1.)

Calcium carbonate (CaCO$_3$), a principal constituent of chemical sedimentary rocks, is formed when calcium hydroxide [Ca(OH)$_2$] neutralizes carbonic acid. Another common salt, ammonium nitrate (NH$_4$NO$_3$), is used as a fertilizer to add nitrogen to the soil. The salts borax (Na$_2$B$_4$O$_7$) and washing soda (Na$_2$CO$_3$) are commonly used in cleansers. The versatile salt called baking soda (NaHCO$_3$) finds uses as an ingredient in baking, as a medicinal antacid, as a deodorizer, and as an ingredient in toothpastes.

**PHOTO 9.3**  These crystallized deposits of salt (sodium chloride) were left behind on the floor of Death Valley, California, when the sun's rays evaporated a pool of briny water containing sodium and chloride ions.

## BOX 9.1

# Ions of the Oceans

On average, about 3.5% by mass of seawater consists of dissolved salts. If Earth's oceans could be completely evaporated, the remaining salts would be plentiful enough to cover the continental landmass to a depth of about 150 m. The abundances of several common seawater ions are tabulated in Table 9.2. More than 60 chemical elements have been identified in seawater, which makes the ocean a veritable storehouse for minerals of many kinds.

Extracting minerals from seawater is not always easy. Sodium chloride, magnesium, and bromine are regularly isolated from seawater, but the more

| **TABLE 9.2** Average Percentage Abundance by Mass of the Most Common Dissolved Ions in Seawater | Ion | Abundance (%) |
|---|---|---|
| | Chlorine ($Cl^-$) | 1.93 |
| | Sodium ($Na^+$) | 1.07 |
| | Sulfate ($SO_4^{2-}$) | 0.27 |
| | Magnesium ($Mg^{2+}$) | 0.13 |
| | Calcium ($Ca^{2+}$) | 0.042 |
| | Potassium ($K^+$) | 0.039 |
| | Bicarbonate ($HCO_3^-$) | 0.014 |
| | Bromide ($Br^-$) | 0.0070 |
| | Strontium ($Sr^{2+}$) | 0.0013 |
| | Fluoride ($F^-$) | 0.0001 |
| | Nitrate ($NO_3^-$) | 0.000035 |

*Note:* Water = 96.5%; all dissolved ions = 3.5%.

# Organic Chemistry

In the study of chemistry, a general distinction is often made between **organic chemistry,** which is nearly all chemistry involving the element carbon, and *inorganic chemistry,* which is the chemistry of everything else. The name *organic* comes from the fact that living organisms on Earth consist largely of carbon compounds. Before the early 1800s, organic chemistry and biochemistry were thought to be virtually synonymous. Today, however, organic compounds are produced by the millions through methods that have nothing to do with life or biochemistry.

Carbon is unique in that its atoms easily bond covalently to other carbon atoms and to several other kinds of atoms as well. Each carbon atom can share four of its outer electrons with nearby atoms. The number and variety of compounds thus formed is rich and diverse. Several million different organic compounds have either been discovered in nature or synthesized by chemists. Organic molecules can be simple and lightweight,

**PHOTO 9.4**   Sulfide minerals are being ejected from this active hydrothermal vent (or "black smoker") on the Pacific Ocean floor.

The dissolved ions in seawater owe their presence to two general processes, both connected to the greater hydrologic cycle diagramed in Chapter 7 (Figure 7.34). In the first process, soluble minerals are leached out of rock and soil by the action of rainfall, surface streams, and groundwater. Sooner or later, nearly all of this mineral-rich water reaches the ocean, carrying dissolved ions with it.

In the second general process, occurrences related to plate tectonics both add to and remove dissolved ions and minerals from the oceans. *Hydrothermal vents* (underwater volcanoes) along the midocean ridges eject superheated water and lava containing numerous minerals from Earth's upper mantle (Photo 9.4). Volcanoes on land spew out lava and ash that increase the mineral content of the land around them. Ions and minerals are removed from seawater at convergence zones where an oceanic plate subducts under another plate. There, mineral-rich seawater and minerals precipitated on the seafloor are drawn back into the upper mantle. Elsewhere and at some future time, this material will emerge on land or under the sea by way of volcanic activity.

valuable metallic minerals, such as gold, are present in such tiny concentrations that they cost far more to extract than they are worth.

| TABLE 9.3 | Element | Valence |
|---|---|---|
| Valences of | Hydrogen (H) | 1 |
| Elements That | Carbon (C) | 4 |
| Commonly | Nitrogen (N) | 3 |
| Participate | Oxygen (O) | 2 |
| in Covalent | Fluorine (F) | 1 |
| Bonding | Silicon (Si) | 4 |
| | Phosphorus (P) | 3 |
| | Sulfur (S) | 2 |
| | Chlorine (Cl) | 1 |

*Note:* With regard to organic chemistry, valence refers to the number of electrons in the outermost part of an atom that can be shared with other atoms. For this reason, the valences given here are also known as "covalences."

like methane ($CH_4$). They can be of moderate weight yet extremely complex, as are many of the molecules that participate in biological processes. Some, like the molecules of plastic materials, are massive, consisting of chains of thousands or millions of repetitious units.

The rules for making common organic compounds, or for deciding which arrangements of carbon and other atoms form possible compounds, are simple. First, you must recognize how many electrons in each atom participate in bonding. This number is also called the *valence* of an atom or an element (see Table 9.3 for the valences of several atoms commonly associated with organic molecules). Second, there should be no unconnected bonds.

With these two rules in mind, we shall explore the incredible versatility of the carbon atom by describing some important categories of organic compounds and by making mention of some of the uses of these chemicals.

## Alkane Hydrocarbons

Compounds containing only hydrogen and carbon are called **hydrocarbons.** Hydrocarbon compounds having carbon atoms linked to each other by single bonds are called **alkanes.** Alkane hydrocarbons are the principal components of natural gas, bottled gas, lighter fluid, gasoline, kerosene, diesel, motor oil, petroleum jelly, paraffin wax, and asphalt. Their source is petroleum (crude oil) and natural gas extracted from within Earth's crust.

The eight simplest (unbranched) alkane hydrocarbons, starting with methane, are organized and illustrated in Figure 9.14. The structural formulas given show only the general arrangement of the atoms in each molecule. Methane's true three-dimensional shape is that of a tetrahedron (a pyramid with four equal faces), with four hydrogen atoms at equal distances from the carbon atom at the center. The longer, heavier unbranched alkanes have their carbon atoms arranged in a zigzag line.

| Structural Formula | Molecular Formula | Name | Freezing Point | Boiling Point |
|---|---|---|---|---|
| 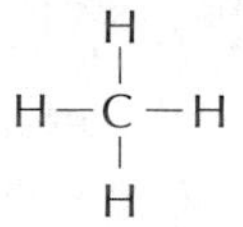 $H-\underset{\underset{H}{\mid}}{\overset{\overset{H}{\mid}}{C}}-H$ | $CH_4$ | Methane (natural gas) | −183 °C | −160 °C |
| 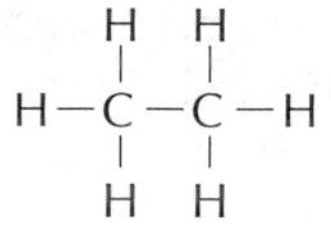 $H-\underset{\underset{H}{\mid}}{\overset{\overset{H}{\mid}}{C}}-\underset{\underset{H}{\mid}}{\overset{\overset{H}{\mid}}{C}}-H$ | $C_2H_6$ | Ethane (natural gas) | −184 °C | −89 °C |
| 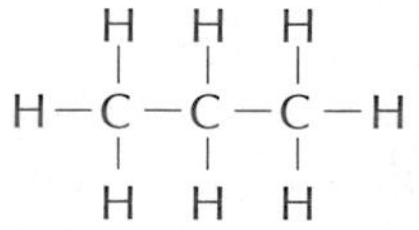 $H-C-C-C-H$ (with H above and below each C) | $C_3H_8$ | Propane (bottled gas) | −188 °C | −42 °C |
| 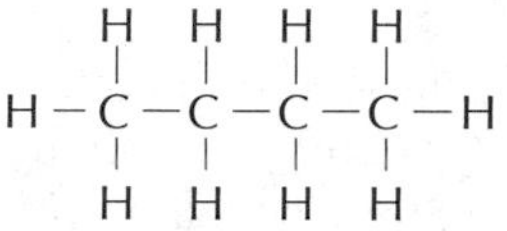 $H-C-C-C-C-H$ | $C_4H_{10}$ | Butane (bottled gas and lighter fluid) | −139 °C | −1 °C |
| 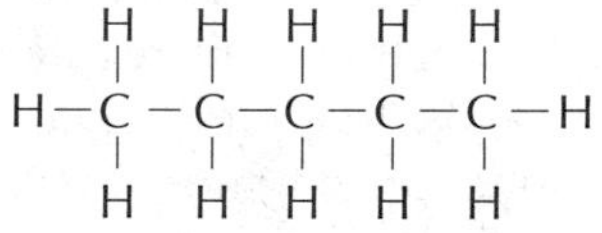 $H-C-C-C-C-C-H$ | $C_5H_{12}$ | Pentane (naphtha) | −130 °C | 36 °C |
| 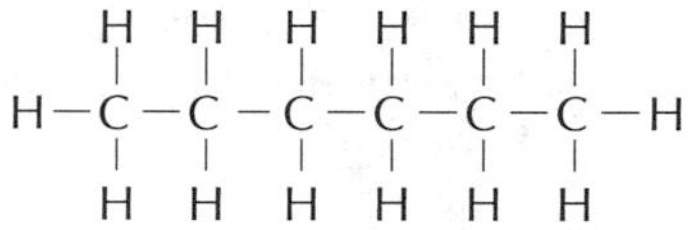 $H-C-C-C-C-C-C-H$ | $C_6H_{14}$ | Hexane (naphtha) | −95 °C | 69 °C |
| 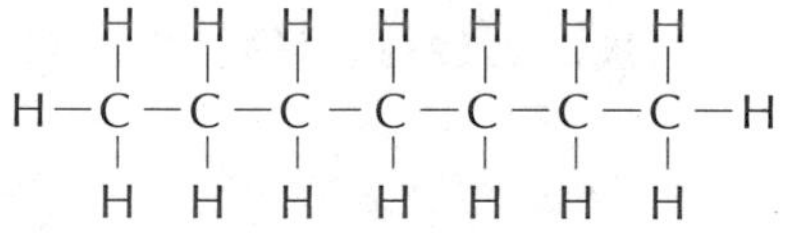 $H-C-C-C-C-C-C-C-H$ | $C_7H_{16}$ | Heptane (component of gasoline) | −91 °C | 98 °C |
| 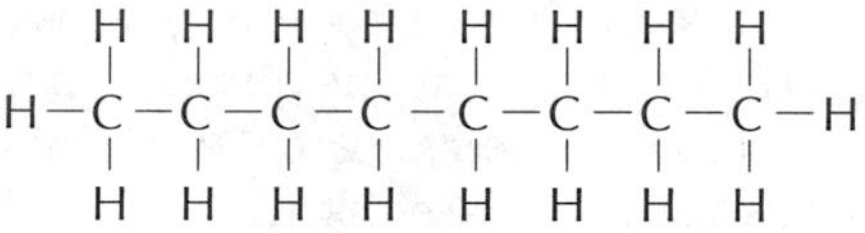 $H-C-C-C-C-C-C-C-C-H$ | $C_8H_{18}$ | Octane (component of gasoline) | −57 °C | 126 °C |

**FIGURE 9.14** The alkane series of hydrocarbons include single-chain molecules of carbon and hydrogen. As the number of carbon atoms (and the weight) of each molecule increases, the freezing and boiling points of the compounds generally increase.

The freezing and boiling points of the various alkanes rise with increasing molecular weight, as suggested by the data given in Figure 9.14. The first four are gases at room temperature; the next six, up to $C_{10}H_{22}$, are volatile (easily evaporated) liquids, and most of them are part of the mixture of hydrocarbons called gasoline. Kerosene and jet fuel, which are somewhat more viscous and oily in feel, generally contain $C_9H_{20}$ through $C_{16}H_{34}$. Diesel fuel, heating oil, and lubricating oil contain alkanes of up to $C_{18}H_{38}$. Semisolid petroleum jelly and paraffin wax contain alkanes of about $C_{20}H_{42}$ and up. Still heavier alkanes, $C_{36}H_{74}$ and up, comprise the sticky tar that glues together sand and gravel in asphalt paving material.

**PHOTO 9.5**   Propane, a gaseous alkane, can be compressed and stored in small bottles (as here) or in larger containers for various household uses. Propane is used for heating and cooking in many rural homes where piped sources of natural gas are not available.

The basic process by which the various components of crude oil are separated is called *fractional distillation.* The crude oil is slowly heated while its vapors are collected and distilled (or condensed) at progressively higher temperatures. Pentane and hexane (which make up naphtha, or cleaning fluid) are usually the first to evaporate and be collected. Next, as the temperature of the remaining mix rises, comes gasoline. Then come kerosene, diesel, and others. The nonvolatile sludge left over at the end is asphalt tar.

Many alkanes contain carbon atoms arranged in some kind of branched structure. A simple example is isobutane, whose molecular formula is the same as "normal" butane but whose structure is fundamentally different. Normal butane and isobutane (see Figure 9.15) are said to be **isomers** of each other, *isomer* meaning "made of the same parts." The simplest alkanes (up to propane) have no possible isomers, but alkanes progressively heavier than butane have isomers that quickly increase in number and complexity. For example, $C_{13}H_{28}$ has 813 theoretically possible isomers. Relatively few of the theoretically possible isomers of the heavier hydrocarbons have actually been synthesized by chemists.

The properties—freezing point, boiling point, density, flammability, and so forth—of the different isomers of a given compound are usually somewhat different. For instance, an isomer of octane called isooctane has its eight carbons arranged in one long chain of five carbon atoms with three side chains, which makes it less volatile and less flammable than regular octane. Isooctane can be added in varying degrees to gasoline fuel to keep the fuel from igniting too quickly in internal combustion engines. Fuels with more isooctane in them (or, more commonly, other compounds that produce the same effect as isooctane) have higher octane ratings and less tendency to cause early ignition in the cylinders, commonly known as "engine knock."

# Saturated and Unsaturated Compounds

The existence of double or triple bonds between carbon atoms in some molecules adds greatly to the richness and variety of organic compounds. The alkanes, having only single bonds between their carbon atoms, are known as **saturated compounds.** The term *saturated* means that no other atoms can be directly added to the carbon atoms of their molecules. Organic compounds having double or triple bonds between carbons are known as

**FIGURE 9.15**   There are two and only two alternative forms, or isomers, of butane, $C_4H_{10}$. Not only are the structures of normal butane (left) and isobutane (right) different, but the physical properties are different. Normal butane boils at −1 °C; isobutane boils at −12 °C. Isobutane is denser by about 3%.

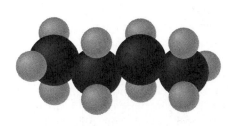

Normal butane
(*n*-butane)

Isobutane

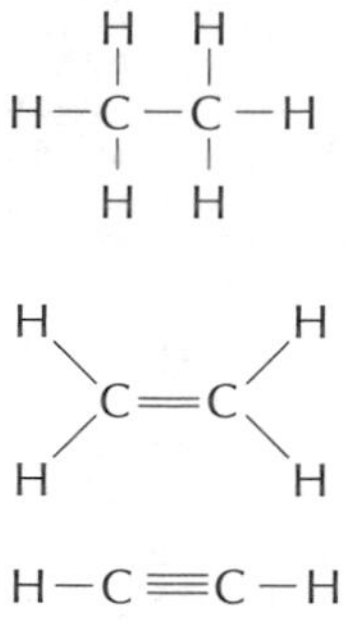

H—C—C—H    $C_2H_6$    Ethane

$C_2H_4$    Ethene

H—C≡C—H    $C_2H_2$    Acetylene

**FIGURE 9.16**   A double bond, like that between the carbons in an ethene molecule, consists of two shared electron pairs. A triple bond, as in acetylene, consists of three shared electron pairs.

**unsaturated compounds.** When the double or triple bonds are broken, more atoms can be added to the molecules of these compounds.

Two examples of unsaturated hydrocarbons, depicted in Figure 9.16, are ethene ($C_2H_4$) and acetylene ($C_2H_2$). Both are more reactive than their alkane counterpart, ethane ($C_2H_6$). Acetylene is used in oxyacetylene welding torches, in which gaseous streams of pure oxygen and acetylene combine to form an extremely hot flame (Photo 9.6). Ethene can react with chlorine, fluorine, and many acids to form a wide variety of saturated compounds.

Under the proper conditions of heat and pressure, and with the help of a catalyst, ethene molecules (as well as certain other unsaturated molecules) can undergo polymerization. Two examples are diagrammed in Figure 9.17. During *polymerization,* many unsaturated molecules, called **monomers,** combine to form saturated molecules made of thousands or even millions of repetitious units. The giant-chain molecules so created are called **polymers,** and the materials they comprise are called *plastics* for their ability to be molded or shaped at high temperatures. In rigid plastics, the polymers often nest against each other in parallel, somewhat like logs

**PHOTO 9.6**   The hot jet of an oxyacetylene torch will cut through steel with ease.

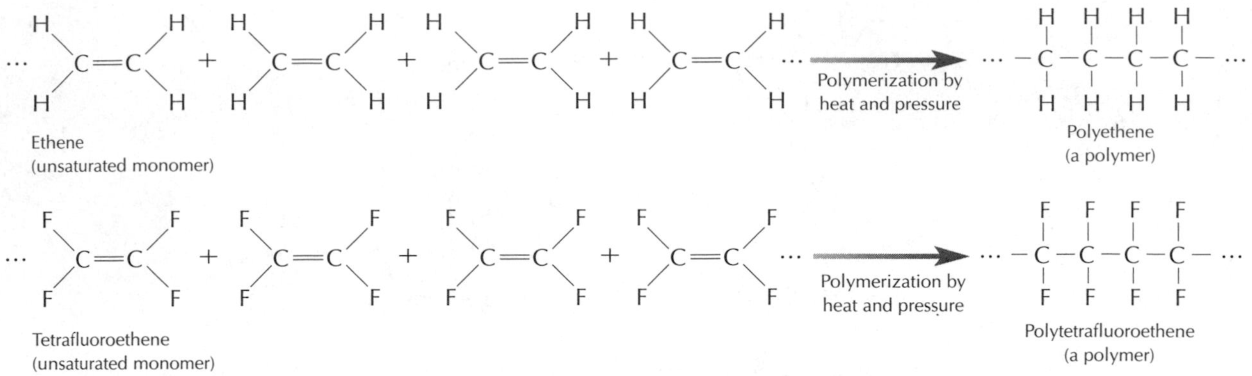

**FIGURE 9.17** Polymerization converts many monomers into polymers consisting of thousands of repeating units. Notice that the 1:2 ratio between carbon and the other type of atom in these examples remains the same after polymerization. Polyethene (or polyethylene) is used as a plastic packaging material. The carbon-fluorine bonds in polytetrafluoroethene (or Teflon) are very strong, so this polymer can resist high temperatures and attack by chemicals; and it remains slippery and quite rigid under pressure. Teflon is used in frying pans and in friction bearings and seals.

floating down a river. In the more flexible plastics, the polymers interweave, like long strands of cooked spaghetti on a plate. Some types of plastic can be molded only once, but most can be melted and reshaped many times.

Common polymer plastics include polyethene and polystyrene (plastic packaging); polyvinyl chloride, or PVC (plastic piping); Lucite and Plexiglas (trade names for a hard plastic often used as a glass substitute); Teflon (the coating on nonstick cooking pans); polyacrylonitrile and polypropylene (textile and carpet fiber); and neoprene (a rubberlike material). Other familiar polymers include nylon and polyester.

**PHOTO 9.7** Polymer plastic sheeting emerging from a processor at a plastic bag factory.

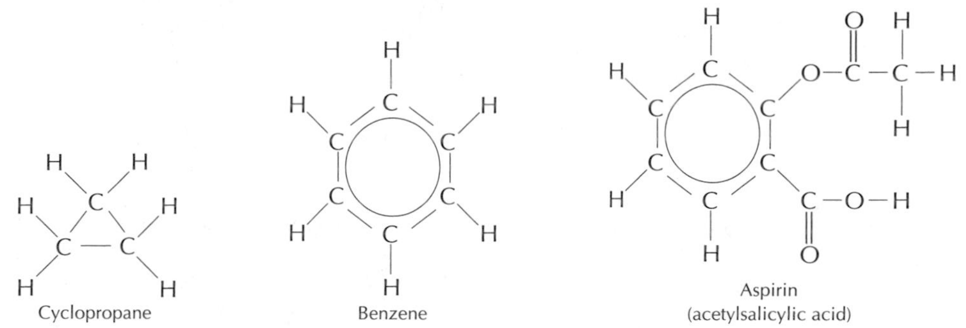

**FIGURE 9.18** Single bonds connect all atoms of a cyclo-propane molecule. An unusual type of covalent bonding connects the six carbon atoms of a benzene molecule: Each carbon atom shares three of its electrons with adjacent carbon and hydrogen atoms, while a total of six electrons (repre-sented by an inner ring in the structural formula) circulate among the carbon atoms in the ring. Molecules of aspirin have a single benzenelike ring of carbon atoms. The number of possible compounds built from one or more carbon rings is virtually endless.

None of the long-chain, synthetic compounds just listed were in use before the 1930s. It is a tribute to modern chemical research and engineering that plastics have by now largely replaced the metal, wood, and other plant materials once found in common consumer products. On the other hand, the easy availability of cheap plastics has led to excessive consumption of products used only briefly and then disposed of. This wastes petroleum (the raw material used in the manufacture of most plastics), wastes energy, and creates the need for more and more landfill space. Fortunately, recycling programs are helping to alleviate the disposal problem.

## Cyclic Structures

In some organic compounds, carbon atoms connect to form a closed, or cyclic, structure. Two examples, cyclopropane and benzene, are diagramed in Figure 9.18. Benzene ($C_6H_6$) is a clear, nonpolar liquid with a strong odor. Benzene and other compounds containing cyclic components of six carbon atoms are called *aromatic compounds* because most of them have distinctive odors. The benzene ring structure itself can act like a building block: it can attach to one or more other benzene rings, and it can form bonds with many other organic molecules.

## Functional Groups

Many organic compounds owe their dominant chemical behavior to the presence of atom groups called *functional groups*. One example is the class of compounds called *alcohols,* identified by the presence of one or more hydroxyl (OH) functional groups. (The hydroxyl group should not be confused with the hydroxide ion, $OH^-$, described earlier.) Two common alcohols, methanol and ethanol, have structures similar to those of the simplest alkanes, methane and ethane (see Figure 9.19), and they can be used as fuel or fuel additives.

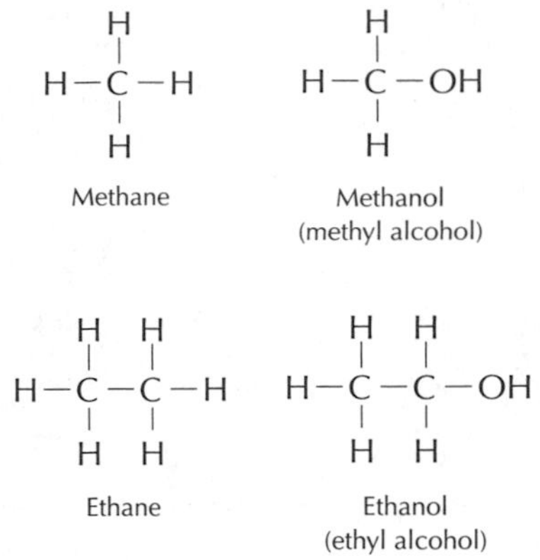

$$
\begin{array}{ccc}
& H & & H \\
& | & & | \\
H-&C&-H \qquad H-&C&-OH \\
& | & & | \\
& H & & H
\end{array}
$$

Methane          Methanol
(methyl alcohol)

$$
\begin{array}{cc}
H\ \ H & H\ \ H \\
|\ \ \ | & |\ \ \ | \\
H-C-C-H & H-C-C-OH \\
|\ \ \ | & |\ \ \ | \\
H\ \ H & H\ \ H
\end{array}
$$

Ethane          Ethanol
(ethyl alcohol)

**FIGURE 9.19**   The replacement of one H atom by an OH group converts methane to methanol and converts ethane to ethanol. Since OH is a commonly recognized functional group of atoms, the line representing the single bond between the O and the H is often omitted in structural formulas.

Ethanol, the "alcohol" in alcoholic beverages, can be manufactured by reacting ethene with water in the presence of a catalyst; but more commonly, it is the result of the fermentation of a sugar solution. Whether it takes place spontaneously in rotting fruit or grain or under controlled conditions in the fermentation tanks of a brewery, the process is sped along by the presence of yeast enzymes acting as a catalyst. One fermentation reaction involves glucose, a sugar. The net reaction that summarizes the complex series of steps associated with fermentation can be written as

$$
\underset{\text{glucose}}{C_6H_{12}O_6} \quad \underset{\substack{\text{catalyzed}\\\text{by yeast}\\\text{enzymes}}}{\rightarrow} \quad \underset{\text{ethanol}}{2C_2H_5OH} \ + \ \underset{\substack{\text{carbon}\\\text{dioxide}}}{2CO_2}
$$

Since chemists deal with dozens of different functional groups, and the compounds formed with the help of each group often number in the dozens or more, it is easy to see how the possibilities of organic chemistry really multiply.

# The Chemistry of Life

On the atomic and molecular levels, there are no fundamental differences between biochemistry—the chemistry of life—and the organic chemistry associated with nonliving matter. Biochemical processes, however, tend to be extremely complex and therefore quite difficult to decipher on the atomic or the molecular level. Nonetheless, a general understanding of biochemistry can be distilled from three facts:

1. The large molecules involved in biochemical processes are polymers. These polymers are made up of monomers that are relatively simple, easily studied, and often easy to synthesize from even simpler chemicals. Plant fiber (cellulose) and starch are polymers of sugar molecules. Proteins are polymers of smaller units called amino acids, which are sometimes referred to as the "building blocks of life."

2. Enzymes (biological catalysts) are important in regulating the rate at which biochemical reactions occur in living organisms. Without enzymes, reactions like photosynthesis and oxidation would occur either too slowly or too rapidly to benefit the organism.

3. The shape of a molecule often determines whether or not it will participate in a biochemical reaction. Today, microbiologists are trying to elucidate the shapes and structures of complex molecules, such as those of enzymes and drugs. With the help of sophisticated computers, researchers can now view and manipulate visual models of these molecules and search for ways to fit them together. Drugs of the future may be selected and designed not by the trial-and-error methods of the past but by computer simulations of their behavior on molecular scales.

Since great complexity often distinguishes the chemistry of life from ordinary organic chemistry, it is fair to ask how this complexity came about. The exact processes leading toward life on Earth may never be entirely deciphered; but a famous experiment, first performed in 1953 by chemists Stanley Miller and Harold Urey, gives some insight into the first small steps. The experiment simulates the action of lightning on Earth's early atmosphere, which may have been a gaseous mixture of water, hydrogen, methane, and ammonia. When electric sparks are introduced into containers filled with gases like these, the gas molecules break apart and recombine to form a variety of organic molecules of considerable complexity. Molecules of certain amino acids (the building blocks of proteins) are often produced.

This experiment and others like it hardly come close to synthesizing life itself, which is a far more complex affair than a mere collection of life's working parts. But the results lead to a possible conclusion that nonliving material—given enough time, a favorable chemical environment, and an abundant influx of energy—could have spontaneously assembled into self-replicating collections of chemicals we would consider to be alive.

## Crystalline Solids

A **crystalline solid** is a substance made of atoms or molecules assembled in some kind of orderly or repeating arrangement. This pattern is often reflected in the shape of the individual crystals, or grains, of the substance. For example, the six-sided structure of a snowflake reflects the underlying hexagonal patterns made by water molecules in the solid state.

All crystalline solids owe their rigidity and their internal symmetry to electromagnetic, or chemical, bonds of some type. For convenience, crystalline solids can be divided into four categories—covalent, ionic, metallic, and molecular—on the basis of the dominant type of bond holding the solid together.

## Covalent Crystals

Covalent bonding in crystals can occur between atoms of the same element, like the carbon atoms of diamonds, or between atoms of different kinds, such as in the networks of silicon and oxygen atoms that comprise many of the minerals in Earth's crust. In diamond, as shown in Figure 9.20(a), each carbon atom is held in place by means of four covalent bonds

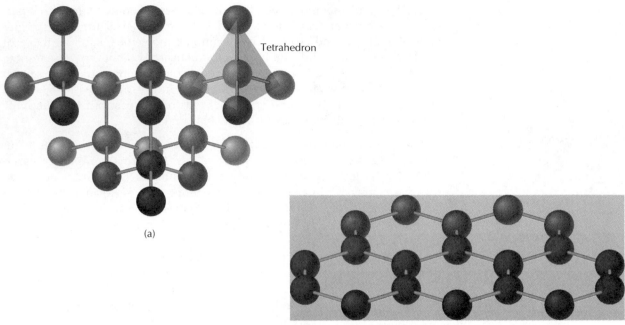

Tetrahedron

(a)

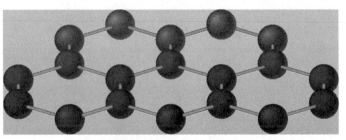

Weak bond between layers

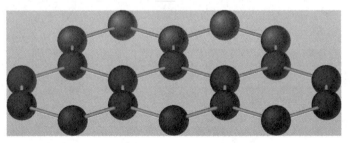

Weak bond between layers

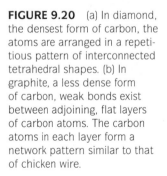

Network of carbon atoms in a single layer

(b)

**FIGURE 9.20** (a) In diamond, the densest form of carbon, the atoms are arranged in a repetitious pattern of interconnected tetrahedral shapes. (b) In graphite, a less dense form of carbon, weak bonds exist between adjoining, flat layers of carbon atoms. The carbon atoms in each layer form a network pattern similar to that of chicken wire.

directed toward four equidistant carbon atoms nearby. Since covalent bonds are generally very strong, covalent crystals are usually very rigid and have high melting points.

Sometimes, rigid covalent bonds within a solid do not apply for all three dimensions. A case in point is graphite, another form of pure carbon, whose structure is shown in Figure 9.20(b). Graphite consists of thin layers of networked carbon atoms that can easily slide over one another. Three

**PHOTO 9.8** This beautiful cut diamond and sooty-looking pile of graphite powder are made up of exactly the same atoms.

electrons from each carbon atom in graphite are shared exclusively with those of its three nearest-neighbor atoms in a common plane, while the fourth electron drifts more or less freely between adjacent layers. The presence of free electrons between the layers weakly binds the layers together. Because the layers can freely slip and slide on microscopic levels, graphite finds good use as a dry lubricant. The relatively large number of free electrons in graphite makes it a good conductor of electricity as well. This property is exploited by test-scoring machines, which can sense the electrical conductivity of pencil marks (pencil "lead" is primarily graphite) applied to the answer sheet.

Recently, more complicated arrangements of carbon atoms have been discovered. These so-called fullerene molecules of pure carbon are described in Box 9.2.

**FIGURE 9.21** Shown here is a tiny portion of the body-centered cubic pattern of ions in a crystal of cesium chloride. Any atom you choose to look at stands at the center of a cube made by eight adjacent atoms.

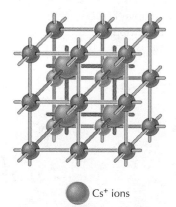

● Cs⁺ ions

● Cl⁻ ions

## Ionic Crystals

As we saw earlier, a face-centered cubic arrangement of ions (as in Figure 9.7) arises when sodium and chlorine combine to form ionically bonded sodium chloride in the solid state. Other geometrical arrangements of ions are possible, such as the "body-centered" cubic arrangement of ions in a cesium chloride crystal (see Figure 9.21). Since ionic bonds are strong, most ionic crystals, like covalent crystals, tend to be rigid and have high melting points.

## Metallic Crystals

Atoms of a metal have one or more outer electrons loosely bound to, and easily lost from, their parent nuclei. In solid metals, these outer electrons skitter about within the lattice of metal ions, behaving much like molecules in an ordinary gas. The negatively charged electron "gas" within a metal acts as a kind of fluid glue that holds together the positively charged metal ions.

The presence of mobile electrons and the relative weakness in the bonds between metal atoms give metals their excellent thermal and electrical conductivity; their shiny, opaque surfaces; their general ability to be bent and shaped; and their ability to combine with other metals in almost any proportion to form an alloy. Some metals are quite soft and have a low melting point, because the cohesive effect of the free electrons in them is relatively feeble. In mercury, which is liquid at room temperature, the

---

**BOX 9.2**

# New Crystal-like Materials

 *Crystalline solids* are made of regularly repeating units—be they atoms molecules, or clusters of molecules—and have definite melting points. *Amorphous* ("without form") *solids,* on the other hand, have no organized internal structure and gradually soften with increasing temperature. Glass is an amorphous solid at room temperature; honey thickens with cold and becomes a not-quite-rigid amorphous solid at very low temperatures.

In the past two decades, scientists have discovered a variety of substances that defy easy classification as either crystalline or amorphous. Discussed here are four new classes of materials: liquid crystals, which have already found important applications in electronic and information technology; and fullerenes, aerogels, and quasicrystals, materials whose future applications are regarded as promising but largely unknown at present.

## Liquid Crystals

Internally, liquid-crystal substances consist of stringy molecules having the ability to align themselves in a common direction (see Figure 9.22). Externally, they can flow just like any other fluid, albeit with some resistance.

The optical characteristics of some liquid crystals are greatly affected by small changes in temperature or by the application of tiny electric currents. These materials have found wide use in a variety of products. Inexpensive liquid-crystal thermometers pressed against a person's skin can now be used to estimate body temperature. A thin layer of liquid crystal sandwiched between conducting surfaces makes up the segments of the familiar block-style numerals of digital wristwatches; when a tiny electric field is applied, molecules in the liquid crystal layer line up in such a way that they can be made to appear opaque, or black.

The liquid-crystal displays (LCDs) found on today's portable or laptop computers are made of hundreds of thousands of *pixels* ("picture elements"),

---

Increasing order ⟶

**FIGURE 9.22** Liquid crystals have molecular arrangements (two of them are shown here) that lie somewhere between the disordered molecular arrangements of liquids and the highly ordered molecular arrangements of crystalline solids.

Liquid          Liquid crystal          Liquid crystal          Crystalline solid

metallic bonds are weak and easily disrupted by internal thermal motions. Tungsten, on the other hand, has strong metallic bonds and possesses the highest melting point of any metal, 3400 °C.

# Molecular Crystals

That bonds can exist between molecules, as well as between individual atoms, is demonstrated by solids such as water ice and carbon dioxide ice (dry ice). These solids are molecular crystals, whose basic units are molecules. In the case of ice, hydrogen bonding tends to link water molecules

each of which can be electronically switched "on" or "off" through instructions sent by the computer's central processing unit. Larger LCD flat screen displays are beginning to replace the bulky vacuum picture tubes traditionally used in desktop computers and television sets.

# Fullerene Molecules

The fullerene family of molecules constitutes a recently recognized third form of pure carbon, after diamond and graphite. Most fullerenes consist of hollow spheres or rounded cages made of 60, 70, and other even-numbered combinations of carbon atoms. $C_{60}$, a soccerball-shaped, almost perfectly round molecule of 60 atoms (see Figure 9.23), bears a strong resemblance to the geodesic domes invented by the engineer-philosopher R. Buckminster Fuller—hence $C_{60}$'s somewhat whimsical common name, buckminsterfullerene, or buckyball, for short. Although the existence of $C_{60}$ was proposed as early as 1966, it was not isolated in quantity until 1990, when researchers succeeded in extracting it from carbon soot, the kind of black soot produced in an ordinary candle flame.

It is likely that fullerenes will not remain mere laboratory curiosities. For instance, when combined with other atoms, $C_{60}$ has demonstrated a variety of electrical characteristics: It can behave as a conductor, insulator, semiconductor, and superconductor. With the help of pulses of light from a laser, researchers can now imprison an atom or molecule within a hollow fullerene cage and then shrink the cage to a minimum size by removing pairs of carbon

atoms from the lattice, a process called "shrink wrapping." One potential application may be the efficient delivery of individual drug molecules, such as those that fight AIDS, into the body. Fullerenelike arrangements of carbon atoms have also been fashioned into tubular, fibrous and sheetlike structures of microscopic size. The new structures possess interesting electronic properties and may be used in the future to produce materials with unparalleled strength. The study of fullerenes may *(continued)*

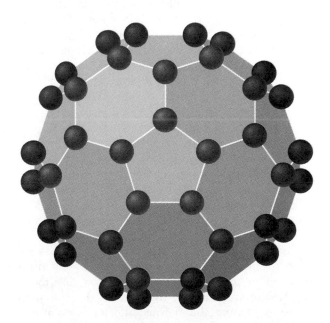

**FIGURE 9.23** The buckyball molecule, $C_{60}$, consists of 60 carbon atoms arranged in a network of 12 pentagons and 20 hexagons. Only one side of the molecule is shown in this illustration.

together in hexagonal patterns. Other molecular solids owe their rigidity to relatively weak forces called *van der Waals forces,* which occur between both polar and nonpolar molecules, as well as between all atoms. Van der Waals forces (which can be explained by the quantum theory of the atom) are so weak that they are often overwhelmed by other forces, such as those associated with ionic and covalent bonds.

Since the bonds associated with molecular crystals are weak and easily broken, molecular solids are soft and have low melting points. Even water ice, a relatively rigid example of a molecular solid, can be easily deformed or melted by applying considerable pressure. We see examples when glaciers flow, when ice-skaters glide (the pressure applied by the thin blades temporarily melts the ice underneath), and when a wire passes through an ice block without cutting it in two (Photo 9.10).

*(Box 9.2 continued)*    well lead to applications as broad and revolutionary as those that resulted from the earlier, pioneering studies of polymer materials.

## Aerogels

Aerogels are the whipped cream of the materials world. These wispy, often transparent solids are made of rigid networks of atoms enclosing innumerable open spaces (or pores). Normally, the pores are filled with air, giving aerogels an appearance likened to solid smoke or frozen mist. Some aerogel materials are rigid enough to support more than a thousand times their own weight without collapsing (see Photo 9.9).

Aerogels made from silica (silicon dioxide), the main ingredient of common sand, have been produced in configurations having densities of just three times that of air. Such lighter-than-a-feather substances can take flight on the slightest breeze. Denser versions of today's aerogels have existed since the early 1930s. Physicists have used them to detain and identify the subatomic fragments that come reeling out of particle accelerator collisions. Space scientists hope to use aerogel cushions attached to orbiting satellites to catch tiny meteoroids without smashing them to powder. Chemists envision loading the porous interiors of aerogels with catalysts to increase the speed and efficiency of reactions that might accelerate fuel and polymer production. A newly developed carbon aerogel has already been used in experimental devices that

**PHOTO 9.9**   The blue haze in this photograph is a silica aerogel, one of the lightest known "solids."

desalinate seawater more economically than conventional methods.

A number of household uses for aerogels have been envisioned, too. A new type of aerogel derived from the organic compounds resorcinol and formaldehyde has thermal-insulating characteristics greatly superior to those of ordinary polyurethane foams. The aerogel's intricate honeycombed structure efficiently traps air and greatly retards convective heat transfer. Insulation materials for homes and for appliances such as water heaters and refrigerators may one day consist of aerogels like these, assuming that cheaper ways to manufacture them can be found. Some aerogels are transparent enough to be used as filler for the

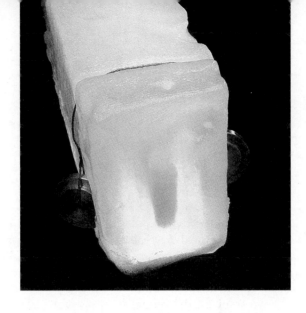

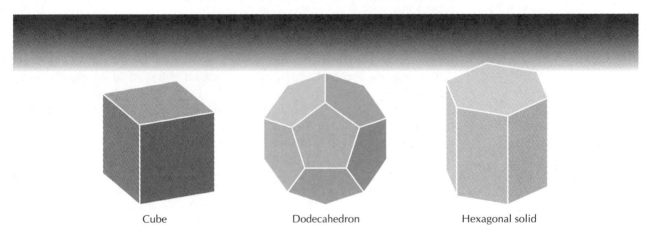

Cube     Dodecahedron     Hexagonal solid

**FIGURE 9.24** Cubes, hexagonal solids, and certain other geometric solids can be packed together with no gaps between units, thus building up the larger units of atoms or molecules we call crystals. Other geometric solids, such as the 12-sided dodecahedron (the near half of which is shown here), cannot fit together in any strictly repetitious pattern, even though each unit has a highly symmetric shape. Materials made of units like these, called quasicrystals, have properties different from those of ordinary crystals.

narrow air spaces between double panes of glass in thermally insulated windows and doors. The widespread application of aerogels in insulation could save energy and replace traditional insulating foams, whose manufacture releases ozone-destroying chlorofluorocarbon (CFC) chemicals to the atmosphere.

## Quasicrystals

When certain metals, such as iron, copper, and aluminum are melted together and quickly cooled, the resulting alloy can solidify into tiny crystal-like grains that cannot fit together in a repeating pattern.

A dodecahedron (see Figure 9.24) is an example of such a grain; there is no way to arrange many of them so that there is no excess space between them. Materials having this kind of structure have been given the name *quasicrystals*.

Applications of quasicrystals are uncertain at present. So far, researchers have been able to synthesize quasicrystal glasses, made almost entirely of metal atoms, which exhibit extraordinary strength and lightness. Other quasicrystal materials may be useful as low-friction coatings.

# Minerals

Earth's *lithosphere,* the crust and the uppermost part of the mantle, is made of **rocks,** which are solid aggregations of one or more minerals. **Minerals,** in turn, are naturally occurring elements or compounds having crystalline structure. Our discussion here centers on minerals; rocks will be discussed in Chapter 16, with an emphasis on the ways they may change over time during the *rock cycle.*

Some minerals can precipitate out of solution near Earth's surface. This happens when calcite (calcium carbonate) crystallizes from mineral-laden water evaporating inside caves. The calcite (a mineral) builds up to form limestone (a type of rock). It also happens when halite (rock salt or sodium chloride) crystallizes from evaporating brackish water.

Most of Earth's minerals, however, crystallize from cooling magma. *Magma* is molten rock beneath Earth's surface. Quick cooling of the magma leads to rocks with tiny (usually microscopic) mineral grains. Gradual cooling usually allows atoms to combine and arrange themselves into geometric patterns having billions or trillions of repeating units. The resulting crystals may be large enough to be visible to the naked eye. Very slow cooling under undisturbed conditions may produce conspicuously large, individual crystals, some of which may be large and beautiful enough to merit being called *gems.*

Dozens of elements participate in the formation of various minerals, and there are many ways in which the same combination of elements can bond together to form various structures. The number of known Earth minerals exceeds 2500, but only about 20 of them are common.

As we see in Table 9.4, the elements iron, oxygen, and silicon account for more than three-fourths of the entire Earth's composition; the crust itself is about three-fourths silicon and oxygen. Since silicon and oxygen are so abundant near Earth's surface, both are involved in the structure of most crustal minerals.

In the chemistry of minerals, silicon plays a similar role to that of carbon in organic chemistry. In nature, silicon never occurs in its semimetallic, elemental state. Rather, it bonds easily to four adjacent atoms or ions of other elements, often oxygen. Unlike carbon, silicon seldom has double or triple bonds, so the number of possible silicon compounds is more limited than those of carbon.

In a wide variety of minerals, silicon and oxygen combine in an approximate ratio of one atom of silicon to two atoms of oxygen. The 1 : 2 ratio is exact for the mineral *silica,* which is also known as quartz or silicon dioxide, $SiO_2$. A multiplicity of minerals contain predominantly silicon and oxygen, plus lesser amounts of metallic atoms; they fall into the class of *silicate minerals.* Silica and silicate minerals make up about 92% of Earth's crust. The remaining 8%, which are *nonsilicates,* includes the carbonates (like calcite), oxides (like the iron or manganese oxides that lend a reddish or dark tinge to some rocks), and the "native" elements (like sulfur, gold, and diamond).

The basic structural unit of silica and the silicates is the silicon-oxygen tetrahedron, illustrated along with some of its variations in Figure 9.25. The tetrahedral units can be linked by means of metal ions; or they can over-

| | Element | Earth's Crust (%) | Whole Earth (%) |
|---|---|---|---|
| **TABLE 9.4** Element Abundance by Mass for Earth | Oxygen | 46.6 | 29.8 |
| | Silicon | 27.7 | 15.6 |
| | Aluminum | 8.1 | 1.5 |
| | Iron | 5.0 | 33.3 |
| | Calcium | 3.6 | 1.8 |
| | Sodium | 2.8 | 0.2 |
| | Potassium | 2.6 | < 0.1 |
| | Magnesium | 2.1 | 13.9 |
| | Nickel | < 0.1 | 2.0 |
| | All others | 1.4 | 1.8 |

lap, with one or more oxygen atoms at the corners being shared with other tetrahedra. Predominantly one-dimensional chain structures, two-dimensional sheets, and three-dimensional networks are all possible as stable structures. The great number and variety of silicate minerals arises because several different kinds of metal atoms can participate in a great variety of structural arrangements.

By bombarding a given mineral sample with X-rays and determining the manner in which the X-rays are diffracted (bent) by the crystal lattice, chemists and physicists can usually discover a mineral's true internal structure. In recent years, this task has been aided by scanning electron microscopes, which produce graphic images of crystal surfaces. Far simpler

**FIGURE 9.25** The silicon-oxygen ($SiO_4$) tetrahedron, with four $O^{2-}$ ions surrounding each $Si^{4+}$ ion, is the basic building block of silica and silicate minerals. In some silicates, such as olivine, the $SiO_4$ tetrahedra are linked by positive metallic ions such as those of iron and magnesium. Other silicate minerals have $SiO_4$ tetrahedra linked in single (shown here) or double chains. One such mineral with a chain structure is asbestos, a substance that has long been used for insulation and fireproofing in buildings. The microscopic fibers of some kinds of asbestos can irritate the lungs and induce lung cancer. In still other silicates, such as mica, the $SiO_4$ tetrahedra are arranged in flat sheets, with metal ions lying between the sheets.

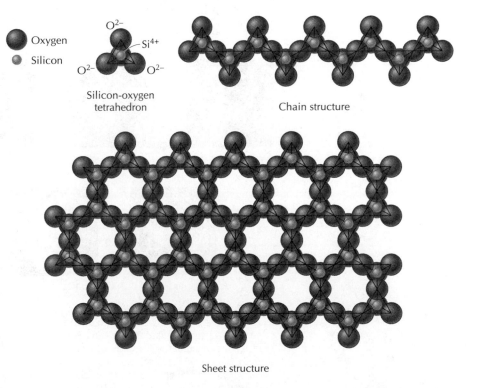

Oxygen
Silicon

$O^{2-}$
$Si^{4+}$
$O^{2-}$   $O^{2-}$

Silicon-oxygen tetrahedron

Chain structure

Sheet structure

methods, involving external characteristics, are helpful in classifying or distinguishing the thousands of different minerals in the field or in the laboratory. Here is a short list of some of the physical properties used by geologists in mineral identification:

• *Color, streak,* and *luster.* Streak is the color of a mineral when crushed to powder form; it may be lighter than the color of a mineral's freshly broken surface. Luster refers to the way a mineral's surface reflects light.

• *Density.* Quartz, for example, has a density of about 2.6 g/cm$^3$, or 2.6 times that of water. Magnetite (magnetic iron ore) has a density of about 5.2 g/cm$^3$.

• *Hardness.* A relative scale of hardness, called the Mohs scale, is often used. Diamond, the hardest mineral of all, is rated 10 on this scale. Talc, a soft, powdery mineral, is rated 1. Quartz rates a 7 on the scale.

• *Crystal form.* When a mineral grain solidifies or precipitates out of solution without being disturbed, its outermost atoms form smooth faces that join each other at sharply defined edges, which creates an overall geometric shape or crystal form. Any mineral can be classified into one of six major crystal groups and numerous subgroups. Two of the simplest subgroups include cubical and tetrahedral forms.

• *Cleavage.* When struck, many minerals have a tendency to split apart along certain planes (see Figure 9.26). Mica, with a sheetlike internal structure, has one cleavage plane. The common mineral feldspar cleaves in two directions, thus forming crystals that can resemble broken, four-sided columns. Calcite, a form of calcium carbonate, cleaves along three planes that are not perpendicular to each other. Some minerals have more than three cleavage planes.

The mining of minerals constitutes a major effort of most industrialized nations. Mineral extraction and processing yields gemstones, a wide

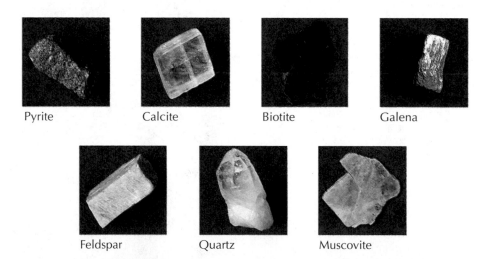

Pyrite    Calcite    Biotite    Galena

Feldspar    Quartz    Muscovite

**PHOTO 9.11**   Several common minerals are shown here. Muscovite (colorless) mica and biotite (dark) mica readily cleave into tablets and sheets. Large sheets of muscovite were once used as window glass by Russians, known as Muscovites—hence the mineral's name. Quartz, one of the harder minerals, does not cleave. The transparent crystals of calcite cleave in three directions. Feldspar, the most common mineral of all, cleaves in two directions. Galena (lead sulfide) cleaves to form cubes with a metallic luster. Pyrite (iron sulfide) is also known as "fool's gold." Small crystals of pyrite may resemble flecks of gold.

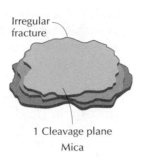

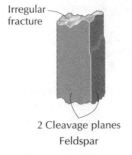

Irregular fracture

Irregular fracture

1 Cleavage plane
Mica

2 Cleavage planes
Feldspar

3 Cleavage planes
Calcite

**FIGURE 9.26** Most but not all minerals exhibit one or more cleavage planes. If broken across (not along) a cleavage plane, a mineral will fracture irregularly.

variety of metals, chemicals for manufacturing, fertilizers for farming, and raw materials such as cement for building.

The study of individual mineral crystals in rocks helps geologists decode the age and origins of the rocks themselves. With radiometric dating techniques (Chapter 12), geologists can determine when a rock or mineral last solidified. Sometimes, the analysis is challenging, as when a sedimentary rock consists of mineral grains of a variety of ages and places of origin. Such a rock could consist of particles swept together by movements of water and winds at different times in the past.

## CHAPTER 9
# Summary

There are only two kinds of electric charge, positive and negative. Bodies of like charge repel each other; bodies of unlike charge attract each other. Positive charge in ordinary matter can be traced to the protons residing in the nuclei of atoms. Negative charge in ordinary matter is carried by electrons. When there are equal numbers of protons and electrons in a body, the body has a net charge of zero. Charged bodies have either an excess or a deficiency of electrons.

Two charged bodies will exert mutual electrostatic forces on each other that are inversely proportional to the square of the distance between the bodies (Coulomb's law). This behavior is reminiscent of the inverse square law governing gravitation. Unlike gravitational forces, which always involve attraction, electrostatic forces involve both attraction and repulsion.

Materials classified as electrical conductors have charged particles that can move freely. In electrical insulators, charged particles can move only with difficulty.

Chemical bonds are essentially combinations of electric forces that hold atoms and molecules together. Ionic bonds feature mutual attraction between ions carrying opposite charges. Covalent bonds feature mutual attraction between atoms that share electrons with each other. A much weaker bond, a hydrogen bond, may exist between molecules that are themselves covalently bonded.

Four types of bonds are responsible for the rigidity of crystalline solids: ionic, covalent, metallic, and molecular bonds. Metallic bonds arise from a "gas" of mobile electrons moving amid metal atoms. These electrons are responsible for the ability of metals to conduct heat and electricity. Molecular bonds involve weak forces between molecules, which constitute the particles of a molecular solid. Molecular solids (such as ice) are somewhat soft and have low melting points.

Electrically polarized molecules (polar molecules), such as those of the polar compound water, tend to attract each other. The tendency for water molecules to stick to each other and to other molecules gives rise to properties such as surface tension and capillary action.

Polar solvents like water will dissolve other polar compounds. Nonpolar solvents, with molecules not electrically polarized, will dissolve other nonpolar substances. Thus, "like dissolves like."

The solubility of solid solutes in liquid solvents tends to increase with temperature. Generally, the more concentrated the liquid solution, the higher is its boiling point and the lower is its freezing point. The solubility of gases in liquids decreases with increasing temperature and increases with increasing pressure.

Aqueous solutions contain ions derived from atoms or atom groups that have lost or gained one or more electrons. Each ion in a solution contributes its own set of properties to the solution. Hydrogen ions in an aqueous solution generate the characteristic properties of acids. Hydroxide ions in solution generate the properties of a base. The neutralization of an acid by a base, or vice versa, is essentially an exothermic reaction between hydrogen and hydroxide ions; water is the product. Neutralization leaves behind certain ions that combine to become a solid salt when the solution is evaporated to dryness.

About 3.5% of the mass of ocean water is dissolved ions, primarily chloride and sodium ions. Ions are added to the ocean by runoff from landmasses and by the injection of minerals through volcanic vents on the ocean floor.

Organic chemistry—the chemistry of carbon—owes its complexity to the ability of carbon to easily bond to itself and to a variety of other nonmetal atoms.

Petroleum and natural gas deposits are the source of the simplest organic compounds, the alkane hydrocarbons, which have single bonds between carbon atoms. The alkanes include a variety of fuels, motor oil, paraffin wax, and other products.

Double or triple bonds can exist between carbon atoms. Cyclic (ringlike) structures, such as benzene, can serve as building blocks for more complex molecules. Functional groups of atoms can be attached to a variety of alkanes and other simple organic molecules to yield many new and different organic compounds. Long chains of carbon atoms, joined to other atoms, form the backbones of polymer molecules, such as those of plastic materials.

Organic chemistry, once thought to be exclusively the chemistry of life, includes biochemistry, but only as a subset.

The silicon-oxygen tetrahedron is the basic structural unit found in most minerals of Earth's crust. These tetrahedra can be linked (often with the help of metal ions) in one-, two-, and three-dimensional patterns, creating a great variety of crystalline forms. Many other external characteristics of a given mineral are associated with the number, arrangement, and bonding of the atoms inside it.

## CHAPTER 9
# Questions

## Multiple Choice

1. Positive charges
   a) always attract other positive charges
   b) always repel other positive charges
   c) attract certain positive charges and repel other positive charges
   d) exert no force on negative charges

2. The electrons in an atom can be thought of as moving around the nucleus, instead of the other way around, because
   a) an electron has more charge than a proton
   b) a proton has more charge than an electron
   c) the electron has more mass than other particles in the atom and therefore more inertia
   d) any nucleon (proton or neutron) has much more mass than any electron and therefore much more inertia.

3. Coulomb's law describes the electrostatic force between two charged bodies. That force depends
   a) only on the charges on the bodies
   b) on the charges and masses of both bodies
   c) on the charges on both bodies and on the distance between the bodies
   d) on the charges and masses of both bodies and on the distance between the bodies

4. Metals are good conductors of electricity because
   a) metal atoms have more electrons circling their nuclei than nonmetals do
   b) metal atoms have many positively charged protons in their nuclei
   c) metal atoms have one or more loosely bound electrons, which can move easily through the metal

d) some metal atoms themselves are charged and can move amid the other atoms of the metal

5. A chemical bond between two atoms or molecules always involves
   a) an attractive electric force
   b) a repulsive electric force
   c) both attractive and repulsive electric forces
   d) strong gravitational attraction

6. Sodium metal and chlorine gas readily combine to form a compound with
   a) ionic bonds between sodium and chlorine atoms
   b) ionic bonds between sodium and chloride ions
   c) covalent bonds between sodium and chlorine atoms
   d) hydrogen bonds between sodium and chlorine atoms

7. Polar molecules
   a) have two opposite magnetic poles
   b) are symmetrical
   c) will not dissolve ionic compounds
   d) are positively charged at one end and negatively charged at the other

8. A polar liquid like water
   a) consists of atoms linked by ionic bonds
   b) will dissolve any other liquid
   c) will dissolve other polar liquids
   d) will dissolve nonpolar liquids

9. Hydrogen bonds are relatively weak chemical bonds that exist between
   a) atoms
   b) ions
   c) electrons
   d) molecules

10. A saturated solution contains
    a) only polar molecules
    b) only nonpolar molecules
    c) the maximum concentration of solute
    d) the maximum concentration of solvent

11. In general, the solubilities of solid solutes and gaseous solutes in a liquid solvent
    a) increase with increasing temperature
    b) decrease with increasing temperature
    c) increase and decrease (solid and gaseous, respectively) with increasing temperature
    d) decrease and increase (solid and gaseous, respectively) with increasing temperature

12. Electrolytes
    a) conduct electricity because they contain dissolved ions
    b) conduct electricity because they contain free electrons
    c) contain more electrons than protons
    d) consist solely of ions

13. The chloride ion ($Cl^-$) in solution is
    a) less reactive than the chlorine molecule ($Cl_2$)
    b) more reactive than the chlorine molecule ($Cl_2$)
    c) as active as the chlorine molecule ($Cl_2$)
    d) not reactive at all

14. An aqueous solution containing a large number of hydroxide ions would
    a) be classified as an acid
    b) be classified as a base
    c) be classified as a nonelectrolyte
    d) taste sour

15. Pure water has a pH of
    a) 0
    b) 2
    c) 7
    d) 10

16. Acid rain falling in heavily polluted regions of the world has had pH values of approximately
    a) 3
    b) 7
    c) 9
    d) 14

17. Acid-base neutralization produces
    a) solutions with a pH of 0
    b) carbon dioxide
    c) nonelectrolyte solutions
    d) water

18. The ions dissolved in the world's oceans
    a) account for over half of the world ocean mass
    b) are primarily hydrogen and hydroxide ions
    c) are primarily sodium and chloride ions
    d) are decreasing in concentration as they combine and form salts that precipitate out on the ocean floor

19. For alkane hydrocarbons, increasing molecular weight (and more carbon atoms) is linked to
    a) higher melting points
    b) higher boiling points
    c) greater numbers of possible isomers
    d) all of the above

20. A saturated organic compound is made of molecules that
    a) must dissolve in water
    b) cannot dissolve in water
    c) have the capability of adding extra atoms to their carbon atoms
    d) cannot have extra atoms added directly to their carbon atoms

21. The conversion of sugar into alcohol and carbon dioxide is called
    a) photosynthesis
    b) distillation
    c) fermentation
    d) oxidation

22. An example of a covalent solid is
    a) ice
    b) diamond
    c) sodium chloride
    d) gold

23. Which of the following kinds of crystals are generally rigid and have high melting points?
    a) ionic and covalent crystals
    b) ionic and metallic crystals
    c) covalent and metallic crystals
    d) ionic and symmetrical crystals

24. Water ice is classified as
    a) a quasicrystal
    b) a covalent crystal
    c) an amorphous crystal
    d) a molecular crystal

25. A large gemstone consisting of a pure mineral could be formed only
    a) underwater
    b) by quick cooling of molten rock
    c) by very slow cooling of molten rock
    d) from molten rock containing carbon

26. The two most abundant elements in Earth's crust are
    a) hydrogen and helium
    b) hydrogen and oxygen
    c) silicon and iron
    d) silicon and oxygen

27. Silica and silicate minerals
    a) make up the majority of Earth's crust
    b) are rich in the element iron
    c) consist of tetrahedral-shaped units containing carbon
    d) consist of silicon, oxygen, and various non-metal atoms

28. Electric currents can be carried by
    a) electrons moving through a metal wire
    b) electrons in substances such as glass and wood
    c) a solution of gasoline and oil
    d) all of the above

# Questions

1. Explain the similarities and differences between gravitational force, as expressed by Newton's law of gravitation, and the electrostatic force, as expressed by Coulomb's law.

2. What is the difference between ionic and covalent chemical bonding?

3. Give some examples of capillary action in the natural world.

4. Why should you never quickly mix together a strong acid and a strong base?

5. Why is organic chemistry (the chemistry of carbon) often so intricate and rich with possibilities?

6. What are isomers?

7. Give some examples of common polymers.

8. What three general characteristics of biochemistry have given us a better comprehension of it?

9. Briefly describe the structures of three forms of carbon: diamond, graphite, and the fullerene family of molecules.

10. What are some common applications of liquid crystals?

11. Describe several ways of distinguishing one mineral from another.

# Problems

1. Two electric charges are separated by 20 cm, and there is a mutual electrostatic force of 1.8 N between them. How much electrostatic force would exist between these charges if the distance were increased to 60 cm?

2. Using the chemical symbols and valences given in Table 9.1, write down the chemical formulas of the following compounds. *Examples*: The compound calcium carbonate is written as

$CaCO_3$, because calcium (Ca) tends to lose two electrons and the atom group carbonate ($CO_3$) tends to pick up two electrons. Therefore, calcium and carbonate pair up one to one. Calcium bromide is written as $CaBr_2$, because each bromine (Br) atom picks up only one electron and two of them are needed to pick up the two electrons from the calcium atom.
a) potassium bromide
b) magnesium oxide
c) magnesium fluoride
d) barium hydroxide

3. Using Table 9.3, decide whether the following compounds are theoretically possible. If the compound is possible, show its structural formula.
a) $C_3H_3$
b) $C_{12}H_{26}$
c) $C_3H_7OH$
d) $C_4F_{10}$

## Questions for Thought

1. What do you think matter in the universe would be like if there were no electrostatic attractions (Coulomb forces) between charged particles?

2. When honey is stored for a long time, it will often crystallize. Bearing in mind that honey is a solution made of sugar (a solute) and water (a solvent) and that sugar's solubility in water increases with temperature, explain (a) why honey crystallizes and (b) what you can do to make the honey clear and free-flowing again.

3. Like land animals, fish breathe oxygen, which they extract from the water they swim in. Most trout thrive in cold, turbulent streams, but they slowly suffocate if the water becomes too warm and stagnant. Why?

4. Do you think that the term *isomer* could refer to two different molecules that are exact mirror images of each other?

5. If life did not arise on Earth owing to natural processes involving chemical compounds native to this planet, can you think of an alternative explanation for its origin?

## Answers to Multiple-Choice Questions

| | | | | | |
|---|---|---|---|---|---|
| 1. b | 2. d | 3. c | 4. c | 5. c | 6. b |
| 7. d | 8. c | 9. d | 10. c | 11. c | 12. a |
| 13. a | 14. b | 15. c | 16. a | 17. d | 18. c |
| 19. d | 20. d | 21. c | 22. b | 23. a | 24. d |
| 25. c | 26. d | 27. a | 28. a | | |

# CHAPTER 10
# Electricity and Magnetism

*The electric generators in this hydroelectric power plant are energy-transforming devices. As falling water pushes against the turbine blades that spin these generators, electrons are pushed through the generator coils. Electric impulses surge down the wires leading from the generators to distant cities and towns where the energy is put to use in a variety of electric and electronic devices.*

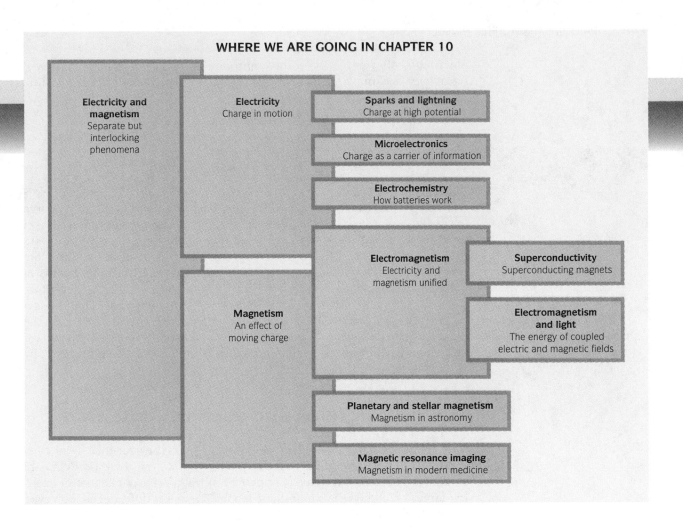

## WHERE WE ARE GOING IN CHAPTER 10

**Electricity and magnetism**
Separate but interlocking phenomena

**Electricity**
Charge in motion

**Sparks and lightning**
Charge at high potential

**Microelectronics**
Charge as a carrier of information

**Electrochemistry**
How batteries work

**Electromagnetism**
Electricity and magnetism unified

**Superconductivity**
Superconducting magnets

**Electromagnetism and light**
The energy of coupled electric and magnetic fields

**Magnetism**
An effect of moving charge

**Planetary and stellar magnetism**
Magnetism in astronomy

**Magnetic resonance imaging**
Magnetism in modern medicine

Electricity and magnetism touch nearly every aspect of our daily lives. Electric motors accomplish much of the work we do in the household; electric lighting banishes the night. We are surrounded at home and at work by an array of information devices—radios, televisions, computers, and others—that operate on basic principles of electromagnetism discovered during the first half of the 19th century.

The practical applications of electricity and magnetism on microscopic and macroscopic scales continue to multiply. Successive generations of computer chips grow in computing power with each new advance in circuit element miniaturization. "Maglev" (magnetically levitated) trains floating on force fields generated by superconducting electromagnets may soon make the leap from the drawing board and test track to widespread use in the United States and other countries.

Our investigation of electricity and magnetism starts with the notion that electric charge can flow from place to place and do work as it goes. Next we move on to magnetism, which arises whenever charges are in motion. We will learn how magnetism, coupled with motion, can produce electricity. After exploring some common and uncommon applications of electricity and magnetism (electromagnetism), we will discover the link between electromagnetism and light.

# Electricity

 Electricity usually refers to *electric current*, which is the movement of electric charge from place to place. In current-carrying metal wires, the charge is carried by swarms of "free" electrons that can more or less freely travel amid the lattice of metal atoms. In electrolyte solutions, dissolved ions are the carriers of charge. Electric currents can also exist in near-vacuum conditions. In the evacuated space inside a television picture tube, beams of electrons flung toward the phosphor screen at the face of the tube constitute an electric current. Even in outer space, currents may flow. The *Voyager* spacecraft discovered that a strong electric current passes between the planet Jupiter and its inner moon Io. The carriers of this current are ionized oxygen and sulfur atoms ejected into space from Io's volcanoes and bent by Jupiter's intense magnetic field.

## Electric Fields and Electric Potential

The immediate cause of any electric current is the presence of an electric field. An **electric field** exists around every charge. Electric fields are similar to gravitational fields (Chapter 5) in that both are capable of exerting forces on bodies that lie in their reach. There are differences, however. Gravitational fields influence *all* material bodies, whether or not they are charged, because every material body has some mass. Electric fields, on the other hand, act only on bodies having electric charge.

Recall from Figure 5.21 how the gravitational field around a large mass acts on a small test mass so as to pull the small mass in the direction of a field line. In an analogous way, we can imagine how an electric field around a body with a relatively large charge on it influences a small test charge placed nearby (see Figure 10.1). The test charge will be pulled along a field line in one direction or the other, depending on the sign of its charge. By common convention, electric field line arrows point away from positive charges and toward negative charges. Therefore, protons and other positive charges tend to move "with" the field lines, and electrons and other negative charges tend to move "against" the field lines.

Unlike gravitational fields, which can only "pull" on a body, electric fields can both "push" and "pull" on a given charge. We see an example in Figure 10.2, where electrons are simultaneously being pushed away from a place of negative charge and being pulled toward a place of positive charge. If these electrons could move without restraint along the field lines, they would pick up speed and gain kinetic energy at the expense of potential energy. The electric field would do work on these electrons, just as Earth's gravitational field does work on masses that accelerate when dropped from a tower. There are several practical ways to exploit the loss of an electron's potential energy as it is pushed and pulled through an electric field. For

**FIGURE 10.1** Like gravitational fields around spherical or point masses, the electric field around a spherical or a point charge can be drawn by using radial lines. The closer together the lines, the stronger the field. Electric field lines are considered to point away from positively charged bodies and toward negatively charged bodies. Therefore, any positive test charge in the field experiences a force parallel to the field lines, and any negative test charge experiences a force antiparallel (opposite) to the field lines. If the object in the middle of this diagram were negatively charged, all arrows shown on this diagram would be reversed.

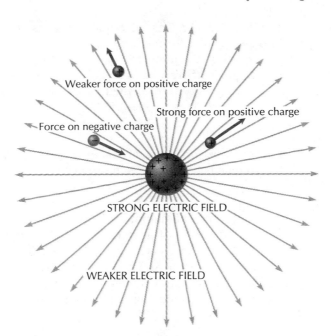

instance, moving electrons can be used to produce heat, light, and mechanical energy.

In Chapter 6, we learned that the gravitational potential energy of a mass depends on its *height* above some chosen reference level. In an analogous way, we can think of a charged particle (typically an electron) as possessing electric potential energy simply by virtue of its *position* in an electric field. If the charged particle is allowed to "fall" through an electric field, it loses some of its electric potential energy, just as a mass does when it falls

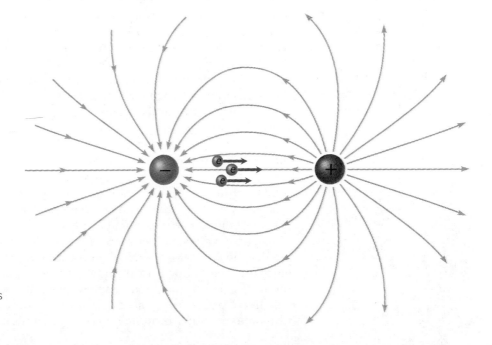

**FIGURE 10.2** Electrons (e) tend to migrate away from a region of negative charge and toward a region of positive charge. As the electrons move from left to right, they lose potential energy in the electric field. The lost potential energy is transformed into kinetic energy or other forms of energy.

to a lower height. That potential energy is not lost; rather, it is transformed into some other form of energy.

In the study of electricity, the analog of height above a reference level is **potential difference,** or *voltage* (*V*). It is defined as the amount of potential energy transformed (work, *W*) when a given amount of charge (*q*) moves through an electric field:

$$V = W/q \qquad\qquad\qquad \text{Equation 10.1}$$

From SI units for work and charge, the joule (J) and the coulomb (C), the unit for voltage becomes the joule per coulomb, or the *volt* (V). By definition,

1 J/C = 1 V

According to Equation 10.1, 1 C of charge moving across a potential difference of 1 V will transform 1 J of potential energy into other forms. If 1 C of charge moves across 2 V, then 2 J of energy are transformed. If 2 C of charge move across a potential difference of 2 V, then 4 J of energy are transformed. In the falling-mass analogy, this is equivalent to saying that the more mass a body has and the farther it can fall, the more work it does when it falls.

Picture a rock poised at the edge of a high cliff ready to fall. We could say that the rock is at a large potential because it is poised at a high place and will give up a lot of potential energy if it falls to the bottom of the cliff. Similarly, any electron or other charged body has a certain potential to move in response to an applied electric field. If the electron cannot move (perhaps because it is stuck inside an electric insulator), then its potential energy remains the same. If it can move (say, through an electric conductor like a metal wire), then work is done. The greater the potential difference (voltage) an electron moves through, the more work it does. Electric sparks and lightning (see Box 10.1) are examples of electrons moving in a very strong electric field and giving up much potential energy as they do.

Imagine that some device (a battery or an electric generator) can maintain a positive charge in one place and a negative charge in another place. A potential difference would then be maintained between these two spots, which we will call terminals. If the terminals are separated by air, no charge flows, because air is an insulator. However, once a wire is connected between the two terminals, free electrons in the metal start to flow through the wire from the negative terminal to the positive terminal. These electrons collide with the fixed atoms in the metal's crystal lattice, the atoms vibrate more vigorously, and the wire gets warm or hot. The greater the voltage between the two terminals, the more violent the collisions are and the faster the thermal energy is produced. If 1 J of thermal energy is transformed for each coulomb of charge (the charge on roughly $10^{19}$ electrons) moving through the wire, then the potential difference between the terminals is 1 V. If 2 J of thermal energy are produced by 1 C of charge, then the potential difference is 2 V; and so on.

A wire connected between the two terminals of an electric battery or an electric generator is part of an **electric circuit:** a looping path that the electrons can follow. Part of the circuit is inside the battery or generator; the larger part of the circuit usually lies outside. The battery or generator acts like a pump that pushes electrons out of one terminal and pulls electrons into the other terminal; there is no gain or loss of electrons in the circuit. In the case of a simple wire circuit, thermal energy is released. This process underlies the operation of electric resistance heating, which is used in electric toasters, stoves, and ovens. (*Note:* Never connect an ordinary wire across the two terminals of a large battery, such as a car battery. Usually, so much current flows through the wire in this "short circuit" that either the wire melts or the battery explodes owing to the heat generated inside.)

Many other kinds of devices capable of various tasks can be part of a circuit. For example, when enough thermal energy is concentrated at a particular spot along a conducting path inside an incandescent lightbulb, some

## BOX 10.1
# Sparks and Lightning

The small sparks of static electricity you have seen and heard crackling during dry weather and the intense flickering and percussive sound of a lightning bolt accompanied by thunder are really the same phenomenon. Both are caused by a separation of charge, which produces an

electric field strong enough to ionize air molecules. Air is normally a good insulator; but when ionized, it gives up electrons that respond freely to the strong push and pull of the electric field. These speeding electrons collide with atoms in the air, giving them extra kinetic energy. This heats and expands the air and stimulates some atoms to release light.

The frictional kinds of charging discussed in the early part of Chapter 9 can give rise to small sparks across strong but very localized electric fields. A hundred thousand volts of potential difference may develop between two points charged by ordinary friction a few centimeters apart. Relatively few electrons need to move in order to equalize the imbalance of charge, so the spark is relatively weak and lasts a very short time. Common static electricity, then, involves high voltage but little current—and therefore not much energy flow.

Lightning is caused by frictional charging on a grand scale. Friction caused by winds within a cloud can strip electrons from water droplets or ice crystals and thereby produce an imbalance of charge. If the imbalance of charge is severe enough between oppositely charged regions, the air in between is ionized and electrons flow quickly, giving up large amounts of potential energy.

A typical lightning bolt, which has a potential difference of hundreds of millions of volts across it, releases about 100 kilowatt-hours (360 million joules) of energy in less than a second. This bolt can flash within a cloud, between parts of separate clouds, or, most spectacularly, between the bottom of a cloud and the ground. The latter type of lightning may occur when the negatively charged bottom of a cloud induces the opposite charge on the ground below (see Figure 10.3).     *(continued)*

**PHOTO 10.1**   Spark plugs initiate the combustion of gasoline and oxygen in the cylinders of older car engines. The 40-kV pulse delivered to this spark plug easily ionizes the air in the gap so that a spark jumps across.

of that thermal energy is converted into light. In fluorescent lights, there is a more direct transformation of electric energy into light, which is why fluorescent lighting yields comparatively more light and less heat than incandescent lighting. Yet another application is the electric motor, in which electrons moving through wire coils exert forces that can do work. But circuits need not consist of a single pathway; for example, a circuit with two lightbulbs in *parallel* with each other (see Photo 10.2) has two separate loops for electrons to follow.

To facilitate further discussion of electric circuits, we now define electric current and resistance.

*(Box 10.1 continued)*    A cloud-to-ground lightning bolt is complex, often involving many quick surges of electrons making their way to the ground through the same ionized column of air. The total charge transferred during a large strike is surprisingly small—only a few coulombs—but during the few microseconds each individual surge is taking place, the current is enormous. Peak temperatures of up to 30,000 K in the ionized channel create the blinding flash. The sudden expansion of air around the visible streak produces the crack of thunder we hear from afar.

**FIGURE 10.3** In this common scenario of cloud-to-ground lightning, the polarization of charge within a cloud repels electrons on the ground below, leaving a positive charge at the surface. During the lightning strike shown, electrons move downward.

The most common earthly targets for cloud-to-ground lightning are things with sharply pointed edges that stand head and shoulders above neighboring objects. There are several reasons for this: Excess charge tends to accumulate on the sharp edges or pointed ends of any conducting or partially conducting object. Most tall objects (be they trees, buildings, or even people standing on flat, open ground) are pointed. Coulomb's law tells us that the smaller the distance between two charged regions—a charged cloud and an object beneath it in this case—the stronger the electric force field. Coulomb's law also explains how a large reservoir of charge can induce a separation of charge on bodies nearby (as detailed in the previous chapter in Figure 9.3).

Metal lightning rods, whose pointed top ends reach above buildings in lightning-prone areas, serve two purposes. First, some of the extra charge accumulating on a lightning rod tip tends to leak into the air above, thereby weakening any electric field between the rod and a charged cloud above it. Second, if a bolt does strike in the immediate area, it will likely hit the rod rather than any part of the building itself. The lower end of a lightning rod is driven deep into the ground, so that when the tip is struck, current flows safely through the rod and down into the ground.

About 100 fatalities are attributed to lightning annually in the United States. Golfers in lightning-prone areas such as Florida are prime targets, since they are often caught out in the open in flat terrain or huddled under trees that may draw lightning.

**PHOTO 10.2**   Electric circuits come in many configurations. In this two-loop *parallel circuit,* each lightbulb is connected directly to the source of power. If one bulb burns out, the other can continue to shine. Christmas tree lights are wired in parallel for this reason. If all the lightbulbs on the string were, instead, arranged in a *series circuit*—one by one along a single loop of wire (like children holding hands in a circle)—then the entire circuit would be vulnerable to the failure of just one lightbulb.

# Electric Current and Resistance

**Electric current** ($I$) is a measure of the rate of flow of electric charge—that is, charge divided by time.

$$I = q/t$$                                                    Equation 10.2

One coulomb of charge flowing in 1 s is, by definition, 1 *ampere* (A), the SI unit of current:

$$1 \text{ C/s} = 1 \text{ A}$$

Amperes are commonly referred to as "amps," and current is loosely referred to as "amperage." Roughly 1 A of current flows through a 100-W lightbulb. Smaller units of current include the milliampere ($1 \text{ mA} = 10^{-3} \text{ A}$) and the microampere ($1 \text{ } \mu\text{A} = 10^{-6} \text{ A}$). The values of current and potential difference are easily measurable in any part of a circuit by instruments called *ammeters* and *voltmeters.* Both are used to sample and measure very small currents diverted from the circuit.

The electric current moving in a wire has a definite direction associated with it. In this book we adopt the convention that the current direction is the same as that of the moving electrons in a circuit—that is, away from a battery's negative terminal and toward the positive terminal. In other texts, an older, opposite convention may be employed: that a positive current flows around the external part of a circuit from the positive terminal to the negative terminal. From a practical standpoint, either convention, when consistently used, can be considered correct.

The **resistance** ($R$) of a circuit element (a wire or some other device) is defined to be the ratio between potential difference across the element and the current through it:

$$R = V/I$$                                                    Equation 10.3

The standard unit for resistance, 1 V/A, is given the name *ohm* (symbolized by the Greek letter $\Omega$):

$$1 \text{ } \Omega = 1 \text{ V/A}$$

In most common metals and a few other nonmetallic conductors as well, the ability to resist current remains fairly constant over a broad range of applied voltage and current. These materials are said to be *ohmic.* For circuit devices made of these materials (generally known as *resistors*), Equation 10.3 becomes what is called **Ohm's law.** Ohm's law can be written in three equivalent forms:

$$I = V/R \qquad V = IR \qquad R = V/I$$

Strictly speaking, Ohm's law works only for resistors kept at a constant temperature. Resistance in ohmic devices generally increases with temperature.

By Ohm's law, the more voltage there is across a resistor, the more current that voltage pushes through the resistor. (Think of voltage as being analogous to the "push" of water pressure in a water hose. The more pressure there is, the greater the flow of water through the hose.) Alternatively, when voltage is kept constant, resistance and current are inversely proportional. That is, the greater the resistance, the smaller the current; and the smaller the resistance, the greater the current. (Again, for the hose analogy, a small-diameter hose, having a lot of resistance to the flow of water, allows relatively little water to flow. A large-diameter hose, having less resistance to the flow of water and fed by the same water pressure, allows more water to flow.) For example, a 1-$\Omega$ resistor passes 1 A of current when connected to a 1-V source of electricity. A 0.5-$\Omega$ resistor passes 2 A when connected to the same source, and a 2-$\Omega$ resistor passes 0.5 A.

Ohm's law is not a fundamental physical law but, rather, a statement of how *most* electrically conducting materials behave when various currents pass through them. Nevertheless, Ohm's law finds frequent use in the analysis of simple and complex circuits.

Many materials do not obey Ohm's law at all. These *nonohmic materials* include ionized gases (for example, the glowing gases in neon signs and fluorescent lights) and semiconductors. Semiconductors are doubly unusual in that their resistance to the flow of current also depends on the current direction. Because of this property, they can be fashioned into devices called *diodes,* which can act as one-way gates in an electric circuit. (Other semiconductor devices called *transistors* are the key elements of today's miniaturized electronic circuits; see Box 10.2.) In the sense that Ohm's law is widely, but not universally, applicable, it is similar to the 10 °C rule (Chapter 6), which applies to most, but not all, chemical reactions.

When we visualize a simple electric circuit, it is often helpful to picture water flowing through a pump and a pipe (see Figure 10.4). The stronger the pump used in a simple hydraulic circuit, the higher the pressure is, and the faster the water tends to flow. The water flow, or current, also depends on the characteristics of the pipes in the system. Smaller-diameter pipes produce more friction and reduce the current. Longer pipes also tend to reduce the current of water. You can check this out the next time you water your lawn. As long as there is adequate pressure and flow in your plumbing system and the faucet is open all the way, short hoses of a given diameter deliver more water than do long hoses of the same diameter.

The electric battery or generator in a circuit is the analog of a pump; the higher the voltage, the greater its ability to push and pull electrons through the circuit. The longer or thinner the wire in an electric circuit, the more resistance it has. The more resistance in a conducting path, the less the electric current is.

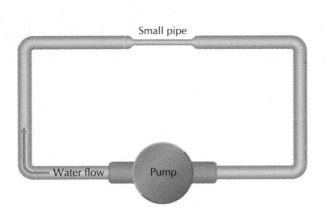

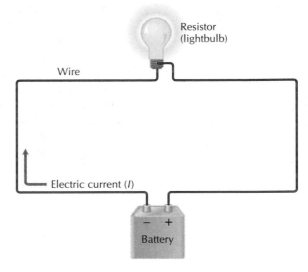

**FIGURE 10.4** By analogy with a closed hydraulic system, electricity in a circuit can be thought of as a current of electrons being pushed around a loop. A resistor (such as the filament of a lightbulb) resists electric current in much the same way that a small pipe restricts the flow of water. In both cases, the restriction of current generates thermal energy. Except when heating is desired, the wires in an electric circuit are designed to have low resistance so that they carry current between circuit elements without much heat loss.

The property of electric resistance is a mixed blessing. Resistance of any kind generates heat. We use resistors to generate heat (as in a toaster) and also to distribute various amounts of current through complex circuits. In many instances, however, we would be better off without the waste heat generated by currents in wires. After all, the primary purpose of wires is to guide electricity from place to place without any losses of energy.

**Problem 10.1**
A current of 1 A is passing through an incandescent lightbulb. There are 100 V of potential difference across the tungsten wire (filament) inside the bulb. Use Ohm's law to find the resistance of the filament.

**Solution**

$$R = V/I$$
$$= (100 \text{ V})/(1 \text{ A})$$
$$= 100 \text{ }\Omega$$

Technically, the resistance of the filament increases somewhat with temperature, so the 100-$\Omega$ value is correct only when the filament is carrying 1 A of current and glowing white hot.

The water flow analogy we are using for circuits illuminates a surprising fact about electric current. Even though an electric *impulse* (any change in current) travels at a substantial fraction of the speed of light from one end of a wire to the other, the moving electrons themselves drift along the wire at a very slow rate: only a few millimeters per second in a typical current-carrying wire! Water behaves similarly: If you force water into one end of a water-filled pipe, water squirts out the other end immediately. Water is virtually incompressible, so it is not necessary for water molecules to traverse the whole length of the pipe for current to flow. Similarly, electrons repel each other and cannot be squeezed together in a wire. When electrons are pushed at one end of a wire, there is an almost immediate response all along the wire.

## Electrical Safety

Ohm's law sheds light on the dangers of electric shock. Human tissues consist of electrolyte solutions capable of conducting some current. Considered as a resistor, your body's interior—from hand to hand, or from hand to foot—has a resistance of about 1000 $\Omega$. Dry skin, however, acts as an insulating layer on the body; its resistance is as much as 100,000 $\Omega$. Wet skin

---

### BOX 10.2
# Microelectronics

The past few decades have seen an exponential growth in the power and capability of electronic circuits and a corresponding shrinkage in the volume occupied by those devices. Today, the circuitry engraved on a single computer chip or microprocessor the size of a fingernail can be vastly more complex and powerful than the earliest electronic computers of the 1940s, which filled entire rooms.

What we refer to as *power* here is simply the ability of a complex electronic circuit to manipulate information in numerical form. Nearly every thing or property of a thing that can be described in words or pictures can also be modeled as a string of numbers. Some things, like money, are already in numerical form. Once reduced to numbers, information can be tabulated, summarized, or processed in some other meaningful way. (This is not to say that numbers can precisely represent every phenomenon; even words and pictures are imperfect at representing many things.)

The faster a microelectronic circuit can manipulate numbers and the more numbers it can manipulate at the same time, the more efficient it can be at its appointed task. The task might be to tabulate payroll figures for a large corporation, guide a spacecraft to its proper target, control a graphic display on a computer screen, or keep track of repetitive pulses inside a digital wristwatch.

Electronic computers are the most versatile information-processing devices ever invented. Although computers can produce net outputs that seem intelligent, the tiny switching units (transistors) that make up the heart of every computer are single-mindedly stupid (see Figure 10.5). Each transistor either lets electrons flow (the "on" state) or prevents electrons from flowing (the "off" state). Consequently, the native language of computers is binary arithmetic, which is a way of manipulating numbers written with only two kinds of digits, ones and zeros.

A circuit combining 16 transistors and 2 resistors is capable of the addition of two binary digits; more complicated circuits are needed for operations like subtraction, multiplication, and division of bina-

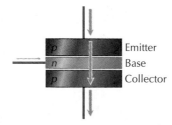

**FIGURE 10.5** A transistor is made of two dissimilar semiconductor materials, each typically silicon or germanium mixed with tiny amounts of another element. In the *pnp* transistor here, the flow of electrons from the *p*-type material of the emitter toward the *p*-type material of the collector (thick gray arrows) can be controlled by the charge applied to the *n*-type material of the base (thin gray arrow). When used in computer circuits, transistors function only as simple switches.

contains dissolved salts (electrolytes) that make it ineffective as an insulator. Small potential differences (such as 1.5 or 6 V) never produce much current in an unprotected body, but potential differences of 100 V or more across a body always do, especially when wet skin is involved.

You would almost certainly be electrocuted if you stood with bare feet on wet concrete (0 V) and with a wet hand grasped a 110-V hot wire. Under those circumstances, the current would be about 0.1 A. Such a current is extremely painful, causes involuntary muscle contractions, disturbs the rhythm of the heart, paralyzes breathing, and is lethal if not interrupted quickly. If your skin were perfectly dry under the same circumstances, the current might be 100 times smaller—not lethal, perhaps, but still very painful.

In order for current to flow in a wire (or in a partial conductor like a human body), there must be complete, or "closed," circuit for the electrons to follow. There must also be a significant potential difference to get the electrons moving. Once moving, those electrons will tend to follow the

ry numbers. More circuitry still is needed to translate computer inputs, such as keystrokes on a keyboard, into binary numbers and to send instructions to output devices such as visual displays, loudspeakers, printers, plotters, and modems.

Today's most powerful home computers contain microprocessors with millions of transistors, each capable of switching on and off millions of times every second. Future generations of microprocessors will be more powerful than today's, not only because they will incorporate more transistors but also because electrical engineers will continue to shrink the size of transistors and connecting wires. The quest for smaller size in microelectronics is important because the speed of electrons traveling from one circuit element to the next is finite (somewhat less than the speed of light). This limits the rate at which a signal can pass from one part of the computer to another. Furthermore, since large packets of electrons are harder to start and stop and move from place to place, efforts are underway to reduce the electric currents flowing between circuit elements as much as possible.

Considerable research is also being devoted to a fundamentally different strategy: optical computing. This approach relies on manipulating pulses of light by means of tiny laserlike devices that would play the same role as transistors do in conventional information-processing circuits. Practical optical computing is probably a long way off, but laser light, at least, is already proving its worth as an effective carrier of information over long distances. Several networks of fiber-optic cables, capable of carrying enormous amounts of information in the form of laser light, are now being strung across North America and around the world.

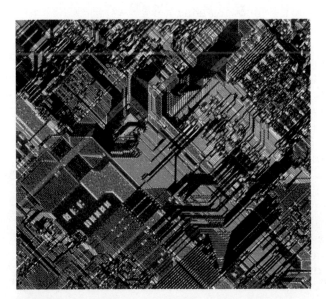

**PHOTO 10.3** Modern microprocessors may contain millions of circuit elements.

path of least resistance. Think of a bird standing on a single, bare-metal, high-voltage wire; it experiences no harm because the resistance of the bird's body, as measured between its two feet, is enormously greater than the resistance of the short segment of wire lying between the bird's feet. The potential difference between the bird's feet is minuscule, and the vast majority of the current passes through the wire and not the bird. If, on the other hand, a bird's wings were to simultaneously come in contact with two wires having thousands of volts of potential difference between them, a large current would leap through the bird and incinerate it. Of course, high-voltage wires are normally strung far enough apart to prevent this type of thing from happening.

Rubber or plastic insulation around wires and other conducting parts of household circuits usually prevent a person from coming in contact with current-carrying pathways. As another safely measure, appliances with metal exterior parts are "case-grounded," which means that a wire connects all structural parts made of metal in the appliance to the ground (Earth itself is essentially uncharged and considered to be at 0 V). Thus, in the event of a short circuit caused by a frayed or broken wire inside, the metal case will remain at the same zero electric potential as the ground. Figure 10.6 shows how grounding works in typical household electric circuits.

## Electric Power

In Chapter 6, we learned that power is the rate at which work is done (or energy transformed). That is,

$$P = W/t = E/t$$

Also, potential difference is defined as the work done per unit charge, and current is defined as charge divided by time:

$$V = W/q \quad \text{and} \quad I = q/t$$

From these three equations, the following equation for electric power can be derived:

$$P = VI \hspace{6cm} \text{Equation 10.4}$$

**Electric power** ($P$), then, is equal to the product of potential difference and current. The SI unit for electric power is the watt (W), just as it is for any other type of power. For instance, 1 A of current flowing through a lightbulb at 100 V releases 100 W of light and heat. Recall from Chapter 6 that 100 W are 100 J of energy released each second.

Remember that power is not energy; it is the rate at which work is done or, equivalently, the rate at which energy is expended. Electric energy can be expended at various rates and for different tasks. When paying an electric bill, for instance, you are purchasing electric energy (and the services that go along with it). You are not paying for power. That is, your monthly bill reflects the total quantity of electric energy you use in that month, not the rate at which you use it at various times.

The typical billing unit for electric energy is the kilowatt-hour (kW·h), which is the equivalent of 1 kW of power used over a period of 1 h. This is

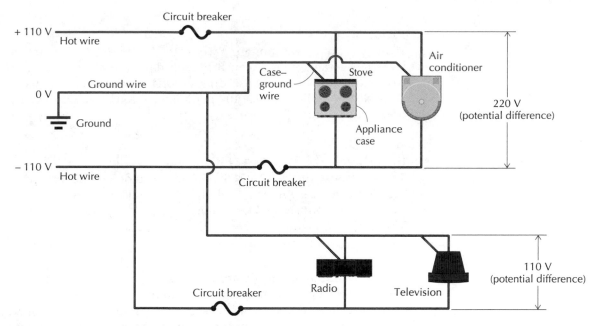

**FIGURE 10.6** Typical household circuits are connected to electric utilities by means of three wires. One wire is grounded, while the other two "hot" wires carry currents that continually alternate in direction (this form of current is called *alternating current,* or ac). The alternating currents in U.S. households produce *average* potential differences of about 110 V relative to the ground wire. Appliances that use large amounts of energy, like electric stoves and air conditioners, are designed to run on ac potential differences of about 220 V, so they are connected across the two hot wires of the circuit. Nearly all other devices run on 110 V; they are connected between the ground wire and either hot wire. For safety reasons, appliances with metal cases or exposed metal parts have a third (case-ground) wire connecting this metal to the ground wire of the circuit. If a hot wire inside an appliance ever breaks or loses its insulation and comes into contact with the case, excess current is immediately diverted along a path through some device (a circuit breaker or fuse) that quickly opens the circuit. For an appliance without a case-ground wire, a person touching a hot-wire case could experience a severe shock if he or she were in electrical contact with the ground at the same time.

the amount of energy expended by a 1000-W toaster if it is used for 1 h; or the amount of energy used by a 100-W (0.1 kW) lightbulb when illuminated for 10 h.

How much energy is a kilowatt-hour in joules? To find out, we shall solve the equation $P = E/t$ for $E$ and work out the result by converting to units of joules and seconds:

$$E = Pt$$
$$= (1 \text{ kW})(1 \text{ h})$$
$$= (1000 \text{ J/s})(3600 \text{ s}) = 3,600,000 \text{ J} = 3.6 \text{ MJ}$$

In the United States, 1 kW·h, or 3.6 MJ, of electric energy can be purchased for roughly 10¢ through electric utilities. The use of a typical 1-kW toaster does not contribute much to a household electric bill, since it is used only briefly and occasionally. Devices that stay on many hours a day (televisions, computers, lighting fixtures, and the like) individually draw less power than a toaster but use much more energy over the long run. A refrigerator, whose compressor motor runs a good fraction of the time, can be a big-ticket item. So, too, are air conditioners and all-electric heating systems in regions of the country that demand their frequent use.

**Problem 10.2**

While you are on vacation for a week, the 500-W (0.5-kW) motor in your refrigerator is on about 8 h each day. The various electric clocks throughout the house draw a total of 25 W continuously. Your electric utility charges 10¢ per kilowatt-hour. What is the cost of electricity during that week?

**Solution**

In one week, there are $7 \times 24$ h = 168 h.
The clocks use (0.025 kW)(168 h) = 4.2 kW·h of electric energy.
The refrigerator is on for a total of $7 \times 8$ h = 56 h.
The refrigerator uses (0.5 kW)(56 h) = 28 kW·h of electric energy.
The total electric energy used is 32.2 kW·h, and the cost is $3.22.

Electric utilities nationwide are encouraging the development and use of energy-saving devices, ranging from compact fluorescent lightbulbs designed to replace ordinary incandescent lamps, to more efficient refrigerators that draw less power. Conservation efforts such as these are reducing or eliminating the need to build more power plants in the future and limiting the consumption of the nonrenewable fuels that keep most of our nation's electric generators spinning.

# Electrochemistry

The wet-cell storage batteries used in cars and the dry-cell batteries used in flashlights are examples of **electrochemical cells.** They are devices that transform chemical energy (electron potential energy) into electric energy by means of exothermic chemical reactions. Electrochemical cells make use of complementary reactions called *oxidation* and *reduction* reactions.

**Oxidation,** in chemistry, is the process by which electrons are lost from an element or compound. The element or compound losing electrons is said to be oxidized. Oxygen often, but not always, plays the role of oxidizer in oxidation reactions. Oxygen in the air oxidizes materials undergoing ordinary combustion. Chlorine, however, can oxidize materials, too: for example, it can oxidize sodium to form sodium chloride.

**Reduction,** in chemistry, is defined as the gain of electrons by an element or a compound. The element or compound receiving electrons is said to be reduced. For example, oxygen itself is reduced when it oxidizes something, and chlorine is reduced when it combines with sodium to form sodium chloride.

If oxidation and reduction reactions are linked so that electrons given up during one reaction can travel through an external circuit and become available as reactants at the site of the other reaction, then both reactions can occur and current will flow through the circuit. An internal connection must be made by means of an electrolyte solution (a liquid or a paste) that allows ions to travel from one reaction to the other. This strategy, outlined in Figure 10.7, is employed in electric batteries.

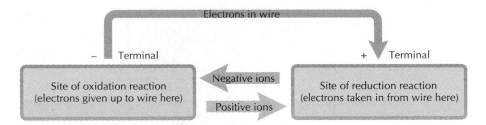

**FIGURE 10.7** In this schematic of a idealized electro-chemical cell, an excess negative charge builds up on the left because electrons are being released by the oxidation reaction there. These electrons travel through a wire to reach the reduction reaction on the right. While the electrons are flowing, positive ions produced on the left and negative ions produced on the right flow through an electrolyte solution between the sites of the two reactions.

Many different pairs of oxidation-reduction reactions are used in various electrochemical cells. In the storage batteries for cars, the reactions take place among plates of lead (Pb) and lead dioxide ($PbO_2$) immersed in a solution of sulfuric acid ($H_2SO_4$), as shown in Figure 10.8. These and other rechargeable batteries make use of reactions that are easily reversible. Energy must be put back into a battery in order to recharge it.

Oxidation-reduction reactions are also used in electrolysis, a method of forcing certain endothermic chemical reactions to occur by means of electricity. A simple example is the electrolysis of molten sodium chloride (Figure 10.9), a process that yields sodium metal and chlorine gas.

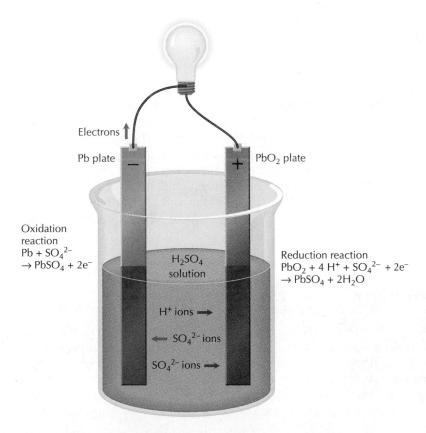

Oxidation reaction
$Pb + SO_4^{2-} \rightarrow PbSO_4 + 2e^-$

Reduction reaction
$PbO_2 + 4H^+ + SO_4^{2-} + 2e^- \rightarrow PbSO_4 + 2H_2O$

**FIGURE 10.8** In this sketch of one cell of a lead-acid storage battery, oxidation and reduction reactions take place whenever electrons are allowed to flow between the negative and positive terminals. In the oxidation reaction on the left, two electrons are given up when one atom of lead and one sulfate ion react to form lead sulfate. After passing through the electric circuit, the two electrons participate in the reduction reaction on the right, which produces lead sulfate and water. As the reactions continue, both plates become coated with lead sulfate, and the acid bath becomes increasingly dilute. Finally, there are not enough ions present in the bath to keep the reactions going, so the battery becomes discharged. Unlike most kinds of dry cells, lead-acid batteries can be recharged by forcing electrons back into the negative terminal and drawing them off at the positive terminal. An external source of electric energy must be used for this process. Freshly charged cells of the lead-acid type provide about 2 V between the two terminals; a series of six such cells make a standard 12-V car battery.

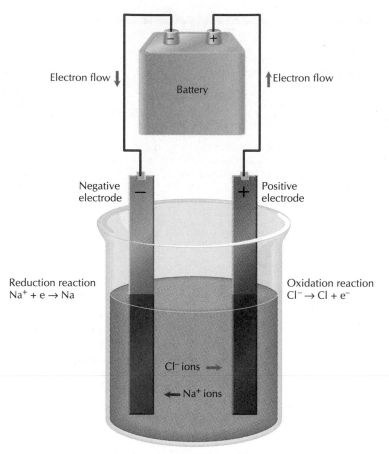

**FIGURE 10.9** When molten sodium chloride is electrolyzed, sodium ions and chloride ions migrate to electrodes kept at negative and positive potentials by an external source. On the negative electrode, each sodium ion gains one electron and becomes a sodium atom. Sodium metal accumulates on the electrode. On the positive electrode, each chloride ion becomes a chlorine atom by losing one electron. Chlorine atoms quickly pair up to form chlorine gas molecules, which bubble up from the solution. The net reaction can be written as $2NaCl \rightarrow 2Na + Cl_2$. (*Note:* This is *not* a suitable home experiment. Chlorine gas can be deadly when inhaled, and sodium metal is highly reactive and dangerous.)

Electron flow    Battery    Electron flow

Negative electrode    Positive electrode

Reduction reaction
$Na^+ + e \rightarrow Na$

Oxidation reaction
$Cl^- \rightarrow Cl + e^-$

$Cl^-$ ions →

← $Na^+$ ions

Beaker of molten sodium chloride

Another practical application of electrolysis is *electroplating,* in which one kind of metal can be deposited on an object made of another metal (base metal). When electrons are forced into a base metal immersed in a solution containing positively charged metal ions, the ions are attracted toward the surface of the base metal, give up their electrons there, and become metal atoms, which coat the base metal. Electroplating by chromium metal is used to protect steel from corrosion and to dress up accessories, as in the "chrome" bumpers and trim of older cars. Cheap metallic jewelry often consists of precious metals such as gold or silver electroplated onto common base metals.

# Magnetism

Like gravity and static electricity, magnetism involves interactions between bodies that are not visibly touching each other. Magnetism's importance goes far beyond the curious behavior of toy magnets. Earth itself is permeated with a magnetic field that stretches far into outer space. Magnetic fields trillions of times stronger than Earth's field exist in the tiny regions of space surrounding protons and electrons. An extremely weak magnetic field suffuses the enormous space of our home galaxy, the Milky Way galaxy. For over a century now, magnetism has figured in the practical generation of electric power. And for the past four decades, magnetic media such as audiotape, videotape, and floppy and hard disks have been used to store ever-greater amounts of information. Despite its intricacies, the study of magnetism can begin with simple magnets and the magnetic fields that surround them.

## How Magnets Work

The magnetic properties of a dark iron-containing mineral called *magnetite* ($Fe_3O_4$), or "lodestone," have been noted since antiquity. Lodestones are natural magnets, capable of attracting nearby iron-containing bodies. The ancient Chinese noticed that a thin sliver of lodestone, when allowed to turn freely in a horizontal plane, has a tendency to align itself along a north-south line. Indeed, you can see this behavior yourself if you suspend an ordinary bar magnet horizontally on a thread (taking care that there are no other magnets or large metal objects nearby). The cause has been recognized for four centuries: Earth acts as if it were a giant magnet itself, with a global magnetic field, or magnetic "sphere of influence," that permeates the planet and its surface (Box 10.3). Since Earth's magnetic field is fixed to the planet's geography, the compass has long been the quintessential tool of navigation, especially at sea, where navigational landmarks are often hard to come by.

An ordinary bar magnet has two different poles, one at each end. By convention, the north-pointing (or north-seeking) end is called the N or *north pole* of the magnet, and the south-seeking end is called the S or *south pole* of the magnet.

Pick up two bar-shaped magnets and tinker with them. In certain configurations, they attract each other; in other configurations, they repel each other. If the ends of each magnet are properly labeled, it does not take long to realize that a repelling force always occurs between magnetic poles of like kind, and an attraction occurs between magnetic poles of unlike kind (see Figure 10.10). The simple rule governing interactions between magnetic poles can be stated as follows: Like poles repel each other; unlike poles attract each other. This behavior is reminiscent of the way electric charges interact, but we must keep in mind that, despite some similarities, *magnetic poles are not the same thing as electric charges.*

Careful experiments involving interactions between long, thin bar magnets whose poles are well separated from each other demonstrate another similarity between electric and magnetic forces: The force of attraction or repulsion between any two magnetic poles depends on the "pole strength"

Like poles repel each other; unlike poles attract each other.

**FIGURE 10.10**   Like poles of magnets repel each other; unlike poles of magnets attract each other.

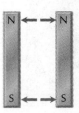 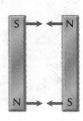

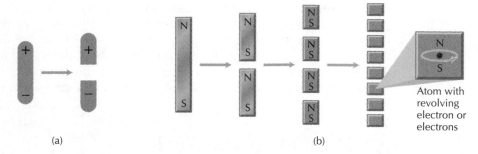

(a)                                              (b)

**FIGURE 10.11** (a) When cut in half, each piece of an electrically polarized object retains its electric charge. (b) When cut in half, a bar magnet is transformed into two separate magnets, each with its own north and south poles. Further subdivision leads to more and more magnets, each having poles with weaker and weaker pole strengths. Even if a magnet is whittled down to the very atoms that make it up, those atoms act as tiny magnets themselves, each with a north and a south pole.

of both poles. Also, the magnitude of the force is inversely proportional to the square of the distance between the two poles. This result recalls Coulomb's law, which governs electrostatic forces between charges.

Here are some of the differences between magnets and charged bodies:

1. Magnets normally are neither charged nor electrically polarized. Magnetic interactions are not caused by an imbalance of electric charge.

2. When a body having some positive or negative charge is sliced in two, the result is simply two different bodies having the same kind of charge. Furthermore, we can isolate the opposite charges on an electrically polarized body by cutting it in half [see Figure 10.11(a)]. If we try to isolate the poles on the ends of a magnet by cutting the magnet in half [Figure 10.11(b)], it never works. We get two new magnets, each with its own set of north and south magnetic poles!

How finely must a magnet be diced before its poles become isolated? As far as we know, it cannot be done, because the atoms that make up magnets behave as tiny magnets themselves. So, too, do the individual protons and electrons that compose those atoms. Some physicists suspect that magnetic *monopoles* (isolated magnetic poles) may exist in the tiniest spatial realms, but so far all attempts to find and isolate such monopoles have been unsuccessful.

**PHOTO 10.4** Magnetic fields easily penetrate most materials, including human tissue.

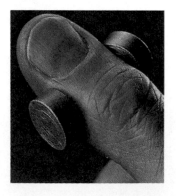

# Magnetic Fields

What really is the cause of the magnetism exhibited by magnets? An important clue is offered in the way that a small magnet behaves when placed near a larger bar magnet or near a current-carrying wire looped into a coil. The small magnet acts as a probe, capable of testing any magnetic properties that might exist in the space surrounding the larger magnet or the coil of wire. (Recall from our discussions so far of gravitational and electric fields how we can map their strength and direction by means of small test masses and charges.) The force field that arises because of the pres-

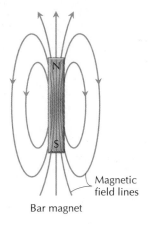

Magnetic
field lines

Bar magnet

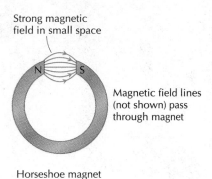

Strong magnetic
field in small space

Magnetic field lines
(not shown) pass
through magnet

Horseshoe magnet

**FIGURE 10.12**   Unlike electric field lines, which start and end at positive and negative electric charges, magnetic field lines always form closed loops. On the outside of a bar magnet, the magnetic field is strongest at the two poles, where the field lines are close together. Inside the magnet, the field is even stronger. The external magnetic field of a bar magnet can be concentrated in a small volume of empty space by bending the magnet into an open-ring or horseshoe shape.

ence of a magnet is called a **magnetic field.** Some typical magnetic field mappings are plotted in Figure 10.12 and pictured in Photo 10.5.

A magnetic field consists of imaginary lines that can be traced with the help of a compass, which is simply a small, thin bar magnet balanced so that it can turn freely. By definition, the direction of a magnetic field line at any point is taken to be the same direction indicated by the N end of a compass needle placed there. Experiment shows that a compass needle behaves similarly around both a bar magnet and a current-carrying helical coil of wire (see Figure 10.13). The compass needle aligns itself on a field line so that its N end is attracted to the S end of the magnet and its S end is attracted to the N end of the magnet. Around the current-carrying coil, the behavior of the compass needle is similar, though the direction indicated by the compass needle can be easily reversed by reversing the current in the coil.

The behaviors illustrated in Figure 10.13 suggest that the same basic mechanism operates in a bar magnet and a coil. We already know that current-carrying wires have moving electrons; so clearly, the magnetism of the coil is due to the motion of electric charges moving in a circular

**FIGURE 10.13**   A small test magnet (a compass needle, for instance) can be used to investigate the magnetic field around a common bar magnet (a) and a helical current-carrying coil of wire (b). In both cases, the compass needle behaves similarly, and it can be used to trace out some of the imaginary "lines of force," or magnetic field lines, that allow us to visualize the magnetic field around each object.

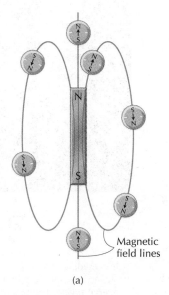

Magnetic
field lines

(a)

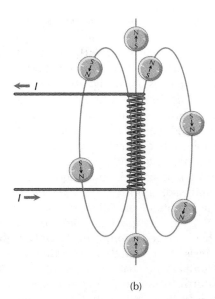

(b)

**PHOTO 10.5** Iron filings can be used as tracers of magnetic field lines. The horseshoe magnet shown here induces magnetism in the tiny slivers of iron, and they in turn align themselves parallel to the magnet's magnetic field.

## BOX 10.3
# Planetary and Stellar Magnetism

Early adherents to Copernicus's heliocentric theory thought that magnetic forces might be the cause of the invisible forces that kept planets and moons moving along their predictable paths. Isaac Newton showed otherwise with his brilliant explanation of universal gravitation. Magnetism's true role in astronomy was revealed much later, when the tools and techniques of modern physics were employed to detect and measure magnetic fields permeating the sun, the planets, and even distant stars.

No sophisticated equipment is needed to detect and map Earth's magnetic field; a compass needle, which turns to align itself with the direction of the field lines, works just fine. The needles in some specialized compasses are designed to swing vertically, thus indicating any *dip* in the direction of the magnetic field lines. A simplified picture of Earth's magnetic field emerges after extensive mapping: Earth behaves magnetically as if a bar magnet were embedded near its center (see Figure 10.14). This imaginary bar magnet is tilted about 12° relative to Earth's spin axis approximately in the direction of the 70° W longitude meridian. Thus, in the United States, at least, compass needles accurately point toward true (or geographic) north only if the compass user is near a path from Minnesota to east Texas. Compass needles point as much as 20° west of true north in Maine and as much as 25° east of true north in Washington State. Users of compasses

**PHOTO 10.6** The *aurora borealis* (northern lights) shown here and the *aurora australis* (southern lights) center on the upper atmospheric regions above Earth's north and south magnetic poles. The auroral glow is caused by solar wind particles colliding with atoms in the upper atmosphere. These particles are steered along Earth's magnetic field lines, which converge toward the planet's magnetic poles.

pattern. Further experiments show that the magnetic forces exerted on the compass needle by the coil increase in proportion to the current in the coil and fall to zero if the current is cut altogether. More detailed experimentation demonstrates that moving charges of any kind, not just electrons moving in a wire, can produce a magnetic field that can, in turn, exert forces on any test magnet lying within the field. These facts suggest the following general rule: Magnetic fields originate from moving charges.

Indeed, all magnetic fields are thought to originate from charged bodies that either move or spin. The magnetism in a magnet is primarily the effect of electrons revolving around atomic nuclei and spinning on their axes. (As we will note later, scientists do not know whether or not electrons actually "spin"; what is certain is that they *act as if* they were tiny spinning particles of negative charge.) Pairs of electrons that spin or revolve in a common direction contribute to a stronger magnetic field. Pairs whose spins or rev-

> Magnetic fields originate from moving charges.

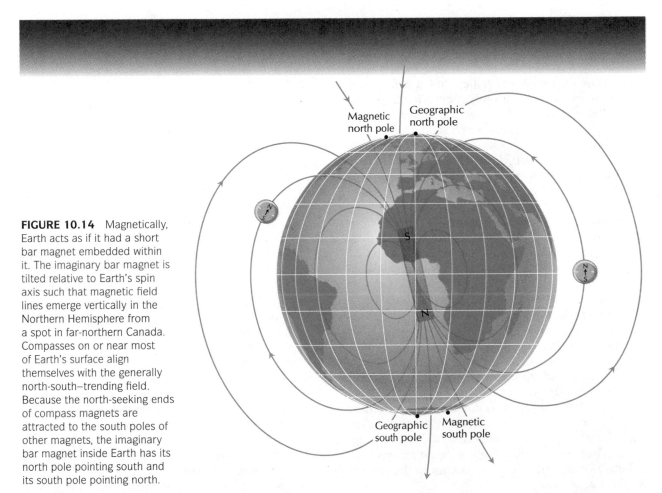

**FIGURE 10.14** Magnetically, Earth acts as if it had a short bar magnet embedded within it. The imaginary bar magnet is tilted relative to Earth's spin axis such that magnetic field lines emerge vertically in the Northern Hemisphere from a spot in far-northern Canada. Compasses on or near most of Earth's surface align themselves with the generally north-south–trending field. Because the north-seeking ends of compass magnets are attracted to the south poles of other magnets, the imaginary bar magnet inside Earth has its north pole pointing south and its south pole pointing north.

need not be confused, since corrections, known as *magnetic declination angles,* are known for virtually every spot on Earth's surface.

That a giant bar magnet does not exist inside Earth is clear. We have evidence from studies of seis-

mic waves that the planet's outer core, at least, is molten. There is also very strong evidence that Earth's magnetic field has reversed itself on a semi-regular basis approximately 300 times in the past 170 million years. When igneous rock    *(continued)*

olutions are opposite cancel each other. Most substances cannot be magnetized, because their atoms have internal fields that tend to cancel each other.

Ordinary permanent magnets are composed of **ferromagnetic material**—various iron, nickel, and cobalt alloys (and certain ceramic materials)—that can retain their magnetism indefinitely. The atoms of these substances act as minuscule magnets, which in turn cluster into microscopic regions called *magnetic domains* where all the atoms are aligned with each other. Each domain acts as a small magnet that is billions of times more strongly magnetized than an individual atom. If many tiny domains are made to align in a common direction inside a given body, or if domains aligned in a particular direction can be made to grow at the expense of other domains (see Figure 10.15), the result is an overall magnetic field that extends well outside the body. One way to organize a ferromagnetic material's domains (or to "magnetize" it) is to expose it to a strong, steady external magnetic field.

---

*(Box 10.3 continued)* cools from a molten state, tiny magnetic grains crystallize and are frozen in alignment with the currently prevailing magnetic field. Radiometric dating (see Chapter 12) has revealed the ages (the time since solidification) of a great many igneous rock formations around the world. The orientation of magnetic grains found in these formations has revealed how Earth's magnetic field has changed in strength and direction through time.

Apparently, Earth's magnetic field stems from its relatively rapid rotation and the presence of an electrically conducting fluid inside. As explained by the leading theory of planetary magnetism, called the *dynamo effect,* Earth's liquid metal outer core is stirred by convection. Earth's relatively rapid rotation forces these currents into circular patterns (in much the same way that the Coriolis force affects circulation patterns in the atmosphere). The circular patterns constitute current loops, and many current loops of similar orientation produce an overall dipole magnetic field similar to that around a coil or a bar magnet. It is difficult to explain how these currents can occasionally reverse themselves and thereby produce reversals in Earth's magnetism, but at least there is a precedent for this type of behavior in what we know occurs in the sun (as we shall soon see).

The planetary dynamo effect seems plausible when we look at the magnetic fields of several of the other planets. (These fields have been measured during spacecraft missions to every planet except Pluto.)

- Mercury, a small and slow-rotating planet with a solid iron core, lacks both qualities that give rise to planetary magnetism: rapid rotation and an electrically conducting *fluid* interior. Its very weak field is possibly a vestige of an earlier magnetic field that was frozen into its solid interior.
- Venus, which is similar to Earth in size and internal structure, has liquid metal in its interior but rotates at the sluggish rate of once per 243 days. Mars rotates only slightly more slowly than Earth; but because Mars is small, its interior has cooled and solidified. Both Venus and Mars have magnetic fields of negligible strength.
- Jupiter's enormously strong magnetic field (ten times stronger at its surface than Earth's field and vastly larger in extent) comes as a surprise. Jupiter is composed primarily of hydrogen and helium, with small traces of metals. Throughout much of its volume, however, Jupiter's hydrogen is compressed to a *liquid metallic state* with a density near that of water. Hydrogen becomes a good conductor when its atoms are squeezed together tightly. Jupiter also spins rapidly, completing one rotation in less than 10 h.
- The other Jovian planets, Saturn, Uranus, and Neptune, share many similarities with Jupiter, including their magnetism. Each has a magnetic field, though somewhat weaker than Jupiter's.

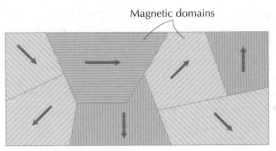

Magnetic domains

Unmagnetized ferromagnetic material

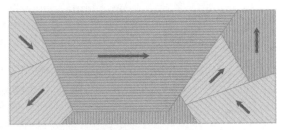

Magnetized ferromagnetic material

**FIGURE 10.15**  When ferro-magnetic materials are magne-tized, domains with orientations similar to that of the external magnetic field grow at the expense of domains not aligned with the external field.

When matter glows, its emitted light can be spectroscopically analyzed to yield information about the strength and direction of any magnetic fields present. For this reason, much is known about our sun's surface magnetism (and, to some extent, the magnetism of other stars). The sun's overall magnetic field turns out to be relatively weak; but here and there, and at certain times, the magnetic field lines concentrate into small volumes of space on or near the sun's photosphere. These regions of strong magnetism (thousands of times stronger than Earth's field) are associated with the dark cool areas on the photosphere called *sunspots* and with *solar prominences,* the great curving columns of gas that arch above some sunspot regions.

Magnetically driven solar activity rises and falls in intensity on the *11-year cycle* mentioned earlier in Chapter 7. The cycle is a bit complicated because the polarity of the magnetic field associated with sunspots reverses itself with each successive cycle. The 11-year cycle is really half of a complete 22-year cycle in which the sun's magnetism "flips" and then "flops" back again. Spectroscopic studies of nearby stars have revealed that the sun is not unique in this kind of cyclical behavior.

A widely supported partial explanation of the cycle invokes the sun's observed differential rota-tion: The sun, being a ball of fluid, rotates so that points on its equator complete one rotation in about 25 days and points near its poles complete one rota-tion in about 30 days. Near the minimum of a solar cycle, the magnetic field lines of the sun form a rel-atively simple pattern. In time, the differentially rotating plasma of the sun distorts and twists these field lines so that eventually they pinch together, erupt through the surface, and produce sunspots and other unusual activity. Eventually, the field lines break and rearrange into a simpler pattern, heralding the return of another minimum in the cycle.

When the cores of certain massive stars collapse to form *neutron stars* a mere 20 km or so across, their magnetic fields collapse as well and undergo a tremendous amplification in intensity. Near a neu-tron star's surface, the field is some $10^{12}$ times stronger than Earth's and $10^{10}$ times stronger than the sun's average field at the surface. Neutron stars typically spin very fast (many times a second) and can emit electrons that get caught on the magnetic field lines anchored to the star itself. Narrow beams of radiation produced by the emerging electrons may sweep across the sky as the neutron star turns. If Earth happens to be in the line of fire of such a beam, we see a burst of light (and other electromag-netic radiation) each time the star spins once. Such neutron stars are known as **pulsars** for their ability to pulse like a rotating lighthouse beacon.

Permanent magnets may not retain their magnetism forever. They can be "demagnetized" through excessive heating, mechanical jarring (for instance, by hammer blows), or exposure to a strong, rapidly fluctuating external magnetic field. These procedures tend to scramble the domains and randomize their orientations.

Pure iron and the softer alloys of iron contain domains that will align easily to an external field and yet spontaneously revert to an unaligned state as soon as the external magnetic field is removed. A paper clip does this when clinging to a strong magnet. The attached clip acts as a magnet,

## BOX 10.4

# Magnetic Resonance Imaging

The technique of *magnetic resonance imaging* (MRI) has recently come into wide use in the health field. Physicians use these images to help diagnose internal injuries or disorders. MRI, which produces computer-generated images of various tissues inside the body, is typically safer and more effective than (and more expensive than) conventional X-raying.

MRI exploits the fact that protons (which are the nuclei of ordinary hydrogen atoms) act as if they are tiny magnets. Protons are a part of nearly every molecule inside the body (think of water, $H_2O$, as an example), and their density at any point in the body depends on the chemical makeup of the tissues present there. Like compass needles, protons have a tendency to align themselves to any magnetic field impressed upon them. Unlike compasses, however, protons also precess (wobble) when subjected to the field. The precession frequency (the rate at which the protons wobble) depends on the strength of the external magnetic field.

Strong magnetic fields are needed to keep protons significantly aligned and precessing, so the MRI unit used for medical imaging typically employs superconducting electromagnets capable of producing fields tens of thousands of times the strength of Earth's field. In order to detect protons, the MRI unit bombards the patient (who is already permeated by the strong magnetic field) with short bursts of radio waves equal in frequency to that of the precessing protons. Each burst temporarily knocks the protons out of alignment, but they quickly spiral back into position, emitting their own weak radio waves as they do so. Detectors surrounding the patient pick up these faint returned signals and a computer processes them, thereby mapping out their points of origin. After synthesizing all the information, the computer constructs, in cross-sectional slices, a map showing the internal structure of the patient's body.

Although not nearly as invasive as X-rays, MRI subjects patients to magnetic fields much stronger than those experienced in everyday life, with possible deleterious effects. A new, totally passive technique called *positron emission tomography* (PET) may supplant MRI in the future. PET scans rely on the detection of incredibly weak signals spontaneously generated by the nuclei of atoms inside body tissues.

**PHOTO 10.7** The technique of magnetic resonance imaging for medical diagnosis requires that a patient remain motionless for up to an hour. The many successive thin sections of the patient's body read by the MRI machine can be assembled into a three-dimensional model by computer.

capable of picking up other clips; but it loses nearly all of its magnetism once it is removed from the strong magnet. When a cylindrical bar of soft iron is placed inside a current-carrying coil, the magnetic field inside the coil induces an alignment of magnetic domains inside the bar, which greatly intensifies the field. But if the current in the coil is stopped, the field collapses. Here, then, is an **electromagnet** that can be turned off and on at will or varied in strength by means of changing the current in the coil!

Electromagnets find hundreds of applications in modern life. Large, crane-mounted electromagnets are used in scrap yards to lift heavy pieces of metal. Small electromagnets can serve mundane uses, such as powering doorbell ringers, buzzers, and loudspeakers; or more exotic uses, such as writing millions of bits of information on the magnetic disks in home computers. The most powerful electromagnets employ coils of superconducting wire (see Box 10.5) capable of supporting large electrical currents without any loss of energy due to heating.

## Electromagnetism

The function of an electromagnet illustrates one linkage between electricity and magnetism: Electricity produces magnetism. Is the reverse true? Will magnetism produce electricity? The answer is yes, but only under certain circumstances. The circumstances involve either a particular kind of motion between charged bodies and a magnetic field, or charged bodies under the influence of a magnetic field that is changing in strength.

## Magnetic Forces on Moving Charges

A nonmoving electric charge in a stationary magnetic field experiences no force. (This observation alone tells us that electric and magnetic fields are distinct from each other.) However, when a charge moves such that it crosses magnetic field lines, the charge experiences a force. Curiously, this force acts neither in the direction the charge is moving nor in the direction of the magnetic field. Rather, it pushes the charge in a *sideways* direction, perpendicular to the magnetic field.

The magnetic force on moving charges is exploited in television picture tubes: An "electron gun" at the back of the tube accelerates electrons and sends a narrow beam of them toward the center of the picture tube screen. Before traveling far, the electron beam is steered horizontally and vertically by rapidly changing currents in a yoke of electromagnets placed around the neck of the tube. The beam, which also varies in intensity, makes hundreds of horizontal scans across the face of the tube, starting at the top and ending at the bottom, every 1/30 second. Thus, 30 new "pictures" are painted every second on the phosphorescent coating on the screen, giving the illusion of a moving picture. In color picture tubes, three beams scan simultaneously to activate red, green, and blue phosphorescent spots on the screen. The next wave of television technology (high-definition television, or HDTV) will feature more scan lines for better spatial resolution and more images per second for smoother motion.

The magnetic force on a moving charge has been harnessed in a number of ways to convert electric energy into mechanical energy. To visualize how this might be done, imagine electrons flowing through a stationary wire above the pole of a magnet (see Figure 10.16). The electrons cross magnetic field lines and therefore are deflected sideways. Since they are confined to the wire and do not normally escape into the air, the entire wire experiences the same sideways push. As it is pushed sideways, the wire may do work by applying a force to something that moves in the direction the wire is pushed. Furthermore, if the current reverses in the wire, the

---

## BOX 10.5
# Superconductivity

Under ordinary conditions, all conductors resist the flow of electrons to some extent. Normally, the free electrons moving about the atomic lattice of the material drift across the material only by means of an applied potential difference. This movement constitutes a current. Once drifting, the electrons bump their way through the vibrating atoms and tend to lose their acquired speeds in collisions. This generates thermal energy, which comes at the expense of electric energy. The situation is analogous to a car moving at constant speed down a flat highway: The car's engine must continually expend energy to overcome air resistance and other forms of friction. So that thermal energy is kept to a minimum, electric wires are usually made of copper and aluminum, because these metals have minimal resistance to the flow of drifting electrons (and they are relatively inexpensive, besides).

In metals, resistance declines as temperature decreases. However, an amazing transformation occurs in some metals when they are chilled to temperatures near absolute zero: Their resistance to dc current disappears completely and they become *superconductors*. The phenomenon of zero resistance, known as **superconductivity,** was discovered in 1911, not long after researchers were able to attain laboratory temperatures below 4 K with the use of liquid helium as a refrigerant. Hundreds of different elements, metal alloys, and compounds are now known to exhibit superconductivity. In every case, the transition between normal conductivity and superconductivity occurs abruptly at a certain *critical temperature,* which is typically very low.

The leading theoretical explanation of super-conductivity relies on a quantum mechanical

description of electrons that pair up and move without any hindrance through the chilled material. From an operational standpoint, a superconducting material can be thought of as one in which no applied potential difference is needed to maintain a given current, once the current has been started. "Supercurrents" in closed loops of superconducting wire have been observed to persist with no measurable loss for periods of years. This is analogous to a car, once accelerated to cruising speed, being able to somehow coast with its engine off down a flat highway forever!

A great deal of electric energy could be saved if electrical transmission lines and wires in electrical devices could be fashioned out of superconducting materials. However, remember that ordinary resistance returns if the wires are not kept chilled below their critical temperature. The maintenance of such low temperatures requires sophisticated refriger-

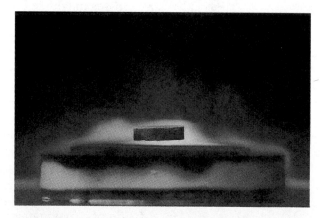

**PHOTO 10.8**    Magnetic levitation is accomplished here by one of the new high-temperature superconductors, a yttrium-barium-copper oxide ceramic. An ordinary magnet floats above the current-carrying superconducting ring below, which is being cooled by liquid nitrogen.

wire moves in the opposite direction. Current-carrying wires interacting with changing magnetic fields can both push and pull.

If current is sent alternately one way and then the other through a wire many times a second, the current is called **alternating current,** or ac. Common household electrical current is ac. One-way current is called **direct current,** or dc; it is used in automobile and flashlight batteries. Alternating current in a wire positioned across a magnetic field will make the wire vibrate at the characteristic frequency of the ac, which is 60 hertz (Hz) for household current in the United States and 50 Hz in some other

ation equipment, which operates on rather large amounts of energy.

Today, superconducting electromagnets are used in a wide variety of applications. Huge currents can flow through the closed coils of these magnets without the deleterious heating effects that plague ordinary electromagnets. Superconducting

**PHOTO 10.9**   This experimental maglev train in Japan uses superconducting coils.

magnets are widely used in MRI devices, in particle accelerators, and in the manufacture of clay-coated paper, to name a few uses.

In one type of high-speed, magnetically levitated (maglev) train (Photo 10.9), superconducting magnets are used on the train's undercarriage to induce currents in rows of closed coils mounted along the guideway underneath. These induced currents exert magnetic forces that lift the moving train a few inches in the air. While floating on this magnetic "cushion," the train is propelled by successively energized coils mounted on the side rails of the guideway. The coils of the superconducting magnets on board the train are chilled to about 4 K by liquid helium, which adds to the complexity and cost of the train.

Research in superconductivity took an exciting turn in the late 1980s with the discovery of a class of superconducting ceramic compounds having much higher critical temperatures than any previously known. Many of these so called high-temperature superconductors, which are made of copper oxides mixed with a variety of rare elements, exhibit superconductivity at temperatures considerably higher than that of liquid nitrogen (77 K), though still well below room temperature (300 K). This is good news, since liquid nitrogen is much easier to prepare and to use as a refrigerant than liquid helium is. The bad news, so far, is that the higher-temperature superconductors are brittle and therefore difficult to form into current-carrying wires. The quest for high-temperature superconductivity continues, with the hope that someday a superconductor of practical use will be found that will operate at or near room temperature.

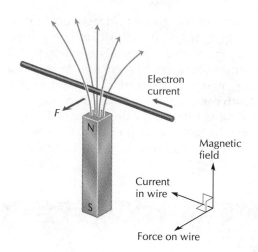

**FIGURE 10.16** When electrons in a wire move to the left across magnetic field lines that point up, the wire experiences a sideways force that tends to push it out of the plane of the page. If the current were reversed, the wire would be pushed into the page. Directly above the north pole of the magnet, the magnetic field, current direction, and force on the wire are mutually perpendicular to each other.

countries. Besides making a primitive buzzer, the wire could conceivably be connected to some mechanism, similar in design to an automotive crankshaft, that could convert the back-and-forth motion of the wire into a rotary motion. The essential point is that electric energy, in the form of electric currents in a wire, can be converted into mechanical energy. In practice, however, a single current-carrying wire placed across the top of a magnet experiences only a small force.

More sophisticated arrangements of wires and magnets are used to fashion practical electric motors. One version, which runs only on ac, consists of freely turning coils of wire (an *armature*) mounted on a shaft and set between the poles of a horseshoe-type magnet (a *field magnet*). Whenever the current in the wires of the armature runs across the magnetic field lines, the armature experiences a *torque*, or twisting force. The rhythmic changes in the ac keep the armature rotating in step with the frequency of the ac: 60 rotations per second if the motor is operating on 60-Hz ac. The greater the number of turns in the armature, the greater the current in the armature; and the stronger the magnets, the greater the torque of the motor.

The simplest dc motors have a switching device (a *commutator*) attached to the motor's shaft that reverses the direction of the current fed to the armature every half a turn. The larger and more complex motors of both types (ac and dc) often use electromagnets (instead of permanent magnets) as field magnets so as to furnish a stronger magnetic field.

The power output of a motor is often limited by the current-carrying capacity of its coils. You have probably noticed how hard-working electric motors tend to get hot or even to burn out from excessive current in the coils. All energized motors dissipate some heat because of resistance in their coils, which somewhat reduces their work output.

## Electromagnetic Induction

In the early 1800s, sustained currents of electricity could be provided only by one source: the electric battery. But a battery is merely a storage device, capable of releasing electric energy at the expense of chemical potential energy. When drained, batteries are dead; and the only way to

**FIGURE 10.17** When a wire is pushed across magnetic field lines and out of the page, as illustrated here, electrons in the wire experience a push to the right, in the same direction as the length of the wire. A current of electrons (an "induced" current) will flow in the wire from left to right as long as there is some continuous path for the electrons to follow (a circuit, not shown here). If the wire is moved into the page, the induced current in the wire flows from right to left.

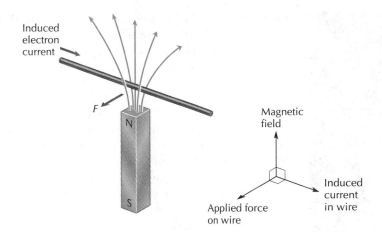

revive them is to pump electric energy back in from some external source. Practical electric energy for the home and industry became possible only after a link between mechanical and electric energy was discovered. This link, called **electromagnetic induction,** is the just reverse process of what we described in our discussion of motors.

Electromagnetic induction works for the same reason that motors work: When electric charges move across magnetic field lines they experience a sideways push. Figure 10.17 shows a wire moving transversely across a magnetic field by virtue of some applied force. As long as the wire continues to "cut" across magnetic field lines (and as long as the wire is attached to some circuit), electrons flow down the length of the wire. Moving the wire once across the field produces a single pulse of dc in the wire. Moving the wire back and forth many times across the field produces ac. We now have a primitive electric generator capable of converting mechanical energy into electric energy. If you compare Figures 10.16 and 10.17 carefully, you will see that the same basic interaction is occurring. What differs is the geometry and, of course, the transformation of energy taking place.

Note that it requires a force not only to start a wire going across magnetic field lines but also to keep it going. Whenever current is being induced in a wire, a resistance to its motion arises quite apart from inertia or friction. The greater the current induced, the more force it takes to push the wire through the field and the more work it takes. This is in accordance with the law of conservation of energy—you cannot get electric energy out of a generator unless an equivalent amount of mechanical energy is expended.

It is instructive at this point to turn back to Chapter 6 and look at Figure 6.15 again. Notice the several pathways by which electric energy can be produced with the help of an electric generator. The mechanical energy that feeds an electric generator can be derived from wind power, water falling from a dam (hydroelectric power), heat engines driven by the burning of fossil fuels, heat-producing nuclear reactions, geothermal heat, or direct solar heating.

The current induced in the simplistic arrangement shown in Figure 10.17 is very feeble. Efficient generators make use of wire coils. When coiled up, many segments of a long wire can cut magnetic field lines at the same

**PHOTO 10.10**   This bicycle generator converts the kinetic energy of the spinning wheel into electric energy, which powers the light. The bicycle rider must pedal harder to turn the generator and illuminate the light.

time, provided that some motion is involved. One way to make a generator (Figure 10.18) is to move a coil toward or away from the pole of a magnet (alternatively, a magnetic pole can be moved toward or away from a coil). Repeated back-and-forth motions of this type will generate ac. Another way is to spin a coil between the opposing poles of a horseshoe magnet so that the wires in the coil cut across the magnetic field lines first one way and then the other. This design, which is typical of today's electric generators, is basically the same as that of a common ac motor. Generators and motors can be used interchangeably in most instances; there are no fundamental differences between them. When you turn the shaft of any motor hooked up to a circuit, you produce electricity in the circuit. Supply electricity (ac or dc, as appropriate) to any generator, and its shaft will turn. This reciprocity, or interchangeability, of function is typical in many practical applications of electromagnetism (see Figure 10.19).

In electromagnetism, as in mechanics, all motion is relative. When electric charges (electrons) in a wire are being deflected near a magnet (as in Figure 10.16), it makes no difference whether the charges move across stationary magnetic field lines or whether magnetic field lines drag across stationary charges. When electromagnetic induction is taking place (Figure 10.17), it makes no difference whether the wires move relative to a stationary magnet or whether the magnetic field moves relative to stationary wires. Einstein's special theory of relativity (Chapter 13) is based in part on the simple fact that motion is relative for all interactions, including electromagnetic interactions. Indeed, it can be shown through special relativity that all magnetic forces are but a relativistic by-product of charges moving relative to each other.

## Changing Magnetic Fields

As shown in Figure 10.18, current can be induced in a coil (which is part of a circuit) when there is relative motion between the coil and a magnet. We can see from this figure that when the coil and magnet are moved farther apart from each other, fewer and fewer magnetic field lines pass through the coil (the field is weaker inside the coil). When the two are moved closer together, more magnetic field lines pass through the coil (the field is stronger inside the coil). Thus, it is really a *change* in the intensity of the field inside the coil that induces current.

Other than physically moving the coil, the magnet, or both, is there a way to change the magnetic field in the coil and thereby induce current? Yes: We can replace the permanent magnet with an electromagnet *driven by a changing current*. A constantly changing current (for instance, ac) in the electromagnet produces a magnetic field that periodically grows and collapses inside the coil. Current is induced in the coil whenever that field is growing or collapsing. Such an arrangement is known as a **transformer.**

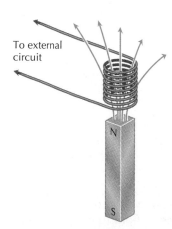

To external circuit

**FIGURE 10.18**   A rudimentary generator can be fashioned out of a wire coil and a permanent magnet. No current will flow to an external circuit unless there is motion between the magnet and the coil. Moving either the coil or the magnet up or down produces a current in the coil. The current induced increases with the number of turns in the coil and with the rate of motion (faster motion causes the magnetic field lines to cut across the wires of the coil more quickly).

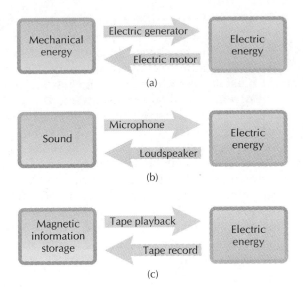

(a)

(b)

(c)

**FIGURE 10.19**   Some symmetries between electromagnetic interactions are indicated here. (a) Electric currents are induced by the mechanical action of a generator, and electric currents fed into a motor produce mechanical energy. (b) In a moving-coil microphone, sound waves hitting a diaphragm produce a vibration that moves a coil near the pole of a stationary magnet. Changing currents are thereby induced in the coil. In a moving-coil loudspeaker, changing currents fed into a coil cause the coil to move back and forth near the pole of a stationary magnet. The coil is attached to a diaphragm that pushes air back and forth, creating sound. (c) A tiny electromagnet is used in the recording head of an audiotape or videotape unit. Changing currents in the electromagnet cause the needlelike grains of magnetic material in the magnetic tape passing by to align into particular patterns and therefore record information. During playback, currents are induced in the coil of the playback head by the magnetic fields of the aligned magnetic grains passing by.

One common transformer design, which features two coils wrapped around a soft iron core, is shown in Figure 10.20. Electric energy is fed into one coil, called the *primary coil,* and is drained from the other, *secondary coil* when it is connected to an external circuit. Transformers transform both ac voltage and current, but only if the number of turns on each coil is different. If there are twice as many turns in the secondary as in the primary (as in Figure 10.20), the voltage is doubled from primary to secondary. This is called a *step-up transformer.* If there are half as many turns in the secondary as in the primary, the voltage is halved. This is called a *step-down transformer.* Any ratio of turns can be used to step up or step down voltage by any amount.

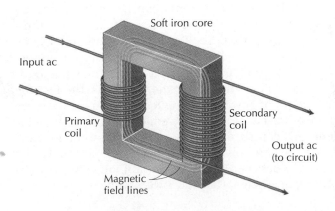

Soft iron core

Input ac

Primary coil

Secondary coil

Output ac (to circuit)

Magnetic field lines

**FIGURE 10.20**   The primary coil of a transformer functions as an electromagnet. When current is fed into the primary coil, the magnetic field lines that arise are guided through a soft iron core through the secondary coil. A constant (dc) current in the primary coil induces no current in the secondary coil, but a rhythmically changing current (such as ac) in the primary induces a similarly changing current in the secondary. Transformers transform nothing if the number of turns in the primary and secondary coils are equal. When the number of turns are unequal, current and voltage differ in each coil.

There is no "free lunch" in transforming voltage. Any increased voltage comes at the expense of a proportionally decreased current. Recall that electric power, the rate at which energy flows, is the product of voltage and current (Equation 10.4). If voltage increases across a transformer, then current must decrease. The power, however, remains the same.

Current and voltage transformations are analogous to the transformations of force and distance in simple machines (see Box 6.1). A machine's mechanical advantage, if it is greater than 1, amplifies force at the expense of the distance through which the force acts. The output work or power of a machine can never be more than the input work or power. Real machines are never perfectly efficient because some mechanical energy is lost to friction, which produces heat. Similarly, the transformation of current and voltage in a transformer is accompanied by resistance heating in the wires and heating caused by small currents in the iron core. These (usually small) losses of power somewhat reduce the current, but not the voltage, coming out of the secondary coil.

**Problem 10.3**

When 1 A of current at 120 V is being fed into the 200-turn primary coil of a given transformer, the 20-turn secondary coil is delivering 9 A of current to an external circuit. What is the voltage across the secondary coil, and how efficient is the transformer?

**Solution**

The ratio of primary to secondary turns is 10:1, so the voltage is stepped down from 120 V to 12 V. The power fed into the primary coil is

$$P = VI = (120 \text{ V})(1 \text{ A}) = 120 \text{ W}$$

The power delivered by the secondary coil is

$$(12 \text{ V})(9 \text{ A}) = 108 \text{ W}$$

The efficiency is

$$(108 \text{ W})/(120 \text{ W}) = 0.9, \text{ or } 90\%$$

Ten percent of the electric energy fed into this transformer is dissipated as heat.

The ease with which voltages and currents can be manipulated by transformers is the reason that most electric power is generated, transmitted, and delivered as ac rather than dc. Typically, generators at large power stations produce ac at several thousand volts. That voltage is immediately stepped up to as much as 500,000 V so that it can be transmitted for long distances through power lines. The higher the voltage is, the less the current is for the same amount of power. Heating losses in power lines are proportional to the square of the current, so any reduction in current yields big dividends in efficiency. Of course, electricity transmitted through wires at half a million volts must be carefully insulated and kept high above the ground by stringing the wires between giant towers. Million-volt transmission lines would be even more efficient, but the towers would then be even larger and more costly.

**PHOTO 10.11** Insulators 2 m long are needed to handle the high voltage entering these transformers at a New Jersey substation.

At the end of long transmission lines, transformers at electrical "substations" step down the voltage to thousands of volts. From there the ac is distributed to smaller substations throughout the city or countryside. After two or three more reductions in voltage, the electric power is reduced to levels safe enough for ordinary uses.

In view of everything we have said so far, the relationship between electricity and magnetism seems to be a reciprocal one: A steady electric current produces a steady magnetic field, and a changing magnetic field can produce electric current. This relationship may seem not quite symmetrical, because a magnetic field must change in order to induce a current, but steady currents give rise to steady magnetic fields. Actually, though, the relationship does become symmetrical when we ignore currents and refer only to fields. Electric fields surround charges; and when charges move, their electric fields move with them. Basically, moving (changing) electric fields produce magnetic fields. Conversely, as we have seen in the case of electromagnetic induction, moving or changing magnetic fields make charges behave as if they were immersed in an electric field. If free to move, these charges constitute an electric current. Essentially, changing magnetic fields produce electric fields.

In the 1860s, the physicist James Clerk Maxwell summed up the nature of electricity and magnetism and the theoretical interconnections between them in succinct form in just four equations. **Maxwell's equations** are as important in the study of electricity and magnetism as Newton's laws are in mechanics. Although highly mathematical, the equations express the following basic ideas:

1. Any two like charges repel each other and any two unlike charges attract each other; they do so with a force whose strength is inversely proportional to the distance between them (Coulomb's law).
2. Magnetic field lines are closed loops; there are no magnetic monopoles.
3. A changing electric field produces a magnetic field.
4. A changing magnetic field produces an electric field.

# Electromagnetism and Light

We have seen that an electric charge moving past a point in space produces a changing electric field, which, in turn, creates a magnetic field. We have also seen the reverse: A changing magnetic field produces an electric field, and that can cause an electric charge to move.

After further study and manipulation of the four equations of electromagnetism named in his honor, Maxwell realized that whenever a charge is accelerated, a rather unusual type of wave should be emitted. This *electromagnetic wave*, as he called it, should consist of coupled, fluctuating electric and magnetic fields moving through space (see Figure 10.21). The fields would propagate outward, feeding off one another as they went. Maxwell calculated how fast these waves should move from known constants associated with electrical and magnetic phenomena, such as the constant $K$ in Coulomb's law. The answer was startling: The speed was just the known speed of light, $c$. Maxwell was the first person to envision the production of radio waves, the first to recognize that electromagnetic radiations of different frequencies are fundamentally the same, and the first to realize that all electromagnetic radiations travel at the same speed.

Maxwell's notions were eventually confirmed. In the 1880s, Heinrich Hertz (whose last name is used as the standard frequency unit) performed experiments that confirmed Maxwell's theoretical predictions. In 1895, Guglielmo Marconi sent telegraph signals by means of radio (then called "wireless") waves through the air for more than a mile; by 1901, Marconi's signals had spanned the Atlantic Ocean. In time, all the other electromagnetic radiations, including light, were shown to be associated with accelerated charges.

Maxwell stuck with the idea that electromagnetic waves, like ordinary waves, must be propagated through *something*—presumably an all-pervading substance in the fabric of space called "ether." Later, Einstein and others showed that there is no ether. Empty space alone can support the propagation of electromagnetic waves.

As we will soon learn in the next chapter, light and the other forms of electromagnetic radiation are only incompletely described as a continuum of waves. On very small scales of distance, all forms of electromagnetic

**FIGURE 10.21** The coupled electric and magnetic fields of an electromagnetic wave oscillate at right angles to each other. When both fields are momentarily at zero strength, they are *changing* most rapidly, and each generates the other. When the fields reach maximum strength, they are momentarily not changing, so they begin to die out and approach zero strength.

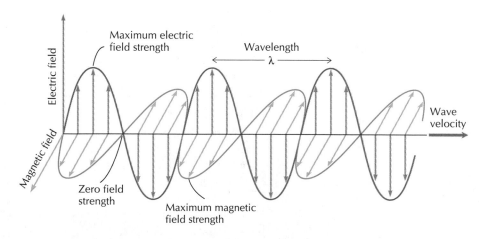

**PHOTO 10.12**  Oscillating electric and magnetic fields, driven by surging electric currents, are radiating at the speed of light from most of these mountaintop antennas. The antennas consisting of straight wires are designed to *broadcast* signals far and wide. Some of the microwave dish antennas (which are enclosed in protective covers) are designed to receive microwave transmissions from distant sources; others are used to transmit microwave signals to specific targets. This particular installation, overlooking San Diego, is used as a relay station for telephone transmissions and for radio networks used by emergency services (fire department, paramedics, and so on).

radiation interact with matter in a *discontinuous* way. It is valid to imagine light as a stream of rapidly oscillating waves and equally valid to think of light as a stream of particlelike concentrations of energy, or *photons*. Either way, electromagnetic waves or photons move through empty space at the same speed, $c$.

# CHAPTER 10
# Summary

Electric fields exist around electric charges. A movable electric charge, such as an electron in a wire, will migrate in response to any electric field present and convert electric potential energy to other forms of energy as it moves. The greater the potential energy given up by a given charge when moving from one point to another, the greater the potential difference, or voltage, is between the two points.

Electric charge will flow through a simple electric circuit as long as a potential difference is maintained between two points in the circuit. Typically, this potential difference is maintained by means of an electric battery or generator.

Electric current in a circuit is a measure of the rate of flow of electric charge. Resistors in a circuit restrict the current and also dissipate heat. The less resistance a resistor has, or the greater the potential difference across it, the greater the current is through it (Ohm's law).

Electric shock can occur when a person becomes part of an electric circuit. People must avoid coming into contact with two points in a circuit with a large potential difference between them.

Electric power, which is equal to voltage times current in a wire or an appliance, is the rate at which electric potential energy is converted into other forms (mechanical work, heat, light, and so on). The typical billing unit for commercial electric energy is the kilowatt-hour, equivalent to 1 kW of power delivered over a period of 1 h.

Electrochemical cells, such as batteries, make use of oxidation reactions (in which electrons are lost) and reduction reactions (in which electrons are gained), which produce electric energy at the expense of chemical potential energy. Electrolysis, which also involves oxidation and reduction reactions, builds up chemical energy at the expense of electric energy.

Semiconductors are materials with properties intermediate between those of conductors and insulators. They can be fashioned into transistors, the miniature switching units that lie at the heart of every computer microprocessor. The development of smaller and more complex microprocessors is leading to computers with increased speed and information-processing power.

Like poles of magnets repel each other; unlike poles attract each other. The concept of a field can be applied to magnets as well as to masses and charges. Unlike gravitational and electric field lines, magnetic field lines always consist of closed loops. Repeatedly dividing a magnet produces more smaller magnets, each of which has two poles and magnetic field lines consisting of closed loops.

Magnetic fields can be produced by electric currents in wires. Indeed, all magnetism arises from moving or spinning charges.

The sun and several of the planets have magnetic fields of significant strength. Earth's magnetic field apparently results from currents stirred by convection in its molten metallic outer core.

Any charge moving across magnetic field lines will experience a force that is exerted at right angles to the velocity of the charge. Electric motors exploit this phenomenon by converting the electric energy of electrons moving through a wire and across a magnetic field into mechanical energy. Electric generators exploit this same phenomenon by inducing an electron current in a wire when the wire is pushed across a magnetic field (electromagnetic induction).

Transformers are electromagnetic devices that change the mix of current and voltage. Increased voltage comes at the expense of decreased current.

When electricity and magnetism are analyzed solely in terms of fields, we can show that a changing electric field produces a magnetic field and that a changing magnetic field produces an electric field. All electromagnetic radiation can be described in terms of coupled, oscillating electric and magnetic fields moving through empty space at the speed of light.

## CHAPTER 10
# Questions

## Multiple Choice

1. When an electron accelerates in response to an electric field, what kind of energy does the electron acquire?
   a) electron potential energy
   b) kinetic energy
   c) rest energy
   d) no energy, since the electron does not gain or lose its charge

2. Electric currents in circuits are analogous to water flowing through pipes in that
   a) voltage can be compared to water pressure
   b) smaller wires resist electricity just as smaller pipes restrict the flow of water through them
   c) much energy can be transferred by small movements of water under great pressure

   and also by electrons moving through a large potential difference
   d) all of the above are true

3. The volt is the standard unit of electric
   a) current
   b) resistance
   c) potential energy
   d) potential difference

4. Ohm's law expresses the relationship between
   a) electric and magnetic forces
   b) electric charge and current
   c) electric power and energy
   d) electric current, voltage, and resistance

5. If the voltage across a resistor is doubled, then the current through the resistor will
   a) double
   b) quadruple

c) remain the same

d) be half as much

6. The intention of case grounding in electrical appliances is to make sure that
   a) the entire appliance maintains a potential of 0 V
   b) a short circuit will never happen inside the appliance
   c) users will always encounter a potential difference of zero when in electrical contact with the outer case and the ground beneath them
   d) none of the above occur

7. Electric power is
   a) the same as electric energy
   b) equivalent to voltage
   c) voltage divided by current
   d) the rate at which electric energy is expended

8. Sparks and lightning bolts must always pass through
   a) negatively charged air
   b) positively charged air
   c) ionized air
   d) moist air

9. Ordinary flashlight batteries rely on paired
   a) endothermic and exothermic reactions
   b) oxidation and reduction reactions
   c) acid-base reactions
   d) electron and proton reactions

10. The magnetic field lines around a magnet indicate
    a) the direction of a magnetic field at points around the magnet
    b) the strength of a magnetic field at points around the magnet
    c) the strength and direction of the magnetic field at various points
    d) none of the above

11. When a permanent magnet is heated or subjected to blows, its magnetic field tends to
    a) become weaker
    b) become stronger
    c) become reversed
    d) be unchanged

12. All magnetism originates in
    a) solid, ferromagnetic material
    b) electrically conducting fluids
    c) moving electric charge
    d) moving magnets

13. A magnetic field exerts no force on
    a) an electric current
    b) a magnet
    c) a stationary electric charge
    d) a ferromagnetic material, as long as it is not magnetized

14. Jupiter's strong magnetic field arises from
    a) the liquid-metallic state of the hydrogen in Jupiter's interior
    b) Jupiter's relatively large metal content
    c) the motion of Jupiter around the sun
    d) Jupiter's immersion in the sun's magnetic field

15. An important factor related to the 11-year (or 22-year) solar activity cycle is the sun's
    a) surface temperature
    b) central temperature
    c) differential rotation
    d) motion around the center of the galaxy

16. A transformer (see Figure 10.20) will not work on dc, because
    a) no dc can flow through the primary coil
    b) no magnetic field is produced by dc in the primary coil
    c) only changing magnetic fields can induce current
    d) magnetic field lines escape the iron core when the magnetic field produced by the primary is constant

17. Alternating current (ac) is commonly used in power grids, because
    a) ac always delivers more voltage
    b) ac always delivers more current
    c) ac can be easily stepped up in voltage for efficient transmission and stepped down in voltage for various practical uses
    d) it is impossible to generate dc at electric power stations

18. Superconducting materials would have a much wider application if
    a) their critical temperatures were as high as room temperature
    b) they were malleable enough to fashion into wires
    c) they could be inexpensively synthesized out of common elements
    d) all of the above were possible

19. According to Maxwell's theoretical findings,
    a) electric and magnetic fields propagate through empty space at the speed of light

b) the presence of an electric field always implies the presence of a magnetic field

c) electric and magnetic fields are identical

d) magnetic field lines always point in the same direction as the electric field that creates them

# Questions

1. If a strong electric field exists at a certain point in space, how could you test for its existence?

2. What is potential difference (voltage), and what is its gravitational analog?

3. Why must a person touch two (not one) conducting objects or surfaces having a potential difference of several tens of volts or more to receive a painful electric shock?

4. What is the difference between a kilowatt and a kilowatt-hour?

5. What chemical changes occur inside a wet-cell storage battery (car battery) when it is being recharged?

6. Why is there a strong technological thrust toward miniaturizing the electronic circuitry used in computer chips?

7. If a strong magnetic field exists at a certain point in space, how could you test for its existence?

8. How could you determine the polarity of a bar magnet whose ends are unmarked?

9. Explain why the imaginary bar magnet (Figure 10.14) inside Earth has its S pole in Earth's Northern Hemisphere and its N pole in Earth's Southern Hemisphere.

10. Strong planetary magnetic fields are associated with what two planetary characteristics?

11. What role does magnetism play in the formation of images on the screen of a television picture tube or computer monitor?

12. How is the fundamental operation of a motor similar to that of a generator? How are they different in their operation?

13. What do transformers transform?

# Problems

1. If 110 J of mechanical work are done and 10 J of waste heat are released by an electric motor when exactly 1 C of charge passes through it, then the motor must be connected to a power source of how many volts?

2. If it takes 2 s for the motor in Problem 1 to deliver 120 J of mechanical work and waste heat, how much electric current is passing through the motor? How much electric power is flowing through the motor during those 2 s?

3. What voltage is needed to produce a current of 4 A in a resistor that is dissipating 24 W of thermal energy?

4. A steam iron draws a current of 15 A when connected to a 120-V power source. What is its resistance?

5. Show that a 110-V potential difference applied across a human body can produce a current inside the body of as much as 0.1 A.

6. Derive Equation 10.4.

7. What is the power rating of an electric motor that draws a current of 5 A when operated at 120 V?

8. A 1-horsepower (rated at 750 W of power consumption) water pump used to circulate and filter water in a backyard swimming pool switches on for an 8-h stint every day. What is the monthly cost of keeping this motor running if the local cost of electricity is 11¢ per kilowatt-hour?

9. If a lightning bolt releases 100 million joules of energy in 1 s across two clouds that differ in potential by 500 million volts, what is the average electric current in the bolt during that 1 s? (*Hint:* you can use Equation 10.1 to calculate the total charge transferred; then use Equation 10.2 to calculate the current.) How many watts of power are released during the 1-s duration of the bolt?

10. The transformer in Figure 10.20 is used in reverse so as to step down an applied voltage of 220 V ac. If 10 A of current are being drawn from the 10-turn coil by an external circuit, how

much current (at what voltage) is flowing through the 20-turn coil? Assume that there are no heating losses.

## Questions for Thought

1. How might you best save electricity in your own household? By curtailing the use of low-power devices used regularly for many hours, or by curtailing the use of power-intensive devices used briefly and intermittently? Explain.

2. A freely moving electron is diverted sideways (perpendicular to its velocity) when it crosses magnetic field lines. Is any work done on that electron? (Please review the discussion of work in Chapter 6 before thinking about this question.)

3. What electromagnetic devices have you used today?

4. How many transformers do you think there are between your home and a nearby electric generating station? If yours is an older neighborhood and the electric utility lines are overhead, you can try to trace the wires from your home to the neighborhood substation, to a larger substation or substations, and ultimately to the generating station. If electric utilities are underground in your area, a call to your power company may be helpful in beginning the trace.

5. Can a step-up transformer used in one circuit be configured in another circuit so as to act as a step-down transformer?

6. If common electrical appliances used superconducting wire, would it be necessary to plug them in to keep them operating?

## Answers to Multiple-Choice Questions

| 1. b | 2. d | 3. d | 4. d | 5. a |
|------|------|------|------|------|
| 6. c | 7. d | 8. c | 9. b | 10. c |
| 11. a | 12. c | 13. c | 14. a | 15. c |
| 16. c | 17. c | 18. d | 19. a | |

# Light and the Atom

*Laser light is characterized by utter conformity: The photons coursing through these beams have virtually the same frequency, wavelength, speed and energy. They also travel "in step" with one another, like soldiers on parade. The production of laser light involves many identical atoms shedding precisely the same amount of energy at just the right moments of time.*

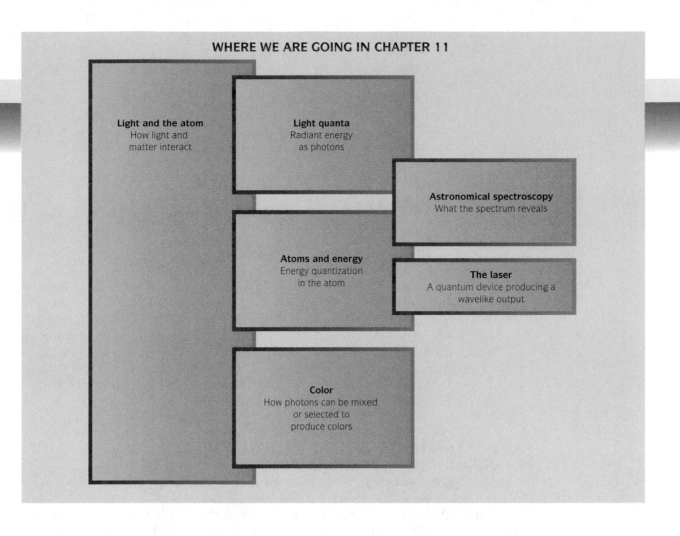

**Light and the atom**
How light and matter interact

**Light quanta**
Radiant energy as photons

**Astronomical spectroscopy**
What the spectrum reveals

**Atoms and energy**
Energy quantization in the atom

**The laser**
A quantum device producing a wavelike output

**Color**
How photons can be mixed or selected to produce colors

What is light? You cannot touch light or even slow it down to "look" at it. Nevertheless, light is one of the most comprehensively studied phenomena in all of nature. There is nothing ambiguous about our measurements of a light beam's velocity and intensity, but other properties of light are not so clear-cut. Many aspects of light seem to be sensitive to the particular experiments we perform to test its nature. In some experiments light behaves as a continuous train of energy-carrying waves, whereas in others light behaves as if it were made of discontinuous packets or particles of energy. These energy packets are called *photons.*

One thing is certain: It is impossible to study light without having it interact with something. In the realm of visible light, that "something" is atoms or molecules: in the eye's retinal cells, in the light-sensitive emulsion of photographic film, or in the light-sensitive electronic elements of a video camera. Light (or any other electromagnetic radiation) makes its presence known only by giving up its energy to matter. Matter, in turn, may give up energy in the form of light.

Two basic ideas constitute the primary focus of this chapter: light's particlelike nature and the particlelike interactions that take place between light and matter. The clear understanding of light's particle nature helped scientists develop the laser, a powerful tool of industry and information technology, and the method of analysis called spectroscopy, whose importance in astronomy is unparalleled.

## Light Quanta

In physics, the word *quantum* (plural, *quanta*) means the smallest possible amount of a quantity. To say that something is "quantized" means that only discrete multiples of the minimum quantity are possible. This idea becomes less mysterious when you think of the way currency is traded. At present in the United States, the smallest unit of currency is the penny. Every transaction involves units that are no smaller than pennies.

When we look at small enough spatial and temporal scales, quantization seems to be the norm in nature. Matter is built up from atoms and subatomic particles with definite masses and charges. And these units themselves have important properties, such as energy and angular momentum, that are quantized as well.

## Origins of Quantum Theory

The idea that light might be quantized, or made of tiny particles, is at least as old as the ancient Greeks. However, by the late 19th century, a slew of experiments had conclusively demonstrated that light travels as a wave and is wavelike in its interactions with material bodies. By the year 1900, a revolution in scientific thought was brewing. It would soon lead to the realization that light has both wavelike properties and particlelike properties. Within three decades, the new ideas about waves and particles had turned physics upside down and the probabilities of quantum mechanics (Chapter 14) began to supplant the certainties of Newtonian physics as the best model of nature's fundamental interactions.

The revolution began with efforts to explain the phenomenon known as *blackbody radiation*. Blackbody radiation is an example of **thermal radiation**—the electromagnetic radiation spontaneously emitted by all bodies above a temperature of absolute zero. The net flow of thermal radiation is outward when an object is surrounded by cooler things and inward when it is surrounded by warmer things. For example, your body emits thermal radiation (specifically, infrared radiation) to its cooler surroundings when you stand outside on a cold night.

**Blackbody radiation** is the thermal radiation emitted by a perfect absorber of electromagnetic radiation. When cool enough not to glow visibly, such a body would appear by human eyes to be perfectly black. (The converse of a blackbody would be a perfectly white body, capable of reflecting all radiation incident on it.) Although blackbodies are theoretical, many objects resemble the ideal blackbody, such as a dark lump of cast iron or a star. (Stars, of course, shine, but they would appear quite dark to human eyes if they were so cool that they did not radiate visible light.)

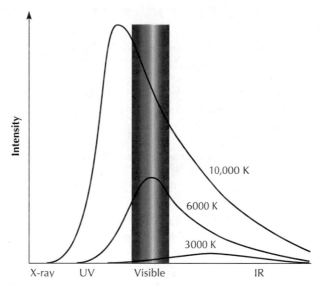

**FIGURE 11.1** The total radiant energy emitted by a blackbody rises with temperature, and the distribution of the energy changes as well. At 3000 K (the approximate temperature of an illuminated incandescent lamp), the peak of the radiation occurs in the infrared (IR). At 6000 K the peak is near the middle of the visible spectrum. At 10,000 K blackbody radiation peaks in the ultraviolet (UV). Star colors, as human eyes perceive them, can be explained by these blackbody radiation curves. Relatively cool stars appear reddish because their radiation *within the visible range* consists of more red than violet and blue. Stars like the sun emit rather evenly across the visible spectrum, so they appear white or yellowish white. The hottest stars, 10,000 K and hotter, emit more strongly in the violet-blue region than in the red region. Their light appears bluish to the eye. The thermal radiation emitted by various bodies in the universe spans a greater range than that shown here. Hot plasma at a temperature of more than 1 million kelvins emits strongly in the X-ray region of the spectrum; cool gases in interstellar space emit in the infrared-to-radio range. Absent all the matter in it, the entire universe radiates as if it were a blackbody at 2.7 K, with its peak radiation in the microwave part of the radio spectrum. This is the so-called 3 K cosmic background radiation—one of the pieces of evidence supporting the big bang theory.

Every blackbody radiates electromagnetic energy over a broad spectrum of frequency or wavelength, and it does so in a way that depends on its temperature. When the body's temperature is raised, the radiation becomes increasingly intense and is spread across the spectrum in such as way that its *peak* intensity is found at ever-shorter wavelengths (see Figure 11.1).

In an effort to explain the theoretical underpinnings of blackbody radiation, the physicist Max Planck proposed (in the year 1900) that the conversion of thermal energy into radiant energy in a radiating body can be thought of as taking place in tiny little bursts. Each burst, or quantum of energy, would be equal to the product of a constant quantity—now called **Planck's constant,** $h$—and the frequency $f$ of the radiation emitted:

$$E = hf \hspace{4cm} \text{Equation 11.1}$$

Planck's constant has a very tiny value, $h = 6.63 \times 10^{-34}$ J·s, so the presumed energy quanta themselves would be very small, too. The conversion of thermal energy to light in, say, the filament of a lightbulb may seem to be continuous over time, but it does indeed take place stepwise by means of a vast number of tiny energy transformations.

Planck himself was not comfortable with the idea of quantum conversions of energy and at the time he did not think that light was anything other than a continuous series of waves. It took a bolder physicist—Albert Einstein—to envision light in a whole new "light."

# The Photoelectric Effect

Einstein proposed in 1905 that light (or more broadly, radiant energy) consists of quanta whose energy are related to their frequency in just the way Planck had described, Equation 11.1. These quanta were later to be named *photons*. In support of his idea, Einstein cited the **photoelectric effect,** earlier discovered by Heinrich Hertz, the man best known for his experimental confirmation of the existence of electromagnetic waves. The photoelectric effect involves the release of electrons from the surface of a metal when it is irradiated with light. The effect can be used to transform light energy into electric energy and is responsible for many important inventions of the 20th century. The photovoltaic solar cells used to power spacecraft, the photocells used in cameras, and the image-producing CCD chips used in home video cameras are all photoelectric devices.

The nature of the photoelectric response is somewhat unusual. The release of electrons does not necessarily depend on the intensity of the light, but it does depend in a strange way on the frequency of the light. For any given metal, a minimum frequency of light is needed to dislodge electrons from its surface. Light having a frequency lower than the minimum required, *no matter how intense it is,* produces absolutely no response. Light having a frequency as high as or higher than that required will release electrons; the higher the frequency of the light, the more average kinetic energy the ejected electrons have. Even very weak light of a high-enough frequency is capable of ejecting some electrons.

Einstein correctly analyzed this behavior as a clear indication that light travels in packets and interacts with matter in short-lived bursts of activity. A packet of light (photon) must have a certain minimum energy (or frequency, according to Equation 11.1) in order to barely succeed in pulling an electron away from the crystal lattice of the metal. If the minimum energy is exceeded, then the ejected electron may have energy to spare. The situation is roughly analogous to a car striking a brick wall (see Figure 11.2). No bricks in the wall will go flying unless the crashing car has sufficient energy to dislodge them.

Einstein explained that because there is absolutely no photoelectric response to light that is below a certain threshhold of frequency (or longer than a certain wavelength), light simply cannot be a *continuous* series of waves. Any continuous disturbance, like ocean waves eroding away sea cliffs, would, over time, disrupt the metal and cause a steady release of electrons.

Today, it is universally accepted that light is quantized into photons. Scientists now utilize light detectors sensitive enough to trap and respond to nearly every photon that comes their way and to distinguish between one photon and another. Precise measurements confirm that photons have the following exact amount of energy:

$$E = hf$$

and

$$E = hc/\lambda \qquad \qquad \text{Equation 11.2}$$

Equation 11.2 comes from the fact that $f = c/\lambda$ (which is the wave equation, Equation 8.1, rewritten for light). We see from these equations

**PHOTO 11.1** This array of photovoltaic cells powers timers and switches and actuates valves in an automatic irrigation system installed for street landscaping near the author's home.

Before                          After

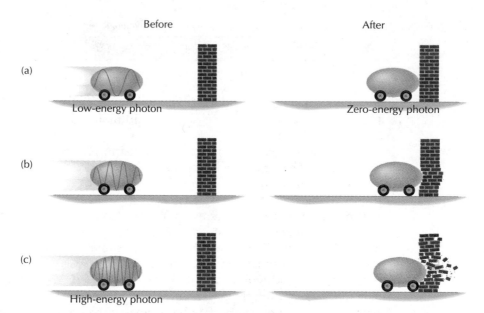

(a)    Low-energy photon                Zero-energy photon

(b)

(c)    High-energy photon

**FIGURE 11.2**  In this analogy of the photoelectric effect, photon "cars" strike a brick wall. (We are using a visual metaphor of wave packets here, although, strictly speaking, the wave nature of light is never exhibited in any phenomenon, such as the photoelectric effect, that demonstrates light's particle nature.) The low-energy, low-frequency photon car in (a) loses all its energy on impact and fails to dislodge any bricks from the surface. The photon car in (b) has higher energy and higher frequency, but not nearly enough energy to dislodge bricks. The highest-frequency photon in (c) has more than enough energy to dislodge bricks and give them a certain nonzero kinetic energy. This analogy of the photoelectric effect is inexact in the sense that each photon absorbed on a metal surface can eject only one electron. Also, keep in mind that all photons travel *at the same speed*. It is the frequency (or wavelength) of each quantum of light that determines its energy. Zero frequency means zero energy. A photon that loses all its energy during an encounter is one that ceases to exist.

that higher-frequency photons carry more energy and, equivalently, that shorter-wavelength photons carry more energy. This relationship, illustrated in Figure 11.3, is important to keep in mind as we proceed through this chapter.

**Problem 11.1**
How much energy is carried by a single photon of red light? (We shall make the calculation easier by assuming that $\lambda = 663$ nm for the red light.)

**Solution**

$$E = hc/\lambda$$
$$= \frac{(6.63 \times 10^{-34}\ \text{J·s})(3 \times 10^{8}\ \text{m/s})}{(6.63 \times 10^{-7}\ \text{m})}$$
$$= 3 \times 10^{-19}\ \text{J}$$

The very small amount of energy carried by each visible-light photon ($3 \times 10^{-19}$ J or thereabouts) means that a vast number of photons must make up the light we see with our eyes. Nearly $10^{19}$ photons are needed to

**FIGURE 11.3** Increased photon energy on the electromagnetic spectrum corresponds to increased frequency and decreased (shorter) wavelength of the light associated with those photons.

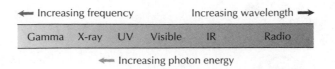

← Increasing frequency          Increasing wavelength →

| Gamma | X-ray | UV | Visible | IR | Radio |

← Increasing photon energy

make the 1 J of radiant energy that is emitted by a 1-W lightbulb in just 1 s. It is not surprising that light seems continuous to our eyes, rather than granular, as it really is.

In no way does this discussion about the particle (photon) nature of light negate the fact that light can also behave as a traveling series of waves. Any experiment designed to test light's particle nature always confirms that light consists of photons. Any experiment designed to test light's wave nature confirms that light travels as waves. However, for any particular event, or for any one kind of experiment, light behaves either as a wave or as a stream of particles, never both at the same time. Light is profoundly strange in this respect, but there are parallels in everyday experience. For example, try looking at a coin with one eye closed. You can examine one side of the coin, then the other. But without some form of trickery (such as using a mirror), you cannot see both sides of the coin at the same time.

## Atoms and Energy

All solids, liquids, and dense gases emit thermal radiation. This radiation is due to continual interactions taking place among the atoms of these substances. These interactions are electromagnetic, and they give rise to emitted photons of many different energies (frequencies). The spectrum of thermal energy is smooth and continuous; it is a **continuous spectrum,** with a peak frequency or wavelength that depends on the temperature of the emitting body (recall the blackbody curves of Figure 11.1, which represent idealized thermal radiation). The peak represents the energy given off by atomic interactions that are most probable in a given body. The less common interactions that are either less energetic or more energetic than the norm produce the "tails" of the energy distribution curves. Most of the radiation flowing through the universe is thermal radiation, because stars are basically thermal radiators, and stars are the biggest contributors of light to the universe.

A very different type of radiation comes from *rarefied gases* (gases of low density), most obviously those that are heated to incandescence. A heated quantity of gas in this condition emits light only at certain discrete frequencies; its spectrum is a series of bright lines superimposed on a dark background—an **emission spectrum.** The particular pattern of bright spectral lines depends on the type of gas: Hydrogen gas produces one pattern, helium gas another, mercury vapor another, and so on.

The same pattern of bright lines emitted by a gas can sometimes be seen as dark lines superimposed on a spectrum that is otherwise continuous. This so-called **absorption spectrum** occurs when light originating from a continuous-spectrum source of light passes through a cooler rarefied gas before reaching the observer. This seemingly contrived situation is actually the normal state of affairs for stars. The hot interior of a star emits

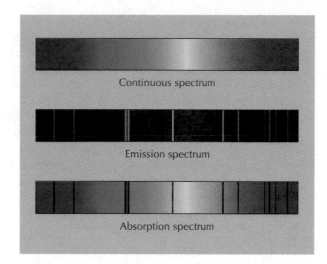

**FIGURE 11.4** A continuous spectrum arises from a hot (or glowing) solid, liquid, or compressed gas. Continuous spectra are emitted by lightbulb filaments, molten metal heated to incandescence, hot lava, and the hot interiors of stars. An emission spectrum (a generic example is shown here) arises from a glowing, rarefied gas. Emission spectra are produced by electrical discharges in neon signs and in the mercury vapor or sodium vapor lamps commonly used for street lighting. *Emission nebulas* (the tenuous clouds of interstellar gas heated or excited by radiation from stars nearby) also produce emission spectra. An absorption spectrum results when light from a continuous-spectrum source passes through a quantity of cooler, rarefied gas. Most stars exhibit absorption spectra. Although photons of all wavelengths stream outward from a star's hot, compressed interior, some of these photons (those of particular wavelengths) are absorbed by the cooler, rarefied gas of the star's photosphere.

a continuous spectrum; but the cooler, rarefied gases in its photosphere (visible surface) *absorb* specific wavelengths. The three types of spectra mentioned—continuous, emission, and absorption—are illustrated and compared in Figure 11.4.

## Explaining Spectral Lines

In a rarefied gas, atoms jostle each other somewhat infrequently, and we can then see the effects of *discrete* interactions that lead to the appearance of discrete spectral lines. These interactions take various forms. Here are three possibilities:

1. An atom can absorb a passing photon and thereby gain energy. This event is accompanied by the disappearance of the photon.
2. An atom can gain energy through collision. Two atoms may collide, or a stray electron may collide with an atom. Some of the kinetic energy possessed by the two particles before collision is transformed into energy stored inside one or both atoms.
3. An atom can lose energy through the creation and emission of a photon.

The first two possibilities are examples of atomic *excitation*. The third represents what we shall call atomic *de-excitation*. What really happens inside an atom during excitation and de-excitation? The physicist Niels Bohr took the first steps in explaining this. His **Bohr model** (1913) of the simplest atom (hydrogen) was groundbreaking in several ways.

Bohr supposed that a hydrogen atom's lone electron whirls around the positively charged nucleus (a single proton) in a circle, much like a planet circling the sun. This is quite plausible, since the electrostatic attraction between the proton and the electron would keep the electron in its place. Bohr proposed that the circling electron, however, unlike a planet, which can circle the sun in any size orbit, can orbit the nucleus only in certain stable orbits having well-defined radii. These orbits are called *stationary states*. The smallest of them, the one corresponding to lowest energy, is called the **ground state.** The larger orbits or stationary states, corresponding to states of higher energy, are called **excited states.** Bohr also proposed that when an electron orbits in any stationary state, *it does not radiate energy*. This was a clear break from Maxwell's finding (Chapter 10) that an orbiting or accelerating charge should radiate away electromagnetic energy. Bohr had to make this assumption; otherwise, the electron in his model would lose all its energy and spiral into the nucleus within a tiny fraction of a second.

The various stationary states of the Bohr hydrogen atom can be labeled with a **quantum number** called $n$. The number $n$, which starts at the ground state with $n = 1$ and continues outward through the various excited states ($n = 2$, $n = 3$, and so on), is simply a code number representing the various discrete *energy levels* of a hydrogen atom that correspond to electron orbits of different size. The jumps in energy from one level to the next are discrete, but they are not all of the same magnitude. In fact, the energy differences between adjacent orbits of higher $n$ value become less and less. Thus, the difference in energy between $n = 1$ and $n = 2$ ($1 \rightarrow 2$) is larger than that between $n = 2$ and $n = 3$ ($2 \rightarrow 3$). The jump $2 \rightarrow 3$ is larger in energy difference than the jump $3 \rightarrow 4$; and so on.

The great utility of the Bohr model is that it explains splendidly the emission spectrum of hydrogen. Of all the elements, hydrogen has the most straightforward spectrum, consisting of a characteristic series of lines that repeat many times throughout the ultraviolet-to-infrared range of the spectrum. The relative simplicity of the hydrogen spectrum is linked to the fact that hydrogen has only one electron, which can jump to one or another of the stationary states prescribed by Bohr's model (see Figure 11.5). A jump to a higher $n$ state is made when the atom absorbs energy (excitation); a jump to a lower $n$ state occurs when the atom loses energy through the emission of a photon (de-excitation).

Let us assume that we have a means of exciting hydrogen atoms. We can do this electrically by sending a high-voltage electric current through a hydrogen-filled *gas discharge tube.* (Fluorescent lights and neon signs are examples of discharge tubes.) The large potential difference applied pushes and pulls electrons through the tube and excites the hydrogen atoms by collision. During the excitation of any hydrogen atom, the electron jumps to $n = 2$ or any higher excited state. (If enough energy is absorbed, the electron may escape from the nucleus entirely, thus ionizing the atom. Ionized hydrogen atoms cannot produce spectral lines.)

Hydrogen atoms (and most other atoms) do not persist in an excited state for long. Within about $10^{-8}$ s (which is a long time compared to the "atomic year," the time of one revolution of the electron, roughly $10^{-16}$ s), the electron spontaneously makes a jump toward a lower-energy stationary state, emitting a photon as it does so. Jumps from higher states down to $n = 1$ are most probable, but other patterns can occur. For example, during

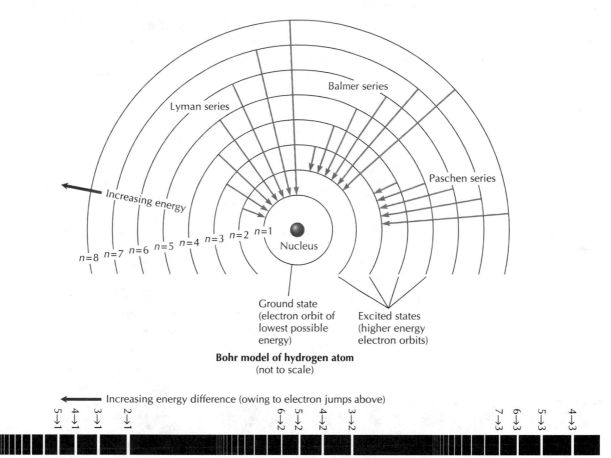

**Bohr model of hydrogen atom**
(not to scale)

**Hydrogen emission spectrum**
(not to scale)

**FIGURE 11.5** The electrons of hydrogen atoms excited by one mechanism or another immediately start cascading toward the ground, or $n = 1$, state. Because all stationary states are quantized (there are no intermediate electron orbits between the integer $n$ orbits), the photon released by any single inward jump of the electron has a specific frequency and wavelength. The Lyman series of spectral lines, which lies exclusively in the ultraviolet, is produced by downward jumps that end at $n = 1$. The Balmer series, with four lines visible to the eye and the rest in the ultraviolet, results from downward jumps ending at $n = 2$. The Paschen series in the infrared is the result of jumps ending at $n = 3$. Other series, associated with jumps down to $n = 4$, 5, 6, and so on, exist in the far infrared (not shown here). Within any given series, the lines become closer together toward the left, because the energy *difference* between adjacent $n$ states becomes less at higher energy levels. Also within each series, the lines become progressively less intense toward the left, because inward jumps from orbits farther and farther out are less and less likely to happen.

de-excitation the jump sequence $6 \rightarrow 4$, then $4 \rightarrow 2$, then $2 \rightarrow 1$ is possible. Three separate photons would be emitted in quick succession during this particular sequence.

Each jump toward a lower-energy stationary state is associated with the emission of a certain quantum of energy (a photon) having a frequency $f = E/h$ and a wavelength $\lambda = hc/E$. The greater the energy difference of the jump, the higher the frequency and the shorter the wavelength of the photon emitted. There are many energy levels ($n$ states) in a hydrogen atom, and a great many possible jumps from one state down to another, so there are a great many lines in the hydrogen spectrum. The hydrogen spectrum appears orderly because the many possible energy jumps within the hydrogen atom can be grouped into a number of separate series, and each series has its counterpart in a separate set of grouped spectral lines. Figure 11.5 makes this connection clear.

# Spectroscopy

With the advance of quantum mechanics in the 1920s and 1930s, more complete models of electron structure were developed. We now regard the Bohr model as being an approximation of electron structure in the hydrogen atom. Still, the model ascribes the correct quantum behaviors to both photons and the energy levels associated with the hydrogen atom. More sophisticated quantum mechanical models were eventually developed to describe electron structure in atoms heavier than hydrogen (helium, lithium, and so on). Because these atoms contain many electrons that can exist at many levels of energy, their spectra are far more complex than that of hydrogen.

An enormously useful aspect of the study of spectra is that atoms of different elements produce utterly different-looking spectra. Even atoms of the *same* element in different ionization states produce different spectra.

## BOX 11.1
## The Laser

The details of the Bohr model may fail to explain some aspects of atomic structure, but the basic idea that an atom can experience jumps from one energy level to another by absorbing or emitting a quantum of energy is central to the understanding of the laser. The acronym **laser** means *l*ight *a*mplification by *s*timulated *e*mission of *r*adiation. The "lasing" process typically involves a resonant cavity that holds a large number of atoms, a large fraction of which are excited to the same energy level and ready to fall into a lower-energy state. The trick is to get these atoms to cooperatively de-excite in exactly the same way and at the proper times. When they do, a coherent beam of light is produced. *Coherent light* consists of photons of virtually identical wavelength, all "marching in step" with each other in a common direction. Typically, a laser produces an exquisitely narrow beam of a pure color that has very little tendency to spread out with distance.

The atoms about to undergo the lasing process are first excited by one of a variety of methods. One way is to use the intense and intermittent flashes of light from a xenon flashtube (of the sort used as flash units in cameras). Another is to supply a steady current of high-voltage electricity.

Not just any atoms will do. The lasing atoms must have a *metastable* (or relatively long-lived) excited state—roughly one-thousandth of a second long—so there is enough time for many atoms in the cavity to reach the same excited state. Once they have, *stimulated emission* can take place: A photon released by one atom undergoing de-excitation quickly triggers another nearby atom to do the same, and that atom's emitted photon repeats the cycle with another nearby atom. A cascade of identical photons ensues (see Figure 11.6).

So that these photons are guided and their effects amplified, the cavity containing the atoms and photons (typically a gas-filled tube) features, at opposite ends, two flat mirrors positioned precisely parallel to each other and separated by an integral number of half wavelengths. Photons traveling along the length of the tube (perpendicular to the mirrors) can then surge back and forth, forming a standing wave of light. (Recall from Chapter 8 how vibrating strings can assume a variety of standing wave patterns such that whole numbers of half wavelengths fit across the two fixed points of the string. Light waves behave similarly.) One of the two mirrors is

**PHOTO 11.2** The glass tubes that make up neon signs are filled with various gases (not just neon) in rarefied states. The color of each glowing tube depends on the mix of gases inside and on the characteristic emission spectrum each gas has. The color for the element neon is a vivid reddish orange.

For example, the spectrum of ordinary or *neutral* helium (helium atoms having their normal complement of two electrons) is different from that of *singly ionized* helium (helium atoms with one electron missing). Furthermore, there is a unique spectrum emitted by each kind of molecule (aggre-

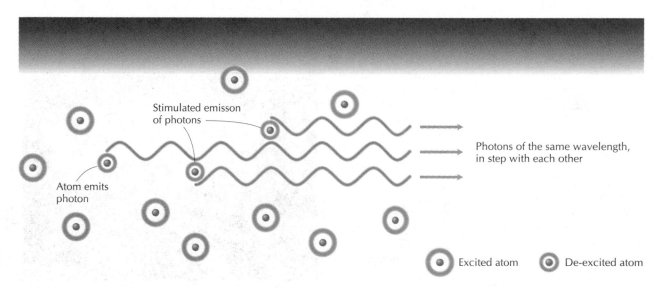

Stimulated emisson of photons

Photons of the same wavelength, in step with each other

Atom emits photon

Excited atom        De-excited atom

**FIGURE 11.6** If many atoms of the same kind can be pumped up to the same excited state, the emission of a photon by one atom can stimulate similar emissions of photons by other atoms nearby. So that light is produced continuously, the de-excited atoms must be excited again by some external mechanism.

partially silvered; it is mostly opaque and reflective, but also slightly transparent. It allows some light to escape in the form of a thin external beam whose radiant energy is extremely concentrated.

The frequency or wavelength of a laser's light is dependent on the energy released by the atoms undergoing their coordinated de-excitations, in accordance with Equation 11.1 or Equation 11.2. Because many kinds of atoms are known to have metastable states, and because the energy levels in those atoms are different, lasers with different wavelengths can be constructed. There are visible-light   *(continued)*

gate of atoms) that is agitated enough to emit radiant energy. The characteristic spectra of atoms and molecules have been compared to fingerprints: The recognition of unique patterns of spectral lines leads to positive identification of the substances responsible for those patterns.

By utilizing **spectroscopy** (or *spectrum analysis*), which is the method of separating wavelengths of light and analyzing the resulting spectrum, scientists can tell much about the physical nature of the source of the light—not just composition, but also temperature, structure, and more. Spectroscopy has long been used by chemists to help identify elements and compounds in samples of unknown composition by heating them in a flame and observing the resulting spectrum. This technique is called *emission spectroscopy*. Geologists use spectroscopic techniques to identify rare elements in rocks and minerals, and microbiologists use them to discover the chemical makeup of biomolecules. Organic molecules, whether or not they are associated with biological organisms, decompose easily when heated and thus cannot be identified through emission spectroscopy. They can, however, be identified through *absorption spectroscopy*. This technique involves transmitting electromagnetic radiation (typically infrared radiation) through a sample of organic material and noting the pattern of missing wavelengths. The missing energy associated with the wavelengths absorbed in the sample produces specific vibrations and rotations in the molecules of the

*(Box 11.1 continued)*    lasers of a variety of colors (so impressively displayed during laser "light shows"), infrared and ultraviolet lasers, and even pulsed X-ray lasers. The latter kind have been under development by the military as a weapon.

Lasers have been put to hundreds of scientific, industrial, and household uses. The common red (helium-neon gas) lasers read digital information from CDs and bar codes. Surveyors use lasers to establish sight lines and for range finding. Modern heavy machinery (everything from farm tractors to tunnel borers) is increasingly being guided along straight lines by reference beams produced by lasers. Physicians use narrowly focused laser beams for delicate eye surgery. More powerful versions of the surgeon's laser scalpel are used as precision cutting tools in factories.

Much effort of late has gone into the development of *microlasers,* miniaturized lasers that find uses in information technology and in basic scientific research that strives to understand the details of atom-photon interaction. Researchers have recently succeeded in producing a detectable beam of laser light using only a handful of photons bouncing back and forth in a 1-mm gap between two tiny, curved mirrors.

**PHOTO 11.3**   Laser beams are used to both write and read information to or from a CD (compact disc). In the writing (mastering) process shown here, a modulated laser beam is used to produce billions of hardened spots on a spinning plastic disc. A metal master disc, pressed from this, is used to manufacture the CD copies. The copies are read by a CD player, which uses a laser and electronic circuitry to decode the information encoded in the pits.

**PHOTO 11.4**   Sophisticated spectrometers like this one are used to determine the chemical makeup of small samples of material. Information encoded in the spectrum of light from the glowing sample shown is decoded with the help of a computer.

sample. Each type of molecule has unique modes of vibration and rotation and therefore produces a unique energy absorption pattern.

In astronomy, spectroscopy has long been an indispensable tool of inquiry. Virtually everything we know about the physical nature of our sun and matter outside our Solar System comes from well over a century's worth of spectroscopic observations performed with the help of telescopes.

## Astronomical Spectroscopy

Light, it has been said, is the "calling card" of astronomy. Virtually all our knowledge of the stars, planets, and galaxies comes from the analysis of light: the intensity and spatial distribution of the light, variations in the intensity of the light (if any), and, most important, the spectrum of light. Spectroscopy has the ability to tease out physical data from nearly any object we observe in the sky, regardless of how far away it is. Light may diminish in intensity as it radiates outward from a source, but the information in it remains.

The tool of astronomical spectroscopy is the *spectrograph,* or *spectrometer,* an instrument (usually attached to a telescope) capable of breaking down light into its component wavelengths. At the heart of a spectrometer lies a glass prism (or series of prisms) or a *diffraction grating* (a device in which diffraction and interference effects are used to spread a beam of light into its various wavelengths). Up until the past decade, nearly all astronomical spectra were recorded for further reference and measurement on film, but now electronic recording is the norm. The old photographic images of spectra, which resemble those depicted in Figure 11.4, may be more intuitively instructive to the average person. But the newer way of

**PHOTO 11.5** This relatively low-resolution spectrogram of the sun shows many dark absorption lines, introduced by atoms of various elements existing in the sun's photosphere. The fact that the sun and nearly all other stars have absorption spectra shows that their interiors are hotter than their surfaces. Some youthful stars have emission lines superimposed on a normal absorption spectrum. These stars are ejecting large quantities of superhot gas into the space around them.

looking at spectra—by means of a graphed line that rises and falls with intensity across the spectrum—is more revealing to the eyes of a specialist.

There has also been a big push over the past two or three decades to gather spectroscopic information over the broadest possible range of wavelengths. Visible-light spectroscopy tells an important, but only small, part of the whole story; therefore, scientists gather and analyze the spectra of everything from radio waves to X-rays and gamma rays from extraterrestrial sources. As mentioned in Chapter 8, Earth's atmosphere absorbs large portions of the electromagnetic spectrum, so today's astronomers make frequent use of specialized telescopes borne on high-altitude balloons or spacecraft to gather the information they want.

## What the Spectrum of a Star Reveals

Astronomical spectroscopy began with the analysis of the copious light of the sun. The first photograph of the solar spectrum was taken in 1843. Before long, astronomers had matched many of the dark spectral lines with corresponding bright spectral lines observed when certain elements are heated to incandescence in earthly laboratories. The spectral fingerprints of hydrogen and several other known elements were recognized in the sun's spectrum early on.

In 1868, a pattern of lines of unknown origin was discovered in the solar spectrum. The substance presumably producing the pattern was given the name helium (after the Greek word *helios,* which means "sun"). For nearly three decades, astronomers were not sure if helium was merely too scarce on Earth to be detected or if it was a substance unique to the sun and perhaps other stars. If the latter were true, it would certainly call into question our ability to understand the nature of the sun's chemical makeup purely by means of observations from Earth. That possibility was put to rest in 1895, when a small amount of helium was found in a sample of radioactive material from Earth's crust. The element helium is, in fact, rare on Earth, though it can be extracted in quantity from certain natural gas wells. Helium today finds many uses, including its use to hoist lighter-than-air blimps. It is, however, a nonrenewable resource, and its extravagant use

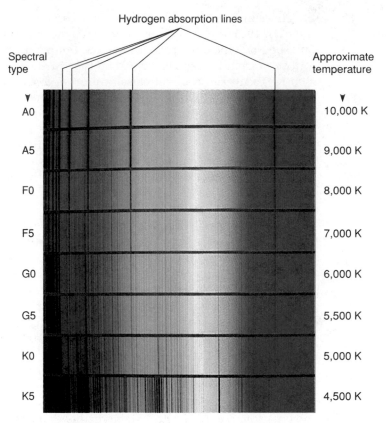

**PHOTO 11.6**  These spectra of eight different stars arranged in order of descending temperature are denoted by spectral type (letter) and decimal subclass (number). For example, a star classified as A5 has spectral features and a surface temperature about midway between that of an A0 star and an F0 star. The most prominent absorption lines exhibited in the spectra of the A-type and F-type stars are those of the Balmer series of hydrogen. Hydrogen is abundant in all stars; however, its ability to absorb light depends on temperature. For this reason, the hydrogen absorption features are very dark in the relatively hot A-type stars and much less prominent in the cooler stars whose spectra are shown here.

in party balloons is perhaps unwise. Once released into the atmosphere, helium is lost for any further use.

Today, tens of thousands of spectral lines have been mapped and measured in the solar spectrum, and about 70 elements have been identified as existing in the sun's photosphere. Hydrogen and helium are by far the most abundant elements in the sun. Together, they make up about 98% of the sun's surface material.

By the earliest years of the 20th century, improvements in photographic materials made feasible the wholesale recording and cataloging of *stellar spectra,* the spectra of stars. Although at that time astronomy was an almost exclusively male pursuit, several women astronomers dominated this rapidly advancing field. Annie Jump Cannon, who was personally responsible for the spectral analysis of hundreds of thousands of stars, is best known for organizing the standard sequence of *spectral types* given to stars. The major spectral types, which are based on the patterns and strengths of spectral lines, are labeled sequentially with the letters O, B, A, F, G, K, and M. Each of these major classes is divided into ten decimal subclasses. Some examples of various spectral types can be seen in Photo 11.6.

The sun is a spectral type G2 star, which is near the middle of the range. In 1925, the astronomer Cecilia Payne succeeded in linking the various spectral types to a scale of surface temperature: Type O stars are the hottest, at greater than 20,000 K, and type M stars are the coolest, at about 3000 K. Today, the spectral classification of a star automatically gives the star's surface temperature to an accuracy of about 1%. The sun's G2 classification corresponds to an average surface temperature of 5800 K.

As the 20th century progressed, more and more clues to the physical characteristics of the stars were found hidden in the details of stellar spectra. In addition to disclosing temperature, structure, and composition, a star's spectrum may reveal something about the pressure and density of the gas at the surface, magnetic fields on the surface, and any line-of-sight motion taking place.

Scientists soon recognized that stars of a given spectral type or temperature may have spectral lines that differ in their sharpness. Those with diffuse spectral lines, indicating greater photospheric pressure and density, were at first identified simplistically as *dwarfs,* because compact size would be a characteristic of stars of higher surface pressure. The sun falls into this general dwarf category, though certain stars, such as the *white dwarfs,* are much smaller still. Stars with sharp spectral lines, indicating very low pressure and density at the surface, were identified as being immensely bloated; we still call them *giants* and *supergiants.*

The presence of a strong magnetic field causes the energy levels of atoms in a gas to split into sublevels so that each of its spectral lines appears (typically) as a closely spaced triplet of lines. This phenomenon, known as the *Zeeman effect,* can be readily demonstrated in the laboratory. It has been used to map out magnetic fields on the sun and to detect strong magnetism in certain stars and clouds of diffuse matter in space.

The Doppler effect in light (Chapter 8) has been used to great advantage in astronomy. If a star is moving toward or away from us, its lines in the visible spectrum shift toward the bluer or redder wavelengths, respectively. The degree of shift is directly related to the star's *radial velocity,* or its velocity in the line of sight. All the spectral lines of the moving star are shifted by the same proportional amount.

As long as the radial velocity of a light source is not large compared with the speed of light, the relationship between Doppler shift and radial velocity is a simple one: The greater the Doppler shift of any given line, the greater the radial velocity. For example, if a particular spectral line appearing at $\lambda = 500$ nm in the spectrum of a nonmoving source of light is observed at $\lambda = 505$ nm in the spectrum of some star, that means the star is receding from us. The 5-nm difference in wavelength is a change of 1%, which means that the star is receding at 1% of the speed of light, or 3000 km/s. If, for another star, the same spectral line is observed to be at $\lambda = 510$ nm, this represents a 2% shift, which means the star is receding at 6000 km/s. For most stars, which have Doppler shifts much smaller than 1%, scientists must measure the spectral lines very precisely in order to derive accurate values for radial velocity.

The Doppler effect is commonly used to decode the motions of binary stars. As two stars orbit each other, the out-and-back component of their motion (in the line of sight) produces Doppler shifts that change with time. This knowledge sometimes allows astronomers to compute the masses of each of the two stars.

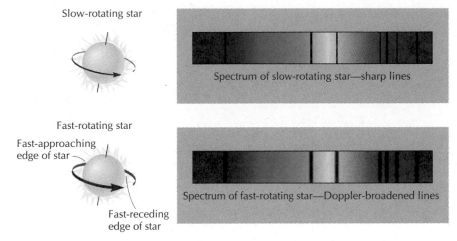

Slow-rotating star

Spectrum of slow-rotating star—sharp lines

Fast-rotating star

Fast-approaching edge of star

Fast-receding edge of star

Spectrum of fast-rotating star—Doppler-broadened lines

**FIGURE 11.7**   The speed of rotation of a star may be discerned from studies of its spectrum. Very little broadening is seen in the spectral lines of a slow-rotating star, because there is little relative movement of different parts of the star. A fast-rotating star may be oriented to our line of sight such that light from its various parts is smeared over a range of wavelengths. Statistical studies of many stars indicate that the hot and typically bright O-, B-, and A-type stars generally rotate relatively quickly; the cooler and generally faint G-, K-, and M-type stars rotate much more slowly. Astronomers believe that many of the slow-rotating stars (the sun included) lost angular momentum to their planets during an early stage in their lives.

Doppler measurements have also allowed astronomers to build up a comprehensive picture of how the sun moves relative to neighboring stars, how billions of stars stream through our home galaxy, how the Milky Way galaxy moves relative to its neighbors, and how the universe as a whole expands. The Doppler effect may also reveal how stars rotate. Rapidly rotating stars may give themselves away by exhibiting blurry, rather than sharp, spectral lines (see Figure 11.7). This *rotational broadening* of a star's spectral lines is distinct from any overall shift in spectral line positions from the star's radial velocity.

## Planetary Spectra

The Solar System's planets shine by the reflected light of the sun, so you might expect that we would learn nothing new by examining a planet's spectrum—it would be an exact copy of the solar spectrum. This is largely true for airless worlds like the moon and Mercury, but not for the other planets. When sunlight passes through the atmosphere of a planet and bounces back toward Earth, new absorption lines may contaminate the originally pristine solar spectrum. These new lines are revealed through careful comparison of both spectra. Long before the era of spacecraft, astronomers had deduced the existence of carbon dioxide on Mars and methane and ammonia on the Jovian planets by using this method.

Spectrometers on board Earth-orbiting spacecraft help in the efforts to map out the resources of our planet's lands and oceans and to monitor changes. Similar instruments on interplanetary space probes have added an important dimension to the photographic information sent back from our planetary neighbors.

# Color

Just as spectra reveal to scientists a wealth of information about the physical nature of light sources, color enriches our lives with information about nearly everything we cast our eyes upon. Color—like musical sounds, fragrant smells, or any other sensation—can give us great aesthetic pleasure as well.

All color sensation arises from only three distinct types of cone-shaped photoreceptors, or *cone cells,* in the retinas of our eyes. Each is sensitive to a wavelength range that is about a third as wide as the spectrum of visible light. The peak response of the three cone-cell types centers on the colors red, green, and blue, which lie at or near the long-wavelength end, the middle part, and the short-wavelength end of the visible spectrum. Amazingly, all the multifarious hues we can distinguish with our brain are produced by no more than three inputs of variable intensity from the retinas of our eyes.

The true nature of color was first discovered by Isaac Newton when he passed a thin slice of sunlight through a prism (as in Photo 11.8), and manipulated various parts of the resulting color spectrum. No matter how hard he tried, he simply could not produce *new* color out of light belonging to any one part of the spectrum. With another prism, Newton combined all the colors produced by the first prism. The combination produced a slice of white light, just like that entering the first prism. Newton's conclusion was that light is a mixture of distinct colors that can be *separated* from each other by refraction or other means but never *changed* in color. This is analogous to strands of a rope that are unraveled but not broken in the middle of the rope.

Newton (like anyone else with normal color vision) saw only six or seven major bands of distinctive color in the spectrum, so he could not

**PHOTO 11.7** The light we see in a flower garden comes from the sun. The color we see and appreciate comes from the ability of the leaves and the flowers to absorb some wavelengths of visible light and reflect the rest.

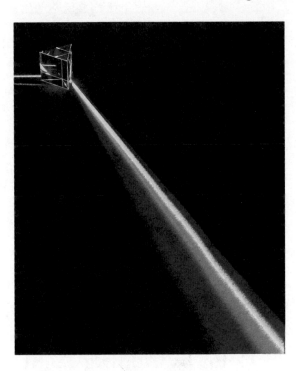

**PHOTO 11.8**  When white light passes through a prism, it breaks up, or disperses, into a spectrum of colors. Violet light is bent (refracted) the most. Red light is bent the least.

prove that light within a single color band is heterogeneous. However, the rope analogy really can be pushed further. Just as the handful of major strands in a hemp rope are made of bundles of smaller, distinct fibers, so, too, are the color bands in the spectrum made of photons with tiny but distinct wavelength differences. Like unbroken fibers in a rope, no photon can be modified or joined with another to assume a new identity.

Nevertheless, we see far more colors with our eyes than the specific bands of color associated with the spectrum. This richness comes about because photons can be mixed and selected in various ways to produce different *sensations* of color. What follows are brief descriptions and examples of photon-mixing and -selecting processes.

## Color by Dispersion

The refraction of light can be explained by means of both a wave model and a particle model for light. Both models show that light with a shorter wavelength should be bent though a prism with a greater angle of deviation. The phenomenon of separating colors or wavelengths through refraction is called **dispersion.**

Prisms or other angular pieces of glass are hardly unique in their ability to disperse light. Rain droplets, which happen to be spherical (or nearly so) as they fall through the air, are capable of bending and internally reflecting sunlight; thus, much of it emerges in a direction almost opposite to the incoming light. Much of the refracted sunlight of a given color emerges at an angle of about 42° with respect to the *antisolar point* (the point 180° away from the sun in the sky, below the horizon in the daytime). Different colors reinforce at slightly different angles from the antisolar

**PHOTO 11.9** The rainbowlike arc lying some 48° above the late-afternoon sun (hidden behind the tree) is not a rainbow. It is a rare "circumzenithal arc" photographed from the author's backyard. The arc is produced by sunlight refracting through the 90° corners between the bottoms and sides of billions of slablike ice crystals floating with their flat surfaces parallel to the ground in the cirrus cloud above. The arc forms part of a small circle centered on the zenith (straight-up point) in the sky.

point, so that a rainbow—a circle of radius 42° centered on the shadow of your head—appears. (To visualize where a rainbow appears, look again at Photo 8.5 in Chapter 8. The photographer stood with his back to the setting sun and aimed the camera approximately toward the antisolar point in order to catch the rainbow amid the raindrops falling on the scene.) Usually, the arc of the rainbow you view lies above the horizon and not much below it, because very few raindrops fall at your feet. A complete circular arc can sometimes be seen, however, in the fine spray of a garden hose when the spray is aimed in a direction away from the sun.

Many rainbowlike phenomena commonly seen in the sky are not caused by rain at all but by the presence of a myriad of tiny, hexagonal ice crystals in intervening cirrus or other high clouds. The large, slightly colored haloes often seen around the sun or the moon prior to approaching storms are caused by light refracting and dispersing through the 60° corners of the ice crystals. Most of the light entering these corners bends through an angle of about 22°, so the ice crystal halo appears as a hazy circle of 22° average radius around either the sun or the moon. In rarer cases, sunlight may refract through the 90° corners that join the sides and the top or bottom of an ice crystal (see Photo 11.9).

# Color by Adding Light

As Newton discovered, an even mixture of all the luminous colors of the spectrum produces white light. Curiously, but not inexplicably, we get the same result when we mix just the colors red, green, and blue. These, of course, are the three colors associated with the peak response of the eye's three different cone cells. They are called the **additive primary colors,** because all possible hues can be generated by adding them in different combinations.

Let us look at some simple combinations of the additive primary colors (see Photo 11.10). Red light added to green light (with no blue light) acti-

**PHOTO 11.10** The white parts of this scene are being illuminated equally by three spotlights of red, green, and blue color. In the shadow of the blue light, but not the green and the red light, the resulting illumination on the screen appears yellow. Blue and yellow added make white, so we say that blue and yellow are complementary colors (in an additive sense). Two other additive complementary pairs are red plus cyan and green plus magenta.

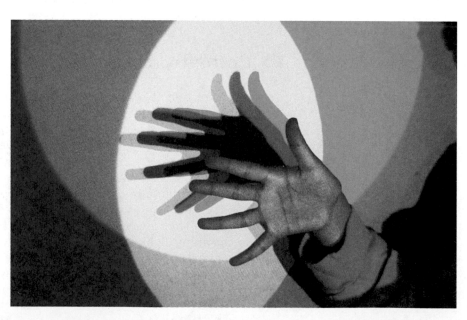

**PHOTO 11.11**  When seen closeup, the glowing image on a TV picture tube breaks up into closely spaced bars (or dots on the face of some tubes) that glow with red, green, and blue light. When all three kinds of stripes or dots are glowing with equal intensity at some spot, the color appears to be white at that spot when seen at normal TV-viewing distance. Unequal intensities among the three kinds of dots or stripes give rise to a wide variety of color hues.

vates only two of the three cone cells and produces the sensation of yellow. Green light plus blue light, again with no other, yields a blue-green hue called *cyan*. Red light plus blue light yields a vivid hue called *magenta*.

Not-so-simple mixtures of various wavelengths in the spectrum can potentially produce all the hues we can perceive. For example, an excess of red in an otherwise even mixture of the three primary colors (or in wavelengths across the whole spectrum) yields a pink hue. Hues can also be synthesized from discrete wavelengths of light, such as those produced in emission spectra. The emission spectrum of hydrogen produces just four narrow lines in the visible spectrum: a strong red line, a weaker line in the blue near the green band, and two fainter violet lines. This mixture creates a strange pink or magentalike hue. The glows of hydrogen and a number of other gases in discharge tubes are pictured in Photo 11.12.

## Color by Subtracting Light

The color we see in an object that does not give off its own light but, instead, is illuminated by white light is a function of how that object subtracts (absorbs) photons of different wavelengths. The absorption can take place during the reflection of light from the object or (as long as the object is at least partially transparent) during the transmission of light through it.

When all wavelengths are reflected in equal proportion from an opaque object, then the object is described as being white or gray. If no light reflects at all, then the object is black. There are no truly white (100% reflectivity) or totally black (zero reflectivity) objects in nature, though carbon soot and snow come close to these ideals. The range between pure black and pure white is filled with various shades of gray. Near the middle of the gray scale are typical concrete surfaces, which reflect 20–50% of the light falling on them. The moon's surface, which looks white when viewed against the dark night sky, is actually a dark gray, averaging 12% reflectivity.

All visible wavelengths are transmitted equally and virtually undiminished through small amounts of transparent (or "colorless") materials such

**PHOTO 11.12**  Energized by high voltage in laboratory discharge tubes, the gases (left to right) hydrogen, helium, neon, and mercury vapor glow. Because photographic film responds somewhat differently than the eye to discrete wavelengths of light, the colors reproduced here do not exactly match the sensation you get when actually viewing the glowing gases.

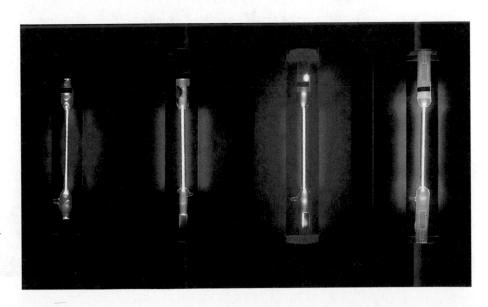

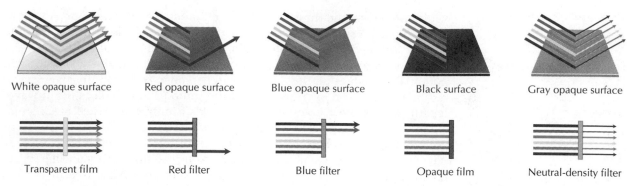

White opaque surface   Red opaque surface   Blue opaque surface   Black surface   Gray opaque surface

Transparent film   Red filter   Blue filter   Opaque film   Neutral-density filter

**FIGURE 11.8** Opaque colored surfaces and semitransparent films and filters change the mix of reflected or transmitted photons by means of selective absorption. Most blue filters and surfaces transmit or reflect a range of wavelengths spanning the blue and violet bands of the visible spectrum.

as air, water, and clear glass. These materials are analogous to white, opaque bodies. Smoked glass, or what photographers call neutral-density filters, can be used to diminish the intensity of transmitted light. They appear gray in color to the eye.

The *selective* absorption or transmission of visible wavelengths by various dyes or pigments produces the colored hues of all the nonglowing objects we see (see Figure 11.8). A red rose does not glow red; rather, it reflects a range of wavelengths in the red part of the spectrum and absorbs the rest. The chlorophyll in the bright green leaves of a tree reflects green and absorbs most other wavelengths. The colored glasses or gels used by stagehands to filter white light and the colored filters used by photographers to achieve certain effects are examples of the selective absorption of light during transmission.

When a painter mixes paints to obtain a new color, the colors reflected by the pigments of those paints do not "add" in the same way that red,

**PHOTO 11.13** The white light incident on the moth was entirely reflected. Some of the same white light incident on the flower was absorbed by molecules in the flower's petals and stamens.

green, blue, or any other colors *already present in light* add. Pigments can only subtract from wavelengths of light that are already there. Red paint, for example, contains pigments that absorb cyan and reflect the rest, which is red light. Three particular colors, it turns out, are all you need to reproduce all possible hues by *subtraction*. These three—yellow, cyan, and magenta—are called the **subtractive primary colors.**

Yellow paint absorbs blue, cyan paint absorbs red, and magenta paint absorbs green. If you mix yellow and cyan paint, all that is left to reflect is green. If you mix cyan paint and magenta paint, only blue reflects. And if you mix yellow paint and magenta paint, only red reflects. Mix all three of the subtractive primaries together as paint and theoretically you get nothing reflected, or black—though the result, in practice, may be a dark brown.

Use a powerful magnifying glass to examine the tiny dots that make up any color photograph in this book. You will see the imprint of yellow, cyan, and magenta ink. You will also see dots of black ink. Black ink is needed to vividly depict blacks and to accurately depict grays when printed on white paper, because the sizes of the printed dots must be adjusted to reflect various amounts (saturations) of particular color combinations. Since the dots normally do not completely overlap, there would be some unwanted reflected light in areas of a picture that are supposed to be dark or black if the three subtractive color dots were used alone. You hold in your hands a "four-color" textbook. Each page received the imprint of at least one ink (black) and as much as four inks in stages when it passed through the printing press.

The *perceived* color of an object depends, of course, on the available light as well as on the pigments present in the object. For example, the reds, whites, and blues in an American flag illuminated by a source of exclusively red light would appear as red, red, and black, respectively. The apparent black color of the blue parts of the flag may seem surprising until you consider that blue pigment absorbs all colors but blue (and usually violet). The perceived colors of many objects may seem strange or even ghastly when illuminated by mercury vapor or sodium vapor street lights. These lamps are much more efficient than incandescent lamps, but their emissions in the visible spectrum are irregularly distributed.

Note that the absorption of light on reflection or transmission is not equivalent to the disappearance of energy. The missing photons give up their energy to the atoms that make up the pigments of the absorbing material, and that energy is soon degraded into thermal energy. Now you know why light-colored objects stay relatively cool in the sun, but dark objects bake.

## Color by Interference

The shimmering colors of a soap bubble, the swirls of color on a film of motor oil floating on a puddle of water, and the iridescent glow of a peacock's feathers owe their pizzazz to interference, along with a pinch of reflection and diffraction. All are described as the effect of the interference of light waves by reflection in thin films.

Illustrated in Figure 11.9 is a film of some transparent material comparable in thickness to the wavelengths of visible light. In nature, any such film is likely to vary in thickness from place to place. Much of the incident

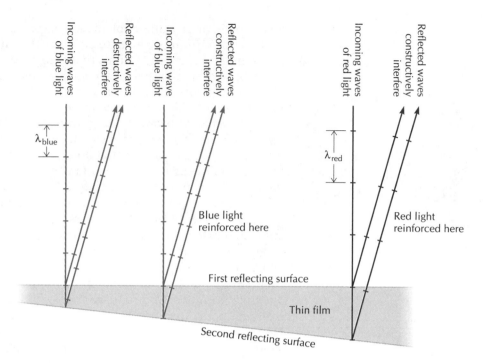

**FIGURE 11.9** A thin film of transparent material, such as oil floating on water or a soap bubble, can produce a double reflection of the light incident on it. Different colors, which have different wavelengths, are reinforced at various spots on a film of variable thickness.

Labels in figure:
Incoming waves of blue light
Reflected waves destructively interfere
Incoming wave of blue light
Reflected waves constructively interfere
Incoming waves of red light
Reflected waves constructively interfere
$\lambda_{blue}$
$\lambda_{red}$
Blue light reinforced here
Red light reinforced here
First reflecting surface
Thin film
Second reflecting surface

**PHOTO 11.14** The colorful interference fringes on this thin film of soapy water are the result of light from two reflections combining constructively and destructively.

light may be transmitted through the film, but some reflects off both surfaces. The reflected light of a single incident ray is now two rays. The two rays, which can be pictured as waves marching side by side, combine and interfere. Constructive interference for any wavelength $\lambda$ occurs when the thickness of the film is equal to $(1/2) \lambda$, $\lambda$, $(3/2) \lambda$, $2 \lambda$, and so on. Destructive interference occurs for intermediate thicknesses. Since the wavelengths of colors vary, different colors are reinforced or diminished upon reflection from different places on a film of variable thickness.

When light is transmitted through a series of many tiny, closely spaced, parallel slits, or when it is reflected from a series of many tiny, closely spaced, parallel, mirrorlike strips, constructive interference takes place for specific wavelengths only in very specific directions. Such a device is called a **diffraction grating,** so named because light diffracts (spreads out) from each slit or strip and because the many slits resemble a grate. A diffraction grating can disperse light. The typical grating in a modern spectrometer contains tens or even hundreds of thousands of microscopically spaced slits. The more slits illuminated by the incoming light, the greater the maximum theoretical resolution (the ability to separate closely spaced wavelengths) in the spectral image. An effect similar to that of a diffraction grating can be seen in white light reflected from a compact disc. The billions of microscopic pits engraved in rows in a CD break up the light waves striking them into a glittering assortment of colorful bands.

# Color by Scattering

Why is the sky blue? Why are clouds white? Why is the sun often reddish at sunrise or sunset? It is not because of pigments. Rather, white light

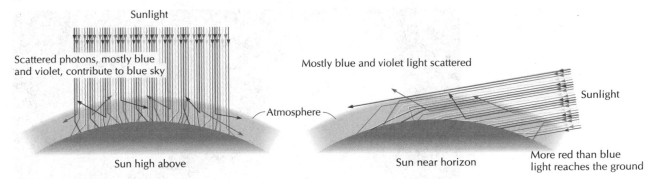

**FIGURE 11.10**  When the sun is high, some of its incoming photons are scattered out of the incident direction. In clear air, more violet and blue is scattered than other colors, and the sky appears a bright blue. When the sun is low, sunlight passes through much more air, and the scattering is more pronounced. Less light reaches the ground, and the light from the sun is reddened because so many violet and blue photons have been scattered away.

is selectively *scattered* among the molecules of air or among the tiny water droplets of a cloud.

**Scattering** takes place as a result of certain interactions between photons and atoms or other small particles of matter. When photons coming from a common direction are absorbed by small particles, some or all of their energy may immediately be reemitted as other photons of like kind that depart from the particle *in random directions*. When the particles are much smaller in size than the wavelength of the incident photons, scattering is infrequent. When particles are as large as or larger than the photon wavelength, scattering is more pronounced.

The microscopic or near-microscopic water droplets of a cloud are much larger than the longest visible wavelength. Therefore, all visible wavelengths are scattered equally and the cloud looks white whenever it is illuminated with white light from the sun. You can also see white-light scattering in the beam of a flashlight passing through a tank of water with a small amount of milk in it. The milk particles are about the same size as water droplets in a cloud.

The nitrogen and oxygen molecules that make up the bulk of the atmosphere are much smaller than visible-light wavelengths. Thus, photons of shorter wavelength (which are more closely matched in size to the particles) have the greater tendency to scatter amid the molecules; the longer-wavelength photons pass right by. Violet photons scatter about ten times more frequently in clear, dry air than red photons do, and ultraviolet photons scatter even more. In the daytime, these scattered photons come to our eyes from directions away from the sun (see Figure 11.10). The admixture of many photons from the violet and blue parts of the spectrum and the scarcer contributions of the other parts create a color we perceive as blue.

The blue of the sky can also be seen in clear air as a blue haze in front of distant mountains. True blue skies are seen only in dry and clean air. Large amounts of water vapor scatter white light and "whiten" the blue sky. Particulate matter (such as the soot particles in smoke and the tiny, floating "aerosol" particles of smog) scatters and absorbs a larger and different fraction of the solar photons. They may paint the sky brown or gray.

**PHOTO 11.15** A star-speckled sky, including the constellation Orion, appears in this scene photographed through the aft flight deck window of the space shuttle *Columbia*.

Sunsets are tinted red for the same reason that the sky is blue. Light transmitted through the atmosphere has more bluish light removed by scattering than reddish light. The lower the sun is in the sky, the more obstacles (air molecules) stand in the path of the incoming sun rays and the dimmer *and* redder the sunlight appears to be.

If there were no air above our heads, there would be no scattering, and the daytime sky above would appear as black as night. For this reason, orbiting astronauts can see the stars in the sky, even with the sun shining, provided that they shield their eyes from any strong direct or reflected sunlight (see Photo 11.15).

**CHAPTER 11**

# Summary

Light can be thought of as a stream of particles, or photons, moving at the speed of light. A certain frequency and wavelength (wave concepts) can be ascribed to each photon, even though it is impossible to test for the existence of the wave properties and the particle properties of light at the same time. Each photon carries a quantum of energy that is proportional to its frequency and inversely proportional to its wavelength. The quantized nature of light and quantum interactions with matter are indicated by several phenomena, including the photoelectric effect and the appearance of discrete bright or dark lines in the spectra of rarefied gases.

The spectrum of hydrogen can be explained in simple terms by the Bohr model. The model assumes that the electron in a hydrogen atom can jump from one discrete orbit, or stationary state, to another, emitting or absorbing energy as it does so. The emitted or absorbed energy may take the form of a photon. In a laser, many atoms are stimulated to release photons of the same energy or wavelength. The photons travel back and forth between two mirrors and emerge from the laser as a thin beam of coherent light.

Each kind of atom or molecule, when energized enough, gives off radiation that has a unique spectrum. Furthermore, the detailed appearance of any spectrum is sensitive to certain physical factors, such as temperature, pressure, and density. Spectroscopy, or spectral analysis, is useful in all the sciences but is especially appropriate in astronomy, where it is has yielded knowledge about the chemical composition, the physical state, and the motion of matter in the stars and on the planets.

The sensory phenomenon of color is related to the response of three different light-sensitive cone cells in the retina of the eye. White light can be described as an even mixture of wavelengths across the visible-light spectrum. Other colors can be obtained in a variety of ways: by separating white light into colors through dispersion, by adding light from different parts of the visible spectrum, by selectively absorbing some wavelengths and reflecting or transmitting the rest, by using interference to selectively reinforce certain wavelengths and diminish others, and by scattering photons amid atoms or other small particles.

## CHAPTER 11
# Questions

## Multiple Choice

1. The formula for the energy involved in quantum interactions between matter and light was first worked out by
   a) Heinrich Hertz, with his discovery of radio waves
   b) Max Planck, to account for the distribution of thermal radiation emitted from light-absorbing bodies
   c) Albert Einstein, who showed that when light strikes a metal surface, it does so in tiny bursts
   d) Niels Bohr, who showed that electron orbits in the hydrogen atom are quantized

2. The temperature of a star (as well as anything resembling a theoretical blackbody) can be estimated by observing its
   a) brightness
   b) color
   c) composition
   d) mass

3. Which of these phenomena specifically demonstrates the existence of light quanta, or photons?
   a) refraction
   b) dispersion
   c) interference
   d) the photoelectric effect

4. A hydrogen atom is in the ground state (lowest-energy state) when its electron
   a) is at rest
   b) travels in the smallest possible orbit around the nucleus
   c) travels in the largest possible orbit around the nucleus
   d) has escaped from the atom

5. Which of the following materials exhibits an emission spectrum in visible light?
   a) solids and liquids heated to incandescence
   b) hot, compressed gases
   c) hot, rarefied gases
   d) a cool, rarefied gas placed in front of a glowing, incandescent lightbulb

6. In the emission spectrum of hydrogen,
   a) the spectral lines are uniformly spaced
   b) all quantum jumps are made from the same electron orbit
   c) all quantum jumps end at the same electron orbit
   d) each bright spectral line is the result of a photon emitted during a particular electron jump from one energy level (orbit) to another

7. Which of the following electron jumps between the quantum states $n$ of a hydrogen atom emits the photon of highest frequency?
   a) $2 \rightarrow 3$
   b) $3 \rightarrow 2$
   c) $4 \rightarrow 3$
   d) $3 \rightarrow 4$

8. The element helium
   a) exists only on Earth and in the sun
   b) is common on Earth and uncommon in the sun
   c) is common in the sun and uncommon on Earth
   d) was the first element to be identified in the sun

9. The standard sequence of stellar spectral type, labeled by the letters O, B, A, F, G, K, and M, is primarily a sequence of
   a) decreasing surface temperature
   b) increasing surface temperature

c) increasing hydrogen abundance

d) increasing helium abundance

10. The spectrum of a star never offers clues to the
    a) existence of a magnetic field on the star
    b) star's rotation
    c) pressure and density on the star's surface
    d) direction of the star in the sky

11. The photons emitted by a laser
    a) have different wavelengths
    b) have different speeds
    c) are coherent
    d) easily diverge from one another

12. Light undergoes dispersion and separates into various colors when it is
    a) refracted through a prism
    b) reflected by a curved surface
    c) diffracted through a small hole
    d) scattered by air molecules

13. A photon of which color has the highest frequency (and shortest wavelength)?
    a) red
    b) blue
    c) yellow
    d) green

14. When magenta light is added to green light, the result is
    a) blue light
    b) red light
    c) orange light
    d) white light

15. If a white spotlight is shined on a gray card in a dark room, the card appears to be
    a) black
    b) white
    c) gray
    d) green

16. Under blue light, the American flag appears to be
    a) all blue
    b) blue and red
    c) white and blue
    d) blue and black

17. Prisms and diffraction gratings are similar in the sense that
    a) both rely on the dispersion of refracted light
    b) both rely on diffraction and interference
    c) neither can be used to reflect light
    d) both are used in spectrometers

18. The clear sky is blue because
    a) blue light is scattered more than red light amid air molecules
    b) red light is scattered more than blue light amid air molecules
    c) the air is being bombarded by cosmic rays from the sun
    d) air reflects the color of the oceans

19. In sunlight or moonlight, clouds appear white because
    a) there is no dispersion when the light refracts through clouds
    b) clouds absorb photons of all wavelengths equally
    c) clouds scatter photons of all wavelengths equally
    d) solid pieces of white material in clouds produce diffuse reflections, like the tiny fibers that make up a white sheet of paper

# Questions

1. In what way or ways are the wave nature and the particle nature of light related?

2. How would you describe the radiation given off by a theoretical blackbody? How is the radiation affected by the temperature of the body?

3. Planck's constant is a very tiny number. What does this say about the energy carried by photons?

4. When light shines on a metal surface, what determines whether or not that light succeeds in ejecting electrons from the metal?

5. What kind of matter can produce a continuous spectrum? An emission spectrum? An absorption spectrum?

6. Explain in your own words how the hydrogen atom works according to the Bohr model.

7. What information can be gleaned by applying the technique of spectroscopy?

8. Why are the visible-light spectra of the planets almost identical in appearance to the sun's spectrum?

9. How are metastable excited states used to advantage in a laser?

10. How was Newton able to decide that the colors of the spectrum must be the result of white light

being separated into components by a prism rather than white light being modified in some way by the prism?

11. How are rainbows and rainbowlike ice crystal effects different? How are they similar?

12. What are the additive primary colors? What are the subtractive primary colors?

13. How does gray fit into the scheme of color?

14. If a thin film has a thickness of a fourth of a wavelength of blue light, what kind of interference (constructive or destructive) takes place when blue light reflects off both sides of the film and the two streams of reflected light combine?

15. If the sun emitted radiation only in the yellow-through-red parts of the visible spectrum, what color would the clear sky be?

## Problems

1. Photons of radio wavelengths individually carry extremely small amounts of energy. Confirm that this is true by calculating the energy carried by a single photon transmitted by a radio station at 100 MHz on the FM dial.

2. Calculate the energy carried by a typical gamma ray photon. You can do this calculation by looking up the range of frequency or wavelength of gamma rays in Figure 8.6 (the electromagnetic spectrum) and then calculating the energy by using Equation 11.1 or Equation 11.2.

## Questions for Thought

1. Can you think of anything else that is quantized (besides money) in our everyday experience?

2. Dark basaltic lava glows a dull red when it emerges from volcanic vents. Can you give a rough estimate of its Kelvin temperature?

3. What kind of spectrum (continuous, emission, or absorption) should the aurora borealis and aurora australis have?

4. Clouds of cold neutral hydrogen and other gases often lie between us (Earth) and many of the stars we see at great distances across the galaxy. How might we detect the presence of that gas? If we can detect it, how might we figure out whether it is moving toward or away from us?

5. How many times today have you interacted with or been affected in some way by a device that utilizes a laser beam?

6. Do you think there is any connection between certain ratios of light wavelengths and pleasing combinations of color, just as there is (for most ears) a connection between simple ratios of sound frequencies and the sensation of pleasing sound?

7. Have you noticed today any examples of color produced by the dispersion of light? Any examples of color by interference?

## Answers to Multiple-Choice Questions

| | | | | |
|---|---|---|---|---|
| 1. b | 2. b | 3. d | 4. b | 5. c |
| 6. d | 7. b | 8. c | 9. a | 10. d |
| 11. c | 12. a | 13. b | 14. d | 15. b |
| 16. d | 17. d | 18. a | 19. c | |

# The Inner Atom

*Sometimes it takes a lot of energy to get a lot of energy. In this experimental particle-beam fusion accelerator at Sandia National Laboratories in New Mexico, a pulsed beam of fast-moving particles triggers nuclear fusion reactions within tiny pellets of deuterium and tritium. During each of the accelerator's "shots," electrical discharges crackle over the surface of the water in which the accelerator is immersed.*

# WHERE WE ARE GOING IN CHAPTER 12

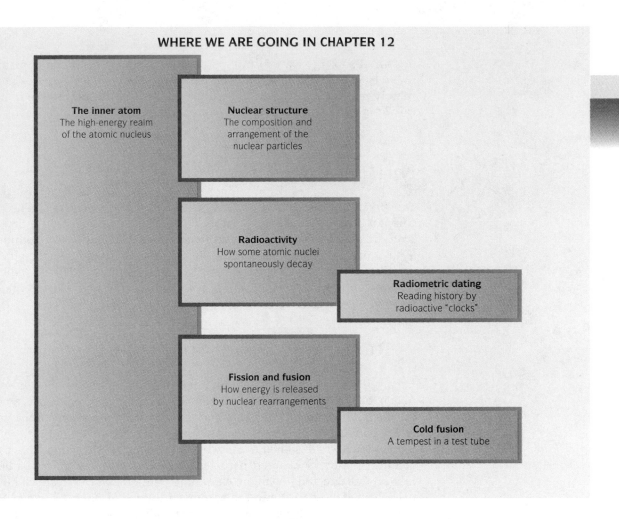

**The inner atom**
The high-energy realm of the atomic nucleus

**Nuclear structure**
The composition and arrangement of the nuclear particles

**Radioactivity**
How some atomic nuclei spontaneously decay

**Radiometric dating**
Reading history by radioactive "clocks"

**Fission and fusion**
How energy is released by nuclear rearrangements

**Cold fusion**
A tempest in a test tube

We now set our sights (or, rather, our minds) on the invisibly small clusters of protons and neutrons that form the nucleus of the atom. Together, these particles are known as **nucleons.** At this level of scale, we must obtain details not by light, whose wavelengths are far too coarse to interact with atomic nuclei, but by speeding subatomic particles of matter and by photons of very tiny wavelength and very large energy (gamma rays).

Interactions between nucleons are governed by forces entirely different from the electromagnetic force that keeps electrons in their atomic orbits. In this chapter, we enter the realm of the strong and weak nuclear forces. Nuclear reactions (which involve changes among nucleons mediated by the nuclear forces) are typically millions of times more energetic than chemical reactions (which involve changes among the atom's electrons mediated by the electromagnetic force).

In the pages ahead, we will review the structure of the inner atom, describe some of the processes governing the radioactive decay of unstable atomic nuclei, and explore an extremely useful application of natural radioactivity, the technique of radiometric dating. To conclude, we will explain the energy-transforming processes of nuclear fission and fusion. In one way or another (as explained in Chapter 6), nuclear processes are ultimately responsible for virtually all the energy that flows through the natural world or is generated for our convenience.

## Nuclear Structure

Our first "windows" on the inner workings of matter in the cosmic zoom of Chapter 2 introduced us to the notion that the atom consists of a tiny, relatively massive, positively charged nucleus. That nucleus is surrounded by a haze of negative charge, which is contributed by lightweight electrons orbiting at relatively large distances from the nucleus. The first indication that the atom might be organized in this way came with the results of an experiment devised by the British physicist Ernest Rutherford in 1911.

## Rutherford Scattering

By the first decade of the 20th century, the atom was visualized as a matrix of positively charged material with tiny nuggets of negative charge (electrons) embedded in it. This model of the atom has been compared to a raisin pudding, with raisins representing electrons. In Rutherford's experiment, which was designed merely to confirm the prevailing model, a narrow beam of positively charged particles known as *alpha particles* (or helium nuclei) was directed toward an exceedingly thin metal foil to check how the particles would scatter amid the atoms of the metal. Like a barrage of pellets shot from a BB gun into a raisin pudding, the alpha particles were expected to shoot right through the foil or be deviated slightly by either the diffuse positive charge or the tiny nuggets of negative charge inside the atom.

The results of the scattering experiment were totally unexpected. The atom is not at all like a raisin pudding but more like a cherry with a small, "hard," positively charged pit in the middle. Most of the alpha particles speeding toward the atoms in the foil plowed through as if no interaction had taken place, like BB pellets passing through the soft fruit of a ripe cherry. Yet a small number were scattered at large angles (some almost straight back), much like BBs ricocheting off a cherry pit (see Figure 12.1).

Two aspects of the experiment amazed Rutherford and his colleagues. First, only a very small fraction of the alpha particles came straight back or nearly so. Thus, the particles (let us call them targets) that reversed the direction of these "backscattered" alpha particles had to be very tiny compared to the space between the targets themselves. Second, the backscattered alpha particles were fast moving and energetic; they had been completely turned around by targets capable of exerting an enormous repulsive force. Therefore (since like charges repel), the targets were positively charged. Also, the targets were probably quite massive—otherwise, they would have been knocked out of the metal and would not have turned back any alpha particles.

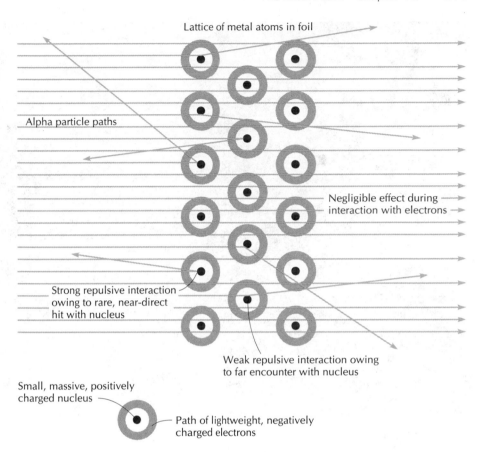

Lattice of metal atoms in foil

Alpha particle paths

Negligible effect during interaction with electrons

Strong repulsive interaction owing to rare, near-direct hit with nucleus

Weak repulsive interaction owing to far encounter with nucleus

Small, massive, positively charged nucleus

Path of lightweight, negatively charged electrons

**FIGURE 12.1** The Rutherford scattering experiment led to the concept that the atom consists of a tiny, dense, positively charged nucleus with lightweight electrons orbiting some distance away.

Today, we recognize that the small targets revealed by the results of Rutherford's scattering experiment are atomic nuclei. A picture was emerging of atoms possessing tiny, dense, massive, positively charged nuclei, having electrons that "orbit" relatively far away. We now know that the atomic electrons do not orbit an atomic nucleus like a planet orbits the sun; but electrons do, in fact, spend much of their time in areas that are about 10,000 times farther away from the nucleus than the radius of the nucleus itself. It is no exaggeration to say that a whole atom is incredibly spacious when compared to just its nucleus.

## The Search for Ultimate Building Blocks

After the Rutherford model of the nucleus was established, further research brought indications that the nucleus consists of a complex aggregate of particles. By the 1930s neutrons and protons had been isolated, and atomic nuclei were visualized as clusters of varying numbers of both particles. As we saw in Chapter 2, the number of protons in the nucleus of an atom (*atomic number*) determines what chemical element that atom belongs to. However, nuclei of the same element can have different masses.

**PHOTO 12.1**   In recent years, particle physicists have boosted the energy released during subatomic particle collisions by bombarding matter with antimatter. Recorded on this detector tracing is the immediate aftermath of two pro- tons colliding with two antiprotons. The spray of particles produced interacts with a strong magnetic field, and thus each one can be analyzed for its mass and charge.

These *isotopes* of a given element have the same number of protons but different numbers of neutrons. In other words, all isotopes of an element have the same atomic number, but each has a different *mass number* (the number of protons plus the number of neutrons).

With the discoveries of protons and neutrons, it was becoming clear that only four fundamental forces were responsible for every kind of interaction: the *strong* and *weak* forces inside the nucleus, and the *electromagnetic* and *gravitational* forces everywhere.

Today, physicists working at the frontiers of knowledge picture the nucleus as a seething mass of ephemeral particles interacting with each other and weaving webs of energy to produce the external properties we associate with protons and neutrons. Much research these days in the field of *particle physics* is concerned with cataloging and explaining the behavior of these particles, either inside or outside the nucleus.

Scientists working in the field of particle physics have compiled a list of hundreds of subnuclear particles—a group so large and diverse that some call it the "particle zoo." Just as a zoologist might organize animals by class, particle physicists have managed to fit all particles of matter into two broad classes: leptons and hadrons. **Leptons,** such as electrons and neutrinos, are lightweight particles that are not affected by the strong force. They do not participate in activities inside the nucleus. **Hadrons**—the heavier protons

**PHOTO 12.2** The Fermi National Accelerator Laboratory (Fermilab) near Chicago, Illinois, remains the premier facility for particle physics research in the United States. The two main "rings" of the machine are 4 mi across. Superconducting magnets accelerate beams of protons and antiprotons in opposite directions in different parts of one ring. With precise timing, both beams are diverted into the second ring, where they collide head-on inside a collision detector. Construction was underway on a similar, but much larger, particle accelerator called the Superconducting Supercollider (SSC) in Texas until funding was withdrawn by Congress in 1994.

and neutrons (nucleons) of atomic nuclei, plus a slew of somewhat lighter, short-lived particles—do respond to the strong force and therefore interact freely inside the nucleus. It now seems certain that hadrons are composed of two or more *quarks,* the fundamental entities introduced at the beginning of the cosmic zoom in Chapter 2.

Other than the protons and neutrons, hadrons tend to exist only for very short intervals and are never seen outside atomic nuclei, except under conditions of extremely high energy. The short-lived hadrons can be created as the result of cosmic ray impacts in the atmosphere, and many more leave tracks in particle accelerator experiments (see Photo 12.1). It is easy to dismiss the study of exotic particles like these as leading to no practical purpose. Yet many exotic particles have been put to use in the medical field (PET, or positron emission tomography, for one) and in other phases of technology. The study of particle physics also gives insight into what was going on during the earliest phases of the big bang, when the whole universe consisted of nothing but interacting exotic particles.

For some time now, the idea of particle interaction has been extended to explain the behavior of the four fundamental forces. In latter-day physics, the notion of a force field (which actually dates from the 19th century) is replaced by mechanisms in which exchanges of *virtual particles* give rise to the appearances of forces. For instance, exchanges of virtual particles called *gluons* between protons and neutrons are what "glue" these particles together in the atomic nucleus. In a less fundamental sense, we would say that the strong force is responsible. Similarly, exchanges of virtual particles called *gravitons* between masses are supposed to produce the attractive forces of gravity.

Physicists may be well on their way to discovering a global theory that would explain all particles and all interactions. It involves simpler schemes of organizing the basic building blocks of matter and a thrust toward unifying the fundamental forces into a single "superforce" that would suffice as a single description of all particle interactions. This goal remains elusive, though—and perhaps it is unattainable.

For the most part in the remainder of this chapter, we are going to revert to the less fundamental but conceptually simpler concepts of proton and neutron interactions. They are sufficient to explain the everyday (though not always apparent) phenomena of radioactivity and nuclear fission and fusion.

# Radioactivity

As mentioned during the early stages of the cosmic zoom in Chapter 2, **radioactivity** is the spontaneous emission of energy from nuclei that are inherently unstable. The earliest studies of radioactivity took place in France near the turn of the 20th century. Henri Becquerel in 1896 discovered that minerals containing very heavy elements such as uranium emit radiation that can make certain materials fluoresce (glow in the dark). By accident, he also discovered that this radiation can penetrate many kinds of matter and expose photographic plates kept in the dark.

Marie Curie and Pierre Curie worked diligently for many years to isolate the agents responsible for the penetrating rays. Many of their key discoveries were made with 8 tons of waste ore, rich in the uranium-bearing mineral *pitchblende,* from Czechoslovakian mines. Using a variety of chemical techniques, they divided the ore into fractions and discarded the portions that were not strongly radioactive. After many stages, several radioactive elements were isolated from each other, including two that were previously unknown, polonium and radium. The Curies became famous

**PHOTO 12.3**   The Curies, Pierre and Marie, in their Paris laboratory around the turn of the 20th century.

when their efforts to understand radioactivity, along with Becquerel's, were acknowledged with a Nobel Prize in 1903. Marie Curie later received another Nobel Prize in 1906 for her contributions to the discoveries of polonium and radium.

# What Is Radioactivity?

The Curies and other pioneers of radioactivity found that chemical manipulation neither increases nor decreases the ability of a given radioactive element to decay. Nor does heating (except, we now know, to millions of degrees) or cooling. The source of radioactive emissions, then, must come from activity within the nucleus, whose composition and structure does not change with temperature or chemical changes.

Further light was shed on the nature of radioactivity when three common types of radioactive emission were distinguished from each other by their ability to penetrate matter and each were analyzed to discover such properties as mass, charge, and energy. These three common radiations were named after the first three letters of the Greek alphabet (alpha, beta, and gamma) according to their penetrating ability.

Alpha emissions or rays, now commonly called **alpha particles,** were recognized as the least penetrating type; a sheet of paper or cardboard suffices to absorb most of them. Beta rays, or **beta particles,** are more strongly penetrating; most can be stopped by a sheet of aluminum. **Gamma rays** are only gradually attenuated by lead, the substance of choice for absorbing all kinds of ionizing radiation (patients wear a lead-filled shield when receiving dental and medical X-rays to block unwanted X-rays).

Figure 12.2 shows what happens to all three kinds of particles when they pass through a magnetic field—a region designed to separate them according to their charge (recall from Chapter 10 the effect of a charge moving across magnetic field lines). The alpha particles are revealed as being positively charged (each carries twice the charge of a proton). They are bundles of two protons and two neutrons, identical to helium nuclei. The beta particles are electrons; they are negatively charged with very low mass. The gamma rays are photons; they are uncharged and unaffected by the magnetic field but very penetrating because of their high energy.

What is the cause of these (and other) radioactive emissions? The answer centers on the strong and weak nuclear forces, which are today under intensive study and still not fully understood. But we can state some generalities. The more massive a nucleus is, the less stable it tends to be. The nature of the strong force is that it pulls together all the nucleons and normally overcomes the tendency of the protons in a nucleus to fly apart owing to electrostatic repulsion. More neutrons are needed as glue in the heavier nuclei, which is why nuclei of elements heavier than calcium, with 20 protons, tend to have more neutrons than protons. Too many neutrons, however, also makes a nucleus unstable.

Within the nuclei themselves, clusters of two protons and two neutrons form particularly stable units. In the case of **alpha decay,** one of these clusters breaks free of the strong force, which rapidly loses its grip toward the edge of a nucleus. When a cluster breaks free, it flies away, propelled by

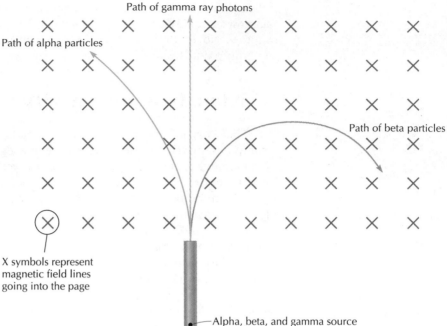

Path of gamma ray photons

Path of alpha particles

Path of beta particles

X symbols represent magnetic field lines going into the page

Alpha, beta, and gamma source

**FIGURE 12.2** When charged particles cross magnetic field lines, they curve sideways, either clockwise or counter-clockwise in this view, depending on their charge. The counter-clockwise-veering beam consists of positively charged and massive alpha particles. The clockwise-veering beam consists of lightweight and negatively charged beta particles (electrons). Gamma rays (photons, actually) have no charge and therefore are not affected by the magnetic field.

electrostatic repulsion, as an alpha particle. The following example of alpha decay is given in chemical notation (recall from Chapter 4 that the subscript before a chemical symbol refers to atomic number and the superscript refers to mass number). It describes a decay in which a common isotope of radium (Ra-226) is transformed into an isotope of radon (Rn-222), accompanied by the emission of an alpha particle (He-4).

$$^{226}_{88}\text{Ra} \rightarrow \, ^{222}_{86}\text{Rn} + \, ^{4}_{2}\text{He}$$

Alpha decay, which is quite common among the heavy elements of Earth's interior, is what has kept Earth hot inside for billions of years. The kinetic energy carried by the fast-moving alpha particles is converted to thermal energy. In addition, each alpha particle captures two electrons after it becomes free, thereby becoming a helium atom. This is where Earth's supply of helium comes from.

**Beta decay** involves the transformation of a neutron into a proton, which can happen in a nucleus having too many neutrons. The result is not just a proton but also an electron, which is necessary to conserve charge: the sum of the charges of one proton (+1) and one electron (−1) equals zero, which is the charge of a neutron. The proton stays put in the nucleus, while the electron is ejected as a beta particle. In the following example of beta decay, a radioactive carbon nucleus (C-14) changes to a stable nucleus of nitrogen (N-14).

$$^{14}_{6}\text{C} \rightarrow \, ^{14}_{7}\text{N} + \text{e}^{-}$$

The weak force, not the strong force, is responsible for beta decay. During beta decay, a tiny amount of energy is missing from the newly created proton and the ejected electron, but it is accounted for by the creation and ejection of an *electron antineutrino*. Neutrinos (the name means "little neutral one") and antineutrinos are chargeless and probably massless particles that are virtually unaffected by ordinary matter. Their existence was not confirmed until 1956.

Gamma ray photons (or sometimes X-ray photons) can be ejected from an unstable nucleus when some kind of spontaneous quantum rearrangement of the protons and neutrons inside the nucleus occurs. This event is called **gamma decay.** These rearrangements are analogous to the quantum jumps that electrons make in the outer atom. The energy emitted by a nuclear quantum jump, however, is thousands or millions of times greater.

## Half-life

All radioactive decay events occur spontaneously and on schedules that can only be described in terms of probability. The "pace" of radioactive decay associated with different radioactive isotopes (*radioisotopes*) may vary greatly, but this pace can be described clearly in terms of a specification called half-life. The **half-life** is the interval of time over which there is an even (50–50) chance that a given nucleus of a radioisotope will decay into something else. Thus, in any given sample of a radioisotope, activity declines with time. Half of the original sample will have decayed away after an initial half-life period. Half of the remaining half (one-fourth) of the original isotope remains and three-fourths is gone after a second half-life. After more and more half-lives pass, less and less of the radioisotope is present, and the activity of the sample keeps declining (see Figure 12.3).

Picture this analogy for exponential decay: Drop a large bunch of coins on a table. Discard all coins that come up tails and repeat the process again with the remaining coins after, say, 1 min. Discard the tails again, and start over in the next minute with the remaining coins. After 1 min, half of your coins remain on the table (strictly speaking, this would only occur on average, after many trials). After 2 min, 1/4 as many coins remain; after 3 min, 1/8 remain; and so on. The coins that you select minute after minute are

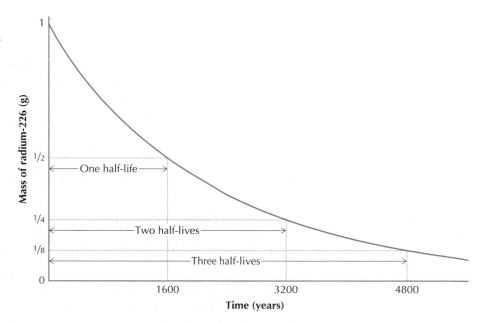

**FIGURE 12.3** All radioisotopes undergo *exponential decay*; an example is shown here for radium-226 decay. Radium-226 ($^{226}_{88}$Ra) has a half-life of approximately 1600 years. It decays into radon-222 ($^{222}_{86}$Rn) by emitting an alpha particle ($^{4}_{2}$He), which reduces its mass number by 4 and its atomic number by 2. After 1 half-life, half of the original sample of radium is gone. The rest has decayed to radon, which itself is radioactive and decays to other isotopes. After 3 half-lives, or about 4800 years, only 1/8 of the original mass of radium remains. After 20 half-lives, or about 32,000 years, only 1/1,000,000 remains.

| TABLE 12.1 Half-lives of Selected Radioisotopes | Isotope | Type of Decay | Half-life |
|---|---|---|---|
| | Hydrogen-3 (tritium) | Beta | 12.3 years |
| | Beryllium-8 | Alpha | $10^{-16}$ s |
| | Carbon-14 | Beta | 5700 years |
| | Cobalt-60 | Beta and gamma | 5.2 years |
| | Iodine-131 | Beta | 8 days |
| | Polonium-213 | Alpha | $4 \times 10^{-6}$ s |
| | Radon-222 | Alpha | 3.8 days |
| | Radium-226 | Alpha | 1620 years |
| | Uranium-235 | Alpha | $7 \times 10^8$ years |
| | Uranium-238 | Alpha | $4.5 \times 10^9$ years |
| | Plutonium-239 | Alpha | 24,000 years |

declining in number, and the half-life of their "decay" is 1 min. Of course, unlike dropped coins, radioactive decay is ongoing; it does not take place in timed episodes.

The half-lives of the different radioisotopes vary greatly. Table 12.1 gives some examples.

## Decay Chains

Most radioactive decay events in nature belong to one of three natural decay sequences that transform isotopes of the heavy elements uranium and thorium into different stable isotopes of the element lead. Uranium and thorium were present in some quantity in the diffuse material that formed the Solar System, but there is less of both of them now and more lead. Figure 12.4 traces the first few steps of one of these series, which begins with the common uranium-238 isotope. Uranium's half-life is 4.5 billion years, nearly the same amount of time as the age of the Solar System, so this chain hardly starts with a bang. Subsequent decays, ranging in half-life from hundreds of thousands of years down to a fraction of a second, comparatively quickly convert the original uranium-238 to lead-206. Energy seeps out of any rock that contains uranium, thorium, or other elements that decay but do not form part of a chain. This energy is keeping Earth (and the other terrestrial planets) hot or molten inside.

## Radiation Benefits and Hazards

The Curies, along with other scientists working with radioactive materials during the early 20th century, were mindful of the ability of radium and other radioactive materials to cure or suppress certain forms of cancer. Radium-containing potions were sold as elixirs for a time, and radium was widely used in cancer therapy until the mid-1950s. Before World War II, radium was also a key ingredient in the paint used to make glow-in-the-dark watch hands and instrument dials.

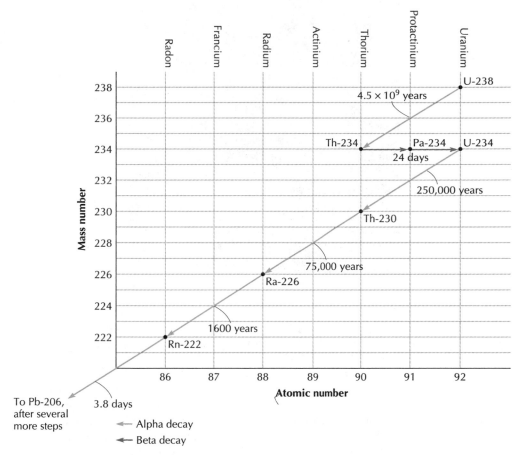

**FIGURE 12.4** Plotted here are the first few steps in the naturally occurring radioactive decay series starting with U-238 and ending with Pb-206. The slowest step (with the longest half-life) by far in this series is the decay of U-238 to Th-234.

By World War II, the harmful effects of radioactive substances were becoming obvious. While exposure to concentrated sources of radiation may preferentially kill cancer cells, it disrupts and kills normal cells, too. People who had taken large amounts of radium-containing potions suffered early deaths as a result. Workers who had handled radium in factories producing fluorescent watch dials died as a result of ingesting or otherwise absorbing tiny amounts of radium.

Today, nonradioactive materials are used in fluorescent paint, and radiation therapy, although widely used to suppress cancer, is recognized as a two-edged sword. Today's radiation therapy involves the irradiation of very specific areas of the body, such as the sites of cancerous tumors. Other, somewhat safer sources of radiation (such as the isotope cobalt-60 and beams of radiation from small particle accelerators) are now commonly used in place of radium.

Oblivious to any connection between radioactivity and ill health, many of the early pioneers of nuclear physics were among the first to succumb to premature death by leukemia and other cancers. Today, the harmful effects of ionizing radiation—either energetic photons (X-rays and gamma rays) or particle-type emissions such as alpha and beta rays—on living biological tis-

**PHOTO 12.4** Today, the hazard of exposure to highly radioactive material is clearly recognized. The technician about to load or unload radioactive material from this vehicle is properly dressed for the job.

# Radiometric Dating

   The universe abounds with obvious and subtle time keeping mechanisms. Electrons revolve, stars spin, and planets both spin and revolve around their home stars in periods of time that remain more or less constant. The study of radioactivity opened up a new method of measuring intervals of past time: **radiometric dating.** This method of marking time relies not on a natural clock that "ticks" uniformly but on one that slows exponentially with time. For this method to work, we must know the half-life of the radioisotope we are using as a clock.

The best-known and most widely applied radiometric-dating method is called *radiocarbon dating.* Every living thing absorbs carbon during its life, in food and in the carbon dioxide of the air. Most of this carbon consists of the stable isotope carbon-12, but a small fraction is radioactive carbon-14, with a half-life of about 5700 years. The carbon-14 is produced through the cosmic ray bombardment of nitrogen-14 atoms in the atmosphere. As C-14 decays (back into N-14, interestingly), it is continually replenished by cosmic ray bombardment at an approximately constant rate. Past changes in the rate are detectable through studies of C-14 content in tree rings, which can be easily dated by counting rings inward from the outermost growth ring of a living tree. Some trees have been alive for thousands of years.

The C-14 content of any living thing is about one atom per trillion C-12 atoms. When an organism dies, no further C-14 can be incorporated into it, and the ratio of C-14 and C-12 inside it exponentially decreases with time. In a determination of the age (time since death) of a piece of bone, wood, or any other long-lasting sample of biological material, the sample is burned to form carbon dioxide, which is then analyzed for its C-14/C-12 ratio. After 5700 years the ratio has fallen to half as much; after 11,400 years the ratio has fallen to a fourth as much; and so on. The method can be extended with considerable accuracy for about five half-lives (some 30,000 years) into the past. Beyond an age of about 100,000 years, so little C-14 remains in the sample that carbon dating is unreliable or impossible. Carbon dating has been a boon to archeological studies, because bones, wood, and cloth (made of either animal hair or plant fiber) can be dated.

Here is an example of how carbon dating can be used. A cave painting contains pigment derived from the crushed leaves of a certain kind of plant. A small sample of the pigment is found to be an eighth as radioactive as a similar sample of pigment extracted from freshly picked leaves of the same kind of plant. To approximate the age of the painting, determine the half-life equivalent of 1/8. One-eighth is equiva-

sue are well understood. On the cellular level, the passage of ionizing radiation creates a trail of ionized atoms, or atoms stripped of one or more electrons. Sometimes, the electrons being ejected carry enough kinetic energy to ionize other nearby atoms. Ionization disrupts the normal chemical processes that keep cells functioning and reproducing properly. If the dose of radiation is small enough, the body can repair the damage done. Larger doses may result in severe cellular damage that cannot be repaired. Sometimes, damaged cells may start to divide uncontrollably, resulting in cancerous growth. (Note that the irradiation of food products by ionizing radiation is thought to pose no threat to people who eat the food. The purpose of this procedure is to kill living microorganisms in the food that would otherwise multiply and shorten its shelf life.)

Of the total dosage of ionizing radiation received by the average individual in the United States, about two-thirds comes from natural sources such as cosmic rays, radioisotopes in rocks and soil, and radioisotopes naturally present in the tissues of the body. Approximately one-third comes primarily from medical and dental X-rays and nuclear medicine. A small

**PHOTO 12.5** Bone material of archeological interest is being dissolved in inorganic solvents prior to being radiocarbon dated. The procedure extracts from the bone amino acids—the actual material dated.

lent to 3 half-lives ($1/2 \times 1/2 \times 1/2 = 1/8$). Each half-life is approximately 5700 years, so the age is about $5700 \times 3 = 17,100$ years.

Radiometric dating has revolutionized geology as well, making possible the conversion of *relative geological time* (discussed in Chapter 3) into *absolute time* measured in years. Geologists, of course, must make use of slow-ticking radioisotopes, with half-lives measured in millions or billions of years. Among the most widely used are potassium-40, with a half-life of 1.25 billion years, and uranium-238, with a half-life of 4.5 billion years. By measuring the relative abundances of isotopes like these, geologists can determine, with reasonable accuracy, the ages of rocks as old as Earth itself.

The ages derived for a rock using radiometric-dating techniques give the time since the rock (or mineral) was last molten—that is, the point in the past when it solidified from magma. The technique will not work for all rocks. For example, the time since a sedimentary rock formed cannot be precisely dated by radiometric means because such a rock is a hodgepodge of previously existing mineral grains that may have solidified at different times.

fraction comes from occupational exposures, consumer products, and fall-out from past nuclear weapons testing, emissions from nuclear power plants, and from other artificial nonmedical sources. These amounts, of course, may vary enormously. Ten or twenty thousand miles of high-altitude airplane travel may result in an extra exposure to cosmic rays equivalent to a normal chest X-ray. Uranium miners, obviously, face a greatly amplified risk of harm from the natural background of radioisotopes in rocks and soil. Many low-level sources of ionizing radiation exist in the world around us, and they are impossible to avoid. What one should try to avoid is *unnecessary* exposure to ionizing radiation.

The radioactive decay of radon, which is part of the U-238 decay chain sketched in Figure 12.4, is of major concern in many parts of the world. Radon arises from uranium, which is present in Earth's surface rocks at an average but highly variable abundance of about 2 g per ton. Uranium-238's decay chain produces thorium, radium, and other elements, which remain sealed in crystalline minerals. The radon formed by the decay of radium, however, is an inert gas (it is in the same family of elements as helium, whose atoms will not bond with anything). Radon near Earth's surface seeps out of the host rock and rises into the atmosphere, where minute quantities of it are inhaled and exhaled by any breathing animal. Radon atoms cannot poison the body in a chemical sense, but their radioactive nuclei, with a half-life of just 3.8 days, may do damage if and when they decay inside a person's lungs. Radon decays to form radioactive polonium. Polonium, which is chemically similar to oxygen, will bond to biological tissue. Polonium's decay leads through a chain of further radioactive "daughters" ultimately to the stable Pb-206 isotope.

In the past, the normal flow of radon gas into the atmosphere was dispersed by winds and weather and caused no serious harm. In modern, tightly sealed houses and buildings, however, radon seepage from rocks or concrete can build up (especially in basements) to concentrations tens or hundreds of times greater than in the air outside. The solution to this potential threat is straightforward: Test the indoor air with a radon kit (available at building supply stores) and install fans or otherwise improve ventilation to reduce the concentration of radon.

# Fission and Fusion

More than a ton of coal is consumed at an electric generating station to supply the yearly electric energy demand of an average household. Only a few grams of uranium fuel suffices for the same task in a nuclear power plant. It would take about 50 million tons of a conventional chemical explosive to equal the energy release of the largest nuclear weapons ever built: a hydrogen bomb with an energy yield of 50 megatons. The processes of fission in a nuclear power plant and fusion in a hydrogen bomb tap into the reservoir of energy stored in the nuclei of atoms. This *nuclear potential energy,* which exists by virtue of the strong nuclear force, outstrips by a factor of many millions the *chemical potential energy* atoms have by virtue of the electromagnetic force acting on their electrons.

Nuclear potential energy is released slowly through radioactivity and more rapidly through the two processes known as fission and fusion. **Nuclear fission** refers to the splitting apart of atomic nuclei. **Nuclear fusion** means that atomic nuclei combine or merge together. Like chemical reactions, which either release energy (exothermic reactions) or absorb energy (endothermic reactions), nuclear reactions can either release or absorb energy. However, except in exotic milieus, such as a star undergoing a supernova explosion, the nuclear reactions around us are exothermic. Nuclear reactions, primarily taking place in stars, are steadily converting nuclear potential energy into radiant energy in our universe today.

## Nuclear Fission

Although nuclear fission events may occasionally take place sporadically on an atom-by-atom basis in nature, the contemporary process of fission utilizes a rapid-fire chain of reactions that evolves much energy. The first controlled chain reaction involving fission took place in 1942 during the Manhattan Project, which was a crash effort the United States launched to develop the atomic bomb for use in the struggle against the Axis powers in World War II. By 1945, the feasibility of such a bomb was confirmed during a successful test of the first nuclear weapon in the New Mexico desert. (The terms *atomic bomb* and *nuclear bomb* are often used interchangeably, but the latter term is more accurate, because the energy released originates from nuclei and has nothing to do with the outer parts of an atom.) The subsequent use of nuclear weapons on the Japanese cities of Hiroshima and Nagasaki at the close of World War II resulted in the deaths of some 100,000 people, brought about the surrender of Japan, and forever changed the world's political landscape, for better or worse.

Like radioactivity, nuclear fission involves the opposing tendencies of a nucleus to fly apart because of the repelling force of its protons and to stay together in the grip of the strong force. In a massive nucleus, such as that of uranium or plutonium, strong-force attraction barely dominates electrostatic repulsion. In some isotopes of heavy elements (the so-called *fissionable* isotopes), the absorption of a single neutron is enough to disrupt the delicate balance. The nucleus splits into two smaller nuclei, which violently push apart, carrying a large amount of kinetic energy. At the same time, gamma rays and two or three individual neutrons are emitted. The neutrons, in turn, may bombard other fissionable nuclei and stimulate them to undergo fission and release still more neutrons. A chain reaction can occur if there are enough fissionable nuclei nearby for the released neutrons to interact with. Figure 12.5 illustrates how a chain reaction can spread among the nuclei of a commonly used fissionable isotope, uranium-235.

Once a chain reaction gets started in a baseball-sized (a few kilograms) chunk of pure U-235, nothing can stop it. The neutrons released by the nuclei undergoing fission increase exponentially and very quickly trigger all the rest of the nuclei to split as well. A tremendous explosion results.

Two things prevent runaway chain reactions from happening in nature. First, the uranium content in rocks and minerals is extremely dilute. In uranium-bearing minerals, uranium atoms exist in chemical com-

**FIGURE 12.5** The absorption of a neutron into a U-235 nucleus causes it to split into two large fragments, along with two (sometimes three, depending on how the nucleus splits) neutrons. Energy is carried off as gamma rays (not shown) and as the kinetic energy of the speeding fragments. If other fissionable nuclei lie nearby, the ejected neutrons can induce them to split apart as well. In a ball of solid U-235 of sufficient mass, the chain reaction multiplies so quickly that nearly all the nuclei are involved within a millionth of a second. The result is a furious explosion.

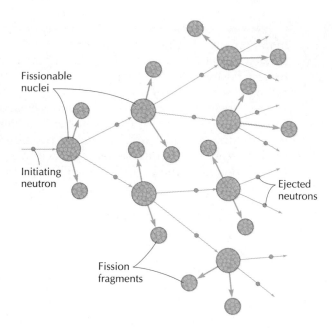

bination with other atoms. Furthermore, only 1 part in 140 of naturally occurring uranium is the isotope U-235. The rest is U-238, which absorbs neutrons but does not undergo fission. Chain reactions cannot be supported in a mass of uranium unless it is artificially enriched in U-235 through the separation and removal of U-238.

Second, pure U-235 (or any other fissionable isotope) must be fashioned into a solid, *critical mass* of compact size in order for a runaway chain reaction to occur. If a chain reaction starts in a piece of U-235 of subcritical mass, too many neutrons reach the surface and escape before they have a chance to initiate many fission events. If that happens, the chain reaction quickly dies out.

The fission bombs of World War II and today are detonated by slamming or compressing together subcritical masses of a fissionable isotope, using conventional explosives such as TNT. When the combined masses of the pieces exceed the critical mass and when the pieces are brought together with just the right timing, the bomb explodes.

Thankfully, the refinement of U-235 from naturally occurring uranium and the technology required to trigger nuclear explosions remain technically difficult and expensive. So far, only a handful of countries have managed to mount successful efforts at building nuclear weapons, and no nuclear explosives have been used in war or conflict since World War II.

A kinder, gentler form of nuclear fission is put to work in conventional (fission) nuclear power plants, in nuclear submarines, and in far-ranging spacecraft. All these devices involve a *nuclear reactor,* whose product is thermal energy. The heat flowing out from a reactor is converted, in one way or another, to electricity.

Typically, the nuclear "fuel" in a power plant reactor consists of U-235-enriched uranium oxide pellets, stacked into thousands of long, thin rods protected by a metal casing. The rods are spaced at certain intervals of distance and immersed in a fluid, usually water. The chain reaction takes place primarily between the rods, not in them, as neutrons are passed from rod to rod. The fluid, which absorbs thermal energy from the fission reactions in the rods, is pumped through a heat exchanger (it is important to keep the radioactive fluid circulating in its own closed system), where it is used to boil water. The boiled water makes steam, which turns a steam turbine. The steam turbine drives an electric generator.

The fluid, which functions like the coolant flowing through the radiator and engine of a car, has another vital role to play. It is used as a *moderator* to regulate the speed of neutrons passing from one rod to another. Slower-moving neutrons are more likely to initiate fission.

Loss of coolant by leakage or evaporation results in faster neutrons and the snuffing out of the chain reaction. Under no circumstances can a reactor undergo a nuclear explosion. Nevertheless, at a critical stage during refueling, the loss of the coolant can result in a *meltdown* in which the highly radioactive fuel melts and may puddle on the bottom of the reactor vessel. The great amount of heat released can cause a steam explosion that can spew much dangerous radiation into the air.

In 1979, a reactor at Three-Mile Island in Pennsylvania underwent a partial meltdown. The result was damage to the reactor and cleanup expenses in the billions of dollars, but no significant quantities of radiation were released to the outside environment. In 1986, a poorly designed reactor at Chernobyl near Kiev, Ukraine, suffered a meltdown of catastrophic proportions. During the resulting steam explosion and fire, large amounts of radiation were released into the atmosphere and carried worldwide on the winds. Tens of thousands of people were forced to evacuate the region surrounding the reactor, and hundreds were killed either quickly or over a period of years as a direct result of radioactive fallout. Thousands more throughout the region are expected to die prematurely by cancers brought about by the contamination of the soil and food supply by radioisotopes.

Nuclear fission is a hard-to-tame beast, and its safety record is checkered. Nuclear power plants throughout the world may run safely, and

**PHOTO 12.6** Nuclear electric generating stations are among the most complex and expensive monuments ever built. Elaborate safety mechanisms are built into most modern nuclear installations. Nevertheless, there always remains the possibility of human error.

**PHOTO 12.7**    The failed 1000-MW fission reactor at Chernobyl stands today under a leaky "sarcophagus," hastily constructed to isolate the worst of the radiation it emits.

nations may maintain control over the use of nuclear weapons. But the mining and refining of uranium always creates health hazards, and the disposal of the radioactive waste generated at nuclear power plants remains problematical. Many argue that these problems are no worse than the hazards involved with the generation of energy from fossil fuels: the environmental degradation of coal mining and oil extraction and the chemical pollution generated by their use. Some harmful health or environmental effects will always be associated with the generation of useful energy by any means, and they are the minimum price we must pay in a technological society. Our goal should always be to minimize these effects.

## Nuclear Fusion

Nuclear fusion floods our Solar System—and most of our universe—with light. The fusion of hydrogen into helium (also called *thermonuclear fusion*, because of the extremely high temperatures needed for it) is the primary energy source of our sun and about 90% of all the stars.

Physicists and astronomers have identified two important thermonuclear fusion processes responsible for the generation of energy in the cores of main sequence (hydrogen-burning) stars. The *proton-proton cycle* (Figure 12.6) is the primary producer of energy in the sun. The *carbon-nitrogen-oxygen* (CNO) *cycle* is of greater importance for stars hotter and more massive than the sun. Both processes accomplish the same thing—conversion of hydrogen into helium—but the CNO cycle involves carbon nuclei, which act as a kind of catalyst to speed up the reaction.

In both reactions, the products include positrons, neutrinos, and gamma rays. The positrons immediately annihilate with any electrons they encounter and become gamma rays. These gamma rays carry away most

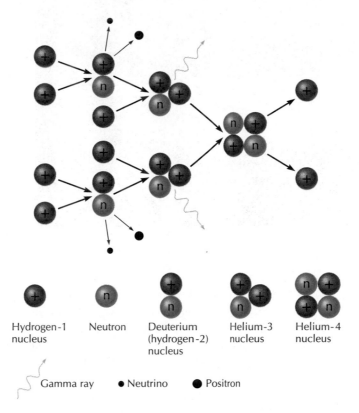

Hydrogen-1    Neutron    Deuterium     Helium-3    Helium-4
nucleus                  (hydrogen-2)  nucleus     nucleus
                         nucleus

Gamma ray    ● Neutrino    ● Positron

**FIGURE 12.6** The proton-proton cycle of fusion reactions taking place in the sun's core features three major steps. In the first—and by far most unlikely—step, two protons (hydrogen-1 nuclei) collide and fuse to form a nucleus of deuterium (hydrogen-2). For any given proton this happens, on average, about once every 14 billion years. During each encounter of this sort, a positron (positively charged electron) is created so as to conserve charge, and a neutrino is emitted. The positron very quickly annihilates with a nearby electron (not shown), producing a gamma ray.  In the second step, a deuterium nucleus fuses with a single proton to form a nucleus of helium-3, an event that occurs within about 6 s. More gamma radiation is emitted. In the third step, two helium-3 nuclei collide to form a single (and stable) nucleus of helium-4. This happens about a million years after helium-3 is formed. Two protons are kicked away in the third step. They can participate in further fusion reactions. Since there are six protons participating in the whole proton-proton chain and the result is a single helium nucleus plus two free protons, the *net* reaction is $4\,{}_1^1\mathrm{H} \rightarrow {}_2^4\mathrm{He}$.

of the energy, which (as detailed in Box 7.3) slowly diffuses toward a star's surface. The neutrinos, which carry away a small fraction of the energy, zip virtually unimpeded through the star and out into space. In stars much more massive than the sun, further fusion reactions involving helium and heavier elements can take place. We shall cover these reactions in the discussion of stellar evolution in Chapter 17.

During fusion reactions (and, to a lesser extent, during fission reactions), the mass of the products is significantly smaller than the mass of the reactants. The lost mass is a consequence of the energy equivalence of mass, in accordance with Einstein's famous equation $E = mc^2$. Each second, the sun "burns" about 600 million metric tons ($6 \times 10^{11}$ kg) of hydrogen fuel into helium "ash," losing some 0.7% of its mass (over 4 million metric tons, or $4 \times 10^9$ kg) as it does so. Using $E = mc^2$ to calculate the energy released by this loss of mass, we get a whopping $3.8 \times 10^{26}$ J. This, remember, is the energy released for only 1 s. Expressed in terms of power, we can say that the sun's radiating power is $3.8 \times 10^{26}$ W.

### Problem 12.1

Calculate how much hydrogen fuel you would need to fulfill all your energy needs for a lifetime if you could convert 0.7% of its mass into energy. Assume that you will demand an average of 10 kW of power over an 80-year span. (This average power takes care of all your electrical and transportation needs and suffices for the manufacture of all the products you use.)

### Solution

In Chapter 6, we noted that 1 kW·h of energy is equivalent to 3.6 million joules. At the rate of 10 kW, $3.6 \times 10^7$ J are used each hour. Let us multiply this figure by the number of hours in an 80-year period:

$$(3.6 \times 10^7 \text{ J/h})(24 \text{ h/day})(365 \text{ days/year})(80 \text{ years})$$
$$= 2.5 \times 10^{13} \text{ J}$$

Now we use $E = mc^2$, solved for $m$, to calculate how much mass must disappear to produce this much energy:

$$m = E/c^2$$
$$= (2.5 \times 10^{13} \text{ J})/(3 \times 10^8 \text{ m/s})^2$$
$$= (2.5 \times 10^{13} \text{ J})/(9 \times 10^{16} \text{ m}^2/\text{s}^2)$$
$$= 0.00028 \text{ kg}$$

A joule is defined as $\text{kg·m}^2/\text{s}^2$, so the mass is expressed in kilograms. The quantity 0.00028 kg (less than 1/3 of a gram) is the amount of mass converted to energy, which is 0.7% (or the fraction 0.007) of the amount of hydrogen fuel needed. The mass of hydrogen is

$$m = (0.00028 \text{ kg})/(0.007)$$
$$= 0.04 \text{ kg}$$

which is about the same as the mass of a large chocolate-chip cookie.

How wonderful it would be if we could conjure up at will the enormous amounts of energy delivered by such paltry amounts of hydrogen. Actually, fusion *is* an accomplished goal here on Earth, but not in a manner most people would consider constructive. The United States detonated the first hydrogen bomb during a test in 1952, and there have been many similar test explosions since then. Hydrogen bombs (or thermonuclear weapons) have a much larger energy yield than the fission bombs described earlier. There is no theoretical upper limit to their size and destructive power. The nuclei fused in thermonuclear weapons are deuterium ($^2_1$H) and tritium

## BOX 12.2
# Cold Fusion

Some nuclear physicists believe there may be a way to avoid the extremely high temperatures needed to trigger fusion among protons. One scheme that has undergone some testing involves the displacement of a normal hydrogen atom's electron by a short-lived, negatively charged hadron called a *muon*. The resulting *muonic atom* is so small compared with a normal hydrogen atom that a passing proton can approach very closely and be caught by the muonic atom's nucleus. Fusion immediately follows, with a consequential release of energy. This and other similar *cold nuclear fusion processes* currently under study may one day produce practical amounts of energy at temperatures of around 1000 K—"hot" compared with room temperature but very "cold" compared with thermonuclear fusion.

These cold fusion schemes are very different from the cold-fusion-in-a-bottle experiments conducted in Utah by two chemists and widely trumpeted in the mass media in 1990. The two claimed they had experimental evidence of fusion taking place at room temperature in an apparatus no more complicated than those typically used in high school chemistry laboratories. The experiment, which involved palladium metal bars in contact with "heavy water" (water whose molecules have at least one deuterium atom instead of a normal hydrogen atom), supposedly produced as products helium, spare neutrons, and energy.

For a short while, the supposed findings caught the public imagination. If such a simple process could be made to yield nuclear energy at room temperature, then any naturally occurring water on Earth could be mined for the heavy-water molecules it contains. Thus, we would have a means of harnessing essentially inexhaustible power.

As it turned out, subsequent efforts by thousands of scientists to reproduce cold-fusion-in-a-bottle failed to confirm the findings of the two chemists. The chemists were apparently misled by small errors in their measurements.

In hindsight, the cold-fusion incident illustrates three truisms about the practice of science.

1. Science is self-correcting. All ideas, sensational or not, are subject to further testing. Sooner or later, erroneous ideas are weeded out by further experiment and observation.

2. Drawing conclusions from scanty data is risky. Scientists, like workers in other professions, may face tremendous pressures to excel. In the sciences, excellence is often synonymous with making discoveries that will lead to important practical applications. The stakes were high in the case of cold fusion, and the temptation to make a premature announcement, especially to the public, was strong.

3. *Peer review* plays an essential role in science. Peer review is the process by which theoretical ideas and experimental results are evaluated by other scientists before they are published in scientific journals. Sensational results announced directly to the mass media without any benefit of peer review are looked upon with doubt by competent scientists. In the case of cold fusion, this doubt led to intense scrutiny and eventually to rejection of the original results (and to embarrassment on the part of the two chemists).

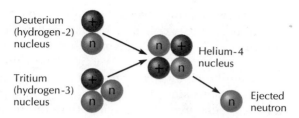

**FIGURE 12.7** Commercial fusion reactors of the future may well use this reaction to produce large amounts of energy. When deuterium and tritium nuclei are brought to a temperature of more than $10^8$ K, they fuse into a helium nucleus and a spare neutron. In a fusion power plant, the kinetic energy of the neutrons would be converted to heat through absorption by a liquid metal, and the heat would drive a steam turbine, as in a conventional power plant.

($_1^3$H), so the most difficult step in solar-style fusion (proton-proton combination) is avoided. Nonetheless, temperatures equal to or greater than the sun's core temperature of 15 million degrees Kelvin are needed for the reaction to start and proceed rapidly. The only way to attain these temperatures quickly is to use an implosion-type fission device. Thermonuclear weapons are thus hybrids, using conventional explosives to trigger a more powerful fission bomb and using the fission bomb to trigger the still more powerful fusion bomb.

The kinder, gentler side of fusion—fusion as a source of power, or controlled fusion—is theoretically promising but fraught with enormous technical difficulties. The challenge is to raise the temperature of small amounts of hydrogen plasma (usually containing much deuterium and tritium) to a temperature of more than 100 million degrees Kelvin and keep it that hot so that fusion into helium can take place over long time periods (see Figure 12.7). Physicists are experimenting with several strategies. One type of experimental reactor confines the superhot plasma with the use of enormously strong magnetic fields. The magnetic fields are also used to heat the plasma. Using such reactors, researchers are approaching the breakeven point in which the energy released from fusion equals the energy necessary to keep the reactor's electromagnets and other components running. So far, tests like these have produced only short-duration bursts of power.

Another strategy in the planning stages would bombard tiny pellets of deuterium and tritium with intense, synchronized laser pulses from many directions. A pellet would explode as it was compressed to a density of around 200 g/cm³, producing much more energy than that contained in the laser beams. A steady stream of pellets would be introduced into the lasers' line of fire (about 5 per second, according to one model on the drawing boards) so as to produce a relatively smooth output of power.

Once the many technical hurdles associated with controlled fusion are overcome, fusion power may become a safe, virtually nonpolluting source of practical energy. Many scientists believe, however, that fusion power plants of the future would likely be much larger than conventional nuclear power plants and would dump much more waste heat into the immediate environment. Thus, thermal pollution might be a significant concern. Fusion research over the past four decades in the United States has been given little emphasis. Unless more funding is devoted to it, the practical utilization of fusion power may have to await the middle of the next century.

# The Iron Plateau

Everywhere in the universe and at every scale of size, matter strives to attain a state of low potential energy. Acted on by gravitational forces over astronomical distances, diffuse matter in space tends to collapse into spherical bodies (stars and planets) that retain as little gravitational potential energy as possible. On Earth, continents shed their mass toward the sea over geologic time, thereby minimizing their gravitational potential energy. In the atmosphere, clouds have a tendency to rid themselves of any electric potential energy they acquire through the phenomenon of lightning. In the atomic realm, any electron acquiring extra potential energy by leaping into a new orbit very quickly loses that energy when it jumps back toward an orbit with less potential energy.

The same striving toward low potential energy exists in the atomic nucleus; but here, it is nuclear potential energy (the potential energy associated with the strong force) that a nucleus tries to shed. Held in the "iron hand" of the strong force, nucleons try to cluster together as closely as possible inside the nucleus. Interestingly, the overall grip of the strong force is significantly stronger not for nuclei that are very light or very massive but for nuclei with masses somewhere between these two extremes.

Coincidentally (in view of the fact that the element iron has traditionally represented strength among materials), one of the most tightly packed nuclei—one whose particles are as tightly bound as they can get—is that of iron-56 ($^{56}_{26}$Fe). Nuclei that are both lighter and significantly heavier than iron-56 have more space between their nucleons and therefore more potential energy per nucleon.

As we have seen, changes (both spontaneous and induced) can occur inside certain nuclei. When heavy nuclei break apart, either incrementally by means of radioactive decay or cataclysmically by means of fission, the total energy that binds together the nucleons in the fragments after the breakup is greater than it was before. Generally, the situation is just the opposite for lightweight nuclei. In most instances, these nuclei strongly resist being broken up. They would have to absorb a large amount of energy in order to do so. However, if these nuclei crash together with sufficent force so that their nucleons can fuse, the total energy that binds them together in the new form is much greater.

The energy that binds nucleons together is called **binding energy.** The *differences* in binding energy are what are responsible for the transformation of nuclear potential energy into other forms during radioactive decay, fission, or fusion. For example, after four hydrogen nuclei fuse to become a helium nucleus in the sun's core, the binding energy associated with the four nucleons has increased. The nucleons end up much closer to each other and caught in the intense grip of the strong force. It would take a lot of energy to pull them away from each other and transform them back into four hydrogen nuclei again. Losses of nuclear potential energy correspond to gains in binding energy. (In the same way, the greater the loss of gravitational potential energy in a book that falls, the more energy it requires to return the book to where it was before it fell.) Under the proper conditions, helium nuclei themselves can fuse with other nuclei to form still heavier

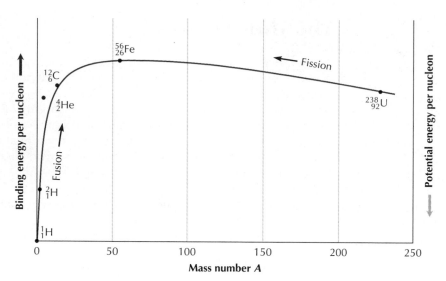

**FIGURE 12.8**  The most stable nuclei have mass numbers near that of iron ($A = 56$), where the binding energy per nucleon is maximized and the nuclear potential energy (the potential energy associated with the strong force) is minimized. When fission occurs, generally for isotopes on the right side of the peak of the curve, the resulting fragment nuclei have more binding energy per nucleon, which means that they have lost nuclear potential energy. When fusion occurs, generally for isotopes on the left side, the resulting fused nucleus has more binding energy per nucleon and less potential energy. Fusion generally evolves more energy than fission, a fact illustrated by the greater steepness of the left part of the generalized curve drawn here.

and more stable (more densely packed) nuclei with even more binding energy. One such fusion reaction, which occurs in older (giant) stars, involves the conversion of three helium nuclei ($_2^4$He) into a single nucleus of carbon ($_6^{12}$C).

Energy and mass, of course, are related by Einstein's formula $E = mc^2$. Thus, any increase in binding energy (accompanied by the same decrease in potential energy) results in a decrease in the mass of all the nucleons involved. The more tightly bound a nucleus is, the greater the binding energy is and the smaller the mass of its particles is.

The general relationship represented by the graph in Figure 12.8 is one of the most important in all of science. It is a plot of the *binding energy per nucleon* (the total binding energy of all nucleons bound together divided by the number of nucleons) versus mass number. On the left side, we see the effect of lightweight nuclei fusing into heavier nuclei. On the right side, we see the effect of heavy nuclei splitting into lighter components. The average curve, which passes through or near all the points representing various elements, attains a plateau at the element iron. Note that, in general, more energy is produced by fusion reactions than by fission reactions.

Iron represents a pinnacle of stability among nuclei. If there were nothing to prevent it, nuclei of various elements would interact freely and quickly transform themselves into iron (and perhaps other stable elements, such as nickel, which are very near iron on the plateau at the top of the curve). Fortunately, as we have already seen, obstacles stand in the way: Only very hot temperatures, such as those found inside stars, can force protons and other lightweight nuclei together. Also, the wholesale splitting apart of heavy nuclei is not spontaneous; it must be induced by artificial means.

## CHAPTER 12
# Summary

The nucleus of the atom contains nucleons—protons and neutrons—bound together by the strong force, the strongest of the four fundamental forces. The electrons of an atom travel at relatively great distances from the nucleus. A fundamental division of matter envisions subnuclear particles such as protons or neutrons (which are hadrons, or heavy particles that respond to the strong force) inside the nucleus and electrons (which are leptons, or lightweight particles that do not respond to the strong force) outside the nucleus.

Not all nuclei are stable; some undergo radioactive decay. All radioactive emissions, which are of several types, can be accounted for by internal changes within the nucleus. All radioactive emissions penetrate matter to some degree and are capable of causing biological damage. Radioactive decay is spontaneous. Half of a sample of a given radioisotope remains undecayed after an amount of time known as its half-life. From the known rates of decay of a number of key radioisotopes, the ages of many organic materials and rocks containing those isotopes can be determined.

Nuclear fission and fusion reactions involve rearrangements of nucleons. Fusion generally produces more energy than fission, though both processes produce millions of times more energy than typical chemical reactions. Fissionable isotopes, which are heavy and marginally stable, can be induced to split apart into smaller fragmentary nuclei, thus releasing much nuclear potential energy at the expense of a small loss of mass. Light nuclei, if brought close enough together by means of high temperature, may fuse and release even more nuclear potential energy, again with the loss of relatively little mass.

Uncontrolled fission and fusion reactions have been successfully exploited in nuclear weapons of two basic types. Controlled fission has been harnessed as a practical source of energy for decades. Controlled fusion—preferable because it promises much higher energy yields and relatively benign environmental consequences—currently remains under development.

## CHAPTER 12
# Questions

## Multiple Choice

1. An atom consists of a
   a) smear of positive charge with embedded electrons
   b) smear of negative charge intermixed with positive charge
   c) small, negatively charged nucleus surrounded at a distance by protons
   d) small, positively charged nucleus surrounded at a distance by electrons

2. The nucleus of an atom
   a) contains about half of its mass
   b) is electrically neutral
   c) contains electrons
   d) can deflect alpha particles that come very near it

3. What force holds together the particles of an atomic nucleus?
   a) the strong force
   b) the weak force
   c) the electromagnetic force
   d) gravitation

4. Quarks do not make up
   a) electrons
   b) protons
   c) neutrons
   d) hadrons

5. Radioactive nuclei do not emit
   a) electrons
   b) protons
   c) alpha particles
   d) gamma rays

6. During what process are electrons emitted by a radioisotope?
   a) alpha decay
   b) beta decay
   c) gamma decay
   d) nuclear fission

7. Which of the following statements is true?
   a) Alpha rays are more easily absorbed by matter than gamma rays.
   b) Alpha particles are negatively charged.
   c) Gamma rays are more easily absorbed by matter than alpha rays.
   d) Gamma rays curve sharply when crossing magnetic field lines.

8. After 4 years have elapsed, 1/16 of a radioisotope remains, while the rest has decayed to something else. Its half-life is
   a) 1 year
   b) 2 years
   c) 4 years
   d) 16 years

9. Nuclear fission can occur in
   a) hydrogen
   b) helium
   c) plutonium
   d) iron

10. In a nuclear power plant, the nuclear reactor supplies
    a) alpha particles
    b) an electric current
    c) mechanical energy
    d) thermal energy

11. A possible fuel for practical fusion reactors is
    a) ordinary hydrogen
    b) deuterium and tritium
    c) uranium
    d) alpha particles

12. After two light nuclei fuse, the binding energy per nucleon is generally
    a) slightly less
    b) much less
    c) greater
    d) the same

## Questions

1. What characteristics distinguish hadrons from leptons? How do both differ from photons?

2. Explain how Rutherford's scattering experiment disproved the notion that the atom is a mishmash of positive and negative charge.

3. What does it mean to say that a substance is radioactive? How does radioactivity differ from any chemical characteristic a substance may have?

4. How can alpha, beta, and gamma rays be distinguished from one another?

5. What keeps a chain reaction going in a fissionable isotope?

6. What are the technical obstacles to be overcome in the quest for practical fusion energy?

7. *Fusion* reactions among light elements are generally exothermic; among heavy elements, they are generally endothermic. *Fission* reactions among heavy elements are generally exothermic; among light elements, they are generally endothermic. Why?

8. What did the 1990 cold-fusion-in-a-bottle experiments illustrate about the workings of science?

## Problems

1. After 5 half-lives, how much of a radioisotope in an initial sample remains? After 10 half-lives, how much remains? After 20 half-lives, how much remains?

2. After 34,000 years, approximately how much of an original sample of 6 milligrams (mg) of carbon-14 remains undecayed?

3. For every gram of hydrogen transformed to helium in the fusion furnace of the sun, how much energy is produced? (*Hint:* Use $E = mc^2$; and remember to use SI units—kg for mass and m/s for the speed of light.)

## Questions for Thought

1. Is it meaningful in any sense to refer to the "half-life" of a human being? Or to the "half-life" of the human race?

2. If the source of Earth's internal heat is radioactive decay, and if Earth's internal heat drives all geologic change, what is likely to happen to Earth in the far future?

3. Plutonium-239, a fissionable isotope, is produced as a by-product in certain nuclear fission reactors known as *breeder reactors*. Breeder reactors have been phased out in the United States. Why is the existence of breeder reactors in other countries considered by many to be hazardous to world peace?

4. Why does radon pose such an insidious threat, in particular, to people living in cold climates?

5. Why is it not practical to use isotopes such as potassium-40 and uranium-238 to determine the ages of "young" rocks (aged a few million years or less)?

## Answers to Multiple-Choice Questions

| 1. d | 2. d | 3. a | 4. a | 5. b | 6. b |
|------|------|------|------|------|------|
| 7. a | 8. a | 9. c | 10. d | 11. b | 12. c |

Every scientific discipline has associated with it one or more overarching concepts, ideas broad and inclusive enough to unify many disparate bits and pieces of knowledge within the discipline itself and often within a larger arena of inquiry. In the five chapters of Part IV, we shall focus on five overarching concepts within the physical sciences: relativity, quantum mechanics, the periodic table of the elements, plate tectonics, and cosmic evolution.

For most of the 20th century, the entire edifice of physics has rested on two theoretical pillars: relativity theory (Chapter 13) and quantum mechanics (Chapter 14). Nearly all physical behavior can be traced to either. Relativity explains much of what goes on in supermacroscopic realms, while the point of view expressed by quantum mechanics is absolutely essential for any real understanding of the world of the very small. Newtonian mechanics, once the overarching concept for all of physics, is now recognized as being subsumed under the grander and more inclusive theories of relativity and quantum mechanics.

The periodic table of the elements (Chapter 15) is an immensely useful organizing concept in the physical sciences. Neither new nor arcane, it is as essential to the practice of chemistry (or any other inquiry involving the properties of matter) as a dictionary and thesaurus are to a writer. The linkage between the periodic table and the quantum mechanical behavior of atoms is an intimate one, which is why our periodic table chapter immediately follows the quantum mechanics chapter.

The theory of plate tectonics (Chapter 16) has revolutionized geology and the other earth sciences just as surely and irrevocably as relativity and quantum mechanics have revolutionized physics. At present, plate tectonics has jelled into a clear picture of how the mostly horizontal movements of landmasses and the seafloor drive geologic changes on Earth's surface. The exact mechanisms that propel these movements are not yet clear, and efforts to understand them continue.

More than a century's worth of astronomical observation and data analysis have contributed to today's "big picture" of the universe, past, present, and future (Chapter 17). The scheme of cosmic evolution described in Chapter 17 begins with the big bang and follows the process by which stars live and die and give up some of their material for future generations of stars. The ultimate fate of the universe, perhaps never to be known with certainty, can at least be projected with some confidence into two known alternatives.

# PART IV

# Overarching Concepts

# CHAPTER 13

# Relativity

*A fun house mirror distorts our view of the world. In a similar but more fundamental way, any two observers moving relative to each other or any two observers in different gravitational environments can never agree on the exact characteristics of anything they are both looking at or measuring. What appears normal to one observer may look distorted to another.*

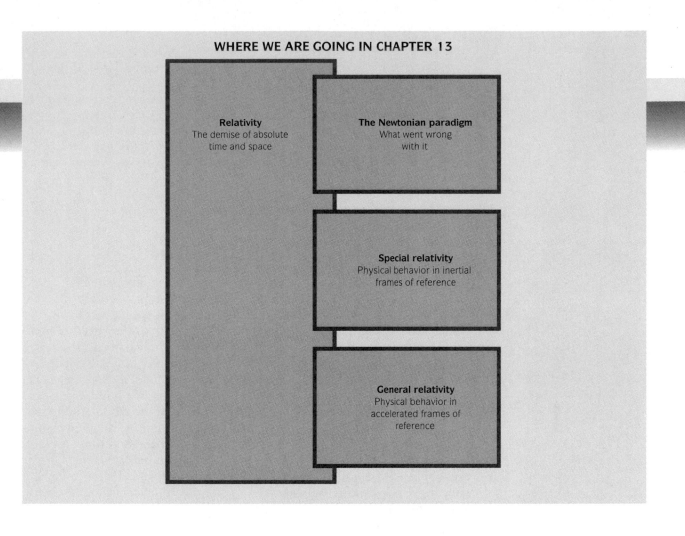

To understand the seemingly topsy-turvy world of Einsteinian relativity, we must realize that *all scientific explanations require context.* As mentioned earlier in this book, science makes no claim to absolute truth. However, it is very much concerned with descriptions of reality that involve measurable and consistent behavior. Sometimes, there is more than one explanation for a given pattern of behavior, each valid within its own context.

As an example, try evaluating the truth of the following two statements: (1) The sun moves around Earth, and (2) Earth moves around the sun.

Within the context of our everyday experience, the first statement is manifestly true. A few hours' observation suffices to show that the sun does indeed swing from east to west across the sky during the day. Also, between daylight periods, the sun somehow arcs under the horizon and appears in the east again at daybreak. Clearly, the sun moves around Earth in a *geocentric* context.

From everything we have learned so far in this book, the second statement is also correct. Relative to the stationary sun (that is, when we adopt a *heliocentric* context), Earth spins on its axis and moves around the sun as well.

Which is the better description of reality? From a scientist's point of view, the second statement is better—or at the very least, more useful. It is a broader and more general description of reality than the first statement. *There is a striving in science toward explanations that are universally applicable,* or at least valid within the broadest and most general context imaginable.

## The Newtonian Paradigm

For most of our everyday lives, we focus our attention on the here and now. We accept the idea that a straight line is the shortest distance between two points. We are accustomed to the effects of moving at speeds (relative to the things around us) that are always very much less than the speed of light. Time flows smoothly, at least as reckoned by the mechanical or electric clocks we use to mark its passage. Whether we give any thought to it or not, the constant 1*g*, downward-pointing gravitational field we live in affects nearly everything we do. From these consistencies, we build a personal "worldview," or *paradigm* (model), of what is real in the world around us.

When Isaac Newton developed his laws of motion, he adopted a similar commonsense paradigm about the nature of space and time. Newton assumed that space is exactly flat and uniform, with no distortions here or anywhere else. He also held that time is absolute: It is like a uniformly running clock whose individual ticks are heard everywhere as being simultaneous.

Newton's laws were applied with spectacular success to a host of problems in physics and astronomy, and for about two centuries (starting around the year 1700) the Newtonian paradigm reigned supreme in science. The laws, as Newton had laid them down and as we introduced them in Chapter 5, seemed to precisely describe everything from the behavior of pendulums and the rise and fall of tides on Earth to the stately movements of the celestial bodies above.

In the hands of Newton and his followers, particularly the physicist-mathematician Pierre-Simon Laplace (1749–1827), the universe began to resemble a perfect—albeit a very complicated—machine consisting of stars, planets, and all kinds of smaller things moving on their own prescribed paths. Scientists believed that if a person could measure with perfect accuracy the attributes of each moving part of this metaphorical machine, then its future destiny could, in principle, be determined for all time.

## What Went Wrong

By the early 20th century, the Newtonian paradigm was in trouble on two fronts. First, there was something wrong with light. Careful measurements of the speed of light were demonstrating that no matter what the motion of the source or the observer of the light, the speed of light in a vac-

**PHOTO 13.1**  Albert Einstein (1879–1955), the founder of relativity theory, also made important contributions to quantum theory.

uum could never be observed as being different from exactly *c*. This result flew in the face of Newton's laws.

To picture this strange situation, imagine a surfer paddling in order to catch up to a wave (symbolizing light, in this case) and ride it. In the Newtonian (commonsense) paradigm, the surfer paddles toward the shore faster and faster, and the waves overtaking and passing him travel slower and slower relative to him. When the surfer succeeds in catching up to a wave, he rides its crest, which means that the speed of the wave is zero relative to him. The speed of light in a vacuum, however, *never changes,* no matter who looks at it or how it is looked at. It is as if a surfer could never catch up to a wave and ride it. No matter how hard the surfer might paddle, the wave (representing light) moving toward the shore would always move past him at the same speed it would if he did not paddle at all. Clearly, a new and dramatically different way of thinking was needed to provide a logical solution to this difficulty. The solution, known as Einstein's *special theory of relativity,* eliminates the Newtonian paradigm of absolute space–absolute time and replaces it with a different set of absolute standards.

Second, the study of light and the atom (as detailed in Chapter 11) was raising serious questions about the notion of a *deterministic* (perfectly predictable) universe. In the realm of the very small, it seemed, probability rather than determinism rules. The solution to this dilemma was to replace the Newtonian paradigm with the theory of quantum mechanics, the subject of the next chapter in this book.

In the jargon used by some historians of science, the overthrow of Newtonian physics by relativity theory and quantum mechanics constitutes two fundamental "paradigm shifts" in science. They were not the first turnarounds in the physical sciences—witness the Copernican revolution—nor are they likely to be the last.

Contemporary physics is built on the cornerstones of relativity (especially the part of relativity called Einstein's *general theory of relativity*) and quantum mechanics. Fundamental physics today is a patchwork: Relativity best describes gravity and large-scale interactions, and quantum mechanics best describes the world of the very small. Efforts by theoretical physicists to unify the two theories continue apace, though progress has been slow.

A third paradigm shift, involving chaos theory, may be afoot in physics today. The hallmark of chaos—sensitive dependence upon initial conditions (see Box 5.4)—argues against the possibility of perfect predictability in a great many systems and strikes another blow against Laplace's notion of determinism. Although studies of chaotic behavior are currently popular, it is too early to judge whether chaos theory should join relativity and quantum mechanics as a fundamental revolution in scientific thought.

## Why Newton's Laws Are Still Useful

Newton's laws are now recognized as being subsumed (or encompassed) by both relativity and quantum mechanics. However, they remain useful when restricted to slow-speed and weak-gravity situations in the macroscopic realm. When ordinary macroscopic behavior is analyzed through both Newtonian and relativistic physics, the results are virtually the same. Also, the confident predictions of Newtonian physics and the probabilistic outcomes of quantum mechanics are essentially in agreement outside the microscopic realm of atoms and molecules.

Newtonian physics remains especially useful because its mathematical formulations are much simpler and easier to use than those of relativity and quantum mechanics. In a similar way, it is usually less complicated to say that the sun "rises" in the sky than to explain what happens in a broader, heliocentric perspective.

For the remainder of this chapter, we will focus our attention on the relativistic paradigm. There are actually two theories of relativity: the special (restricted) theory, proposed by Einstein in 1905, and the general (unrestricted) theory, proposed by Einstein in 1915. We will describe them separately.

# Special Relativity

The special theory of relativity is concerned with frames of reference that are *not* experiencing any acceleration. They are known as *inertial* reference frames. The dependence upon inertial reference frames is what makes this theory special, or restricted.

The two basic *postulates* (assumptions) of the special theory of relativity are simple. The consequences (when viewed through the prism of common sense) are not, as we shall soon see. Postulate 1 says that all physical laws are consistent within all inertial frames of reference. Postulate 2 says that the speed of light in empty space is the same for all observers.

Postulate 1 seems reasonable. If you were playing a game of billiards on a flat table inside an airplane moving at constant speed in a perfectly straight line, the behavior of the rolling and colliding billiard balls would be exactly the same as it would be if the table were at rest on the ground. In other words, motion (in mechanics, at least) is relative. This much was recognized by both Galileo and Newton, the founders of the science of mechanics.

Recall from Chapter 10 how electromagnetic interactions are relative as well. It matters not a whit whether you thrust a magnet into a stationary coil or thrust a coil toward a stationary magnet; the resulting electromagnetic induction in the coil is the same. If you perform the same experiments on board the inertial frame of a cruising airplane, you get exactly the same results.

Numerous experiments have shown that it is not just the laws of mechanics (such as conservation of momentum) that are the same within all inertial reference frames. All physical laws are the same, including those dealing with electromagnetic phenomena (such as electricity, magnetism, and light) and nuclear phenomena.

Postulate 2 is counterintuitive—nonsense, if you will. This simple example will illustrate how "nonsensical" it seems: Imagine a spaceship (see Figure 13.1) moving away from Earth at half the speed of light, or $0.5c$. The astronaut, Joe, inside the ship, aims a headlight forward and a taillight rearward and measures the speed of the light leaving these two light sources. His measurements show that both light beams are departing his ship at exactly the speed of light ($1.0c$).

**Postulate 1**

All physical laws are consistent within all inertial frames of reference.

**Postulate 2**

The speed of light in empty space is the same for all observers.

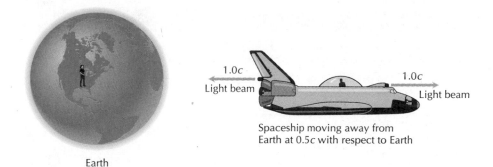

Earth

Spaceship moving away from
Earth at 0.5c with respect to Earth

**FIGURE 13.1**   Joe, aboard a spaceship moving at half the speed of light away from Earth, measures the light leaving his ship fore and aft. He finds that in both cases the light travels at exactly the speed of light. Jane, on Earth, sees the light from the taillight moving toward her at precisely $1.0c$, and light from the headlight moving away from her at precisely $1.0c$. Why?

What does Jane, an observer on Earth, see? According to Newtonian physics (or common sense), she should see the light from the taillight approaching her at $0.5c$ and the light from the headlight moving away from her at $1.5c$. According to Postulate 2 in special relativity, however, she should measure the light from the taillight approaching her at $1.0c$ and the light from the headlight moving away at $1.0c$. Which version is right? The Newtonian prediction *seems* right; but the verdict, as revealed by a host of experiments designed to test the behavior of light, is that the relativistic prediction *is* right. The speed of light in empty space is exactly the same no matter how a light source moves and no matter how the light-measuring apparatus moves.

Comparisons like this, which pit the Newtonian paradigm against the relativistic paradigm, pose a profound paradox. The paradox can be resolved by assuming that time and space (length), which Newton took to be absolute for all observers, *are not the same* for all observers. The fundamental properties of length and time (and, as it turns out, mass as well) that seem to be so unchangeable in our everyday experience are *relative*. They have values that depend on the relative motion between the observer and what is being observed.

Before we resolve the spaceship paradox illustrated in Figure 13.1, let us look into some of the specific consequences of special relativity, which necessarily stem from the two postulates. We will merely describe these consequences, not derive them. Derivations can be found in upper-level college physics textbooks.

## Length Contraction

When a body moves past you in uniform motion, it is shorter, in the direction it is traveling, by a factor that depends on its speed. Alternatively, when you move uniformly past a body, that body is shorter, in the same direction you are traveling, by the same speed-dependent factor. The faster the relative motion, the greater the relativistic **length contraction.**

In other words, it is impossible to specify, without ambiguity, the spatial dimensions of anything unless the role of the observer is recognized. If

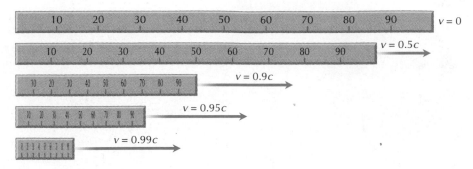

**FIGURE 13.2** A meterstick contracts in the direction it is traveling by an amount that depends on its speed relative to the observer. When not moving, the meterstick is exactly 1 m long—as long as it can ever be. When moving at half the speed of light relative to an observer, it is 83 cm long for that observer. At a relative speed of 0.99c, the meterstick is contracted to approximately 1/7 its uncontracted length. Further increases in speed give rise to even greater contractions. The contracted length can be calculated by the formula $L = L_0\sqrt{1 - v^2/c^2}$, where $L$ is the contracted length when in motion, $L_0$ is the length at rest, $v$ is the relative speed, and $c$ is the speed of light.

an observer is not moving relative to a body, then that body has the maximum possible size. If there is relative motion, then the body being observed is contracted (shortened) in the same direction as the relative motion. For most ordinary physics, which involves speeds much less than light's speed, this is a moot point. The speed-dependent factor mentioned earlier is almost exactly 1 for ordinary speeds. For example, the contraction of a meterstick moving at 1/10,000 of the speed of light (typical of current interplanetary spacecraft) is a mere 5 parts in a billion. Thus, someone aboard the *Voyager* spacecraft whizzing past a meter stick at 30 km/s would notice that it is only 0.999999995 m long, which is not much different from exactly 1 m long.

A meterstick moving at half the speed of light (0.5c), however, is about 83% as long as it would be if it were stationary, or 83 cm long (see Figure 13.2). What would an observer moving alongside that "contracted" meterstick see? Since there would be no relative motion between that observer and the meterstick, the meterstick would have its normal length of 1 m. One meter, which is the so-called *proper length* of a meterstick, is the maximum length a meterstick can have. Table 13.1 gives several examples of contracted length for various speeds approaching c.

| TABLE 13.1 | Relativity of Length, Time, and Mass | | |
|---|---|---|---|
| Speed of Meterstick Clock, or Mass Relative to Observer | Observed Length of a 1-m Stick | Observed Rate at Which a Clock Ticks | Observed Mass of an Object with a Rest Mass of 1 kg |
| 0 | Exactly 1 m | Normal rate | Exactly 1 kg |
| 0.5c | 0.83 m | 0.83 × normal rate | 1.2 kg |
| 0.9c | 0.43 m | 0.43 × normal rate | 2.3 kg |
| 0.95c | 0.31 m | 0.31 × normal rate | 3.2 kg |
| 0.99c | 0.14 m | 0.14 × normal rate | 7.1 kg |
| c | Zero | Clock is frozen in time | Infinite mass |

Note that it is not just material bodies that contract in the direction of travel. Space itself contracts when there is relative motion. Thus, distances between objects moving toward or away from each other are smaller, too, in the same way that a meterstick is shorter when there is relative motion between it and the observer.

Another important point to keep in mind is that the dimensions of an object look perfectly normal to a person not moving relative to that object. A witch riding a broomstick at nearly the speed of light sees nothing amiss with her broomstick. Length contraction does not turn a grape whizzing through space at some very high speed relative to Earth into grape jam.

## Time Dilation

When a clock moves past you (or you move past a clock), the clock "ticks" slower as you observe it. The **time dilation** ("stretching" of time) of relative motion is very slight unless the clock is (or you are) moving at a substantial fraction of the speed of light.

When a clock moves at 0.5$c$ (relative to some observer), the clock ticks about 83% as fast as it would if the clock were stationary relative to the same observer. A clock moving at 99% of the speed of light (0.99$c$), ticks about 14% (1/7) as fast. These and other examples are listed in Table 13.1.

We are not talking merely about mechanical and electric clocks here. Recall from Chapter 3 that time is what a clock measures. Any repetitive series of events, contrived or natural, can serve as a clock. Time itself is slowed, or dilated, in any reference frame moving uniformly relative to an observer. Time "flows" fastest when there is no relative motion between clock and observer. This time, the time kept by a clock observed at rest, is called *proper time.*

The subject of time dilation often brings up the issue of time travel. In point of fact, just by standing still with clock in hand you cannot help but travel through time, from past to future, at the normal rate: the proper time, according to your clock. The theory of relativity has nothing to say about traveling *backward* in time, and the future-pointing "arrow of time" discussed in Chapter 3 seems to prohibit it, anyway.

Relativity theory does, however, permit more rapid rates of time travel into the future under certain circumstances. As an example, let us imagine an astronaut traveling away from Earth at 99% of the speed of light toward a star 10 LY away. According to clocks on Earth being watched by people on Earth, the astronaut's one-way journey takes slightly longer than 10 years. (The journey would take exactly 10 years if the spacecraft could move at exactly the speed of light instead of at 0.99$c$.)

Since there is a relative speed of 0.99$c$ between Earth and the traveling astronaut, the astronaut looks back at Earth and sees Earth clocks ticking 1/7 as fast as his clock. According to his reading of Earth clocks, the one-way trip to the star takes only 1/7 of 10 years, which is about 1.4 years.

What we have not considered yet is how the spacecraft's own clock appears to the astronaut inside. It keeps proper time for the astronaut, who is stationary with respect to it. In other words, the astronaut perceives his clock as ticking at its normal rate. You might think, therefore, that the trip

**FIGURE 13.3** In this space-time diagram, a hypothetical astronaut, moving at nearly the speed of light, completes a 20 LY journey, out and back. By the astronaut's clock, the journey takes approximately 2.8 years. By Earth clocks, the same journey takes approximately 20 years. Time dilation effects, and the apparent paradoxes they produce, are pronounced only at speeds close to the speed of light.

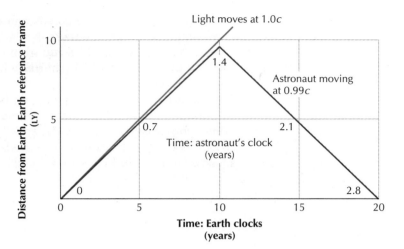

takes around 10 years by the astronaut's clock because the star is 10 LY away and the ship is traveling at very nearly the speed of light.

Amazingly, the astronaut does indeed reach the star in just 1.4 years, as reckoned by both the clock he uses on board *and* by his reading of clocks on Earth. How? Because the distance to the star, *as measured by the astronaut,* is no longer 10 LY. The astronaut is moving at 0.99c relative to the star and directly toward it, so the distance to that star, in the astronaut's own reference frame, is contracted by 1/7! (If this is confusing, just remember that it is not just metersticks that experience length contraction; space does, too, in the direction of travel.)

If, after reaching the star, the astronaut could instantly turn around and return to Earth at the same 0.99c speed, the elapsed time shown on the spacecraft clocks would register a total of 2.8 years for the round-trip. That 2.8 years is the proper time for the astronaut, who remained in the same reference frame as his clock for the whole journey. Clocks on Earth, viewed by those who stayed behind on Earth, would have registered a total of just over 20 years for the astronaut's round-trip journey. That 20 years is the proper time for the people on Earth (see Figure 13.3). What this means, literally, is that if the astronaut had left a twin sibling behind on Earth, the stay-at-home twin would have aged about 17 more years than the traveling twin upon completion of the journey. This seemingly impossible situation is known as the *twin paradox,* though it is not a paradox within the relativistic paradigm. Relativity permits alternative paths through time for observers in reference frames with different histories.

Anyone attempting an actual experiment in space travel at the speed outlined above would face enormous technical and practical difficulties. (The fastest any astronaut has ever traveled relative to Earth is about 20,000 times slower than the speed of light. These speeds were experienced by the Apollo astronauts traveling to and from the moon.) Yet similar experiments that test the validity of relative time have been performed for decades with tools such as atomic clocks and particle accelerators. So far, every experiment devised and performed to test the validity of time dilation has confirmed its existence.

When two identical atomic clocks are synchronized and one of them is flown around the world, the traveling clock loses a tiny, but measurable, amount of time in comparison to the stay-at-home clock. When *muons*—subatomic particles whose average lifetimes are just 2 ms when measured at rest in the laboratory—are accelerated to a speed of 0.99*c* in particle accelerators, they take about 7 times longer to decay and therefore move 7 times farther around the accelerator rings than they would otherwise be expected to travel. Some muons are created naturally when cosmic ray particles collide with atomic nuclei in the upper atmosphere. These freshly created muons streak toward Earth at speeds of around 0.998*c,* and some are picked up by muon detectors on the surface. Were it not for the pronounced effect of time dilation on these fast muons (a factor of around 15, which makes them last about 30 ms), virtually no muons would be recorded by the detectors. They would decay too quickly for appreciable numbers of them to reach the ground.

The length contraction consequence of Einstein's special theory indicates that if a body is observed to be moving at exactly the speed of light, it is contracted to zero length in the direction of travel. The time dilation consequence says that a body observed moving at *c* does not age at all because time comes to an absolute standstill. These strange characteristics are valid for photons, which have no mass and do move at exactly the speed of light. But they do not apply to material bodies (bodies with mass) because of another consequence of the special theory, the relativity of mass.

## Relativity of Mass

A moving body has a greater mass than a body at rest. When at rest, a body has the least possible mass. This mass is called its *rest mass.* We are using the word *mass* here to mean the body's resistance to a change in its motion: its inertia. When a mass moves (relative to something else, of course), it does not somehow acquire new atoms or molecules; it simply has a greater tendency to resist acceleration.

Mass increases are small for speeds that are not large compared with the speed of light. In fact, the mass *increase* is proportional to the *decreasing* length of the moving body and to the *decreasing* rate at which moving clocks tick. For example, a body with a rest mass of 1 kg moving at 0.99*c* will have a "moving" (relativistic) mass of about 7 kg (see Table 13.1 and Figure 13.4).

At speeds very nearly equal to *c,* the mass increase is drastic. At exactly *c,* any material body, whatever its rest mass, would have infinite mass. It would take an infinitely large force to accelerate even the smallest mass to the speed of light. The speed of light is thus the ultimate, and not quite attainable, speed limit for material bodies.

## Resolving the Paradox

We are now ready to resolve at least one aspect of the paradox presented in Figure 13.1. Joe, with his meterstick and clock, measures the speed of the light leaving his spaceship by means of gauging how far the light travels over a known interval of time. The formula $v = d/t$ gives the speed of light for him, which is 300,000 km/s, or 1.0*c.*

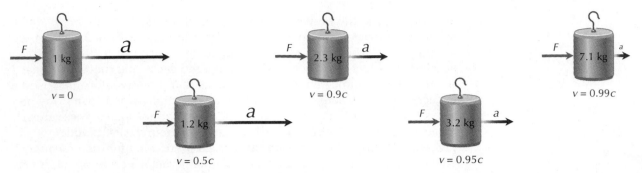

**FIGURE 13.4**   The faster a body moves, the more difficult it is to change its motion (accelerate it). The rest mass of the body shown is 1 kg. At half the speed of light, this body acts as if it had a mass of 1.2 kg. For the same applied force $F$, it accelerates a little less than it would when at rest. At $0.99c$, the body acts as if it had a mass of 7.1 kg, and it accelerates about 1/7 as much in response to the same applied force $F$. The moving mass of a body moving at any speed can be calculated by the formula $m = m_0/\sqrt{1 - v^2/c^2}$.

Recall that Joe's spaceship is moving at $0.5c$ relative to Earth. At that speed, space is contracted by 83% and clocks tick 83% as fast as they would at rest, as noted in Table 13.1. Jane, on Earth, watches Joe measuring the light; but for her, Joe's meterstick is only 83% as long as hers, and Joe's clock is ticking only 83% as fast as hers does. Because both distance and time are altered by the same factor, the ratio $d/t$ (the speed of the light measured in Joe's reference frame by Jane) has the same value of $1.0c$!

What happens when the light from the taillight of the spacecraft actually reaches Jane? Jane's clock is running about 1.2 times faster than Joe's clock as seen by her, but at the same time her meterstick is 1.2 times longer than his as she sees it. Again, the values of $d$ and $t$ are different, but the ratio $d/t$ is the same. That ratio is precisely the speed of light.

Here is the consequence of this line of thinking: The properties of space and time, which we formerly assumed were constant for all observers, are not actually constant. The speed of light, which we formerly assumed would depend on motion, does not depend on motion; it has an absolutely constant value. No matter how an observer moves or how a source of light moves, and no matter what disagreements different observers may have about the values of length and time when they look into various reference frames, light always leaves any source at 300,000 km/s and arrives at any observer at 300,000 km/s.

Note that the constant speed of light is not the only consistency associated with the special theory of relativity. Recall the first postulate, which speaks of the consistency of physical laws in all inertial reference frames. Thus, even though the measured values of space, time, mass, and so on are different for different observers moving relative to each other, we do not end up with an assortment of inconsistent phenomena. A runner who wins a race in one reference frame does not lose the race in another, despite the fact that observers in different reference frames may disagree on the length of the course and how much time it took the winner to complete the course.

Four hundred years ago, few people believed that Earth could be spinning and revolving. No one could feel these motions. Nevertheless, the heliocentric model gained acceptance on grounds other than direct experience. Today, time-lapse photography obtained from spacecraft shows planets spinning on their axes and revolving around the sun. We also have the eyewitness accounts of astronauts who have journeyed to the moon and observed a spinning Earth. The heliocentric model, originally posed on theoretical grounds, was finally confirmed by direct experience.

If we, as humans, could travel through our surroundings at nearly the speed of light, then the various distortions of space and time predicted by special relativity would presumably become visually apparent. Unfortunately, humans cannot use direct experience to confirm this theory, because the technical hurdles involved in accelerating humans to relativistic speeds are immense. Still, special relativity has gained nearly universal acceptance among scientists. Special relativity predicts many strange, but testable, consequences; and most of those consequences have been thoroughly tested and confirmed. One of the these consequences, which at first glance seems to have nothing to do with the two postulates of special relativity, is that energy is equivalent to mass times the speed of light squared, or $E = mc^2$. This relationship, perhaps the most important in all of science, has profoundly changed the world.

# General Relativity

**Principle of Equivalence**
All physical laws are consistent within any two reference frames undergoing the same acceleration.

In Einstein's general theory of relativity, the restriction of inertial reference frames is removed. General relativity deals with frames of reference undergoing acceleration. The general theory rests on one simple assumption, known as the principle of equivalence: All physical laws are consistent within any two reference frames undergoing the same acceleration.

An "accelerated" reference frame can be one that is under the influence of a gravitational field, such as the 1g environment we experience on Earth; or it can be produced by the application of some force that has nothing to do with gravity. Regarding the first instance, we on Earth do live in an accelerated reference frame. If a trapdoor in a castle opens under the feet of a prisoner, there is no longer any force holding the prisoner up. The prisoner accelerates toward the floor of the dungeon. Regarding the second instance, a drag racer experiences the effect of acceleration when his car is burning rubber on the track. This acceleration (which is parallel to the ground) is distinct from the acceleration associated with Earth's gravity.

A spaceship, far away from any star or planet and drifting along with its engines turned off, is an example of an inertial frame of reference. Everyone on board is weightless. If the engines are turned on so that the ship accelerates at 9.8 m/s², however, then everyone on board falls toward the rear of the ship and experiences an "artificial gravity" of 1g, just like that on Earth's surface. The constantly applied force of the rocket engines produces this effect. The principle of equivalence says that all physical laws are the same in these two accelerated reference frames, regardless of the cause of the acceleration.

# Consequences of the Principle of Equivalence

So far, so good; general relativity sounds simple. But let us make some comparisons and draw some conclusions about the nature of light as seen in these accelerated reference frames. In Figure 13.5(a), an astronaut inside a spaceship accelerating upward at 10 m/s$^2$ aims a flashlight beam out a porthole on the ship's side. (For simplicity, we are making the approximation $1g \approx 10$ m/s$^2$). The observers are exerting downward forces on the floor (back end) of the spaceship, because the floor is forcing them to go 10 m/s faster with every passing second. Each feels an apparent downward force equal to his or her normal weight on Earth. By very carefully observing the light racing out from the flashlight at 300,000 km/s, or $3 \times 10^8$ m/s, they notice that the beam dips slightly. After 1 s of travel, the beam has dropped 5 m behind. After 2 s, the beam has fallen behind a total of 20 m; after 3 s, it is 45 m behind; and so on. This has happened because the spaceship is accelerating: After 1 s, the spaceship is 5 m farther ahead than it would have been if the engines had been off. After 2 s, the ship is 20 m farther ahead; and so on. Thus, in the accelerated reference frame of the astronauts, the path of the light is curved.

Since, according to Einstein, all $1g$ reference frames are equivalent, light should behave in the same way on Earth. In Figure 13.5(b), we imagine a flat Earth with a constant gravitational field of $1g$ on its surface. Observers standing there exert downward forces on the ground, just as observers on the spacecraft exert "downward" forces on the floor of the spaceship. (We must imagine a flat Earth for our example, because light would move very quickly away from the curved surface of the real Earth. We are also ignoring the effect of atmospheric refraction by assuming that there is no atmosphere and that the light is traveling through empty space.)

Let us say that one of our flat-Earth observers aims a flashlight horizontally so that its beam travels parallel to the ground. The beam starts off parallel, but it does not remain so. In fact, light moving in this $1g$ environment drops 5 m in the first second, a total of 20 m in the next second, and so on, as the principle of equivalence suggests. The observers conclude that light travels on a curved path in a gravitational field and that it must somehow feel gravitational force. (This is one way of looking at this phenomenon. Another way is to suppose that the presence of Earth's mass slightly curves the space around it and that light merely follows this curvature.)

By following other lines of reasoning, we can show that time *slows down* in an accelerated reference frame, regardless of whether the acceleration is due to gravity or to artificial means. This is another *time dilation* effect distinct from the one already discussed in connection with special relativity.

Einstein went further than merely thinking of the light curvature and time dilation effects as being separate curiosities. He conceived of space and time as being interconnected as *space-time,* and he imagined that space-time would be "flat" in the absence of any mass and "curved" in the presence of mass.

Technically, there is no truly flat space-time anywhere in the universe, because the universe has plenty of matter (mass) in it. In a hypothetical flat

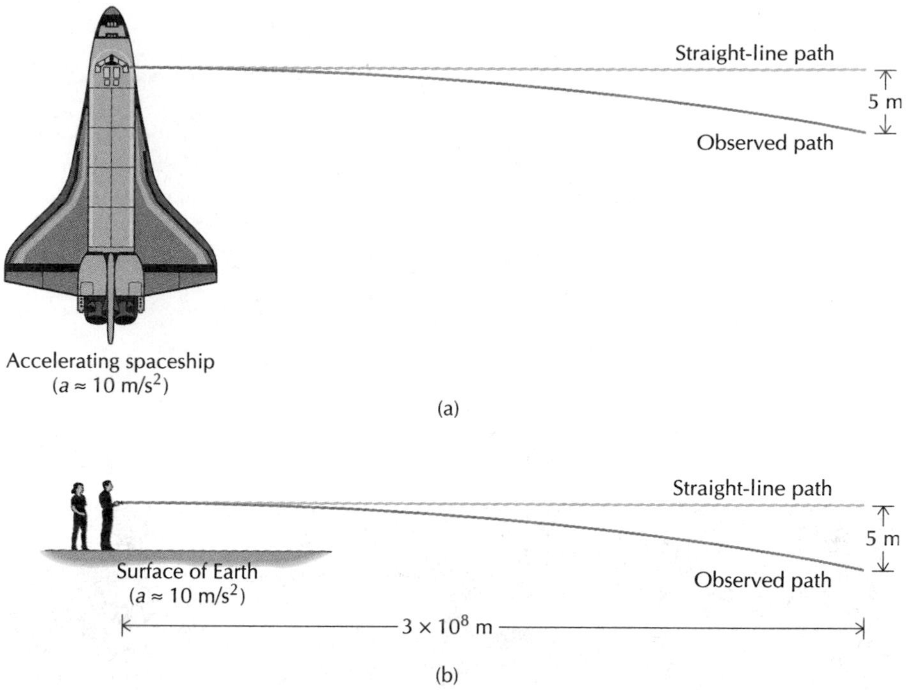

**FIGURE 13.5** As an observer accelerating at $1g$ from spacecraft propulsion sees it, light from a flashlight moves outward 300 million meters in 1 s and also falls behind its intended path by 5 m in the same time (a). According to the principle of equivalence, light should fall in exactly the same way for an observer in the $1g$ environment at Earth's surface (b). By looking at the behavior of the light beams alone, a person cannot tell what is causing the $1g$ acceleration.

(virtually empty) space, however, light would always travel on perfectly straight lines between any two points and there would be no distortions in space or time other than those made possible by special relativity.

The greater the mass present, the greater the density of that mass is; and the closer you are to that mass, the stronger the curvature of space-time is. We can think of light as tracing the contours of space-time, naturally following its curvature at any point. Matter is also influenced by the curvature of space-time. Since most matter normally moves at speeds much less than the speed of light, the paths of material bodies—for example, planets revolving around the central mass that is our sun—are much more sharply curved than the paths of light beams that share the same space-time. Also, the greater the curvature of space-time at a given point, the more slowly clocks will tick at that point.

Einstein's conclusion was that gravity is just an expression of curved space-time around massive bodies. In his way of thinking, a massive body curves the space and alters the time around itself, and both light and material bodies passing through that curved space-time respond to the curvature.

Although curved space-time, which features four dimensions (three of space and one of time) can be described mathematically, it cannot be completely visualized in our minds. We can gain some understanding of its nature, however, by considering an analog that makes use of just two dimensions of space: the curved, two-dimensional space that forms the surface of a sphere.

On the surface of a sphere, the shortest distance between two points is not a straight line but an arc of a great circle (any transcontinental aircraft flying above Earth's surface is likely to be on this kind of route for reasons of efficiency). It is not difficult to imagine several *different* paths of the same length between two points on a sphere, provided those points are

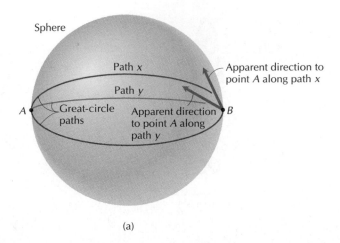

(a)

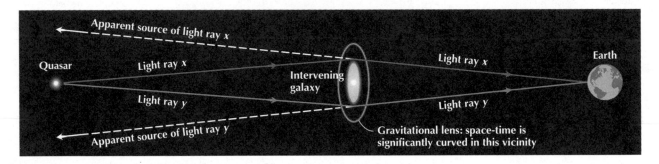

(b)

**FIGURE 13.6**    On the surface of a sphere (a curved, two-dimensional space), there can be multiple paths of minimum distance between two points on opposite sides of the sphere. In the three-dimensional space of the universe, space-time is essentially flat except near massive bodies, where it can be curved in a way *analogous* to the curved space on the surface of a sphere. The curvature of space-time around a galaxy can bend light rays from a quasar (a bright, pointlike source of light) in such a way that observers on Earth see two or more images of the same quasar. If a spherical or disk-shaped galaxy is aligned just right between a quasar and Earth, observers on Earth may see the light from the quasar distorted into a ring surrounding the galaxy. This effect has been called an "Einstein ring."

carefully chosen, as illustrated in Figure 13.6(a). In a similar way, massive galaxies curve the space-time around them and therefore bend the light passing near them. Thus, if the geometry is just right, as illustrated in Figure 13.6(b), the light moving out from a distant, bright quasar can take two or more paths around a massive galaxy. If these paths, or rays, of light converge on Earth, we will see two or more images of the same quasar. Galaxies, or other massive objects that affect light in this fashion are called *gravitational lenses.* Thousands of gravitational lenses have been discovered so far. The most famous is an "Einstein cross" (see Photo 13.2): four images of the same quasar symmetrically surrounding the image of an intervening galaxy.

The curvature of space-time is very slight on or near the surfaces of planets and normal stars. For the sun, the change in direction of light passing just above its surface is just 1.7 arcseconds (1.7″), a mere 1/2000 of a degree. As in the example in Figure 13.6(b), rays of light from a distant star passing around the sun bend inward. This makes the star appear to be slightly farther away from the sun in the sky than it is. Astronomers detected this tiny

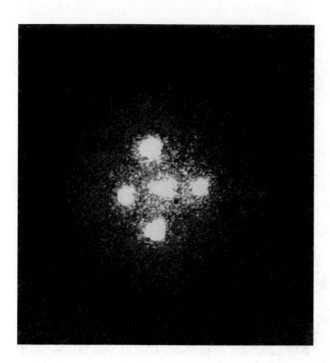

**PHOTO 13.2**   In this Hubble telescope view of the object known as G2237 + 0305 (a code number denoting its right ascension and declination coordinates in the sky), a quasar some 8 billion light-years away is being multiply imaged by an intervening galaxy some 400 million light-years distant. The arrangement of four quasar images surrounding a single galaxy image is known as an "Einstein cross." A similar though very precise alignment between a distant quasar and a closer galaxy may produce an "Einstein ring" in which the quasar's light is spread into a thin, concentric ring.

but measurable effect during a total solar eclipse in 1919, just four years after Einstein published his general theory. They were able to photograph several stars in the sky behind the blotted-out sun, and the positions of these stars were shifted by the amount predicted by general relativity. This was a great triumph for the theory, and it helped make Einstein famous.

In a contemporary test of general relativity, two identical atomic clocks were placed at different heights in the same tall building. Even though the clocks were perfectly synchronized when they were together, the clock nearest Earth's surface ticked just a bit more slowly when the two clocks were separated. The curvature of space-time, which gives rise to time dilation in the general theory, is just a little greater in the slightly stronger gravitational field closer to Earth.

The bending of a light beam by 1.7″, or comparative measurements of time that differ by parts per billion, are hardly spectacular demonstrations (though they convince most physicists that general relativity must be right). Our Solar System has no examples of bodies sufficiently massive and dense to severely warp space and time. Things get very interesting, though, near *neutron stars* and *black holes,* both of which are massive and enormously compressed astronomical objects. The gravitational field on or near a typical neutron star (billions of *g*'s) is so intense that a light beam directed at a slight upward angle will actually return to the surface, much as a football thrown during a pass will arc back to the ground. A passing light beam may be deflected through an angle of many degrees. Time is also significantly affected near a neutron star. A clock resting on the surface, as observed by someone far away from the neutron star, would seem to tick much more slowly than it would if it were far from the neutron star. Black holes have such a strong gravitational field inside their boundary, called the *event horizon,* that no light and no material body can ever escape to the outside. Inside the event horizon, space-time folds in on itself.

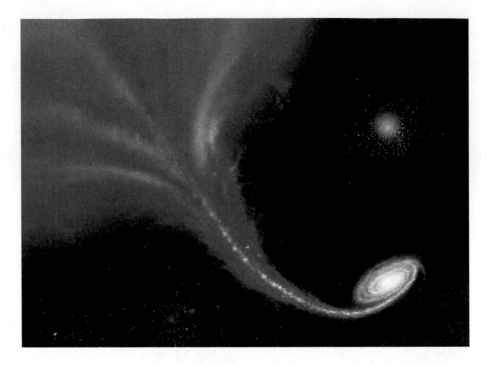

**PHOTO 13.3**  In this artist's conception of a close binary system containing a black hole, matter from the surface of a red-giant star collects into an inward-spiraling accretion disk before falling into the black hole itself. Near the black hole's event horizon at the center of the disk, the material heats up to $10^6$ K or more and emits X-rays.

Black holes and neutron stars are generally very compact in size—no more than tens of kilometers in diameter—and are capable of severely altering space-time only in regions relatively close to them. If our sun collapsed (hypothetically) to form a neutron star or a black hole, Earth's orbital path would not be changed at all. The curvature of space-time associated with the sun's presence 1 astronomical unit (AU) away would be exactly the same no matter how small the sun might shrink.

No one has ever seen a neutron star or a black hole up close, but the idea that they exist is strongly supported by most physicists and astronomers who study them. For one thing, the general theory of relativity makes such objects theoretically possible. The predicted consequences of the general theory have been confirmed in every area of experimental testing devised so far, so it is not unreasonable to try to push the theory by extrapolating to conditions of very strong gravity. For another thing, neutron stars and black holes can, in many cases, be invoked to explain astronomical phenomena that cannot be explained in any other known way.

In some instances, a suspected black hole is paired with another, more or less normal star in a *close binary system*: any two bodies of stellar mass that circle each other at close range. The black hole exerts a tremendous differential gravitational force (or tidal force) across the normal star and warps it into a balloonlike shape (see Photo 13.3). If the black hole is close enough, it pulls gaseous material away from the star's surface. The gas collects into a thin *accretion disk* around the black hole (which arises because the two objects are revolving around each other) and spirals inward. The gas becomes very compressed as it converges on the comparatively tiny black hole at the center, and the temperature rises accordingly. Just before the gas (a plasma by then) disappears beneath the event horizon, never to be seen again, its temperature reaches millions of degrees—more than hot

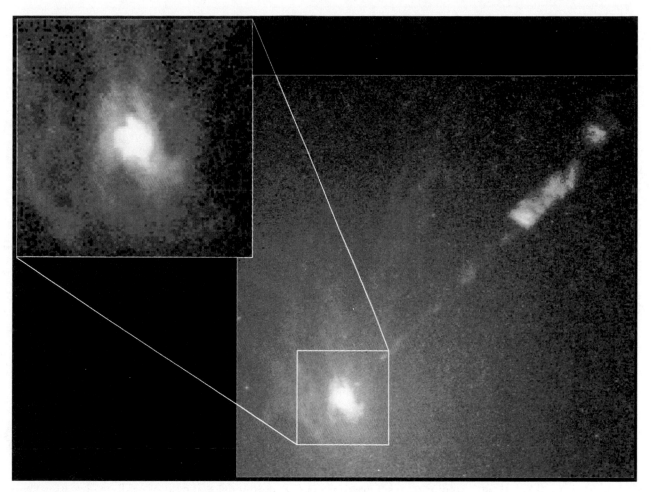

**PHOTO 13.4** In 1994, the Hubble space telescope photographed a spiral-shaped disk of hot gases (inset) in the core of the active galaxy M87, which lies at the center of the Virgo cluster of galaxies some 40 million light-years distant from us. Doppler measurements of the rapidly moving material in the disk suggest that a supermassive black hole, with a mass of more than 1 billion suns, lies at its center. As material spirals in and piles up along the inner edge of the disk (just outside the black hole), some of it is thrown outward in a brilliant jet, visible as a diagonal streak in the larger photograph. The existence of this jet has been known for decades.

enough to emit X-rays. These X-rays can be detected by X-ray telescopes in Earth orbit, and calculations show that they originate from a point in space close to a dark (invisible) and very massive object orbiting the star. The analysis of the light from the star by optical telescopes reveals that the star itself revolves around the dark object, whose mass can be estimated. After all the observational facts are in hand, the simplest way to explain the identity of the dark object is to assume that it is a black hole.

Astrophysicists think they are seeing the same kind of thing happening on a much larger scale within the nuclei (cores) of many galaxies, including our own. There, they think, *supermassive black holes* lurk. These objects could be the result of millions of smaller black holes coalescing over time within the relatively dense, central region of a galaxy. The Hubble space telescope has provided us with the first close-up photograph of an accretion disk surrounding a suspected, supermassive black hole (Photo 13.4).

The infall of diffuse material into supermassive black holes may well explain the abnormal luminosity of certain galaxy nuclei known as *active galactic nuclei* (AGNs). The most familiar AGNs are *quasars.* All that is needed to power a typical quasar, which emits about ten times as much energy as an average whole galaxy, is just one solar mass of material per year falling into a supermassive black hole.

One curious observed property of quasars, which are the brightest of the AGNs, is that none are closer than about 2 billion light-years. Considering the look-back time, which is no less than about 2 billion years for quasars, quasars may be a thing of the past: Presumably, billions of years ago, supermassive black holes in galactic nuclei had more abundant sources of mass around them to capture than they do now. This material would have consisted of diffuse gas, dust, and whole stars, which would be torn apart by tidal forces if they strayed too close to the accretion disk. With the accretion of any new material, a black hole grows in size and mass, but its growth will slow after it has ingested everything within easy reach. Quite possibly, the supermassive black holes in some of our neighboring galaxies may have acted as quasars billions of years ago. They have since settled down, starved for the inflowing material that makes them shine so brightly.

General relativity theory has illuminated the grand canvas of cosmology as well as the minuscule world of black holes. The big bang theory was conceived, in part, on the basis of general relativity. General relativity also provides tools by which we may eventually determine the ultimate fate of the universe: open, meaning the universe will expand forever, and closed, meaning the universe will eventually collapse. It says that the curvature of space in the universe as a whole should be positive if the universe is closed and negative if the universe is open. In a positively curved space, light rays that start out parallel eventually converge, just as parallel lines originating at one point on a sphere (a special, two-dimensional case of positive curvature) will converge somewhere else on the sphere. In a negatively curved space, parallel light rays will eventually diverge. Some characteristics of flat and curved spaces are demonstrated in Figure 13.7.

Astronomers have followed several avenues of inquiry that exploit the geometrical properties associated with curved space, but there have been no definitive conclusions. If anything, the studies have shown that on very large scales space is somewhere near to being flat and that the universe's expansion is thus close to the boundary between being open and being closed.

An interesting possibility is associated with the universe having an overall positive curvature. The universe could be finite in volume yet unbounded, in the same way that the surface of a sphere is finite in area yet unbounded. Unbounded means that by traveling in a given direction you will never reach an edge or escape from the universe. This possibility is even more intriguing when you consider that the whole universe would then act as a black hole of immense proportions—not even light could ever escape from it!

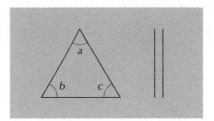

Flat space
(like the surface of a tabletop)
$a + b + c = 180°$
parallel lines remain parallel

**FIGURE 13.7**  In these two-dimensional analogs of three-dimensional space, the type and the amount of curvature can be deduced by certain geometrical properties, two of which are given here.

Sphere: positive curvature
(like the surface of a ball)
$a + b + c > 180°$
parallel lines converge

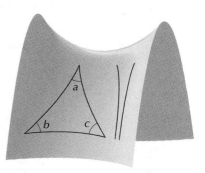

Saddle surface: negative curvature
(like the surface of a Pringles potato chip
$a + b + c < 180°$
parallel lines diverge

# CHAPTER 13
## Summary

Einstein's special and general theories of relativity have superseded Newton's laws as the most inclusive and accurate descriptions of the macroscopic world.

According to the special theory of relativity, the laws of physics are consistent within all inertial reference frames, and the speed of light in empty space is constant regardless of the motion of the source or the observer of the light. Thus, length, time, mass, and certain other properties can be different for observers in relative motion. These aspects of the special theory must be invoked when dealing with phenomena involving very high speeds.

According to the general theory of relativity, all frames of reference undergoing the same acceleration are equivalent to each other. The general theory of relativity describes gravity as the effect of curved space-time. Massive bodies create distortions in space-time, and light and matter alike respond to these distortions as they move through space-time. General relativity provides the framework for understanding such astronomical phenomena as neutron stars, black holes, and quasars.

## CHAPTER 13
# Questions

## Multiple Choice

1. Within the Newtonian paradigm,
   a) the speed of light in a vacuum is constant
   b) the force of gravity becomes zero in space
   c) times and distances are always the same for all observers
   d) times and distances can be different for different observers

2. Within the relativistic paradigm (Einstein's relativity theory),
   a) space is flat and time flows uniformly
   b) the laws of physics depend on the observer's location and speed
   c) times and distances are always the same for all observers
   d) times and distances can be different for different observers

3. The proper length of a given baseball bat is
   a) the minimum possible length that bat can have
   b) its average length as measured by many observers
   c) the bat's length when the bat is at rest relative to the observer
   d) the bat's length when it moves at the speed of light

4. Einstein's theories of relativity
   a) are subject to further testing
   b) are not subject to further testing because they have been proved
   c) are impossible to test in principle
   d) have not been tested yet because of a lack of adequate equipment

5. According to the general theory of relativity,
   a) time passes faster in frames of reference that are accelerated in some way
   b) gravitation acts only on bodies that are moving with respect to some mass
   c) gravitation is an effect of curved space-time
   d) gravitation works the same as in Newton's theory

6. When a beam of starlight passes near a massive object (like the sun), it is
   a) deflected slightly toward the massive object
   b) deflected slightly away from the massive object
   c) not deflected at all
   d) always absorbed by the massive object

7. If a precision clock is operating atop Mount Everest and another identical clock is running in the city of Kathmandu, Nepal, in the foothills below the mountain,
   a) the Mount Everest clock ticks very slightly faster
   b) the Mount Everest clock ticks very slightly slower
   c) the Mount Everest clock ticks much slower
   d) both clocks tick the same

8. Black holes emit no light, but it is believed they have been detected in
   a) close binary systems
   b) the center of the Milky Way galaxy
   c) in the nuclei of certain other galaxies outside our own galaxy
   d) all of the above

9. Quasars
   a) typically emit less energy than the sun
   b) reside at the centers of some planetary (solar) systems
   c) lurk at the center of every galaxy
   d) may be caused by mass falling through an accretion disk into a supermassive black hole

10. If two parallel beams of light never intersect and never diverge from each other, then the space they are traveling through is
    a) flat
    b) positively curved
    c) negatively curved
    d) saddle-shaped

## Questions

1. How is the Newtonian paradigm (on which Newtonian physics is based) similar to the geocentric paradigm held by the ancient Greeks?

2. What are the two postulates of special relativity?

3. What things are "constant" in the special theory of relativity? What things are "relative" in special relativity?

4. In what sense or senses is it possible to "travel" through time?

5. How does time dilation work in inertial reference frames? In accelerated reference frames?

6. Why is it not possible for a mass to travel at the speed of light relative to something else?

7. What does the principle of equivalence mean in general relativity?

8. If a clock were resting on Jupiter, where the gravitational acceleration is about $2.5g$, would it tick faster or slower than an identical clock on Earth? Assume that the rates of both clocks are being observed by someone well above the surface of each planet.

9. Why are black holes said to be black?

## Problems

1. Assume you are in a spacecraft moving uniformly at 99% of the speed of light. How long would a meterstick be if it were on board next to you? How fast would a clock tick if it were next to you? If you had a 1-liter (1-kg) carton of milk on board, what would its mass be?

2. Assume you are on Earth watching a spacecraft moving by uniformly at 99% of the speed of light. How long would a meterstick on board the spacecraft be as you would see it? How fast would you observe a clock on board to be ticking? If there was a 1-liter (normally 1-kg) carton of milk on board, what would its mass be as measured in a reference frame fixed to Earth?

## Questions for Thought

1. When Einstein was a young man, he wondered what a ticking clock would look like if he could move away from the clock at the speed of light. How would the clock appear in those circumstances? According to special relativity, can a person travel at light's speed away from anything?

2. Three historical paradigm shifts were mentioned in this chapter, leading to the Copernican (heliocentric) model, relativity, and quantum mechanics. Can you think of any other paradigm shifts in either the natural sciences (such as physics and biology) or in the human sciences (such as psychology and sociology)?

3. In the twin paradox, one twin is thought of as leaving Earth, traveling outward at close to the speed of light, and returning to Earth. We learned that the traveling twin ages less than his stay-at-home sibling. Since we have learned that motion is relative, would it not be equivalent to say that from the traveling twin's point of view, it was Earth that moved away from the traveling twin and later returned? If so, that would throw a monkey wrench in our argument that the twins must have different ages when they get back together. There is, however, a fundamental difference in the experiences of the two twins. What do you think this difference is?

4. One of the effects of general relativity has been expressed as "matter tells space how to curve and curved space tells matter how to move." Is this another example of symmetry in nature?

5. Often, the light emitted by a quasar being "lensed" into two or more images by an intervening galaxy varies in brightness in an irregular fashion. Sometimes, the changes in light intensity are not synchronized; brightness fluctuations taking place in one image will be followed by the same fluctuations in another image of the same quasar after a certain period of time. Give two reasons for this phenomenon.

## Answers to Multiple-Choice Questions

1. c    2. d    3. c    4. a    5. c    6. a
7. a    8. d    9. d    10. a

# CHAPTER 14
# Quantum Mechanics

*Using either a ramp or a set of stairs, you can accomplish the same task—increasing your gravitational potential energy. Any combination of large and small steps will take you up a ramp. On a staircase, however, you move upward by means of incremental changes. In the spatial realm of atoms and subatomic particles, continuity disappears and every change occurs in a stepwise fashion.*

## WHERE WE ARE GOING IN CHAPTER 14

**Quantum mechanics**
Physical behavior in the world of the very small

**Wave-particle duality**
Both light and matter exhibit wave and particle behaviors

**The uncertainty principle**
It is impossible to know everything about a particle

**Quantum numbers**
Four code numbers are used to symbolize an atomic electron's state

Welcome to the quantum zone. You hop in your car one morning and start heading north toward school. For some reason, the smooth acceleration you are accustomed to during the trip does not happen. The speedometer needle locks on 10 mi/h, then suddenly jumps to 20 mi/h, then to 30 mi/h, and so on as you depress the accelerator pedal. The scenery drifts by at 10 mi/h, then suddenly at double that rate, then suddenly at triple that rate, and so forth.

Another strange thing happens. There is no way you can get out of first gear when you are going 10 mi/h. After the jump to 20 mi/h, however, you have a choice of two gears, first and second. After another leap to 30 mi/h, three gears are available. You notice that the needle on your car's tachometer (engine-revolutions-per-minute gauge) jumps from one specific value to another specific value whenever you change gears but you do not change the car's speed. You find that the crankshaft of your engine can turn only at specific rates (or revolutions per minute) and that no intermediate rates of engine speed are possible.

And here's one more strange thing. At 10 mi/h, your car acts as if it were affixed to trolley tracks in the roadway; your car can move only north or south. At 20 mi/h, your car is able to turn, but only very sharply, onto any street going east or west. From those streets, abrupt turns onto any north-south street are possible. At 30 mi/h, your options are broadened; you can now move in any of 8 directions: north, northeast, east, southeast, south, southwest, west, and northwest. The wheels of your car simply refuse to spin in all other directions besides these 8 points on the compass. At 40 mi/h, there are 16 directions to choose from.

There is more strangeness in the quantum zone. Your car itself is a kind of "fuzzy beast." It is like a long series of waves rippling down the road. Waves are difficult to locate precisely in space, so it is hard to say just where your car is at any moment within the train of waves. It might be on or near the leading edge, on the trailing edge, or somewhere in the middle of the wave train.

You are approaching a traffic light that is changing from yellow to red, and you are probably exceeding the speed limit, too. A traffic officer on the curb, seeking to enrich the coffers of the municipal treasury, wants to issue you two tickets: one for speeding and one for running a red light. Alas, he finds he cannot gather definitive evidence for both of these infractions at the same time. He has to settle for one only. If he measures the speed of the entire wave train and that speed is in excess of the speed limit, then he has caught you speeding. However, at the same time, he cannot know whether or not you ran a red light because he does not know exactly where your car is within the wave train. On the other hand, if the officer is able to precisely locate your car entering the intersection on a red light, then he really does not know your car's speed. To determine speed (the distance between two positions divided by time), he needs a later measurement of your car's position, and by that time, your car could be located somewhere else (ahead of or behind the place where it was before) within the wave train.

These imaginary quantum zone experiences, which are based loosely on the behavior of electrons, may give you some insight into several aspects of quantum mechanical theory. Together, they are a far-from-complete or even accurate analogy of what really goes on inside the atom—but then again, quantum mechanics by its very nature is abstract. It always resists being compared to ordinary experience. Nonetheless, some of these imaginary experiences should make sense by the time you finish this chapter.

Quantum mechanics is a mathematical description of the behavior of particles far too small to see. Time and time again, it has proved its worth through the success of its predictions. As such, scientists accept its validity. We need not fully understand something (in an experiential sense) to make use of it. In fact, well over half of the world's entire economy is linked to inventions that have stemmed from the applied principles of quantum mechanics.

Let us now investigate, in a more formal manner, the aspects of quantum theory that pertain to tiny bits of matter. (Recall that we have already discussed quantum theory as it pertains to light in Chapter 11.)

## Wave-Particle Duality

Light has a *wave-particle duality*: its behavior incorporates wavelike features and particlelike features. Might not pieces of matter, which we have traditionally thought of as particles, act like waves? If so, then a certain kind of symmetry would encompass both matter and pure energy (for example, light). In other words, both would exhibit wave-particle duality.

## Matter Waves

These thoughts and some of their implications were first expressed by the French physicist Louis de Broglie in his doctoral dissertation, completed in 1923. Mindful of Einstein's conclusions that mass and light energy are equivalent ($E = mc^2$) and that there is an intimate relationship between space and time, de Broglie deduced that particles of matter could indeed be thought of as waves of matter, or **matter waves.** De Broglie predicted that the matter waves associated with a moving particle of matter should have a wavelength of

$$\lambda = h/p = h/mv \qquad \text{Equation 14.1}$$

where $h$ is Planck's constant ($6.63 \times 10^{-34}$ J·s), and $p$ is the particle's momentum (which, as we learned in Chapter 5, is the particle's mass multiplied by its velocity, or $mv$).

Equation 14.1 says that a particle of matter somehow propagates as moving waves with a wavelength that depends on both the mass and the speed of the moving particle. To give ourselves an inkling of what this means, let us calculate the wavelength of a large (definitely macroscopic) "particle," such as a typical adult human, when that person is walking slowly down the street. We will assume the person has a mass of 66.3 kg and saunters along at 1 m/s:

$$\lambda = h/mv$$

$$= (6.63 \times 10^{-34} \text{ J·s})/(66.3 \text{ kg})(1 \text{ m/s})$$

$$= 10^{-35} \text{ m}$$

This wavelength is absurdly tiny, and we could never hope to measure it. Even the wavelength of a typical gamma ray (about $10^{-12}$ m) or the diameter of a proton or a neutron (about $10^{-15}$ m) is enormously large compared to this matter wavelength of $10^{-35}$ m. It is obvious that no wave behavior can be detected in a "particle" as big as a person. However, if we choose something with a much smaller mass—say, an electron moving at a speed of $10^7$ m/s, which is about how fast atomic electrons move—we find that the electron's matter wave wavelength is comparable to the size of an atom. If they exist, matter waves must be crucial to the understanding of atomic structure.

The meaning of matter waves is difficult to interpret, but the matter (or de Broglie) wavelengh $\lambda$ of a particle can be somewhat concretely described as being related to the size of the region of influence of the particle. For example, slow-moving neutrons are more likely to be captured by fissionable nuclei (as we mentioned in connection with nuclear fission in Chapter 12) than fast-moving neutrons. The slow neutrons have a longer λ and act as if they were "bigger."

# Matter Waves and the Bohr Atom

It was not clear in the beginning just what matter waves might be, but de Broglie went ahead and applied the matter-as-moving-waves idea to the electrons of Bohr's model of the hydrogen atom (the Bohr model was explained in Chapter 11).

The speed of the electron in each of the Bohr orbits is known. For each orbit the speed is just sufficient to keep the electron orbiting around the proton in the nucleus to which it is attracted. (Similarly, a planet must move around the sun at a particular speed if it is to remain in the same orbit.) The radii of the various Bohr orbits were known as well; they were related to the potential energies of the electron at different distances from the nucleus. Since these orbits are all circles, the distance around the rim of each orbit (circumference) is just $2\pi r$.

Knowing the mass of the electron and its speed in each of the Bohr orbits, de Broglie used Equation 14.1 to calculate the electron's wavelength in each case. The amazing result, schematically illustrated in Figure 14.1, is that the electron's matter waves fit in a standing wave pattern around the circumference of each orbit. In other words, the electron, as a traveling wave, reinforces itself (experiences constructive interference) every time it completes an orbit. One complete wave fits into the first Bohr orbit ($n = 1$, the ground state). Two complete waves fit into the second Bohr orbit ($n = 2$, the "first excited state"). Three complete waves fit into the $n = 3$ orbit, and so on. By this way of thinking, the electron cannot exist in any orbit other than those described by $n = 1$, 2, 3, and so on, because those orbits would involve a mismatch of wave "crests" and "troughs" as the wave repeats its path around the orbit. These situations of destructive interference make it impossible for an electron to persist in intermediate orbits.

The electron "waves" sketched in Figure 14.1 resemble the fundamental and harmonic frequencies of a vibrating string (see Figure 8.21), except that they are wrapped around circles. It seems that in the hydrogen atom, the electron can exist in only one mode of vibration at a time; and it may jump from one discrete mode (or orbit) to another, absorbing or giving up energy as it does.

The perfect fit of hypothetical matter waves around the circumference of electron orbits was either an incredible coincidence or a true description of what really happens. Soon, a host of experiments were demonstrating that matter waves are indeed real. When a beam of X-rays is sent through a thin metal foil, many are diffracted (bent) when passing through the crystalline lattice of metal atoms (see Photo 14.1). The X-rays emerge in a geometrical pattern somewhat similar to that produced by light when it reflects from a diffraction grating or the pitted surface of a compact disc. These X-rays are simply exhibiting normal wave behavior. When electrons, having a matter wavelength equal to that of the X-rays, are used in the same experiments, the diffraction effect is exactly the same. Somehow, the

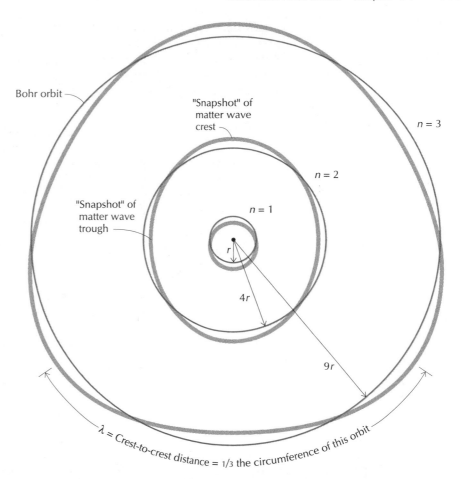

Bohr orbit

"Snapshot" of matter wave crest

*n* = 3

*n* = 2

*n* = 1

"Snapshot" of matter wave trough

*r*

4*r*

9*r*

λ = Crest-to-crest distance = 1/3 the circumference of this orbit

**FIGURE 14.1** The three inner-most Bohr orbits, drawn to scale here, have radii (and circumferences) proportional to $n^2$. One complete matter wave, having one "crest" and one "trough," fits around the *n* = 1 orbit and closes in on itself, thereby constructively interfering with itself. (Notice that there is just one crest and one trough around the *n* = 1 orbit.) Two complete matter waves fit around the *n* = 2 orbit; three complete matter waves fit around the *n* = 3 orbit; and so on. The larger the orbit, the slower the electron moves and the longer is its wavelength according to Equation 14.1.

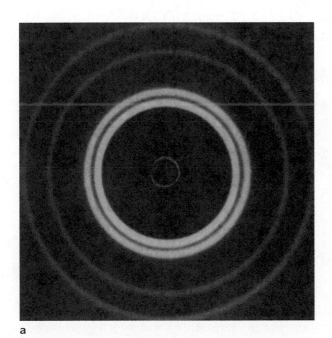

a

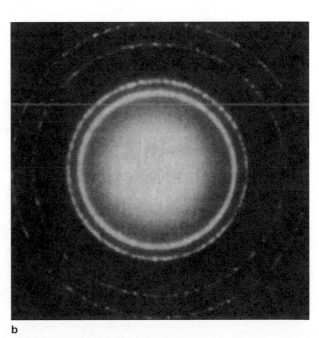

b

**PHOTO 14.1** When X-ray photons (a) and electrons (b) pass through the same thin film of metal, the diffraction patterns they produce are almost identical.

electrons act just like waves. The wave nature of speeding electrons has been put to good use in *electron microscopes* (Box 2.1), which use beams of electrons instead of light waves. The faster the electrons move in a beam, the shorter their wavelength is and the better they are able to produce high-resolution images of tiny objects.

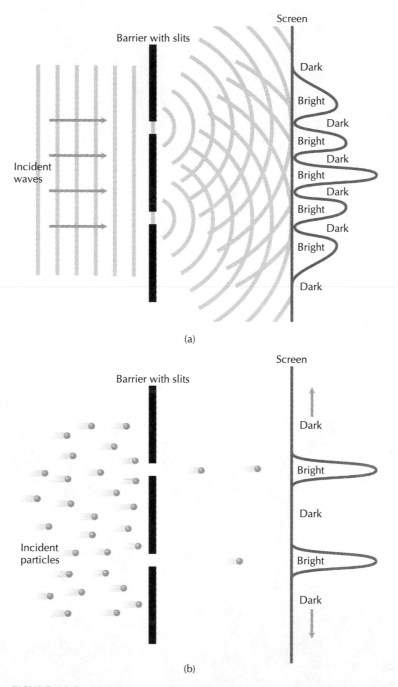

**FIGURE 14.2** (a) When waves diffract through two slits and interfere, a pattern of many interference fringes appears on the screen. The waves must be of relatively large wavelength compared with the slit width in order to diffract significantly. (b) Particles, assuming that they have a wavelength too small to diffract significantly, have trajectories that carry them to only two spots on the screen.

# Two-Slit Experiments

There is another even simpler way to check whether electrons are indeed waves with a requisite matter wavelength. Electrons of the same speed (and therefore the same wavelength) may exhibit interference patterns when projected through two closely spaced holes or parallel slits onto a phosphor screen. These patterns, consisting of alternating bright and dark bands (which represent where many and few electrons have struck the screen, respectively), are like those of interfering mechanical or electromagnetic waves (see Figure 8.20). It is worth going over this two-source interference phenomenon in some detail, since it illustrates several things about wave-particle duality. (Refer to Figure 14.2 as we go along.)

Incident light waves striking a barrier perforated by two small slits [Figure 14.2(a)] diffract around the edges of both slits, producing, in effect, two new sources of light of the same wavelength. These waves combine and interfere, producing an interference fringe pattern.

If a beam of very fast electrons, having negligibly small wavelengths, is directed toward the same slits and a phosphor screen [Figure 14.2(b)], very little diffraction occurs, and only two bright spots appear on the screen. If, however, the electrons are slowed down so that their wavelengths are the same as those of the light waves, an interference fringe pattern appears that is identical to the one produced by the light waves. Clearly, both photons and electrons share the same wave behaviors when their wavelengths are equal.

It is also clear, in a "wave" way of thinking, that in order for waves to interfere, they must combine and occupy the same space *at the same time*. In the quantum mechanical world, this time constraint is not necessary. Even when electrons (or photons, it matters not) are doled out one by one on their parallel trajectories toward the two slits, an interference pattern builds on the screen. Such a slow-forming fringe pattern can be captured by time-lapse photography, and we see the result of such an experiment in Photo 14.2.

One interpretation of this strange behavior is that a single particle such as an electron or a photon can "split" itself, like a river current bending around a rock, and interfere with itself after recombining on the far side of the slit. But what happens if you cover up one slit and allow individual photons or electrons to pass only through the other slit? The result is that each acts just like an *undivided* particle that passes through the single slit and

after 100 electrons

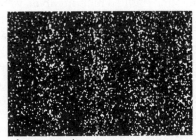

after 3000 electrons

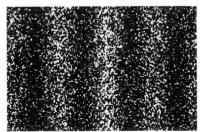

after 70,000 electrons

**PHOTO 14.2**   These photographs show the pattern made by same-speed electrons after various numbers pass through two slits. The same interference pattern would have built up on the photographic plates if photons (of the same wavelength) had been used instead of electrons.

arrives at the screen at a single instant of time. The wavelike interference fringes disappear as well.

In this and every other quantum experiment performed so far, small things act exactly like particles when subjected to experiments designed to test for particle properties. These same things behave exactly like waves when wave properties are tested for. According to one way of thinking, somehow the small entity "knows" which of its two behaviors are being tested for, and it behaves accordingly.

Our discussion at this point could easily launch into realms of speculation and philosophy. Suffice it to say that at the quantum level, tiny units of matter or energy (photons) are neither solely particles nor solely waves. They are something else entirely, exhibiting both wave and particle properties. Unfortunately, we cannot conjure up in our minds a visual picture of wave-particle duality, nor can we find for it analogies that are very satisfactory. In this sense, though, the character of the quantum world is not unique. How would you attempt to describe, say, the color of ultraviolet light? What analogy from everyday life can possibly explain everything about the wave nature of light, let alone its particle nature?

## The Uncertainty Principle

**Uncertainty Principle**
It is impossible to know both the exact position and the exact momentum of a particle at the same time.

If, on the quantum level, matter is significantly smeared out into waves, then there must be a substantial amount of imprecision involved in measurements made at very small scales. Werner Heisenberg, one of the many contributors to the theory of quantum mechanics, proposed in 1927 that it is impossible to know both the exact position and the exact momentum of a particle at the same time. Heisenberg's mathematical statement of this postulate, which is called the uncertainty principle, can be written as

$$(\Delta p)(\Delta x) \approx h \qquad \text{Equation 14.2}$$

where $p$ is the momentum of a particle, $x$ represents its position, and $h$ is Planck's constant. The quantity $\Delta p$ represents the uncertainty in the measurement of momentum, and $\Delta x$ represents the uncertainty in the measurement of position. This state of affairs was alluded to in the imaginary tale that opened this chapter: The traffic officer could not issue two tickets for two infractions at the same time—one based on speeding (which involves knowing the car's momentum or speed precisely) and the other based on running a red light (which involves measuring the car's position precisely).

Two things are important to keep in mind about the uncertainty principle. One is that a relatively good knowledge of a particle's momentum (or essentially, velocity, as long as the particle's mass is constant) implies a relatively poor knowledge of the particle's position at the same time. The opposite is true as well. A relatively good knowledge of a particle's position implies a relatively poor knowledge of the particle's momentum at the same time.

The second important thing to realize is that the two uncertainties, multiplied together, are approximately equal to $h$, which has an incredibly small value. Thus, any trade-offs between $\Delta p$ and $\Delta x$ are significant only for particles with tiny momenta and tiny sizes. These are the same particles (particularly atoms and subatomic particles) for which all the other quantum effects are significant. Because the value of $h$ is very small, quantum uncertainty has almost no effect on macroscopic bodies. Only if $h$ (which, like $c$, the speed of light, is a fundamental constant of nature) were very much larger than it is would it be impossible to simultaneously get good measurements of both the speed and the position of a car nearing an intersection.

To get some feeling for how the uncertainty principle works for small particles, imagine trying to simultaneously locate and measure the momentum of some very small particle. To "see" this particle, you must send at least one other "particle" (a photon of light or an electron) that will bounce off the particle you are trying to observe and then come back to you. Photons carry momentum, and certainly moving electrons do, too. So when the probing photon or electron returns to you, it has already interacted with and disturbed the particle you are trying to observe in a way that cannot be known unless you succeed in locating the particle a second time. The very act of observing the particle disturbed it in an unpredictable way. The returning electron or photon told you the particle's position at the moment of encounter, but the particle skittered away at an unknown speed and direction after the encounter, and that made it impossible for you to determine the particle's momentum.

Some interesting analogies have been drawn between the study of social behavior and the uncertainty principle. For example, suppose that an anthropologist parachutes into a jungle village whose inhabitants have had no previous contact with the modern world. By the very act of the anthropologist's godlike arrival and her interaction with the villagers, the culture might very well change in an unpredictable way. It is therefore quite likely that the anthropologist will never be able to formulate an objective and accurate description of all aspects of that culture as it was before she arrived.

Despite such comparisons, it is important to realize that only at the quantum level does the act of making a physical measurement *significantly* affect the object being measured. Even with this restriction in mind, however, the philosophical implications of the uncertainty principle are enough to destroy Laplace's notion of the universe running like a perfect machine. If it is, in principle, impossible to measure all the properties of a particle or system of particles at any one instant of time, then it is impossible to make confident predictions concerning all later instants of time.

## Quantum Numbers

In the older (pre–quantum mechanics, or classical physics) way of picturing the atom, an electron moving about the nucleus is described as having energy, momentum, and a certain orbit (path of travel). In the newer quantum mechanical way of picturing the atom, the uncertainty principle comes into play, and our knowledge of an electron's position and movement lacks the crispness and apparent certainty with which we view the larger world. In the quantum world, an atomic electron still has energy and

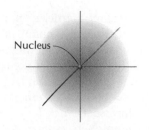

Nucleus

**FIGURE 14.3** The probability of finding a ground-state ($n = 1$) electron around the nucleus of a hydrogen atom is mapped by the shading in this illustration. The electron "probability cloud" is spherical and is centered on the nucleus.

momentum, but these quantities can change only in discrete steps; they are quantized just as everything else is in the realm of the very small.

We may be uncertain about exactly where an atomic electron is and where it is headed at any instant of time, but we can certainly identify trends, or *probabilities* associated with its position. These probabilities were given a firm mathematical description in the work of Erwin Schrödinger, another giant in the development of quantum mechanics. *Schrödinger's wave equation* (which because of its mathematical complexity will remain unwritten here) describes particles, such as electrons, as waves occupying all three dimensions of space and (if need be) changing with time.

If an electron in an atom remains in a stationary state, then its wave-like nature does not change with time, and Schrödinger's wave equation can be used to describe the probability of its position in three-dimensional space. For an electron in the ground state ($n = 1$) of a hydrogen atom, the electron ends up "looking" as if it occupied a fuzzy sphere around the nucleus (see Figure 14.3). In other words, rather than following the simplistic circular path of a Bohr model orbit, the electron follows some unknown random path at any instant. Over time, however, the pattern of probability of the electron's position (called the electron's *probability cloud*) spreads out radially from the nucleus so that the negative charge carried by the electron occupies a fuzzy sphere. If you choose to measure the exact position of the electron at some instant, it will most likely be found at or near the same distance from the nucleus as the radius of the $n = 1$ Bohr orbit. (Of course, then you don't know its momentum very well. You are uncertain of its speed and direction.) There is also some chance that your instantaneous positional measurement might locate the electron inside or outside the most probable radius.

The number $n$, which increases in integer steps in a hydrogen atom (as well as in all other atoms), denotes different quantized energy states in which an electron may be found when it is bound to an atom. It can be thought of as a code number symbolizing those various energy states. It is also used as a multiplier in formulas that determine just how much overall potential energy the electron has in each state and how far the electron is (on average) from the nucleus in each state.

It turns out that electrons, when bound in some way to atoms, possess properties apart from overall potential energy, which is symbolized by $n$. These other properties, which involve the electron's angular momentum, are only three in number. (Recall that angular momentum is a quantity associated with bodies that revolve or spin. An atomic electron can be thought of as capable of doing both.) Each of these other three properties is also quantized. Thus, a *complete* description of an atomic electron's state of being (its "quantum state") can be symbolized by a total of four code numbers, called **quantum numbers.** We will now briefly describe the meaning of each of these quantum numbers.

## The Four Quantum Numbers

The quantum number $n$, which is called the *principal quantum number*, describes the electron's average distance from the nucleus as well as its overall potential energy. Within a given quantum state $n$, however, the exact potential energy of the electron may vary somewhat owing to different possible values of its angular momentum. As we have seen, $n$ takes on integer values: $n = 1, 2, 3, \ldots$. The overall effect of $n$ is on the size of the electron probability cloud; the larger $n$ is, the farther out from the nucleus an electron is most likely to be found.

The quantum number $l$, which is called the *orbital angular momentum quantum number*, describes the magnitude of the electron's angular momentum as it moves about the nucleus. It, too, is quantized. (Macroscopically, this would be like a bicycle wheel that could spin only at certain speeds with no intermediate speeds possible.) The possible angular momentum magnitudes are limited in a way that depends on the electron's $n$ state; this is codified by the rule $l = 0, 1, 2, \ldots, n - 1$. In other words, if an electron is in the $n = 1$ state, then $l = 0$ is the only possible angular momentum the electron can have—which is to say that its angular momentum is zero. That particular state is represented by the electron occupying a spherical fuzzy cloud, as in Figure 14.3. An $l = 0$ electron has not stopped moving; rather, it moves around the nucleus in all possible directions, so that its angular momentum sums to zero. In the $n = 2$ state, however, an electron may assume either of two possible angular momentum magnitude states, $l = 0$ (zero angular momentum) and $l = 1$ (a nonzero angular momentum). The overall effect of $l$ is on the shape of the electron probability cloud. An electron with $l = 0$ forms a spherical cloud, an electron with $l = 1$ forms a dumbbell-shaped cloud, and electrons with higher $l$ values form probability clouds of greater complexity.

Recall from Chapter 5 (and Figure 5.11 specifically) that angular momentum is a vector that has direction as well as magnitude. The quantum number $m_l$, the *orbital magnetic quantum number*, determines the quantized direction of the electron's angular momentum as it moves about the nucleus. (Macroscopically, this would be like the axis of a spinning bicycle wheel that could be oriented only in specific directions.) The term *magnetic* is used to describe $m_l$ because an electron with an angular momentum, like a bunch of electrons circulating in a coil, acts just like a tiny bar magnet. The possible values of $m_l$ depend on $l$ in this manner: $-l, \ldots, -2, -1, 0, +1, +2, \ldots, +l$. If $l = 0$, then $m_l = 0$ only. If $l = 1$, then the possible values of $m_l$ are $-1$, $0$, and $+1$. If $l = 2$, then there are five possible values of $m_l$, namely, $-2, -1, 0, +1, +2$. The overall effect of $m_l$ is the orientation of the shape of the electron probability cloud. The spherical ($l = 0$) shape has only one orientation. The dumbbell ($l = 1$) shape can be oriented in three different ways. The shapes associated with $l = 2$ can be oriented in five different ways.

The fourth and final quantum number $m_s$, the *spin magnetic quantum number*, is a description of an atomic electron's spin. Regardless of what other quantum numbers a given electron has, it either spins in one particular direction or in the opposite direction. These spin states are denoted $m_s = +1/2$ and $m_s = -1/2$. There is some question as to whether or not electrons actually spin, because we cannot see them spinning. But that is really beside the point, because even nonmoving electrons act like very weak bar magnets that either point "up" or "down." This is exactly how they would behave if they were spheres of negative charge spinning either one way or the other.

# The Exclusion Principle

As we saw in Chapter 11, the electron in a hydrogen atom has a strong tendency to fall into its lowest energy state. For the quantum numbers $n$, $l$, $m_l$, and $m_s$ this state could either be 1, 0, 0, +1/2 or 1, 0, 0, −1/2. In the excited states ($n > 1$), the possibilities for different sets of quantum numbers increase, because the possible values of $m_l$ depend on the value of $l$, and because the values of $l$ depend, in turn, on $n$. In Figure 14.4, we see that there are 2 unique sets of the four quantum numbers for $n = 1$, and 8 unique sets for $n = 2$. Similarly, there are 18 unique sets for $n = 3$, 32 for $n = 4$, and even more for higher $n$'s.

So far, we have limited our discussion of atomic electrons to the single electron in a hydrogen atom. What about all the other atoms that make up the chemical elements heavier than hydrogen. They have more than one electron. (Uranium, in fact, has 92!) Where do these many electrons fit? Do they all crowd into the same $n = 1$ probability cloud associated with the hydrogen atom's ground state? Or do they somehow "stack up" in some sort of organized pattern? The chemical behavior of the various elements, which is widely divergent in most instances, suggests that electrons arrange themselves differently in the atoms belonging to different elements.

In 1925 Wolfgang Pauli found the key to electron arrangements in multi-electron atoms. His exclusion principle says that no two electrons in the same atom can have the same set of four quantum numbers. This means that there is a *unique set* of attributes for every electron in a given atom; no two electrons in a given atom can behave in exactly the same way. So how do the electrons arrange themselves? Just remember that each of the electrons in an undisturbed (unexcited) atom strives always toward a state of lower energy. Let us proceed, starting with the hydrogen atom.

Hydrogen's single electron has either a 1, 0, 0, +1/2 or a 1, 0, 0, −1/2 quantum state. (There is a minuscule energy difference between the two states, so both are about equally possible.)

An atom of helium normally has two electrons. Both occupy the $n = 1$ state, or (to use a bit of jargon) *shell*. Their quantum states are described by 1, 0, 0, +1/2 and 1, 0, 0, −1/2. This shell is spherical and fuzzy, and the two electrons that occupy it never collide with each other, as particles would do, because both electrons are waves spread out around the helium atom's nucleus. The two electrons can share exactly the same space because (in accordance with the exclusion principle) there is something different about them: They have opposite spins, or $m_s$ values.

An atom of the next element, lithium, has three electrons. The inner two electrons have quantum states just like those of helium, but the third electron cannot fit into the $n = 1$ shell because that would violate the exclusion principle. We say that for lithium (as for helium) the $n = 1$ shell is filled to capacity, or "closed." Instead, the third electron occupies the next shell out, $n = 2$, and its quantum state is normally 2, 0, 0, +1/2 or 2, 0, 0, −1/2 (these two states have the lowest energy among the eight possible states that electrons can have within the $n = 2$ shell). Earlier, we learned that the $n = 2$ state is an excited state for a hydrogen atom. In a lithium atom, the third electron is not in an excited state because its energy is

---

**Exclusion Principle**

No two electrons in the same atom can have the same set of four quantum numbers.

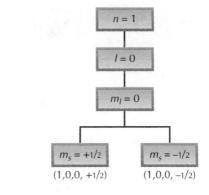

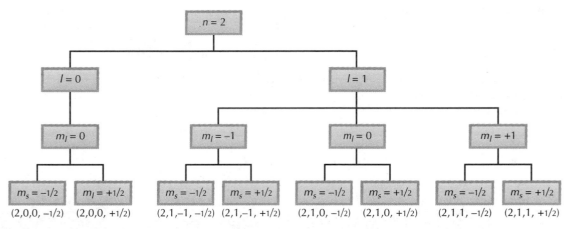

**FIGURE 14.4**  From the rules for quantum numbers, only two unique sets of quantum numbers can be found for $n = 1$. Only eight unique sets of quantum numbers can be found for $n = 2$.

minimized; it cannot lose energy by jumping to the $n = 1$ shell because that shell is closed. If an unexcited lithium atom is disturbed in some way, however, any of its electrons may jump to some higher energy state, in which case the atom is "excited."

Beryllium, the fourth element, has its inner two electrons in the $n = 1$ shell. Its two other electrons in the $n = 2$ shell have the quantum numbers 2, 0, 0, $+1/2$ and 2, 0, 0, $-1/2$.

The fifth through tenth elements are boron, carbon, nitrogen, oxygen, fluorine, and neon. Their electrons progressively fill up the $n = 2$ shell, which, as we saw in Figure 14.4, can contain a maximum of 8 electrons. Neon, with 10 electrons in all, has 2 in its $n = 1$ shell and 8 in its $n = 2$ shell. Both shells in the neon atom are closed, so the next element, sodium, with one more than 10 electrons, must have its 11th electron in the next shell out, $n = 3$.

If it seems to you that the electrons in more and more complex atoms arrange themselves in an orderly and repetitious manner, you are right. This repetition is reflected in the *periodic table of the elements,* the subject of the next chapter.

---

CHAPTER 14
# Summary

Light and matter exhibit both wavelike properties and particlelike properties. This behavior is called wave-particle duality.

A particle of matter has a wavelength that increases with decreasing momentum (mass and speed). Because the wavelength of a particle also depends on Planck's constant $h$, which has an extremely small value, only particles with very small momenta exhibit measurable wavelike properties. In the atomic realm, the wavelike properties of particles are significant.

The wavelike properties of tiny particles can be tested by subjecting them to the same experiments that have been used to confirm the wave nature of light. Particles and photons behave similarly when their wavelengths are the same.

It is impossible to know both the exact position and the exact momentum of a particle at the same time. The relative uncertainty involved is large only for small particles.

Because uncertainty is so prevalent in the realm of the atom, atomic electrons are best described as behaving like standing waves positioned around the nucleus. Each electron in an atom can be completely described by a set of four quantum numbers. Furthermore, each electron in an atom differs from all the others in the sense that no two electrons in the same atom can have exactly the same set of four quantum numbers.

---

CHAPTER 14
# Questions

## Multiple Choice

1. According to quantum mechanics,
   a) only tiny, slow-moving particles exhibit wave behavior
   b) only charged particles exhibit wave behavior
   c) no particle ever exhibits wave behavior
   d) all particles exhibit wave behavior

2. According to quantum mechanics, a hydrogen atom's electron
   a) circles the nucleus like a planet circles the sun
   b) radiates outward from the nucleus like a wave
   c) fills a shell-like cloud of probable positions, centered on the nucleus
   d) spends much of its time inside the nucleus

3. Describing a moving body as a series of moving "matter waves" is legitimate because
   a) it is based upon common sense
   b) matter waves are visible in microscopes
   c) any kind of matter can be made to vibrate

   d) theory agrees with experiment and observations of behavior in the atomic realm

4. What experiment or experiments demonstrate that tiny particles, such as electrons, have a wave nature?
   a) the two-slit experiment
   b) the behavior of electrons in an electron microscope
   c) X-ray diffraction experiments involving thin metal foils
   d) all of the above

5. According to the uncertainty principle, it is impossible to precisely know a particle's
   a) mass
   b) position
   c) momentum
   d) position and momentum at the same time

6. Heisenberg's uncertainty principle is most significant for
   a) particles with very small mass or momentum
   b) particles with very large mass or momentum

c) particles that do not spin

d) charged particles

7. The fact that no two electrons in an atom can have the same set of quantum numbers is known as

a) the exclusion principle

b) Archimedes' principle

c) the principle of equivalence

d) the uncertainty principle

8. Each electron in an atom can be characterized (and distinguished from other electrons in the same atom) by

a) its mass and charge

b) its energy and spin

c) its set of four quantum numbers

d) none of the above

9. Two electrons can peacefully coexist in the same $n = 1$ shell of a helium atom because

a) the electrons are so small that they can never collide

b) their matter waves never intersect

c) they have opposite spins

d) they both have the same angular momentum

## Questions

1. If all bodies, small or large, are supposed to possess a wave nature and have a matter wavelength that can be calculated, why can't we visibly observe the wave effects that a moving person has?

2. What does the term *wave-particle duality* mean?

3. When an electron is moving around the nucleus of a hydrogen atom in the $n = 9$ state, what is the ratio between the circumference of that Bohr orbit and the matter wavelength of that electron?

4. In your own words, explain what each of the four quantum numbers represents.

## Problems

1. This chapter stated that an electron moving at $10^7$ m/s has a matter wavelength that is about equivalent to the size of an atom. Confirm that this is true: Calculate the electron's wavelength and compare it with the size of a typical atom (which is pictured in the cosmic zoom feature of Chapter 2).

2. Using the rules for quantum numbers, confirm that as many as 18 electrons will fit into the $n = 3$ shell of an atom. Do the same for the $n = 4$ shell.

3. Assume that the constant $h$ is 1 J·s (instead of its actual value, which is very small). If you walk along at a speed of 1 m/s, and you know that figure with an uncertainty of 10% (or 0.1 m/s), then how uncertain are you about your position at that same instant of time?

## Questions for Thought

1. When electrons strike the phosphor screen of a TV picture tube, they are moving very much faster than $10^7$ m/s. Do these electrons act more like waves than particles, or more like particles than waves?

2. What happens to the wavelength of a particle of matter when its speed (or momentum) drops to zero? Does this seem to say something related to the uncertainty principle?

3. When I was in high school in the 1960s, the textbook used in my physics course was *Physics, An Exact Science.* How appropriate do you think this title is in view of what you have read in this chapter? In what sense (if any) is physics, or any science, "exact"?

4. It is sometimes risky to use an analogy from another academic discipline when attempting to explain a difficult concept in a scientific field like physics. An analogy is never completely equivalent to the phenomenon it sheds light on. Do you think the anthropological analogy (that is, the anthropologist who studies a primitive culture in the jungle and therefore disturbs the culture) used for the uncertainty principle is appropriate?

5. Quantum mechanics, as it is currently understood, calls into question what *understanding* really means. How would you define the term *understanding*?

## Answers to Multiple-Choice Questions

1. d   2. c   3. d   4. d   5. d   6. a

7. a   8. c   9. c

# The Periodic Table

*Grandma's cookies were never quite like this. As the cookies grow in size, left to right and row by row, they fall into color patterns. The chemical elements are organized in a similar way on the periodic table. Their atomic numbers increase one by one, left to right and row by row. Elements appearing in any given column have at least one, and often several, properties in common.*

**The periodic table**
How the chemical
elements are organized

**Elements organized by
electron structure**
How the atoms of physics relate
to the elements of chemistry

**Elements surveyed**
Unity within the diverse
groups of the periodic table

In 1869, a Russian chemistry professor by the name of Dimitri Mendeleev was busy compiling information on the properties of the chemical elements for inclusion in a new textbook. After laying out the elements in the order of their *atomic weight* or *atomic mass* (a measure of the average mass of their atoms), Mendeleev noticed a striking trend. With increasing atomic mass, the physical and chemical properties of the various elements changed one after the other, but in a periodic fashion. The term *periodic* means "recurring" or "repetitious." In other words, Mendeleev noticed that there were repetitions of series of elements with similar sequences of properties.

For example, Mendeleev recognized regularity in the chemical and physical behavior of elements we now call *alkali metals* (lithium, sodium, potassium, and others). Each of these elements, when in its elemental state, is a soft metal that reacts vigorously with water. When lithium, sodium, and potassium are chemically combined with the element chlorine, the resulting ionic compounds are similar as well: Each consists of a one-to-one ratio of alkali metal ions to chloride ions (LiCl, NaCl, and KCl). In the arrangement of increasing atomic masses, the element beryllium follows

**PHOTO 15.1** The intrepid and versatile Dimitri Mendeleev (1834–1907) making a balloon ascension to study a solar eclipse in 1887.

immediately after lithium, magnesium immediately follows sodium, and calcium immediately follows potassium. These other elements, Mendeleev noted, fall into a different, and again fairly homogeneous, category called the *alkaline earths*. When alkaline earth elements are combined with chlorine, the result is a group of similar compounds having a one-to-two ratio of alkaline earth ions to chloride ions ($BeCl_2$, $MgCl_2$, $CaCl_2$, and so on).

Mendeleev also recognized that some elements with quite dissimilar physical properties could still have enough in common chemically to justify classifying them in the same group. One example is carbon and silicon. Both can react with oxygen to form compounds that are *dioxides*: one atom of carbon or silicon for every two atoms of oxygen. Although the formulas are similar ($CO_2$ and $SiO_2$), the two compounds differ markedly in their physical characteristics. At normal temperature and pressure, carbon dioxide is a heavier-than-air gas, and silicon dioxide (silica or quartz) is a hard, crystalline solid.

To illustrate the periodic properties of the elements effectively, Mendeleev arranged all of the 63 elements known in his time into a table having 12 horizontal rows. Elements having similar properties were made to fall into 8 vertical columns of the table. Mendeleev's table was the precursor of the modern **periodic table of the elements.** Just by glancing at the table, a chemist could recognize connections and similarities of behavior—some that were previously unrecognized—among the known elements. Even more important, the table allowed a glimpse into the future. There were certain obvious gaps in the periodic pattern, places where (if the table had any real meaning) new elements should lie. Confident in the efficacy of his table, Mendeleev boldly predicted the atomic masses and other properties of several of the missing elements. Within a few years, many of the missing elements were isolated from natural sources, and their measured properties agreed quite closely with Mendeleev's predictions.

One such gap was filled in Mendeleev's lifetime by the discovery of the element designated germanium. Germanium, like silicon, is a semiconductor. Mendeleev's predicted properties for this missing element, which he called eka-silicon because of its presumed place on the table right under silicon, match the observed properties of germanium almost exactly.

Also during Mendeleev's lifetime, elements belonging to an entirely new category, the *noble gases* (helium, neon, argon, and more), were discovered. Although Mendeleev had not predicted them, the noble gases fit perfectly as an addition to his periodic scheme. After Mendeleev's death in 1907, the table grew further with the discoveries of elements heavier than uranium (element number 92). Today, the periodic table remains one of the most useful tools in science.

# Elements Organized by Electron Structure

The structure of the original Mendeleev-style table has evolved over the years into one that is even more informative. The modern periodic table consists of 7 horizontal rows, or periods, rather than Mendeleev's 12 rows. These periods contain varying numbers of elements, from few elements in the top periods to many elements in the bottom periods. According to the

most common depictions of the table, there are 18 vertical columns, or groups, each containing elements with similar properties.

At this point, you should glance at the modern, complete periodic table, Figure 15.7 or the back endpaper of this book, to get a feeling for how the elements are arranged. (You will understand the table better after finishing this chapter.) The modern periodic table has its elements sequentially arranged by *atomic number* (number of protons), rather than by atomic mass—a subtle but important distinction we shall touch upon later in this chapter.

## Regularity Recognized

There is nothing particularly mysterious about how the modern periodic table works. For example, examine the pattern of numbers on the following table.

| 1 | 2 | 3 | 4 | 5 | 6 | 7 | 8 | 9 | 10 |
|---|---|---|---|---|---|---|---|---|----|
| 11 | 12 | 13 | 14 | 15 | 16 | 17 | 18 | 19 | 20 |
| 21 | 22 | 23 | 24 | 25 | 26 | 27 | 28 | 29 | 30 |

This purely mathematical "periodic table of whole numbers" echoes one key feature of the chemical periodic table. The numbers increase by one for each box to the right. New rows (or periods) start after every ten numbers, because all numbers in a given vertical column have something in common—they share the same last digit. When listed in this way, whole numbers fall into a periodic and predictable pattern.

In the modern periodic table, the whole numbers, which are written above each chemical symbol, denote the atomic number of each element. An element's atomic number refers specifically to the number of protons in the nucleus of each atom of that element, and it also refers to the number of electrons normally present in the space surrounding the nucleus (this is true as long as the atom is neither ionized nor chemically combined with another atom or atoms).

Unfortunately (for chemistry students and professors alike), the periodic table of the elements is not simple. The table—indeed, chemistry itself—can be compared to the English language, in that they both have a maddening trait: Each is well organized yet riddled to some degree with inconsistencies or anomalies. "Proper" English has certain rules and consistencies, but they are often broken. (Recall the spelling rule "i before e, except after c.") Anomalies in chemistry are associated with deviations from simple patterns; they may be difficult to understand, but they are not random or inexplicable.

Periodic trends in chemistry, as demonstrated by the periodic table, are imperfectly regular. For one thing, the periods (horizontal rows) do not include the same number of elements. The seven periods from top to bottom on the table have 2, 8, 8, 18, 18, 32, and 32 elements, respectively, although the last period currently contains an empty region of massive elements yet to be discovered or synthesized. Elements in any given group

(vertical column) of the table exhibit strong similarities in chemical behavior. Across each period (moving horizontally across the table), the elements exhibit chemical and physical properties that change in some regular way.

Despite the obvious, regular trends associated with it, though, the periodic table conceals many surprises. (These surprises give researchers something to do and make the study of chemistry fascinating and challenging for the astute student.) Each chemical element, no matter how similar to other elements of the same group, is unique in at least some of its properties. These properties are sought after by chemists who wish to combine the various elements into novel compounds and by physicists and engineers who look for advantageous properties associated with different materials.

Until the 20th century, no one had any clear idea about why the various elements exhibit a periodicity of properties when listed according to atomic mass and why the periodic table should work as an organizing tool. Quantum mechanics (Chapter 14) changed that. By the 1920s, quantum mechanics was providing a satisfactory, albeit abstract and mathematical, description of how electrons behave when they are bound to the nuclei of atoms. It turns out that *the patterns made by the outermost atomic electrons are strongly correlated to the physical properties and chemical behaviors of the elements made of those atoms.*

In the next few pages, we shall build the beginnings of the periodic table, not by means of organizing the elements by their properties, as Mendeleev did it, but with the modern knowledge that the electrons of elements of greater and greater atomic number accumulate in ever-larger and more complex shell-like structures. Later in this chapter, we shall examine several of the major groups of elements, discuss some similarities among them, and tell about some of their uses.

In a nutshell, the periodic table explains *how* elements are related to each other, and the quantum mechanical model of the atom goes a long way toward explaining *why* elements behave as they do. In Chapter 5, we encountered something similar with regard to the paths of the planets around the sun. Kepler's laws (Box 5.2) explain *how* the planets orbit the sun, and Newton's laws explain *why* planets move in the way they do in terms of a more general understanding of motion and gravity.

## Electron Shell Models

From the discussion of the four electron quantum numbers in Chapter 14, let us summarize some important facts about the patterns made by an atom's electrons. (For review, you may find it helpful to refer to Table 15.1, which lists the four quantum numbers associated with each electron of an atom.)

1. A series of **shells** (shell-like structures) represented by the quantum number $n$, having values of 1, 2, 3, ..., exists around the nucleus. With increasing $n$, the shells increase in size, and in general, any electrons in them have higher energies.

2. Shells can contain electrons that orbit the nucleus in various spatial patterns that chemists call **orbitals.** The shapes of these orbitals are rather simple for low values of $n$ and more complex for higher values of $n$. Within

| **TABLE 15.1** Atomic Electron Quantum Numbers Summarized | Name and Symbol | Values | Meaning |
| --- | --- | --- | --- |
| | Principal quantum number, $n$ | 1, 2, 3, ... | The electron's overall energy (and the overall size of its region of influence around the nucleus) |
| | Orbital angular momentum quantum number, $l$ | 0, 1, 2, ..., $(n-1)$ | The electron's angular momentum magnitude as it moves around the nucleus (relates to the shape of an electron's orbital) |
| | Orbital magnetic quantum number, $m_l$ | 0, $\pm 1$, $\pm 2$, ..., $\pm l$ | The electron's angular momentum direction as it moves around the nucleus (relates to the orientation of an electron's orbital) |
| | Spin magnetic quantum number, $m_s$ | $\pm 1/2$ | Either of two possible states of electron spin |

any given shell, all orbitals that have the same value of $l$ (and similar shapes, as well) are grouped into a **subshell.** The number of subshells in a shell corresponds to the possible values of the quantum number $l$ for that shell. There is one subshell for the $n = 1$ shell ($l = 0$ only). There are two subshells for the $n = 2$ shell ($l = 0$ and $l = 1$), in accordance with the rules for quantum numbers. Three subshells lie in the $n = 3$ shell, and so on. To get some sense of what orbitals look like, examine Figure 15.1. The overall

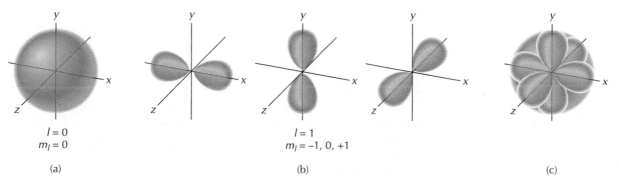

$l = 0$
$m_l = 0$

(a)

$l = 1$
$m_l = -1, 0, +1$

(b)

(c)

**FIGURE 15.1**   In these diagrams, electron orbitals are shown as three-dimensional surfaces centered around the nucleus (omitted for clarity), which lies at the center of the $x$-, $y$-, and $z$-axes. In each drawing, the volume enclosed by the surface represents the region where the electrons (as many as two electrons may inhabit the same orbital) have a 90% probability of being present. (a) The single orbital associated with the $l = 0$ subshell is spherical. (b) The three orbitals associated with the $l = 1$ subshell have a dumbbell shape, and their axes are mutually perpendicular. Notice that there is a "node," or a place where electrons in an $l = 1$ orbital spend no time. This strange behavior can be understood by realizing that the patterns made by electrons in an orbital are analogous to standing waves in a string (for example, note the nodes visible in the vibrating strings of Figure 8.21 and Photo 8.8). A node in a vibrating string represents a point of zero amplitude, whereas a node in an orbital represents a point at which the probability of finding an electron is zero. (c) In a multielectron atom, many orbitals overlap and form a complex pattern. In an atom like neon, all orbitals of the subshells present are filled. The combined pattern looks much like a sphere. To a first approximation, atoms like helium, neon, and others with filled subshells act like tiny spheres.

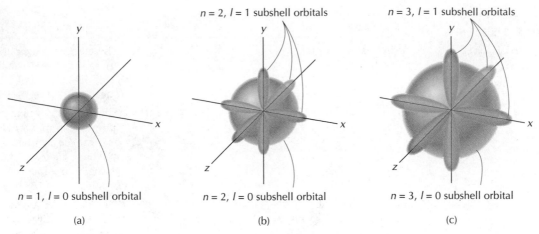

$n = 2, l = 1$ subshell orbitals

$n = 3, l = 1$ subshell orbitals

$n = 1, l = 0$ subshell orbital

$n = 2, l = 0$ subshell orbital

$n = 3, l = 0$ subshell orbital

(a)

(b)

(c)

**FIGURE 15.2** (a) The $n = 1$ shell has only one subshell ($l = 0$), with a single spherical orbital. (b) The $n = 2$ shell has two subshells, one with a single spherical orbital and the other ($l = 1$) with three dumbbell-shaped orbitals. (c) The $n = 3$ shell has three subshells, only two of which ($l = 0$ and $l = 1$) are shown here. The third subshell ($l = 2$), consisting of five orbitals of even more complex shape, is omitted for clarity.

Note from the entire sequence that, in general, orbitals are larger with increasing $n$. Also keep in mind that orbitals can contain no more than two electrons, and that many orbitals can overlap each other in the manner of many waves overlapping. In this illustration, the dumbell-shaped ($l = 1$) orbitals are distorted in shape—made thinner than they really are—so that we can visualize their orientations.

shapes of orbitals in the $l = 0$ and $l = 1$ subshells are shown. With increasing $l$, the orbitals become more and more complex in shape.

3. The number of orbitals in a given subshell corresponds to the possible values of the quantum number $m_l$ for that subshell. The possible values of $m_l$ go up in steps of two for each higher value of $l$. Thus, all $l = 0$ subshells have only one orbital; all $l = 1$ subshells have three orbitals; all $l = 2$ subshells have five orbitals; and so on. Within a given shell, electrons in the orbitals of subshells with higher $l$ values have somewhat higher energies (strictly speaking, this is true only for atoms with more than one electron).

4. In undisturbed atoms of higher atomic number (and having more and more electrons), the added electrons tend to occupy orbitals in which they have the least energy possible.

5. No two electrons in the same atom can have the same set of quantum numbers (exclusion principle). Because of this feature and the one given in item 4, the electrons of an atom fill up the lower-energy orbitals of an atom first and then begin filling up the higher-energy orbitals.

6. Each orbital can contain no more than two electrons. If two electrons are present in an orbital, then they have opposite spins ($m_s = \pm 1/2$).

The diagrams of Figure 15.2 illustrate how the electron orbitals in the first three shells are arranged around the nucleus. The $n = 1$ shell has one subshell, because only one value of $l$ is possible for $n = 1$. The $n = 2$ shell has two subshells; $n = 3$ has three subshells; and so on.

In undisturbed atoms having more and more electrons, the electrons fill the $n = 1$ shell first, then the $n = 2$ shell (its $l = 0$ subshell first, then its $l = 1$ subshell). After the first two shells are filled, the $n = 3$ shell starts filling, first its $l = 0$ subshell, then its $l = 1$ subshell. At this point, the nice, simple pattern of orbital electron filling runs into an anomaly. A pattern still exists for the more massive elements, though it is a bit more complicated. Any further addition of electrons leads not to filling up the $l = 2$ subshell in the $n = 3$ shell, as you might guess, but rather to filling the lowest-energy orbital of the $n = 4$ shell, its $l = 0$ subshell. Only after this subshell is filled do electrons start filling the $n = 3$, $l = 2$ subshell. At last, the $n=3$ shell is filled.

Next, electrons fill the $n = 4$, $l = 1$ subshell. After that, yet another anomaly occurs. The $n = 5$, $l = 0$ subshell starts filling next. This happens after only the first two of the four subshells of the $n = 4$ shell have been filled.

Why do electrons in the $n = 3$ and higher shells, seeking the lowest possible energy, leapfrog to the next shell and "prematurely" start filling the inner orbitals of the higher shells? They behave this way primarily because the energy differences between shells get less and less with higher $n$—so much so that the energy differences between subshells of the same shell can overwhelm the energy differences between the shells themselves. Figure 15.3 shows the relative energy of all the subshells belonging to the $n = 1$ through $n = 4$ shells, along with the lowest-energy subshell of the $n = 5$ shell. We will use this figure to construct periods 1 through 4 of the periodic table, and to start period 5. As we start this process, refer to Figure 15.4, which is an abbreviated version of the modern periodic table containing the first 20 elements. Beyond element 20 (calcium), we will switch to the complete periodic table, Figure 15.7.

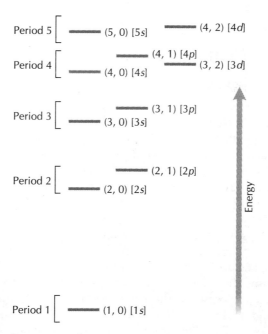

**FIGURE 15.3** The relative energy of electrons in various subshells of multielectron atoms is graphically illustrated here. Two notations are given for each subshell: in paren-theses are the quantum numbers $n$ and $l$, and in brackets is the standard notation used by chemists to denote orbitals.

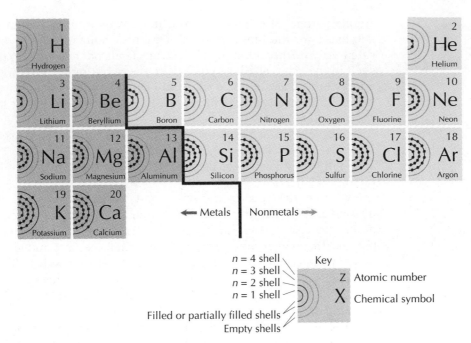

**FIGURE 15.4** In this abbreviated periodic table, elements of increasing atomic number are arranged according to their electron structure. Two electrons fill the $n = 1$ shell, eight electrons fill the $n = 2$ shell, and eight electrons fill the $n = 3$ shell. (Technically, at argon the eight electrons of the $n = 3$ shell fill up only the inner two subshells.) The physical and chemical properties of the elements shown are strongly correlated to the number of electrons in the outermost shell of each element's atoms. Thus, the vertical-column families, or *groups,* consist of similar elements. (*Note:* The half circles used to illustrate the shells associated with quantum number $n$ are purely schematic. The dots represent only the number of electrons belonging in each shell.)

# Constructing the Periodic Table

**FIGURE 15.5** In a hydrogen molecule, the two shared electrons form a new egg-shaped *molecular orbital* in which both electrons stay mostly inside the boundary shown here and yet spend more time between the two hydrogen nuclei than elsewhere.

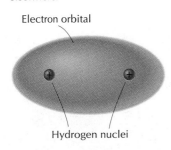

Electron orbital

Hydrogen nuclei

Hydrogen (a one-proton atom) in the ground state has one electron in its $n = 1$ shell. Helium (a two-proton atom) has two electrons (of opposite spins) that fill its $n = 1$ shell to completion. Helium has a *closed-shell* electron configuration that is very stable. Helium is inert to chemical activity; its atoms will not bond to anything. A hydrogen atom, on the other hand, has an outer shell that is partially filled with electrons: an *open-shell* electron configuration. A hydrogen atom can bond with another hydrogen atom (see Figure 15.5) or with a different atom, thus achieving a kind of pseudo-closed-shell configuration.

The mathematical solutions of quantum mechanics can explain how the two electrons of a hydrogen molecule can exist in the same common space around both nuclei (two protons), with both electrons spending more time between the protons than elsewhere. The concentrated negative charge between the protons holds the two nuclei, and therefore the entire molecule, together. This type of electron-sharing arrangement, as we learned in Chapter 9, is a covalent bond. Similar solutions of quantum mechanical equations show how covalent bonds can exist between a hydrogen atom and any other atom capable of sharing electrons. For example, in the ammonia molecule ($NH_3$), six electrons are shared in three different

molecular orbitals: three electrons from the single N atom and one each from the three H atoms. In the methane molecule ($CH_4$), eight electrons are shared in four different molecular orbitals: four electrons from the single C atom and one each from the four H atoms.

As noted above, helium's electrons fill its $n = 1$ shell to completion. Compared with hydrogen's nucleus of one proton, helium's nucleus has two protons (a nuclear charge of $+2$) and twice the attractive force on each of its two electrons. However, the repulsion of helium's two electrons from each other is an even stronger influence. The repulsion pushes the electrons farther out from the nucleus. Thus, the helium atom is somewhat larger than the hydrogen atom. But, helium atoms are still quite small, especially when compared with the diatomic or triatomic molecules of gases such as nitrogen ($N_2$), oxygen ($O_2$), water vapor ($H_2O$), and carbon dioxide ($CO_2$). This explains why helium balloons deflate more quickly than air-filled balloons. The stretched rubber of a balloon is microscopically porous; nearly any gas particle can escape, but small ones escape faster.

After hydrogen and helium, the next element is lithium, whose atoms have three protons and three electrons. Lithium's third electron tends to settle into the innermost subshell of the $n = 2$ shell (the $l = 0$ subshell, which has a single spherical orbital). That single electron "feels" an attractive force from the $+3$ charge of the nucleus, but it is also pushed away to a somewhat lesser degree by the two electrons occupying the smaller spherical orbital of the $n = 1$ shell underneath it. As a result of this *electron shielding* by the inner electrons, lithium's outer electron is weakly bound to the nucleus. Its orbit is relatively large, and not much energy is required to pull the electron away from the nucleus—that is, to ionize the atom. The energy required to do this is called the atom's *ionization energy*. We say that lithium's size is relatively large, that its ionization energy is relatively small, and that it is chemically very active. Lithium has a valence of $+1$ (see Table 9.1), because only one of its electrons can be lost to another atom during a chemical reaction. When it loses an electron, the lithium atom becomes a lithium *ion,* with a closed-shell electron configuration like that of helium. It is much less chemically active in that state.

Hydrogen has a unique electron configuration. It is the only element with only one electron in its outer (and only) shell and lacking one electron to fill its outer shell at the same time. Still, hydrogen has more in common with lithium than any of the next few elements of increasing atomic number. This is the first instance of repetition on the periodic table. Lithium, therefore, is placed below hydrogen on the periodic table, and helium lies somewhere to the right of hydrogen in period 1. Lithium begins period 2.

Beryllium, the next element, of atomic number 4, has four electrons; the outermost two electrons are in the $n = 2$ shell. These electrons tend to fill up the single spherical orbital belonging to the $l = 0$ subshell. But as it happens, the slight energy available through thermal activity at room temperature is enough to "promote" one of those electrons into the next subshell ($l = 1$). This "promotion" process is none other than excitation by collision—like that of the collisional excitations we discussed in Chapter 11 in connection with the hydrogen spectrum, only less energetic. The promoted electron fills one of the three dumbbell-shaped orbitals in the $l = 1$ subshell. Because the two electrons in the beryllium atom's $n = 2$ shell no longer "close" its inner subshell, beryllium is capable of chemical bonding. Furthermore, these two electrons, held in the grip of a nucleus with a $+4$ positive charge (compared with lithium's $+3$ nuclear charge), orbit in a tighter pattern. The beryllium atom is somewhat smaller than the lithium atom, and its ionization energy is somewhat greater. Beryllium normally has a valence of $+2$, with two electrons to donate to other atoms, and it is somewhat less reactive than lithium.

The next element is boron, with 5 electrons. Again, ordinary thermal motion is sufficient to distribute its three outer-shell electrons into three different orbitals. On the periodic table, Boron lies just right of the stairstep line indicating the division between metals and nonmetals. (Technically, boron is a semimetal, with characteristics intermediate between metals and nonmetals.) Boron has a fair ability to attract electrons from other atoms and lure them into covalent bonds. *Electronegativity* is the term chemists use to identify this property. The more electronegative an atom is, the more readily it can accept or share electrons from other atoms. Electronegativity generally increases to the right along the periods of the periodic table.

Carbon follows next, with 6 electrons. With the influx of a little energy, the outer four of these electrons typically distribute themselves one each into the four different orbitals of the $n = 2$ shell. In that state, carbon participates in covalent bonding by sharing its four electrons with other atoms. Each covalent bond involves an overlap between the orbital of one atom and the orbital of another atom. Each atom contributes a single electron, so two electrons are shared in each molecular orbital (recall the simplest example of this, the diatomic hydrogen molecule in Figure 15.5).

In Window C of the cosmic zoom in Chapter 2, we saw a visualization of the electron orbitals of a carbon atom when it is bonded in some simple fashion to other atoms. Those orbitals, which are not the same as those of lone carbon atoms, are "hybridized" versions of the original $l = 0$ and $l = 1$ orbitals. Basically, the four pairs of electrons in carbon's four hybridized orbitals repel each other and thus tend to spread as far apart from each other as possible. This phenomenon is reflected in the crystal lattice of diamond (depicted in Figure 9.20), where carbon atoms are bonded in a tetrahedral pattern.

As we approach the end of period 2, we note that as a general rule, *when two or more orbitals of equal energy are available* (as they are when the orbitals belong to the same subshell), *the electrons do not pair up in an orbital until each orbital has one.* These electrons may be compared to strangers boarding a bus: Individuals (at least in the United States) tend not to double up on the seats until each seat has at least one passenger.

Beyond the element carbon are nitrogen, oxygen, fluorine, and neon, with 7, 8, 9, and 10 electrons, respectively. Several general trends (along with some exceptions) continue in this sequence of elements. The outer electrons, acted on by stronger and stronger positive nuclear charges, cluster ever closer to the nucleus and are held with a tighter electrostatic grip. The atomic size decreases and the ionization energy increases. The outer orbitals with unpaired electrons gain electrons one by one in the sequence until every orbital in the $n = 2$ shell is filled.

The nitrogen atom, lacking three electrons and having three unpaired electrons in its outer orbitals, readily links to other atoms by means of three covalent bonds. Oxygen, with two unpaired electrons, has a valence of $-2$ and can bond by accepting two electrons or by sharing two electron pairs. Oxygen has large electronegativity (second only to fluorine) and therefore a great ability to oxidize (take electrons from) other elements.

Fluorine, the most electronegative atom of all, easily establishes a single bond with almost any other atom on the periodic table. Its valence is $-1$. When paired with lithium, the electron lost from the lithium atom spends so much time near the fluorine atom that the bond is considered ionic. (We should expect this from our study of chemical bonding in Chapter 9. Lithium, being a metal, bonds ionically with fluorine, a nonmetal.) In this state both atoms are ions—the lithium ion having a $+1$ charge and the fluoride ion having a $-1$ charge.

After fluorine, which is the most ferociously reactive nonmetal of all, period 2 ends with a crashing bore, the inert element neon. At last, with neon, the $n = 2$ shell is closed. According to the exclusion principle, no more electrons can fit into it. Neon is similar to helium, so helium is properly placed above neon on the periodic table. The closed-shell electron configurations of both helium and neon preclude any further additions (or even sharing) of electrons, so both have zero electronegativity and both are chemically inert.

We are now ready to build period 3. This is easy because the sequence of $l = 0$ and $l = 1$ subshell filling by electrons is the same as for period 2! Sodium lies below lithium, magnesium lies below beryllium, aluminum lies below boron, and so on, because the electron configurations in the outer shells of each pair are similar. The electrons are now filling the larger $n = 3$ shell, so the atoms in period 3 are somewhat bigger overall, but the same systematic trends we saw in period 2 hold. Two of these systematic trends—atomic radius and ionization energy—are illustrated for our abbreviated version of the periodic table in Figure 15.6.

Period 3 ends with argon, which closely resembles helium and neon in properties and structure, except that the subshells $l = 0$ and $l = 1$ in shell $n = 3$ are closed at this point, not the whole shell. The $l = 2$ subshell has yet to be filled.

Next comes period 4 and the first anomaly referred to earlier. In period 4, the innermost ($l = 0$) subshell of the $n = 4$ shell quickly fills with just two electrons, at the element calcium. (This is exactly where our abbrevi-

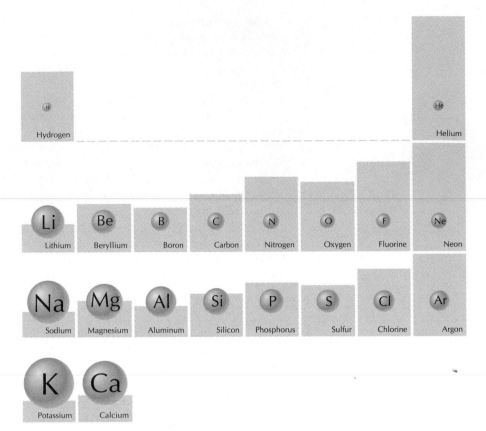

**FIGURE 15.6** This graphic display of an abbreviated periodic table illustrates two characteristics of atoms of the first 20 elements: The relative atomic radius is proportional to the size of each atom as drawn here. The relative ionization energy is proportional to the height of each rectangle. Notice how atoms get smaller across any given period (from the left side to the right side) and generally larger for successive periods (from the top to the bottom). Ionization energy generally increases across a period and decreases in successive periods. These trends continue through the parts of the complete periodic table not included here.

ated table, Figure 15.4, ends. From now on refer to the complete periodic table, Figure 15.7.) After calcium, and for the next 10 elements in the series, the electrons go back to filling the third and outermost ($l = 2$) subshell of the $n = 3$ shell. There are five orbitals in this subshell, and thus 10 electrons are needed to fill them. The ten elements in this intermediate series are known as *transition metals*. All have different properties because of the differing numbers of electrons in their outer ($l = 2$) subshell, but they also have many similarities because each has two electrons in an outer ($n = 4$) shell that is far from being filled.

The transition metals (of period 4 and those of the next periods, 5 and 6) include the familiar elements iron (Fe), copper (Cu), silver (Ag), gold (Au), and mercury (Hg). Transition metals may easily lose anywhere from no electrons (gold, for example) to as many as several. Iron can lose either two or three electrons (its valence is either $+2$ or $+3$); and in its different states of ionization, iron has quite different properties and uses. The ferrous ($Fe^{2+}$) ion is a component of hemoglobin, which transports oxygen in the blood; the ferric ($Fe^{3+}$) ion is a component of rust.

The transition metals of period 4 end with zinc. At the next element, gallium, the filling of the $n = 4$ shell resumes with a sequence just like that of the latter part of period 3, which is right above it on the periodic table. The last element in period 4 is krypton (no relation to the fictitious "kryptonite" of *Superman* fame). Like argon, krypton achieves a configuration of two closed subshells, not a closed-shell configuration. Like helium, neon, and argon, which belong to the same group, krypton strongly shuns any form of chemical combination.

# Periodic Table of the Elements

| | I (1) | II (2) | | | | | | | | | | | | III (13) | IV (14) | V (15) | VI (16) | VII (17) | VIII (18) |
|---|---|---|---|---|---|---|---|---|---|---|---|---|---|---|---|---|---|---|---|
| 1 | 1 **H** 1.008 | | | | | | | | | | | | | | | | | | 2 **He** 4.003 |
| 2 | 3 **Li** 6.941 | 4 **Be** 9.012 | | | | | | | | | | | | 5 **B** 10.81 | 6 **C** 12.01 | 7 **N** 14.01 | 8 **O** 16.00 | 9 **F** 19.00 | 10 **Ne** 20.18 |
| 3 | 11 **Na** 22.99 | 12 **Mg** 24.31 | (3) | (4) | (5) | (6) | (7) | (8) | (9) | (10) | (11) | (12) | | 13 **Al** 26.98 | 14 **Si** 28.09 | 15 **P** 30.97 | 16 **S** 32.06 | 17 **Cl** 35.45 | 18 **Ar** 39.95 |
| 4 | 19 **K** 39.10 | 20 **Ca** 40.08 | 21 **Sc** 44.96 | 22 **Ti** 47.90 | 23 **V** 50.94 | 24 **Cr** 52.00 | 25 **Mn** 54.94 | 26 **Fe** 55.85 | 27 **Co** 58.93 | 28 **Ni** 58.7 | 29 **Cu** 63.55 | 30 **Zn** 65.38 | | 31 **Ga** 69.72 | 32 **Ge** 72.59 | 33 **As** 74.92 | 34 **Se** 78.96 | 35 **Br** 79.90 | 36 **Kr** 83.80 |
| 5 | 37 **Rb** 85.47 | 38 **Sr** 87.62 | 39 **Y** 88.91 | 40 **Zr** 91.22 | 41 **Nb** 92.91 | 42 **Mo** 95.94 | 43 **Tc** 98.91 | 44 **Ru** 101.1 | 45 **Rh** 102.9 | 46 **Pd** 106.4 | 47 **Ag** 107.9 | 48 **Cd** 112.4 | | 49 **In** 114.8 | 50 **Sn** 118.7 | 51 **Sb** 121.8 | 52 **Te** 127.6 | 53 **I** 126.9 | 54 **Xe** 131.3 |
| 6 | 55 **Cs** 132.9 | 56 **Ba** 137.3 | 57* **La** 138.9 | 72 **Hf** 178.5 | 73 **Ta** 180.9 | 74 **W** 183.9 | 75 **Re** 186.2 | 76 **Os** 190.2 | 77 **Ir** 192.2 | 78 **Pt** 195.1 | 79 **Au** 197 | 80 **Hg** 200.6 | | 81 **Tl** 204.4 | 82 **Pb** 207.2 | 83 **Bi** 209.0 | 84 **Po** (210) | 85 **At** (210) | 86 **Rn** (222) |
| 7 | 87 **Fr** (223) | 88 **Ra** 226.0 | 89** **Ac** (227) | 104 (261) | 105 (262) | 106 (263) | 107 (262) | 108 (265) | 109 (266) | 110 | 111 | | | | | | | | |

Atomic number → 11
Symbol → **Na**
Atomic mass → 22.99

Atomic masses are based on carbon-12. Numbers in parentheses are mass numbers of most stable or best known isotopes of radioactive elements.

**Transition Metals**

Period

## Inner Transition Metals

| *Lanthanide series 6 | 58 **Ce** 140.1 | 59 **Pr** 140.9 | 60 **Nd** 144.2 | 61 **Pm** (145) | 62 **Sm** 150.4 | 63 **Eu** 152.0 | 64 **Gd** 157.3 | 65 **Tb** 158.9 | 66 **Dy** 162.5 | 67 **Ho** 164.9 | 68 **Er** 167.3 | 69 **Tm** 168.9 | 70 **Yb** 173.0 | 71 **Lu** 175.0 |
|---|---|---|---|---|---|---|---|---|---|---|---|---|---|---|
| **Actinide series 7 | 90 **Th** 232.0 | 91 **Pa** 231.0 | 92 **U** 238.0 | 93 **Np** 237.0 | 94 **Pu** (244) | 95 **Am** (243) | 96 **Cm** (247) | 97 **Bk** (247) | 98 **Cf** (251) | 99 **Es** (252) | 100 **Fm** (257) | 101 **Md** (258) | 102 **No** (259) | 103 **Lr** (260) |

☐ Metals   ☐ Semimetals (metalloids)   ☐ Nonmetals   ☐ Noble gases

**FIGURE 15.7**  The most common depiction of the periodic table arranges elements into 7 periods and 18 groups. An older, simplified Roman-numeral designation may be used to group the elements that are not transition metals.

Period 5 repeats the pattern of period 4, only now it is shells $n = 5$ and $n = 4$ that are filling up. Again, there is a series of transition metals in between the two-element groups on the left and the six-element groups on the right.

In period 6, another anomaly occurs after barium (Ba). Things get very complicated as different subshells in different, higher shells start to fill at different rates. A simpler pattern is restored at thallium (Tl), when the $l = 1$ subshell of shell $n = 6$ starts filling in the same way that other $l = 1$ subshells filled previously in the periods directly above period 6.

There are so many elements in period 6 between barium and thallium that a subset of them, called the *lanthanides* (an "inner" set of transition metals all very similar to the element lanthanum, atomic number 57), are shown as a "footnote" in the table pictured in Figure 15.7. A similar thing happens in period 7, with another set of inner transition metals, the *actinides,* occupying the space of the element actinium, atomic number 89. Several of the lanthanides (or *rare earth* elements) find use as phosphors in the manufacture of television screens and computer monitors. The actinides, including thorium and uranium, are more useful for their nuclear properties, as we have noted in Chapter 12, than for their chemical properties. The heavier actinides, in particular, have unstable nuclei. Most have been artificially synthesized in small quantities through nuclear reactions.

Currently, the periodic table ends in period 7 near element 111, though efforts continue to make still heavier elements in the hope that there is an "island of nuclear stability" somewhere ahead on the series. Names have been proposed for the known elements beyond 103, but none has been made official at the time of this writing.

By now it should be clear why elements on the periodic table are listed in order of atomic number rather than in order of atomic mass (as in Mendeleev's scheme). *The chemical properties of an element depend on outer-electron structure,* and that structure depends on the number of electrons present. In atoms (which are uncharged) the number of electrons present equals the atomic number. Atomic mass, on the other hand, depends on the total mass of the protons *and* neutrons in the nucleus. Most of the elements have more than one common isotope, and it is the average mass of different isotopes (as they occur in nature) that determines the atomic mass listed for a given element. Average atomic mass *generally* increases with atomic number among the elements, but there are several exceptions. Argon, for example, has a greater atomic mass than potassium, yet its obvious place on the table (on the basis of both its behavior and its electron structure) is before potassium.

We shall now survey some of the categories and groups of elements in the periodic table, and we will comment on the properties and uses of some of them.

# Elements Surveyed

For the 18 elements appearing in each of the periods 4 and 5, the groups of the periodic table can be labeled 1 through 18 from left to right. An older usage is to label the 8 groups that appear in periods 2 and 3 with roman numerals I through VIII. Both designations appear on our complete periodic table in Figure 15.7.

Before we survey some of these groups, let us look into the distinction between metals, nonmetals, and the semimetals (metalloids)—the "fence-sitting" elements near the stairstep border on the periodic table. *Metals* are malleable (subject to being bent or shaped to some degree as a solid), are ductile (capable of being stretched to some extent as a solid), are opaque to light, and possess a metallic luster (shiny appearance). As noted in previous chapters, metals are good conductors of both thermal and electric energy. Under normal temperatures and pressures, the *nonmetal* elements are either gaseous and transparent under normal conditions or brittle solids (bromine is a liquid, however). Nonmetals are good thermal and electric insulators. As noted in Chapter 9, metal atoms easily give up electrons, and nonmetal atoms readily accept extra electrons or share electrons. The *semimetals (metalloids)* have properties intermediate between those of metals and nonmetals. Silicon and germanium are fashioned into semiconductors for use in all sorts of electronic devices including computers (see Box 10.2). Germanium, with properties more metallic than those of silicon, can be used to make faster electronic circuits. Silicon chips prevail, however, largely because silicon is abundant in nature and germanium is rare and expensive.

# Group I: The Alkali Metals

Imagine spreading some lithium, sodium, or potassium metal on your toast in the morning. These and the other *alkali metals* are soft enough to be cut with a knife (rather like pats of butter served on ice). Each of these elements is also dangerously reactive, especially when exposed to water, so you would not want to even touch the stuff, let alone expose it to any moisture in a loaf of bread. The alkali metals are so reactive that their bare surfaces corrode almost instantly from the oxygen in the air. Thus, in nature they are found in ionic compounds (or as ions dissolved in water) but never in an elemental state. Chemical activity among the alkali metals increases as one moves downward through the periods (lithium, sodium, potassium, and so on).

Alkali metals tend to lose one electron and form ions with a single positive charge ($Li^+$, $Na^+$, $K^+$). Substantial concentrations of sodium and potassium ions dissolved in the body's fluids are essential to the proper functioning of the body's cells. We lose certain amounts of both ions when

**PHOTO 15.3**  Shown here are three alkali metals in their elemental state: lithium rods and lumps of potassium and sodium.

**PHOTO 15.4**  Fireworks occur when potassium reacts with water, because hydrogen gas, evolved from the reaction, catches fire.

we perspire heavily (the salty taste of perspiration is due mainly to sodium); thus, we must replace them in the food or drinks we consume. Most diets contain plenty of sodium chloride but not always enough compounds that contain potassium. Bananas are an excellent source of potassium. Alkali metal ions also play an important role in the chemistry of the minerals that make up rocks. Many of the common silicate minerals have alkali metal ions bound up amid their crystal lattices.

Hydrogen, though certainly not a metal under normal conditions, lies above the alkali metals on the periodic table primarily because it has just one electron in its outermost (and only) shell. When extremely compressed, however, hydrogen acts like a liquid metal. The deep interior regions of the planets Jupiter and Saturn are composed mostly of hydrogen in this liquid metallic state.

## Group II: The Alkaline Earths

The *alkaline earth* elements are metals that, like the alkali metals, tend to form compounds easily. In compounds or when dissolved, these elements exist as ions that have lost two electrons ($Mg^{2+}$, $Ca^{2+}$, and so on). The word *earth* in their name refers to the tendency of their compounds to not melt or change when exposed to fire. Compounds of the alkaline earths comprise many common minerals, such as calcite, or calcium carbonate ($CaCO_3$), magnesite ($MgCO_3$), and gypsum ($CaSO_4 \cdot 2H_2O$). Like the alkali metals, the alkaline earths are not found in the elemental (metallic) state in nature, but they can be extracted from compounds by means of endothermic reactions. Fine threads of magnesium metal, which quickly react with oxygen and evolve light when energized by electricity, were commonly used for camera flashbulbs up until about two decades ago.

Calcium and magnesium ions are important biologically. Calcium ions build bones, and the magnesium ion is the key component in chlorophyll—the molecule in green plants that helps transform sunlight and carbon dioxide into water and sugar.

## Groups III and IV

Groups III and IV contain elements (such as carbon, aluminum, and silicon) that are at the forefront of modern applications and high-technology research. High-tech does not necessarily mean uncommon, however. Silicon and aluminum are the second and third most abundant elements in Earth's crust. They, along with oxygen (the most abundant element in the crust), form the bulk of the silica and silicate minerals we discussed in Chapter 9. The great abundance of silicon and its metalloid properties makes it the material of choice for semiconductor circuits (refer to Box 10.2).

Aluminum in the elemental state is a soft, lightweight, silvery metal with little strength. When mixed with small amounts of copper, magnesium, zinc, or other elements, however, the resulting aluminum alloys exhibit nearly all the properties metals are valued for. These alloys can be bent or cast into nearly any shape and rolled into thick plates or into thin foils. The more rigid aluminum alloys, which have a large strength-to-mass ratio, are used in everything from lawn chair frames to aircraft frames. Aluminum (as well as copper) is commonly used for wiring, and aluminum's excellent reflecting characteristics make it the material of choice for metallic mirror coatings. Aluminum never occurs in its elemental state in nature, though it can be extracted economically from an ore called *bauxite*. The recycling of aluminum scrap metal saves over 90% of the energy required to make aluminum from bauxite.

Clay, which consists of aluminum silicate minerals, can be heated to very high temperatures to form extremely hard, water-resistant, and heat-resistant ceramic materials. Some modern ceramics, which may be partly composed of various transition metals, are harder than any metal or metallic alloy. Yet the brittleness of ceramic materials remains a problem. Effort is currently being devoted to improving ceramics so that they can be used as more effective or less expensive substitutes for metals and their alloys.

Carbon, as we discussed in Chapter 9, is the central element of organic chemistry and biochemistry, though its inorganic uses are increasing. For example, new lightweight, strong, carbon-fiber structural elements are beginning to appear in products ranging from sporting equipment to bridges. The newly discovered molecular forms of pure carbon called *buckyballs* or *fullerenes* (Box 9.2) are under intense study. These exotic forms of carbon may have many future applications unforeseen at present.

## Group V: The Nitrogen Group

Molecular nitrogen ($N_2$) is the principal constituent of the air we breathe. Bonded to hydrogen in the form of ammonia ($NH_3$), nitrogen is also an essential plant nutrient. Only blue-green algae and certain plants (legumes) have the ability to transform molecular nitrogen into ammonia through their roots, a process called *nitrogen fixation*. Today, much of the

**PHOTO 15.5** Nitrogen- and phosphorus-rich fertilizers, along with imported water, have turned some desert regions into agricultural "breadbaskets."

world's crops are grown with the help of ammonia-based fertilizer, which is artificially synthesized in huge quantities. Remember, though, that any such artificial synthesis requires a huge input of energy. Phosphorus (another nitrogen group element), when combined with oxygen in phosphates, is also an important component of many artificial fertilizers. It would be impossible to adequately feed the current human population without the added boost in food production brought about by the application of plant fertilizers.

## Group VI: The Oxygen Group

Oxygen, in the air and cycled through our bodies and through plant organisms, is a common currency of life. Oxygen makes up nearly half of Earth's crust by mass and also comprises the majority of the mass of water ($H_2O$) in its oceans. Oxygen, released by plants over eons, is what makes our atmosphere unique among the atmospheres of all the other known planets. The oxygen group elements participate in reactions by gaining two electrons or by forming covalent bonds involving two pairs of shared electrons. Note that the oxygen atoms in Earth's crust and oceans exist in chemical combination with other atoms. The oxygen gas we need to breathe is a different substance. Atmospheric oxygen is in a "free" state, existing almost entirely as the diatomic gas $O_2$.

The abundance of oxygen gas in the atmosphere is due almost entirely to photosynthesis taking place in living plant material, as discussed in Chapters 3 and 6. A very small fraction of the $O_2$ in the atmosphere is converted into ozone, $O_3$, which is present in small but significant amounts in the stratosphere, high above Earth's surface.

Sulfur, another element in the oxygen group, is extensively mined and used to produce sulfuric acid, $H_2SO_4$, the most common product of the chemical industry. Sulfuric acid is utilized in the refining of metals and in the manufacture of a wide variety of substances.

## Group VII: The Halogens

The *halogens*, whose name means "salt forming," readily combine with the metals of groups I and II to form compounds known as *salts*. Halogen atoms lack only one electron to complete a shell or a subshell, and as a group, they are the most active nonmetals. Chemical activity among the halogens increases as one moves through the group toward fluorine at the top. At room temperature, fluorine and chlorine are gases, bromine is liquid, and iodine is solid. (Astatine, like many elements in the higher periods, is a rare, radioactive solid.) The halogens are also known for their distinctive colors (see Photo 15.6).

As elements, the halogens consist of diatomic molecules ($F_2$, $Cl_2$, $Br_2$); but in chemical combination, the halogens become ions with a single negative charge ($F^-$, $Cl^-$, $Br^-$). In their elemental state, the halogens are poisonous. Fluorine and chlorine are deadly when inhaled in rather small concentrations. Chlorine gas was the original chemical weapon of World War I. Small amounts of chlorine are used to disinfect drinking water, and both chlorine and bromine find use as disinfectants in swimming pools and spas.

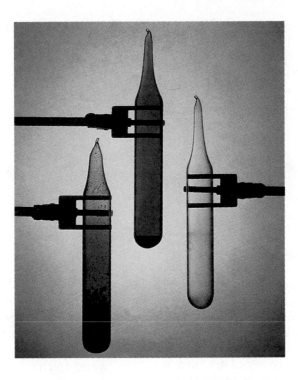

**PHOTO 15.6** Three halogen elements are pictured here. Iodine, a grayish black solid, readily sublimes (evaporates) to a violet-tinted vapor. Bromine, a reddish brown liquid, readily forms a vapor of the same color. Chlorine is a greenish yellow gas at room temperature.

## Group VIII: The Noble (Inert) Gases

Since some of the elements of group VIII have been shown to form compounds with great difficulty, they are not entirely inert and are thus often referred to as *noble gases*. Helium, neon, argon, and krypton are the primary gases used in "neon" decorative lighting. Argon is used to fill incandescent lightbulbs, and xenon is used in gas discharge tubes capable of short-duration bursts of light (electronic-flash units).

Although the noble gases are chemically nondescript, they do exhibit a large variation in at least one physical property: Their densities increase with increasing period. When kept at the same temperature and pressure, helium is about a seventh as dense as air, and xenon is about five times denser than air. You've heard of a "lead balloon"? Fill up a balloon with xenon and that is (with some exaggeration) what you have.

## CHAPTER 15
## Summary

When the chemical elements are listed in order of atomic number, elements with similar physical and chemical properties recur at intervals of 2, 8, 8, 18, 18, 32, and 32 elements. The periodic table arranges these seven clusters of elements into seven horizontal rows, or periods. Elements of similar kind, based on physical and chemical properties, appear in the vertical columns, or groups, of the table.

The arrangement of the elements on the periodic table may be arrived at by recognizing similarities and trends of behavior among the elements. These similarities and trends are a direct consequence of the electron structures of the atoms of each element.

The quantum numbers describing the atomic electron states, along with the exclusion principle, can be used to construct the periodic table. As different shells and subshells of electron structure are filled, in atoms of higher and higher atomic number, new periods (or new sequences within periods) start.

Along with showing groups of similar elements arranged in the vertical columns, the periodic table separates metals and nonmetals, and it shows trends of atomic size and chemical activity within the metal and nonmetal categories. The periodic trends and behavior patterns exhibited throughout the table are not perfectly regular.

## CHAPTER 15
# Questions

## Multiple Choice

1. Mendeleev's periodic table was considered a significant description of patterns and trends among the elements because
   a) elements in the rows of the table exhibited regular patterns of change
   b) elements in a given column exhibited similar properties
   c) empty spaces on the table were later filled in by the subsequent discovery of new elements
   d) of all of the above

2. The form of the periodic table can be explained by the
   a) application of Newtonian mechanics to the atom
   b) application of quantum mechanics to the atom
   c) fact that atoms approximately double in size after every period
   d) fact that atoms of the elements, with few exceptions, become more massive as their atomic number increases

3. The start of each period on the periodic table corresponds to the filling of
   a) a new (formerly empty) shell of quantum number $n$
   b) a new subshell of quantum number $l = 1$ or higher
   c) a new subshell of quantum number $l = 2$
   d) any new subshell

4. The start of the transition metals in periods 4 and 5 corresponds to the filling of
   a) a new (formerly empty) shell of quantum number $n$
   b) a new subshell of quantum number $l = 1$ or higher
   c) a new subshell of quantum number $l = 2$
   d) any new subshell

5. All of the noble gases
   a) are totally incapable of forming compounds
   b) may be used to catalyze (speed up) combustion reactions
   c) exhibit brilliant colors at room temperature
   d) have in common a closed-outer-shell or closed-outer-subshell electron configuration

6. In an undisturbed atom of any kind, the electrons tend to
   a) orbit in a spherical pattern
   b) occupy orbitals in which they have the least energy possible
   c) fall into the nucleus owing to electrostatic attraction
   d) occupy all the orbitals available in the atom until all are filled with one electron

7. Carbon has four bonds (or four electrons to share) because
   a) its shells and subshells contain exactly four electrons
   b) its outer shell has orbitals that contain four electrons
   c) its outermost orbital is shaped like a cloverleaf and has a total of four electrons in it
   d) of none of the above

8. Excluding the noble gases, elements with greater ionization energy tend to be
   a) active metals (such as the alkali metals)
   b) active nonmetals (such as the halogens)
   c) relatively inactive metals
   d) elements of high atomic number

9. Iron, gold, silver, and zinc are all examples of
   a) alkali metals
   b) alkaline earth metals
   c) transition metals
   d) inner transition metals

10. The atoms of group I (the alkali metals)
    a) easily lose the one electron in their outermost shell

b) lack only one electron to complete a shell or a subshell

c) will combine with nearly every other kind of atom

d) are among the smallest of every period

11. Alkaline earth (group II) elements play important roles in mineral and biological chemistry when they are
    a) in atomic form
    b) in ionic form
    c) ionized at high temperatures
    d) all of the above

12. Nitrogen and phosphorus compounds are major constituents of
    a) the air
    b) the majority of the minerals that make up rocks
    c) human bones
    d) plant fertilizers

13. Oxygen and sulfur are related in that
    a) they are in the same period of the periodic table
    b) they are in the same group of the periodic table
    c) each can lose two electrons
    d) each forms compounds only with difficulty

14. The halogen (group VII) atoms
    a) easily lose the one electron in their outermost shell
    b) lack only one electron to complete a shell or a subshell
    c) will combine only with metal atoms
    d) are among the largest of every period

## Questions

1. What did Dimitri Mendeleev discover about the elements when they were listed in order of their atomic masses?

2. In as few words as possible, describe the connection between the behavior of atomic electrons (as described by quantum mechanics) and the physical properties and chemical behavior of the elements.

3. Using the idea of electrons inhabiting orbitals, describe briefly how atoms can participate in covalent bonding.

4. Why is it that the number of elements in each successive period of the periodic table (2, 8, 8,

18, 18, 32, 32) does not completely match the number of electrons that can fit in successive shells of increasing quantum number $n$ (2, 8, 18, 32, and so on) that we saw at the end of the previous chapter?

5. On the basis of its electron structure, why is it that hydrogen is chemically active and helium is not?

6. Why is the lithium atom quite large in size and its ionization energy relatively low?

7. What is implied by the fact that elements such as fluorine and oxygen have large electronegativities?

## Questions for Thought

1. Some periodic tables place hydrogen directly above fluorine. Why do you think they do? (A look at Figures 15.4 and 15.6 may be instructive.)

2. From the trends of the known elements exhibited in the periodic table, where do you think element 119 might be placed, assuming it can ever be synthesized? What are some of the chemical properties it would be expected to have?

3. An ongoing squabble among chemists concerns the proposed names of elements 104 and higher. Some think that it is fine to name them after living physicists and chemists. Others would rather commemorate the dead. What do you think is appropriate?

4. From the brief descriptions of periodic table groups given in the last section of this chapter, how do you think our world (Earth and the life on it) would function if it were lacking the elements of any one of these groups?

5. Why do you think that the noble gases increase in density with increasing period? (*Hint:* Refer to Avogadro's law in Chapter 4.)

## Answers to Multiple-Choice Questions

| | | | | | |
|---|---|---|---|---|---|
| 1. d | 2. b | 3. a | 4. c | 5. d | 6. b |
| 7. b | 8. b | 9. c | 10. a | 11. b | 12. d |
| 13. b | 14. b | | | | |

# Plate Tectonics

*To the slow and relentless beat of geologic time, continents move across the face of Earth, oceans expand and shrink, and mountain ranges rise and fall. These changes, of planet-wide importance, are accompanied by local upheavals such as earthquakes and volcanic eruptions on land and under the sea. The theoretical mechanism explaining all of these changes is known as plate tectonics.*

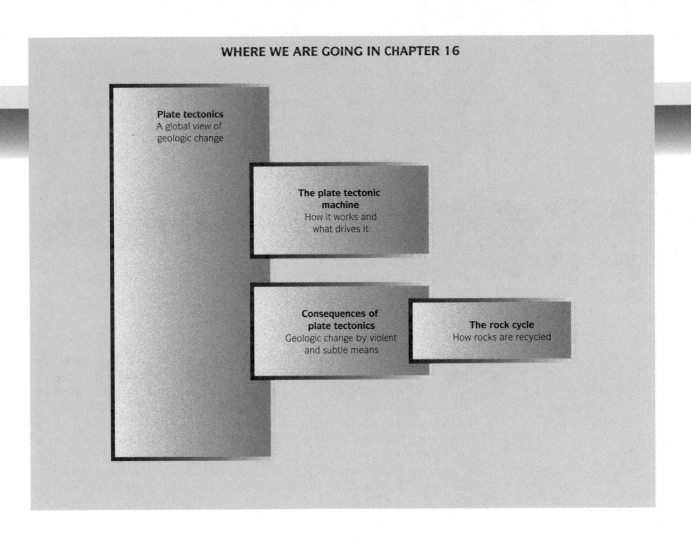

Earth spins once a day and revolves once a year—despite the feeling we have that we and Earth itself cannot possibly be moving. Einstein's experimentally correct theories of relativity posit that time is relative, yet the notion of relative time still boggles the mind. It is quite sensible to think that the more carefully we measure something, the more we learn about it; but quantum mechanics says, "it ain't necessarily so."

Revolutions in thought like these have occurred many times in the history of science. More often than not, these revolutions have involved paradigm shifts in which new, and often counterintuitive, notions led to broader and more enlightened perspectives on the world.

In the earth sciences (particularly geology, oceanography, and paleontology), the most recent revolution in thought took place in the 1960s. Prior to that decade, most scientists accepted a fixed-Earth model in which our world's geography was presumed to remain roughly constant over time. By the end of that decade, the new model of **plate tectonics** had swept away the old model, relegating it to the metaphorical "trash can" (refer to Figure 1.4) filled with old ideas. In the plate tectonic model, continents (or pieces of continents) drift thousands of miles; entire oceans appear and disappear over geologic spans of time. The continents do not simply roam the oceans like aimlessly drifting rafts; they belong to one or more *lithospheric plates*—sections of Earth's thin, rocky shell (the lithosphere) that move relative to each other and interact in several possible ways.

# The Plate Tectonic Machine

The ascent of the plate tectonic model was not simply a matter of one philosophical viewpoint gaining favor over another. It was supported by a great deal of evidence gained by meticulous observations and measurements.

## Historical Origins

The concept of the continents (not plates) moving and rearranging themselves is not new at all. The shapes of the continents, even as depicted on the earliest maps of the world, suggested to some observers how they could be fitted together like rough jigsaw puzzle pieces. Perhaps all were joined in a single large landmass at some time in the distant past. The idea was bolstered in 1912 when Alfred Wegener published his comprehensive theory of *continental drift* (Figure 16.1). Wegener drew on a large amount of geological, paleontological, and climatological data to support his theory. He noted that when the opposing coastlines of continents such as South America and Africa are compared, there is a continuity of structures such as rock strata and mountain ranges. He realized that many of the same extinct life-forms recorded in fossils of great age are common to places on

300 million years ago

200 million years ago

100 million years ago

Present

**FIGURE 16.1** In this modern interpretation of continental drift, we see how Pangaea, an aggregate of earlier continents, came together about 200 million years ago and then broke apart into the present continents.

**FIGURE 16.2** This shaded relief map of the seafloor reveals the midocean ridges—which together constitute the world's largest chain of mountains by far—and deep trenches, many of which lie near the edges of the continents.

widely spaced continents, thus indicating that those continents were close together or in contact at one time.

Wegener also noted evidence of past glaciation in parts of continents that are not now very close to either polar region. He felt that these regions could have drifted from polar latitudes to the warmer areas on the globe where they presently lie. Wegener's supporters later introduced further evidence based on the extensive glaciation that had taken place on parts of South America, Africa, Australia, and India hundreds of millions of years ago. Along with Antarctica, these once-glaciated parts could be made to fit together over the south polar region. A picture was emerging of continents once joined that are now drifting apart from one another.

Although rich in circumstantial evidence, Wegener's continental drift theory lacked an explanation of a plausible mechanism, a driving force, that could set the continents moving and keep them moving. It also failed to explain how continents could either plow through or somehow slide over the oceanic crust, which was known to be solid and rigid. Because of these critical deficiencies, the majority of geologists and other earth scientists (before the 1960s) regarded the continental drift hypothesis as unorthodox and even outlandish.

The seeds of the plate tectonic revolution were planted shortly after World War II as oceanographers began mapping the seafloor and preparing detailed topographic maps. By the 1960s it was clear that the entire globe is girdled by submarine mountain chains, called *midocean ridges* (see Figure 16.2). The most prominent of these, the Mid-Atlantic Ridge, runs down the middle of the Atlantic Ocean, almost exactly halfway between the two American continents and the joined continents of Europe and Africa. The mapping program also revealed long and deep trenches on the ocean floor, often located either near the margins of continents or alongside island chains of volcanic origin. Both the ridges and the trenches are the sites of intense geologic activity, such as earthquakes and volcanic eruptions. It was clear that an inclusive picture of geologic change on Earth would have to include mechanisms that alter the ocean floor as well as the landmasses.

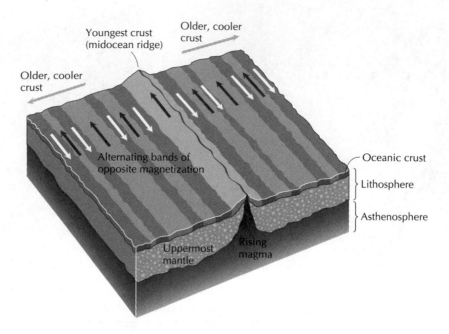

**FIGURE 16.3**   As magma rises up along the axis of the mid-ocean ridges, it solidifies into new oceanic crust and also builds the solid-rock portion of the upper mantle. Taken together, these two sections (the lithosphere) are less than 100 km thick. The rising magma pushes outward on both sides, causing a rift. The lithosphere on both sides of the rift moves outward like a conveyor belt on a softer, underlying layer of the mantle known as the *asthenosphere*.

# Divergent Boundaries

Studies of the seafloor rocks on either side of the midocean ridges revealed several remarkably regular patterns, three of which are detailed here and illustrated in Figure 16.3. Radiometric dating of rocks both near and far from the ridges showed that their ages (the time since they were last molten) increase with increasing distance from the axis. Rocks very near the ridges were found to be not only the youngest of all but also *very* young. These young rocks are also warmer than the rocks found at larger distances from the axes of the ridges.

Another pattern of regularity is related to the orientation of the tiny, elongated grains of magnetic materials (such as *magnetite*), which form when molten rock from Earth's interior (called *magma* whenever it lies below Earth's surface) rises to the seafloor and crystallizes (solidifies) on contact with the cold seawater. Like compass needles, these magnetic grains point either north or south in accordance with the prevailing magnetic field at the time they were "frozen" into the crystallizing rock. As we learned in Chapter 10, Earth's magnetic field reverses its polarity every half million years or so and has been doing that for at least hundreds of millions of years. Detailed magnetic mappings of the seafloor on both sides of any seafloor ridge show a symmetric pattern of alternating stripes of "normal" magnetic orientation and "reversed" magnetic orientation. From this evidence, geologists were compelled to conclude that new seafloor is created at rifts that run along the midocean ridges and moves outward on both sides like a conveyor belt. This phenomenon is known as **seafloor spreading,** and it takes place at these ridges. Because the plates are separating, or diverging, these ridges are called *divergent-plate boundaries*.

In some instances, the rifting associated with seafloor spreading occurs on land. The intensely volcanic island of Iceland lies on the northern end of the Mid-Atlantic Ridge, and part of eastern Africa is the site of rifting that may eventually cause the ocean to invade the continent's interior.

If new seafloor is being created in some areas, it is reasonable that some-where else it is being destroyed—or somehow recycled back into Earth's inte-rior. These places, it turns out, include the deep oceanic trenches shown in Figure 16.2. They are also the sites of volcanic activity. They exist along what are called *convergent-plate boundaries*.

## Convergent Boundaries

The outward-moving edge of a lithospheric plate must interact in some way with plates adjacent to it. It may crumple, sink under, or slide laterally against the edge of a neighboring plate. The first two of these possible inter-actions take place at the convergent-plate boundaries.

When neither plate carries continental crust, **subduction**—the sinking of one plate beneath another—occurs at a convergent boundary. Subduc-tion produces a deep furrow in the ocean floor, or a *deep ocean trench*. If the melting edge of the subducted plate produces enough magma, the magma may rise and create a chain of volcanic islands. The Japanese and Philippine islands were created in just this way.

Where an oceanic plate collides with the edge of a plate occupied by a continent, the result again is subduction. The thinner, denser oceanic plate invariably bends under and melts beneath the margin of the thicker, less dense, and high-riding continental crust. The result is much volcanic activ-ity a short distance inland from the continent's coastline: a volcanic moun-tain range. The Andes Mountains along the west coast of South America (see Figure 16.4) and the Cascade Range along the west coast of North America are prime examples of this process.

A rarer occurrence is that of two continent-bearing plates colliding with each other. For over 200 million years, the plate that bears the former con-tinent of India has been heading north toward the plate bearing Europe and Asia. About 50 million years ago, India began to collide with what is now Tibet in Asia. Parts of the southern margin of Asia buckled upward to form the Himalayas, the loftiest mountain range on Earth. The collision contin-ues today, building the Himalayas higher even as erosion tries to tear them down.

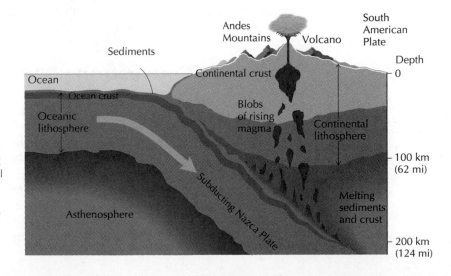

**FIGURE 16.4** This cross-sectional view through South America's west coast shows the oceanic Nazca Plate subducting under a slab of thick continental lithosphere. Magma rising from the melting oceanic crust and sediments gradually rises to the surface and builds the Andes Mountains. (Adapted from Garrison, 1993)

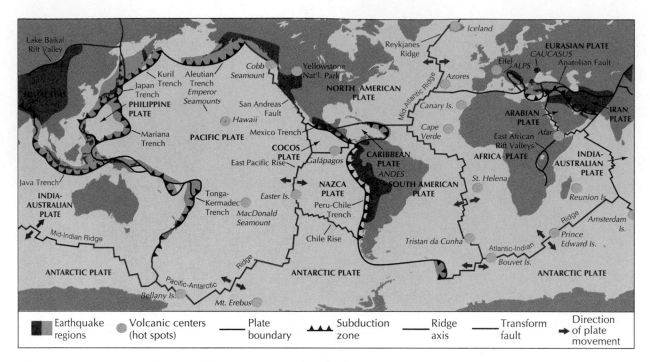

**FIGURE 16.5** This world map shows the major lithospheric plates and the directions of relative plate movement. Also shown are the locations of the principal *hot spots,* where magma rises to the surface in chimneylike columns from deep within Earth. Plate boundaries are nearly always the sites of volcanic activity, earthquakes, or both. (From Garrison, 1993)

# Transform Boundaries

In some places, the edge of one plate merely slides against the edge of another. This sliding, however, is rarely smooth. Typically, the plate edges remain stuck for a while (sometimes decades or centuries) and then unpredictably lurch, releasing energy in the form of seismic waves. California's San Andreas Fault, the world's most famous *transform fault,* lies at the transform boundary of the Pacific Plate and the North American Plate.

Convergent-, divergent-, and transform-plate boundaries throughout the world are illustrated on Figure 16.5. This figure also delineates the major lithospheric plates and a number of "hot spots," which will be discussed later.

# Mantle Convection—The Driving Force?

Although the theory of plate tectonics successfully and comprehensively explains how Earth's lithospheric plates move and interact with each other, geologists are less certain about why plates move in the first place. The question is hard to answer because we cannot see in any direct way what is going on under Earth's crust.

Some aspects of this issue are clear: It takes a lot of force to get a plate moving, and a lot of work to move a plate through some distance. Plate movements are largely horizontal, so gravity cannot be the primary source of the energy required. Winds and ocean currents are far too weak to exert

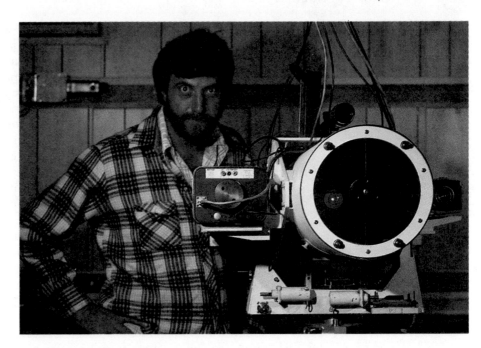

**PHOTO 16.1** This two-color laser is used to monitor movements along the San Andreas Fault at Parkfield in central California, where horizontal shifts of about 1 m occur every 30 years or so.

any significant force on plates. We must look to Earth's interior to identify the source, and what we find there is plenty of heat. Earth's interior is hot (on the order of 4000 K) because of thermal energy generated by the decay of radioactive isotopes trapped inside the mantle and core. In a general sense, then, we can think of plate motion as being the output of a giant heat engine, which runs on thermal energy produced by natural radioactivity.

But what are the cogs and the wheels of this metaphorical heat engine? Convection currents are a plausible answer. If heat energy is unequally distributed at any particular depth, then the hotter material at that depth would have a tendency to expand and rise. The upward motion of these plumes would set up convection currents, like those sketched in Figure 16.6. Seafloor spreading could occur where hot, upwelling plumes of magma break through the thin oceanic crust, solidify, and push apart sections of the lithosphere along a broad front. Elsewhere, subduction could be aided or driven by plate material that sinks to replace the upwelling plumes.

**FIGURE 16.6** This simplified model of mantle convection assumes that there are differences in temperature at the same depth deep within the Earth. It also assumes that the rising mantle material does not cool so fast that it loses its buoyancy and fails to reach the surface.

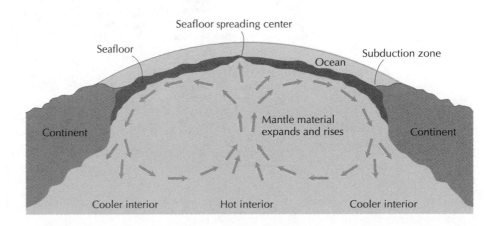

This general idea, which has been tacitly accepted by most geologists for some time, is now being supported and sharpened by a host of recent discoveries about Earth's internal structure and the composition of the material coming up through the plumes. It now seems probable that at least some of the rising plumes originate from as far as 3000 km down, at the remote boundary between the outer, liquid-metal core and the rocky mantle lying above it. Studies of earthquake-generated seismic waves, which travel through all parts of Earth's interior, have allowed geophysicists to indirectly glimpse the structure of that boundary. Rather than being smooth and static, as previously thought, the core-mantle boundary is now recognized as bumpy and turbulent. The whole region seethes with complex chemical and physical interactions, making it quite possibly the most dynamic layer anywhere in or on the planet. These interactions involve thermal energy moving out through the core and heating parts of the lowermost mantle. The heated parts move upward in the form of hot, buoyant plumes that reach the lithosphere after tens or hundreds of millions of years. There, some break through at what are called *hot spots,* causing some or all of the volcanic activity that pushes aside preexisting crust and sets the plates moving.

Some 60 mantle plumes have been identified so far. Some rise to the sites of seafloor spreading along plate boundaries; others emerge somewhere in the middle of existing plates. On land, a rising plume may produce a succession of volcanoes, like those in and near Yellowstone National Park and along the eastern Sierra Nevada of California that blanketed vast areas of North America with lava and ash hundreds of thousands of years ago. These were titanic blowouts, since enormous pressure had built up under the thick continental crust before it could be released. Under the sea, the crust is much thinner and hot-spot eruptions are much less violent. Under seafloor crust that is drifting owing to plate motion, a stationary plume can cause a chain of volcanic islands, like today's Hawaiian Islands, which were created (and continue to be created) in exactly this manner (see Figure 16.7).

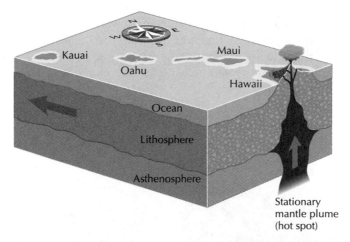

**FIGURE 16.7**    The roughly linear chain of the Hawaiian Islands are expressions of hot-spot volcanism modified by the Pacific Plate drifting over a stationary plume of upwelling magma. Kauai is the oldest of the major Hawaiian Islands. Ever since it drifted from the plume millions of years ago, its profile has been reduced by erosion to a fraction of its original stature. Maui, the second-highest island in the chain, only recently drifted away from the hot spot. Currently, the plume is busy building the Big Island of Hawaii, though undersea volcanic activity off the southeast shore of the island hints of a new island that will rise there sometime in the future.

**PHOTO 16.2** This computer-generated perspective of the surface of Venus, derived from radar measurements made by the spacecraft *Magellan*, shows Maat Mons, the largest volcano on Venus. Maat Mons rises 8 km above the surrounding terrain, about as high as a similar volcanic structure on Earth, Hawaii's Big Island, stands above the Pacific Ocean floor.

Recent analysis of the material emerging from mantle plumes has revealed the unexpected presence of carbon and oxygen, presumably of biological origin. This finding confirms the idea that slabs of ocean crust, along with their burden of biologically rich sediment built up over millions of years, sink and melt into the mantle at subduction zones, ride convection currents downward, and eventually rise to the surface in some other part of the world.

It is interesting to compare Earth's internal and external geologic activity with that of the other terrestrial planets (plus the moon). Both the moon and Mercury show hard, crater-pocked surfaces that have changed little over the past 3 billion years. Volcanic activity, faulting, and other indications of geologic vitality have virtually ceased on both of these bodies because their small size has allowed them to radiate away most of their internal heat. The surfaces of these geologically dead worlds preserve much of the early phase of their existence, when they were bombarded by impacting objects.

Venus, a near-twin of Earth with regard to mass, size, and internal composition, exhibits a broad and still-puzzling array of geological activity. There is much evidence of volcanic activity (see Photo 16.2), probably precipitated by upwellings similar to Earth's mantle plumes; but there are also strange downwellings in which the parts of its crust apparently get sucked straight downward. Mantle convection within Venus may resemble that within Earth, but its surface expression is different. There is little on Venus that suggests a mechanism of large, rigid plates bumping and interacting with one another.

These differences are not surprising in view of the fact that the surfaces and atmospheres of the two planets are vastly different. Venus has no oceans, although its pizza-oven-hot (750 K) atmosphere exerts the same pressure at the surface as Earth's oceans do at a depth of nearly a kilometer. The seething-hot rock at Venus's surface is more susceptible to deformation than Earth's colder, more brittle crustal rock. Rotation may play a

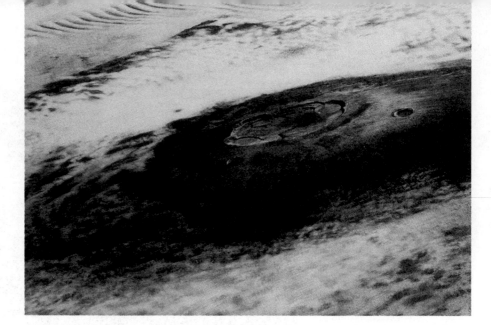

**PHOTO 16.3** Olympus Mons, the largest volcano and the highest elevation on Mars, pokes through thin clouds in this shot taken by the *Viking* spacecraft orbiter.

role, too. Venus hardly rotates at all (its "day" is 243 of our days), which means that the movement of material within it could not be influenced in any significant way by the Coriolis effect.

The surface geology of Mars gives us a snapshot of a dying world. Mars is larger than the moon and Mercury but smaller than Earth. Its core, quite possibly the major source of its internal heat, is proportionally smaller than Earth's core. Mars shows much evidence of geologic upheaval in the distant past, but that activity has apparently slowed to a crawl today. Mars has several large, dome-shaped volcanic mountains, all probably dormant within the past few centuries. The largest of these, Olympus Mons (Photo 16.3), resembles the Big Island of Hawaii (the largest volcano on Earth), but it is more than two times higher and some six times wider at its base. The absence of plate tectonics on Mars apparently has allowed a moderately active mantle plume inside Mars to gradually build a huge dome in the same spot on the crust. If the crust were moving in a plate tectonic manner, then a row of smaller volcanoes, like that of the Hawaiian chain on Earth, would have been produced instead of one large dome.

## Consequences of Plate Tectonics

Plate tectonic cycles involve the creation of mostly seafloor crust (and an underlying layer of solid mantle) at the divergent boundaries, island formation and mountain building of various kinds at the convergent boundaries, and other changes associated with motions on the transform boundaries. Plate tectonics (the word *tectonic* is derived from a Greek word meaning "builder") has built the continents and oceanic basins as we see them today and will profoundly affect Earth's future geography. Without the constructive effects engendered by plate movements (or any other manifestation of mantle convection), the destructive forces of erosion would dominate geologic change on Earth. Over time, the continents would slowly wash into the sea, perhaps to disappear completely within a billion years or so. In the next few pages, we will focus on earthquakes and volcanoes (processes directly connected to plate tectonic mechanisms) and also detail the processes of weathering and erosion, which tear down, rather than build up, surface features on our planet. Finally, we will look at the processes of sedimentation and metamorphism, which would not occur if our planet was purely static.

# Earthquakes

Seismic activity occurs mostly along plate boundaries where the edges of the lithospheric plates either converge or slide sideways against each other. On or near a convergent-plate boundary, where rocks are subject to compression, seismic movements are most likely to occur along thrust faults [Figure 16.8(a)]. A *thrust fault* has an inclined surface along which one mass of rock is thrust above another by means of compressive stress. Thrust faulting on a grand scale produced the jagged Teton range in Wyoming and the Sierra Nevada of California (Photo 16.4). Thrust faulting on a slightly less dramatic scale is pushing mountains up north of Los Angeles. The enormously destructive 1994 Northridge quake, centered in the densely populated San Fernando Valley north of Los Angeles, raised parts of the valley and the surrounding mountains about half a meter.

Thrust faulting also occurs beneath the ocean floor, as evidenced by the devastating earthquake that hit Mexico City in 1985. The seismic waves that rocked the city originated from sudden earth movements at a subduction zone, hundreds of kilometers to the west. There, the oceanic Cocos Plate is diving below the offshore margin of the North American Plate. Similarly, Japan's catastrophic Kobe earthquake in 1995 resulted from movement on a thrust fault just offshore. Another similar situation exists at a subduction zone off the coast of extreme northern California, Oregon, and Washington, where the remaining sliver of a once larger oceanic plate, the Juan de Fuca Plate, slips underneath the margin of the North American Plate. Although no large earthquakes have occurred in historical times in this region, smaller quakes have, and a large one may be imminent.

Compression in Earth's crust does not always give rise to fitful episodes like the earthquakes we are describing. In softer sedimentary rock, the stress is often relieved by a process of slow and gradual crumpling called

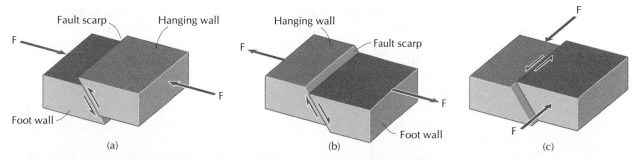

**FIGURE 16.8** (a) A thrust fault has an inclined surface that is subject to compressive stress; one rock mass (the *hanging wall*) overrides the other (the *foot wall*). (b) A normal fault has an inclined surface that is subject to tension; as the two opposing rock masses pull apart, one sinks below the other. (c) On a strike-slip fault, two opposing rock masses slide against each other primarily in a horizontal direction. Vertical movement along a fault exposed at Earth's surface, as in (a) and (b), can create a fault scarp, or cliff. In time, erosion tends to round off fault scarps, making them appear less distinct.

**PHOTO 16.4** From Owens Valley in eastern California, the east side of the Sierra Nevada rises abruptly some 3 km, a result of normal faulting.

*folding*. The parallel ridges and valleys that characterize the Appalachian Mountains are a reflection of many parallel folds that underlie much of the eastern United States. The folds originated nearly 300 million years ago, when that portion of the Earth's crust was under great compressional stress. The effects of more recent folding can be seen in much of the Rocky Mountains, where folds are commonly exposed in eroded sedimentary strata.

Another kind of fault can develop in places where a rock formation, or even a large section of the lithosphere or crust, is subject to tension. This is a *normal fault* [Figure 16.8(b)]. When two masses of rock pull apart from one another, typically on an inclined surface, one will have a tendency to slip downward. Normal faulting on a grand scale is responsible for the topography of the "basin and range" country that covers much of Nevada and Utah and parts of southern Idaho and eastern California. The region is dominated by towering mountain ranges interspersed with broad, valley-like troughs. The troughs are underlain by sections of crust that have broken from their moorings and gradually slipped downward. The mountain ranges are blocks that have gradually risen alongside the troughs. These upward and downward movements are thought to be a consequence of the stretching and thinning of the crust in that region. In California's Death Valley, normal faulting produced total vertical movements exceeding 7 km. Over time, though, while the mountains bordering Death Valley have been rising, the valley itself has been sinking and collecting sediments eroded and washed down from those same mountains. Thus, the difference in elevation between the topmost valley sediments and the summits of the mountains now averages not the original 7 km but, rather, 2–3 km, which is still impressive.

Tension at great depths in the lithosphere may have been responsible for triggering powerful earthquakes in places we might least expect it—like the middle of the continental North American Plate. An earthquake that severely shook the area around New Madrid, Missouri, in 1814 was powerful enough to temporarily reverse the flow of the Mississippi River in that region and to ring church bells on the East Coast. Such events are believed to be rel-

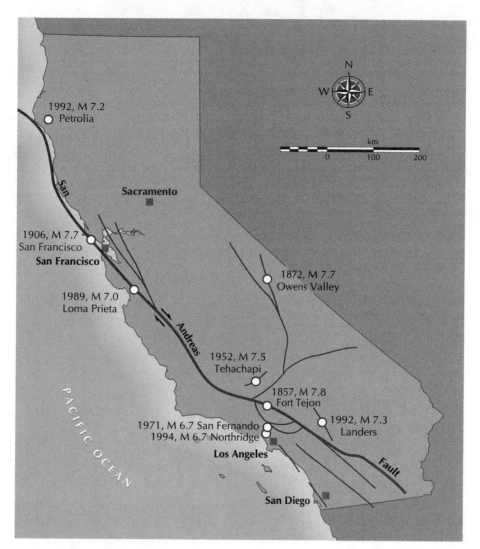

**FIGURE 16.9** This map of California shows several of the state's most significant faults and the dates, moment magnitudes (M), and names of several important California earthquakes of the past 150 years. The San Andreas Fault traces the principal boundary between the Pacific Plate (on the left) and the North American Plate (on the right). It is a right-lateral transform fault, which means that an observer on either side of the fault facing the other side sees motion to the right whenever tectonic movement is taking place. About half of the roughly 6 cm per year of relative motion between the two plates takes place on smaller faults parallel to the San Andreas, some of which are shown here. At the current overall rate of relative plate movement, about 6 m per century, present-day Los Angeles will stand abreast of present-day San Francisco 10 million years from now.

atively rare, but if a similar event occurred today, it would likely cause widespread destruction over an area stretching from Memphis to St. Louis.

On or near a plate boundary, where sideways motion is predominant, *strike-slip faults* or *transform faults* develop [Figure 16.8(c)]. Strike-slip faults on land are found in relatively few areas, but coastal California has several. Here, the North American and Pacific plates interact by sliding horizontally against each other. The San Andreas Fault, the major division between the two plates, extends more than 1000 km and passes through or near the state's largest urban areas (Figure 16.9). The structure of the plate

**PHOTO 16.5** Sudden, lurching movements have been taking place along the San Andreas Fault (at upper right) for millions of years. On this particular section of the fault, displacements of 5 to 10 m (about 15 to 30 ft) take place every two to four centuries. Rock formations on one side of the San Andreas Fault can be matched with identical formations on the opposite side hundreds of kilometers away. The fault itself can be traced for more than two-thirds the length of California.

boundary in California is actually very complex, with up to half of all the movement taking place off the main San Andreas Fault on a series of paralleling fault strands.

## Volcanic Activity

Volcanic eruptions occur when magma rising from somewhere in the mantle pushes through fissures in the solid rock above and reaches the surface. On the seafloor-spreading centers, the heat from the rising magma is immediately quenched by the great mass of cold water above. The magma, which comes from deep in the mantle, is dark and dense. It solidifies quickly into a *basaltic* (dark and dense) *crust* which becomes new seafloor.

At hot spots, located either on land or under the sea, a large amount of material (lava) may accumulate to form a large, gradually sloping **shield volcano.** The island of Hawaii is the largest shield volcano on Earth, both in its height (about 9 km as measured from the ocean floor) and its breadth. The magma that feeds a shield volcano is relatively fluid (not viscous or sticky) and does not contain a large amount of dissolved gas. Thus, it rises with little resistance through the fissures and erupts at the surface in a relatively nonviolent way (see Photo 16.6). The solidified lava forms a dark volcanic rock called *basalt.*

Volcanoes that arise on convergent boundaries are typically very different in character. They may erupt violently at intervals of thousands of years or more, spewing hot gases and "ash" (pulverized rock) far and wide. Quite commonly, these violent episodes are interspersed with periods of relative calm in which lava seeps from the fissures and rebuilds the damage done by the violent explosions. The result is a relatively steep-sided,

**PHOTO 16.6** An eruption at Mauna Loa, the large volcanic *caldera* (crater) atop the Big Island of Hawaii. In recent years, more frequent eruptions have been taking place along the flank of the island at Kiluea.

cone-shaped mountain called a **stratovolcano** (or composite volcano). A cross section cut into a stratovolcano reveals the alternating layers of ash and lava deposited on the mountain's flanks.

Mount Saint Helens and Mount Rainier in the Cascade Range of Washington State and Mount Fuji in Japan are prime examples of stratovolcanoes. Mount Saint Helens (Photo 16.7) lost its former symmetrical shape in 1980 when a large eruption tore off about 1 km$^3$ of mass from the mountain's top and north flank. Much larger explosions have occurred in rather recent historic times. In 1883, for example, the Indonesian island of Krakatoa nearly disappeared in a series of eruptions that pulverized some 20 km$^3$ of solid material and flung it into the sky.

**PHOTO 16.7** A giant ash cloud emerges from Mount Saint Helens during the catastrophic eruption of May 22, 1980. The ash was propelled as high as 18 km within 8 min. The blast devastated some 500 km$^2$ of land, much of it richly forested before the eruption.

Stratovolcanoes are explosive because the magma rising at the convergent boundaries is quite different from the dark and dense basaltic magma mentioned earlier in connection with seafloor spreading and shield volcanoes. It typically contains a fair portion of silica-rich minerals, which form a more viscous melt, plus plenty of dissolved gases, including water (steam). Near the surface, this "sticky" type of magma tends to form a plug that seals the throat of the volcano. Pressure from the rising magma builds until, like a ticking time bomb, the volcano explodes.

Where does the silica-rich and gas-rich magma come from? Another look at Figure 16.4 reveals a possible source. The blobs of magma rising from the edge of a subducting plate contain not only basaltic seafloor but also seafloor sediments and very likely some ocean water. When the blobs rise into the solid layer of continental crust (which is relatively rich in silica), they become enriched with silica as they melt their way through toward the surface.

The rock, or lava, associated with the surface expression of volcanic activity comes in three principal types. *Rhyolite,* which is relatively rich in light-colored minerals such as quartz (silica) and feldspar, and *andesite,* which consists of primarily feldspar and secondarily various ferromagnesian (dark, iron- and magnesium-rich) minerals, are associated with stratovolcanoes. *Basalt,* which contains some feldspar but is especially rich in ferromagnesian minerals, is associated with shield volcanoes and newly formed seafloor crust. All three belong to a category called **extrusive igneous rocks** (to *extrude* means to "push out," and *igneous* means "arising from fire"). Since extrusive igneous rocks are created from magma that cools quickly, they are invariably fine-grained, consisting of microscopic mineral crystals.

In addition to extrusive igneous rocks, there are many common examples of **intrusive igneous rock.** These rocks are formed by rising magma that intrudes and pushes aside previously existing rock but fails to reach the surface. Three principal types of intrusive igneous rock are *granite, diorite,* and *gabbro.* Each has the same composition as rhyolite, andesite, and basalt, respectively, since they form from the same types of magma. Because intrusive rocks cool much more slowly than the extrusive (volcanic) rocks, they can form larger crystals and so are invariably coarse-grained (their crystals are easily visible to the naked eye). Granite and diorite have a salt-and-pepper-like appearance, with predominantly light-colored minerals. Gabbro has a pepper-and-salt-like appearance, with dark flecks of ferromagnesian minerals mostly evident. The whole family of intrusive igneous rocks are sometimes called "granitic" rocks for their granitelike appearance. For comparison, igneous rocks of both the intrusive and extrusive types are pictured in Photo 16.8.

When a large blob of intrusive magma collects underground and fails to reach the surface, it eventually crystallizes to form a *pluton* (see Figure 3.23 in Chapter 3). Long chains of plutons may form a *batholith.* Plutons and batholiths are typically a little less dense and more buoyant than the rocks that surround them. In time (typically tens of millions of years) they work their way to the surface as the overlying layers of rock are eroded away. The cores of the great mountain chains of the West Coast (the Cascade Range excepted) consist of granitic rocks belonging to roughly 100-million-year-old batholiths that have at last risen to poke through the surface.

Granite    Diorite    Gabbro

Rhyolite    Andesite    Basalt

**PHOTO 16.8**   The intrusive, or granitic, rocks at top are divided into specific types by mineral content. Granite is generally the lightest in color; gabbro, which is sometimes referred to as "black granite," is the darkest. The extrusive, or volcanic, rocks at bottom form a similar sequence of types based on mineral content. Extrusive rocks are generally composed of very small mineral grains, but the andesite rock shown here contains some visible feldspar crystals. It is known as *andesite porphyry.*

# Weathering and Erosion

Weathering and erosion are the processes by which rocks are broken down and the resulting debris carried away. **Weathering** refers to the small-scale disintegration of rocks. *Chemical weathering* takes place when oxygen (or other gases in the atmosphere), acidic rain, or groundwater breaks down mineral grains on the surfaces of rocks. *Mechanical weathering* takes place when, for example, water in tiny cracks expands when freezing and pushes apart pieces of rock, or when the growing roots of

**PHOTO 16.9**   As a result of weathering, these boulders, once joined into a single rock mass, have been rounded and separated from each other over a period of many thousands of years.

plants exert strong enough forces to disrupt weak rocks. Weathering produces unconsolidated material, which becomes soil if it is mixed with decaying plant matter. Once it is broken down to small pieces, the debris from rocks can be picked up and transported from one place to another by various agents such as moving water, wind, or moving ice. This large-scale movement of rock material is known as **erosion.** Let us examine, in some detail, the principal agents of erosion.

### Erosion by Moving Water

Water moving in streams and rivers is the most obvious agent of erosion. The steeper the slope, the faster the water moves within a given drainage and the more easily it can pick up and move solid materials. During floods, the fast-moving water may have enough momentum to pick up boulders and roll them downstream. The rolling rocks and suspended particles scrape away at the bottom and banks of a stream course, doing far more excavation work than the moving water itself.

A stream originating in the mountains typically starts out with a relatively large *gradient* (rate of elevation loss), but the stream loses its ability to erode once it emerges upon the flatter terrain below. Larger rocks are the first to settle out. Successively smaller rocks and sand particles settle out as the stream slackens further. The finest suspended particles, of silt or clay, are carried farthest downstream. In high-volume rivers, such as the Missouri and the Mississippi, large amounts of silt can be transported all the way to the ocean. If you look at the type of debris deposited along the banks of a stream after a flood recedes, you can get a good idea of how fast the flood waters were moving and how effective they were at transporting material.

We can broadly classify rivers or streams as being *young, mature,* and *old* (see Figure 16.10). Young streams, typically found at higher elevations in the mountains, are characterized by V-shaped valleys or canyons. Over time, they carve deeper channels, and debris accumulating in the channel from the side walls gets swept downstream. Usually, a young stream is also geologically young, but there are exceptions. The Grand Canyon of the Colorado River in northern Arizona has managed to keep its youthful profile for several million years because the entire Colorado plateau that surrounds it has been gradually uplifted during that same time by geologic

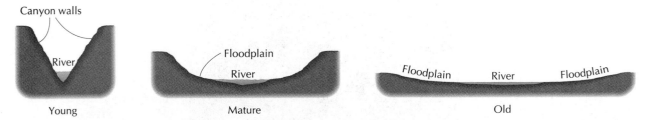

**FIGURE 16.10**  Young watercourses tend to carve out V-shaped canyons or ravines. Mature watercourses may flow through U-shaped canyons or steep-sided valleys, and often meander from side to side within them. Old watercourses are shallow, are relatively slow moving, and readily meander across their gently sloping floodplains. Each of these cross-sectional diagrams has been vertically exaggerated for clarity.

forces. The uplift has proceeded at a pace about as fast as the river has been able to cut downward.

Most of the cutting and widening of the Grand Canyon has taken place not steadily but during great floods that recur at intervals of hundreds or thousands of years. During these catastrophic events, immense quantities of sediment wash down the side canyons and slopes into the main channel. After each flood, the Colorado River may spend decades or centuries removing and transporting the accumulated sediment farther downstream.

At the mature stage, a stream or river will have cut itself down to such a low level that its erosive and transport abilities are close to being in balance. At any given time, a deep layer of sediment lies beneath the flowing water. Most of the erosion now takes place not on the bottom but on the sides of the channel, and the profile of the valley or canyon containing the stream may become U-shaped. When not flooding, the stream may meander back and forth along the valley's flat floor, or *floodplain*. During a flood, a silt-laden stream may spill over its banks onto the floodplain beside it and enrich the plain with a fresh coating of soil. It is no accident that floodplains are commonly utilized for agricultural purposes.

In old age, a stream becomes very wide and shallow. It may meander almost aimlessly across a broad floodplain through a flat or gently rolling landscape, typically near sea level. The stream's lower branches, or tributaries, become fewer in number but bigger, as the watershed divides (ridges) between former tributaries are erased by erosion. The mature, lower ends of the world's great rivers carry great loads of sediment to the ocean, where much of it drops out of suspension near the shoreline and forms a *delta*. The Mississippi River alone discharges about 500 million tons of sediments into the Gulf of Mexico every year.

## Erosion by Wind

Winds sweeping across deserts and other dry lands are capable of transporting large amounts of fine material over distances of hundreds or thousands of miles, sometimes from continent to continent. It was just this process that created the deep, fertile topsoil, known as *loess,* that now covers much of the central United States, central and eastern Europe, and eastern China. An important source of the American loess was a vast area of mud left behind as glaciers receded from the northern United States and Canada at the close of the last ice age. After the mud was dried and pulverized, strong winds blew the mineral grains southward onto what became the prairies of the central United States.

Sand particles, picked up by strong winds, can erode and carve any hard object, as anyone caught in a sandstorm while driving on a desert highway knows. The world's arid regions are full of wind-carved topographical features.

On Mars, a planet whose surface has been devoid of liquid water throughout nearly all of its geologic history, erosion by dry winds has produced a gigantic canyon called Valles Marineris (Photo 16.10). The canyon may have started as a long and narrow rift, perhaps caused by tectonic activity, roughly 4 billion years ago. Although tectonic activity presumably waned on Mars over the ensuing eons, sand-laden winds continued to chip

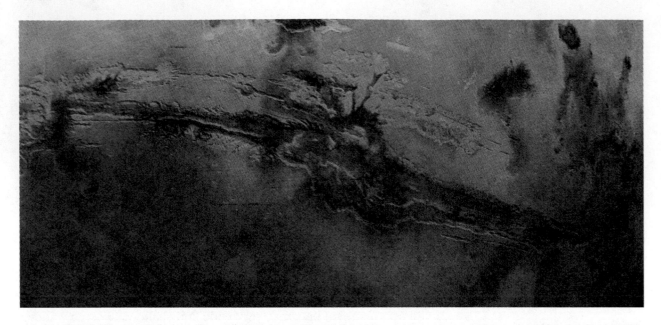

**PHOTO 16.10** If brought to Earth, Valles Marineris on Mars would stretch across the width of the continental United States. The entire Grand Canyon of Arizona would fit into one of the "small" tributary canyons at the top or bottom.

away at the feet of the clifflike canyon walls, causing their collapse. The rift gradually became wider and deeper as winds carried the collapsed debris away.

## Erosion by Moving Ice

Glaciers, which are essentially rivers of ice, are the predominant erosive force in many cold regions on Earth. Today, glacial erosion on a large scale is taking place only in the polar regions. There, *continental glaciers,* or ice caps, up to 4 km thick sit atop the continent of Antarctica in the south and the large island of Greenland in the north. These sheets of ice slowly flow outward toward the sea and grind away at the rock underneath, picking up sediment, transporting it to the sea, and dumping it there.

On a smaller scale, *valley glaciers,* which are confined to mountain valleys, are responsible for carving the jagged profiles of the Alps, the Alaskan coastal ranges, the Cascade Range, and other cold-climate mountain ranges throughout the world.

During the ice ages, glacial erosion helped carve large portions of North America into the terrains we see today. Parts of Wisconsin, for example, are full of peculiar, parallel ridges and valleys that are the result of the undersides of glaciers scraping across the landscape. New York's Hudson River valley is a glacier-carved feature. Large boulders found in what is today's Central Park in Manhattan were pushed there by a southward-advancing glacier more than 10,000 years ago and marooned when the glacier retreated.

## Mass Wasting

The downslope movement of rocks or unconsolidated material under the influence of gravity is prosaically called *mass wasting.* Although mass wasting usually occurs with the help of running water or some other transport medium, it may also occur by gravity alone.

**PHOTO 16.11** The Lamplugh Glacier of Alaska sweeps down from the coastal mountains into Glacier Bay.

**PHOTO 16.12** These *sea stacks* (the Twelve Apostles rock formations off the Victoria coast in Australia) exemplify the erosive power of the waves. Note the horizontal sedimentary strata in the stacks, which match that in the coastal cliffs.

Fast or sudden types of mass wasting include landslides and mudflows. Landslides, which commonly occur in mountainous areas, may be triggered by the weathering and weakening of rock debris or soil on steep slopes and by violent earthquakes. Mudflows are aided by the presence of water, which fills the spaces between the debris particles and allows them to slide against each other more easily. Catastrophic mudflows are often associated with the eruptions of stratovolcanoes. When volcanic ash absorbs water from thunderstorms or melting ice on the volcano slopes, the resulting mixture becomes a dense fluid that can travel long distances down the river valleys below. This happened during the eruption of Mount Saint Helens in 1980 and is sure to happen again someday when other Cascade Range volcanoes, such as Mount Rainier near Seattle, Washington, reawaken from their current state of slumber.

Less dramatic but still effective over long time periods are the slow forms of mass wasting such as *creep*. Creep involves the imperceptibly slow movements of weathered debris down a slope. Tilting fences or walls anchored in unconsolidated sediment or soil often indicate the slow but sure impact of creep.

## Other Erosional Agents

The action of waves and ocean currents against coastlines is a relatively minor agent of erosion, though it is of great concern to coastal residents. On the gently shelving Atlantic and Gulf coasts of the United States, waves and currents easily move sand and silt from place to place. During calm periods, much of this material collects in long sandbars—called *barrier islands* when they poke above the surface—paralleling the coastline. Barrier islands are ephemeral features, subject to destruction by storms.

On the geologically young West Coast of North America, the continent's abrupt edge is subject to continual battering by the ocean waves. This wave action steadily nibbles away at the coastline, but steady uplift and mountain building (consequences of the plate tectonic interactions taking place there) ensure that the edge of the continent endures.

Much of the rain and snow falling on land sinks below the surface to become groundwater. Groundwater moves slowly through porous strata and more rapidly through caverns. Its erosive effects come mainly from its ability to dissolve certain kinds of rock, notably limestone, as it flows by. In some areas underlain by limestone, sinkholes may occur when the ceilings of underground caverns collapse.

## Sedimentation

Sedimentation—the filling in of low spots by deposits of sediment—depends indirectly on the ongoing constructive forces of plate tectonics and more directly on the destructive forces of erosion. As we have seen, waterborne sediments can be sorted according to particle size along a stream or river. Further sorting may take place when the sediments reach the sea, by means of wave action or ocean currents. Some water-carried sediment winds up in desert valleys or basins, which have no outlet to the sea. The wind sorts sediments (the smallest and lightest particles carried by moving air can travel higher and farther), and glaciers gouge out particles both large and small and carry them downward.

When loose or "unconsolidated" sediments are covered by later deposits and subjected to considerable pressure (a common occurrence), the individual rocks or mineral grains may become cemented together in a

**PHOTO 16.13** This cross-bedded sandstone, exposed in Zion National Park, Utah, was formed from successive layers of windblown drifts of sand. Most sandstones are of marine origin, arising from the lithification of sand deposited offshore at continental margins. This particular sandstone is of nonmarine origin.

**PHOTO 16.14** Stalactites (upper) and stalagmites (lower) arise from the dripping of mineral-laden water from the ceiling of a cavern. Both are classified as sedimentary rock.

slow process called *lithification.* The result is a family of rather weak, layered (stratified) rocks called **sedimentary rocks.** These rocks can be classified by particle size. *Conglomerate,* made of smooth pebbles of various sizes, and *breccia,* made of irregular fragments, are examples of rocks made of the largest particle sizes. *Sandstone* is typically made of particles up to 2 mm across. Sandstone can result from deposits of windblown sand particles as well as from particles transported by water. *Shale* is made of finer particles still, typically much less than 0.1 mm across.

Sediments can also be deposited whenever mineral-laden water evaporates and leaves solid minerals behind. In limestone caverns, precipitates can accumulate to form needlelike stalactites and stalagmites and a variety of other interesting cave decorations. These, too, are considered to be a type of sedimentary rock.

Some sedimentary rocks are of biological origin. Coal is the chemically altered remains of plant life. Some kinds of limestone are derived from the shells and other remains of marine animals that sunk to the ocean floor and were buried long ago. The White Cliffs (chalk cliffs) of Dover, England, are a well-known example. Several types of sedimentary rock are shown in Photo 16.15.

Shale

Sandstone

Conglomerate

Halite

Limestone and Lignite

**PHOTO 16.15** The sedimentary rocks at the left have formed from fragments of preexisting rocks. The very fine mineral grains in shale were deposited as silt in quiet water. Sandstone consists of sand grains cemented together. The sample shown here, from the Grand Canyon, is tinted red by iron oxide. Conglomerate contains water-worn pebbles cemented together with finer-grained material. The sedimentary rock halite, or rock salt, is an example of chemically precipitated rock. It formed when salt-laden water evaporated from an ancient sea. The sample of white limestone shown is a nearly pure form of crystallized calcium carbonate. The very dark-colored rock in front of the limestone sample is lignite, a low grade of coal formed from the carbon of decayed plants.

# Metamorphism

Rocks may undergo a change more radical than that of lithification. When subjected to a very large amount of compression, heat, or both (usually for a long period of time), any preexisting rock may be transformed into a **metamorphic rock** (*metamorphic* means "changed in form"). Rocks are commonly metamorphosed into new forms when magma intrudes nearby, typically at or near plate boundaries. Metamorphism involves physical and chemical transformations in a rock short of melting the rock. If melting takes place, then the rock is no longer a rock; it is magma.

During metamorphism, heat and pressure form new minerals in the rock, change the rock's texture, and generally make the rock harder and more cohesive. For example, limestone can be transformed into *marble,* a metamorphic rock well known for its rigidity and permanence. Sandstone, which is composed of primarily quartz grains cemented together, can be metamorphosed into *quartzite.* Shale can metamorphose into *slate,* and slate can undergo further metamorphism to become *schist.* Both slate and schist retain the stratified texture of shale and can be split into layers. Granite and several other kinds of igneous and sedimentary rock can be metamorphosed into a compressed rock called *gneiss.* The mineral grains in gneiss become elongated and separated into thin layers; the exposed surfaces of

## BOX 16.1
# The Rock Cycle

The theory of plate tectonics broadly paints a unified picture of how continents drift, how oceans widen, how continents converge, and why violent events such as earthquakes and volcanic eruptions occur. In a similar manner, the rock cycle serves as a unified conceptual scheme of how Earth's rocky material evolves from one basic kind to another while participating in geologic change. The major patterns of change in the rock cycle are illustrated in Figure 16.11.

The rock cycle can begin with any type of rock, though it originally began with magma (molten rock), the ancestor of all rocks on Earth except for meteorites. Wherever rising and cooling magma solidifies, it forms (by definition) igneous rock. Let us follow one possible scenario in the life of a mass of igneous rock by following arrows counterclockwise around the rock cycle diagram of Figure 16.11.

When and if igneous rock is exposed at or near the surface, it becomes weathered and eroded. Gravity acts on the fragmented remains, eventually

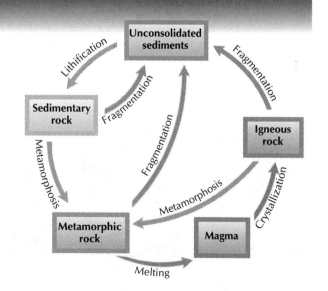

**FIGURE 16.11**   This simplified illustration of the rock cycle shows several important pathways between the various forms that rocky material can assume on Earth. Some processes are omitted for simplicity. Most of the processes mentioned here take place on scales of geologic time, though some (the cooling of magma into volcanic rock, for example) make take place in short-duration bursts.

Gneiss          Quartzite          Marble          Slate          Schist

**PHOTO 16.16** Each of these metamorphic rocks is an altered form of some previous type of rock. Imperfect foliation (layering) characterizes gneiss, which can derive from granite and a number of other rocks. Quartzite is metamorphosed sandstone with grains so firmly united that the rock can break through the grains rather than around them. Marble, the metamorphic equivalent of limestone, is distinguished by its purer color and more lively sparkle. Slate is shale metamorphosed by pressure. It has a tendency to split easily. Slate can be further metamorphosed into schist.

gneiss have a banded appearance. Several samples of metamorphic rock are pictured in Photo 16.16

As we have seen, the mechanism of plate tectonics—aided by the weather and the pull of gravity—is implicated, directly or indirectly, in nearly all the local geologic changes occurring on Earth. In Box 16.1, we examine another, broader consequence of plate tectonics: the rock cycle.

depositing them in low basins on land or on the seafloor. Lithification, leading to sedimentary rock, occurs where the sediments are squeezed by the weight of overlying layers. Extreme pressure and heat applied to sedimentary rock, perhaps in a zone where plates are colliding, may transform these layered rocks into "metasedimentary," or metamorphosed sedimentary rocks. If such metamorphic rocks are never exposed at the surface but, instead, sink downward into a subduction zone, they will eventually melt to form magma. This magma, if caught in a mantle convection current, may be brought to the surface, where it will again crystallize into igneous rock.

Dozens of scenarios are possible within the framework of the rock cycle. Some involve the cycles in which material stays near the surface and is heavily influenced by weather and gravity; others involve material sinking deep into Earth's mantle.

Although nearly all rocks change with time, the rock cycle at any given time favors the igneous type over the sedimentary and metamorphic types. Except for sediments accumulated on it, all oceanic crust is born of upwelling magma that solidifies primarily along the midocean ridges. The continents are typically composed of igneous rock, interspersed here and there with metamorphic rock. If we were to survey the kinds of rock *exposed on the surface* of the crust, however, we would find that sedimentary rock, along with unconsolidated sediments, are by far the most common. This material forms a thin veneer over the mostly igneous crust below.

The life of sedimentary rock is generally quite short. On land it is easily eroded back into sediment. Most of the sediment that accumulates on the ocean floor (along with the oceanic crust below it) ends up descending into the mantle within a few hundred million years. The oldest unaltered rocks on Earth are igneous and metamorphic. Roughly 4 billion years old, they have been very resistant to erosion and are associated with pieces of continental crust that have somehow escaped the ravages of the rock cycle.

# CHAPTER 16
# Summary

In the 1960s, a fixed-Earth model of nonmoving continents and static oceans was overthrown and replaced by the theory of plate tectonics. According to plate tectonics, the Earth's hard outer shell of rock (lithosphere) is divided into large and small slabs, or plates, which move relative to each other. These plates may include oceanic crust, continental crust, or both. Early and inconclusive evidence for plate tectonics included the idea that if the continents could be fitted together like jigsaw puzzle pieces into a larger whole, many structures such as mountain ranges and rock formations would match each other across the continental boundaries. More recent and conclusive evidence included the discovery of midocean ridges and deep-ocean trenches. Detailed observations showed that these were boundaries along which the seafloor either pulls apart (diverges) or comes together (converges).

At divergent-plate boundaries, which typically coincide with midocean ridges, magma wells up from the mantle below to form new seafloor. Older seafloor is pushed aside and moves outward. At convergent-plate boundaries, where two plates collide, subduction (the sliding of one plate under another) typically occurs. This is accompanied by volcanic and seismic activity. At transform-plate boundaries, the two plate edges slide laterally, a process that triggers earthquakes but does not produce volcanic activity.

Slow convection taking place within Earth's mantle is thought to drive the plate tectonic mechanism. Mantle convection also apparently occurs inside Venus and to some extent inside Mars.

Seismic activity takes place on thrust faults and normal faults, producing mostly vertical movement, and on strike-slip faults, which produces mostly horizontal movement.

Relatively nonviolent volcanic activity takes place at the divergent-plate boundaries and at hot spots on the oceanic crust associated with plumes of rising magma (mantle plumes). Violent volcanic eruptions typically occur on or near the subduction zones that lie along convergent-plate boundaries. Even more violent, but much less frequent, are the eruptions that occur when a mantle plume melts through continental crust.

When magma cools and crystallizes, it forms igneous rock. Igneous rocks can be either extrusive (volcanic and quick cooling) or intrusive (slow cooling). Weathering breaks down rocks into smaller particles (sediment). Erosion—by means of agents such as moving water, wind, moving ice, and gravity—picks up sediment, transports it to lower elevations, and deposits it on land or offshore. Buried sediments may be transformed into sedimentary rock through the slow process of lithification.

All types of rock, if subjected to enough heat and pressure, can be transformed into metamorphic rocks or else melted to form magma. The rock cycle summarizes how rocks can change from one basic type to another in a variety of ways. At any given moment, sediments and sedimentary rock form a thin veneer on most of Earth's surface, but igneous rock makes up the bulk of the crust, especially under the oceans.

## CHAPTER 16
# Questions

## Multiple Choice

1. Earth's lithospheric plates
   a) move up and down in response to swelling and shrinkage of Earth's mantle
   b) move horizontally, bumping and grinding against each other
   c) are the roots of the continents, which plow through oceanic crust at the rate of about 2 inches per year
   d) are stacked up in layers, much like sedimentary rock layers but on a much larger scale

2. Subduction in geology means the
   a) flow of water through porous rocks underground
   b) erosion of volcanic islands by wind, rain, and ocean waves
   c) melting of continental crust under a lithospheric plate containing ocean bottom
   d) process in which the edge of one lithospheric plate is forced under another plate

3. The oceanic crust
   a) is solid and thick relative to the continental crust
   b) is partly molten
   c) has rifts through which magma rises, solidifies, and spreads out
   d) is flat and filled with a nearly uniform layer of sediment

4. Earthquakes and volcanoes most frequently occur
   a) at plate boundaries
   b) near the equator
   c) near the poles
   d) near the midpoints of the plates

5. Compared with the continents, the ocean floors are
   a) much younger
   b) much older
   c) about the same age
   d) in some places older and in others younger

6. The deep-ocean trenches are the sites of
   a) seafloor spreading
   b) hot-spot volcanic activity
   c) subduction
   d) strike-slip faults

7. The landmass of India
   a) has always been part of Asia
   b) lies upon a rift that will eventually split it apart
   c) is colliding with the Asian landmass
   d) is moving away from the Asian landmass

8. Strike-slip faults (or transform faults) are those in which
   a) one block of land is squeezed above another block
   b) one block of land drops below another block
   c) one block of land slides horizontally against another block
   d) a block of land is folded upward owing to compression

9. Steep, symmetrical volcanoes with composite layers of ash and lava are called
   a) stratovolcanoes
   b) shield volcanoes
   c) cinder cones
   d) hot spots

10. The Hawaiian Islands
    a) are part of a sunken continent
    b) float on the surface of the ocean
    c) are located on a convergent-plate boundary
    d) are volcanic peaks

11. In their late, mature stages, river valleys feature
    a) narrow, V-shaped cross sections
    b) fast-moving water
    c) narrow floodplains
    d) meandering watercourses

12. Sedimentary rock includes
    a) rocks made of bits and pieces of preexisting rocks
    b) minerals precipitated out of solution in groundwater
    c) chemically altered organic plant or animal material
    d) all of the above

13. Rocks formed by magma cooling and crystallizing are called
    a) igneous rocks
    b) sedimentary rocks
    c) metamorphic rocks
    d) metasedimentary rocks

14. A common example of a light-colored, intrusive igneous rock is
    a) granite
    b) shale
    c) gneiss
    d) basalt

15. A common example of a dark, extrusive igneous rock is
    a) granite
    b) slate
    c) gneiss
    d) basalt

16. Which rock types can be transformed into metamorphic rock?
    a) igneous
    b) sedimentary
    c) metamorphic
    d) all of the above

## Questions

1. What evidence did Wegener present in support of his theory of continental drift? What supporting evidence did his theory lack?

2. What evidence indicates that seafloor spreading is occurring?

3. What are the distinctions between the divergent-, convergent-, and transform-plate boundaries? What kinds of phenomena occur at each boundary?

4. What are mantle plumes? How might they relate to the mechanism that moves the lithospheric plates?

5. Geologically speaking, how much alike are Earth and Venus? How much alike are Earth and Mars?

6. In what manner do landmasses shift along thrust faults? Along normal faults? Along transform (strike-slip) faults?

7. Why are stratovolcanoes capable of violent eruptions but shield volcanoes are not?

8. How does weathering differ from erosion?

9. List and summarize the major agents of erosion.

10. Describe the three general stages of river development.

11. What is the general relationship between altitude and the amount of sediment likely to be deposited on a given patch of land?

12. How could you distinguish sandstone from shale?

13. How might you distinguish limestone from marble?

14. Which rock type is most abundant by mass throughout Earth's crust? Which rock type covers most of Earth's surface?

## Questions for Thought

1. In what ways do you think (or do you know) that past tectonic activity has created or modified the landforms, or the underlying geology, of your geographic region?

PHOTO 16.17

2. What kind of rocks (if any) are exposed in your geographic region? (Field studies may help answer this question, or you may want to review geologic maps at your library.)

3. The rock pictured in Photo 16.17 is one of the famous "moving rocks" of Racetrack Playa in Death Valley National Park, California. What agent or force could have caused the rocks to move? (*Hint: Playas,* or dry lakes filled with very fine sediment, occasionally fill with runoff from storms. When dry, their surfaces are almost perfectly flat.)

4. Take a close look at any nearby river, stream, or arroyo and attempt to classify it as young, mature, or old. Try to identify the extent of the floodplain, if any, that flanks this watercourse.

5. Compare and contrast the convection mechanisms that transport heat in Earth's atmosphere and the convection in Earth's mantle that presumably drives plate tectonics.

6. Can you think of any rock-transforming processes not illustrated in Figure 16.11 that belong to the rock cycle?

## Answers to Multiple-Choice Questions

1. b   2. d   3. c   4. a   5. a   6. c
7. c   8. c   9. a   10. d   11. d   12. d
13. a   14. a   15. d   16. d

# CHAPTER 17
# Cosmic Evolution

*In star death lies the seeds of future generations of stars and planets. The expanding gas filaments of the Vela nebula, a supernova remnant some 1500 LY away, are enriching the tenuous interstellar medium with elements such as carbon, oxygen, nitrogen, and iron—the raw material of terrestrial-type planets. A tiny pulsar has been found within the Vela nebula. It is the collapsed core of a star that exploded some 12,000 years ago.*

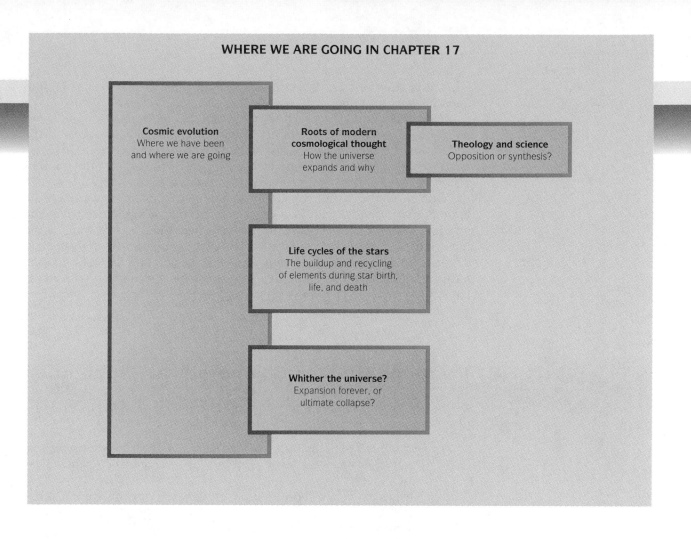

When we try to pick out anything by itself, we find it hitched to everything else in the universe.

These insightful words, written by explorer and naturalist John Muir a century ago, have proven to be remarkably prophetic. Muir's "universe"—the interconnected web of living and nonliving things on Earth coupled with energy from the sun—pales in comparison to the larger universe we comprehend today. We are now aware that the atoms within each of us have a history stretching back billions of years. Before Earth was formed, these atoms were adrift in interstellar space; and before that, they were inside one or more stars.

In the universe of our present comprehension, stars of the past have synthesized most of the elements of which we (human beings) are made. As is sometimes pointed out, we are truly "star stuff." Billions of years hence, when heat from the swelling sun incinerates the inner planets, the atoms that make up Earth will return to deep space. Millions or perhaps billions of years after that, these same atoms may join with other material to form new stars and possibly new planets.

This grand scenario, which we will elaborate on in this chapter, is not mere speculation. It is built on the great astronomical discoveries of the 20th century. Like pieces of a puzzle being fitted together, these discoveries have been integrated to build a reasonably consistent and coherent vision of the universe's origin, evolution, and future.

Our review of cosmic evolution begins with modern *cosmology*, the branch of astronomy (with firm connections to physics) that describes the origin and subsequent evolution of the universe. Next, we shall follow the life cycles of stars, including our own, the sun. Our existence on this planet depends on the stability of the sun, and both the sun and Earth are mortal. Last, by means of recognizing present cosmological trends, we will speculate on the ultimate fate of the universe. Will the universe exist forever, or will it have a definite end?

# Roots of Modern Cosmological Thought

 During the early part of the 20th century, three things about our particular place and stature in the universe became apparent:

1. Our sun is not in the center of our galaxy, and furthermore, the sun is a quite ordinary star among billions of other stars belonging to the same galaxy.
2. Our galaxy is not the only one; it coexists with billions of other galaxies.
3. The universe as a whole is expanding.

We will not concern ourselves with the details of the first discovery. Today, two well-established facts are that our Solar System lies some 10 kiloparsecs (kpc), or about 30,000 LY, from the galaxy's center and that the galaxy contains at least $10^{11}$ stars (recall Windows P and Q in the cosmic zoom of Chapter 2). From these facts we can conclude that we are not located at the center of anything and that there is nothing special or unusual about the region of the universe we inhabit.

The second discovery made us realize that the universe is far larger than we once thought possible. Prior to the 1920s, astronomers were quite familiar with a class of fuzzy, spiral-shaped objects known as *spiral nebulas*. For the most part, they were regarded as being whirlpools of luminous gas located within the Milky Way galaxy. During the 1920s, several lines of inquiry produced enough evidence to convince astronomers that spiral nebulas were really immense aggregates of stars outside our own galaxy; they were galaxies in their own right. (Galaxies did, and still do, look fuzzy through a telescope because they are so far away. With modern equipment, it is possible to partially "resolve"—to pick out individual stars in—the nearer galaxies.) The term *spiral galaxy* eventually replaced the name *spiral nebula*. It soon became clear that many other fuzzy-looking objects in the sky that do not necessarily have a spiral structure are galaxies, too.

The earliest hints of the third discovery came from a spectroscopic study of several galaxies begun in 1912, more than a decade before the true nature of galaxies was generally accepted. (Spectroscopy was discussed in

Chapter 11. It involves the separation of light into component wavelengths, or colors, and the analysis of the resulting spectral line patterns.) Today, obtaining the spectrum of a galaxy is routine; but in the early years of this century, spectra could be obtained only by painstaking effort, often involving photographic exposures of as much as 40 hours over several nights. The early studies, and all similar investigations since, have shown that nearly all galaxy spectra have *red shifts;* that is, the spectral features are shifted toward longer (or "redder") wavelengths. A red shift, according to the Doppler effect (Chapter 8), means recession; the faster a light source is receding, the greater is its red shift.

## The Nature of the Expansion

The apparent trend of galaxy recession was confirmed by astronomer Edwin Hubble and others in the 1920s and early 1930s. By then, astronomical equipment and measuring techniques had improved, and researchers could estimate the distances to various galaxies, as well as measure their recession speeds, with some degree of confidence. Hubble came up with an important generalization: *Galaxies in every direction are receding, and the farther a galaxy is from us, the faster it moves away.* The implication is that the universe, which is populated by billions of galaxies, is itself expanding. This idea remains a central axiom of cosmology today.

The relationship between the speed of recession and the distance for a galaxy is called the **Hubble law.** It can be written as

$$v = Hd$$

Equation 17.1

**PHOTO 17.1**   Edwin Hubble (1889–1953), photographed here at the 48-inch Schmidt telescope on Palomar Mountain, California. Hubble's legacy lives on in the telescope that bears his name, the Hubble space telescope.

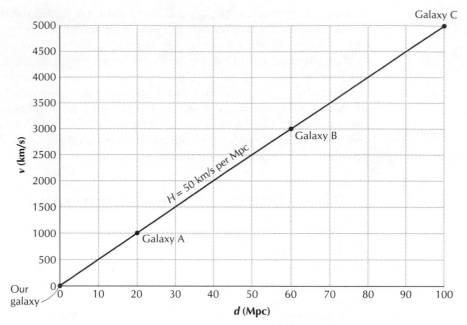

**FIGURE 17.1**    The Hubble law in graphical form shows the straight-line (proportional) relationship between velocity of recession (*v*) and distance (*d*). The actual data points for galaxies (which are not shown, except for some sample data points) fall on or close to the diagonal line. The slope of the diagonal line, called the Hubble constant, depends largely on how the distances to galaxies are calibrated. The value *H* = 50 km/s per Mpc is assumed here. If an alternative value of *H* = 100 km/s per Mpc were used, the diagonal line would be steeper, which would imply that the universe is expanding twice as fast. The Hubble law has been extended to distances and velocities much larger than those indicated on this graph.

where *v* is the speed of recession, *d* is the distance, and *H* is the *Hubble constant,* the constant by which you multiply *d* to get *v*.

Although it can be expressed in a tidy mathematical form, the Hubble law is not a fundamental physical law. It does not necessarily embody an exact relationship between speed and distance. As a "law," Hubble's law is on a par with Ohm's law in that it describes a trend rather than an inflexible rule.

The constant *H* is essentially a measure of how fast the universe is expanding. Estimates for *H* depend upon current calibrations of the distance scale for galaxies, and the competing methods for calibrating this scale yield somewhat different values for *H*. By one set of estimates, galaxies that are receding at a velocity of 50 km/s are thought to be 1 megaparsec (1 Mpc = $10^6$ pc) away. Consequently, *H* is 50 km/s per Mpc. Other estimates say that galaxies receding at 50 km/s are as little as 0.5 Mpc away. If this is true, then *H* could be as much as 100 km/s per Mpc. (In this chapter, we will adopt and make further use of the value *H* = 50 km/s per Mpc.)

The Hubble law can be clearly visualized in graphical form, as in Figure 17.1. Independent measurements of distance and velocity for various galaxies show that a galaxy estimated to be 10 Mpc away recedes at 500 km/s. One that is 20 Mpc away recedes at 1000 km/s. When the data for *v* and *d* for a large number of galaxies are plotted, virtually all the points fall close to the diagonal line shown. The scatter of points can be explained by uncertainties in the measurements of galaxy distance and by the fact that the expansion is known to be not completely uniform (galaxies are moving

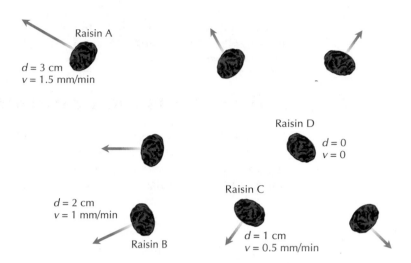

**FIGURE 17.2** In a small portion of an infinite, expanding raisin bread loaf, raisins are pulling apart from one another. For an observer on raisin D (who is zero distance from raisin D and moves at zero speed with respect to raisin D), all other raisins recede at velocities proportional to their distances. The constant of proportionality describing the ratio between velocity and distance for these raisins (analogous to the Hubble constant) would be 0.5 mm/min per cm.

relative to each other and therefore have their own random motions superimposed on the general expansion). The fact that the diagonal line passes through the origin (0, 0) of the graph is significant. The origin is "us"—the observers on planet Earth in the Milky Way galaxy. We are zero distance from ourselves; we are not receding from ourselves.

The mechanics of the expanding universe are often explained in terms of a somewhat humorous analogy: the expanding raisin bread model. Imagine mixing many raisins into a lump of dough with lots of yeast in it. Make sure that no two raisins are touching each other, and bake at 350 °F. What happens? The dough rises (expands) and the raisins move apart from each other.

Imagine an infinitely large, expanding raisin bread loaf (Figure 17.2), with the raisins representing galaxies and the dough representing the space among the galaxies. Any two raisins a short distance apart move apart slowly, but raisins far apart recede faster. If you could place yourself on a raisin, *any* raisin, you would see *all* the other raisins receding from you.

The question naturally arises, "Where is the center of all this expansion?" With regard to the universe, this question cannot be answered; indeed, the question may be moot. In every direction we have looked so far, galaxies recede faster and faster the farther we see. If an edge to the universe exists somewhere out there, we have certainly not seen it yet. If, on the other hand, the universe really is infinite in extent and has no outer edge, it would be impossible to pick out a center. Another possibility is that the space in the universe might be curved in such a way that it is finite in extent but *unbounded*. Unbounded means that you could never come to an edge or boundary, no matter how far you could see or how far you could travel. This kind of universe is analogous to the idealized, two-dimensional, curved space that makes up Earth's surface. A traveler confined to Earth's surface will never reach an edge, nor can any center be found on Earth's surface either.

The raisin bread analogy goes further. The raisins themselves do not expand in size; neither do many of the basic units of matter in the universe. Stars, planets, and galaxies are not, in any overall sense, expanding. Galaxies move about within clusters of galaxies somewhat like bees in swarm, and the clusters of galaxies within superclusters of galaxies do much the same, but there is not much overall expansion within these large units of matter. The universe appears to have a more or less uniform expansion only on the very largest scales.

## How Old Is the Universe?

*When* did the universe start expanding? We can try to answer this question by "running the clock backward." If, in the raisin bread analogy, we imagine time being reversed, we would see the raisins getting closer and closer. A space-time diagram of our raisin loaf is given in Figure 17.3. From the information for various raisins given in Figure 17.2, all the raisins should have been together 20 minutes before present. Ten minutes ago, the raisins were half as far apart as they are at present.

By means of a similar extrapolation, using the Hubble law, we can imagine what the universe was like long ago. The result, shown for several sample galaxies in Figure 17.4, indicates that all the galaxies were together some 20 billion years ago. This figure is presumably the "age" of the universe: the time since the big bang, the event that is presumed to have initiated the expansion.

The figure of 20 billion years depends on several assumptions. It is based on $H = 50$ km/s per Mpc. It is also based on a universe that has expanded at a constant rate during the entire time since the expansion started. A constant expansion rate is unrealistic, however, because the attractive force of gravity, which pulls on everything, has undoubtedly been slowing the rate of expansion. If, indeed, the universe was expanding somewhat *faster* in the past than it is now, then our estimate for the age of the universe must be less than 20 billion years, perhaps 15 billion years. Fifteen billion years, an oft-quoted figure for the universe's age, is a compromise based on somewhat uncertain data from a variety of sources. It is quite consistent with present estimates of the ages of Earth and the other bodies of the Solar System (4.6 billion years old) and just barely consistent with the ages of the oldest stars in the Milky Way galaxy (roughly 12–14 billion years old).

As just described, the evidence is quite compelling that we live in a universe that is expanding. But what causes or caused the expansion? One

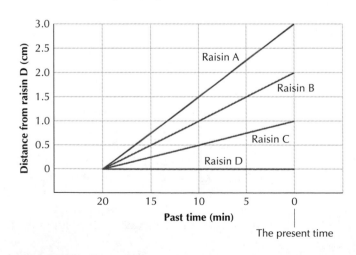

**FIGURE 17.3** This space-diagram for the raisin bread illustrated in Figure 17.2 traces the history (measured in units of past time) of the distance between raisin D and raisins A, B, and C. The extrapolation back into time assumes that the raisins have not changed their speed or direction since they began separating.

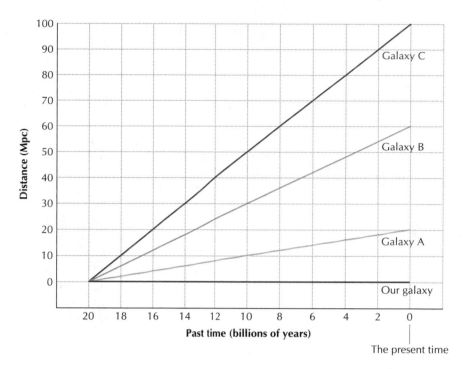

**FIGURE 17.4**  If the present rate of galaxy recession (assumed to be 50 km/s for every 1 Mpc of distance) is extrapolated into the past, then all galaxies were together 20 billion years ago. The past distances of three galaxies (the galaxies plotted in Figure 17.1) relative to our galaxy are graphed here.

straightforward way of answering this question is to invoke a description of a universe that was very dense, very hot, and rapidly expanding during the earliest moments of its existence. This "exploding" state of affairs is known as the big bang. It is also reasonable to assume that after the big bang the universe kept expanding by virtue of its own momentum, as it continues to do today. The discussion on the following pages builds on what we have already learned about the big bang in Step 1 of Chapter 3's timeline.

Note that the big bang theory, as it is presently framed, cannot answer *why* the big bang itself occurred. The answer to that question may forever lie beyond what science can tell us (see Box 17.1). Nonetheless, there is plenty of evidence to support the plausibility of a big bang origin.

## Support for the Big Bang Theory

The notion of a universe expanding from a very dense, hot state dates back to at least 1928, when Georges Lemaître pictured an initial, massive "primeval atom" exploding and giving rise to an immense number of stars and planets including a "well-cooled cinder" that happens to be Earth. In the late 1940s and the 1950s, George Gamow, Ralph Alpher, Robert Herman, and others developed a big bang model (as it became popularly known) that featured details of hydrogen and helium production in the earliest seconds and minutes of the universe's existence. Their model also predicted that the photons released at the *dawn of light* (see Chapter 3's timeline), some 1 million years after the big bang, should still be sailing through space today, filling the universe with a uniform radiation. Their calculations showed that by now these photons should have been stretched in wavelength (and therefore lowered in energy) by a factor of about 1000. In other words, their effective temperature (recall the discussion of *blackbody radiation* in Chapter 11) would have cooled from a fiery 3000 K at $t \approx 1$ million years to a distinctly chilly 3 K (only 3 degrees above absolute zero) today.

Photons emitted from bodies at 3 K span a range of wavelengths in the microwave part of the electromagnetic spectrum, so it is there that we would search for the telltale remnants of radiation from the early universe. In the early 1960s, a group from Princeton University was engaged in building a large antenna and a sensitive receiver to do just that, but they were beaten to the task by two physicists working on an unrelated project. Arno Penzias and Robert Wilson were using a large, horn-shaped microwave antenna to try to track down the source of some static that was interfering with telephone signals sent by means of microwaves. In 1965 they detected, by accident, exactly the sort of cosmic radiation the Princeton group was searching for. Apart from all other sources of static, their antenna was picking up "background radiation" from every direction in the sky.

Today we call this radiation the **cosmic background radiation** (CBR). Amazingly enough, anyone with a television set can easily detect it. Just tune the set to any empty channel, preferably in the UHF range (channels 14–83). About 1 or 2% of the random specks of "snow" you see displayed

## BOX 17.1
# Theology and Science

Is theology opposed to science? Are science and religion mutually exclusive, covering separate realms? Or do they complement each other?

Cosmology, the scientific study of the universe at large, provides a fertile testing ground for questions like these. Both cosmologists (astronomers and physicists) and theologians seek answers to some of the same major questions, like how and why the universe came into being.

Cosmologists attempt to trace effects and causes back to the earliest moments of the universe. Cosmological hypotheses are, in principle, scientifically testable, because evidence of some of the earliest happenings in the universe exists even today. Quantities we can measure, such as the ratios of various elements in the universe and the cosmic background radiation, suggest the existence of past physical causes.

Cosmologists would hit a brick wall, though, if they tried to answer questions like "Who created the universe?" and "What is the purpose of the universe and, for that matter, our own existence?" These questions cannot be scientifically tested, but they are fair game for philosophers and theologians.

Clearly, theology and science approach creation from different perspectives. But these perspectives need not be antagonistic. In fact, most modern religious thinkers see little or no conflict between the notion of a big bang creation and the existence of God or some higher being.

In Eastern religions (Buddhism, Hinduism, Taoism, and others), for example, there is no distinction between the creator and the created. God and nature are the same thing. This pantheistic point of view, which also characterizes the worldviews of many aboriginal cultures, does not conflict with the big bang, or any other scientific explanations of early origins.

In Western religious tradition, God has often been seen as creating the physical world and then standing apart from it. An alternative to this is a *process model* of creation, favored by many theologians today. The deity, although separate, is considered to be still at work in the physical world today (indeed, throughout all time and space).

The process model meshes well with the notion of a universe that starts with a primordial big bang and then develops further. Supporters of the process model believe that the findings of modern cosmology and modern physics are rich sources of evidence for divine design: Nature is full of remarkable coinci-

on your screen are caused by CBR photons picked up by your television antenna or your cable TV system.

As measured from space by the COBE (*Cosmic Background Explorer* satellite)—the latest in a succession of instruments designed to measure CBR characteristics—the radiation is found to correspond to a body radiating energy at a temperature of 2.735 K. This radiation (which does not include the energy given off by stars, planets, and other bodies formed since the big bang) fills the universe almost perfectly uniformly. The existence of the CBR strongly supports the notion of a "hot big bang." Its near-perfect uniformity is consistent with the idea that the big bang took place simultaneously for every part of the universe we observe today and that all parts of the universe were essentially in contact with each other at the earliest moments of time.

Figure 17.5 schematically represents the universe's past history as we understand it today. The four axes, which should not be construed as indicating any particular direction in space, chart various changes since the earliest moment we can (in principle) observe: the dawn of light. Notice that as we look out farther, we are also looking back into time. This *lookback time* is a window into the past, and it works for anything we see as far

dences that conspire together to make a universe that not only survives for a long time but also produces complex biological systems including intelligent life. These coincidences include certain "constants" that govern the strength of the four fundamental forces and laws of nature that allow matter, energy, space, and time to interact so as to produce a universe that is not totally chaotic. A religious person can point to this evidence and say that is God at work. (An atheist would point to the same evidence and say that is natural law; an agnostic would simply say, "I don't know.")

In today's world, science and religion need not be antagonistic as long as they are free of political overtones. Unfortunately, in parts of the United States today, efforts are being made by certain ideological groups to force the teaching of an account of creation, often called "scientific creationism," in elementary and high school science courses. Based on a literal and very narrow interpretation of certain passages in the Christian Bible, scientific creationism asserts that (1) everything in the universe, including life, was created suddenly thousands of years ago, not millions or billions of years ago; (2) Earth's geologic record is almost entirely explained in terms of catastrophic events, especially a worldwide flood; and (3) all living organisms (including humans) were created simultaneously, with changes in the descendents of those organisms being minor.

Scientific creationism is not a part of contemporary scientific thought, and it is hardly a part of mainstream theological thought, either. Its assertions are utterly refuted by the work of the vast majority of all scientists, many of whom adhere to Judeo-Christian religious beliefs. Virtually everything in this textbook, which represents a summary of current scientific knowledge, argues against scientific creationism.

Scientific creationism was long ago rejected by the selection process intrinsic to the scientific method. From a scientific point of view, it remains a historical curiosity, on a par with the notion of a flat Earth: a reasonable worldview for someone who takes into account a tiny subset of knowledge about Earth and the universe and is ignorant of all the rest.

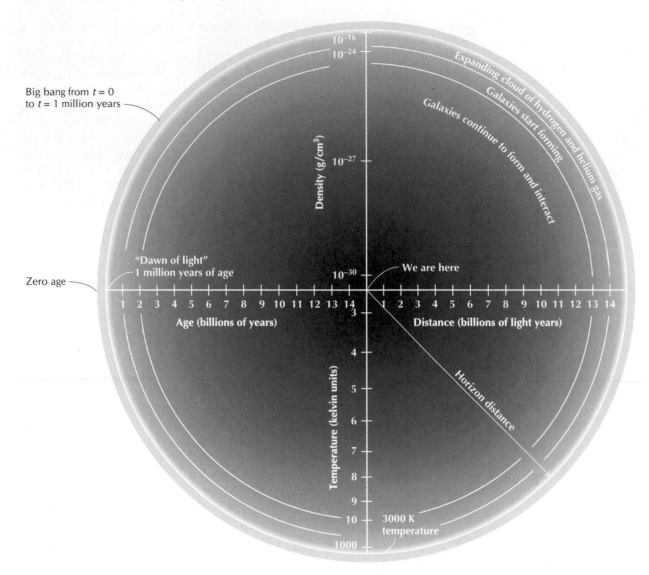

Big bang from $t = 0$
to $t = 1$ million years

Zero age

"Dawn of light"
1 million years of age

We are here

Density (g/cm$^3$)

$10^{-16}$
$10^{-24}$

$10^{-27}$

$10^{-30}$

Expanding cloud of hydrogen and helium gas

Galaxies start forming

Galaxies continue to form and interact

Age (billions of years)

1  2  3  4  5  6  7  8  9  10  11  12  13  14

Distance (billions of light years)

1  2  3  4  5  6  7  8  9  10  11  12  13  14

Temperature (kelvin units)

3
4
5
6
7
8
9
10
1000

Horizon distance

3000 K
temperature

**FIGURE 17.5**  This schematic history traces events and changing conditions in the observable universe. From our present perspective (here and now), at the center of the diagram, we look out to greater and greater distances in space. We also look further back into the past to see the universe at younger and younger ages. With present telescopes, we are seeing newly formed or forming galaxies at estimated distances of as much as 13–14 billion light-years. In principle, we could never see anything in the universe younger than the dawn of light (roughly 15 billion years ago, at $t = 10^6$ years), because the universe was opaque before that time. For up to a billion years after the dawn of light, matter in the universe remained diffuse; it had not yet collected into stars and galaxies that could emit large amounts of light. Astronomers have not spotted such matter yet. Omitted from this diagram are the photons that make up the cosmic background radiation. They were released from all parts of the universe at the dawn of light. These photons, now shifted to microwave wavelengths, fill the universe uniformly. We detect them coming from all directions, not from any particular sources of matter.

back as the dawn of light. Any facts about still earlier times must be deduced by extrapolating backward in time from the dawn of light. For that, we are guided by the known laws of physics, which presumably were the same then as now. We are also guided by the results of high-energy particle accelerator experiments, which tell us how matter and energy interact under very hot (high-energy) conditions.

**PHOTO 17.2** This Hubble space telescope high-resolution image of a rich cluster of galaxies several billion light-years away shows us how galaxies looked when the universe was about two-thirds of its present age. Apparently, spiral galaxies were more abundant in the past in clusters such as these than they are now. Presumably, many spiral galaxies have disappeared or been altered in form through mergers and disruptions, some of which are evident here.

Calculations based on the results of particle accelerator experiments show that after the earliest half hour or so of the universe's existence, nearly all of the ordinary matter (protons, neutrons, and electrons) in it should have consisted of hydrogen nuclei (about three-fourths of the total by mass) and helium nuclei (about one-fourth of the total by mass). Large numbers of lightweight electrons accounted for almost all the remaining small fraction of the mass of ordinary matter. The electrons were later captured by the hydrogen and helium nuclei as soon as the universe had cooled to a sufficiently low temperature (about 3000 K) at the dawn of light.

Today, when astronomers measure the abundances of the elements in both the stars and the interstellar matter of the universe, they find a roughly 75%/25% (by mass) mix of hydrogen and helium. This agreement of observation and theory is taken to be additional strong support in favor of the big bang theory.

## Difficulties with the Big Bang Theory

The supporting evidence mentioned so far—the expansion of the universe, the existence of the CBR, and the abundance ratios of the light elements hydrogen and helium—confirms the general idea of an expanding universe originating from an extremely hot, dense state. This standard model of a big bang universe has successfully passed through the "loop" of the scientific method (Figure 1.4) many times. The standard model, however, runs into difficulties in several ways.

One difficulty is called the *flatness problem*. On the basis of Einstein's general theory of relativity, the universe would be expected to have some kind of overall curvature (like the curved-space examples in Figure 13.7), either positive or negative, depending on how much matter is present in the universe and how fast the universe expands. A positive curvature implies a closed universe (one that will expand to a maximum size and then contract). A negative curvature implies an open universe (one that will expand forever).

A truly "flat" universe, one that is delicately balanced between the closed and open alternatives, is possible but highly unlikely.

The space we observe in our present universe, when measured over large enough scales of distance, seems indeed to be remarkably flat. That is, any light beam traverses a straight or almost straight path over very long distances in space, despite some kinks in its direction over shorter distances. The implication, if the universe is very close to being flat, is that the early expansion was vigorous enough to result in a universe that doesn't briefly expand and then quickly collapse, but not so vigorous as to result in a universe that expands too quickly for matter to congeal into large structures such as galaxies, stars, and planets. According to the standard big bang model, this just-right expansion rate would be evidence of a very special and overwhelmingly unlikely condition on the state of the early universe.

Another difficulty is the *horizon problem.* The astounding uniformity of the CBR implies that the universe was almost perfectly homogeneous in the earliest stages of the big bang. It seems as if all parts of the universe began expanding in a perfectly uniform manner at exactly the same time. Yet when we look in opposite directions to very distant regions of the universe, we realize these regions are too far apart, and have always been too far apart, to ever have "communicated" with each other by the fastest known means—at the speed of light.

A way of solving these difficulties has come, again, by way of discoveries in high-energy physics (particle physics). High-energy particle collisions involve enormously large amounts of energy being released in very tiny volumes of space. Quantum mechanical interactions predominate in these tiny domains, and it is not unusual during these experiments for small quantities of matter to be "created" out of a supply of pure energy. Since the universe was presumably also very tiny at the earliest moments of its existence, any theory attempting to describe its state then would have to involve quantum mechanical interactions. One idea posits that the protouniverse spontaneously appeared from a perfect vacuum, albeit one with a lot of energy in it. Curiously, this idea, which obviously has no parallel in our world of ordinary experience, violates none of the known laws of physics.

From these considerations, a modification of the big bang theory, called the *inflationary model,* has been proposed. It seems to solve both the flatness and the horizon problems. The inflationary model deals with the earliest moments of the universe's existence (up to $t \approx 10^{-30}$ s), during which the tiny protouniverse presumably "inflated" tremendously under a short-lived, repulsive force enormously stronger than any force in existence today. This phase of inflation, which was distinct and much more vigorous than the normal big bang expansion that followed, allowed regions that were in contact with each other to balloon outward so that there could be no further contact by means of anything traveling at the speed of light or slower. This model implies that the entire universe is much more vast than we can see over our current horizon distance (as shown on the outer rim of the circle in Figure 17.5) of about 15 billion light-years. It also implies that although space in the universe may have had some strong overall curvature in the beginning, inflation "ironed" out the curvature. The part of the universe we are capable of seeing today, out to about 15 billion light-years, may be like a small, circular patch on the surface of an immense expanding balloon. Just as an observer on the roof of the tallest building in Peoria, Illinois, sees nothing but an almost perfectly flat horizon all around, we in our provincial little patch of the universe get the impression that space is essentially flat around us.

The inflation scenario, though speculative, is supported by many astronomers and physicists as a working model of the universe's earliest history. Others, in the minority, feel that inflation, or even the big bang itself, never happened. The universe, some say, makes sense in terms of a steady-state model or some other model featuring the continuous creation of matter and energy. According to some individuals, the preference of Western scientists toward models featuring a definite beginning (like the big bang theory) is a consequence of their inability to accept infinite time or cycles that repeat themselves endlessly. Perhaps time and space really are unbounded; perhaps there was no beginning to time and space, nor will there be an end.

It is instructive to read encyclopedia accounts of a century ago on any subject within science and to realize how naive or misled researchers were on certain subjects. This does not mean that the process of science was a failure a century ago. Science permits dissent. Occasionally, a growing movement of dissent overwhelms previous theories and modes of thought (as was certainly the case for plate tectonics in the earth sciences). In this way, science is self-correcting. We do not know what those future encyclopedias will say, but if the past is any guide, our perspectives then will surely be broader and clearer than they are now.

# Life Cycles of the Stars

Stars, like living organisms on Earth, undergo several major changes during their lifetimes. The metaphor cannot be pushed very far, however. For one thing, the "birth," "midlife," and "death" of stars takes place during periods of millions or billions of years, not mere days or years. Also, unlike the life cycles of elephants and ants, for instance, the more massive a star, the more ephemeral is its existence. The most massive stars shine brightly for less than a million years, whereas the least massive stars will glow dimly for periods much longer than the current age (about 15 billion years) of the universe. Our survey of stellar evolution starts with star birth, continues through midlife (the *main sequence* phase) and ends with star death, which may or may not involve a catastrophic transition known as a *supernova.*

One way of tracing a star's evolution in time is to plot its *evolutionary track*—a path that represents how it changes over time—on a graph of fundamental importance in astronomy: the *Hertzsprung-Russell diagram (H-R diagram)*. The diagram is a plot of luminosity versus surface temperature (or, alternatively, spectral type) for stars. When the data points for a large number of stars are plotted on an H-R diagram (as in Figure 17.6), several trends are apparent. About 90% of the stars (the normal, or main sequence, stars) form a prominent sequence of points across the graph. Most of the other points fall into other recognizable areas. Clustered near the lower left (low luminosity, high temperature) corner of the graph are points associated with what are called *white-dwarf* stars. Scattered along the top and

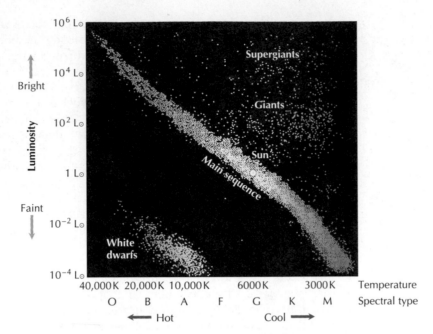

**FIGURE 17.6** This schematic H-R diagram for a large sample of stars in our region of the galaxy illustrates that a large majority of all stars (about 90%) belong to a main sequence in which luminosity is correlated to surface temperature. Within the main sequence itself, there are far more cool and dim stars than there are hot and bright stars. The sun, with a temperature of 5800 K and a luminosity of 1 solar luminosity unit (1 L⊙), lies near the middle of the main sequence. About 9–10% of the stars plotted are *white dwarfs,* which are hot and faint. Although hot stars tend to be bright, white dwarfs are dim only because they are very small. A typical white dwarf is about twice as hot as the sun, about 1/100 the diameter of the sun, and about 1/1000 as bright as the sun. Less than 1% of all stars are giants and supergiants: stars that are both bright and large. The typical giant star (a *red giant*) is somewhat cooler than the sun, roughly 100 times the sun's diameter, and some hundreds or thousands of times brighter than the sun. Supergiant stars are bigger and brighter than giants.

extending to the right are points associated with the bright and, for the most part, relatively cool *giant* and *supergiant* stars. As their names suggest, white dwarfs are quite small (about the size of Earth), and supergiants are extremely large. White dwarfs are not uncommon. One, in fact, orbits around Sirius—the brightest star of the night sky and one of the sun's close neighbors (see Window P in Chapter 2). Supergiants are exceedingly rare; one of the largest known, which is Betelgeuse in the constellation Orion, would fill up the inner Solar System out to the orbit of Mars if it were to stand in the sun's place.

Within the main sequence itself, the position of a star's data point turns out to be strongly correlated to mass. Stars having slightly less than 1/10 the sun's mass anchor the lower right end of the sequence. Stars of roughly 50 solar masses belong to the uppermost extremity of the sequence. There is a vast difference in the relative number of stars on various parts of the main sequence. The cool, dim main sequence stars (plotted at the lower right) are enormously abundant. The hot, bright main sequence stars (plotted at the upper left) are very rare.

An astronomer looking at an H-R diagram plotted for a given sample or population of stars sees not just a static display of the characteristics of all the stars included in the sample but, rather, an instantaneous "snapshot" of moments in the lifetimes of those stars. For instance, the sun, currently a main sequence star with a data point near the middle of the diagram, will one day be a giant star. Later, it will evolve into a white dwarf star.

When projected forward or backward in time, the sun's (or any other star's) data point can be imagined as "moving" on the diagram in response to slow changes in luminosity and surface temperature. The changes that stars go through, as commonly traced by evolutionary tracks, occur at different rates: Some take place in relatively short intervals of cosmic time (as little as hundreds or thousands of years for some stars), and others take place over periods of millions of billions of years.

## Star Birth

Each year, one or more "new" stars appear in the Milky Way galaxy. More accurately put, more than a million *protostars* (stars in the making) in our galaxy are undergoing the slow process of contracting from clouds of interstellar gas and dust during any 1-million-year-long period of time. More than 90% of these protostars are less massive than our sun. They are destined to become cooler and dimmer than the sun. The relatively rare stars that accumulate more mass than the sun in their formative stages are destined to become hot and bright. These stars, although few in number, contribute most of the light generated in our galaxy. Star formation has been taking place in our galaxy for more than 10 billion years, though its pace has probably slowed somewhat over time.

The birth of a star takes at least thousands of times longer than a human lifetime, so no one has ever watched it happen in "real time." Nevertheless, we can observe snapshots of the process—but not without difficulty, since protostars tend to hide inside cocoons of the opaque, dusty material from which they form. Fortunately, though, powerful theoretical tools and techniques help astronomers describe the birth stage and other phases of stellar evolution. Starting with known or assumed values of a star's (or a protostar's) mass, temperature, gas pressure, and density, scientists can develop mathematical models that will predict how a star will change over time.

Within our galaxy, at least, stars are most likely to form within the giant molecular clouds of gas and dust that populate the galaxy's disk component. In these clouds, which are relatively cold and dense compared to the interstellar material outside, even denser regions, of varying size and mass, tend to develop. Under the proper conditions, these denser clouds, or "condensations," may start collapsing under their own gravity.

Astronomers believe that in most, if not all, cases, some external mechanism is needed to trigger the collapse of the condensations. Shock waves rippling through the interstellar medium from a nearby supernova, added pressure within the molecular cloud caused by the relatively sudden birth of luminous stars within the same cloud, and pressure waves moving slowly outward through the galaxy's disk have all been explored as possible mechanisms.

As gravity squeezes a condensation, pressure and temperature rise. Gravitational potential energy is given up in favor of thermal energy as material falls inward toward the center of the condensation. As a result, heat (infrared) is radiated. The condensation is now a protostar.

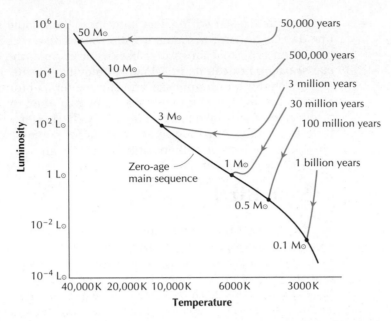

**FIGURE 17.7**    The evolutionary tracks (green) on this H-R diagram represent how surface temperature and luminosity change with time in protostars contracting toward the main sequence. (Note that the arrowed paths shown here do not represent stars traveling in any way through space; they represent changing physical characteristics for stars of different mass.) A low-mass protostar decreases in luminosity even as it heats up, because the area of its surface is being reduced very quickly. As indicated by the times given, the more massive a protostar, the shorter the protostar stage of its life. One-solar-mass (1 M☉) protostars (the protosun, for example) take about 30 million years to contract. Once a star reaches the zero-age main sequence, it is no longer a protostar but is a full-fledged main sequence star. Notice how the masses of the main sequence stars are correlated to their position on the graph.

If the original condensation was rotating even slowly, during contraction it spins faster and some of its material may settle into a disk surrounding the protostar. This disk may later evolve into a system of revolving planets around the forming star (the process outlined as the origin of our Solar Sytem in Chapter 3). Some condensations with a large amount of angular momentum may evolve to become binary or multiple stars. Scientists believe that stars in these configurations are less likely to possess planetary systems.

From theoretical calculations, the protostars themselves evolve in the manner summarized by the evolutionary tracks shown in Figure 17.7. After the initial contraction from gas and dust, protostars are large, softly glowing objects with relatively cool (compared to most stars) surfaces. Normally, we do not actually see these protostars because they are enveloped in dust. Only later, when they become hotter, can they expel their dusty cocoons and become apparent at visible-light wavelengths. For this reason, protostars do not show up on most H-R diagrams.

High-mass protostars contract quickly. As they do so, their surface temperatures increase, but their luminosities do not change much. Lower-mass protostars, on the other hand, take a long time to contract, and they decrease in brightness as they age.

Regardless of how a protostar's surface characteristics change, the temperature and pressure in the interior keep increasing. Eventually, sufficient pressure and temperature (about 10 million K) are achieved in the core of a protostar to trigger the onset of hydrogen fusion. As the nuclear fires turn on, the added pressure balances gravity and the star stops contracting. This event marks the star's "arrival" on the H-R diagram at the *zero-age main sequence* and the beginning of the star's long existence as a stable, main sequence star. Every main sequence star, the sun included, is powered by the fusion of hydrogen into helium.

## Life on the Main Sequence

The zero-age main sequence curve in Figure 17.7 represents the luminosity and temperature characteristics of stars of differing mass as they begin their main sequence life. The most massive main sequence stars (which are also exceedingly rare) are hot and shine very brightly—nearly a million times the sun's luminosity. Stars of much less mass than the sun are thousands of times less luminous than the sun. There is a good reason for this enormous variation in luminosity. The more massive a star, the higher its interior temperature. Fusion reactions are very sensitive to temperature, so even a modest increase in core temperature promotes a much greater output of energy. On a star's suface, this means a somewhat higher temperature and also a much greater luminosity.

As time goes on (billions of years, in the sun's case), a dense core of helium begins to accumulate in the center of a main sequence star. This helium is the "ash" from hydrogen's "burning" (*burning* here refers to nuclear fusion reactions, *not* chemical reactions). Since the core is not nearly hot enough to undergo further nuclear reactions, it is crushed ever smaller by the weight and pressure of the layers above it—and its temperature rises. This causes the hydrogen layers above to burn just a little faster. This process is ever-accelerating and leads to an increase in the luminosity of the star. Slowly at first, then more and more rapidly, the star brightens and its surface expands. On the H-R diagram, this brightening (sometimes accompanied by slight cooling on the surface) translates into a star moving up and to the right from its initial position on the zero-age main sequence curve. This process explains why the main sequence found on random sample H-R diagrams like that of Figure 17.6 is somewhat broad: At any given time, some stars have recently arrived on the main sequence, but others that have been main sequence stars for a long time have experienced some change in their luminosity and temperature.

The sun itself is about halfway through its tenure as a main sequence star. Upon its arrival on the main sequence some 4.6 billion years ago, it was about 70% as luminous as it is today. Some 5 billion years from now it will be several times brighter and well on its way, up and to the right on the H-R diagram, toward the domain of the giant stars.

## Life as a Giant, and Beyond

Like race cars with powerful engines and large fuel tanks that run out of gas quickly, massive main sequence stars don't remain the same for long. They quickly use up their hydrogen fuel (in as little as a million years for the most massive stars) and evolve into supergiant stars. Main sequence stars of low mass, on the other hand, are like small economy cars with

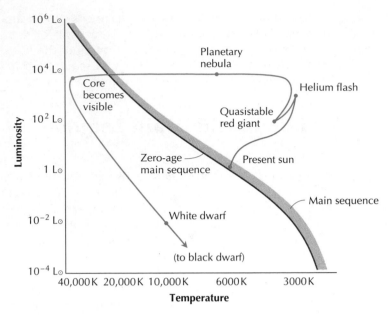

**FIGURE 17.8** This simplified evolutionary track (green line) for the sun traces its changing luminosity and surface temperature over the next several billion years on the H-R diagram. Slowly at first, then more and more rapidly, the sun departs the main sequence and evolves toward the red-giant region at the upper right of the diagram. (Always keep in mind that any "movements" made by stars on the diagram represent changing physical characteristics and not actual movements in space.) Some 6 billion years from now, helium ignites (begins fusing) in the core, a sudden event known as the "helium flash." Afterward, the sun settles into a quasistable state in which the supply of helium in its core steadily fuses into carbon and oxygen. After roughly a billion years in that state, nearly all the helium is exhausted, and the superhot carbon-oxygen core ejects the sun's outer layers of gas in the form of an expanding bubble of gas—a planetary nebula. This stage lasts only thousands of years, rather than millions or billions of years. As the expanding gas thins, it reveals the naked core remnant of the sun, which by then has a surface temperature of around 40,000 K. Since no further nuclear reactions can take place in it, the core can only get cooler and less luminous as it radiates thermal energy into space. After many millions of years, it is a white-dwarf star: still quite hot and somewhat luminous. After a virtually unlimited amount of time, it fades into a cold and nonluminous "black dwarf."

somewhat small gas tanks, small engines having phenomenal gas economy, and long ranges between gas stops. These stars are barely simmering, with core temperatures just high enough for hydrogen fusion to work. Some may persist as main sequence stars for upwards of a trillion years. The sun falls in between these extremes, with a main sequence lifetime of about 10 billion years.

To see what "life" is like at the giant (or supergiant) phase and beyond, let us follow the evolution of the sun over the next several billion years. The following scenario (which is also plotted in Figure 17.8) would be about the same for most other stars, too, as long as their masses do not exceed several times the sun's mass. (The rate that a star changes depends on its mass. The more massive a star, the more quickly it goes through all of its evolutionary phases.)

By about 6 billion years from now, the sun, somewhat cooler and bloated on the outside, shines as much as 1000 times brighter. Internally, it is extremely hot: nearing 100 million K in the helium ash core. Eventually, at a critical moment (estimated to last minutes—truly an eyeblink on the astronomical time scale), the rising temperature causes the compressed helium to explosively ignite. The event is known as the *helium flash.*

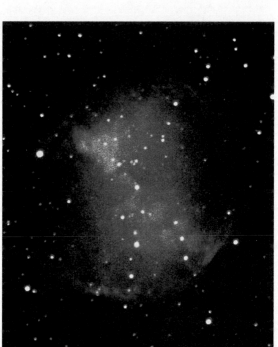

**PHOTO 17.3** The Dumbbell nebula, one of the best-known planetary nebulas in the sky, consists of gas expanding away from the central star (the small blue star in the middle) at about 27 km/s. The central star started ejecting this gas some 50,000 years ago. The Dumbbell nebula is unusual in that the gas was not ejected symmetrically.

Curiously, virtually all the energy released by the helium flash is transformed into the work needed to expand the helium core. Subsequently, the sun settles down into a quasistable state in which its primary source of energy is helium burning: the steady fusion of helium nuclei into carbon nuclei and lesser amounts of oxygen nuclei. At the same time, some of the remaining hydrogen in a shell surrounding the core continues to fuse into helium. During this quasistable phase, which lasts about a billion years (only one-tenth as the long as the previous, stable main sequence phase), the sun remains in the giant-star region of the H-R diagram.

Eventually, a dense, hot core of carbon-oxygen ash begins accumulating at the sun's center, just as helium accumulated in the earlier, main sequence sun. Temperatures much hotter than 100 million kelvin are needed to fuse the carbon and oxygen to other elements. Only in stars much more massive than the sun can this happen, however.

In time, the quasistable phase comes to an end. The shell of burning helium, which surrounds the carbon-oxygen core, develops instabilities. Internal explosions called *thermal pulses* occur again and again, causing the sun to expand and contract and thereby become brighter and dimmer, more or less rhythmically. For a time, the sun becomes a variable star, somewhat like the Cepheid variables mentioned in Box 8.1. Relatively quickly, the energy released by the thermal pulses ejects the outer, nonburning layers of the sun in the form of an expanding cloud of diffuse gases called a *planetary nebula* (see Photo 17.3). After as little as a thousand years, these gases become thin and transparent enough that the light from the now-exposed core of the sun shines through. It is now the "central star" of the nebula, and it is still very hot and bright. If any stage in the life of a sunlike star can be called "death," this is the beginning of it. By then, all that is left in this dense core (nearly a million times denser than the present sun) is carbon and oxygen, which is not hot enough outside or inside to participate in any further nuclear reactions. Like a glowing coal in a fireplace, this core can only get cooler and dimmer as time goes on—relatively quickly at first, then more slowly. On the H-R diagram (Figure 17.8), it slides toward the white-dwarf region and on farther to become a cold and nonluminous *black dwarf.* Such is the ultimate fate of our sun and the majority of stars in the sky.

It may be tempting to think of the previous stellar evolution scenario as some kind of fanciful projection, but it is supported by an abundance of astrophysical evidence. For example, when H-R diagrams of star clusters are plotted, we see reflected in the hundreds or thousands of plotted points snapshots in the life histories of each of the plotted stars. In clusters, all stars begin their lives at about the same time. Massive stars evolve quickly from the upper main sequence toward the giant or supergiant regions on the diagram; the less massive, more slowly evolving stars remain on the main sequence for a longer time. When nearly all members of a cluster are main sequence stars, we know that cluster is relatively young (typically tens of millions of years old). When no stars brighter than the sun appear on the main sequence, but instead a number of them appear in the giant region of the diagram, we know that the cluster is very old (about 10 billion years old). We know these ages because stars of 1 solar mass last about 10 billion years in the main sequence phase; after that, they become giants.

# Exploding Stars and Nucleosynthesis

In a star much more massive than the sun (an upper main sequence star), further fusion reactions can take place as it evolves toward the upper middle or upper right portion of the H-R diagram, the region of supergiant stars. These reactions take place sequentially, resulting in the synthesis of nuclei heavier than carbon and oxygen. By the time it starts accumulating iron in its core, the star has a structure resembling that of an onion, with light elements on the outside and heavier elements toward the center (see Figure 17.9). At the star's center, the temperature is approaching an incredible 1 billion kelvin. It is also, at that time, rapidly heading for a crisis, because the iron nucleus lies at the pinnacle of stability on the binding energy curve for nuclear reactions (refer to Figure 12.8 in Chapter 12) and therefore will not spontaneously participate in fusion reactions to form still heavier elements. Instead, more and more iron becomes concentrated at the center. When enough iron accumulates, a critical point is reached in which nothing can stop the force of gravity from suddenly collapsing the core. Within a time period shorter than a second, the core region, which is about the size of Earth, catastrophically implodes to form a sphere some 20 km across—a neutron star. Coupled with this implosion is a titanic explosion that results when the star's outer layers rush inward and pile up against the newly collapsed, rigid core. This external explosion is one form of a supernova.

So far we have learned that the primordial material of the universe, hydrogen and helium, is converted to the heavier elements carbon and oxygen in aging stars of sunlike mass. We have further learned that massive

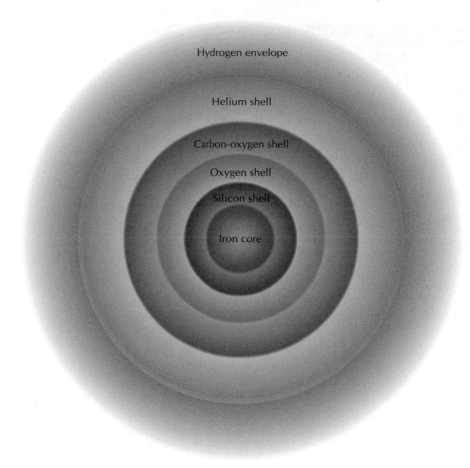

Hydrogen envelope

Helium shell

Carbon-oxygen shell

Oxygen shell

Silicon shell

Iron core

**FIGURE 17.9** An aging, massive star develops layers of successively heavier elements. One idealized model of a star about to undergo a supernova explosion is shown here. When enough silicon fuses into iron, the accumulated iron in the core suddenly collapses under its own weight and the weight of the star's outer layers.

stars, when they age, go on to synthesize elements as heavy as iron in their cores. When a supernova outburst takes place, a final burst of nucleosynthesis takes place, creating (either instantaneously or by slower radioactive decay) an enormous variety of elements and isotopes. This newly synthesized debris, as well as some of the star's original material, is propelled outward into the spaces between the stars and mixes in with the material already filling those spaces. The admixture of primordial gases (hydrogen and helium) and the newer heavier elements formed in the stars becomes the raw material for future generations of stars and planets. The death of old stars helps provide for the life of new stars and the planets that will likely form around some of them.

In our galaxy, the earliest generation of stars condensed from primordial hydrogen and helium. The massive, short-lived stars among these, enriched with newly synthesized elements, exploded and seeded the galaxy with material that would help form future stars. Successive generations of stars repeated this process, further enriching the interstellar medium with heavy elements. Some 5 to 10 billion years after this process started in our galaxy, interstellar material condensed into what became our Solar System.

Just before the Solar System formed, the heavy-element content of the interstellar medium in our part of the galaxy was about 2%. This percentage is reflected in the overall composition of the Solar System today. The sun, which has a bit less than 99.9% of the total mass of the Solar System, has a

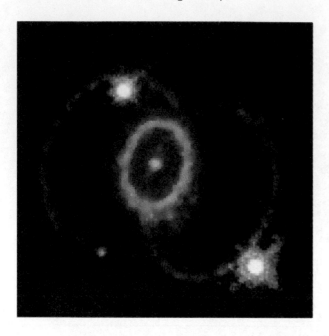

**PHOTO 17.4**   Supernova 1987A, in the dwarf galaxy called the Large Magellanic Cloud some 160,000 ʟʏ away from Earth, is offering the current generation of astronomers a rare chance to follow the expansion of a supernova's remnant material in the years and decades following the supernova itself. The triple-ring structure surrounding the site of the explosion, shown in this 1994 Hubble space telescope photograph, has yet to be explained. Supernovas occur so rarely that only three have been seen in our Milky Way galaxy during the past 1000 years.

surface composition of 73% hydrogen, 25% helium, and about 2% of all the other elements. The two most massive planets, Jupiter and Saturn, have a similar breakdown of elements. (As noted in Chapter 3, Earth and the other terrestrial planets wound up with mostly heavy elements and little hydrogen and helium because they condensed in a warm region near the protosun. Close to the protosun, the light elements and compounds never condensed, and they were swept outward by the solar wind.)

## Neutron Stars and Black Holes

One loose end remains in our story of stellar evolution. What becomes of the cores of massive stars when they implode? These remnants are mostly neutron stars; less commonly (perhaps extremely rarely), they are black holes. If the remnant is a neutron star, then it is a sphere roughly 20 km across having a mass somewhere between 1.4 and about 3 solar masses and a density of trillions of times that of the sun. Freshly formed neutron stars typically spin furiously fast as well, several times or tens of times each second, because the angular momentum of a star's core is conserved as it collapses (recall how bodies that shrink inward tend to spin faster).

Neutron stars are commonly found in or near the centers of expanding debris clouds called *supernova remnants*. Quite often, the strong magnetic fields anchored in these spinning neutron stars cause intense beams of radiation to sweep around the sky like the rotating beam of a lighthouse. When Earth happens to lie in the path of one of these beams, we see a flash every time the neutron star turns. This neutron star is then called a *pulsar*. The Crab nebula (pictured in Chapter 5), one of the nearest, youngest, and most famous of supernova remnants, harbors such a pulsar. Pulsars do not last forever. The energy they radiate is derived almost entirely from their angular momentum. Like the frictional forces that gradually slow a rotating bicycle wheel, pulsars spin down gradually—a process that takes millions of years or more.

If enough mass is tied up in the collapsing core of a star, theory predicts that the compressive effects of the gravitational field, which become ever greater as the core gets smaller, cannot be halted by any known force. The matter in the collapsed core shrinks toward zero volume and infinite density and disappears from sight as soon as the space around it becomes so sharply bent that it curves in on itself. This is a **black hole.** It emits no light (or anything else) to the outside, but it may make its presence known by its gravitational effect on any neighboring masses.

White dwarfs (on their way to becoming black dwarfs), neutron stars, and black holes are the three possible "end points" of a star's evolution. As a star ages, however, it returns some of its material to interstellar space, which helps perpetuate the lives of future stars.

## Whither the Universe?

It is an expression of natural human curiosity to wonder where we came from and where we are going in this universe of ours. From a scientific perspective, the first part has already been answered, however tentatively, by the big bang theory and the subsequent evolution of the stars and planets in the universe as traced by astronomers. What might we say about the far future?

First, let us consider just our immediate surroundings. Life as we know it will probably be extinguished from our planet when the steadily brightening sun eventually (don't hold your breath) boils away the oceans of our planet around 2 billion years from now. During the later stages of the sun's giant phase, billions of years hence, the sun may expand to a radius of 1 astronomical unit (AU) or more, enveloping Earth. Our planet's crust and mantle will likely be evaporated. As Earth plows through the sun's outermost layers, the frictional drag will pull Earth into the sun's inner regions. Our planet (and its neighbors) will be completely vaporized. The vaporized remnants will eventually be pushed outward into space during the sun's planetary nebula phase. Then they will condense to form tiny particles of interstellar dust.

When we make projections for the universe as a whole, we normally adopt the quite reasonable assumption that gravity is the only large-scale force currently acting on our expanding universe. (Other forces were obviously at work during the big bang.) With this in mind, there are only two possibilities:

1. The universe is open. It will expand forever, though at an ever-slowing rate because of gravity.
2. The universe is closed. Gravity will succeed in slowing the universe to a halt at some future time; thereafter, the universe will contract.

How may we decide whether the universe is open or closed? The analogy of a rocket fired directly away from Earth is a good one. Moving at less than Earth's escape speed, the rocket goes up and falls back. If it moves at greater than the escape speed, the rocket does not return, even though it

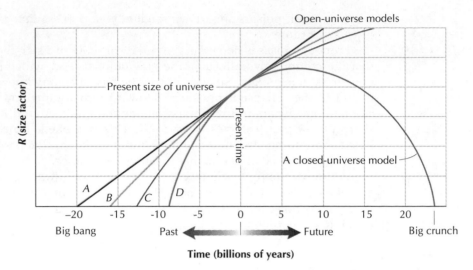

**FIGURE 17.10**   In this graph of the size of the universe (*R*) versus time, four model universes are plotted. Straight line *A* represents a universe with no matter in it. Since there is no mass in this universe to slow the expansion, the expansion is constant and the universe is open. Curve *B* represents a more realistic universe with matter in it, but the density is low. It, too, is open. Curve *C* represents a universe having exactly the critical density. It is open, but just barely. Curve *D* represents a universe densely populated with matter. After its expansion is halted by gravity, it will contract. All of these models are based on an assumed present Hubble constant of 50 km/s per Mpc. If the Hubble constant proves to be larger than this, then each model universe is younger, and more matter would be needed to "close" the universe.

will always be subject to the Earth's far-reaching, weaker-with-distance gravitational pull.

Extend this line of thinking to two bodies (galaxies, perhaps) moving away from each other. The more massive the bodies and the closer they are to each other, the greater the relative speed necessary for them to "escape" each other. At less than escape speed, the bodies will fall back and collide with each other.

Now think of a very large number of bodies undergoing mutual recession in such a way that they all originated from the same point. If we could somehow measure the mass of all the bodies and their distances from one another, we could determine the escape speed for the system—that is, the minimum speed at which the bodies would continue to separate from each other indefinitely.

The expansion of the universe can be described in this line of thinking. The quantity we need to know is density. If the average density of the universe is less than a value called the *critical density,* then the present rate of expansion of the universe is sufficient to ensure that expansion will continue forever. If the density of the universe is exactly equal to the critical density, then the universe will stop expanding, but only after an infinite amount of time has passed. The universe would then be *marginally open.* If the universe is more dense than the critical density, then it will surely reach a maximum size and then begin to contract. In this closed universe, the contraction would presumably lead to a "big crunch." The three alternatives (which really boil down to two, open and closed) are schematically plotted in Figure 17.10.

To decide whether the universe is open or closed, we need to know (1) how dense the universe is and (2) how fast the universe (both space and the matter in it) is expanding. The Hubble constant gives us the latter; however, there is much disagreement at present about the true value of the Hubble constant. New measurements of galaxy distances made with the help of the Hubble space telescope seem to indicate a value of $H \approx 80$ km/s per Mpc, instead of the earlier, widely accepted value of $H \approx 50$ km/s per Mpc. (The new measurements are quite puzzling, because they seem to indicate a "younger" universe of some 10 billion years. In such a universe there would not be enough time to account for the existence of stars that are calculated to have been in existence for nearly 15 billion years. It is entirely possible that the seeds of this current controversy will lead to fundamentally new ways of modeling the universe.)

The average density of the universe is notoriously difficult to determine as well. It is based on measurements of mass and volume, and there are rather large uncertainties associated with both. The volume of any large part of the universe depends on measurements of galaxy distances and indirectly on the Hubble constant, which (as we just pointed out) is not known with certainty. To complicate matters further, estimates of the amount of mass in the universe vary greatly. Taking into account all the ordinary or *visible matter* (the matter that gives off light or other electromagnetic radiation) in the universe or in any representative part of it, we come to the conclusion that the universe has only about 1% of the matter needed to close it. That is, the actual density is about 1/100 of the critical density, and therefore, we live in an open universe.

However, there is abundant evidence that the universe contains enormous amounts of *dark matter* lurking amid all the visible matter, such as stars, glowing nebulas, and other sources of light or other radiations we can detect with telescopes. A case in point is the massive but unseen coronas that belong to our galaxy (as mentioned in Chapter 2) and other galaxies. The existence of these coronas is inferred from the fact that the outer stars of these galaxies are moving far faster than they should unless there exist great reservoirs of mass either superimposed on the galaxies or surrounding them. More evidence is found in the fact that member galaxies within a cluster of galaxies typically move relative to each other at unexpectedly rapid speeds. There simply must be much more mass contained in the clusters than we can see, because the gravitational forces causing this rapid motion are large.

No one knows whether dark matter comes in the form of vast numbers of black holes, failed stars too dim to see, or ghostly subatomic particles capable of exerting gravitational force. Perhaps there are several forms of dark matter. The identification and quantitative measurement of dark matter is one of the central problems in astronomy today. Estimates of the density of the universe and, thus, knowlege of the ultimate fate of the universe largely depend upon the ability of astronomers and physicists to solve the problem of dark matter.

For many people, the issues of dark matter and the ultimate fate of the universe belong to a speculative realm with no connection to their personal life. But there will always be others, especially scientists on the cutting edge of discovery, whose inquiring minds simply want to know.

# CHAPTER 17
# Summary

Current evidence strongly suggests that galaxies in every direction are receding from us. All galaxies are observed to recede at rates that are roughly proportional to their distances, a relationship known as the Hubble law. The Hubble law indicates that the universe as a whole is expanding and that it was smaller in the past than it is now. The Hubble constant, a measure of how fast the universe is expanding today, is estimated as having a value in the range of 50–100 km/s of recession speed per megaparsec (Mpc) of distance.

The existence of the cosmic background radiation (CBR) and the observed relative abundances of hydrogen and helium in the universe today support the big bang theory: that the universe expanded from a very small, dense, high-temperature state. The estimated age of the universe (time since the big bang) hinges primarily on the true value of the Hubble constant. If the Hubble constant is 50 km/s per Mpc, then the age is probably near 15 billion years.

The Hertzsprung-Russell (H-R) diagram, a plot of luminosity versus surface temperature for stars, may be used to categorize stars into groups and also to visualize the changing conditions of stars as they evolve from one stage to another in their lives. Most stars in the universe are members of the main sequence; their data points fall upon a well-populated curve stretching from the hot, bright region of the H-R diagram to the cool, faint region. Lesser numbers of stars are classified as white dwarfs (small in size, relatively hot in temperature, and faint in luminosity) or as giants and supergiants (large, bright, and relatively cool).

Stars are born within diffuse gas-and-dust regions in space called giant molecular clouds. Condensations within these clouds gravitationally collapse to form protostars, and protostars generate radiant heat and then light as they get hotter through further contraction. Protostars eventually stabilize and become main sequence stars, which are stable stars that generate their energy by hydrogen fusion. The sun has been a main sequence star for nearly 5 billion years now.

As stars age, they swell in size, gradually becoming luminous giant or supergiant stars. Inside giants that have evolved from sunlike main sequence stars, helium begins fusing into carbon and oxygen, but no further nuclear reactions take place. These stars are destined to become white dwarf stars. The relatively rare supergiant stars are, for the most part, massive stars that have evolved from the upper main sequence. Nuclear reactions in the cores of these stars continue to produce elements heavier than carbon and oxygen. Many of these stars undergo a supernova explosion and, in so doing, scatter newly synthesized heavy elements into interstellar space. New stars coalesce from clouds of primordial hydrogen and helium enriched by the heavy elements contributed by supernovas.

The fact that the universe is expanding leads to two possible scenarios of its future: The universe may be open, which means it will expand forever; or it may be closed, which means it will eventually stop expanding and subsequently contract. The density of matter in the universe is a crucial factor in deciding between these outcomes. This density is difficult to estimate largely because of the presence of large amounts of dark matter, which cannot be detected by conventional means because it emits no electromagnetic radiation. When and if the characteristics and quantity of dark matter in the universe are determined, the ultimate fate of the universe will be easier to predict.

## CHAPTER 17
# Questions

## Multiple Choice

1. The Hubble law states that
   a) galaxies are receding from us with velocities directly proportional to their distances
   b) the farther away a galaxy is, the younger it appears
   c) the farther away a galaxy is, the older it appears
   d) the farther away a galaxy is, the faster it rotates

2. If galaxy X is four times more distant than galaxy Y, then according to the Hubble law, galaxy X recedes
   a) two times faster than galaxy Y
   b) four times faster than galaxy Y
   c) one-half as fast as galaxy Y
   d) one-quarter as fast as galaxy Y

3. The cosmic background radiation
   a) indicates that the universe will expand forever
   b) comes from a single direction that indicates the center of the universe
   c) is the remnant radiation left by a supernova
   d) supports the notion that the universe started with a big bang

4. A universe that is flat, or has no overall curvature, is one that
   a) is static
   b) is closed
   c) is marginally open
   d) has a finite lifetime

5. The inflationary model
   a) explains why the universe is expanding today
   b) covers specific events that took place in the first tiny fraction of the first second of the big bang
   c) predicts that the universe will eventually collapse in on itself
   d) would be contradicted by the discovery of large amounts of dark matter

6. The Hertzsprung-Russell (H-R) diagram is fundamentally a plot of
   a) luminosity versus size for stars
   b) mass versus temperature for stars
   c) luminosity versus temperature for stars
   d) mass versus luminosity for stars

7. The main sequence represents stars
   a) of the same mass
   b) of the same luminosity
   c) of the same temperature
   d) whose sole source of energy is the fusion of hydrogen into helium

8. Stars lying above and to the right of the main sequence on the H-R diagram are relatively
   a) large and faint
   b) cool and faint
   c) large and bright
   d) small and bright

9. A protostar's mass determines
   a) the length of time it will take the protostar to evolve to the main sequence
   b) how bright the protostar will be once it becomes a stable star on the main sequence
   c) the temperature of the protostar's surface, once it arrives on the main sequence
   d) all of the above

10. Stars of relatively large mass
    a) have relatively short lives
    b) have relatively long lives
    c) are more common than stars of small mass
    d) always end their lives as white dwarfs

11. For a star having about the same mass as the sun, which is the correct sequence of evolutionary stages?
    a) protostar, main sequence star, giant star, supernova, neutron star
    b) protostar, main sequence star, supergiant star, supernova, black hole
    c) protostar, main sequence star, giant star, planetary nebula, white dwarf, black dwarf
    d) protostar, giant star, main sequence star, nova, white dwarf

12. What is the sun's estimated lifetime on the main sequence?
    a) 4.6 million years
    b) 5 billion years
    c) 10 billion years
    d) 15 billion years

13. Elements heavier than iron on the periodic table are primarily produced in stars in which evolutionary stage?
    a) protostar
    b) main sequence

c) giant
d) supernova

14. Pulsars spin very rapidly because
 a) of their huge size
 b) of the way they conserved angular momentum as they shrank toward a very compact size during the supernova phase
 c) their gravitational fields are enormously strong
 d) the stars they formed from were spinning just as fast

15. In a closed universe,
 a) expansion will continue forever
 b) no additional new stars will form
 c) the universe will neither expand nor collapse
 d) expansion will stop and the universe will ultimately collapse

16. To determine whether the universe is open or closed, one must accurately measure the universe's
 a) size
 b) mass
 c) volume
 d) average density

# Questions

1. Many things about the history of the universe are illustrated in Figure 17.5. How and why is the age of a galaxy as we see it related to that galaxy's distance? How and why is the overall temperature and density of any part of the universe related to its distance from us?

2. How have discoveries in the field of particle physics (high-energy physics) shed light on the workings of the very early universe?

3. What is the significance of an evolutionary track on the H-R diagram?

4. Briefly describe how a star is born within a giant molecular cloud.

5. Why is it that protostars cannot be seen clearly?

6. What is the significance of the *zero-age main sequence* on the H-R diagram? How does it differ from the *main sequence*?

7. Why do scientists believe that many of the atoms contained in our bodies were at one time inside one or more stars?

8. What do white dwarfs, neutron stars, and black holes have in common with each other? How are they different from each other?

9. What is dark matter, and why is it so important to identify and measure it?

10. In this chapter, we have used chemical terms such as *burning, ignite,* and *ash* in their metaphorical sense to describe *nuclear,* rather than *chemical* (combustion), reactions. From what you have learned in earlier chapters about reactions involving hydrogen, can you describe the essential differences between hydrogen burning in a chemical sense and hydrogen burning in a nuclear sense?

# Problems

1. According to the Hubble law (Equation 17.1), what should be the distance of a galaxy that recedes at 850 km/s? What should be the velocity of a galaxy that is 200 Mpc distant? (Assume that the Hubble constant is 50 km/s per Mpc.)

2. If the Hubble constant is 50 km/s per Mpc, what is the maximum distance we can theoretically see within our universe at the present time? Assume that the expansion rate of the universe has not changed over time. (*Hint:* We could not see anything moving away from us at faster than the speed of light—if, indeed, anything does.)

# Questions for Thought

1. Will the temperature of the cosmic background radiation ever reach 0 K if the universe expands forever? Can this happen if the universe expands to a certain maximum size and then contracts?

2. Do you think the big bang theory is a better example of the workings of the scientific method than, say, the development of the Copernican model or (in the medical field) the detective work that was involved in the discovery of insulin?

3. Since helium is an inert gas belonging to the noble gas group of the periodic table, how can it possibly "ignite" in a giant star and form carbon and oxygen?

4. If our universe is, in fact, an open one, what do you think it will be like many hundreds of billions of years from now? Consider such issues as the average density of the universe and the relative amounts of interstellar matter, ordinary stars, and "dead" stars (such as white dwarfs, neutron stars, and black holes) that might be contained in it.

## Answers to Multiple-Choice Questions

| | | | | | |
|---|---|---|---|---|---|
| 1. a | 2. b | 3. d | 4. c | 5. b | 6. c |
| 7. d | 8. c | 9. d | 10. a | 11. c | 12. c |
| 13. d | 14. b | 15. d | 16. d | | |

# EPILOGUE
# Our Fragile Earth

The broad perspectives of space and time that have pervaded much of this book illustrate how the human mind, with the help of research tools, can reach across great distances in space and across long intervals of time. Modern science has given us another, more introspective outlook on the world. About 25 years ago, several Apollo space missions carried astronauts to the moon and back. Through the eyes of the Apollo astronauts and through the exquisite photographs they brought back, humanity could, for the first time, witness in full color and great detail our planet's bubblelike surface. These photographs still have impact. Earth, in all its delicate glory, appears as an unbounded but clearly *finite* space, an island of life in a sea of blackness.

There is no question that Earth, considered as a big hunk of rock and metal, will endure for billions of years to come. What is in question is how rich in resources and how biologically diverse our planet will be in the short term. Each year, more forests, grasslands, and wetlands disappear. Deserts continue to overtake formerly productive farmlands. Human-induced erosion scours away vital topsoils. Earth's atmosphere, streams, soils, and groundwater become more polluted with toxic wastes. The rapid extraction and burning of fossil fuels and the wholesale burning of large tracts of tropical forest significantly alter the chemistry of the atmosphere.

Every day, biologists estimate, several dozen species of plants, animals, or insects become extinct as a direct or an indirect outcome of human activities. Even when we recognize that Earth today harbors literally millions of species, the rate of loss (tens of thousands of species annually) is staggering.

Earth has never been a static planet, nor (as we have seen in Chapter 3) is it a stranger to catastrophic events. What is unusual about our particular place in Earth history is that the current rate of human-induced biological alteration rivals episodes of mass extinction by *natural* causes Earth suffered in the far-distant past. From a geological or an astronomical perspective, today's biosphere is being radically transformed in an eyeblink.

**PHOTO E.1**   Visible and pungent forms of air pollution emanate from sources like this paper mill in Washington State. Other forms of air pollution, such as carbon monoxide, are colorless and odorless yet just as harmful to the biosphere.

**PHOTO E.2**   Systematic deforestation in tropical Brazil is evident in this false-color image obtained by the *Landsat* satellite. Areas of dense vegetation are coded red; blue lines indicate slash-and-burn activity.

A major cause of our problems can be traced to the explosion of world population, which totals about 5.7 billion people today. The world population is growing by approximately 1.7% every year and is expected to reach 6 billion or more by the year 2000. Looking at population figures of the past, we may think that explosive growth has sneaked up on us: In 1850, the world population was just 1 billion.

**PHOTO E.3**   Incredibly crowded cities such as Tokyo are the present culmination of a trend that has continued for thousands of years. Even larger, more dense megacities may house the majority of the world's citizens in the 21st century.

The current 1.7% annual growth rate translates to a yearly increase in world population of about 100 million per year (0.1 billion per year). On the basis of a continued 1.7% growth, coupled with an estimated 2000 population of 6 billion, is it valid to predict that Earth will have roughly 7 billion people in the year 2010, 8 billion in the year 2020, 9 billion in the 2030, and 10 billion in the year 2040? The answer is no. Even these grim numbers fall considerably short of actual population projections—assuming the annual growth rate remains 1.7%. The 1.7% increase each year refers to a population base that itself increases by 1.7% each year. With every passing year, the factor 1.7% multiplies a larger number, and the population added each year grows ever larger.

Whenever a quantity such as population, or money in the bank, grows at a fixed positive percentage per year, we say that its growth is exponential. The following relationship reveals the power of exponential growth over time.

doubling time in years ≈ 70/annual % increase

By dividing 1.7 into 70, we obtain the current doubling time of the world population: approximately 40 years. Thus, the population of 6 billion in the year 2000 will balloon to about 12 billion in the year 2040, not the 10 billion of our previous, erroneous estimate. If we extrapolate along similar lines, Earth will harbor 24 billion people by about 2080 and 48 billion by about 2120. Should these trends continue for 500 years (the year 2500), the world population would reach more than 30 trillion. At this point all of humanity, standing shoulder to shoulder, would fill every spot of dry land on Earth. (The effect of exponential growth and its attendant doubling time is graphically illustrated in Figure E.1.)

It is plainly absurd that 30 trillion people could reside on this planet. There is obviously something wrong with the assumption that Earth's population today can continue to grow at an annual rate of 1.7% for centuries to come. Long before the year 2500, humanity will either voluntarily gain

**PHOTO E.5** If humanity can keep the lid on population growth, there will always be enough space for people to live and breathe.

the noble goal of environmental protection. I believe, however, that what everyone must begin to share is a common vision of our planet as a finite assemblage of matter, organized in a unique way and gravely susceptible to human actions that do not consider future consequences. The ethic of domination over nature at any cost—which has been hastened by the development of modern science and technology and practiced with single-minded determination by some industrialized societies for two centuries—simply must change.

I also believe that no single, enlightened ethic or set of attitudes can replace the old ethic of unbridled growth and exploitation. There are many paths toward the goal of sustaining a healthy planet. Some people have adopted a purely human-centered "spaceship Earth" point of view: Earth is an intricate machine needing the utmost care and attention if it is to continue to nurture human civilization. Others have adopted a similar, somewhat broader perspective: All living species merit our attention and protection, because our welfare ultimately hinges on their good health.

Many people in the environmental movement have embraced an even stronger "ecocentric" (life-centered) viewpoint in which all life—not just human life—has intrinsic value. According to this way of thinking, humans are not necessarily the pinnacle of evolution. Rather, we are a part of life on Earth, perhaps not even essential to its well-being. Thus, we are obligated to learn about the rest of nature, to discover how we as humans fit into it, and to model our lives after the economical and sustainable mechanisms that nature has already evolved.

The natural sciences, and the technology that flows from them, are powerful tools of change. They have influenced nearly every aspect of our lives, and they have presided over all the human-engineered processes that have so radically transformed the surface of our planet over the past century. Advancement of scientific knowledge will surely continue. What we do with that knowledge is up to us, the leaders and citizens of the present and future world. It behooves us to make sure that the science we study, and the technology it engenders, is applied in ways beneficial to humans and all the rest of nature.

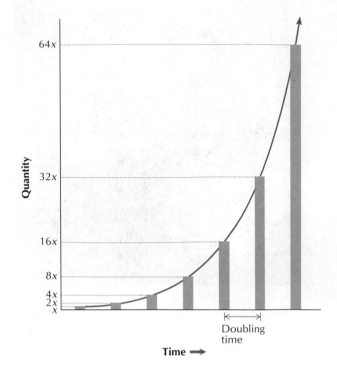

**FIGURE E.1** The rate at which exponential growth takes place can be characterized by an interval of time called the doubling time. The exponential growth over six successive doubling times of an initial quantity $x$ is 2 to the 6th power, or a factor of 64.

control of its exploding population growth or else face widespread famin disease, or other agents that will halt any further increases in population cause a population crash.

Even if we discount the population problem, another exponential-growt issue looms in our future: the ever-increasing (and already staggering) rat at which Earth's natural resources are being consumed. In many instance this rate far outstrips the rate of population growth. The industrialize countries of the world have had the lead in resource consumption although many fast-developing countries, such as China, are now catching up. In some parts of the world, the exploitation of nonrenewable resources such as fossil fuels and minerals is proceeding at truly frightening rates of growth. (Even a mere 7% annual growth in the use of a resource, which may seem slight at first glance, is explosive over the lifetime of a typical human being: From the previous formula, $70/7 \rightarrow 10$ years, which means that the use would double in one decade, quadruple in two decades, and increase eightfold in three decades.)

It seems that the answer to our problems should lie in the prudent use of Earth's renewable resources—wood from trees that can replace themselves in healthy forests, freshwater that cleanses itself as it participates in the hydrologic cycle, and crops that rely on the regenerative ability of the soil that sustains them. Unfortunately, in most parts of the world, the exploitation of these resources has been far too rapid. Wood becomes a non-renewable resource when trees of the tropical forests are slashed and burned, and the thin soils that supported them are exhausted by a few years of intensive agriculture. Water ceases to be reusable when it is poisoned by toxic pollution.

How, then, can we humans extricate ourselves from our worsening global predicament? Hundreds (or even thousands) of distinct actions can support

# APPENDIX A

# Powers of Ten and Scientific Notation

Many of the numbers used to express quantitites in physical science are either very large or very small. For convenience, these numbers may be written in scientific (powers-of-ten) notation. For example, the approximate mass of the planet Jupiter can be written as either 2,000,000,000,000,000,000,000,000,000 kilograms (kg) or $2 \times 10^{27}$ kg. Which of those expressions would you prefer to write?

Here are some of the powers of ten.

$10^{-3} = 0.001$ or $1/1000$     (one-thousandth)

$10^{-2} = 0.01$ or $1/100$

$10^{-1} = 0.1$ or $1/10$

$10^{0} = 1$     (one)

$10^{1} = 10$

$10^{2} = 100$

$10^{3} = 1000$     (one thousand)

$10^{4} = 10,000$

$10^{5} = 100,000$

$10^{6} = 1,000,000$     (one million)

$10^{7} = 10,000,000$

$10^{8} = 100,000,000$

$10^{9} = 1,000,000,000$     (one billion)

$10^{10} = 10,000,000,000$

$10^{11} = 100,000,000,000$

$10^{12} = 1,000,000,000,000$     (one trillion)

To express numbers that are not exactly equal to any of the powers of ten, we can use a power of ten multiplied by an ordinary number. Here are some examples.

$$682 = 6.82 \times 10^2$$

$$0.05 = 5 \times 10^{-2}$$

$$300,000 = 3 \times 10^5$$

$$0.0000000345 = 3.45 \times 10^{-8}$$

$$87,209,000,000,000 = 87.209 \times 10^{12}$$

There is little or no advantage to writing the first three numbers in scientific notation. The last two examples illustrate the effective use of scientific notation's shorthand. The last example, $87.209 \times 10^{12}$, is written in nonstandard form. The number multiplying the power of ten is usually written as more than 1 and less than 10. The nonstandard $87.209 \times 10^{12}$ can be converted into the standard form by dividing the number 87.209 by 10 and by simultaneously multiplying $10^{12}$ by 10. The result is $8.7209 \times 10^{13}$. For all such conversions, use this rule: The power of ten is increased by one for every place the decimal point is shifted to the left, and it is decreased by one for every place the decimal point is shifted to the right.

Multiplying or dividing powers of ten uses the following simple rules: When multiplying powers of ten, add the exponents. When dividing powers of ten, subtract the exponent of the power of ten in the denominator from the exponent of the power of ten in the numerator. Here are some examples.

$$(10^2)(10^3) = 10^5$$

$$(10^8)(10^2)(10^{10}) = 10^{20}$$

$$(10^5)(10^{-9}) = 10^{-4}$$

$$10^{12}/10^5 = 10^7$$

$$10^{10}/10^{-4} = 10^{14}$$

When multiplying or dividing two numbers in scientific notation, treat the powers of ten separately from the ordinary numbers that multiply them. Here are two examples.

$$(2 \times 10^{14})(8 \times 10^{11}) = (2 \times 8) \times (10^{14})(10^{11})$$
$$= 16 \times 10^{25}$$
$$= 1.6 \times 10^{26}$$

$$(2 \times 10^{14})/(8 \times 10^{11}) = (2/8) \times (10^{14}/10^{11})$$
$$= 0.25 \times 10^3$$
$$= 2.5 \times 10^2$$
$$= 250$$

# Math Refresher

Physical laws are often most clearly and succinctly stated in the form of algebraic equations, or formulas. Thus, some knowledge of basic algebra is needed for an appreciation of fundamental aspects of physics and other branches of physical science.

In algebra, equations contain symbols, or variables (typically represented by italicized Roman letters or Greek letters), that represent quantities having no specified values. A formula having two or more variables in it reveals certain general relationships between the variables. If the values for all but one variable in a formula are known or specified, then the value of the remaining variable can be obtained by calculation. As an example, we shall look at one form of Newton's second law of motion, which relates the quantities force ($F$), mass ($m$), and acceleration ($a$).

$$F = ma$$

This formula says that a body with a mass $m$ will accelerate, or change its motion, in response to a force applied to it. More specifically, the formula says that force equals mass multiplied by acceleration. The same formula could be written as either $F = m \times a$ or $F = (m)(a)$, but the form $F = ma$ is simple and conventional.

When none of the three variables in $F = ma$ is specified, their relationship is complex. But simplicity emerges when one variable is held constant. Keep in mind that each variable in the formula represents some physical property or phenomenon. Let us hold $m$ constant—say by experimenting with a single body of matter that has a constant mass. In this case, force and acceleration are *proportional* (or *directly proportional*) to each other. Another way of describing this is by the notation

$$F \propto a \qquad \text{(Read "$F$ is proportional to $a$.")}$$

According to the rules of algebra, one side of an equation can be multiplied (or divided) by any number as long as the other side is multiplied (or divided) by that same number. If the force applied to a given mass is doubled, then the product $ma$ must also double. Since we are assuming that the mass cannot change, only the acceleration can be different. The acceleration must be twice as much as it was before the change in the force took place.

To illustrate another way the formula can work, let us assume that a constant force can be applied to bodies of different mass. The left side of the equation has a constant $F$; therefore, the product $ma$ must also be constant. Individually, $m$ and $a$ can be different, however. If $m$ is multiplied by any factor, $a$ must be divided by the same factor. When force is held constant, mass and acceleration are *inversely proportional* to each other.

Frequently, when using formulas, we wish to calculate the value of a particular variable when we already know the values of the other variables in the formula. For this situation, it is always useful to isolate the unknown quantity to be calculated on one side of the equation. The unknown quantity to be isolated should be positive in sign and to the first power—not squared or cubed, for instance. When we isolate the unknown quantity in this way, we say we have *solved* the formula for the quantity we want to calculate.

If in our formula $F = ma$ we are interested in calculating $a$ based on known values of $F$ and $m$, the following steps should be taken.

$F = ma$

$F/m = ma/m$     (Both sides are divided by $m$.)

$F/m = a$     ($m/m$ on the right equals 1; the two $m$'s cancel.)

$a = F/m$     (The formula is written in conventional form, with the unknown quantity on the left.)

Recalling our earlier assumption that $F$ is constant, we can see that any *increase* in $m$ leads to a proportionate *decrease* in $a$, and vice versa.

The results of a law such as $F = ma$ may also be presented graphically. Let us assume that in a series of experiments using a given mass subject to various forces, we get the values of force and acceleration listed in the table at left. The newton (N) is the name of the SI unit for force. Reduced to fundamental units, $1\ N = 1\ kg \cdot m/s^2$. The SI unit for acceleration is meters per second squared ($m/s^2$). The data may be summarized graphically as shown in Figure B.1.

The fact that the straight line on the graph passes through the data points and the origin tells us that force and acceleration are proportional to each other for the constant mass we are using. The value of that mass is none other than the slope of the line. The slope is easily computed as any change on the line along the $y$-axis (force axis) divided by the corresponding change on the line along the $x$-axis (acceleration axis). The value of $m$ may also be computed from any one of the four trials summarized in the table using the $F = ma$ formula solved for $m$. Using the last entry in the table ($F = 11\ N = 11\ kg \cdot m/s^2$, and $a = 22\ m/s^2$), we get

$m = F/a$

   $= (11\ N)/(22\ m/s^2)$

   $= 0.5\ kg$

| $F$ (N) | $a$ (m/s$^2$) |
|---------|---------------|
| 1       | 2             |
| 3       | 6             |
| 6       | 12            |
| 11      | 22            |

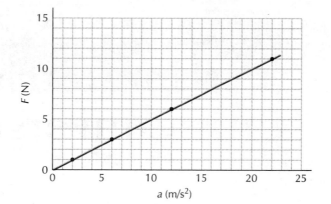

**FIGURE B.1** The graphical depiction of $F = ma$ on a Cartesian (ordinary) graph is a straight line.

Unlike $F = ma$ for Newton's second law, the complicated formulas associated with some physical laws and relationships have graphs that are quite complex. In fact, some laws and relationships cannot easily be modeled by mathematical formulas; from experimental data, though, they can be displayed graphically and are best understood in that way.

# British/Metric Unit Conversions

## Length

| British | → | Metric | Metric | → | British |
|---------|-----|--------|--------|-----|---------|
| Multiply | By | To Get | Multiply | By | To Get |
| Inches (in.) | 2.54 | Centimeters (cm) | Centimeters (cm) | 0.394 | Inches (in.) |
| Feet (ft) | 30.48 | Centimeters (cm) | Centimeters (cm) | 0.0328 | Feet (ft) |
| Inches (in.) | 0.0254 | Meters (m) | Meters (m) | 39.37 | Inches (in.) |
| Feet (ft) | 0.3048 | Meters (m) | Meters (m) | 3.2808 | Feet (ft) |
| Miles (mi) | 1.61 | Kilometers (km) | Kilometers (km) | 0.621 | Miles (mi) |

## Area

| British | → | Metric | Metric | → | British |
|---------|-----|--------|--------|-----|---------|
| Multiply | By | To Get | Multiply | By | To Get |
| Square inches ($in^2$) | 6.45 | Square centimeters ($cm^2$) | Square centimeters ($cm^2$) | 0.155 | Square inches ($in^2$) |
| Square feet ($ft^2$) | 0.0929 | Square meters ($m^2$) | Square meters ($m^2$) | 10.76 | Square feet ($ft^2$) |

## Volume

| British | → | Metric | | Metric | → | British |
|---|---|---|---|---|---|---|
| Multiply | By | To Get | | Multiply | By | To Get |
| Cubic inches (in$^3$) | 16.4 | Cubic centimeters or milliliters (cm$^3$ or ml) | | Cubic centimeters or milliliters (cm$^3$ or ml) | 0.061 | Cubic inches (in$^3$) |
| Cubic feet (ft$^3$) | 0.0283 | Cubic meters (m$^3$) | | Cubic meters (m$^3$) | 35.3 | Cubic feet (ft$^3$) |
| Quarts (qt) | 0.946 | Liters (= 1000 cm$^3$) | | Liters (= 1000 cm$^3$) | 1.06 | Quarts (qt) |
| Gallons (gal) | 3.786 | Liters (= 1000 cm$^3$) | | Liters (= 1000 cm$^3$) | 0.264 | Gallons (gal) |
| Cubic feet (ft$^3$) | 28.3 | Liters (= 1000 cm$^3$) | | Liters (= 1000 cm$^3$) | 0.0353 | Cubic feet (ft$^3$) |

## Speed

| British | → | Metric | | Metric | → | British |
|---|---|---|---|---|---|---|
| Multiply | By | To Get | | Multiply | By | To Get |
| Feet per second (ft/s) | 0.305 | Meters per second (m/s) | | Meters per second (m/s) | 3.281 | Feet per second (ft/s) |
| Miles per hour (mi/h) | 0.447 | Meters per second (m/s) | | Meters per second (m/s) | 2.237 | Miles per hour (mi/h) |
| Miles per hour (mi/h) | 1.61 | Kilometers per hour (km/h) | | Kilometers per hour (km/h) | 0.621 | Miles per hour (mi/h) |
| Miles per hour (mi/h) | 0.000447 | Kilometers per second (km/s) | | Kilometers per second (km/s) | 2237 | Miles per hour (mi/h) |

## Force

| British | → | Metric | | Metric | → | British |
|---|---|---|---|---|---|---|
| Multiply | By | To Get | | Multiply | By | To Get |
| Pounds (lb) | 4.45 | Newtons (N) | | Newtons (N) | 0.225 | Pounds (lb) |

## Work and Energy

| British | → | Metric | | Metric | → | British |
|---|---|---|---|---|---|---|
| Multiply | By | To Get | | Multiply | By | To Get |
| Foot-pounds (ft·lb) | 1.355 | Joules (J) | | Joules (J) | 0.738 | Foot-pounds (ft·lb) |
| British thermal units (Btu) | 1054 | Joules (J) | | Joules (J) | 0.000949 | British thermal units (Btu) |
| British thermal units (Btu) | 252 | Calories (cal) | | Calories (cal) | 0.00397 | British thermal units (Btu) |

### Power

| British | $\rightarrow$ | Metric | Metric | $\rightarrow$ | British |
|---|---|---|---|---|---|
| Multiply | By | To Get | Multiply | By | To Get |
| Horsepower (hp) | 746 | Watts (W) | Watts (W) | 0.00134 | Horsepower (hp) |
| British thermal units per hour (Btu/h) | 0.293 | Watts (W) | Watts (W) | 3.41 | British thermal units per hour (Btu/h) |

### Pressure

| British | $\rightarrow$ | Metric | Metric | $\rightarrow$ | British |
|---|---|---|---|---|---|
| Multiply | By | To Get | Multiply | By | To Get |
| Pounds per square inch (lb/in$^2$, or psi) | 6895 | Pascals (Pa) | Pascals (Pa) | 0.000145 | Pounds per square inch (lb/in$^2$, or psi) |
| Pounds per square inch (lb/in$^2$, or psi) | 0.0690 | 1 bar = 1000 millibars (1000 mb) | 1 bar = 1000 millibars (1000 mb) | 14.5 | Pounds per square inch (lb/in$^2$, or psi) |

## Temperature

Use the following formulas to convert between the Fahrenheit (F), Celsius (C), and Kelvin (K) temperature scales.

$$F = \tfrac{9}{5} C + 32$$

$$C = \tfrac{5}{9}(F - 32)$$

$$K = C + 273.15$$

$$C = K - 273.15$$

# Physical Constants and Data

## Physical Data

| Quantity | Symbol | Value |
| --- | --- | --- |
| Acceleration due to gravity at Earth's surface | $g$ | $9.81 \text{ m/s}^2$ |
| Universal constant of gravitation | $G$ | $6.67 \times 10^{-11} \text{ N·m}^2/\text{kg}^2$ |
| Coulomb's law constant | $K$ | $9.00 \times 10^9 \text{ N·m}^2/\text{C}^2$ |
| Proton mass | $m_p$ | $1.672 \times 10^{-27} \text{ kg}$ |
| Neutron mass | $m_n$ | $1.675 \times 10^{-27} \text{ kg}$ |
| Electron mass | $m_e$ | $9.11 \times 10^{-31} \text{ kg}$ |
| Electron charge | $e$ | $-1.60 \times 10^{-19} \text{ C}$ |
| Proton charge | | $+1.60 \times 10^{-19} \text{ C}$ |
| Speed of light in a vacuum | $c$ | $3.00 \times 10^8 \text{ m/s}$ |
| Speed of sound (in air at 20 °C and 1 atmosphere) | | $343 \text{ m/s}$ |
| Standard atmospheric pressure (1 atm) | | $1.01 \times 10^5 \text{ Pa, or } 1.01 \text{ bar}$ |
| Planck's constant | $h$ | $6.625 \times 10^{-34} \text{ J·s}$ |

## Earth and Solar System Data

| Quantity | Value |
| --- | --- |
| Earth's mass | $5.98 \times 10^{24} \text{ kg}$ |
| Earth's average radius | $6.38 \times 10^6 \text{ m}$ |
| Earth's average density | $5.52 \text{ g/cm}^3$ |
| Average Earth-sun distance (= 1 AU) | $1.50 \times 10^{11} \text{ m}$ |
| Mass of the moon | $7.36 \times 10^{22} \text{ kg}$ |
| Radius of the moon | $1.74 \times 10^6 \text{ m}$ |
| Average Earth-moon distance | $3.84 \times 10^8 \text{ m}$ |
| Mass of the sun | $1.99 \times 10^{30} \text{ kg}$ |
| Radius of the sun | $6.96 \times 10^8 \text{ m}$ |
| Mass of Jupiter | $1.90 \times 10^{27} \text{ kg}$ |
| Equatorial radius of Jupiter | $7.14 \times 10^7 \text{ m}$ |

# Answers to Odd-Numbered Questions and Problems

## Chapter 1

### Questions

1. Scientific explanations of all kinds must be capable of being tested by means of experiment or through further observation.

### Problems

1. There are $10^9$ nanometers in a meter, and $10^3$ meters in a kilometer. Therefore,

$$1 \text{ km} = (10^9 \text{ nm/m})(10^3 \text{ m/km})$$
$$= \mathbf{10^{12} \text{ nm}}$$

## Chapter 2

### Questions

1. When a sphere is quadrupled in radius (or diameter), its surface area becomes $4^2 = 16$ times larger, and its volume becomes $4^3 = 64$ times larger.

3. One, two, and three coordinates. (For Lake Michigan, it is important to specify depth.)

5. RA = 0 h 45 m; Dec = $+42°$

7. Isotopes are atoms of the same element that have nuclei of different masses. All isotopes of the same element have the same number of protons. They differ only in the number of neutrons. Isotopes of the same element have similar chemical behavior, but differ in their nuclear properties.

9. Gravity's effect is cumulative; every mass attracts every other mass. In the earthly environment we live in, Earth's great mass (and our relative proximity to it) results in the noticeable weights that objects have. Electromagnetic forces may be inherently stronger than gravitational forces, but these forces cancel out almost completely for matter in bulk. This is because large objects tend to contain equal or nearly equal amounts of positive and negative charge. Objects with large amounts of *net* charge are rare.

11. Entropy is a measure of the amount of disorder or randomness in a physical system.

13. Earth is one of the inner (terrestrial) planets orbiting the sun. The solar system (including Earth) lies nearer the visible edge of the Milky Way Galaxy than its center. The Milky Way Galaxy lies at one end of the Local Group of galaxies, which is one among many clusters of galaxies in the Local Supercluster of galaxies. The Local Supercluster probably lies along a galaxy-rich sheet that forms the wall of at least one "cosmic void." The universe contains per-

haps thousands of superclusters arranged in a bubble-and-void pattern.

## Problems

1. If a sphere increases in linear size by a factor of 10, its volume becomes $10^3 = 1000$ times larger. The sun's diameter compared to Jupiter's diameter is **10** times larger.

3. $40,000 \text{ km}/360° = $ **111 km** per degree of latitude, or of longitude at Earth's equator. Longitude meridians get closer together as they approach the poles, so the distance per degree of longitude is less at latitudes other than $0°$.

5. The speed of the radio pulse is 300,000 km/s. The one-way time is 500 s. The distance to Mars is calculated by $distance = speed \times time = $ **150,000,000 km**. This happens to be very nearly equal to **1 AU**.

7. The 250-million-year period expressed in seconds is $7.89 \times 10^{15}$ s.

$$Distance = speed \times time$$
$$= (250 \text{ km/s})(7.89 \times 10^{15} \text{ s})$$
$$= 1.97 \times 10^{18} \text{ km}$$
$$= 1.97 \times 10^5 \text{ LY}$$

which is the length of one complete orbit. The radius of that orbit is $3.14 \times 10^4$ LY, or **31,400 LY**. This is consistent with the dimensions of Window Q.

# Chapter 3

## Questions

1. Yes. In the Northern Hemisphere, Polaris and circumpolar stars can be seen above the north horizon. In the Southern Hemisphere circumpolar stars are seen above the south horizon.

3. This is due to the fact that the moon, a sphere, is being lit by a distant light source, which is the sun. Try lighting up a golf ball or other small sphere by a flashlight placed obliquely behind it. The curved sliver of light on the ball, which looks like a crescent, always has cusps that point away from wherever the flashlight appears to be.

5. The sun's uncovered disk is nearly a million times brighter than the corona which becomes visible during a total solar eclipse. The corona is

safe enough to look at directly, but even a small sliver of the sun is thousands of times brighter, requiring the use of a special optical filter. Anyone unsure of how to observe a solar eclipse properly should do so with supervision by an astronomy club, an academic observatory, or an experienced veteran of previous eclipses.

7. Correlation is the process of matching sequences of rock known to exist in one geographic area with similar sequences in other areas. It is used to put local sequences together to get an overall picture of Earth's surface in both space and time.

9. Primarily, Galileo's observations demonstrated that Venus has a complete range of phases, which shows that it cannot always lie between the sun and Earth; and that Jupiter's four bright moons circle Jupiter is a way that is very similar to the behavior of the planets around the sun in the heliocentric model. Galileo's observations also helped discredit some of Aristotle's ideas: Heavenly bodies seen through a telescope were not as perfectly spherical, unblemished, and unearthly as they were once thought to be.

## Problems

1. In Chapter 2 we learned that $distance = speed \times time$. We also know that the speed of light is $3 \times 10^8$ m/s.

$$distance = (3 \times 10^8 \text{ m/s})(3 \times 10^{-15} \text{ s})$$
$$= 9 \times 10^{-7} \text{ m}$$

This is approximately $10^{-6}$ m, which is a millionth of a meter, or **1 $\mu$m**.

3. If $distance = speed \times time$, then $time = distance/speed$. If a 1-cm-long bullet appears as a blur 2 cm long, then it has moved a distance of 1 cm, or 1/100 m.

$$time = (1/100 \text{ m})/(500 \text{ m/s})$$
$$= 1/50,000 \text{ s}$$

In decimal form, this is **0.00002 s**.

5. At the mid-latitude of the United States, each degree of longitude corresponds to about $1200 \text{ km}/15 = 80$ km of east-west distance. This is equivalent to a time difference of 1/15 hour, or 4 minutes. Town A is $240 \text{ km}/80 \text{ km} = 3°$ of longitude west of Town B, so its sunrises occur **12 minutes later**.

# Chapter 4

## Questions

1. Water is not a mixture because it cannot be separated into simpler components by *physical* means (such as temperature and/or pressure changes). If water is not a mixture, then it must be either an element or a compound. Water can be separated into the elements hydrogen and oxygen by *chemical* means (electrolysis, for example), therefore water is a compound. Furthermore, water is a compound composed of hydrogen and oxygen in a definite proportion: 1 part hydrogen to 8 parts oxygen by mass, or 2 atoms of hydrogen to every 1 atom of oxygen.

3. In the simplest sense, when a chemical equation is balanced, the number of atoms of each type must be the same on both sides of the equation. When an equation is balanced, the total mass and total charge of the reactants and the products are equal.

5. In a given compound, two or more elements are combined in a definite (not variable) proportion *by mass*. According to atomic theory, this is explained by the fact that atoms of different kinds bind together in a definite ratio to form a specific compound.

## Problems

1. By writing the balanced equation

$$3H_2 + N_2 \rightarrow 2NH_3$$

we find that three molecules of hydrogen combine with one molecule of nitrogen to form two molecules of ammonia. Equal volumes of gases (nitrogen, hydrogen and ammonia, in this case) contain the same number of molecules, so **two** volumes of ammonia are formed.

3. Table 4.1 reveals that nitrogen atoms are 14 times as massive as hydrogen atoms. Since both are diatomic, $N_2$ molecules are 14 times as massive as $H_2$ molecules. The balanced equation $3H_2 + N_2 \rightarrow 2NH_3$ tells us that three $H_2$ molecules combine with one $N_2$ molecule to form two molecules of ammonia. The ratio of mass between $3H_2$ and $N_2$ is 3/14. There are 28 g of $N_2$, so the mass of $H_2$ must be $(3/14)(28) = $ **6 g**.

# Chapter 5

## Questions

1. Instantaneous speed is the ratio of distance and time for a particular, infinitesimally small interval of time. Average speed is the ratio of distance and time over a time interval that is not necessarily short.

3. Inertia is an inherent property of mass. It is manifested as an apparent resistance to any change in motion (or change in velocity).

5. The angular momentum vector **l** points "down," out from the south pole of Venus, when the right hand rule is applied. Venus and Earth have opposite spins.

7. The weight of an object is the result of a gravitational interaction between Earth's fixed mass and the presumably fixed mass of the object. This weight, however, depends on the distance between the two masses, which is not necessarily fixed.

9. Chaos refers to situations in which an arbitrarily small change in initial conditions can produce relatively large and unpredictable results in the detailed behavior of a system of interacting bodies.

## Problems

1. When each of the entries in Table 5.1 is multiplied by the number of seconds in a year ($3.2 \times 10^7$), the results are
   a) snail: **$3.2 \times 10^5$ m** (320 km!)
   b) walking human: **$4.8 \times 10^7$ m** (48,000 km, around the world and then some)
   c) cruising auto: **$9.6 \times 10^8$ m** (over twice the distance to the moon)
   d) jet aircraft: **$9.6 \times 10^9$ m** (about 1/16 the distance to the sun)
   e) light: **$9.6 \times 10^{15}$ m** (one light year)

3. After each passing second, the car is moving 2 m/s faster, so it takes **5 s** to go 10 m/s faster than it was going in the first place. This is true only if the car accelerates *in the forward direction*.

5. $\mathbf{F} = m\mathbf{a} = (900 \text{ kg})(3 \text{ m/s}^2) = $ **2700 N**, in the same direction as **a**.

7. Most adult humans have an Earth-weight of around 500 to 800 N.

9. Kepler's third law:

$$P^2 = R^3$$
$$P^2 = 4^3 = 64$$
$$P = \textbf{8 years}$$

# Chapter 6

## Questions

1. Energy can be thought of as the ability to do work. All types of energy can do work. But energy occurs in many different forms and comes from many sources: kinetic, gravitational potential, nuclear, chemical and solar (to name a few).

3. *Mechanical energy* includes the kinetic and potential energies of a system of bodies, typically assumed to be of larger than molecular size. (By contrast, *thermal energy* includes the kinetic and potential energies of the individual atoms or molecules in a macroscopic body.)

5. As she climbs to the top of the tower, she increases her *GPE*. When she jumps, her *KE* increases from zero at the same rate her *GPE* decreases (assuming negligible air resistance). At 30 m below the top her *GPE* continues to decrease, but at a slower rate. This is because her *KE* starts transforming into elastic *PE* that is stored in the bungee cord. As she reaches her maximum distance down, *GPE* is minimized, *KE* is momentarily zero, and the bungee cord stores maximum elastic *PE*. After oscillating up and down a few times, she comes to a state of rest in which her *KE* is zero, *GPE* is less than it was when she jumped, and some elastic *PE* is stored in the bungee cord. During the whole process, some energy was dissipated as thermal energy (mostly in the cord), otherwise she would never stop oscillating.

7. Chemical reaction rates can be increased or decreased by changes in temperature, by changes in the concentration or surface area of one or more of the reactants, and (for many reactions) by using catalysts.

## Problems

1. The work is the same, **20 J**, as the same task is accomplished. However, the force required in the second case is twice as great.

3. Assuming a mass of 60 kg (132 lb.) and a speed of 6 m/s (a 13.4 mi/h sprint),

$$KE = 1/2 \, mv^2$$
$$= (1/2)(60 \text{ kg})(6 \text{ m/s})^2$$
$$= 1080 \text{ J}$$

$GPE = mgh$, so $h = GPE/mg$. It is assumed that $KE = GPE$, so $GPE = 1080$ J

$$h = GPE/mg$$
$$= 1080 \text{ J}/(60 \text{ kg})(10 \text{ m/s}^2)$$
$$= 1.8 \text{ m}$$

This height of 1.8 m ($\approx$ 6 ft) is not very impressive, but remember that pole vaulters also use their arms to push themselves upward.

5. The output work done is

$$W = Fd$$
$$= (5000 \text{ N})(0.5 \text{ m})$$
$$= 2500 \text{ J}$$

Ideally, the input work would be the same, but because the efficiency is only 20%, the input work is really 5 times greater, or 12,500 J.

$$F = W/d$$
$$= 12,500 \text{ J}/100 \text{ m}$$
$$= \textbf{125 N}$$

# Chapter 7

## Questions

1. Temperature is a measure of the average *KE* of particles in a body. Temperature can be used to determine which way heat flows from one body to another. A body's thermal energy, or heat energy, is its total internal energy; that is, the kinetic and potential energies of all of its particles. Heat can be thought of as thermal energy in motion from place to place.

3. When two bodies are in contact with one another, their molecules undergo collisions. Faster moving molecules (those associated with a higher temperature) tend to transfer their momentum to slower moving molecules; thus, the spontaneous, net flow of heat is always from the higher-temperature body to the lower-temperature body.

5. The more volatile a substance, the weaker are the forces between its molecules.

7. Since the corona has an extremely low density, relatively few particles are present in it. However, these particles do have a very high average *KE*, and temperature depends on the average *KE*.

9. Pressure is force per area. A modest force applied over a small area can result in a large pressure.

11. *Conduction:* The molecule-to-molecule transfer of thermal energy. Conduction is slow, and usually important only in solids. *Convection:* The transfer of thermal energy by the large-scale motion of molecules in a fluid, primarily as a result of temperature or pressure differences in the fluid. *Radiation:* The transfer of energy from one place to another by electromagnetic waves, which travel at the speed of light and do not need any medium to propagate through.

13. During the day, the land heats up faster than the water. The air over the land expands and rises, giving birth (under relatively undisturbed conditions) to a convection cycle in which a sea breeze blows. At night, the land cools faster than the water. If the land's temperature becomes less than the water temperature, the convection cycle can reverse, producing a land breeze.

15. Cloud formation requires: (1) a source of water vapor, (2) a sufficient drop in the temperature (to the dew point or lower), and (3) condensation nuclei in the air.

## Problems

1. The maximum temperature difference of 50 °C produces a change in length of approximately 0.01% × 5 = 0.05%, or 0.0005. The change in length across 100 m is $(0.0005)(100\text{ m}) = 0.05\text{ m} = \textbf{5 cm}$. (The expansion gaps in many bridges are designed to be about this wide.)

3. The answer to Problem 2 is $t = 375$ s. The total energy required to raise the water's temperature 75 °C to the boiling point is 75 kcal. Converting to joules, $(75,000\text{ cal})(4.186\text{ J/cal}) = 314,000\text{ J}$. We now use $P = E/t$ to find the power:

$$P = E/t$$
$$= 314,000\text{ J}/375\text{ s}$$
$$= \textbf{837 W}$$

which is nearly a kilowatt.

5. Pressure $= F/A$
$$= 20,000\text{ N}/20\text{ m}^2$$
$$= 1000\text{ N/m}^2$$
$$= \textbf{1 kPa}$$

7. The answer is that pressure doubles to **2 bars**. The ideal gas law, expressed as $PV/T = $ constant, assumes that the mass of gas present is constant. Doubling the mass of the gas, while keeping the volume and temperature the same, doubles the constant in the formula, and also doubles $P$. Twice as many molecules moving inside the same container with the same speed will strike the walls of the container twice as often and impart twice the force. When force on the same area is doubled, the pressure also doubles.

# Chapter 8

## Questions

1. They are inversely related: $f \propto 1/\lambda$

3. Transverse waves involve disturbances at right angles to the direction of propagation. Longitudinal waves involve disturbances that go parallel to the direction of propagation.

5. Astronomers like to speak of the "visible window" and the "radio window," which refer to the two wavelength ranges for which Earth's atmosphere is quite transparent.

7. Specular reflection (from smooth surfaces) preserves any image or detailed information in the incident light. Diffuse reflection erases the same. Both types obey the law of reflection.

9. The Doppler effect applies only to motion in the line of sight (along the line from source to observer). It gives no information about motion perpendicular to the line of sight. Doppler radar measures only the line-of-sight component of the velocity.

## Problems

1. The period of the waves is 4 s, so the frequency is $(1)/(4\text{ s})$, or 0.25 Hz, which can be written as $0.25\text{ s}^{-1}$.

$$\lambda = v/f$$
$$= (2\text{ m/s})/(0.25\text{ s}^{-1})$$
$$= \textbf{8 m}$$

3. The sun is 15 magnitudes, or $5 + 5 + 5$ magnitudes, brighter than the full moon. Each interval of 5 magnitudes represents a brightness ratio of 100. So the sun is $100 \times 100 \times 100 = \mathbf{10^6 \ times}$ as bright.

# Chapter 9

## Questions

1. Both are inverse-square laws involving the interaction of two bodies. Electrostatic force, however, is much stronger and can be both attractive and repulsive, whereas gravity is only attractive.

3. A striking example is a large tree (say, a 100-m tall coast redwood in California) which uses capillary action to move many cubic meters of water a day from its roots to its leaves or needles, all with no moving parts! Animals also have complex capillary systems.

5. The keystone element of organic chemistry, carbon, has an almost unique ability to bond to many other atoms and to form long chains with itself and other atoms. This allows a great variety of carbon-containing compounds to form, including some containing millions of atoms. (Silicon, similar to carbon in some ways, tends not to form double and triple bonds with other atoms, so silicon-based chemistry is much more limited.)

7. Some common examples of polymers include the plastics: polyethene, polystyrene, polypropylene, polyvinylchloride (PVC), Teflon, nylon, and polyester. Some biological molecules, such as starch and protein, are polymers made of rather complex monomers.

9. Diamond is a covalent crystal in which each atom is strongly bonded to four neighboring carbon atoms. Diamond is the most compact and rigid form of carbon. In graphite, one of the four bonds is weaker; this results in layers of carbon atoms that can slide past one another. As a result, graphite is relatively soft, weakly conducts electricity, and can be used as a dry lubricant. Buckyballs and other fullerene molecules are relatively large, hollow molecules that have structures resembling geodesic domes. A lot of interesting chemistry is likely to come from further studies of the fullerenes.

11. Field tests include color, streak, luster, hardness, crystal form, and cleavage. Lab work includes measurements of density, and sophisticated X-ray diffraction tests which can reveal internal structure.

## Problems

1. By the inverse-square relationship embodied by Coulomb's law, a tripling of the distance between the charges leads to an electrostatic force that is 1/9 as strong, or **0.2 N**.

3. a)  $C_3H_3$    No such compound

   b)  $C_{12}H_{26}$  Yes, this is an alkane:

   c) $C_3H_7OH$  Yes, this is an alcohol:

   d) $C_4F_{10}$    Yes:

# Chapter 10

## Questions

1. By noting the force the field exerts on a charged object.

3. A potential difference is the cause of current, just as a difference in height is required for a mass to fall. The greater the potential difference across two parts of a body, the greater the current through that body, and the more damage the current does if that body is a human body. Touching two metal surfaces having the same potential difference is like stepping from one plank to another at the same height. Touching two metal surfaces with a large potential differ-

ence between them is like falling from a high plank to a low one.

5. Electrons are forced into the battery on the negative side where they combine with lead ions in lead sulfate to plate metallic lead on the negative terminal and release sulfate ions into the aqueous solution. On the other, positive terminal, electrons are pulled off, breaking down the lead sulfate into ions and increasing the acid concentraton. The reactions at the terminals are of the oxidation and reduction type; during the charging process they are endothermic.

7. A current-bearing wire or a magnet would experience a force due to the magnetic field.

9. Since, by convention, we have labelled the north-seeking end of a compass needle its "N" pole, then that end of the needle would be attracted to the "S" pole of a hypothetical bar magnet embedded inside Earth. This hypothetical S pole lies in the Northern Hemisphere.

11. Magnetic forces, set up by coils along the neck of the picture tube or monitor, steer the electron beam rapidly across the face of the tube and more slowly downward to "paint" a picture on the phosphorescent screen on the tube's face.

13. Transformers transform voltage and current. Ideally, the product of these two things, which is power, remains the same during the transformation process.

## Problems

1. The total energy expended is

$$110 \text{ J} + 10 \text{ J} = 120 \text{ J}$$

and that is equal to the amount of electric energy inputted. From Equation 10.1,

$$
\begin{aligned}
V &= W/q \\
&= 120 \text{ J}/1 \text{ C} \\
&= 120 \text{ J/C} \\
&= \textbf{120 V}
\end{aligned}
$$

3. Use Equation 10.4, $P = VI$.

$$
\begin{aligned}
V &= P/I \\
&= 24 \text{ W}/4 \text{ A} \\
&= \textbf{6 V}
\end{aligned}
$$

5. Use Ohm's law and the fact that the "wet skin" resistance of the human body is about 1000 $\Omega$.

$$
\begin{aligned}
I &= V/R \\
&= 110 \text{ V}/1000 \text{ }\Omega \\
&= \textbf{0.11 A}
\end{aligned}
$$

7. $P = VI = (120 \text{ V})(5 \text{ A}) = \textbf{600 W}$

9. From Equation 10.1,

$$
\begin{aligned}
q &= W/V \\
&= (10^8 \text{ J})/(5 \times 10^8 \text{ V}) \\
&= 0.2 \text{ C}
\end{aligned}
$$

From Equation 10.2,

$$
\begin{aligned}
I &= q/t \\
&= 0.2 \text{ C}/1 \text{ s} \\
&= \textbf{0.2 A}
\end{aligned}
$$

From Equation 10.4,

$$
\begin{aligned}
P &= VI \\
&= (5 \times 10^8 \text{ V})(0.2 \text{ A}) \\
&= \textbf{10}^8 \textbf{ W} \text{ (or 100 MW)}
\end{aligned}
$$

[As a check: $P = W/t = (10^8 \text{ J})/(1 \text{ s}) = 10^8$ W] This shows that in lightning, lots of power is delivered by relatively little current.

# Chapter 11

## Questions

1. Light, an "entity" if you will, has both a particle and a wave nature and these must be related. The frequency of a light wave, $f$, is related to the energy of a photon, $E$, by $E = hf$, for example.

3. Plank's constant is so very tiny that even when it is multiplied by the frequency of, say, a visible light photon in the equation $E = hf$, $E$ is still a tiny fraction of 1 J.

5. "Hot" solids, liquids, and compressed gases produce a continuous spectrum. Heated or excited rarefied gases produce an emission spectrum. A relatively cool gas will absorb certain frequencies from light passing through it, thereby producing an absorption spectrum.

7. The electronic structure of atoms and molecules can be studied, as well as the chemical make-up

of any substance producing electromagnetic radiation strong enough to detect. The latter is true whether the sample is in an earthly laboratory or out in space any arbitrary distance away.

9. An electron at a metastable excited energy level in an atom "stores" energy for a relatively long time until it falls back to a lower energy level and releases a photon of a specific frequency. Passing photons of the same frequency can trigger the early release of photons from many atoms in these metastable states and trigger a cascade of virtually identical photons. The result is coherent radiation.

11. Rainbows are caused by light refracting and dispersing through spherical water droplets; most ice-crystal effects are caused by light refracting through hexagonal ice-crystals (some involve light reflecting off the smooth sides of these crystals). Rainbows are seen generally "away" from the sun in the sky; most ice-crystal displays occur in parts of the sky near the sun.

13. When the reflection (or absorption) of light on a surface is equal for all wavelenths of visible light, then that surface can be characterized as "gray" on a scale that extends between zero reflectance (black) and 100% reflectance (white).

15. Yellow photons, having a higher frequency than red photons, would scatter more, and the sky would appear yellow with a reddish tinge.

## Problems

1. $E = hf$

   $= (6.63 \times 10^{-34} \text{ J·s})(10^8 \text{ s}^{-1})$

   $= \textbf{6.63} \times \textbf{10}^{-26} \textbf{ J}$

(This energy lies at the bottom of the energy-range illustration in Chapter 6, Figure 6.2.)

# Chapter 12

## Questions

1. Leptons are particles of extremely low mass unaffected by the strong force, while hadrons are more massive and feel the strong force. Both types of particles have rest mass (though neutrinos, which are leptons, have very little or no

mass). Photons have no rest mass and travel only at the speed of light.

3. A radioactive substance has an unstable nucleus which will spontaneously decay and emit particles (or gamma radiation). Radioactivity is affected neither by physical changes (temperature, pressure, magnetic fields, etc.) nor by chemical changes; it is a phenomenon associated only with the nucleus of the atom.

5. The neutrons released by a fission reaction can induce nearby nuclei to also undergo fission. If the number of neutron encounters with nearby nuclei are frequent enough, the rate of the chain reaction can build.

7. The basic reason is that iron nuclei, and those of elements near iron in atomic mass, are most stable—they have the greatest possible binding energy and the least possible nuclear *PE*. If the statements in the question were not true, one could get energy out of heavy element fusion, then fuse the lighter products to produce more energy, then get still more energy out of fission again. This sounds very attractive, but it violates the law of conservation of energy.

## Problems

1. After 5 half lives, $(1/2)^5 = \textbf{1/32}$
   After 10 half lives, $(1/2)^{10} \approx \textbf{1/1000}$
   After 20 half lives, $(1/2)^{20} \approx \textbf{1/1,000,000}$

3. Remember that only 0.007 of the original 1 g is converted to energy by $E = mc^2$. This is 0.007 g, or 0.000007 kg.

   $E = mc^2$

   $= (7 \times 10^{-6} \text{ kg})(3 \times 10^8 \text{ m/s})^2$

   $= (7 \times 10^{-6} \text{ kg})(9 \times 10^{16} \text{ m}^2/\text{s}^2)$

   $= \textbf{6.3} \times \textbf{10}^{11} \textbf{ J}$

# Chapter 13

## Questions

1. In our own experience, space seems to be flat and time, as measured by clocks, seems to flow uniformly. This Newtonian way of looking at things is not unlike the geocentric idea of the Greeks, in which Earth is stationary and unmoving because it seems that way. Both the Newtonian and geocentric paradigms were ultimately

overthrown and replaced by more general concepts about the universe.

3. The speed of light in a vacuum is constant, while measurements of space, time, and mass are relative.

5. Time never flows faster than the proper time measured by an observer who stays in the same inertial reference frame. Clocks slow down (time is dilated) for frames of reference that are in uniform motion with respect to that same observer. The faster the relative motion, the greater the time dilation. A similar time dilation effect occurs in accelerated frames of reference. The greater the acceleration (or apparent gravity) in a reference frame, the slower a clock in it ticks with respect to an identical clock in an inertial (nonaccelerated) reference frame.

7. The principle of equivalence says that physical laws are consistent within any two reference frames that have the same acceleration. A rocket accelerating at 1 $g$ in space and Earth's gravitational field of 1 $g$ of acceleration are entirely equivalent regarding the laws of physics, according to Einstein.

9. The escape velocity for anything "under" the event horizon of a black hole is greater than the speed of light, so no light emitted inside the black hole escapes. Since no light from the inside escapes, and no light can reflect and come back out either, the black hole appears perfectly black and invisible.

## Problems

1. No fancy computations are needed for this problem. Since you are in the same reference frame as the meter stick, clock, and 1-kg mass, they have their proper (normal) characteristics.

# Chapter 14

## Questions

1. The De Broglie wavelength, or matter wavelength is so very small ($\lambda \approx 10^{-35}$ m) that there is no way to observe it or its effects.

3. In the $n = 9$ state, 9 wavelengths fit around the circumference of the Bohr orbit. The Bohr orbits

turned out to be a simplification of the orbitals that electrons actually occupy in the hydrogen atom, but there do exist relationships between $n$ and the orbitals that make sense.

## Problems

1. Use Equation 14.1.

$$\lambda = h/mv$$
$$= \frac{(6.63 \times 10^{-34})}{(9.1 \times 10^{-31} \text{ kg})(10^7 \text{ m/s})}$$
$$= \mathbf{7.3 \times 10^{-11} \text{ m}}$$

which is about the same as the radius of a typical atom.

3. Use Equation 14.2, $(\Delta mv)(\Delta x) \approx h$. For an assumed mass of 100 kg and $h = 1$ J·s

$$\Delta x = h/\Delta mv$$
$$= (1 \text{ J·s})/(100 \text{ kg})(0.1 \text{ m/s})$$
$$= 0.1 \text{ m}$$

The uncertainty in your position is roughly 10 cm.

# Chapter 15

## Questions

1. Mendeleev found that the physical and chemical characteristics of the elements repeated in a periodic fashion.

3. By sharing electrons, a combination of atoms can have a more stable, lower energy set of orbitals than the individual atoms would have. During covalent bonding, atomic orbitals change to molecular orbitals.

5. Hydrogen has an incomplete electron shell; it can participate in covalent bonding with another atom. Helium's electrons are in a closed-shell configuration, a stable arrangement that makes helium aloof to chemical bonding of any kind. Hydrogen can increase its stability by forming bonds; helium cannot.

7. An atom with a large electronegativity has a strong ability to attract electrons from other atoms for bonding. Fluorine and oxygen, which are on the right side of the periodic table, tend to form bonds with elements on the left side—

especially those that have low ionization energies and can lose their electrons easily.

# Chapter 16

## Questions

1. In support of continental drift, Wegener noted the apparent "fit" between the edges of today's separated continents, and the continuity, across these edges, of large geologic structures, fossil patterns, and glaciation patterns. He could not come up with a plausible mechanism to explain how continents could actually move across distances of thousands of miles.

3. The edges of lithospheric plates pull apart from each other, collide with each other, and move sideways (parallel to Earth's surface) with respect to each other at the divergent, convergent, and transform boundaries, respectively. At the divergent boundaries, which primarily occur under the oceans, volcanic activity (accompanied by earthquakes) produces ridges, rift valleys, and volcanic islands. At most convergent boundaries, subduction occurs, accompanied by thrust faulting and volcanism. On the ocean floor, this results in deep ocean trenches and volcanic island chains. Along the edges of continents, subduction gives rise to volcanic mountain ranges which lie just inland from the coast. Faulting and folding, producing mountain ranges (such as the Himilayas) occurs when two continental landmasses collide along a convergent boundary. Along transform boundaries, strike-slip faulting produces earthquakes and not much volcanism.

5. Venus is geologically active like Earth, but possibly due to the absence of liquid water and other factors, the nature of its surface geology is very different. Mars seems to have had an active geology like Earth's at one time, but its interior has cooled, and there is relatively little activity occurring there now.

7. Stratovolcanoes feature "sticky" magma which tends to plug up the throat of the volcano for rather long intervals of time. Gases dissolved in this magma build up pressure until "she blows." Shield volcanoes have a less viscous magma and relatively little dissolved gas. More frequent eruptions relieve the pressure.

9. Flowing water, moving air (wind), and moving ice (glaciers) are usually cited as the primary agents of erosion. Mass wasting (due to gravity), wave action, and the passage of groundwater may also be considered "agents."

11. The lower the altitude, the more the accumulated sediment.

13. The marble has a more pronounced crystalline structure, is purer in color, and is "harder." (Hardness is technically a mineral property, and marble is a rock.)

# Chapter 17

## Questions

1. Since light travels at a finite speed, "far away" means "long ago." Since most galaxies are presumed to have formed at about the same time early in the history of the universe, it is generally true that a galaxy's age, as we see it, is inversely related to its distance. As time goes on, the universe is becoming less dense and cooler overall, so as we look farther out into space and farther back in time, we see the universe's contents as being denser and hotter.

3. An evolutionary track shows how the luminosity and surface temperature of a large mass of gas (protostar or star) changes as time goes on.

5. Protostars are surrounded by clouds of gas and dust, the material from which they are forming. The dust is opaque to light, though some radiation—infrared radiation—from the forming star does get through.

7. Hydrogen and helium atoms were formed as a result of the big bang, but the heavier elements had to be synthesized by slow nuclear processes in stars and rapid nuclear reactions during supernova explosions. Our sun (and its planets) must have incorporated some of this heavier material.

9. Dark matter is "stuff" of unknown composition in interstellar or intergalactic space that exerts gravitational pull, yet does not emit any radiation we can detect with our present equipment. Dark matter in the universe has probably played a pivotal role in the organization of visible matter on large scales (galaxies, clusters of galaxies, bubbles and voids). A correct estimate of the

total mass or the average density of dark matter would help us determine whether the universe will continue to expand, or at some point in time begin to contract.

## Problems

1. $d = v/H$

    $= (850 \text{ km/s})/(50 \text{ km/s per Mpc})$

    $= \textbf{17 Mpc,}$ or about 55 million LY

$v = Hd$

    $= (50 \text{ km/s per Mpc})(200 \text{ Mpc})$

    $= \textbf{10,000 km/s}$

which is 1/30 the speed of light.

# Glossary

**Absolute Zero**  The lowest conceivable temperature, corresponding to a halt in the random motions of particles in a substance.

**Absorption Spectrum**  The pattern of spectral lines that appears as dark lines superimposed on an otherwise continuous spectrum.

**Acceleration**  A change in velocity per change in time.

**Accuracy**  How close a measurement of a quantity comes to the accepted or true value of that quantity.

**Acid**  A substance that produces positively charged hydrogen ions when dissolved in water.

**Additive Primary Colors**  Red, green, and blue; the three colors associated with the peak response of the eye's three different retinal cone cells and from which all possible hues can be generated by adding them in different combinations.

**Adhesion**  The tendency of unlike molecules to attract each other.

**Alkane**  The alkane hydrocarbons are those with only single covalent bonds between the carbon atoms.

**Alpha Decay**  When a cluster of 2 protons and 2 neutrons within a nucleus breaks free of the strong force and flies away as an alpha particle, we say that nucleus has undergone alpha decay.

**Alpha Particle**  The nucleus of a helium atom (2 protons and 2 neutrons). An alpha particle radiated from a radioactive source is known as an *alpha ray*.

**Alternating Current** (**ac**)  An electric current that alternates in direction, typically with at least 50 or 60 complete cycles per second.

**Altitude**  In the astronomical sense, the vertical angle of an object in the sky relative to zero altitude at the horizon.

**Amplitude**  For oscillating bodies, the maximum displacement from the equilibrium position. For waves, the maximum departure of a wave disturbance from its average position.

**Angular Momentum** The quantity of angular motion of a body moving sideways or obliquely past a given point or axis.

**Apparent Magnitude** The observed brightness of an object in the sky as expressed on a particular logarithmic scale.

**Apparent Solar Time** See **Solar Time.**

**Asteroid** A rocky or metallic body, up to a few hundred kilometers in diameter, left over from the formation of the solar system. Most asteroids orbit the sun in the *asteroid belt* which lies between the orbits of Mars and Jupiter.

**Astronomical Unit (AU)** The average distance between Earth and the sun, equivalent to approximately 93 million miles or 150 million kilometers.

**Astronomy** The scientific study of the universe and its contents.

**Atom** The smallest unit of an element that can exist alone or combined with other atoms.

**Atomic Nucleus** The central part of an atom, containing a cluster of protons and neutrons.

**Atomic Number** The number of protons in an atomic nucleus.

**Autumnal Equinox** See **Equinox.**

**Base** A substance that produces negatively charged hydroxide ions when dissolved in water.

**Beta Decay** The transformation of a neutron into a proton inside an atomic nucleus, which also creates an electron that is immediately ejected as a beta particle.

**Beta Particle** A fast-moving, high-energy electron released from a radioactive nucleus, also known as a *beta ray.*

**Big Bang Theory** The leading scientific explanation of the universe's earliest history. The theory supposes that the universe began in an extremely hot and dense state, and that it has been expanding and cooling ever since.

**Binary Star** Two stars that revolve around one another.

**Binding Energy** The energy required to separate the particles of an atomic nucleus.

**Blackbody Radiation** The thermal radiation emitted by a perfect absorber of electromagnetic radiation.

**Black Hole** A concentration of mass so dense that neither light nor matter can escape from it.

**Body Wave** Seismic wave energy generated by earthquakes traveling through Earth's interior. Body waves are segregated into two types: *P waves* and *S waves.*

**Bohr Model** Physicist Niels Bohr's model of the atom, in which he proposed that electrons tend to move around the nucleus only in certain stable orbits, and that any change in an electron's orbit results in the absorption or emission of a certain amount of energy.

**Capillary Action** The ability of polar liquids to pull themselves through small tubes, such as capillaries in the body.

**Catalyst** A chemical that either speeds up or slows down the rate of a reaction.

**Celestial Equator** The great circle on the celestial sphere that lies directly above Earth's equator.

**Celestial Meridian** The celestial arc running between the north and south celestial poles and through the zenith, marking the division between the east and west hemispheres of the sky.

**Celestial Poles (North and South)** The points on the celestial sphere directly above Earth's north and south poles.

**Celestial Sphere** The totality of the sky, above and below the horizon. Earth lies at the center of this imaginary sphere, on which are attached all visible celestial bodies.

**Chaos** Situations in which the behavior of interacting bodies depends sensitively on the initial conditions and on the environment in which the interactions are taking place.

**Charge (Electric)** An inherent property of matter, and of only two types, *positive* and *negative.*

**Chemical Bond** The combination of electromagnetic forces that holds atoms, ions, or groups of atoms together.

**Chemical Change** A change in which a pure substance or substances (elements and/or compounds) are transformed into one or more different pure substances.

**Chemical Equation** A concise description of a chemical reaction.

**Chemical Formula** A combination of chemical symbols and subscript numbers that describes the number of atoms of each element present in a molecule, or the ratio between atoms of different elements in an ionic compound.

**Chemical Reaction** The transformation of matter by means of making or breaking chemical bonds.

**Chemical Symbol** A one- or two-letter abbreviation for an element.

**Chemistry** The study of the composition and properties of matter, and of transformations of matter (especially chemical changes).

**Cohesion** The tendency for like molecules to attract each other.

**Comet** An icy, dusty body, a few kilometers in diameter or less, that typically orbits the sun in a large and elongated elliptical orbit. A comet passing near the sun may become visible and form a tail.

**Compound** A pure substance that can be decomposed into simpler substances through a chemical change.

**Condensation Theory** The modern explanation of how a mass of diffuse interstellar gas and dust drew together to form the sun and its family of planets and other smaller bodies.

**Conduction (Thermal)** See **Thermal Conduction.**

**Conductor (Electrical Conductor)** A material through which electrons move easily and electric current flows without much resistance.

**Constellation** A grouping of stars in a common region of the sky, imagined to represent a mythological character, animal, or object.

**Constructive Interference** See **Interference.**

**Continuous Spectrum** The smooth spectrum of thermal energy, with a peak frequency or wavelength that depends on the temperature of the emitting body.

**Convection** The transfer of thermal energy in a heated fluid by means of currents.

**Correlation (Geologic)** The matching and comparison of sequences of rock in different places to establish relative geological age.

**Cosmic Background Radiation (CBR)** An almost perfectly uniform microwave radiation originating from all parts of the sky taken to be strong evidence for a *Big Bang* origin of our universe.

**Coulomb's Law** The mathematical relationship between charge, distance, and electric force between two charged bodies.

**Covalent Bond** A chemical bond formed by the sharing of a pair of electrons between atoms. *Covalent compounds* consist of atoms held together by one or more covalent bonds.

**Crystalline Solid** A substance made of atoms or molecules assembled in some kind of orderly or repeating arrangement.

**Dark Matter** Material in outer space, which has remained undetectable by any direct means and appears to be exerting large gravitational forces on luminous objects we can see.

**Decibel Scale** A logarithmic scale of sound intensity or loudness.

**Declination** A celestial coordinate analogous to latitude on Earth.

**Density** Mass divided by volume; a measure of the compactness of a given quantity of matter.

**Destructive Interference** See **Interference.**

**Diffraction** The bending of a wave around an obstacle.

**Diffraction Grating** A device with many tiny, closely spaced, parallel slits or grooves from which light spreads out into a spectrum.

**Diffuse Reflection** Haphazard scattering of waves or rays that are reflected off rough surfaces.

**Direct Current (DC)** One-way electric current.

**Dispersion** The phenomenon of separating colors or wavelengths through refraction.

**Diurnal Circle** The apparent path followed by a celestial object as a consequence of Earth's rotation during one day.

**Doppler Effect** The apparent change in wave frequency owing to relative motion between a wave source and an observer.

**Ecliptic** The apparent path of the sun relative to the stars on the celestial sphere, arising because of Earth's motion around the sun.

**Ecliptic Plane** The plane defined by Earth's revolution around the sun.

**Elastic Collision** A collision in which kinetic energy is conserved.

**Electric Circuit** A looping path that electrons can follow.

**Electric Current** A measure of the rate of flow of electric charge; charge divided by time.

**Electric Field**   The force field that arises because of the presence of a charged body.

**Electric Power**   The product of potential difference (voltage) and current.

**Electrochemical Cell**   A device that transforms chemical energy into electric energy by means of exothermic chemical reactions.

**Electrolyte**   A substance that can dissociate (break up) into ions when mixed with water. *Nonelectrolyte* substances do not dissociate in water.

**Electromagnet**   A magnet formed by a coil of wire that is energized only when an electric current flows through the wire.

**Electromagnetic Force**   The attractive or repulsive force associated with bodies that possess electric charge.

**Electromagnetic Induction**   The production of a voltage or a current in a wire when an external magnetic field around the wire changes.

**Electromagnetic Wave**   A wave consisting of oscillating electric and magnetic fields coupled to each other and moving through space at the speed of light. Also known as *electromagnetic radiation*.

**Electron**   A negatively charged subatomic particle that is one of the three basic constituents of an atom. Electrons exist in *orbitals* around atomic nuclei, though they may break free and travel from place to place as *electric current*.

**Element**   A pure substance that cannot be chemically decomposed. Also, a substance consisting solely of atoms having the same number of protons.

**Elliptical Galaxy**   A galaxy having an overall elliptical shape and a smooth appearance with no apparent internal structure.

**Emission Spectrum**   A pattern of bright spectral lines superimposed on a dark background.

**Endothermic Reaction**   Any chemical (or nuclear) reaction that absorbs energy, typically resulting in lower temperatures around the site of the reaction.

**Energy**   A quantity always associated with change in a physical system; the capacity to do work.

**Entropy**   A measure of the disorder in a physical system.

**Equinox**   The two times during the year when the sun lies over Earth's equator and the lengths of daylight and night are equal. The vernal (spring) equinox occurs on or about March 21. The autumnal (fall) equinox occurs on or about September 23.

**Erosion**   The movement of surface material on a planet from one place to another (generally down a slope) caused by agents such as moving water, ice, and wind.

**Excited State**   The stationary states of an atomic electron associated with levels of energy higher than that of the *ground state*.

**Exothermic Reaction**   Any chemical (or nuclear) reaction that gives off energy, typically as heat and/or light.

**Extrusive Igneous Rock**   See **Igneous Rock.**

**Ferromagnetic Material**   Substances such as iron, nickel, and cobalt alloys, plus some ceramics, that can maintain their magnetism indefinitely.

**Force**   In physics, force means any agent of change.

**Frequency**   The number of waves that pass a given point in 1 second of time.

**Fundamental Forces** (or *Fundamental Interactions*)   The four forces that control all natural processes; the strong, weak, electromagnetic, and gravitational forces.

**Fundamental Properties**   Properties that cannot be expressed in simpler terms; the four fundamental properties are length, time, mass, and charge.

**Gamma Decay**   The ejection of a gamma ray photon from an unstable nucleus, caused by a spontaneous rearrangement of the protons and neutrons inside the nucleus.

**Gamma Ray**   A highly penetrating photon; radiation belonging to the short-wavelength, high-frequency end of the electromagnetic spectrum.

**Geocentric Model**   An ancient model of the cosmos in which a spherical, motionless Earth lies at the center of everything else.

**Geology**   The study of Earth's surface and its interior.

**Globular Cluster**   A dense accumulation of tens or hundreds of thousands of stars, typically spherical in shape.

**Gravitational Force** (gravitation, or gravity)   The attractive force that exists between all bodies that possess mass.

**Gravitational Wave**   A wave theoretically produced in empty space by any accelerating mass.

**Greenhouse Effect**   The effect of gases (known as greenhouse gases) that trap radiant energy near Earth's surface. The present increasing concentrations of greenhouse gases in our atmosphere may lead to planet-wide temperature increases in the future.

**Gregorian Calendar**   Our modern calendar, which keeps the days of the year synchronized with the seasons to an accuracy of one day in 3300 years.

**Ground State**   The state of lowest possible energy for an atomic electron, typically associated with the smallest possible orbit the electron can have.

**Hadron**   A subatomic particle that responds to the strong force. Hadrons include protons, neutrons, and a slew of other, short-lived particles.

**Half-life**   The interval of time over which there is a 50-50 chance that a given radioactive nucleus will undergo some kind of change and decay into something else.

**Heliocentric Model**   Generally, Copernicus's model of the cosmos, in which the sun lies at the center of a system of orbiting planets, and Earth turns on an axis and revolves around the sun.

**Hubble Law**   The relationship between distance and recession velocity for many galaxies; in general, the farther away a galaxy is, the faster it recedes.

**Hydrocarbon**   A compound containing only hydrogen and carbon.

**Hydrogen Bond**   A chemical bond existing between a hydrogen atom in a polar molecule and the negatively charged portion of another polar molecule.

**Hypothesis**   A provisional explanation often used as a guide for further scientific investigation.

**Ideal Gas Law**   A relationship between pressure, volume, and temperature which is very nearly true for gases that are not too dense or cold.

**Igneous Rock**   A rock formed from crystallized magma, or molten rock. When magma cools and crystallizes deep underground it forms *intrusive igneous rock*, which typically has an assortment of relatively large (macroscopic) mineral crystals. When magma reaches the surface it cools to form *extrusive igneous rock* (volcanic rock) which has small, often microscopic mineral crystals.

**Inelastic Collision**   A collision in which kinetic energy is not conserved.

**Inertia**   The natural tendency of a massive body to remain in the same state of motion if no forces are applied to it.

**Infrared Radiation**   Electromagnetic radiation that is adjacent to red, visible light on the electromagnetic spectrum.

**Insulator (Electrical)**   A material that greatly resists the flow of electric charge through it.

**Interference**   A superposition of two or more waves. *Constructive interference* occurs when two waves combine so as to result in a wave with a larger amplitude. *Destructive interference* occurs when two waves combine to produce a wave with a smaller amplitude.

**International System of Units (SI)**   The modern form of the metric system, with certain agreed upon conventions, used by scientists worldwide.

**Intrusive Igneous Rock**   See **Igneous Rock.**

**Ion**   An atom, or a cluster of atoms, that possesses a net charge.

**Ionic Bond**   The chemical bond that occurs when electrons are lost by atoms of one element and gained by atoms of another element.

**Ionizing Radiation**   Any radiation, by fast-moving particles or by high-frequency photons, that is energetic enough to ionize atoms or molecules.

**Irregular Galaxy**   A galaxy whose shape and structure has no apparent order or regularity.

**Isomer**   A molecule that has exactly the same number and type of atoms as another molecule, but with a different structure.

**Isotopes**   Atoms whose nuclei have the same number of protons but different numbers of neutrons.

**Jovian Planets**   The large, gas-rich outer planets, namely Jupiter, Saturn, Uranus, and Neptune.

**Kepler's First Law**   The rule that states that planets move around the sun in orbits of elliptical shape, with the sun at one focus of each ellipse.

**Kepler's Second Law**   The rule that states that the line joining a planet and the sun sweeps out equal areas in space in equal intervals of time.

**Kepler's Third Law**   The rule that states that the squares of the periods of the planets are proportional to the cubes of the average radii of their orbits.

**Kilogram**   The base SI unit for mass.

**Kinetic Energy**   The energy possessed by a body by virtue of its motion.

**Laser**   A device that produces a coherent beam of electromagnetic radiation, with waves "in step" with one another.

**Latent Heat of Fusion**   The thermal energy stored in a liquid that is released when the liquid freezes into a solid, and absorbed when the solid melts.

**Latent Heat of Vaporization**   The thermal energy stored in a gas that is released when the gas condenses into a liquid, and absorbed when the liquid evaporates.

**Latitude**   An angular measurement specifying the north-south position of a point on Earth's surface.

**Leap Year**   A year of 366 days, occurring every 4 years, that adjusts for the fact that there are approximately $365\frac{1}{4}$ days in an astronomical year.

**Length**   A measurement of distance between two points in space.

**Length Contraction**   The relativistic effect of a body's decrease in length in the dimension that is parallel to the body's motion.

**Lepton**   A subatomic particle of small mass (such as an electron) that does not respond to the strong force and therefore cannot be bound into a nucleus.

**Light-year (LY)**   The distance traveled by light in one year of time, equivalent to approximately 6 trillion miles or 10 trillion kilometers.

**Linear Momentum**   A body's mass multiplied by its velocity; a measure of the body's "quantity of motion."

**Lithification**   The process by which unconsolidated sediments are transformed into sedimentary rock.

**Lithosphere**   Earth's entire crust along with the uppermost rigid layer of the mantle underneath. The entire lithosphere is solid and brittle, and is broken into many *lithospheric plates* which creep horizontally and interact with each other in the processes associated with *plate tectonics*.

**Local Group**   A cluster of about 30 galaxies, of which the Milky Way and Andromeda galaxies are a part.

**Local Supercluster**   A cluster of about 100 clusters of galaxies, including the Local Group and the Virgo cluster.

**Longitude**   An angular measurement specifying the east-west position of a point on Earth's surface.

**Longitudinal Wave**   A wave that oscillates in the direction parallel to the direction the wave is moving.

**Lunar Eclipse**   The event in which Earth's shadow partially or completely envelops the moon.

**Magnetic field**   The force field that arises because of the presence of a magnet or a moving charge or charges.

**Mass**   A fundamental property of matter that expresses its quantity.

**Mass Number**   The number of nucleons in the nucleus of an atom.

**Matter**   Tangible things that have mass and occupy space.

**Matter Waves**   Abstract waves, associated with the probability of a particle's position at different parts of space, that are significant only for particles of very small mass.

**Maxwell's Equations**   Four equations, from the work of physicist James Clerk Maxwell, that summarize the interconnections between electrical and magnetic phenomena.

**Mean Solar Time**   See **Solar Time.**

**Mechanical Energy**   The sum of the kinetic and gravitational potential energies possessed by every part of a system of interacting bodies.

**Mechanical Wave**   A wave that propagates only through matter, not empty space.

**Meridian (Celestial)**   See **Celestial Meridian.**

**Metal**   In chemistry, a substance with atoms that lose electrons during chemical reactions.

**Metamorphic Rock**   A preexisting rock that is transformed through extreme compression, heat, or both.

**Meteorology**   The study of Earth's atmosphere, weather, and climate.

**Meter**   The base SI unit for length or distance.

**Microwave**   A radio wave of relatively short wavelength.

**Milky Way**   A faintly luminous band of light encircling the celestial sphere, often visible on dark, clear nights.

**Milky Way Galaxy**   The disklike, spiral galaxy to which our sun belongs.

**Mineral**   In the geologic sense, a naturally occurring element or compound having a crystalline structure.

**Mixture**   Two or more elements or compounds mixed together that can be separated by physical means.

**Model**   In the scientific sense, a metaphor, or a simplified version, of a complicated scientific explanation that brings out its basic character.

**Molecule**   An aggregate of atoms having a net charge of zero. If an aggregate of atoms (or a single atom) has a net charge, it is known as an *ion*.

**Monomer**   An unsaturated molecule that may link together with other, similar molecules to form a *polymer*.

**Natural Science**   The study of the entire physical world, both nonliving and living. Natural science deals with phenomena that can be observed and measured in some concrete way.

**Neutron**   A subatomic particle having no charge. Neutrons exist in the nucleus of every atom except in the nucleus of the simplest isotope of hydrogen.

**Nonmetal**   In chemistry, a substance with atoms that may gain electrons during chemical reactions or may share electrons with other nonmetal atoms.

**Nuclear Fission**   The splitting apart of atomic nuclei.

**Nuclear Fusion**   The combining or merging of atomic nuclei.

**Nucleons**   The protons and neutrons that are present within an atomic nucleus.

**Nucleus**   See **Atomic Nucleus.**

**Ohm's Law**   The relationship between current, voltage, and resistance applicable to certain materials or devices in an electric circuit.

**Open Cluster**   A group of stars loosely bound to each other by gravity.

**Optical Fiber**   A long, cylindrical thread of transparent material such as glass or plastic, with smooth walls, capable of transmitting light from place to place.

**Orbital**   A spatial pattern, visualized or drawn for an electron in an atom or a molecule, that suggests where the electron spends most of its time.

**Organic Chemistry**   Nearly all the chemistry associated with the element carbon. *Inorganic chemistry* is the chemistry of everything else.

**Oxidation**   The process by which electrons are lost from an element or a compound. The recipient of these lost electrons is often, but not always, atoms of the element oxygen.

**Parsec (pc)**   A unit of distance defined as 206,265 astronomical units or approximately 3.26 light-years.

**Period**   The time necessary for any repetitive process to complete one cycle; the time interval between the passage of one wave and the next wave that follows.

**Periodic Table of the Elements**   An organizing scheme by which all the chemical elements are grouped by similar physical characteristics and chemical behaviors.

**Photoelectric Effect**   The release of electrons from a metal surface irradiated with light.

**Photon**   A packet of pure energy, large numbers of which are associated with light and every other type of electromagnetic radiation.

**Physical Change**   Any alteration of a substance (involving changes of temperature, phase, or magnetization, for example) that does not change the identity of the substance.

**Physical Science**   The academic disciplines concerned with the study of the natural world, generally without emphasis on living matter. Physics, chemistry, astronomy, geology, and meteorology are major subdivisions within the physical sciences.

**Physics**   The study of the fundamental aspects of nature, such as matter, energy, motion, and force.

**Planck's Constant**   A fundamental constant that relates the energy and frequency of photons, and establishes certain relationships between energy and time, and position and momentum in the atomic and subatomic realm.

**Plate (Lithospheric or Tectonic)**   See **Lithosphere.**

**Plate Tectonics**   The theory that Earth's large and small lithospheric plates move horizontally over our planet's soft interior and interact with each other in a variety of ways.

**Pole (Celestial)**   See **Celestial Poles.**

**Polymer**   A substance made of molecules with a chain-like structure consisting of many repetitious units called *monomers*.

**Potential Difference** (or *Voltage*)   The amount of energy that is transformed when a particular amount of charge moves through an electric field.

**Potential Energy**   The energy possessed by a body by virtue of its position in a force field.

**Power**   The rate at which work is done; equivalently, the rate at which energy is transformed.

**Precision**   The extent to which repeated measurements of the same quantity are consistent.

**Pressure**   The ratio between force and the area over which the force is applied.

**Principle** (or *Law*) In the scientific sense, the product of a hypothesis that survives testing through experimentation or observation.

**Principle of Superposition** The concept that in an undisturbed sequence of geologic deposits, each layer has been superimposed on top of a layer that is older than itself.

**Principle of Uniform Change** The concept that many relatively small geologic changes, occurring over very long spans of time, are sufficient to explain Earth's past and present landscape.

**Proton** A positively charged subatomic particle. One or more protons exist in the nucleus of every atom.

**Pulsar** A rapidly spinning neutron star that emits regularly timed pulses of electromagnetic radiation.

**Pure Substance** A substance made up of atoms or molecules with a definite (not variable) composition.

**P Wave** A longitudinal seismic wave that propagates like a low-frequency sound wave through both the solid and liquid parts of Earth's interior.

**Quantum Number** A number that symbolizes some discrete state of an atomic electron. Only four quantum numbers, describing energy, momentum, and magnetic properties, are needed for the description of how each electron behaves in an atom.

**Quark** A tiny particle which is among the smallest and most fundamental entities believed to exist in nature; protons and neutrons are made of three quarks each.

**Radar Ranging** (or *Radar*) A method of determining distance by reflecting a radio pulse off any hard-surfaced body such as a planet or the moon, and timing the return of the pulse.

**Radiation** On the subject of heat, radiation means the transfer of energy by means of electromagnetic waves. In nuclear physics, the term radiation includes the emission of particles from atomic nuclei.

**Radioactivity** The spontaneous emission of particles or energy from nuclei that are inherently unstable.

**Radiometric dating** A method of dating in geology that uses the known rate of decay of certain radioactive isotopes to determine the ages of rock formations or the minerals in them. When the same method is used for the radioactive carbon present in tiny quantities in materials of biological origin, the technique is called *radiocarbon dating*.

**Radio Waves** Electromagnetic waves in the lowest frequency, longest wavelength category.

**Real Image** An image produced by the convergence of light rays from a common source on a point in space.

**Reduction** In chemistry, the gain of electrons by an element or a compound.

**Reflection** See **Diffuse Reflection** and **Specular Reflection.**

**Reflector Telescope** A telescope whose primary element is a large, concave mirror (*primary mirror*) that reflects and converges rays of light toward a point in front of the mirror.

**Refraction** The bending of a wave by virtue of a change in its speed.

**Refractor Telescope** A telescope whose primary element is a large lens (*objective*) that refracts, or bends, incoming light so that it comes to a focus at some distance behind the lens.

**Resistance (Electrical)** The measure of how much a circuit element can resist the flow of electricity; the ratio between voltage across the circuit element and the current through it.

**Resonance** The amplification of vibrations within an object due to the fact that the object has a natural frequency of vibration that matches (or is a multiple of) that of an external source of vibration.

**Rest Energy** (or *Mass Energy*) The energy equivalent of mass, a consequence of the principles set forth by the special theory of relativity.

**Right Ascension** A celestial coordinate analogous to longitude on Earth.

**Rock** A solid aggregation of one or more *minerals*.

**Saturated Compounds** Compounds whose molecules have only single bonds between their carbon atoms.

**Scalar** A quantity that expresses magnitude only, and not direction.

**Scattering** The emission of light in random directions which interacts with particles of matter. In nuclear physics, scattering also applies to the behavior of streams of subatomic particles undergoing interactions with nuclei (such as in Rutherford scattering).

**Scientific Method** The methodology by which scientists gather and synthesize information, and attempt to discover and describe regularity in nature.

**Seafloor Spreading**   The spreading of lithospheric plates due to magma rising from Earth's interior, which occurs along midocean ridges.

**Second**   The base SI unit for time.

**Sedimentary Rock**   A rock derived from particles cemented together or precipitated from a solution of minerals.

**Semiconductor**   A material that conducts electricity with an ability somewhere between that of conductors and insulators.

**Shell**   In the atom, a group of orbitals with electrons that have similar overall energy.

**Shield Volcano**   A large, gradually sloping volcano formed by many relatively nonviolent flows of basaltic lava issuing from one or more vents.

**Sidereal Time**   Time as reckoned by the diurnal motions of the stars. In one sidereal day, Earth rotates once with respect to the stars.

**Solar Eclipse**   The event in which the moon passes partially or wholly in front of the sun as seen from Earth.

**Solar System**   The sun, the nine planets, the moons orbiting the planets, and all other bodies subject to the sun's gravitational influence: that is, comets, asteroids, and debris such as interplanetary dust.

**Solar Time**   Time as reckoned by the diurnal motion of the sun in the sky. *Mean solar time* is based on the sun's average periodic movement around the sky (a day, or 24 hours).

**Solstice**   The times during the year when the sun lies farthest north or south of the celestial equator. During the *summer solstice*, on or about June 21, the sun's rays illuminate Earth's northern hemisphere for more than half of each 24-hour day. During the *winter solstice*, on or about December 22, the sun's rays illuminate Earth's southern hemisphere for more than half of each 24-hour day.

**Solubility**   A measure of the amount of solute that will dissolve in a solvent at a given temperature and pressure.

**Solute and Solvent**   Respectively, the larger and smaller quantities of a substance that, when mixed together, made a solution.

**Solution**   Different substances so uniformly mixed that they cannot be separated by mechanical means.

**Space**   A fundamental attribute of the physical world; generally associated with three mutually perpendicular directions, or dimensions.

**Specific Heat Capacity**   A measure of a substance's ability to store thermal energy.

**Spectroscopy**   The method of separating wavelengths of light and analyzing the resulting spectrum.

**Specular Reflection**   Wave reflection that occurs for surfaces that are smooth compared to the wavelength of the waves being reflected.

**Speed**   The standard measure of how quickly a body moves through space; defined by distance divided by time.

**Spiral Galaxy**   A disk-shaped galaxy having a spiral-like structure inside.

**Standard Time**   The time setting within any standard time zone that is equal to the mean solar time on that zone's standard meridian of longitude.

**Standing Wave**   An interference pattern in which waves apparently stand still.

**Stellar Parallax**   The geometric method by which astronomers measure the distances to nearby stars.

**Stratovolcano (or Composite Volcano)**   A steep sided, cone-shaped volcano formed by alternating layers of ash and lava, often subject to violent eruptions.

**Strong Force**   The strongest of the four fundamental forces of nature, responsible for binding together protons and neutrons in atomic nuclei, but ineffective outside the nucleus.

**Subduction**   The sinking of one lithospheric plate beneath another when the two plates collide.

**Subshell**   In the atom, a group of orbitals with electrons that have similar overall energy and angular momentum magnitude.

**Subtractive Primary Colors**   The only three colors needed to reproduce all possible hues by subtraction: yellow, cyan, and magenta.

**Summer Solstice**   See **Solstice.**

**Superconductivity**   Zero resistance to electric current.

**Superconductor**   A material (typically chilled to an extremely low temperature) that behaves as a perfect conductor, having no resistance at all to the flow of electrons.

**Surface Tension**   The tendency of the molecules within a liquid to attract each other forming a surface that slightly resists penetration; a characteristic that is somewhat pronounced in polar liquids such as water.

**Surface Wave**   A seismic wave that travels along the ground. They are responsible for most of the damage to buildings during major earthquakes.

**S Wave**   A transverse seismic wave that propagates only through rigid portions of Earth's interior.

**Temperature**   A measure of the average kinetic energy possessed by a substance's constituent particles.

**Terrestrial Planets**   The smaller, rocky-metallic inner planets, namely Mercury, Venus, Earth, and Mars.

**Theory**   A detailed explanation covering some general aspect of nature.

**Thermal Conduction**   The transfer of thermal energy within a substance without any net motion of the substance's particles.

**Thermal Energy**   The sum of the internal potential and kinetic energies of the constituent particles of a body.

**Thermal Radiation**   The electromagnetic radiation spontaneously emitted by all bodies having a temperature above absolute zero.

**Time**   A fundamental, but abstract attribute of the physical world that is related to change.

**Time Dilation**   The relativistic effect of time passing slower when there is relative motion between the observer and what is being observed.

**Transformer**   An electrical device that either increases voltage (with an accompanying decrease in current), or decreases voltage (with an accompanying increase in current).

**Transverse Wave**   A wave that oscillates in a direction perpendicular to the direction the wave is moving.

**Trigonometric Parallax**   A geometrical method of determining distance by relying on observations of the same object made from two different places.

**Tsunami**   A seismic wave that travels along the ocean surface, a "tidal wave."

**Ultraviolet Radiation**   Electromagnetic radiation that is adjacent to violet, visible light on the electromagnetic spectrum.

**Unit**   A well-defined and agreed-upon value of a measurable property.

**Unsaturated Compound**   Organic compounds whose molecules have at least one double or triple bond between carbons.

**Vector**   A quantity that expresses both magnitude and spatial direction.

**Velocity**   Both a body's speed and its direction of motion.

**Vernal Equinox**   See **Equinox.**

**Virtual Image**   An image produced by rays of light entering your eye (from a mirror or lens) but impossible to project onto a screen.

**Visible Light**   The relatively small range of electromagnetic energy to which the human eye responds.

**Wavelength**   The distance between any part of a wave and the same part of the next wave that follows.

**Weak Force**   One of the four fundamental forces of nature. The weak force can sometimes disrupt particles within atomic nuclei, but is ineffective outside the nucleus.

**Weathering**   The small-scale disintegration of rocks.

**Weight**   The measure of gravitational force on a body.

**Winter Solstice**   See **Solstice.**

**Work**   The force exerted on a body multiplied by the distance over which the force acts.

**X-ray**   A high-frequency, short-wavelength type of electromagnetic radiation that can pass through many materials that are opaque to light.

**Zodiac**   The belt centered on the ecliptic in which the moon and planets as we see them from Earth are found.

# Index

# Credits

This page constitutes an extension of the copyright page. We have made every effort to trace the ownership of all copyrighted material and to secure permission from copyright holders. In the event of any question arising as to the use of any material, we will be pleased to make the necessary corrections in future printings. Thanks are due to the following authors, publishers, and agents for permission to use the material indicated.

## Photographs

**Preface: viii:** NASA. **viii:** © Julian Baum/SPL/Photo Researchers. **viii:** © NCSA, University of Illinois/SPL/ Photo Researchers.
**Table of contents: xvii:** © Jerry Schad. **xx:** © Dr. Jeremy Burgess/SPL/Photo Researchers. **xxiv:** © Clive Freeman/ Biosym Technologies/SPL/Photo Researchers. **xxvii:** © Paul Conklin/PhotoEdit.
**Part I: 3:** © Jerry Schad.
**Chapter 1: 4:** © Michael Abbey/Photo Researchers. **6:** © Jerry Schad. **13:** National Institute of Standards and Technology, U.S. Department of Commerce. **16:** © Bill Sanderson/ SPL/Photo Researchers
**Chapter 2: 20:** NASA. **22L:** © Dr. Dennis Kunkel/ Phototake. **22 M:** © Yoav Levy/Phototake. **22R:** NASA. **26:** © 1993 Richard Megna/Fundamental Photographs. **29:** © Dan McCoy/ Rainbow. **32:** © Jerry Schad. **34:** © David Nunuk/ SPL/Photo Researchers. **35:** © Jerry Schad. **42:** © Michael L. Abramson/ Time Magazine. **47:** © Manfred Kage/Peter Arnold, Inc. **53:** © Jerry Schad. **56:** NASA. **60:** NASA. **61:** © Jerry Schad. **62:** © Phillipe Plailly/ Eurelios/SPL/ Photo Researchers. **71:** © Tony Hallas/SPL/Photo Researchers.
**Chapter 3: 78:** © 1990 Robert Mathena/Fundamental Photographs. **81:** © Culver Pictures. **82:** © Gerard Lacz/ Peter Arnold, Inc. **96:** © Jerry Schad. **97:** © Tom Brewster/ Phototake. **98:** © Jerry Schad. **99:** © Jerry Schad. **104:** © Gregory Sams/SPL/Photo Researchers. **106:** © Philippe Plailly/ SPL/Photo Researchers. **107:** © Patrice Loiez, CERN/SPL/ Photo Researchers. **109:** © U. S. Navy/SPL/Photo Researchers. **110:** © Dr. Jeremy Burgess/SPL/Photo Researchers. **111:** © Space Telescope Science Institute/NASA/SPL. Photo Researchers. **112:** © NASA/SPL/Photo Researchers. **113:** © Royal Observatory/SPL/Photo Researchers. **114:** © Royal Observatory, Edinburgh/SPL/Photo Researchers. **115:** NASA. **116:** © Tony Hallas/SPL/Photo Researchers. **118:** © JPL/Phototake. **119:** © A.S.P./Science Source/Photo Researchers. **120:** © Francois Gohier/Photo Researchers. **121,T:** © Rev. Ronald Royer/SPL/Photo Researchers. **|121,B:** © Account Phototake/ Phototake. **123:** © John Reader/SPL/Photo Researchers. **126:** © Francis Leroy/ SPL/Photo Researchers. **127:** Emory Kristof National Geographic. **128:** © CDC/Science Source/ Photo Researchers. **129:** © R. Noonan/Photo Researchers. **130:** © Jerry Schad. **131:** © Nuridsany et Perennou/Photo Researchers. **132:** © Francis Leroy, Biocosmos/SPL/Photo Researchers. **133:** © M. K. Rasmussen. **134:** © Jerry Schad. **135:** © Bill Bachman/Photo Researchers. **138:** © Francois Gohier/Photo Researchers. **139:** © Julian Baum/SPL/Photo Researchers. **140:** © John Reader/SPL/Photo Researchers. **141:** © Tom McHugh/Photo Researchers. **142:** © Gianni Tortoli/Photo Researchers. **143:** © James A. Sugar/Black Star. **144:** © Photo Researchers. **145:** © The Granger Collection,

New York. **146:** © Jean-Loup Charmet/SPL/Photo Researchers. **147:** © NCSA, University of Illinois/SPL/Photo Researchers. **150:** © SPL/ Photo Researchers. **156:** © Jerry Schad.
**Chapter 4: 158:** © 1992 Richard Megna/Fundamental Photographs. **160:** © James Prince/Photo Researchers. **162:** © 1991 Kristen Brochmann/Fundamental Photographs. **165,L:** © James Sugar/Black Star. **165 R:** © 1987 Ken Kay/Fundamental Photographs. **170:** © Royal Observatory, Edinburgh/SPL/Photo Researchers. **171:** .NASA. **172,T:** © Photo Researchers. **172,B:** © Jean-Loup Charmet/ SPL/Photo Researchers.
**Part II: 181:** © Dr. Jeremy Burgess/SPL/Photo Researchers.
**Chapter 5: 182:** © James Sugar/Black Star. **193:** © Stock Montage. **195:** © 1994 Richard J. Wainscoat and John Kormendy, Institute for Astronomy, University of Hawaii. **197:** © Jerry Schad. **198:** © Dan McCoy/Rainbow. **199,L:** © Jerry Schad. **199,R:** © Jerry Schad. **205:** NASA. **211:** © Jerry Schad. **212:** © Vandystadt/Photo Researchers. **214:** © Henry Gorskinsky/Peter Arnold, Inc. **225,T:** © Jeff Greenberg/Photo Researchers. **225,B:** © Jeff Greenberg/ Photo Researchers. **229:** © JPL/Black Star. **230:** NASA. **231:** © Rick Falco/Black Star. **232:** NASA. **233:** © 1992 Diane Schiumo/Fundamental Photographs. **234:** © Scott Camazine/ Photo Researchers. **235:** © A.S.P./Science Source/Photo Researchers.
**Chapter 6: 242:** © Wesley Bocxe/Photo Researchers. **246:** © Vandystadt/Photo Researchers. **249:** © Carl Purcell/ Photo Researchers. **263:** © Jonathan Nourok/ PhotoEdit. **264,L:** © 1991 Richard Megna/Fundamental Photographs. **264,R:** © 1991 Richard Megna/Fundamental Photographs. **265:** © Jerry Schad. **266:** © 1993 Richard Megna/ Fundamental Photographs. **267:** Lunar and Planetary Institute. **268:** NASA. **270:** © Jerry Schad. **272,T:** © 1994 Richard Megna/Fundamental Photographs. **272,B:** © 1994 Richard Megna/Fundamental Photographs. **275:** © Allen Green/ Photo Researchers. **277:** © Bob Fraser/Black Star. **279:** © Hank Morgan/Photo Researchers. **280,L:** © Dan McCoy/Rainbow. **280,R:** © Kevin Schafer/Peter Arnold, Inc. **281:** © Jerry Schad.
**Chapter 7: 286:** © Daedelus Enterprises/Peter Arnold, Inc. **290:** © Jerry Schad. **293:** NASA. **294,L:** NASA. **294,R:** © Charles Mason/Black Star. **295, T:** © Jerry Schad. **295,BL:** © 1993 Richard Megna/Fundamental Photographs. **295,BR:** © 1993 Richard Megna/Fundamental Photographs. **303:** © Jerry Schad. **304:** © Richard Folwell/SPL/Photo Researchers. **310:** © Jerry Schad. **314:** © Carl Purcell/Photo Researchers. **315:** © 1968 Fundamental Photographs. **316:** © Tony Freeman/PhotoEdit. **322:** NASA. **327:** NASA. **328:** © Claude Charlier/Black Star. **329:** © Fred K. Smith. **331,R:** © Jerry Schad. **331,L:** © Jerry Schad. **332,TL:** © Pekka Parviainen/SPL/Photo Researchers. **332,TR:** © Pekka Parviainen/SPL/Photo Researchers. **332,BL:** © Gene E. Moore/Phototake. **332,BR:** © Jerry Schad. **333,BL:** © Jerry Schad. **333,TL:** © Howie Bluestein/ Photo Researchers. **333,TR:** © Jerry Schad. **333,BR:** © Jerry Schad. **335:** © Jerry Schad. **340:** © Alvis Uptis/The Image Bank.
**Chapter 8: 352:** © 1992 Martin Bough/Fundamental Photographs. **354:** © Jerry Schad. **368:** © Jerry Schad. **369:** © Jerry Schad. **372:** © Jerry Schad. **373:** © Doug Johnson/ SPL/Photo Researchers. **375:** © Jerry Schad. **376:** © Will & Deni McIntyre/Photo Researchers. **379,T:** © 1991 Richard Megna/Fundamental Photographs. **379,B:** © UPI/Bettmann. **380:** © Leonard Lessin/Peter Arnold, Inc. **382:** © Tim Kelley/Black Star. **383,T:** © Erich Schrempp/Photo Researchers. **383,B:** © Jerry Schad. **385:** © Krafft/ Explorer/ Photo Researchers. **387:** © 1990 Fundamental Photographs.

## Illustrations